Facilities Maintenance & Repair Costs with RSMeans data

Brian Adams, Senior Editor

2019
26th annual edition

Chief Data Officer
Noam Reininger

Engineering Director
Bob Mewis *(1, 3, 4, 5, 11, 12)*

Contributing Editors
Brian Adams *(21, 22)*
Christopher Babbitt
Sam Babbitt
Michelle Curran
Matthew Doheny *(8, 9, 10)*
John Gomes *(13, 41)*
Derrick Hale, PE *(2, 31, 32, 33, 34, 35, 44, 46)*
Michael Henry

Joseph Kelble *(14, 23, 25)*
Charles Kibbee
Gerard Lafond, PE
Thomas Lane *(6, 7)*
Jake MacDonald
Elisa Mello
Michael Ouillette *(26, 27, 28, 48)*
Gabe Sirota
Matthew Sorrentino
Kevin Souza
David Yazbek

Product Manager
Andrea Sillah

Production Manager
Debbie Panarelli

Production
Jonathan Forgit
Mary Lou Geary
Sharon Larsen
Sheryl Rose

Data Quality Manager
Joseph Ingargiola

Technical Support
Kedar Gaikwad
Todd Klapprodt
John Liu

Cover Design
Blaire Collins

Data Analytics
Tim Duggan
Todd Glowac
Matthew Kelliher-Gibson

Numbers in italics are the divisional responsibilities for each editor. Please contact the designated editor directly with any questions.

RSMeans data from Gordian
Construction Publishers & Consultants
1099 Hingham Street, Suite 201
Rockland, MA 02370
United States of America
1.800.448.8182
RSMeans.com

Copyright 2018 by The Gordian Group Inc.
All rights reserved.
Cover photo © iStock.com/tverkhovinets

Printed in the United States of America
ISSN 1074-0953
ISBN 978-1-946872-57-9

Gordian's authors, editors, and engineers apply diligence and judgment in locating and using reliable sources for the information published. However, Gordian makes no express or implied warranty or guarantee in connection with the content of the information contained herein, including the accuracy, correctness, value, sufficiency, or completeness of the data, methods, and other information contained herein. Gordian makes no express or implied warranty of merchantability or fitness for a particular purpose. Gordian shall have no liability to any customer or third party for any loss, expense, or damage, including consequential, incidental, special, or punitive damage, including lost profits or lost revenue, caused directly or indirectly by any error or omission, or arising out of, or in connection with, the information contained herein. For the purposes of this paragraph, "Gordian" shall include The Gordian Group, Inc., and its divisions, subsidiaries, successors, parent companies, and their employees, partners, principals, agents and representatives, and any third-party providers or sources of information or data. Gordian grants the purchaser of this publication limited license to use the cost data

contained herein for purchaser's internal business purposes in connection with construction estimating and related work. The publication, and all cost data contained herein, may not be reproduced, integrated into any software or computer program, developed into a database or other electronic compilation, stored in an information storage or retrieval system, or transmitted or distributed to anyone in any form or by any means, electronic or mechanical, including photocopying or scanning, without prior written permission of Gordian. This publication is subject to protection under copyright law, trade secret law and other intellectual property laws of the United States, Canada and other jurisdictions. Gordian and its affiliates exclusively own and retain all rights, title and interest in and to this publication and the cost data contained herein including, without limitation, all copyright, patent, trademark and trade secret rights. Except for the limited license contained herein, your purchase of this publication does not grant you any intellectual property rights in the publication or the cost data.

| 0309 | $864.99 per copy (in United States)
Price is subject to change without prior notice. |

Related Data and Services

Our engineers recommend the following products and services to complement *Facilities Maintenance & Repair Costs with RSMeans data:*

Annual Cost Data Books
2019 Facilities Construction Costs with RSMeans data
2019 Electrical Costs with RSMeans data
2019 Mechanical Costs with RSMeans data
2019 Plumbing Costs with RSMeans data

Reference Books
Facilities Planning and Relocation
Life Cycle Costing for Facilities
Cost Planning & Estimating for Facilities Maintenance
Estimating Building Costs
RSMeans Estimating Handbook
Building Security: Strategies & Costs

Seminars and In-House Training
Facilities Maintenance & Repair Estimating
Unit Price Estimating
Training for our online estimating solution
Mechanical & Electrical Estimating
Scheduling with MSProject for Construction Professionals

RSMeans data Online
For access to the latest cost data, an intuitive search, and an easy-to-use estimate builder, take advantage of the time savings available from our online application. To learn more visit: RSMeans.com/2019online.

Enterprise Solutions
Building owners, facility managers, building product manufacturers, and attorneys across the public and private sectors engage with RSMeans data Enterprise to solve unique challenges where trusted construction cost data is critical. To learn more visit: RSMeans.com/Enterprise.

Custom Built Data Sets
Building and Space Models: Quickly plan construction costs across multiple locations based on geography, project size, building system component, product options, and other variables for precise budgeting and cost control.

Predictive Analytics: Accurately plan future builds with custom graphical interactive dashboards, negotiate future costs of tenant build-outs, and identify and compare national account pricing.

Consulting
Building Product Manufacturing Analytics: Validate your claims and assist with new product launches.

Third-Party Legal Resources: Used in cases of construction cost or estimate disputes, construction product failure vs. installation failure, eminent domain, class action construction product liability, and more.

API
For resellers or internal application integration, RSMeans data is offered via API. Deliver Unit, Assembly, and Square Foot Model data within your interface. To learn more about how you can provide your customers with the latest in localized construction cost data visit: RSMeans.com/API.

Table of Contents

Foreword

The Value of RSMeans data from Gordian

Since 1942, RSMeans data has been the industry-standard materials, labor, and equipment cost information database for contractors, facility owners and managers, architects, engineers, and anyone else that requires the latest localized construction cost information. More than 75 years later, the objective remains the same: to provide facility and construction professionals with the most current and comprehensive construction cost database possible.

With the constant influx of new construction methods and materials, in addition to ever-changing labor and material costs, last year's cost data is not reliable for today's designs, estimates, or budgets. Gordian's cost engineers apply real-world construction experience to identify and quantify new building products and methodologies, adjust productivity rates, and adjust costs to local market conditions across the nation. This adds up to more than 22,000 hours in cost research annually. This unparalleled construction cost expertise is why so many facility and construction professionals rely on RSMeans data year over year.

About Gordian

Gordian originated in the spirit of innovation and a strong commitment to helping clients reach and exceed their construction goals. In 1982, Gordian's chairman and founder, Harry H. Mellon, created Job Order Contracting while serving as chief engineer at the Supreme Headquarters Allied Powers Europe. Job Order Contracting is a unique indefinite delivery/indefinite quantity (IDIQ) process, which enables facility owners to complete a substantial number of repair, maintenance, and construction projects with a single, competitively awarded contract. Realizing facility and infrastructure owners across various industries could greatly benefit from the time and cost saving advantages of this innovative construction procurement solution, he established Gordian in 1990.

Continuing the commitment to providing the most relevant and accurate facility and construction data, software, and expertise in the industry, Gordian enhanced the fortitude of its data with the acquisition of RSMeans in 2014. And in an effort to expand its facility management capabilities, Gordian acquired Sightlines, the leading provider of facilities benchmarking data and analysis, in 2015.

Our Offerings

Gordian is the leader in facility and construction cost data, software, and expertise for all phases of the building life cycle. From planning to design, procurement, construction, and operations, Gordian's solutions help clients maximize efficiency, optimize cost savings, and increase building quality with its highly specialized data engineers, software, and unique proprietary data sets.

Our Commitment

At Gordian, we do more than talk about the quality of our data and the usefulness of its application. We stand behind all of our RSMeans data—from historical cost indexes to construction materials and techniques—to craft current costs and predict future trends. If you have any questions about our products or services, please call us toll-free at 800.448.8182 or visit our website at gordian.com.

How the Cost Data Is Built: An Overview

A Complete Reference for the Facility Manager

This data set was designed as a complete reference and cost data source for facility managers, owners, engineers, contractors, and anyone who manages real estate. A complete facilities management program starts with a building audit. The audit describes the structures, their characteristics and major equipment, and apparent deficiencies. An audit is not a prerequisite to using any data contained in this data set, but it does provide an organizational framework. A list of equipment and deficiencies forms the basis for a repair budget and preventive maintenance program. An audit also provides facility size and usage criteria on which the general maintenance (cleaning) program will be based. Together, the five sections of this data set provide a framework and definitive data for developing a complete program for all facets of facilities maintenance.

Even if you have a regular building audit and preventive maintenance program in place, this information will be useful.

For the occasional user, the data set provides:

- a reference for time and material requirements for maintenance & repair tasks to fill in where your organization lacks its own historical data;
- preventive maintenance (PM) checklists with labor-hour standards that can be used to benchmark your program;
- a quick reference for general maintenance (cleaning) productivity rates to analyze proposals from outside contractors;
- an explanation of Life Cycle Cost analysis as a tool to assist in making the correct budget decisions; and
- data to validate line items in budgets.

For the frequent user, the data set supplies all of the above plus:

- guidance for getting the audit under way;
- a foundation for zero-based budgeting of maintenance & repair, preventive maintenance, and general maintenance;
- cost data that can be used to estimate preventive maintenance, deferred maintenance (repairs), and general maintenance (cleaning) programs; and
- detailed task descriptions for establishing a complete PM program.

How to Use the Cost Data: The Details

This section contains an in-depth explanation of how the data is arranged and how you can use it to determine a reliable maintenance & repair cost estimate. It includes information about how we develop our cost figures and how to completely prepare your estimate.

Section 1: Maintenance & Repair (M&R)

The Maintenance & Repair section is a listing of common maintenance tasks performed at facilities. The tasks include removal and replacement, repair, and refinishing. This section is organized according to the UNIFORMAT II system.

The purpose of Section 1 is to provide cost data and an approximate frequency of occurrence for each maintenance & repair task. This data is used by facility managers, owners, plant engineers, and contractors to prepare estimates for deferred maintenance projects. The data is also used to estimate labor-hours and material costs for in-house staff. The frequency listing facilitates preparation of a zero-based budget.

Under the System Description column, each task, including its components, is described. In some cases additional information (such as "2% of total roof area") appears in parentheses. The next column, Frequency, refers to how often one should expect, and therefore estimate, that this task will have to be performed. Labor-hours have been enhanced, where appropriate, by 30% to include setup and delay time. In some cases, the list of steps comprising the task includes most of the potential repairs that could be made on a system. For those tasks, individual items, as appropriate, can be selected from the all-encompassing list.

Maintenance & Repair

B30 ROOFING					B3013	Roof Covering					
B3013 105			Built-Up Roofing								
System Description		Freq. (Years)	Crew	Unit	Labor Hours	2019 Bare Costs				Total In-House	Total w/O&P
						Material	Labor	Equipment	Total		
0600	Place new BUR membrane over existing	20	G-5	Sq.							
	Set up, secure and take down ladder				.020		.90		.90	1.31	1.59
	Sweep / spud ballast clean				.500		22.50		22.50	33	40
	Vent existing membrane				.130		5.85		5.85	8.50	10.35
	Cut out buckled flashing				.024		1.08		1.08	1.58	1.92
	Install 2 ply membrane flashing				.027	.42	1.20		1.62	2.21	2.65
	Install 4 ply bituminous roofing				3.733	145	157	34	336	425	495
	Reinstall ballast				.381		17.15		17.15	25	30.50
	Clean up				.390		17.60		17.60	25.50	31
	Total				5.205	145.42	223.28	34	402.70	522.10	613.01
0700	Total BUR roof replacement	28	G-1	Sq.							
	Set up, secure and take down ladder				.020		.90		.90	1.31	1.59
	Sweep / spud ballast clean				.500		22.50		22.50	33	40
	Remove built-up roofing				2.500		104		104	135	167
	Remove insulation board				1.026		43		43	55	69
	Remove flashing				.024		1.08		1.08	1.58	1.92
	Install 2" perlite insulation				.879	110	40		150	179	208
	Install 2 ply membrane flashing				.027	.42	1.20		1.62	2.21	2.65
	Install 4 ply bituminous membrane				2.800	145	118	25.50	288.50	360	420
	Clean up				1.000		45		45	65.50	79.50
	Total										
B3013 120			Modified Bituminous / Therm								
System Description		Freq. (Years)	Crew	U							
0100	Debris removal by hand & visual inspection	1	2 ROFC	M.S							
	Set up, secure and take down ladder										
	Pick up trash / debris & clean up										
	Visual inspection										
	Total										

The Construction Specifications Institute (CSI) and Construction Specifications Canada (CSC) have produced the 2016 edition of MasterFormat®, a system of titles and numbers used extensively to organize construction information.

All unit prices in the RSMeans cost data are now arranged in the 50-division MasterFormat® 2016 system.

Section 2: Preventive Maintenance (PM)

The Preventive Maintenance section provides the framework for a complete PM program. The establishment of a program is frequently hindered by the lack of a comprehensive list of equipment, actual PM steps, and budget documentation. This section fulfills all these needs. The facility manager, plant engineer, or owner can use these schedules to establish labor-hours and a budget. The maintenance contractor can use these PM checklists as the basis for a comprehensive maintenance proposal or as an estimating aid when bidding PM contracts. Copies or print-outs of individual sheets can also be distributed to maintenance personnel to identify the procedures required. The hours listed are predicated on work being performed by experienced technicians familiar with the PM tasks and equipped with the proper tools and materials.

The PM section lists the tasks and their frequency, whether weekly, monthly, quarterly, semi-annually, or annually. The frequency of those procedures is based on non-critical usage (e.g., in "normal use" situations, not facilities such as surgical suites or computer rooms that demand absolute adherence to a limited range of environmental conditions). The labor-hours to perform each item on the checklist are listed in the next column. Beneath the table is cost data for the PM schedule. The data is shown both annually and annualized to provide the facility manager with an estimating range. If all tasks on the schedule are performed once a year, the *annually* line should be used in the PM estimate. The *annualized* data is used when all items on the schedule are performed at the frequency shown.

Section 3: General Maintenance (GM)

This section provides labor-hour estimates and costs to perform day-to-day maintenance. The data is used to estimate cleaning times, compare and assess estimates submitted by maintenance companies, or to budget in-house staff. The information is divided into *Interior* and *Exterior* maintenance. A common maintenance laborer (*Clam*) is used to perform these tasks.

Section 4: Facilities Audits

The facilities audit is generally the basis from which the maintenance & repair, preventive maintenance, and general maintenance estimates are prepared. The audit provides the following:

1. A list of deficiencies for the maintenance & repair estimate

2. A list of equipment and other items from which to prepare a preventive maintenance estimate

3. A list of facilities' usage and size requirements to prepare a general maintenance estimate

The Introduction to Facilities Audits in this section explains the rationale for the audit and the steps required to complete it successfully.

Preventive Maintenance

D30 HVAC			D3025 130	Boiler, Hot Water, Oil/Gas/Comb.					
PM Components			Labor-hrs.	W	M	Q	S	A
PM System D3025 130 1950								
Boiler, hot water; oil, gas or combination fired, up to 120 MBH								
1	Check combustion chamber for air or gas leaks.		.077					√
2	Inspect and clean oil burner gun and ignition assembly, where applicable.		.658					√
3	Inspect fuel system for leaks and change fuel filter element, where applicable.		.098					√
4	Check fuel lines and connections for damage.		.023	√	√	√	√
5	Check for proper operational response of burner to thermostat controls.		.133		√	√	√
6	Check and lubricate burner and blower motors.		.079			√	√	√
7	Check main flame failure protection and main flame detection scanner on boiler equipped with spark ignition (oil burner).		.124		√	√	√	√
8	Check electrical wiring to burner controls and blower.		.079					√
9	Clean firebox (sweep and vacuum).		.577					√
10	Check operation of mercury control switches (i.e., steam pressure, hot water temperature limit, atomizing or combustion air proving, etc.).		.143		√	√	√	√
11	Check operation and condition of safety pressure relief valve.		.030	√	√	√	√
12	Check operation of boiler low water cut off devices.		.056	√	√	√	√
13	Check hot water pressure gauges.		.073	√	√	√	√
14	Inspect and clean water column sight glass (or replace).		.127	√	√	√	√
15	Clean fire side of water jacket boiler.		.433					√
16	Check condition of flue pipe, damper and exhaust stack.		.147		√	√	√
17	Check boiler operation through complete cycle, up to 30 minutes.		.650					√
18	Check fuel level with gauge pole, add as required.		.046	√	√	√	√
19	Clean area around boiler.		.066	√	√	√	√
20	Fill out maintenance checklist and report deficiencies.		.022	√	√	√	√
Total labor-hours/period					.710	1.069	1.069	3.641
Total labor-hours/year					5.678	2.138	1.069	3.641

			Cost Each					
			2019 Bare Costs				Total	Total
Description		Labor-hrs.	Material	Labor	Equip.	Total	In-House	w/O&P

vi

This section also provides forms for listing all the organization's facilities and a separate detailed form for the specifics of each facility. Checklists are provided for major building components. Each checklist should be accompanied by a blank audit form. This form is used to record the audit findings, prioritize the deficiencies, and estimate the cost to remedy these deficiencies.

Reference Section

This section includes information on Equipment Rental Costs, Crew Listings, Travel Costs, Reference Tables, Life Cycle Costing, and a listing of abbreviations.

Equipment Rental Costs: This section contains the average costs to rent and operate hundreds of pieces of construction equipment.

Crew Listings: This section lists all the crews referenced. For the purposes of this data set, a crew is composed of more than one trade classification and/or the addition of power equipment to any trade classification. Power equipment is included in the cost of the crew. Costs are shown both with bare labor rates and with the installing contractor's overhead and profit added. For each, the total crew cost per eight-hour day and the composite cost per labor-hour are listed.

Travel Costs: This chart provides labor-hour costs for round-trip travel between a base of operation and the project site.

Reference Tables: In this section, you'll find reference tables, explanations, and estimating information that support how we develop the unit price data, technical data, and estimating procedures.

Life Cycle Costing: This section provides definitions and basic equations for performing life cycle analyses. A sample problem is solved to demonstrate the methodology.

Abbreviations: A listing of the abbreviations used throughout this data set, along with the terms they represent, is included in this section.

The Design Assumptions of this Data Set

This data set is designed to be as easy to use as possible. To that end, we have made certain assumptions and limited its scope.

1. Unlike any other data set, *Facilities Maintenance & Repair Costs with RSMeans data* was designed for estimating a wide range of maintenance tasks in diverse environments. Due to this diversity, the level of accuracy of the data is +/− 20%.

2. We have established material prices based on a national average.

3. We have computed labor costs based on a 30-city national average of union wage rates.

4. Except where major equipment or component replacement is described, the projects in this data set are small. Therefore, material prices and labor-hours have been enhanced to reflect the increased costs of small-scale work.

5. The PM frequencies are based on non-critical applications. Increased frequencies are required for critical environments.

General Maintenance

01 93 Facility Maintenance

01 93 04 — Landscaping Maintenance

01 93 04.15 Edging	Crew	Daily Output	Labor-Hours	Unit	Material	2019 Bare Costs Labor	Equipment	Total	Total In-House	Total Incl O&P
0010 **EDGING**										
0020 Hand edging, at walks	1 Clam	16	.500	C.L.F.		15.40		15.40	20.00	25
0030 At planting, mulch or stone beds		7	1.143			35		35	46.00	56.50
0040 Power edging, at walks		88	.091			2.80		2.80	3.65	4.51
0050 At planting, mulch or stone beds	▼	24	.333	▼		10.25		10.25	13.35	16.55
01 93 04.20 Flower, Shrub and Tree Care										
0010 **FLOWER, SHRUB & TREE CARE**										
0020 Flower or shrub beds, bark mulch, 3" deep hand spreader	1 Clam	100	.080	S.Y.	3.55	2.46		6.01	7.10	8.40
0030 Peat moss, 1" deep hand spreader		900	.009		4.88	.27		5.15	5.75	6.55
0040 Wood chips, 2" deep hand spreader		220	.036	▼	1.63	1.12		2.75	3.25	3.85
0050 Cleaning		1	8	M.S.F.		246		246	320.00	395
0060 Fertilizing, dry granular, 3#/M.S.F.		85	.094		1.14	2.90		4.04	5.00	6.10
0070 Weeding, mulched bed		20	.400			12.30		12.30	16.05	19.85
0080 Unmulched bed		8	1	▼		31		31	40.00	49.50
0090 Trees, pruning from ground, 1-1/2" caliper		84	.095	Ea.		2.93		2.93	3.82	4.73
0100 2" caliper		70	.114			3.52		3.52	4.58	5.65
0110 2-1/2" caliper		50	.160			4.93		4.93	6.40	7.95
0120 3" caliper	▼	30	.267			8.20		8.20	10.70	13.25
0130 4" caliper	2 Clam	21	.762			23.50		23.50	30.50	38
0140 6" caliper		12	1.333			41		41	53.50	66
0150 9" caliper		7.50	2.133			65.50		65.50	85.50	106
0160 12" caliper	▼	6.50	2.462			76		76	98.50	122
0170 Fertilize, slow release tablets	1 Clam	100	.080		2.32	2.46		4.78	5.75	6.85
0180 Pest control, spray		24	.333		.27	10.25		37.25	42.00	50.50

How to Use the Cost Data: The Details

What's Behind the Numbers? The Development of Cost Data

RSMeans data engineers continually monitor developments in the construction industry in order to ensure reliable, thorough, and up-to-date cost information. While overall construction costs may vary relative to general economic conditions, price fluctuations within the industry are dependent upon many factors. Individual price variations may, in fact, be opposite to overall economic trends. Therefore, costs are constantly tracked and complete updates are performed yearly. Also, new items are frequently added in response to changes in materials and methods.

Costs in U.S. Dollars

All costs represent U.S. national averages and are given in U.S. dollars. The City Cost Index (CCI) with RSMeans data can be used to adjust costs to a particular location. The CCI for Canada can be used to adjust U.S. national averages to local costs in Canadian dollars. No exchange rate conversion is necessary because it has already been factored in.

G The processes or products identified by the green symbol in our publications have been determined to be environmentally responsible and/or resource-efficient solely by RSMeans data engineering staff. The inclusion of the green symbol does not represent compliance with any specific industry association or standard.

Material Costs

RSMeans data engineers contact manufacturers, dealers, distributors, and contractors all across the U.S. and Canada to determine national average material costs. If you have access to current material costs for your specific location, you may wish to make adjustments to reflect differences from the national average. Included within material costs are fasteners for a normal installation. RSMeans data engineers use manufacturers' recommendations, written specifications, and/or standard construction practices for the sizing and spacing of fasteners. Adjustments to material costs may be required for your specific application or location. The manufacturer's warranty is assumed. Extended warranties are not included in the material costs. **Material costs do not include sales tax.**

Labor Costs

Labor costs are based upon a mathematical average of trade-specific wages in 30 major U.S. cities. The type of wage (union, open shop, or residential) is identified on the inside back cover of printed publications or selected by the estimator when using the electronic products. Markups for the wages can also be found on the inside back cover of printed publications and/or under the labor references found in the electronic products.

- If wage rates in your area vary from those used, or if rate increases are expected within a given year, labor costs should be adjusted accordingly.

Labor costs reflect productivity based on actual working conditions. In addition to actual installation, these figures include time spent during a normal weekday on tasks, such as material receiving and handling, mobilization at the site, site movement, breaks, and cleanup.

Productivity data is developed over an extended period so as not to be influenced by abnormal variations and reflects a typical average.

Equipment Costs

Equipment costs include not only rental but also operating costs for equipment under normal use. The operating costs include parts and labor for routine servicing, such as the repair and replacement of pumps, filters, and worn lines. Normal operating expendables, such as fuel, lubricants, tires, and electricity (where applicable), are also included. Extraordinary operating expendables with highly variable wear patterns, such as diamond bits and blades, are excluded. These costs are included under materials. Equipment rental rates are obtained from industry sources throughout North America—contractors, suppliers, dealers, manufacturers, and distributors.

Rental rates can also be treated as reimbursement costs for contractor-owned equipment. Owned equipment costs include depreciation, loan payments, interest, taxes, insurance, storage, and major repairs.

Equipment costs do not include operators' wages.

Equipment Cost/Day—The cost of equipment required for each crew is included in the Crew Listings in the Reference Section (small tools that are considered essential everyday tools are not listed out separately). The Crew Listings itemize specialized tools and heavy equipment along with labor trades. The daily cost of itemized equipment included in a crew is based on dividing the weekly bare rental rate by 5 (number of working days per week), then adding the hourly operating cost times 8 (the number of hours per day). This Equipment Cost/Day is shown in the last column of the Equipment Rental Costs in the Reference Section.

Mobilization, Demobilization—The cost to move construction equipment from an equipment yard or rental company to the job site and back again is not included in equipment costs. Mobilization (to the site) and demobilization (from the site) costs can be found in the Unit Price Section. If a piece of equipment is already at the job site, it is not appropriate to utilize mobilization or demobilization costs again in an estimate.

Overhead and Profit

Total Cost including O&P for the installing contractor is shown in the last column of the Unit Price and/or Assemblies. This figure is the sum of the bare material cost plus 10% for profit, the bare labor cost plus total overhead and profit, and the bare equipment cost plus 10% for profit. Details for the calculation of overhead and profit on labor are shown on the inside back cover of the printed product and in the Reference Section of the electronic product.

General Conditions

Cost data in this data set are presented in two ways: Bare Costs and Total Cost including O&P (Overhead and Profit). General Conditions, or General Requirements, of the contract should also be added to the Total Cost including O&P when applicable. Costs for General Conditions are listed in Division 1 of the Unit Price Section and in the Reference Section.

General Conditions for the installing contractor may range from 0% to 10% of the Total Cost including O&P. For the general or prime contractor, costs for General Conditions may range from 5% to 15% of the Total Cost including O&P, with a figure of 10% as the most typical allowance. If applicable, the Assemblies and Models sections use costs that include the installing contractor's overhead and profit (O&P).

Factors Affecting Costs

Costs can vary depending upon a number of variables. Here's a listing of some factors that affect costs and points to consider.

Quality—The prices for materials and the workmanship upon which productivity is based represent sound construction work. They are also in line with industry standard and manufacturer specifications and are frequently used by federal, state, and local governments.

Overtime—We have made no allowance for overtime. If you anticipate premium time or work beyond normal working hours, be sure to make an appropriate adjustment to your labor costs.

Productivity—The productivity, daily output, and labor-hour figures for each line item are based on an eight-hour work day in daylight hours in moderate temperatures and up to a 14' working height unless otherwise indicated. For work that extends beyond normal work hours or is performed under adverse conditions, productivity may decrease.

Size of Project—The size, scope of work, and type of construction project will have a significant impact on cost. Economies of scale can reduce costs for large projects. Unit costs can often run higher for small projects.

Location—Material prices are for metropolitan areas. However, in dense urban areas, traffic and site storage limitations may increase costs. Beyond a 20-mile radius of metropolitan areas, extra trucking or transportation charges may also increase the material costs slightly. On the other hand, lower wage rates may be in effect. Be sure to consider both of these factors when preparing an estimate, particularly if the job site is located in a central city or remote rural location. In addition, highly specialized subcontract items may require travel and per-diem expenses for mechanics.

Other Factors—

- season of year
- contractor management
- weather conditions
- local union restrictions
- building code requirements
- availability of:
  - adequate energy
  - skilled labor
  - building materials
- owner's special requirements/restrictions
- safety requirements
- environmental considerations
- access

Unpredictable Factors—General business conditions influence "in-place" costs of all items. Substitute materials and construction methods may have to be employed. These may affect the installed cost and/or life cycle costs. Such factors may be difficult to evaluate and cannot necessarily be predicted on the basis of the job's location in a particular section of the country. Thus, where these factors apply, you may find significant but unavoidable cost variations for which you will have to apply a measure of judgment to your estimate.

Rounding of Costs

In printed publications only, all unit prices in excess of $5.00 have been rounded to make them easier to use and still maintain adequate precision of the results.

How Subcontracted Items Affect Costs

A considerable portion of all large construction jobs is usually subcontracted. In fact, the percentage done by subcontractors is constantly increasing and may run over 90%. Since the workers employed by these companies do nothing else but install their particular products, they soon become experts in that line. As a result, installation by these firms is accomplished so efficiently that the total in-place cost, even with the general contractor's overhead and profit, is no more, and often less, than if the principal contractor had handled the installation. Companies that deal with construction specialties are anxious to have their products perform well and, consequently, the installation will be the best possible.

Contingencies

The allowance for contingencies generally provides for unforeseen construction difficulties. On alterations or repair jobs, 20% is not too much. If drawings are final and only field contingencies are being considered, 2% or 3% is probably sufficient and often nothing needs to be added. Contractually, changes in plans will be covered by extras. The contractor should consider inflationary price trends and possible material shortages during the course of the job. These escalation factors are dependent upon both economic conditions and the anticipated time between the estimate and actual construction. If drawings are not complete or approved, or a budget cost is wanted, it is wise to add 5% to 10%. Contingencies, then, are a matter of judgment.

Important Estimating Considerations

The productivity, or daily output, of each craftsman or crew assumes a well-managed job where tradesmen with the proper tools and equipment, along with the appropriate construction materials, are present. Included are daily set-up and cleanup time, break time, and plan layout time. Unless otherwise indicated, time for material movement on site (for items that can be transported by hand) of up to 200' into the building and to the first or second floor is also included. If material has to be transported by other means, over greater distances, or to higher floors, an additional allowance should be considered by the estimator.

While horizontal movement is typically a sole function of distances, vertical transport introduces other variables that can significantly impact productivity. In an occupied building, the use of elevators (assuming access, size, and required protective measures are acceptable) must be understood at the time of the estimate. For new construction, hoist wait and cycle times can easily be 15 minutes and may result in scheduled access extending beyond the normal work day. Finally, all vertical transport will impose strict weight limits likely to preclude the use of any motorized material handling.

The productivity, or daily output, also assumes installation that meets manufacturer/designer/ standard specifications. A time allowance for quality control checks, minor adjustments, and any task required to ensure proper function or operation is also included. For items that require connections to services, time is included for positioning, leveling, securing the unit, and making all the necessary connections (and start up where applicable) to ensure a complete installation. Estimating of the services themselves (electrical, plumbing, water, steam, hydraulics, dust collection, etc.) is separate.

In some cases, the estimator must consider the use of a crane and an appropriate crew for the installation of large or heavy items. For those situations where a crane is not included in the assigned crew and as part of the line item cost,

then equipment rental costs, mobilization and demobilization costs, and operator and support personnel costs must be considered.

Labor-Hours

The labor-hours expressed in this publication are derived by dividing the total daily labor-hours for the crew by the daily output. Based on average installation time and the assumptions listed above, the labor-hours include: direct labor, indirect labor, and nonproductive time. A typical day for a craftsman might include but is not limited to:

- Direct Work
  - Measuring and layout
  - Preparing materials
  - Actual installation
  - Quality assurance/quality control
- Indirect Work
  - Reading plans or specifications
  - Preparing space
  - Receiving materials
  - Material movement
  - Giving or receiving instruction
  - Miscellaneous
- Non-Work
  - Chatting
  - Personal issues
  - Breaks
  - Interruptions (i.e., sickness, weather, material or equipment shortages, etc.)

If any of the items for a typical day do not apply to the particular work or project situation, the estimator should make any necessary adjustments.

Final Checklist

Estimating can be a straightforward process provided you remember the basics. Here's a checklist of some of the steps you should remember to complete before finalizing your estimate.

Did you remember to:

- factor in the City Cost Index for your locale?
- take into consideration which items have been marked up and by how much?
- mark up the entire estimate sufficiently for your purposes?
- read the background information on techniques and technical matters that could impact your project time span and cost?
- include all components of your project in the final estimate?
- double check your figures for accuracy?
- call RSMeans data engineers if you have any questions about your estimate or the data you've used? Remember, Gordian stands behind all of our products, including our extensive RSMeans data solutions. If you have any questions about your estimate, about the costs you've used from our data, or even about the technical aspects of the job that may affect your estimate, feel free to call the Gordian RSMeans editors at 1.800.448.8182.

Included with Purchase

RSMeans Data Online
Free Trial

rsmeans.com/2019freetrial

Maintenance & Repair

Table of Contents

Table of Contents (cont.)

Table of Contents (cont.)

How to Use the Maintenance & Repair Assemblies Cost Tables

The following is a detailed explanation of a sample Maintenance & Repair Assemblies Cost Table. Next to each bold number that follows is the described item with the appropriate component of the sample entry following in parentheses.

Total system costs, as well as the individual component costs, are shown. In most cases, the intent is for the user to apply the total system costs.

However, changes and adjustments to the components or partial use of selected components is also appropriate. In particular, selected equipment system tables in the mechanical section include complete listings of operations that are meant to be chosen from rather than used in total.

B30 ROOFING — B3013 — Roof Covering

B3013 105 ❶ Built-Up Roofing

System Description	Freq. (Years)	Crew	Unit	Labor Hours	2019 Bare Costs Material	2019 Bare Costs Labor	2019 Bare Costs Equipment	2019 Bare Costs Total ❼	Total In-House ❽	Total w/O&P ❾
0600 Place new BUR membrane over existing	20	G-5	Sq.							
Set up, secure and take down ladder				.020		.90		.90	1.31	1.59
Sweep / spud ballast clean				.500		22.50		22.50	33	40
Vent existing membrane				.130		5.85		5.85	8.50	10.35
Cut out buckled flashing				.024		1.08		1.08	1.58	1.92
Install 2 ply membrane flashing				.027	.42	1.20		1.62	2.21	2.65
Install 4 ply bituminous roofing				3.733	145	157	34	336	425	495
Reinstall ballast				.381		17.15		17.15	25	30.50
Clean up				.390		17.60		17.60	25.50	31
Total				5.205	145.42	223.28	34	402.70	**522.10**	**613.01**
0700 Total BUR roof replacement ❷	28 ❸	G-1 ❹	Sq. ❺							
Set up, secure and take down ladder				.020		.90		.90	1.31	1.59
Sweep / spud ballast clean				.500		22.50		22.50	33	40
Remove built-up roofing				2.500		104		104	135	167
Remove insulation board				1.026		43		43	55	69
Remove flashing				.024		1.08		1.08	1.58	1.92
Install 2" perlite insulation				.879	110	40		150	179	208
Install 2 ply membrane flashing				.027	.42	1.20		1.62	2.21	2.65
Install 4 ply bituminous membrane				2.800	145	118	25.50	288.50	360	420
Clean up				1.000		45		45	65.50	79.50
Total ❻				8.776	255.42	375.68	25.50	656.60	**832.60**	**989.66**

B3013 120 Modified Bituminous / Thermoplastic

System Description	Freq. (Years)	Crew	Unit	Labor Hours	2019 Bare Costs Material	2019 Bare Costs Labor	2019 Bare Costs Equipment	2019 Bare Costs Total	Total In-House	Total w/O&P
0100 Debris removal by hand & visual inspection	1	2 ROFC	M.S.F.							
Set up, secure and take down ladder				.052		2.34		2.34	3.41	4.14
Pick up trash / debris & clean up				.327		14.70		14.70	21.50	26
Visual inspection				.327		14.70		14.70	21.50	26
Total				.705		31.74		31.74	**46.41**	**56.14**
0200 Non - destructive moisture inspection	5	2 ROFC	M.S.F.							
Set up, secure and take down ladder				.052		2.34		2.34	3.41	4.14
Infrared inspection of roof membrane				2.133		96		96	140	170
Total				2.185		98.34		98.34	**143.41**	**174.14**

❶ **System/Line Numbers (B3013 105 0700)**
Each Maintenance & Repair Assembly has been assigned a unique identification number based on the UNIFORMAT II classification system.

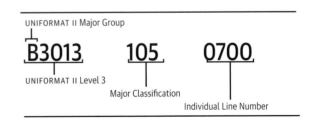

UNIFORMAT II Major Group
B3013 **105** **0700**
UNIFORMAT II Level 3
Major Classification
Individual Line Number

② System Description

A one-line description of each system is given, followed by a description of each component that makes up the system. In selected items, a percentage factor for planning annual costs is also given.

③ Frequency (28)

The projected frequency of occurrence in years for each system is listed. For example, **28** indicates that this operation should be budgeted for once every 28 years. If the frequency is less than 1, then the operation should be planned for more than once a year. For example, a value of **.5** indicates a planned occurrence of 2 times a year.

④ Crew (G-1)

The "Crew" column designates the typical trade or crew used for the task, although other trades may be used for minor portions of the work. If the task can be accomplished by one trade and requires no power equipment, that trade and the number of workers are listed (for example, "4 Rofc"). If the task requires a composite crew, a crew code designation is listed (for example, "G-1"). You'll find full details on all composite crews in the Crew Listings.

- For a complete list of all trades utilized in this data set, as well as their abbreviations, see the inside back cover of the data set.

Crews - Maintenance

Crew No.	Bare Costs		In-House Costs		Incl. Subs O&P		Cost Per Labor-Hour		
Crew G-1	Hr.	Daily	Hr.	Daily	Hr.	Daily	Bare Costs	In House	Incl. O&P
1 Roofer Foreman (outside)	$47.05	$ 376.40	$68.50	$ 548.00	$83.10	$ 664.80	$42.14	$61.36	$74.44
4 Roofers Composition	45.05	1441.60	65.60	2099.20	79.60	2547.20			
2 Roofer Helpers	33.85	541.60	49.30	788.80	59.80	956.80			
1 Application Equipment		181.15		181.15		199.26			
1 Tar Kettle/Pot		178.20		178.20		196.02			
1 Crew Truck		154.35		154.35		169.79	9.17	9.17	10.09
56 L.H., Daily Totals		$2873.30		$3949.70		$4733.87	$51.31	$70.53	$84.53

⑤ Unit of Measure for Each System (Sq.)

The abbreviated designation indicates the unit of measure, as defined by industry standards, upon which the price of the component is based. In this example, "Cost per Square" is the unit of measure for this system or "assembly."

⑥ Labor-Hours (8.776)

The "Labor-Hours" figure represents the number of labor-hours required to perform one unit of work. To calculate the number of labor-hours required for a particular task, multiply the quantity of the item by the number of labor-hours shown. For example:

Quantity	X	Productivity Rate	=	Duration
30 Sq.	X	8.766 Labor-Hours/Sq.	=	263 Labor-Hours

⑦ Bare Costs:
Material (255.42)

This figure for the unit material cost for the line item is the "bare" material cost with no overhead and profit allowances included. *Costs shown reflect national average material prices for the current year and generally include delivery to the facility. Small purchases may require an additional delivery charge. No sales taxes are included.*

Labor (375.68)

The unit labor costs are derived by multiplying bare labor-hour costs for Crew G-1 and other trades by labor-hour units. In this case, the bare labor-hour cost for the G-1 crew is found in the Crew Section. (If a trade is listed or used for minor portions of the work, the hourly labor cost—the wage rate—is found on the inside back cover of this data set).

Equipment (25.50)

Equipment costs for each crew are listed in the description of each crew. The unit equipment cost is derived by multiplying the bare equipment hourly cost by the labor-hour units.

Total (656.60)

The total of the bare costs is the arithmetic total of the three previous columns: material, labor, and equipment.

Material	+	Labor	+	Equipment	=	Total
$255.42	+	$375.68	+	$25.50	=	$656.60

⑧ Total In-House Costs (832.60)

"Total In-House Costs" include suggested markups to bare costs for the direct overhead requirements of in-house maintenance staff. The figure in this column is the sum of three components: the bare material cost plus 10% (for purchasing and handling small quantities); the bare labor cost plus workers' compensation and average fixed overhead (per the labor rate table on the inside back cover of this data set or, if a crew is listed, from the Crew Listings); and the bare equipment cost.

⑨ Total Costs Including O&P (989.66)

"Total Costs Including O&P" include suggested markups to bare costs for an outside subcontractor's overhead and profit requirements. The figure in this column is the sum of three components: the bare material cost plus 25% for profit; the bare labor cost plus total overhead and profit (per the labor rate table on the inside back cover of the printed product or the Reference Section of the electronic product, or, if a crew is listed, from the Crew Listings); and the bare equipment cost plus 10% for profit.

A1033 110   Concrete, Unfinished

	System Description	Freq. (Years)	Crew	Unit	Labor Hours	2019 Bare Costs				Total In-House	Total w/O&P
						Material	Labor	Equipment	Total		
0010	**Minor repairs to concrete floor**	15	1 CEFI	S.F.							
	Repair spall				.612		25.50	3.59	29.09	37	45
	Repair concrete flooring				.061	3.08	2.99		6.07	7.20	8.55
	Total				.673	3.08	28.49	3.59	35.16	44.20	53.55
0020	**Replace unfinished concrete floor**	75	2 CEFI	C.S.F.							
	Break up concrete slab				1.664		74	22.16	96.16	120	143
	Load into truck				1.950		84	34.50	118.50	147	173
	Rebar #3 to #7				.139	10.75	7.60		18.35	21.50	25.50
	Edge form				.693	.80	34		34.80	45	56
	Slab on grade 4″ in place with finish				1.577	120	73.50	.75	194.25	227	267
	Expansion joint				.003	.05	.13		.18	.23	.28
	Cure concrete				.109	6.30	4.50		10.80	12.80	15.15
	Total				6.135	137.90	277.73	57.41	473.04	573.53	679.93

For customer support on your Facilities Maintenance & Repair Costs with RSMeans data, call 800.448.8182.

7

B10 SUPERSTRUCTURE — B1013 — Floor Construction

B1013 510 — Exterior Concrete Stairs

	System Description	Freq. (Years)	Crew	Unit	Labor Hours	2019 Bare Costs				Total In-House	Total w/O&P
						Material	Labor	Equipment	Total		
1010	**Repair exterior concrete steps**	30	1 CEFI	S.F.							
	Repair spalls in steps				.104	13.10	5.10		18.20	21	24.50
	Total				.104	13.10	5.10		18.20	**21**	**24.50**
1020	**Replace exterior concrete steps**	75	1 CEFI	S.F.							
	Remove concrete				.520		21.50	3.05	24.55	31.50	38.50
	Install forms				.328	.87	16.50		17.37	22.50	27.50
	Pour concrete				.567	3.12	24	.63	27.75	35.50	43.50
	Total				1.416	3.99	62	3.68	69.67	**89.50**	**109.50**
1030	**Refinish exterior concrete steps**	3	1 PORD	S.F.							
	Prepare surface				.016	.08	.69		.69	.89	1.11
	Refinish surface, brushwork, 1 coat				.013	.08	.54		.62	.79	.96
	Total				.029	.08	1.23		1.31	**1.68**	**2.07**
2030	**Refinish exterior concrete landing**	3	1 PORD	S.F.							
	Prepare surface				.016	.08	.69		.69	.89	1.11
	Refinish surface, brushwork, 1 coat				.013	.08	.54		.62	.79	.96
	Total				.029	.08	1.23		1.31	**1.68**	**2.07**

B1013 520 — Exterior Masonry Steps

	System Description	Freq. (Years)	Crew	Unit	Labor Hours	2019 Bare Costs				Total In-House	Total w/O&P
						Material	Labor	Equipment	Total		
1010	**Repair exterior masonry steps**	4	1 BRIC	S.F.							
	Remove damaged masonry				.074		3.05		3.05	3.97	4.92
	Clean area				.019	.05	.95		1	1.31	1.60
	Replace with new masonry				.236	3.53	11.05		14.58	18.45	22.50
	Total				.329	3.58	15.05		18.63	**23.73**	**29.02**

For customer support on your Facilities Maintenance & Repair Costs with RSMeans data, call 800.448.8182.

B10 SUPERSTRUCTURE | B1013 | Floor Construction

B1013 530 Exterior Wood Stairs

	System Description	Freq. (Years)	Crew	Unit	Labor Hours	2019 Bare Costs				Total In-House	Total w/O&P
						Material	Labor	Equipment	Total		
1030	**Refinish exterior wood steps**	3	1 PORD	S.F.							
	Prepare surface				.016		.69		.69	.89	1.11
	Wash / dry surface				.001		.05		.05	.07	.08
	Refinish surface, brushwork, 1 coat				.013	.08	.54		.62	.79	.96
	Total				**.030**	**.08**	**1.28**		**1.36**	**1.75**	**2.15**
1040	**Replace exterior wood steps**	30	2 CARP	S.F.							
	Remove stairs (to 6' wide)				.100		4.10		4.10	5.35	6.65
	New stringers (to 5 risers)				.048	.83	2.48		3.31	4.14	5.05
	Risers				.083	.76	4.30		5.06	6.45	7.90
	Treads				.123	2.10	6.35		8.45	10.60	12.90
	Total				**.354**	**3.69**	**17.23**		**20.92**	**26.54**	**32.50**
2030	**Refinish exterior wood landing**	3	1 PORD	S.F.							
	Prepare surface				.016		.69		.69	.89	1.11
	Wash / dry surface				.001		.05		.05	.07	.08
	Refinish surface, brushwork, 1 coat				.013	.08	.54		.62	.79	.96
	Total				**.030**	**.08**	**1.28**		**1.36**	**1.75**	**2.15**

B1013 540 Exterior Wood Railing

	System Description	Freq. (Years)	Crew	Unit	Labor Hours	2019 Bare Costs				Total In-House	Total w/O&P
						Material	Labor	Equipment	Total		
1010	**Replace exterior wood balustrade**	20	1 CARP	L.F.							
	Remove damaged balusters and handrail				.087		3.56		3.56	4.64	5.75
	Install new balusters and handrail				.473	24	24.50		48.50	58	69.50
	Total				**.559**	**24**	**28.06**		**52.06**	**62.64**	**75.25**
1030	**Refinish exterior wood balustrade**	7	1 PORD	L.F.							
	Prepare balusters and handrails for painting				.052		2.25		2.25	2.90	3.59
	Wash / dry surface				.001		.05		.05	.07	.08
	Paint balusters and handrails, brushwork, primer + 1 coat				.089	.90	3.84		4.74	5.95	7.30
	Total				**.142**	**.90**	**6.14**		**7.04**	**8.92**	**10.97**

For customer support on your Facilities Maintenance & Repair Costs with RSMeans data, call 800.448.8182.

9

B1013 550 | Exterior Metal Stairs

	System Description	Freq. (Years)	Crew	Unit	Labor Hours	2019 Bare Costs Material	2019 Bare Costs Labor	2019 Bare Costs Equipment	2019 Bare Costs Total	Total In-House	Total w/O&P
1010	**Repair exterior metal steps**	15	1 SSWK	S.F.							
	Remove damaged stair tread				.099		5.55		5.55	7.50	9.30
	New stair tread				.104	48.05	5.86		53.91	60.50	70
	Total				**.203**	**48.05**	**11.41**		**59.46**	**68**	**79.30**
1040	**Replace exterior metal steps**	40	2 SSWK	S.F.							
	Remove stairs				.168		9.45		9.45	12.80	15.70
	New stairs				.239	153.01	13.30	7.29	173.60	194	221
	Total				**.408**	**153.01**	**22.75**	**7.29**	**183.05**	**206.80**	**236.70**
1050	**Refinish exterior metal steps**	3	1 PORD	S.F.							
	Prepare metal railing for painting				.019		.83		.83	1.06	1.32
	Wash / dry surface				.001		.05		.05	.07	.08
	Paint metal steps, brushwork, primer + 1 coat				.015	.11	.67		.78	.98	1.20
	Total				**.036**	**.11**	**1.55**		**1.66**	**2.11**	**2.60**
2030	**Refinish exterior metal landing**	3	1 PORD	S.F.							
	Prepare metal railing for painting				.019		.83		.83	1.06	1.32
	Wash / dry surface				.001		.05		.05	.07	.08
	Paint metal landing, brushwork, primer + 1 coat				.015	.11	.67		.78	.98	1.20
	Total				**.036**	**.11**	**1.55**		**1.66**	**2.11**	**2.60**

B1013 560 | Exterior Metal Railing

	System Description	Freq. (Years)	Crew	Unit	Labor Hours	2019 Bare Costs Material	2019 Bare Costs Labor	2019 Bare Costs Equipment	2019 Bare Costs Total	Total In-House	Total w/O&P
1010	**Replace exterior metal hand rail**	30	1 SSWK	L.F.							
	Remove damaged railings				.087		3.56		3.56	4.64	5.75
	Install new metal railings				.163	37	9.20	.49	46.69	53.50	62
	Total				**.250**	**37**	**12.76**	**.49**	**50.25**	**58.14**	**67.75**
1030	**Refinish exterior metal hand rail**	7	1 PORD	L.F.							
	Prepare metal railing for painting				.019		.83		.83	1.06	1.32
	Wash / dry surface				.001		.05		.05	.07	.08
	Paint metal railing, brushwork, primer + 1 coat				.015	.11	.67		.78	.98	1.20
	Total				**.036**	**.11**	**1.55**		**1.66**	**2.11**	**2.60**

For customer support on your Facilities Maintenance & Repair Costs with RSMeans data, call 800.448.8182.

B1013 570 — Exterior Wrought Iron Balustrade

	System Description	Freq. (Years)	Crew	Unit	Labor Hours	2019 Bare Costs				Total In-House	Total w/O&P
						Material	Labor	Equipment	Total		
1010	**Repair exterior wrought iron balustrade**	20	1 SSWK	L.F.							
	Cut / remove damaged section				.325		13.35		13.35	17.40	21.50
	Reset new section				.867	95.50	48.50		144	171	200
	Total				1.192	95.50	61.85		157.35	188.40	221.50
1030	**Refinish exterior wrought iron balustrade**	7	1 PORD	L.F.							
	Prepare metal balustrade for painting				.019		.83		.83	1.06	1.32
	Wash / dry surface				.001		.05		.05	.07	.08
	Paint metal balustrade, brushwork, primer + 1 coat				.015	.11	.67		.78	.98	1.20
	Total				.036	.11	1.55		1.66	2.11	2.60

B1013 580 — Exterior Metal Fire Escapes

	System Description	Freq. (Years)	Crew	Unit	Labor Hours	2019 Bare Costs				Total In-House	Total w/O&P
						Material	Labor	Equipment	Total		
1010	**Refinish exterior fire escape balcony, 2' wide**	7	1 PORD	L.F.							
	Prepare metal fire escape balcony for painting				.100		4.33		4.33	5.55	6.90
	Brush on primer coat				.100	2.57	4.32		6.89	8.40	10.10
	Brush on finish coat				.100	2.28	4.32		6.60	8.10	9.75
	Total				.300	4.85	12.97		17.82	22.05	26.75
1020	**Replace exterior fire escape balcony, 2' wide**	25	2 SSWK	L.F.							
	Remove metal fire escape balcony				.500		28		28	38	46.50
	Install new metal fire escape balcony				2.000	60	112		172	218	262
	Total				2.500	60	140		200	256	308.50
2010	**Refinish exterior fire escape stair and platform**	7	1 PORD	Flight							
	Prepare fire escape stair and platform for painting				2.400		104		104	134	166
	Brush on primer coat				2.400	61.56	103.68		165.24	201	243
	Brush on finish coat				2.400	54.72	103.68		158.40	194	234
	Total				7.200	116.28	311.36		427.64	529	643
2020	**Replace exterior fire escape stair and platform**	25	2 SSWK	Flight							
	Remove metal fire escape stair and platform				14.995		840		840	1,150	1,400
	Install new metal fire escape stair and platform				59.925	1,075	3,350		4,425	5,725	6,950
	Total				74.920	1,075	4,190		5,265	6,875	8,350

B1013 580   Exterior Metal Fire Escapes

System Description	Freq. (Years)	Crew	Unit	Labor Hours	2019 Bare Costs				Total In-House	Total w/O&P
					Material	Labor	Equipment	Total		
3010 Refinish exterior fire escape ladder	7	1 PORD	V.L.F.							
Prepare metal fire escape ladder for painting				.020		.87		.87	1.11	1.38
Brush on primer coat				.020	.34	.87		1.21	1.49	1.81
Brush on finish coat				.020	.30	.87		1.17	1.44	1.76
Total				.060	.64	2.61		3.25	4.04	4.95
3020 Replace exterior fire escape ladder	25	2 SSWK	V.L.F.							
Remove metal fire escape ladder				.250		14		14	19	23.50
Install new metal fire escape ladder				.500	41.50	28		69.50	83.50	98.50
Total				.750	41.50	42		83.50	102.50	122

B1013 910   Exterior Steel Decking

System Description	Freq. (Years)	Crew	Unit	Labor Hours	2019 Bare Costs				Total In-House	Total w/O&P
					Material	Labor	Equipment	Total		
0020 Replace exterior steel decking	30	2 SSWK	S.F.							
Remove decking				.011	2.78	.66	.14	3.58	4.11	4.73
Assemble and weld decking				.011	2.78	.66	.14	3.58	4.11	4.73
Total				.023	5.56	1.32	.28	7.16	8.22	9.46

B1013 920   Exterior Metal Grating

System Description	Freq. (Years)	Crew	Unit	Labor Hours	2019 Bare Costs				Total In-House	Total w/O&P
					Material	Labor	Equipment	Total		
0010 Repair exterior metal floor grating - (2% of grating)	10	1 SSWK	S.F.							
Misc. welding of floor grating				.348	1.17	20	4.16	25.33	33	39.50
Total				.348	1.17	20	4.16	25.33	33	39.50
0020 Replace exterior metal floor grating	30	2 SSWK	S.F.							
Remove grating				.011	2.78	.66	.14	3.58	4.11	4.73
Assemble and weld grating				.059	19.75	3.36	.18	23.29	26.50	30.50
Total				.071	22.53	4.02	.32	26.87	30.61	35.23

System Description	Freq. (Years)	Crew	Unit	Labor Hours	Material	Labor	Equipment	Total	Total In-House	Total w/O&P
					2019 Bare Costs					
2010 Replace precast concrete coping, 12" wide	50	2 BRIC	L.F.							
Set up and secure scaffold, 3 floors to roof				.429		22.17		22.17	29	36
Remove coping				.050		2.05		2.05	2.68	3.31
Install new precast concrete coping				.229	26	10.45		36.45	42.50	49.50
Remove scaffold				.429		22.17		22.17	29	36
Total				1.137	26	56.84		82.84	103.18	124.81
2020 Replace limestone coping, 12" wide	50	2 BRIC	L.F.							
Set up and secure scaffold, 3 floors to roof				.429		22.17		22.17	29	36
Remove coping				.050		2.05		2.05	2.68	3.31
Install new limestone coping				.178	16.80	8.15		24.95	29	34.50
Remove scaffold				.429		22.17		22.17	29	36
Total				1.086	16.80	54.54		71.34	89.68	109.81
2030 Replace marble coping, 12" wide	50	2 BRIC	L.F.							
Set up and secure scaffold, 3 floors to roof				.429		22.17		22.17	29	36
Remove coping				.050		2.05		2.05	2.68	3.31
Install new marble coping				.200	19.35	9.15		28.50	33.50	39
Remove scaffold				.429		22.17		22.17	29	36
Total				1.108	19.35	55.54		74.89	94.18	114.31
2040 Replace terra cotta coping, 12" wide	50	2 BRIC	L.F.							
Set up and secure scaffold, 3 floors to roof				.429		22.17		22.17	29	36
Remove coping				.050		2.05		2.05	2.68	3.31
Install new terra cotta coping				.200	8.70	9.15		17.85	21.50	26
Remove scaffold				.429		22.17		22.17	29	36
Total				1.108	8.70	55.54		64.24	82.18	101.31

For customer support on your Facilities Maintenance & Repair Costs with RSMeans data, call 800.448.8182.

13

B2013 109   Concrete Block

System Description	Freq. (Years)	Crew	Unit	Labor Hours	2019 Bare Costs				Total In-House	Total w/O&P
					Material	Labor	Equipment	Total		
1010 Repair 8" concrete block wall - 1st floor	25	1 BRIC	S.F.							
Set up, secure and take down ladder				.130		6.65		6.65	8.75	10.80
Remove damaged block				.183		7.52		7.52	9.80	12.10
Wire truss reinforcing				.005	.29	.27		.56	.66	.79
Install block				.111	2.79	5.20		7.99	9.90	11.95
Total				.429	3.08	19.64		22.72	**29.11**	**35.64**
1020 Replace 8" concrete block wall - 1st floor	60	2 BRIC	C.S.F.							
Set up and secure scaffold				.750		38.75		38.75	50.50	62.50
Remove damaged block				4.571		188		188	245	305
Wire truss reinforcing				.260	14.25	13.25		27.50	33	39.50
Install block				11.111	279	520		799	990	1,200
Remove scaffold				.750		38.75		38.75	50.50	62.50
Total				17.442	293.25	798.75		1,092	**1,369**	**1,669.50**
1030 Waterproof concrete block wall - 1st floor	10	1 ROFC	C.S.F.							
Steam clean				2.971		123	17	140	179	218
Spray surface, silicone				.260	40	12		52	61	71
Total				3.231	40	135	17	192	**240**	**289**
2020 Replace 8" concrete block wall - 2nd floor	60	2 BRIC	C.S.F.							
Set up and secure scaffold				1.500		77.50		77.50	101	125
Remove block				4.571		188		188	245	305
Wire truss reinforcing				.260	14.25	13.25		27.50	33	39.50
Install block				11.111	279	520		799	990	1,200
Remove scaffold				1.500		77.50		77.50	101	125
Total				18.942	293.25	876.25		1,169.50	**1,470**	**1,794.50**
2030 Waterproof concrete block wall - 2nd floor	10	G-8	C.S.F.							
Position aerial lift truck, raise & lower platform				.250		11.25	21	32.25	39.50	43
Steam clean				11.765		483	110	593	750	900
Spray surface, silicone				.260	40	12	22	74	85	95
Total				12.275	40	506.25	153	699.25	**874.50**	**1,038**

B2013 109 Concrete Block

System Description	Freq. (Years)	Crew	Unit	Labor Hours	2019 Bare Costs				Total In-House	Total w/O&P
					Material	Labor	Equipment	Total		
3020 Replace 8" concrete block wall - 3rd floor	60	2 BRIC	C.S.F.							
Set up and secure scaffold				1.500		77.25		77.25	101	125
Remove damaged block				4.571		188		188	245	305
Wire truss reinforcing				.260	14.25	13.25		27.50	33	39.50
Install block				11.111	279	520		799	990	1,200
Remove scaffold				1.500		77.25		77.25	101	125
Total				18.942	293.25	875.75		1,169	1,470	1,794.50
3030 Waterproof concrete block wall - 3rd floor	10	G-8	C.S.F.							
Position aerial lift truck, raise & lower platform				.300		13.50	25	38.50	47.50	51.50
Steam clean				11.765		483	110	593	750	900
Spray surface, silicone				.260	40	12	22	74	85	95
Total				12.325	40	508.50	157	705.50	882.50	1,046.50

B2013 114 Concrete Block Wall, Painted

System Description	Freq. (Years)	Crew	Unit	Labor Hours	2019 Bare Costs				Total In-House	Total w/O&P
					Material	Labor	Equipment	Total		
1030 Point & refinish block wall - 1st floor	25	1 BRIC	C.S.F.							
Set up and secure scaffold				.750		38.75		38.75	50.50	62.50
Steam clean				2.600		107		107	139	172
Clean & point block				3.924		200		200	264	325
Paint surface, brushwork, 1 coat				1.040	19.50	45.50		65	80.50	96
Remove scaffold				.750		38.75		38.75	50.50	62.50
Total				9.064	19.50	430		449.50	584.50	718
2030 Point & refinish block wall - 2nd floor	25	1 BRIC	C.S.F.							
Set up and secure scaffold				1.500		77.50		77.50	101	125
Steam clean				2.600		107		107	139	172
Clean & point block				3.924		200		200	264	325
Paint surface, brushwork, 1 coat				1.040	19.50	45.50		65	80.50	96
Remove scaffold				1.500		77.50		77.50	101	125
Total				10.564	19.50	507.50		527	685.50	843

For customer support on your Facilities Maintenance & Repair Costs with RSMeans data, call 800.448.8182.

15

B2013 114 Concrete Block Wall, Painted

	System Description	Freq. (Years)	Crew	Unit	Labor Hours	2019 Bare Costs Material	Labor	Equipment	Total	Total In-House	Total w/O&P
3020	**Point & refinish block wall - 3rd floor**	25	1 BRIC	C.S.F.							
	Set up and secure scaffold				2.250		116.25		116.25	151	188
	Steam clean				2.600		107		107	139	172
	Clean & point block				3.924		200		200	264	325
	Paint surface, brushwork, 1 coat				1.040	19.50	45.50		65	80.50	96
	Remove scaffold				2.250		116.25		116.25	151	188
	Total				12.064	19.50	585		604.50	785.50	969

B2013 118 Terra Cotta

	System Description	Freq. (Years)	Crew	Unit	Labor Hours	2019 Bare Costs Material	Labor	Equipment	Total	Total In-House	Total w/O&P
1020	**Replace terra cotta wall - 1st floor**	50	2 BRIC	C.S.F.							
	Set up and secure scaffold				.750		38.75		38.75	50.50	62.50
	Remove terra cotta				1.040		43	6.10	49.10	63	76
	Replace terra cotta				10.400	565	487		1,052	1,275	1,500
	Remove scaffold				.750		38.75		38.75	50.50	62.50
	Total				12.940	565	607.50	6.10	1,178.60	1,439	1,701
2010	**Replace terra cotta wall - 2nd floor**	50	2 BRIC	C.S.F.							
	Set up and secure scaffold				1.500		77.50		77.50	101	125
	Remove terra cotta				1.040		43	6.10	49.10	63	76
	Replace terra cotta				10.400	565	487		1,052	1,275	1,500
	Remove scaffold				1.500		77.50		77.50	101	125
	Total				14.440	565	685	6.10	1,256.10	1,540	1,826
2020	**Replace terra cotta wall - 3rd floor**	50	2 BRIC	C.S.F.							
	Set up and secure scaffold				2.250		116.25		116.25	151	188
	Remove terra cotta				1.040		43	6.10	49.10	63	76
	Replace terra cotta				10.400	565	487		1,052	1,275	1,500
	Remove scaffold				2.250		116.25		116.25	151	188
	Total				15.940	565	762.50	6.10	1,333.60	1,640	1,952

For customer support on your Facilities Maintenance & Repair Costs with RSMeans data, call 800.448.8182.

B20 EXTERIOR CLOSURE — B2013 — Exterior Walls

B2013 119 — Clay Brick

	System Description	Freq. (Years)	Crew	Unit	2019 Bare Costs					Total In-House	Total w/O&P
					Labor Hours	Material	Labor	Equipment	Total		
1010	**Repair brick wall - 1st floor**										
	Set up, secure and take down ladder	25	1 BRIC	S.F.	.130		6.65		6.65	8.75	10.80
	Remove damage				.451		18.46		18.46	24	30
	Install brick				.236	3.53	11.05		14.58	18.45	22.50
	Total				.817	3.53	36.16		39.69	51.20	63.30
1020	**Point brick wall - 1st floor**										
	Set up and secure scaffold	25	2 BRIC	C.S.F.	1.333		57.50		57.50	74.50	92
	Steam clean				2.971		123	17	140	179	218
	Clean & point brick				10.000	4	510		514	675	835
	Remove scaffold				1.333		57.50		57.50	74.50	92
	Total				15.638	4	748	17	769	1,003	1,237
1030	**Waterproof brick wall - 1st floor**										
	Steam clean	10	1 ROFC	C.S.F.	2.971		123	17	140	179	218
	Spray surface, silicone				.260	40	12		52	61	71
	Total				3.231	40	135	17	192	240	289
2010	**Replace brick wall - 2nd floor**										
	Set up and secure scaffold	75	2 BRIC	C.S.F.	1.500		77.50		77.50	101	125
	Remove damage				10.401		426		426	555	690
	Install brick				23.636	353	1,105		1,458	1,850	2,250
	Remove scaffold				1.500		77.50		77.50	101	125
	Total				37.037	353	1,686		2,039	2,607	3,190
2020	**Point brick wall - 2nd floor**										
	Set up and secure scaffold	25	2 BRIC	C.S.F.	1.500		77.50		77.50	101	125
	Steam clean				2.971		123	17	140	179	218
	Clean & point brick				10.000	4	510		514	675	835
	Remove scaffold				1.500		77.50		77.50	101	125
	Total				15.971	4	788	17	809	1,056	1,303
2030	**Waterproof brick wall - 2nd floor**										
	Position aerial lift truck, raise & lower platform	10	G-8	C.S.F.	.250		11.25	21	32.25	39.50	43
	Steam clean				11.765		483	110	593	750	900
	Spray surface, silicone				.260	40	12	22	74	85	95
	Total				12.275	40	506.25	153	699.25	874.50	1,038

For customer support on your Facilities Maintenance & Repair Costs with RSMeans data, call 800.448.8182.

17

B2013 119 | Clay Brick

System Description	Freq. (Years)	Crew	Unit	Labor Hours	2019 Bare Costs				Total In-House	Total w/O&P
					Material	Labor	Equipment	Total		
3010 Replace brick wall - 3rd floor	75	2 BRIC	C.S.F.							
Set up and secure scaffold				1.500		77.25		77.25	101	125
Remove damage				10.401		426		426	555	690
Install brick				23.636	353	1,105		1,458	1,850	2,250
Remove scaffold				1.500		77.25		77.25	101	125
Total				37.037	353	1,685.50		2,038.50	2,607	3,190
3020 Point brick wall - 3rd floor	25	2 BRIC	C.S.F.							
Set up and secure scaffold				1.500		77.25		77.25	101	125
Steam clean				2.971		123	17	140	179	218
Clean & point brick				10.000	4	510		514	675	835
Remove scaffold				1.500		77.25		77.25	101	125
Total				15.971	4	787.50	17	808.50	1,056	1,303
3030 Waterproof brick wall - 3rd floor	10	G-8	C.S.F.							
Position aerial lift truck, raise & lower platform				.300		13.50	25	38.50	47.50	51.50
Steam clean				11.765		483	110	593	750	900
Spray surface, silicone				.260	40	12	22	74	85	95
Total				12.325	40	508.50	157	705.50	882.50	1,046.50

B2013 121 | Clay Brick, Painted

System Description	Freq. (Years)	Crew	Unit	Labor Hours	2019 Bare Costs				Total In-House	Total w/O&P
					Material	Labor	Equipment	Total		
1010 Repair painted brick wall - 1st floor	25	1 BRIC	S.F.							
Set up, secure and take down ladder				.130		6.65		6.65	8.75	10.80
Remove damage				.451		18.46		18.46	24	30
Install brick				.236	3.53	11.05		14.58	18.45	22.50
Paint surface, brushwork, 1 coat				.010	.20	.46		.66	.81	.96
Total				.827	3.73	36.62		40.35	52.01	64.26

For customer support on your Facilities Maintenance & Repair Costs with RSMeans data, call 800.448.8182.

B20 EXTERIOR CLOSURE   B2013   Exterior Walls

B2013 121   Clay Brick, Painted

System Description	Freq. (Years)	Crew	Unit	Labor Hours	2019 Bare Costs				Total In-House	Total w/O&P
					Material	Labor	Equipment	Total		
1030 Point brick wall - 1st floor	25	1 BRIC	C.S.F.							
Set up and secure scaffold				.500		25.75		25.75	33.50	42
Steam clean				2.971		123	17	140	179	218
Clean & point brick				10.000	4	510		514	675	835
Paint surface, brushwork, 1 coat				1.040	19.50	45.50		65	80.50	96
Remove scaffold				.500		25.75		25.75	33.50	42
Total				15.011	23.50	730	17	770.50	1,001.50	1,233
2010 Replace brick wall - 2nd floor	75	1 BRIC	C.S.F.							
Set up and secure scaffold				1.500		77.50		77.50	101	125
Remove damage				10.401		426		426	555	690
Install brick				23.636	353	1,105		1,458	1,850	2,250
Paint surface, brushwork, 1 coat				1.040	19.50	45.50		65	80.50	96
Remove scaffold				1.500		77.50		77.50	101	125
Total				38.077	372.50	1,731.50		2,104	2,687.50	3,286
2020 Point and refinish brick wall - 2nd floor	25	1 BRIC	C.S.F.							
Set up and secure scaffold				1.500		77.50		77.50	101	125
Steam clean				2.971		123	17	140	179	218
Clean & point brick				10.000	4	510		514	675	835
Paint surface, brushwork, 1 coat				1.040	19.50	45.50		65	80.50	96
Remove scaffold				1.500		77.50		77.50	101	125
Total				17.011	23.50	833.50	17	874	1,136.50	1,399
3010 Replace brick wall - 3rd floor	75	1 BRIC	C.S.F.							
Set up and secure scaffold				1.500		77.25		77.25	101	125
Remove damage				10.401		426		426	555	690
Install brick				23.636	353	1,105		1,458	1,850	2,250
Paint surface, brushwork, 1 coat				.010	.20	.46		.66	.81	.96
Remove scaffold				1.500		77.25		77.25	101	125
Total				37.047	353.20	1,685.96		2,039.16	2,607.81	3,190.96

For customer support on your Facilities Maintenance & Repair Costs with RSMeans data, call 800.448.8182.

19

B2013 121 Clay Brick, Painted

	System Description	Freq. (Years)	Crew	Unit	Labor Hours	2019 Bare Costs Material	2019 Bare Costs Labor	2019 Bare Costs Equipment	2019 Bare Costs Total	Total In-House	Total w/O&P
3020	**Point and refinish brick wall - 3rd floor**	25	1 BRIC	C.S.F.							
	Set up and secure scaffold				1.500		77.25		77.25	101	125
	Steam clean				2.971		123		140	179	218
	Clean & point brick				10.000	4	510	17	514	675	835
	Paint surface, brushwork, 1 coat				1.040	19.50	45.50		65	80.50	96
	Remove scaffold				1.500		77.25		77.25	101	125
	Total				17.011	23.50	833	17	373.50	1,136.50	1,399

B2013 123 Field Stone

	System Description	Freq. (Years)	Crew	Unit	Labor Hours	2019 Bare Costs Material	2019 Bare Costs Labor	2019 Bare Costs Equipment	2019 Bare Costs Total	Total In-House	Total w/O&P
1010	**Repair field stone wall - 1st floor**	25	1 BRIC	S.F.							
	Set up, secure and take down ladder				.390		16.25		16.25	21	26
	Remove damaged stones				.173		7.20	1.02	8.22	10.50	12.70
	Reset stones				.416	21	21	4.83	46.83	56	65.50
	Total				.979	21	44.45	5.85	71.30	87.50	104.20
1020	**Point stone wall, 8" x 8" stones - 1st floor**	25	2 BRIC	C.S.F.							
	Set up and secure scaffold				1.333		57.50		57.50	74.50	92
	Steam clean				2.971		123		140	179	218
	Clean & point stonework joints				5.714	74	292		366	465	570
	Remove scaffold				1.333		57.50		57.50	74.50	92
	Total				11.352	74	530	17	621	793	972
2020	**Point stone wall, 8" x 8" stones - 2nd floor**	25	2 BRIC	C.S.F.							
	Set up and secure scaffold				1.500		77.50		77.50	101	125
	Steam clean				2.971		123		140	179	218
	Clean & point stonework joints				5.714	74	292		366	465	570
	Remove scaffold				1.500		77.50		77.50	101	125
	Total				11.685	74	570	17	661	846	1,038

For customer support on your Facilities Maintenance & Repair Costs with RSMeans data, call 800.448.8182.

B2013 123 Field Stone

System Description	Freq. (Years)	Crew	Unit	Labor Hours	2019 Bare Costs				Total In-House	Total w/O&P
					Material	Labor	Equipment	Total		
3020 Point stone wall, 8" x 8" stones - 3rd floor	25	2 BRIC	C.S.F.							
Set up and secure scaffold				1.500		77.25		77.25	101	125
Steam clean				2.971		123	17	140	179	218
Clean & point stonework joints				5.714	74	292		366	465	570
Remove scaffold				1.500		77.25		77.25	101	125
Total				11.685	74	569.50	17	660.50	846	1,038

B2013 140 Glass Block

System Description	Freq. (Years)	Crew	Unit	Labor Hours	2019 Bare Costs				Total In-House	Total w/O&P
					Material	Labor	Equipment	Total		
1020 Replace glass block wall - 1st floor	75	2 BRIC	C.S.F.							
Set up and secure scaffold				.750		38.75		38.75	50.50	62.50
Remove blocks				5.200		216	31	247	315	380
Replace blocks				45.217	2,350	2,100		4,450	5,375	6,400
Remove scaffold				.750		38.75		38.75	50.50	62.50
Total				51.917	2,350	2,393.50	31	4,774.50	5,791	6,905
2020 Replace glass block wall - 2nd floor	75	2 BRIC	C.S.F.							
Set up and secure scaffold				1.500		77.50		77.50	101	125
Remove blocks				5.200		216	31	247	315	380
Replace blocks				45.217	2,350	2,100		4,450	5,375	6,400
Remove scaffold				1.500		77.50		77.50	101	125
Total				53.417	2,350	2,471	31	4,852	5,892	7,030
3020 Replace glass block wall - 3rd floor	75	2 BRIC	C.S.F.							
Set up and secure scaffold				2.250		116.25		116.25	151	188
Remove blocks				5.200		216	31	247	315	380
Replace blocks				45.217	2,350	2,100		4,450	5,375	6,400
Remove scaffold				2.250		116.25		116.25	151	188
Total				54.917	2,350	2,548.50	31	4,929.50	5,992	7,156

For customer support on your Facilities Maintenance & Repair Costs with RSMeans data, call 800.448.8182.

21

B2013 141 Aluminum Siding

System Description	Freq. (Years)	Crew	Unit	Labor Hours	2019 Bare Costs				Total In-House	Total w/O&P
					Material	Labor	Equipment	Total		
1010 Replace aluminum siding - 1st floor	35	1 CARP	C.S.F.							
Set up, secure and take down ladder				.520		26.80		26.80	35	43.50
Remove damaged siding				1.434		59		59	77	95
Install siding				5.368	185	274		459	555	665
Total				7.322	185	359.80		544.80	667	803.50
1020 Refinish aluminum siding - 1st floor	20	1 PORD	C.S.F.							
Set up, secure and take down ladder				1.300		56		56	72.50	90
Prepare surface				1.159		50		50	65	80
Refinish surface, brushwork, primer + 1 coat				1.231	13	53		66	83	101
Total				3.690	13	159		172	220.50	271
2010 Replace aluminum siding - 2nd floor	35	1 CARP	C.S.F.							
Set up and secure scaffold				1.500		77.50		77.50	101	125
Remove damaged siding				1.434		59		59	77	95
Install siding				5.368	185	274		459	555	665
Remove scaffold				1.500		77.50		77.50	101	125
Total				9.802	185	488		673	834	1,010
2020 Refinish aluminum siding - 2nd floor	20	1 PORD	C.S.F.							
Set up and secure scaffold				1.500		77.50		77.50	101	125
Prepare surface				1.159		50		50	65	80
Refinish surface, brushwork, primer + 1 coat				1.231	13	53		66	83	101
Remove scaffold				1.500		77.50		77.50	101	125
Total				5.390	13	258		271	350	431
3010 Replace aluminum siding - 3rd floor	35	1 CARP	C.S.F.							
Set up and secure scaffold				2.250		116.25		116.25	151	188
Remove damaged siding				1.434		59		59	77	95
Install siding				5.368	185	274		459	555	665
Remove scaffold				2.250		116.25		116.25	151	188
Total				11.302	185	565.50		750.50	934	1,136

For customer support on your Facilities Maintenance & Repair Costs with RSMeans data, call 800.448.8182.

	System Description	Freq. (Years)	Crew	Unit	Labor Hours	2019 Bare Costs				Total In-House	Total w/O&P
						Material	Labor	Equipment	Total		
3020	**Refinish aluminum siding - 3rd floor**	20	1 PORD	C.S.F.							
	Set up and secure scaffold				2.250		116.25		116.25	151	188
	Prepare surface				1.159		50		50	65	80
	Refinish surface, brushwork, primer + 1 coat				1.231	13	53		66	83	101
	Remove scaffold				2.250		116.25		116.25	151	188
	Total				6.890	13	335.50		348.50	450	557
4120	**Spray refinish aluminum siding - 1st floor**	20	1 PORD	C.S.F.							
	Set up, secure and take down ladder				1.300		56		56	72.50	90
	Prepare surface				1.159		50		50	65	80
	Protect/mask unpainted surfaces				.031	.30	1.30		1.60	2	2.48
	Spray full primer coat				.444	9	20		29	38	45.50
	Spray finish coat				.500	9	22		31	42	50.50
	Total				3.434	18.30	149.30		167.60	219.50	268.48
4220	**Spray refinish aluminum siding - 2nd floor**	20	2 PORD	C.S.F.							
	Set up and secure scaffold				1.500		77.50		77.50	101	125
	Prepare surface				1.159		50		50	65	80
	Protect/mask unpainted surfaces				.031	.30	1.30		1.60	2	2.48
	Spray full primer coat				.444	9	20		29	38	45.50
	Spray finish coat				.500	9	22		31	42	50.50
	Remove scaffold				1.500		77.50		77.50	101	125
	Total				5.134	18.30	248.30		266.60	349	428.48
4320	**Spray refinish aluminum siding - 3rd floor**	20	2 PORD	C.S.F.							
	Set up and secure scaffold				2.250		116.25		116.25	151	188
	Prepare surface				1.159		50		50	65	80
	Protect/mask unpainted surfaces				.031	.30	1.30		1.60	2	2.48
	Spray full primer coat				.444	9	20		29	38	45.50
	Spray finish coat				.500	9	22		31	42	50.50
	Remove scaffold				2.250		116.25		116.25	151	188
	Total				6.634	18.30	325.80		344.10	449	554.48

B2013 142 | Steel Siding

System Description	Freq. (Years)	Crew	Unit	Labor Hours	2019 Bare Costs				Total In-House	Total w/O&P
					Material	Labor	Equipment	Total		
1010 Replace steel siding - 1st floor	35	1 CARP	C.S.F.							
Set up, secure and take down ladder				.520		26.80		26.30	35	43.50
Remove damaged siding				1.434		59		59	77	95
Install siding				5.266	175	269		444	535	645
Total				7.220	175	354.80		529.30	647	783.50
1020 Refinish steel siding - 1st floor	20	1 PORD	C.S.F.							
Set up, secure and take down ladder				1.300		56		56	72.50	90
Prepare surface				1.159		50		50	65	80
Refinish surface, brushwork, primer + 1 coat				1.231	13	53		66	83	101
Total				3.690	13	159		172	220.50	271
2010 Replace steel siding - 2nd floor	35	1 CARP	C.S.F.							
Set up and secure scaffold				1.500		77.50		77.50	101	125
Remove damaged siding				1.434		59		59	77	95
Install siding				5.266	175	269		444	535	645
Remove scaffold				1.500		77.50		77.50	101	125
Total				9.700	175	483		658	814	990
2020 Refinish steel siding - 2nd floor	20	1 PORD	C.S.F.							
Set up and secure scaffold				1.500		77.50		77.50	101	125
Prepare surface				1.159		50		50	65	80
Refinish surface, brushwork, primer + 1 coat				1.231	13	53		66	83	101
Remove scaffold				1.500		77.50		77.50	101	125
Total				5.390	13	258		271	350	431
3010 Replace steel siding - 3rd floor	35	1 CARP	C.S.F.							
Set up and secure scaffold				2.250		116.25		116.25	151	188
Remove damaged siding				1.434		59		59	77	95
Install siding				5.266	175	269		444	535	645
Remove scaffold				2.250		116.25		116.25	151	188
Total				11.200	175	560.50		735.50	914	1,116

For customer support on your Facilities Maintenance & Repair Costs with RSMeans data, call 800.448.8182.

B2013 142 Steel Siding

	System Description	Freq. (Years)	Crew	Unit	Labor Hours	Material	Labor	Equipment	Total	Total In-House	Total w/O&P
							2019 Bare Costs				
3020	**Refinish steel siding - 3rd floor**	20	1 PORD	C.S.F.							
	Set up and secure scaffold				2.250		116.25		116.25	151	188
	Prepare surface				1.159		50		50	65	80
	Refinish surface, brushwork, primer + 1 coat				1.231	13	53		66	83	101
	Remove scaffold				2.250		116.25		116.25	151	188
	Total				6.890	13	335.50		348.50	450	557
4120	**Spray refinish steel siding - 1st floor**	20	1 PORD	C.S.F.							
	Set up, secure and take down ladder				1.300		56		56	72.50	90
	Prepare surface				1.159		50		50	65	80
	Protect/mask unpainted surfaces				.031	.30	1.30		1.60	2	2.48
	Spray full primer coat				.444	9	20		29	38	45.50
	Spray finish coat				.500	9	22		31	42	50.50
	Total				3.434	18.30	149.30		167.60	219.50	268.48
4220	**Spray refinish steel siding - 2nd floor**	20	2 PORD	C.S.F.							
	Set up and secure scaffold				1.500		77.50		77.50	101	125
	Prepare surface				1.159		50		50	65	80
	Protect/mask unpainted surfaces				.031	.30	1.30		1.60	2	2.48
	Spray full primer coat				.444	9	20		29	38	45.50
	Spray finish coat				.500	9	22		31	42	50.50
	Remove scaffold				1.500		77.50		77.50	101	125
	Total				5.134	18.30	248.30		266.60	349	428.48
4320	**Spray refinish steel siding - 3rd floor**	20	2 PORD	C.S.F.							
	Set up and secure scaffold				2.250		116.25		116.25	151	188
	Prepare surface				1.159		50		50	65	80
	Protect/mask unpainted surfaces				.031	.30	1.30		1.60	2	2.48
	Spray full primer coat				.444	9	20		29	38	45.50
	Spray finish coat				.500	9	22		31	42	50.50
	Remove scaffold				2.250		116.25		116.25	151	188
	Total				6.634	18.30	325.80		344.10	449	554.48

For customer support on your Facilities Maintenance & Repair Costs with RSMeans data, call 800.448.8182.

25

B2013 143 Masonite Panel, Sealed

	System Description	Freq. (Years)	Crew	Unit	Labor Hours	2019 Bare Costs Material	2019 Bare Costs Labor	2019 Bare Costs Equipment	2019 Bare Costs Total	Total In-House	Total w/O&P
1010	**Replace hardboard panels - 1st floor**	12	1 CARP	C.S.F.							
	Set up, secure and take down ladder				.520		26.80		26.80	35	43.50
	Remove damaged panel				1.434		59		59	77	95
	Cut new hardboard panel to fit				2.773	268	143		411	480	565
	Total				4.727	268	228.80		496.80	592	703.50
2010	**Replace hardboard panels - 2nd floor**	12	1 CARP	C.S.F.							
	Set up and secure scaffold				1.500		77.50		77.50	101	125
	Remove damaged panel				1.434		59		59	77	95
	Cut new hardboard panel to fit				2.773	268	143		411	480	565
	Remove scaffold				1.500		77.50		77.50	101	125
	Total				7.207	268	357		625	759	910
3010	**Replace hardboard panels - 3rd floor**	12	1 CARP	C.S.F.							
	Set up and secure scaffold				2.250		116.25		116.25	151	188
	Remove damaged panel				1.434		59		59	77	95
	Cut new hardboard panel to fit				2.773	268	143		411	480	565
	Remove scaffold				2.250		116.25		116.25	151	188
	Total				8.707	268	434.50		702.50	859	1,036

B2013 145 Wood, Clapboard Finished, 1 Coat

	System Description	Freq. (Years)	Crew	Unit	Labor Hours	2019 Bare Costs Material	2019 Bare Costs Labor	2019 Bare Costs Equipment	2019 Bare Costs Total	Total In-House	Total w/O&P
1010	**Replace & finish wood clapboards - 1st floor**	25	1 CARP	C.S.F.							
	Set up, secure and take down ladder				.520		26.80		26.80	35	43.50
	Remove damage				2.600		134		134	175	217
	Replace wood clapboards				2.711	340.77	140		480.77	555	650
	Refinish surface, brushwork, primer + 1 coat				2.000	21	87		108	134	164
	Total				7.831	361.77	387.80		749.57	899	1,074.50

B2013 145 Wood, Clapboard Finished, 1 Coat

	System Description	Freq. (Years)	Crew	Unit	Labor Hours	2019 Bare Costs				Total In-House	Total w/O&P
						Material	Labor	Equipment	Total		
2010	**Replace & finish wood clapboards - 2nd floor**	25	1 CARP	C.S.F.							
	Set up and secure scaffold				1.500		77.50		77.50	101	125
	Remove damage				2.600		134		134	175	217
	Replace wood clapboards				3.524	443	182		625	725	850
	Refinish surface, brushwork, primer + 1 coat				2.000	21	87		108	134	164
	Remove scaffold				1.500		77.50		77.50	101	125
	Total				11.124	464	558		1,022	1,236	1,481
3010	**Replace & finish wood clapboards - 3rd floor**	25	1 CARP	C.S.F.							
	Set up and secure scaffold				2.250		116.25		116.25	151	188
	Remove damage				2.600		134		134	175	217
	Replace wood clapboards				3.489	438.61	180.20		618.81	715	840
	Refinish surface, brushwork, primer + 1 coat				2.000	21	87		108	134	164
	Remove scaffold				2.250		116.25		116.25	151	188
	Total				12.589	459.61	633.70		1,093.31	1,326	1,597

B2013 147 Wood Shingles, Unfinished

	System Description	Freq. (Years)	Crew	Unit	Labor Hours	2019 Bare Costs				Total In-House	Total w/O&P
						Material	Labor	Equipment	Total		
1010	**Repair wood shingles - 1st floor**	12	1 CARP	S.F.							
	Set up, secure and take down ladder				.052		2.68		2.68	3.50	4.34
	Remove damaged shingles				.039		1.61		1.61	2.09	2.58
	Replace shingles				.056	2.59	2.90		5.49	6.65	7.90
	Total				.147	2.59	7.19		9.78	12.24	14.82
1020	**Replace wood shingles - 1st floor**	40	2 CARP	C.S.F.							
	Set up and secure scaffold				.750		38.75		38.75	50.50	62.50
	Remove damaged shingles				2.600		107		107	139	172
	Replace shingles				4.324	199	223		422	510	610
	Remove scaffold				.750		38.75		38.75	50.50	62.50
	Total				8.424	199	407.50		606.50	750	907

For customer support on your Facilities Maintenance & Repair Costs with RSMeans data, call 800.448.8182.

27

B2013 147 Wood Shingles, Unfinished

	System Description	Freq. (Years)	Crew	Unit	Labor Hours	2019 Bare Costs Material	Labor	Equipment	Total	Total In-House	Total w/O&P
2010	**Replace wood shingles - 2nd floor**	40	1 CARP	C.S.F.							
	Set up and secure scaffold				1.000		51.50		51.50	67.50	83.50
	Remove damaged shingles				2.600		107		107	139	172
	Replace shingles				4.324	199	223		422	510	610
	Remove scaffold				1.000		51.50		51.50	67.50	83.50
	Total				8.924	199	433		632	784	949
3010	**Replace wood shingles - 3rd floor**	40	1 CARP	C.S.F.							
	Set up and secure scaffold				1.500		77.25		77.25	101	125
	Remove damaged shingles				2.600		107		107	139	172
	Replace shingles				4.324	199	223		422	510	610
	Remove scaffold				1.500		77.25		77.25	101	125
	Total				9.924	199	484.50		683.50	851	1,032

B2013 148 Wood Shingles, Finished

	System Description	Freq. (Years)	Crew	Unit	Labor Hours	2019 Bare Costs Material	Labor	Equipment	Total	Total In-House	Total w/O&P
1020	**Replace & finish wood shingles - 1st floor**	45	1 CARP	C.S.F.							
	Set up, secure and take down ladder				1.500		77.50		77.50	101	125
	Remove damage				2.600		107		107	139	172
	Replace shingles				4.324	199	223		422	510	610
	Refinish surface, brushwork, primer + 1 coat				2.000	21	87		108	134	164
	Total				10.424	220	494.50		714.50	884	1,071
1030	**Refinish wood shingles - 1st floor**	5	1 PORD	C.S.F.							
	Set up and secure scaffold				.750		38.75		38.75	50.50	62.50
	Prepare surface				1.667		72		72	93	115
	Refinish surface, brushwork, primer + 1 coat				2.000	21	87		108	134	164
	Remove scaffold				.750		38.75		38.75	50.50	62.50
	Total				5.167	21	236.50		257.50	328	404

For customer support on your Facilities Maintenance & Repair Costs with RSMeans data, call 800.448.8182.

B2013 148 | **Wood Shingles, Finished**

	System Description	Freq. (Years)	Crew	Unit	Labor Hours	Material	Labor	Equipment	Total	Total In-House	Total w/O&P
							2019 Bare Costs				
2020	**Refinish wood shingles - 2nd floor**	5	1 PORD	C.S.F.							
	Set up and secure scaffold				1.500		77.50		77.50	101	125
	Prepare surface				1.667		72		72	93	115
	Refinish surface, brushwork, primer + 1 coat				2.000	21	87		108	134	164
	Remove scaffold				1.500		77.50		77.50	101	125
	Total				6.667	21	314		335	429	529
3020	**Refinish wood shingles - 3rd floor**	5	1 PORD	C.S.F.							
	Set up and secure scaffold				2.250		116.25		116.25	151	188
	Prepare surface				1.667		72		72	93	115
	Refinish surface, brushwork, primer + 1 coat				2.000	21	87		108	134	164
	Remove scaffold				2.250		116.25		116.25	151	188
	Total				8.167	21	391.50		412.50	529	655
4120	**Spray refinish wood siding - 1st floor**	5	1 PORD	C.S.F.							
	Set up, secure and take down ladder				1.300		56		56	72.50	90
	Prepare surface				1.159		50		50	65	80
	Protect/mask unpainted surfaces				.031	.30	1.30		1.60	2	2.48
	Spray full primer coat				.410	12	18		30	36	43
	Spray finish coat				.410	17	18		35	42	49.50
	Total				3.310	29.30	143.30		172.60	217.50	264.98
4220	**Spray refinish wood siding - 2nd floor**	5	2 PORD	C.S.F.							
	Set up and secure scaffold				1.500		77.50		77.50	101	125
	Prepare surface				1.159		50		50	65	80
	Ptotect/mask unpainted surfaces				.031	.30	1.30		1.60	2	2.48
	Spray full primer coat				.410	12	18		30	36	43
	Spray finish coat				.410	17	18		35	42	49.50
	Remove scaffold				1.500		77.50		77.50	101	125
	Total				5.010	29.30	242.30		271.60	347	424.98

For customer support on your Facilities Maintenance & Repair Costs with RSMeans data, call 800.448.8182.

29

B2013 148 Wood Shingles, Finished

	System Description	Freq. (Years)	Crew	Unit	Labor Hours	Material	Labor	Equipment	Total	Total In-House	Total w/O&P
							2019 Bare Costs				
4320	**Spray refinish wood siding - 3rd floor**	5	2 PORD	C.S.F.							
	Set up and secure scaffold				2.250		116.25		116.25	151	188
	Prepare surface				1.159		50		50	65	80
	Protect/mask unpainted surfaces				.031	.30	1.30		1.60	2	2.48
	Spray full primer coat				.410	12	18		30	36	43
	Spray finish coat				.410	17	18		35	42	49.50
	Remove scaffold				2.250		116.25		116.25	151	188
	Total				6.510	29.30	319.80		349.10	447	550.98

B2013 150 Fiberglass Panel, Rigid

	System Description	Freq. (Years)	Crew	Unit	Labor Hours	Material	Labor	Equipment	Total	Total In-House	Total w/O&P
							2019 Bare Costs				
1010	**Replace fiberglass panels - 1st floor**	9	1 CARP	C.S.F.							
	Set up, secure and take down ladder				.520		26.80		26.30	35	43.50
	Remove damaged fiberglass panel				1.434		59		59	77	95
	Install new fiberglass panel				4.727	395	241		636	745	875
	Total				6.681	395	326.80		721.80	857	1,013.50
2010	**Replace fiberglass panels - 2nd floor**	9	1 CARP	C.S.F.							
	Set up and secure scaffold				1.500		77.50		77.50	101	125
	Remove damaged fiberglass panel				1.434		59		59	77	95
	Install new fiberglass panel				4.727	395	241		636	745	875
	Remove scaffold				1.500		77.50		77.50	101	125
	Total				9.161	395	455		850	1,024	1,220
3010	**Replace fiberglass panels - 3rd floor**	9	1 CARP	C.S.F.							
	Set up and secure scaffold				2.250		116.25		116.25	151	188
	Remove damaged fiberglass panel				1.434		59		59	77	95
	Install new fiberglass panel				4.727	395	241		636	745	875
	Remove scaffold				2.250		116.25		116.25	151	188
	Total				10.661	395	532.50		927.50	1,124	1,346

For customer support on your Facilities Maintenance & Repair Costs with RSMeans data, call 800.448.8182.

B20 EXTERIOR CLOSURE		B2013	Exterior Walls

B2013 152 Synthetic Veneer Plaster

	System Description	Freq. (Years)	Crew	Unit	Labor Hours	2019 Bare Costs				Total In-House	Total w/O&P
						Material	Labor	Equipment	Total		
1030	Refinish veneer plaster - 1st floor	10	1 BRIC	C.S.F.							
	Set up and secure scaffold				.750		38.75		38.75	50.50	62.50
	Pressure wash surface				.260		11	2	13	17	20
	Refinish surface				1.270	163	55		218	250	292
	Remove scaffold				.750		38.75		38.75	50.50	62.50
	Total				3.030	163	143.50	2	308.50	368	437
2020	Refinish veneer plaster - 2nd floor	10	1 BRIC	C.S.F.							
	Set up and secure scaffold				1.500		77.50		77.50	101	125
	Pressure wash surface				.260		11	2	13	17	20
	Refinish surface				1.270	163	55		218	250	292
	Remove scaffold				1.500		77.50		77.50	101	125
	Total				4.530	163	221	2	386	469	562
3020	Refinish veneer plaster - 3rd floor	10	1 BRIC	C.S.F.							
	Set up and secure scaffold				2.250		116.25		116.25	151	188
	Pressure wash surface				.260		11	2	13	17	20
	Refinish surface				1.270	163	55		218	250	292
	Remove scaffold				2.250		116.25		116.25	151	188
	Total				6.030	163	298.50	2	463.50	569	688

B2013 156 Exterior Soffit Gypsum Board, Painted

	System Description	Freq. (Years)	Crew	Unit	Labor Hours	2019 Bare Costs				Total In-House	Total w/O&P
						Material	Labor	Equipment	Total		
1030	Refinish exterior painted gypsum board - 1st floor soffit	5	1 PORD	C.S.F.							
	Set up and secure scaffold				.750		38.75		38.75	50.50	62.50
	Prepare surface				1.159		50		50	65	80
	Paint surface, ro ler, primer + 1 coat				1.231	12	53		65	82	100
	Remove scaffolc				.750		38.75		38.75	50.50	62.50
	Total				3.890	12	180.50		192.50	248	305

For customer support on your Facilities Maintenance & Repair Costs with RSMeans data, call 800.448.8182.

31

B2013 156 Exterior Soffit Gypsum Board, Painted

	System Description	Freq. (Years)	Crew	Unit	Labor Hours	2019 Bare Costs Material	2019 Bare Costs Labor	2019 Bare Costs Equipment	2019 Bare Costs Total	Total In-House	Total w/O&P
1040	**Replace exterior painted gypsum board - 1st floor soffit**	25	2 CARP	C.S.F.							
	Set up and secure scaffold				.750		38.75		38.75	50.50	62.50
	Remove old exterior gypsum board				2.737		112		112	146	181
	Install new exterior gypsum board				3.467	71	179		250	310	380
	Paint surface, roller, primer + 1 coat				1.231	12	53		65	82	100
	Remove scaffold				.750		38.75		38.75	50.50	62.50
	Total				8.935	83	421.50		504.50	639	786
2010	**Replace exterior painted gypsum board - 2nd floor soffit**	25	1 CARP	C.S.F.							
	Set up and secure scaffold				1.500		77.50		77.50	101	125
	Remove old exterior gypsum board				2.737		112		112	146	181
	Install new exterior gypsum board				3.467	71	179		250	310	380
	Paint surface, roller, primer + 1 coat				2.462	24	106		130	164	200
	Remove scaffold				1.500		77.50		77.50	101	125
	Total				11.666	95	552		647	822	1,011
2020	**Refinish exterior painted gypsum board - 2nd floor soffit**	5	1 PORD	C.S.F.							
	Set up and secure scaffold				1.500		77.50		77.50	101	125
	Prepare surface				1.159		50		50	65	80
	Paint surface, roller, primer + 1 coat				1.231	12	53		65	82	100
	Remove scaffold				1.500		77.50		77.50	101	125
	Total				5.390	12	258		270	349	430
3010	**Replace exterior painted gypsum board - 3rd floor soffit**	25	1 CARP	C.S.F.							
	Set up and secure scaffold				2.250		116.25		116.25	151	188
	Remove old exterior gypsum board				2.737		112		112	146	181
	Install new exterior gypsum board				3.467	71	179		250	310	380
	Paint surface, roller, primer + 1 coat				3.693	36	159		195	246	300
	Remove scaffold				2.250		116.25		116.25	151	188
	Total				14.397	107	682.50		789.50	1,004	1,237
3020	**Refinish exterior painted gypsum board - 3rd floor soffit**	5	1 PORD	C.S.F.							
	Set up and secure scaffold				2.250		116.25		116.25	151	188
	Prepare surface				1.159		50		50	65	80
	Paint surface, roller, primer + 1 coat				1.231	12	53		65	82	100
	Remove scaffold				2.250		116.25		116.25	151	188
	Total				6.890	12	335.50		347.50	449	556

For customer support on your Facilities Maintenance & Repair Costs with RSMeans data, call 800.448.8182.

B2013 157 — Overhang, Exterior Entry

	System Description	Freq. (Years)	Crew	Unit	Labor Hours	2019 Bare Costs Material	Labor	Equipment	Total	Total In-House	Total w/O&P
1030	**Refinish wood overhang**	5	1 PORD	S.F.							
	Prepare surface				.016		.69		.69	.89	1.11
	Wash / dry surface				.001		.05		.05	.07	.08
	Refinish surface, brushwork, primer + 1 coat				.020	.21	.87		1.08	1.34	1.64
	Total				.037	.21	1.61		1.82	2.30	2.83

B2013 158 — Recaulking

	System Description	Freq. (Years)	Crew	Unit	Labor Hours	2019 Bare Costs Material	Labor	Equipment	Total	Total In-House	Total w/O&P
1010	**Recaulk expansion & control joints**	20	1 BRIC	L.F.							
	Set up, secure and take down ladder				.130		6.65		6.65	8.75	10.80
	Cut out and recaulk, silicone				.067	.70	3.40		4.10	5.25	6.45
	Total				.197	.70	10.05		10.75	14	17.25
2010	**Recaulk door**	20	1 BRIC	L.F.							
	Set up, secure and take down ladder				.130		6.65		6.65	8.75	10.80
	Cut out and recaulk, silicone				.067	.70	3.40		4.10	5.25	6.45
	Total				.197	.70	10.05		10.75	14	17.25
3010	**Recaulk window, 4′ x 6′ - 1st floor**	20	2 BRIC	Ea.							
	Set up and secure scaffold				1.333		68		68	89.50	111
	Cut out and recaulk, silicone				1.600	16.80	81.60		98.40	126	154
	Remove scaffold				1.333		68		68	89.50	111
	Total				4.267	16.80	217.60		234.40	305	376
3020	**Recaulk window, 4′ x 6′ - 2nd floor**	20	2 BRIC	Ea.							
	Set up and secure scaffold				1.500		77.50		77.50	101	125
	Cut out and recaulk, silicone				1.600	16.80	81.60		98.40	126	154
	Remove scaffold				1.500		77.50		77.50	101	125
	Total				4.600	16.80	236.60		253.40	328	404
3030	**Recaulk window, 4′ x 6′ - 3rd floor**	20	2 BRIC	Ea.							
	Set up and secure scaffold				1.500		77.25		77.25	101	125
	Cut out and recaulk, silicone				1.600	16.80	81.60		98.40	126	154
	Remove scaffold				1.500		77.25		77.25	101	125
	Total				4.600	16.80	236.10		252.90	328	404

For customer support on your Facilities Maintenance & Repair Costs with RSMeans data, call 800.448.8182.

33

B2013 310 Wood Louvers and Shutters

System Description	Freq. (Years)	Crew	Unit	Labor Hours	2019 Bare Costs				Total In-House	Total w/O&P
					Material	Labor	Equipment	Total		
1010 Repair wood louver in frame 1st floor	6	1 CARP	Ea.							
Set up, secure and take down ladder				.001		.06		.06	.08	.10
Remove louver				.473		19.40		19.40	25.50	31.50
Remove damaged stile				.078		4.05		4.05	5.30	6.55
Remove damaged slat				.039		2		2	2.60	3.22
Install new slat				.077	.82	4		4.82	6.10	7.45
Install new stile				.157	2.37	8.10		10.47	13.15	16.05
Reinstall louver				.075		3.86		3.86	5.05	6.25
Total				.900	3.19	41.47		44.56	57.78	71.12
1030 Refinish wood louver - 1st floor	5	1 PORD	Ea.							
Set up and secure scaffold				.444		19.20		19.20	25	30.50
Prepare louver for painting				.434		18.72		18.72	24	30
Paint louvered surface (to 16 S.F.), brushwork, primer + 1 coat				.800	2.27	34.50		36.77	47	58.50
Remove scaffold				.444		19.20		19.20	25	30.50
Total				2.123	2.27	91.62		93.89	121	149.50
1040 Replace wood louver - 1st floor	45	1 CARP	Ea.							
Set up and secure scaffold				.444		23		23	30	37
Remove louver				.473		19.40		19.40	25.50	31.50
Install new wood louver				.520	147	26.75		173.75	197	227
Remove scaffold				.444		23		23	30	37
Total				1.882	147	92.15		239.15	282.50	332.50
2010 Repair wood louver - 2nd floor	6	1 CARP	Ea.							
Set up and secure scaffold				.889		46		46	60	74
Remove louver				.473		19.40		19.40	25.50	31.50
Remove damaged stile				.078		4.05		4.05	5.30	6.55
Remove damaged slat				.039		2		2	2.60	3.22
Install new slat				.077	.82	4		4.82	6.10	7.45
Install new stile				.157	2.37	8.10		10.47	13.15	16.05
Reinstall louver				.075		3.86		3.86	5.05	6.25
Remove scaffold				.889		46		46	60	74
Total				2.677	3.19	133.41		136.60	177.70	219.02

For customer support on your Facilities Maintenance & Repair Costs with RSMeans data, call 800.448.8182.

	System Description	Freq. (Years)	Crew	Unit	Labor Hours	2019 Bare Costs				Total In-House	Total w/O&P
						Material	Labor	Equipment	Total		
2030	**Refinish wood louver - 2nd floor**	4	1 PORD	Ea.							
	Set up and secure scaffold				.889		38.50		38.50	49.50	61.50
	Prepare louver for painting				.434		18.72		18.72	24	30
	Paint louvered surface (to 16 S.F.), brushwork, primer + 1 coat				.800	2.27	34.50		36.77	47	58.50
	Remove scaffold				.889		38.50		38.50	49.50	61.50
	Total				3.012	2.27	130.22		132.49	170	211.50
2040	**Replace wood louver - 2nd floor**	45	1 CARP	Ea.							
	Set up and secure scaffold				.889		46		46	60	74
	Remove louver				.473		19.40		19.40	25.50	31.50
	Install new wood louver				.520	147	26.75		173.75	197	227
	Remove scaffold				.889		46		46	60	74
	Total				2.771	147	138.15		285.15	342.50	406.50
3010	**Repair wood louver - 3rd floor**	6	1 CARP	Ea.							
	Set up and secure scaffold				1.333		69		69	89.50	111
	Remove louver				.473		19.40		19.40	25.50	31.50
	Remove damaged stile				.078		4.05		4.05	5.30	6.55
	Remove damaged slat				.039		2		2	2.60	3.22
	Install new slat				.077	.82	4		4.82	6.10	7.45
	Install new stile				.157	2.37	8.10		10.47	13.15	16.05
	Reinstall louver				.075		3.86		3.86	5.05	6.25
	Remove scaffold				1.333		69		69	89.50	111
	Total				3.565	3.19	179.41		182.60	236.70	293.02
3030	**Refinish wood louver - 3rd floor**	4	1 PORD	Ea.							
	Set up and secure scaffold				1.333		57.50		57.50	74.50	92
	Prepare louver for painting				.434		18.72		18.72	24	30
	Paint louvered surface (to 16 S.F.), brushwork, primer + 1 coat				.800	2.27	34.50		36.77	47	58.50
	Remove scaffold				1.333		57.50		57.50	74.50	92
	Total				3.901	2.27	168.22		170.49	220	272.50
3040	**Replace wood louver - 3rd floor**	45	1 CARP	Ea.							
	Set up and secure scaffold				1.333		69		69	89.50	111
	Remove louver				.473		19.40		19.40	25.50	31.50
	Install new wood louver				.520	147	26.75		173.75	197	227
	Remove scaffold				1.333		69		69	89.50	111
	Total				3.660	147	184.15		331.15	401.50	480.50

For customer support on your Facilities Maintenance & Repair Costs with RSMeans data, call 800.448.8182.

35

B2013 320 Aluminum Louver

	System Description	Freq. (Years)	Crew	Unit	Labor Hours	2019 Bare Costs				Total In-House	Total w/O&P
						Material	Labor	Equipment	Total		
1030	**Refinish aluminum louver - 1st floor**	5	1 PORD	Ea.							
	Set up and secure scaffold				.444		19.20		19.20	25	30.50
	Prepare louver for painting				.434		18.72		18.72	24	30
	Paint louvered surface (to 16 S.F.), brushwork, primer + 1 coat				.800	2.27	34.50		36.77	47	58.50
	Remove scaffold				.444		19.20		19.20	25	30.50
	Total				2.123	2.27	91.62		93.89	121	149.50
1040	**Replace aluminum louver - 1st floor**	60	1 CARP	Ea.							
	Set up and secure scaffold				.444		23		23	30	37
	Remove louver				.473		19.40		19.40	25.50	31.50
	Install new louver				1.783	261	108.60		369.60	425	495
	Remove scaffold				.444		23		23	30	37
	Total				3.145	261	174		435	510.50	600.50
2020	**Refinish aluminum louver - 2nd floor**	12	1 PORD	Ea.							
	Set up and secure scaffold				.889		46		46	60	74
	Prepare louver for painting				.434		18.72		18.72	24	30
	Paint louvered surface (to 16 S.F.), brushwork, primer + 1 coat				.800	2.27	34.50		36.77	47	58.50
	Remove scaffold				.889		46		46	60	74
	Total				3.012	2.27	145.22		147.49	191	236.50
2040	**Replace aluminum louver - 2nd floor**	60	1 CARP	Ea.							
	Set up and secure scaffold				.889		46		46	60	74
	Remove louver				.473		19.40		19.40	25.50	31.50
	Install new louver				1.783	261	108.60		369.60	425	495
	Remove scaffold				.889		46		46	60	74
	Total				4.034	261	220		481	570.50	674.50
3030	**Refinish aluminum louver - 3rd floor**	5	1 PORD	Ea.							
	Set up and secure scaffold				1.333		57.50		57.50	74.50	92
	Prepare louver for painting				.434		18.72		18.72	24	30
	Paint louvered surface (to 16 S.F.), brushwork, primer + 1 coat				.800	2.27	34.50		36.77	47	58.50
	Remove scaffold				1.333		57.50		57.50	74.50	92
	Total				3.901	2.27	168.22		170.49	220	272.50

For customer support on your Facilities Maintenance & Repair Costs with RSMeans data, call 800.448.8182.

B20 EXTERIOR CLOSURE | B2013 | Exterior Walls

B2013 320 | Aluminum Louver

System Description	Freq. (Years)	Crew	Unit	Labor Hours	2019 Bare Costs				Total In-House	Total w/O&P
					Material	Labor	Equipment	Total		
3040										
Replace aluminum louver - 3rd floor	60	1 CARP	Ea.							
Set up and secure scaffold				1.333		69		69	89.50	111
Remove louver				.473		19.40		19.40	25.50	31.50
Install new louver				1.783	261	108.60		369.60	425	495
Remove scaffold				1.333		69		69	89.50	111
Total				4.922	261	266		527	629.50	748.50

B2013 330 | Steel Louver

System Description	Freq. (Years)	Crew	Unit	Labor Hours	2019 Bare Costs				Total In-House	Total w/O&P
					Material	Labor	Equipment	Total		
1030										
Refinish steel louver - 1st floor	5	1 PORD	Ea.							
Set up and secure scaffold				.444		19.20		19.20	25	30.50
Prepare louver for painting				.434		18.72		18.72	24	30
Paint louvered surface (to 16 S.F.), brushwork, primer + 1 coat				.800	2.27	34.50		36.77	47	58.50
Remove scaffold				.444		19.20		19.20	25	30.50
Total				2.123	2.27	91.62		93.89	121	149.50
1040										
Replace steel louver - 1st floor	40	1 CARP	Ea.							
Set up and secure scaffold				.444		23		23	30	37
Remove louver				.473		19.40		19.40	25.50	31.50
Install new louver				1.783	261	108.60		369.60	425	495
Remove scaffold				.444		23		23	30	37
Total				3.145	261	174		435	510.50	600.50
2030										
Refinish steel louver - 2nd floor	5	1 PORD	Ea.							
Set up and secure scaffold				.889		38.50		38.50	49.50	61.50
Prepare louver for painting				.434		18.72		18.72	24	30
Paint louvered surface (to 16 S.F.), brushwork, primer + 1 coat				.800	2.27	34.50		36.77	47	58.50
Remove scaffold				.889		38.50		38.50	49.50	61.50
Total				3.012	2.27	130.22		132.49	170	211.50

For customer support on your Facilities Maintenance & Repair Costs with RSMeans data, call 800.448.8182.

37

	System Description	Freq. (Years)	Crew	Unit	Labor Hours	Material	Labor	Equipment	Total	Total In-House	Total w/O&P
2040	**Replace steel louver - 2nd floor**	40	1 CARP	Ea.							
	Set up and secure scaffold				.889		46		46	60	74
	Remove louver				.473		19.40		19.40	25.50	31.50
	Install new louver				1.783	261	108.60		369.60	425	495
	Remove scaffold				.889		46		46	60	74
	Total				4.034	261	220		481	570.50	674.50
3030	**Refinish steel louver - 3rd floor**	5	1 PORD	Ea.							
	Set up and secure scaffold				1.333		57.50		57.50	74.50	92
	Prepare louver for painting				.434		18.72		18.72	24	30
	Paint louvered surface (to 16 S.F.), brushwork, primer + 1 coat				.800	2.27	34.50		36.77	47	58.50
	Remove scaffold				1.333		57.50		57.50	74.50	92
	Total				3.901	2.27	168.22		170.49	220	272.50
3040	**Replace steel louver - 3rd floor**	40	1 CARP	Ea.							
	Set up and secure scaffold				1.333		69		69	89.50	111
	Remove louver				.473		19.40		19.40	25.50	31.50
	Install new louver				1.783	261	108.60		369.60	425	495
	Remove scaffold				1.333		69		69	89.50	111
	Total				4.922	261	266		527	629.50	748.50

38

For customer support on your Facilities Maintenance & Repair Costs with RSMeans data, call 800.448.8182.

B2023 102 — Steel Frame, Operating

System Description	Freq. (Years)	Crew	Unit	Labor Hours	2019 Bare Costs				Total In-House	Total w/O&P
					Material	Labor	Equipment	Total		
1010 Replace glass - 1st floor (1% of glass)	1	1 CARP	S.F.							
Remove damaged glass				.052		2.13		2.13	2.78	3.44
Install new glass				.043	6.45	2.15		8.60	9.90	11.50
Total				.095	6.45	4.28		10.73	12.68	14.94
1020 Repair 3'-9" x 5'-5" steel frame window - 1st floor	20	1 CARP	Ea.							
Set up and secure scaffold				.444		23		23	30	37
Remove window gaskets				.250		12.90		12.90	16.85	21
Install new gaskets				.564	115.54	29.16		144.70	165	191
Misc. hardware replacement				.333	76	17.20		93.20	106	123
Remove scaffold				.444		23		23	30	37
Total				2.037	191.54	105.26		296.80	347.85	409
1030 Refinish 3'-9" x 5'-5" steel frame window - 1st floor	5	1 PORD	Ea.							
Set up, secure and take down ladder				.267		11.52		11.52	14.85	18.30
Prepare window for painting				.473		20.50		20.50	26.50	32.50
Paint window, brushwork, primer + 1 coat				1.000	2.06	43.50		45.56	58	71.50
Total				1.739	2.06	75.52		77.58	99.35	122.30
1040 Replace 3'-9" x 5'-5" steel frame window - 1st floor	45	1 CARP	Ea.							
Set up and secure scaffold				.444		23		23	30	37
Remove window				.650		26.50		26.50	35	43
Install new window with frame & glazing (to 21 S.F.)				2.078	1,250	116		1,366	1,525	1,750
Install sealant				.764	8.44	39.06		47.50	60.50	74
Remove scaffold				.444		23		23	30	37
Total				4.381	1,258.44	227.56		1,486	1,680.50	1,941
1050 Finish new 3'-9" x 5'-5" steel frame window - 1st floor	45	1 PORD	Ea.							
Set up, secure and take down ladder				.267		11.52		11.52	14.85	18.30
Prepare window for painting				.236		10.25		10.25	13.15	16.25
Paint window, brushwork, primer + 1 coat				1.000	2.06	43.50		45.56	58	71.50
Total				1.503	2.06	65.27		67.33	86	106.05

For customer support on your Facilities Maintenance & Repair Costs with RSMeans data, call 800.448.8182.

39

B2023 102 | **Steel Frame, Operating**

System Description	Freq. (Years)	Crew	Unit	Labor Hours	2019 Bare Costs Material	Labor	Equipment	Total	Total In-House	Total w/O&P
2010 Replace glass - 2nd floor (1% of glass)	1	1 CARP	S.F.							
Set up and secure scaffold				.889		46		46	60	74
Remove damaged glass				.052		2.13		2.13	2.78	3.44
Install new glass				.043	6.45	2.15		8.60	9.90	11.50
Remove scaffold				.889		46		46	60	74
Total				1.873	6.45	96.28		102.73	132.68	162.94
2020 Repair 3'-9" x 5'-5" steel frame window - 2nd floor	20	1 CARP	Ea.							
Set up and secure scaffold				.889		46		46	60	74
Remove window gaskets				.250		12.90		12.90	16.85	21
Install new gaskets				.564	115.54	29.16		144.70	165	191
Misc. hardware replacement				.333	76	17.20		93.20	106	123
Remove scaffold				.889		46		46	60	74
Total				2.925	191.54	151.26		342.80	407.85	483
2030 Refinish 3'-9" x 5'-5" steel frame window - 2nd floor	5	1 PORD	Ea.							
Set up and secure scaffold				.889		38.50		38.50	49.50	61.50
Prepare window for painting				.473		20.50		20.50	26.50	32.50
Paint window, brushwork, primer + 1 coat				1.000	2.06	43.50		45.56	58	71.50
Remove scaffold				.889		38.50		38.50	49.50	61.50
Total				3.251	2.06	141		143.06	183.50	227
2040 Replace 3'-9" x 5'-5" steel frame window - 2nd floor	45	1 CARP	Ea.							
Set up and secure scaffold				.889		46		46	60	74
Remove window				.650		26.50		26.50	35	43
Install new window with frame & glazing (to 21 S.F.)				2.078	1,250	116		1,366	1,525	1,750
Install sealant				.764	8.44	39.06		47.50	60.50	74
Remove scaffold				.889		46		46	60	74
Total				5.270	1,258.44	273.56		1,532	1,740.50	2,015
2050 Finish new 3'-9" x 5'-5" steel frame window - 2nd floor	45	1 PORD	Ea.							
Set up and secure scaffold				.889		38.50		38.50	49.50	61.50
Prepare window for painting				.236		10.25		10.25	13.15	16.25
Paint window, brushwork, primer + 1 coat				1.000	2.06	43.50		45.56	58	71.50
Remove scaffold				.889		38.50		38.50	49.50	61.50
Total				3.014	2.06	130.75		132.81	170.15	210.75

B20 EXTERIOR CLOSURE — B2023 — Exterior Windows

B2023 102 — Steel Frame, Operating

System Description	Freq. (Years)	Crew	Unit	Labor Hours	Material	Labor	Equipment	Total	Total In-House	Total w/O&P
3010 Replace glass - 3rd floor (1% of glass)	1	1 CARP	S.F.							
Set up and secure scaffold				.123		6.35		6.35	8.30	10.25
Remove damaged glass				.052		2.13		2.13	2.78	3.44
Install new glass				.043	6.45	2.15		8.60	9.90	11.50
Remove scaffold				.123		6.35		6.35	8.30	10.25
Total				.341	6.45	16.98		23.43	29.28	35.44
3020 Repair 3'-9" x 5'-5" steel frame window - 3rd floor	20	1 CARP	Ea.							
Set up and secure scaffold				1.333		69		69	89.50	111
Remove window gaskets				.250		12.90		12.90	16.85	21
Install new gaskets				.564	115.54	29.16		144.70	165	191
Misc. hardware replacement				.333	76	17.20		93.20	106	123
Remove scaffold				1.333		69		69	89.50	111
Total				3.814	191.54	197.26		388.80	466.85	557
3030 Refinish 3'-9" x 5'-5" steel frame window - 3rd floor	5	1 PORD	Ea.							
Set up and secure scaffold				1.333		57.50		57.50	74.50	92
Prepare window for painting				.473		20.50		20.50	26.50	32.50
Paint window, brushwork, primer + 1 coat				1.000	2.06	43.50		45.56	58	71.50
Remove scaffold				1.333		57.50		57.50	74.50	92
Total				4.139	2.06	179		181.06	233.50	288
3040 Replace 3'-9" x 5'-5" steel frame window - 3rd floor	45	1 CARP	Ea.							
Set up and secure scaffold				1.333		69		69	89.50	111
Remove window				.650		26.50		26.50	35	43
Install new window with frame & glazing (to 21 S.F.)				2.078	1,250	116		1,366	1,525	1,750
Install sealant				.764	8.44	39.06		47.50	60.50	74
Remove scaffold				1.333		69		69	89.50	111
Total				6.159	1,258.44	319.56		1,578	1,799.50	2,089
3050 Finish new 3'-9" x 5'-5" steel frame window - 3rd floor	45	1 PORD	Ea.							
Set up and secure scaffold				1.333		57.50		57.50	74.50	92
Prepare window for painting				.236		10.25		10.25	13.15	16.25
Paint window, brushwork, primer + 1 coat				1.000	2.06	43.50		45.56	58	71.50
Remove scaffold				1.333		57.50		57.50	74.50	92
Total				3.903	2.06	168.75		170.81	220.15	271.75

For customer support on your Facilities Maintenance & Repair Costs with RSMeans data, call 800.448.8182.

41

B2023 103 | **Steel Frame, Fixed**

System Description	Freq. (Years)	Crew	Unit	Labor Hours	2019 Bare Costs				Total In-House	Total w/O&P
					Material	Labor	Equipment	Total		
1010 Replace glass - 1st floor (1% of glass)	1	1 CARP	S.F.							
Remove damaged glass				.052		2.13		2.13	2.78	3.44
Install new glass				.043	6.45	2.15		8.60	9.90	11.50
Total				.095	6.45	4.28		10.73	12.68	**14.94**
1020 Repair 2'-0" x 3'-0" steel frame window - 1st flr.	20	1 CARP	Ea.							
Set up and secure scaffold				.444		23		23	30	37
Remove window gaskets				.250		12.90		12.90	16.85	21
Install new gaskets				.308	63	15.90		78.90	90	104
Remove scaffold				.444		23		23	30	37
Total				1.447	63	74.80		137.80	166.85	**199**
1030 Refinish 2'-0" x 3'-0" st. frame window - 1st flr.	5	1 PORD	Ea.							
Set up, secure and take down ladder				.267		11.52		11.52	14.85	18.30
Prepare window for painting				.473		20.50		20.50	26.50	32.50
Paint window, brushwork, primer + 1 coat				1.000	2.06	43.50		45.56	58	71.50
Total				1.739	2.06	75.52		77.58	99.35	**122.30**
1040 Replace 2'-0" x 3'-0" steel frame window - 1st flr.	45	1 CARP	Ea.							
Set up and secure scaffold				.444		23		23	30	37
Remove window				.650		26.50		26.50	35	43
Install new window with frame & glazing (2'-0" x 3'-0")				2.078	214	116		330	395	460
Install sealant				.764	8.44	39.06		47.50	60.50	74
Remove scaffold				.444		23		23	30	37
Total				4.381	222.44	227.56		450	550.50	**651**
2010 Replace glass - 2nd floor (1% of glass)	1	1 CARP	S.F.							
Set up and secure scaffold				.044		2.30		2.30	2.99	3.70
Remove damaged glass				.052		2.13		2.13	2.78	3.44
Install new glass				.043	6.45	2.15		8.60	9.90	11.50
Remove scaffold				.044		2.30		2.30	2.99	3.70
Total				.184	6.45	8.88		15.33	18.66	**22.34**

For customer support on your Facilities Maintenance & Repair Costs with RSMeans data, call 800.448.8182.

B20 EXTERIOR CLOSURE | B2023 | Exterior Windows

B2023 103 | Steel Frame, Fixed

	System Description	Freq. (Years)	Crew	Unit	Labor Hours	2019 Bare Costs				Total In-House	Total w/O&P
						Material	Labor	Equipment	Total		
2020	**Repair 2'-0" x 3'-0" steel frame window - 2nd flr.**	20	1 CARP	Ea.							
	Set up and secure scaffold				.889		46		46	60	74
	Remove window gaskets				.250		12.90		12.90	16.85	21
	Install new gaskets				.308	63	15.90		78.90	90	104
	Remove scaffold				.889		46		46	60	74
	Total				2.335	63	120.80		183.80	226.85	273
2030	**Refinish 2'-0" x 3'-0" steel frame window - 2nd flr.**	5	1 PORD	Ea.							
	Set up and secure scaffold				.889		38.50		38.50	49.50	61.50
	Prepare window for painting				.473		20.50		20.50	26.50	32.50
	Paint window, brushwork, primer + 1 coat				1.000	2.06	43.50		45.56	58	71.50
	Remove scaffold				.889		38.50		38.50	49.50	61.50
	Total				3.251	2.06	141		143.06	183.50	227
2040	**Replace 2'-0" x 3'-0" steel frame window - 2nd flr.**	45	1 CARP	Ea.							
	Set up and secure scaffold				.889		46		46	60	74
	Remove window				.650		26.50		26.50	35	43
	Install new window with frame & glazing (2'-0" x 3'-0")				2.078	214	116		330	395	460
	Install sealant				.764	8.44	39.06		47.50	60.50	74
	Remove scaffold				.889		46		46	60	74
	Total				5.270	222.44	273.56		496	610.50	725
3010	**Replace glass - 3rd floor (1% of glass)**	1	1 CARP	S.F.							
	Set up and secure scaffold				.123		6.35		6.35	8.30	10.25
	Remove damaged glass				.052		2.13		2.13	2.78	3.44
	Install new glass				.043	6.45	2.15		8.60	9.90	11.50
	Remove scaffold				.123		6.35		6.35	8.30	10.25
	Total				.341	6.45	16.98		23.43	29.28	35.44
3020	**Repair 2'-0" x 3'-0" steel frame window - 3rd flr.**	20	1 CARP	Ea.							
	Set up and secure scaffold				1.333		69		69	89.50	111
	Remove window gaskets				.250		12.90		12.90	16.85	21
	Install new gaskets				.308	63	15.90		78.90	90	104
	Remove scaffold				1.333		69		69	89.50	111
	Total				3.224	63	166.80		229.80	285.85	347

For customer support on your Facilities Maintenance & Repair Costs with RSMeans data, call 800.448.8182.

43

B20 EXTERIOR CLOSURE · B2023 · Exterior Windows

B2023 103 Steel Frame, Fixed

	System Description	Freq. (Years)	Crew	Unit	Labor Hours	Material	Labor	Equipment	Total	Total In-House	Total w/O&P
3030	**Refinish 2'-0" x 3'-0" steel frame window - 3rd floor**	5	1 PORD	Ea.							
	Set up and secure scaffold				1.333		57.50		57.50	74.50	92
	Prepare window for painting				.473		20.50		20.50	26.50	32.50
	Paint window, brushwork, primer + 1 coat				1.000	2.06	43.50		45.56	58	71.50
	Remove scaffold				1.333		57.50		57.50	74.50	92
	Total				4.139	2.06	179		181.06	233.50	288
3040	**Replace 2'-0" x 3'-0" steel frame window - 3rd floor**	45	1 CARP	Ea.							
	Set up and secure scaffold				1.333		69		69	89.50	111
	Remove window				.650		26.50		26.50	35	43
	Install new window with frame & glazing (2'-0" x 3'-0")				2.078	214	116		330	395	460
	Install sealant				.764	8.44	39.06		47.50	60.50	74
	Remove scaffold				1.333		69		69	89.50	111
	Total				6.159	222.44	319.56		542	669.50	799

B2023 110 Aluminum Window, Operating

	System Description	Freq. (Years)	Crew	Unit	Labor Hours	Material	Labor	Equipment	Total	Total In-House	Total w/O&P
1010	**Replace glass - 1st flr. (1% of glass)**	1	1 CARP	S.F.							
	Remove damaged glass				.052		2.13		2.13	2.78	3.44
	Install new glass				.043	6.45	2.15		8.60	9.90	11.50
	Total				.095	6.45	4.28		10.73	12.68	14.94
1020	**Repair 3' x 4' aluminum window - 1st floor**	20	1 CARP	Ea.							
	Set up and secure scaffold				.444		23		23	30	37
	Remove window gaskets				.250		12.90		12.90	16.85	21
	Install new gaskets				.431	88.20	22.26		110.46	126	146
	Misc. hardware replacement				.333	76	17.20		93.20	106	123
	Remove scaffold				.444		23		23	30	37
	Total				1.903	164.20	98.36		262.56	308.85	364

For customer support on your Facilities Maintenance & Repair Costs with RSMeans data, call 800.448.8182.

B20 EXTERIOR CLOSURE B2023 Exterior Windows

B2023 110 Aluminum Window, Operating

System Description	Freq. (Years)	Crew	Unit	Labor Hours	2019 Bare Costs				Total In-House	Total w/O&P
					Material	Labor	Equipment	Total		
1030 Replace 3' x 4' aluminum window - 1st floor	50	1 CARP	Ea.							
Set up and secure scaffold				.444		23		23	30	37
Remove window				.650		26.50		26.50	35	43
Install new window with frame & glazing (to 12 S.F.)				2.078	500	116		616	710	820
Install sealant				.555	6.13	28.39		34.52	44	54
Remove scaffold				.444		23		23	30	37
Total				4.173	506.13	216.89		723.02	849	991
2010 Replace glass - 2nd flr. (1% of glass)	1	1 CARP	S.F.							
Set up and secure scaffold				.044		2.30		2.30	2.99	3.70
Remove damaged glass				.052		2.13		2.13	2.78	3.44
Install new glass				.043	6.45	2.15		8.60	9.90	11.50
Remove scaffold				.044		2.30		2.30	2.99	3.70
Total				.184	6.45	8.88		15.33	18.66	22.34
2020 Repair 3' x 4' aluminum window - 2nd floor	20	1 CARP	Ea.							
Set up and secure scaffold				.889		46		46	60	74
Remove window gaskets				.250		12.90		12.90	16.85	21
Install new gaskets				.431	88.20	22.26		110.46	126	146
Misc. hardware replacement				.333	76	17.20		93.20	106	123
Remove scaffold				.889		46		46	60	74
Total				2.792	164.20	144.36		308.56	368.85	438
2030 Replace 3' x 4' aluminum window - 2nd floor	50	1 CARP	Ea.							
Set up and secure scaffold				.889		46		46	60	74
Remove window				.650		26.50		26.50	35	43
Install new window with frame & glazing (to 12 S.F.)				2.078	500	116		616	710	820
Install sealant				.555	6.13	28.39		34.52	44	54
Remove scaffold				.889		46		46	60	74
Total				5.062	506.13	262.89		769.02	909	1,065
3010 Replace glass - 3rd floor (1% of glass)	1	1 CARP	S.F.							
Set up and secure scaffold				.123		6.35		6.35	8.30	10.25
Remove damaged glass				.052		2.13		2.13	2.78	3.44
Install new glass				.043	6.45	2.15		8.60	9.90	11.50
Remove scaffold				.123		6.35		6.35	8.30	10.25
Total				.341	6.45	16.98		23.43	29.28	35.44

For customer support on your Facilities Maintenance & Repair Costs with RSMeans data, call 800.448.8182.

45

B2023 110 Aluminum Window, Operating

	System Description	Freq. (Years)	Crew	Unit	Labor Hours	2019 Bare Costs				Total In-House	Total w/O&P
						Material	Labor	Equipment	Total		
3020	**Repair 3' x 4' aluminum window - 3rd floor**	20	1 CARP	Ea.							
	Set up and secure scaffold				1.333		69		69	89.50	111
	Remove window gaskets				.250		12.90		12.90	16.85	21
	Install new gaskets				.431	88.20	22.26		110.46	126	146
	Misc. hardware replacement				.333	76	17.20		93.20	106	123
	Remove scaffold				1.333		69		69	89.50	111
	Total				3.681	164.20	190.36		354.56	427.85	512
3030	**Replace 3' x 4' aluminum window - 3rd floor**	50	1 CARP	Ea.							
	Set up and secure scaffold				1.333		69		69	89.50	111
	Remove window				.650		26.50		26.50	35	43
	Install new window with frame & glazing (to 12 S.F.)				2.078	500	116		616	710	820
	Install sealant				.555	6.13	28.39		34.52	44	54
	Remove scaffold				1.333		69		69	89.50	111
	Total				5.950	506.13	308.89		815.02	968	1,139

B2023 111 Aluminum Window, Fixed

	System Description	Freq. (Years)	Crew	Unit	Labor Hours	2019 Bare Costs				Total In-House	Total w/O&P
						Material	Labor	Equipment	Total		
1010	**Replace glass - 1st floor (1% of glass)**	1	1 CARP	S.F.							
	Remove damaged glass				.052		2.13		2.13	2.78	3.44
	Install new glass				.043	6.45	2.15		8.60	9.90	11.50
	Total				.095	6.45	4.28		10.73	12.68	14.94
1015	**Replace glass - 1st floor, 1" insulating panel with heat reflective glass**	30	2 GLAZ	S.F.							
	Remove damaged glass				.052		2.13		2.13	2.78	3.44
	Tinted insulated glass				.213	24	10.60		34.60	40	47
	Total				.265	24	12.73		36.73	42.78	50.44
1020	**Repair 2'-0" x 3'-0" aluminum window - 1st floor**	20	1 CARP	Ea.							
	Set up and secure scaffold				.444		23		23	30	37
	Remove window gaskets				.250		12.90		12.90	16.85	21
	Install new gaskets				.308	63	15.90		78.90	90	104
	Remove scaffold				.444		23		23	30	37
	Total				1.447	63	74.80		137.80	166.85	199

B2023 111 | **Aluminum Window, Fixed**

	System Description	Freq. (Years)	Crew	Unit	Labor Hours	2019 Bare Costs				Total In-House	Total w/O&P
						Material	Labor	Equipment	Total		
1030	**Replace 2'-0" x 3'-0" aluminum window - 1st floor**	50	1 CARP	Ea.							
	Set up and secure scaffold				.444		23		23	30	37
	Remove window				.650		26.50		26.50	35	43
	Install new window with frame & glazing (2'-0" x 3'-0")				2.078	214	116		330	395	460
	Install sealant				.555	6.13	28.39		34.52	44	54
	Remove scaffold				.444		23		23	30	37
	Total				4.173	220.13	216.89		437.02	534	631
2010	**Replace glass - 2nd floor (1% of glass)**	1	1 CARP	S.F.							
	Set up and secure scaffold				.889		46		46	60	74
	Remove damaged glass				.052		2.13		2.13	2.78	3.44
	Install new glass				.043	6.45	2.15		8.60	9.90	11.50
	Remove scaffold				.889		46		46	60	74
	Total				1.873	6.45	96.28		102.73	132.68	162.94
2020	**Repair 2'-0" x 3'-0" aluminum window - 2nd floor**	20	1 CARP	Ea.							
	Set up and secure scaffold				.889		46		46	60	74
	Remove window gaskets				.250		12.90		12.90	16.85	21
	Install new gaskets				.308	63	15.90		78.90	90	104
	Remove scaffold				.889		46		46	60	74
	Total				2.335	63	120.80		183.80	226.85	273
2030	**Replace 2'-0" x 3'-0" aluminum window - 2nd floor**	50	1 CARP	Ea.							
	Set up and secure scaffold				.889		46		46	60	74
	Remove window				.650		26.50		26.50	35	43
	Install new window with frame & glazing (2'-0" x 3'-0")				2.078	214	116		330	395	460
	Install sealant				.555	6.13	28.39		34.52	44	54
	Remove scaffold				.889		46		46	60	74
	Total				5.062	220.13	262.89		483.02	594	705
3010	**Replace glass - 3rd floor (1% of glass)**	1	1 CARP	S.F.							
	Set up and secure scaffold				.123		6.35		6.35	8.30	10.25
	Remove damaged glass				.052		2.13		2.13	2.78	3.44
	Install new glass				.043	6.45	2.15		8.60	9.90	11.50
	Remove scaffold				.123		6.35		6.35	8.30	10.25
	Total				.341	6.45	16.98		23.43	29.28	35.44

For customer support on your Facilities Maintenance & Repair Costs with RSMeans data, call 800.448.8182.

47

B2023 111 Aluminum Window, Fixed

	System Description	Freq. (Years)	Crew	Unit	Labor Hours	Material	Labor	Equipment	Total	Total In-House	Total w/O&P
							2019 Bare Costs				
3020	**Repair 2'-0" x 3'-0" aluminum window - 3rd floor**	20	1 CARP	Ea.							
	Set up and secure scaffold				1.333		69		69	89.50	111
	Remove window gaskets				.250		12.90		12.90	16.85	21
	Install new gaskets				.308	63	15.90		78.90	90	104
	Remove scaffold				1.333		69		69	89.50	111
	Total				3.224	63	166.80		229.80	285.85	347
3030	**Replace 2'-0" x 3'-0" aluminum window - 3rd floor**	50	1 CARP	Ea.							
	Set up and secure scaffold				1.333		69		69	89.50	111
	Remove window				.650		26.50		26.50	35	43
	Install new window with frame & glazing (2'-0" x 3'-0")				2.078	214	116		330	395	460
	Install sealant				.555	6.13	28.39		34.52	44	54
	Remove scaffold				1.333		69		69	89.50	111
	Total				5.950	220.13	308.89		529.02	653	779

B2023 112 Wood Frame, Operating

	System Description	Freq. (Years)	Crew	Unit	Labor Hours	Material	Labor	Equipment	Total	Total In-House	Total w/O&P
							2019 Bare Costs				
1010	**Replace glass - 1st floor (1% of glass)**	1	1 CARP	S.F.							
	Remove damaged glass				.052		2.13		2.13	2.78	3.44
	Install new glass				.043	6.45	2.15		8.60	9.90	11.50
	Total				.095	6.45	4.28		10.73	12.68	14.94
1020	**Repair 2'-3" x 6'-0" wood frame window - 1st flr.**	15	1 CARP	Ea.							
	Set up and secure scaffold				.444		23		23	30	37
	Remove putty				.667		34.50		34.50	45	55.50
	Install new putty				.333	6.30	17.20		23.50	29.50	36
	Misc. hardware replacement				.333	76	17.20		93.20	106	123
	Remove scaffold				.444		23		23	30	37
	Total				2.222	82.30	114.90		197.20	240.50	288.50
1030	**Refinish 2'-3" x 6'-0" wood frame window - 1st flr.**	4	1 PORD	Ea.							
	Set up, secure and take down ladder				.267		11.55		11.55	14.85	18.45
	Prepare window for painting				.473		20.50		20.50	26.50	32.50
	Paint window, brushwork, primer + 1 coat				1.000	2.06	43.50		45.55	58	71.50
	Total				1.739	2.06	75.55		77.61	99.35	122.45

For customer support on your Facilities Maintenance & Repair Costs with RSMeans data, call 800.448.8182.

B2023 112 Wood Frame, Operating

	System Description	Freq. (Years)	Crew	Unit	Labor Hours	2019 Bare Costs				Total In-House	Total w/O&P
						Material	Labor	Equipment	Total		
1040	**Replace 2'-3" x 6'-0" wood frame window - 1st flr.**	40	1 CARP	Ea.							
	Set up and secure scaffold				.444		23		23	30	37
	Remove window				.473		19.40		19.40	25.50	31.50
	Install new window with frame & glazing (to 14 S.F.)				1.143	340	59		399	450	520
	Install sealant				.625	6.90	31.95		38.85	49.50	60.50
	Remove scaffold				.444		23		23	30	37
	Total				3.130	346.90	156.35		503.25	585	686
2010	**Replace glass - 2nd floor (1% of glass)**	1	1 CARP	S.F.							
	Set up and secure scaffold				.044		2.30		2.30	2.99	3.70
	Remove damaged glass				.052		2.13		2.13	2.78	3.44
	Install new glass				.043	6.45	2.15		8.60	9.90	11.50
	Remove scaffold				.044		2.30		2.30	2.99	3.70
	Total				.184	6.45	8.88		15.33	18.66	22.34
2020	**Repair 2'-3" x 6'-0" wood frame window - 2nd flr.**	15	1 CARP	Ea.							
	Set up and secure scaffold				.889		46		46	60	74
	Remove putty				.667		34.50		34.50	45	55.50
	New putty				.333	6.30	17.20		23.50	29.50	36
	Misc. hardware replacement				.333	76	17.20		93.20	106	123
	Remove scaffold				.889		46		46	60	74
	Total				3.111	82.30	160.90		243.20	300.50	362.50
2030	**Refinish 2'-3" x 6'-0" wood frame window - 2nd flr.**	5	1 PORD	Ea.							
	Set up and secure scaffold				.889		38.50		38.50	49.50	61.50
	Prepare window for painting				.473		20.50		20.50	26.50	32.50
	Paint window, brushwork, primer + 1 coat				1.000	2.06	43.50		45.56	58	71.50
	Remove scaffold				.889		38.50		38.50	49.50	61.50
	Total				3.251	2.06	141		143.06	183.50	227
2040	**Replace 2'-3" x 6'-0" wood frame window - 2nd flr.**	40	1 CARP	Ea.							
	Set up and secure scaffold				.889		46		46	60	74
	Remove window				.473		19.40		19.40	25.50	31.50
	Install new window with frame & glazing (to 14 S.F.)				1.143	340	59		399	450	520
	Install sealant				.625	6.90	31.95		38.85	49.50	60.50
	Remove scaffold				.889		46		46	60	74
	Total				4.018	346.90	202.35		549.25	645	760

For customer support on your Facilities Maintenance & Repair Costs with RSMeans data, call 800.448.8182.

49

B20 EXTERIOR CLOSURE — B2023 — Exterior Windows

B2023 112 — Wood Frame, Operating

System Description	Freq. (Years)	Crew	Unit	Labor Hours	Material	Labor	Equipment	Total	Total In-House	Total w/O&P
3010 Replace glass - 3rd floor (1% of glass)	1	1 CARP	S.F.							
Set up and secure scaffold				.123		6.35		6.35	8.30	10.25
Remove damaged glass				.052		2.13		2.13	2.78	3.44
Install new glass				.043	6.45	2.15		8.60	9.90	11.50
Remove scaffold				.123		6.35		6.35	8.30	10.25
Total				.341	6.45	16.98		23.43	29.28	35.44
3020 Repair 2'-3" x 6'-0" wood frame window - 3rd flr.	15	1 CARP	Ea.							
Set up and secure scaffold				1.333		69		69	89.50	111
Remove putty				.667		34.50		34.50	45	55.50
Install new putty				.333	6.30	17.20		23.50	29.50	36
Misc. hardware replacement				.333	76	17.20		93.20	106	123
Remove scaffold				1.333		69		69	89.50	111
Total				4.000	82.30	206.90		289.20	359.50	436.50
3030 Refinish 2'-3" x 6'-0" wood frame window - 3rd flr.	5	1 PORD	Ea.							
Set up and secure scaffold				1.333		57.50		57.50	74.50	92
Prepare window for painting				.473		20.50		20.50	26.50	32.50
Paint window, brushwork, primer + 1 coat				1.000	2.06	43.50		45.56	58	71.50
Remove scaffold				1.333		57.50		57.50	74.50	92
Total				4.139	2.06	179		181.06	233.50	288
3040 Replace 2'-3" x 6'-0" wood frame window - 3rd flr.	40	1 CARP	Ea.							
Set up and secure scaffold				1.333		69		69	89.50	111
Remove window				.473		19.40		19.40	25.50	31.50
Install new window with frame & glazing (to 14 S.F.)				1.143	340	59		399	450	520
Install sealant				.625	6.90	31.95		38.85	49.50	60.50
Remove scaffold				1.333		69		69	89.50	111
Total				4.907	346.90	248.35		595.25	704	834

For customer support on your Facilities Maintenance & Repair Costs with RSMeans data, call 800.448.8182.

B2023 113 Wood Frame, Fixed

System Description	Freq. (Years)	Crew	Unit	Labor Hours	2019 Bare Costs				Total In-House	Total w/O&P
					Material	Labor	Equipment	Total		
1010										
Replace glass - 1st floor	1	1 CARP	S.F.							
Remove damaged glass				.052		2.13		2.13	2.78	3.44
Install new glass				.043	6.45	2.15		8.60	9.90	11.50
Total				.095	6.45	4.28		10.73	12.68	14.94
1020										
Repair 3'-6" x 4'-0" wood frame window - 1st flr.	20	1 CARP	Ea.							
Set up and secure scaffold				.444		23		23	30	37
Remove putty				.667		34.50		34.50	45	55.50
New putty				.333	6.30	17.20		23.50	29.50	36
Remove scaffold				.444		23		23	30	37
Total				1.889	6.30	97.70		104	134.50	165.50
1030										
Refinish 3'-6" x 4'-0" wood frame window - 1st flr.	5	1 PORD	Ea.							
Set up, secure and take down ladder				.444		19.20		19.20	25	30.50
Prepare window for painting				.473		20.50		20.50	26.50	32.50
Paint window, brushwork, primer + 1 coat				1.000	2.06	43.50		45.56	58	71.50
Total				1.917	2.06	83.20		85.26	109.50	134.50
1040										
Replace 3'-6" x 4'-0" wood frame window - 1st flr.	40	1 CARP	Ea.							
Set up and secure scaffold				.444		23		23	30	37
Remove window				.473		19.40		19.40	25.50	31.50
Install new window with frame & glazing (to 14 S.F.)				1.733	435	89.50		524.50	595	690
Install sealant				.625	6.90	31.95		38.85	49.50	60.50
Remove scaffold				.444		23		23	30	37
Total				3.720	441.90	186.85		628.75	730	856
2010										
Replace glass - 2nd floor (1% of glass)	1	1 CARP	S.F.							
Set up and secure scaffold				.044		2.30		2.30	2.99	3.70
Remove damaged glass				.052		2.13		2.13	2.78	3.44
Install new glass				.043	6.45	2.15		8.60	9.90	11.50
Remove scaffold				.044		2.30		2.30	2.99	3.70
Total				.184	6.45	8.88		15.33	18.66	22.34

For customer support on your Facilities Maintenance & Repair Costs with RSMeans data, call 800.448.8182.

51

B2023 113 | **Wood Frame, Fixed**

System Description	Freq. (Years)	Crew	Unit	Labor Hours	2019 Bare Costs Material	Labor	Equipment	Total	Total In-House	Total w/O&P
2020 Repair 3'-6" x 4'-0" wood frame window - 2nd flr.	20	1 CARP	Ea.							
Set up and secure scaffold				.889		46		46	60	74
Remove putty				.667		34.50		34.50	45	55.50
Install new putty				.333	6.30	17.20		23.50	29.50	36
Remove scaffold				.889		46		46	60	74
Total				2.778	6.30	143.70		150	194.50	239.50
2030 Refinish 3'-6" x 4'-0" wood frame window - 2nd flr.	5	1 PORD	Ea.							
Set up and secure scaffold				.889		38.50		38.50	49.50	61.50
Prepare window for painting				.473		20.50		20.50	26.50	32.50
Paint window, brushwork, primer + 1 coat				1.000	2.06	43.50		45.55	58	71.50
Remove scaffold				.889		38.50		38.50	49.50	61.50
Total				3.251	2.06	141		143.06	183.50	227
2040 Replace 3'-6" x 4'-0" wood frame window - 2nd flr.	40	1 CARP	Ea.							
Set up and secure scaffold				.889		46		46	60	74
Remove window				.473		19.40		19.40	25.50	31.50
Install new window with frame & glazing (to 14 S.F.)				1.733	435	89.50		524.50	595	690
Install sealant				.625	6.90	31.95		38.85	49.50	60.50
Remove scaffold				.889		46		46	60	74
Total				4.609	441.90	232.85		674.75	790	930
3010 Replace glass - 3rd floor (1% of glass)	1	1 CARP	S.F.							
Set up and secure scaffold				.123		6.35		6.35	8.30	10.25
Remove damaged glass				.052		2.13		2.13	2.78	3.44
Install new glass				.043	6.45	2.15		8.60	9.90	11.50
Remove scaffold				.123		6.35		6.35	8.30	10.25
Total				.341	6.45	16.98		23.43	29.28	35.44
3020 Repair 3'-6" x 4'-0" wood frame window - 3rd flr.	20	1 CARP	Ea.							
Set up and secure scaffold				1.333		69		69	89.50	111
Remove putty				.667		34.50		34.50	45	55.50
Install new putty				.333	6.30	17.20		23.50	29.50	36
Remove scaffold				1.333		69		69	89.50	111
Total				3.667	6.30	189.70		196	253.50	313.50

For customer support on your Facilities Maintenance & Repair Costs with RSMeans data, call 800.448.8182.

B20 EXTERIOR CLOSURE | B2023 | Exterior Windows

B2023 113 | Wood Frame, Fixed

System Description	Freq. (Years)	Crew	Unit	Labor Hours	Material	Labor	Equipment	Total	Total In-House	Total w/O&P
					2019 Bare Costs					
3030										
Refinish 3'-6" x 4'-0" wood frame window - 3rd flr.	5	1 PORD	Ea.							
Set up and secure scaffold				1.333		57.50		57.50	74.50	92
Prepare window for painting				.473		20.50		20.50	26.50	32.50
Paint window, brushwork, primer + 1 coat				1.000	2.06	43.50		45.56	58	71.50
Remove scaffold				1.333		57.50		57.50	74.50	92
Total				4.139	2.06	179		181.06	233.50	288
3040										
Replace 3'-6" x 4'-0" wood frame window - 3rd flr.	40	1 CARP	Ea.							
Set up and secure scaffold				1.333		69		69	89.50	111
Remove window				.473		19.40		19.40	25.50	31.50
Install new window with frame & glazing (to 14 S.F.)				1.733	435	89.50		524.50	595	690
Install sealant				.625	6.90	31.95		38.85	49.50	60.50
Remove scaffold				1.333		69		69	89.50	111
Total				5.498	441.90	278.85		720.75	849	1,004

B2023 116 | Aluminum Frame Storm Window

System Description	Freq. (Years)	Crew	Unit	Labor Hours	Material	Labor	Equipment	Total	Total In-House	Total w/O&P
					2019 Bare Costs					
1010										
Replace glass - (1% of glass)	1	1 CARP	S.F.							
Remove glass				.052		2.13		2.13	2.78	3.44
Install glass / storm window				.173	10.15	8.60		18.75	22.50	26.50
Total				.225	10.15	10.73		20.88	25.28	29.94
1030										
Refinish 3'-0" x 4'-0" aluminum frame storm window - 1st floor	10	1 PORD	Ea.							
Set up and secure scaffold				.444		19.20		19.20	25	30.50
Prepare window frame surface				.473		20.50		20.50	26.50	32.50
Refinish window frame surface, brushwork, primer + 1 coat				.615	1.08	26.50		27.58	35.50	44
Remove scaffold				.444		19.20		19.20	25	30.50
Total				1.977	1.08	85.40		86.48	112	137.50
1040										
Replace 3'-0" x 4'-0" aluminum frame storm window - 1st floor	50	1 CARP	Ea.							
Set up and secure scaffold				.444		23		23	30	37
Remove old window				.385		15.80		15.80	20.50	25.50
Install new window (to 12 S.F.)				.743	125	38.50		163.50	188	218
Remove scaffold				.444		23		23	30	37
Total				2.017	125	100.30		225.30	268.50	317.50

For customer support on your Facilities Maintenance & Repair Costs with RSMeans data, call 800.448.8182.

53

B2023 116 Aluminum Frame Storm Window

System Description	Freq. (Years)	Crew	Unit	Labor Hours	2019 Bare Costs				Total In-House	Total w/O&P
					Material	Labor	Equipment	Total		
2030 Refinish 3'-0" x 4'-0" aluminum frame storm window - 2nd floor	10	1 PORD	Ea.							
Set up and secure scaffold				.889		38.50		38.50	49.50	61.50
Prepare window frame surface				.473		20.50		20.50	26.50	32.50
Refinish window frame surface, brushwork, primer + 1 coat				.615	1.08	26.50		27.58	35.50	44
Remove scaffold				.889		38.50		38.50	49.50	61.50
Total				2.866	1.08	124		125.08	161	199.50
2040 Replace 3'-0" x 4'-0" aluminum frame storm window - 2nd floor	50	1 CARP	Ea.							
Set up and secure scaffold				.889		46		46	60	74
Remove old window				.385		15.80		15.30	20.50	25.50
Install new window (to 12 S.F.)				.743	125	38.50		163.50	188	218
Remove scaffold				.889		46		46	60	74
Total				2.906	125	146.30		271.30	328.50	391.50
3030 Refinish 3'-0" x 4'-0" aluminum frame storm window - 3rd floor	10	1 PORD	Ea.							
Set up and secure scaffold				1.333		57.50		57.50	74.50	92
Prepare window frame surface				.473		20.50		20.50	26.50	32.50
Refinish window frame surface, brushwork, primer + 1 coat				.615	1.08	26.50		27.58	35.50	44
Remove scaffold				1.333		57.50		57.50	74.50	92
Total				3.755	1.08	162		163.08	211	260.50
3040 Replace 3'-0" x 4'-0" aluminum frame storm window - 3rd floor	50	1 CARP	Ea.							
Set up and secure scaffold				1.333		69		69	89.50	111
Remove old window				.385		15.80		15.80	20.50	25.50
Install new window (to 12 S.F.)				.743	125	38.50		163.50	188	218
Remove scaffold				1.333		69		69	89.50	111
Total				3.795	125	192.30		317.30	387.50	465.50

B2023 117 Steel Frame Storm Window

System Description	Freq. (Years)	Crew	Unit	Labor Hours	2019 Bare Costs				Total In-House	Total w/O&P
					Material	Labor	Equipment	Total		
1010 Replace glass - (1% of glass)	1	1 CARP	S.F.							
Remove glass				.052		2.13		2.13	2.78	3.44
Install glass / storm window				.173	10.15	8.60		18.75	22.50	26.50
Total				.225	10.15	10.73		20.88	25.28	29.94

For customer support on your Facilities Maintenance & Repair Costs with RSMeans data, call 800.448.8182.

	System Description	Freq. (Years)	Crew	Unit	Labor Hours	2019 Bare Costs				Total In-House	Total w/O&P
						Material	Labor	Equipment	Total		
1030	**Refinish 3'-0" x 4'-0" steel frame storm window - 1st floor**	5	1 PORD	Ea.							
	Set up and secure scaffold				.444		19.20		19.20	25	30.50
	Prepare window frame surface				.473		20.50		20.50	26.50	32.50
	Refinish window frame surface, brushwork, primer + 1 coat				.615	1.08	26.50		27.58	35.50	44
	Remove scaffold				.444		19.20		19.20	25	30.50
	Total				1.977	1.08	85.40		86.48	112	137.50
1040	**Replace 3'-0" x 4'-0" steel frame storm window - 1st floor**	25	1 CARP	Ea.							
	Set up and secure scaffold				.444		23		23	30	37
	Remove old window				.385		15.80		15.80	20.50	25.50
	Install new window (to 12 S.F.)				.743	125	38.50		163.50	188	218
	Remove scaffold				.444		23		23	30	37
	Total				2.017	125	100.30		225.30	268.50	317.50
2030	**Refinish 3'-0" x 4'-0" steel frame storm window - 2nd floor**	5	1 PORD	Ea.							
	Set up and secure scaffold				.889		38.50		38.50	49.50	61.50
	Prepare window frame surface				.473		20.50		20.50	26.50	32.50
	Refinish window frame surface, brushwork, primer + 1 coat				.615	1.08	26.50		27.58	35.50	44
	Remove scaffold				.889		38.50		38.50	49.50	61.50
	Total				2.866	1.08	124		125.08	161	199.50
2040	**Replace 3'-0" x 4'-0" steel frame storm window - 2nd floor**	25	1 CARP	Ea.							
	Set up and secure scaffold				.889		46		46	60	74
	Remove old window				.385		15.80		15.80	20.50	25.50
	Install new window (to 12 S.F.)				.743	125	38.50		163.50	188	218
	Remove scaffold				.889		46		46	60	74
	Total				2.906	125	146.30		271.30	328.50	391.50
3030	**Refinish 3'-0" x 4'-0" steel frame storm window - 3rd floor**	5	1 PORD	Ea.							
	Set up and secure scaffold				1.333		57.50		57.50	74.50	92
	Prepare window frame surface				.473		20.50		20.50	26.50	32.50
	Refinish window frame surface, brushwork, primer + 1 coat				.615	1.08	26.50		27.58	35.50	44
	Remove scaffold				1.333		57.50		57.50	74.50	92
	Total				3.755	1.08	162		163.08	211	260.50

For customer support on your Facilities Maintenance & Repair Costs with RSMeans data, call 800.448.8182.

55

B2023 117 Steel Frame Storm Window

	System Description	Freq. (Years)	Crew	Unit	Labor Hours	2019 Bare Costs				Total In-House	Total w/O&P
						Material	Labor	Equipment	Total		
3040	**Replace 3'-0" x 4'-0" steel frame storm window - 3rd floor**	25	1 CARP	Ea.							
	Set up and secure scaffold				1.333		69		69	89.50	111
	Remove old window				.385		15.80		15.30	20.50	25.50
	Install new window (to 12 S.F.)				.743	125	38.50		163.50	188	218
	Remove scaffold				1.333		69		69	89.50	111
	Total				3.795	125	192.30		317.50	387.50	465.50

B2023 119 Wood Frame Storm Window

	System Description	Freq. (Years)	Crew	Unit	Labor Hours	2019 Bare Costs				Total In-House	Total w/O&P
						Material	Labor	Equipment	Total		
1010	**Replace glass - (1% of glass)**	1	1 CARP	S.F.							
	Remove glass				.052		2.13		2.13	2.78	3.44
	Install new float glass				.043	6.45	2.15		8.60	9.90	11.50
	Total				.095	6.45	4.28		10.73	12.68	14.94
1020	**Repair 3'-0" x 5'-0" wood frame storm window**	10	1 CARP	Ea.							
	Remove putty				.667		34.50		34.50	45	55.50
	Install new putty				.333	6.30	17.20		23.50	29.50	36
	Total				1.000	6.30	51.70		58	74.50	91.50
1030	**Refinish 3'-0" x 5'-0" wood frame storm window**	5	1 PORD	Ea.							
	Prepare window frame surface				.473		20.50		20.50	26.50	32.50
	Refinish window frame surface, brushwork, primer + 1 coat				.615	1.08	26.50		27.58	35.50	44
	Total				1.088	1.08	47		48.08	62	76.50
1040	**Replace 3'-0" x 5'-0" wood frame storm window - 1st floor**	40	1 CARP	Ea.							
	Set up and secure scaffold				.444		23		23	30	37
	Remove old window				.473		19.40		19.40	25.50	31.50
	Install new window (to 15 S.F.)				.693	174	35.80		209.80	238	275
	Remove scaffold				.444		23		23	30	37
	Total				2.055	174	101.20		275.20	323.50	380.50

56

B2023 119 Wood Frame Storm Window

	System Description	Freq. (Years)	Crew	Unit	Labor Hours	2019 Bare Costs				Total In-House	Total w/O&P
						Material	Labor	Equipment	Total		
2040	**Replace 3'-0" x 5'-0" wood frame storm window - 2nd floor**	40	1 CARP	Ea.							
	Set up and secure scaffold				.889		46		46	60	74
	Remove old window				.473		19.40		19.40	25.50	31.50
	Install new window (to 15 S.F.)				.693	174	35.80		209.80	238	275
	Remove scaffold				.889		46		46	60	74
	Total				2.944	174	147.20		321.20	383.50	454.50
3040	**Replace 3'-0" x 5'-0" wood frame storm window - 3rd floor**	40	1 CARP	Ea.							
	Set up and secure scaffold				1.333		69		69	89.50	111
	Remove old window				.473		19.40		19.40	25.50	31.50
	Install new window (to 15 S.F.)				.693	174	35.80		209.80	238	275
	Remove scaffold				1.333		69		69	89.50	111
	Total				3.833	174	193.20		367.20	442.50	528.50

B2023 122 Vinyl Clad Wood Window, Operating

	System Description	Freq. (Years)	Crew	Unit	Labor Hours	2019 Bare Costs				Total In-House	Total w/O&P
						Material	Labor	Equipment	Total		
1030	**Repl. 3'-0" x 4'-0" vinyl clad window - 1st flr.**	60	1 CARP	Ea.							
	Set up and secure scaffold				.444		23		23	30	37
	Remove window				.473		19.40		19.40	25.50	31.50
	Install new window (3'-0" x 4'-0")				1.156	405	59.50		464.50	525	605
	Remove scaffold				.444		23		23	30	37
	Total				2.518	405	124.90		529.90	610.50	710.50
2030	**Repl. 3'-0" x 4'-0" vinyl clad window - 2nd flr.**	60	1 CARP	Ea.							
	Set up and secure scaffold				.889		46		46	60	74
	Remove window				.473		19.40		19.40	25.50	31.50
	Install new window (3'-0" x 4'-0")				1.156	405	59.50		464.50	525	605
	Remove scaffold				.889		46		46	60	74
	Total				3.407	405	170.90		575.90	670.50	784.50

For customer support on your Facilities Maintenance & Repair Costs with RSMeans data, call 800.448.8182.

57

B2023 122 Vinyl Clad Wood Window, Operating

System Description	Freq. (Years)	Crew	Unit	Labor Hours	2019 Bare Costs Material	2019 Bare Costs Labor	2019 Bare Costs Equipment	2019 Bare Costs Total	Total In-House	Total w/O&P
3030										
Repl. 3'-0" x 4'-0" vinyl clad window - 3rd flr.	60	1 CARP	Ea.							
Set up and secure scaffold				1.333		69		69	89.50	111
Remove window				.473		19.40		19.40	25.50	31.50
Install new window (3'0" x 4'0")				1.156	405	59.50		464.50	525	605
Remove scaffold				1.333		69		69	89.50	111
Total				4.295	405	216.90		621.90	729.50	858.50

B2023 123 Vinyl Clad Wood Window, Fixed

System Description	Freq. (Years)	Crew	Unit	Labor Hours	2019 Bare Costs Material	2019 Bare Costs Labor	2019 Bare Costs Equipment	2019 Bare Costs Total	Total In-House	Total w/O&P
1030										
Repl. 4'-0" x 4'-0" vinyl clad window - 1st flr.	60	1 CARP	Ea.							
Set up and secure scaffold				.444		23		23	30	37
Remove window				.473		19.40		19.40	25.50	31.50
Install new window (4'-0" x 4'-0")				1.733	530	89.50		619.50	700	805
Remove scaffold				.444		23		23	30	37
Total				3.095	530	154.90		684.90	785.50	910.50
2030										
Repl. 4'-0" x 4'-0" vinyl clad window - 2nd flr.	60	1 CARP	Ea.							
Set up and secure scaffold				.889		46		46	60	74
Remove window				.473		19.40		19.40	25.50	31.50
Install new window (4'-0" x 4'-0")				1.733	530	89.50		619.50	700	805
Remove scaffold				.889		46		46	60	74
Total				3.984	530	200.90		730.90	845.50	984.50
3030										
Repl. 4'-0" x 4'-0" vinyl clad window - 3rd flr.	60	1 CARP	Ea.							
Set up and secure scaffold				1.333		69		69	89.50	111
Remove window				.473		19.40		19.40	25.50	31.50
Install new window (4'-0" x 4'-0")				1.733	530	89.50		619.50	700	805
Remove scaffold				1.333		69		69	89.50	111
Total				4.873	530	246.90		776.90	904.50	1,058.50

For customer support on your Facilities Maintenance & Repair Costs with RSMeans data, call 800.448.8182.

B2023 150 **Glass Block, Fixed**

	System Description	Freq. (Years)	Crew	Unit	Labor Hours	2019 Bare Costs				Total In-House	Total w/O&P
						Material	Labor	Equipment	Total		
1020	**Repair window - 1st floor (2% of glass)**	8	1 BRIC	S.F.							
	Cut out damaged block				.433		18	2.55	20.55	26	32
	Replace glass block				.452	23.50	21		44.50	53.50	64
	Total				.886	23.50	39	2.55	65.05	79.50	96
1030	**Replace glass block window - 1st floor**	75	1 BRIC	S.F.							
	Set up, secure and take down ladder				.267		13.60		13.60	17.95	22
	Demolish glass block				.074		3.05		3.05	3.97	4.92
	Replace glass block				.452	23.50	21		44.50	53.50	64
	Total				.793	23.50	37.65		61.15	75.42	90.92
2020	**Repair window - 2nd floor (2% of glass)**	8	1 BRIC	S.F.							
	Set up and secure scaffold				.889		45.50		45.50	60	74
	Cut out damaged block				.433		18	2.55	20.55	26	32
	Replace glass block				.452	23.50	21		44.50	53.50	64
	Remove scaffold				.889		45.50		45.50	60	74
	Total				2.663	23.50	130	2.55	156.05	199.50	244
2030	**Replace glass block window - 2nd floor**	75	1 BRIC	S.F.							
	Set up and secure scaffold				.889		45.50		45.50	60	74
	Demolish glass block				.074		3.05		3.05	3.97	4.92
	Replace glass block				.452	23.50	21		44.50	53.50	64
	Remove scaffold				.889		45.50		45.50	60	74
	Total				2.304	23.50	115.05		138.55	177.47	216.92
3020	**Repair window - 3rd floor (2% of glass)**	8	1 BRIC	S.F.							
	Set up and secure scaffold				1.333		68		68	89.50	111
	Cut out damaged block				.433		18	2.55	20.55	26	32
	Replace glass block				.452	23.50	21		44.50	53.50	64
	Remove scaffold				1.333		68		68	89.50	111
	Total				3.552	23.50	175	2.55	201.05	258.50	318
3030	**Replace glass block window - 3rd floor**	75	1 BRIC	S.F.							
	Set up and secure scaffold				1.333		68		68	89.50	111
	Demolish glass block				.074		3.05		3.05	3.97	4.92
	Replace glass block				.452	23.50	21		44.50	53.50	64
	Remove scaffold				1.333		68		68	89.50	111
	Total				3.193	23.50	160.05		183.55	236.47	290.92

For customer support on your Facilities Maintenance & Repair Costs with RSMeans data, call 800.448.8182.

59

B2023 160 Aluminum Shutter

	System Description	Freq. (Years)	Crew	Unit	Labor Hours	2019 Bare Costs				Total In-House	Total w/O&P
						Material	Labor	Equipment	Total		
1030	**Refinish aluminum shutter - 1st floor**	5	1 PORD	Ea.							
	Set up and secure scaffold				.444		19.20		19.20	25	30.50
	Prepare shutter for painting				.434		18.72		18.72	24	30
	Paint shutter surface (to 16 S.F.), brushwork, primer + 1 coat				.800	2.27	34.50		36.77	47	58.50
	Remove scaffold				.444		19.20		19.20	25	30.50
	Total				2.123	2.27	91.62		93.89	**121**	**149.50**
1040	**Replace aluminum shutter - 1st floor**	60	1 CARP	Ea.							
	Set up and secure scaffold				.444		23		23	30	37
	Remove shutter				.260		13.50		13.50	17.50	22
	Install new shutter				.520	100	26.75		126.75	145	168
	Remove scaffold				.444		23		23	30	37
	Total				1.669	100	86.25		186.25	**222.50**	**264**
2030	**Refinish aluminum shutter - 2nd floor**	5	1 PORD	Ea.							
	Set up and secure scaffold				.889		38.50		38.50	49.50	61.50
	Prepare shutter for painting				.434		18.72		18.72	24	30
	Paint shutter surface (to 16 S.F.), brushwork, primer + 1 coat				.800	2.27	34.50		36.77	47	58.50
	Remove scaffold				.889		38.50		38.50	49.50	61.50
	Total				3.012	2.27	130.22		132.49	**170**	**211.50**
2040	**Replace aluminum shutter - 2nd floor**	60	1 CARP	Ea.							
	Set up and secure scaffold				.889		46		46	60	74
	Remove shutter				.260		13.50		13.50	17.50	22
	Install new shutter				.520	100	26.75		126.75	145	168
	Remove scaffold				.889		46		46	60	74
	Total				2.558	100	132.25		232.25	**282.50**	**338**
3030	**Refinish aluminum shutter - 3rd floor**	5	1 PORD	Ea.							
	Set up and secure scaffold				1.333		57.50		57.50	74.50	92
	Prepare shutter for painting				.434		18.72		18.72	24	30
	Paint shutter surface (to 16 S.F.), brushwork, primer + 1 coat				.800	2.27	34.50		36.77	47	58.50
	Remove scaffold				1.333		57.50		57.50	74.50	92
	Total				3.901	2.27	168.22		170.49	**220**	**272.50**

For customer support on your Facilities Maintenance & Repair Costs with RSMeans data, call 800.448.8182.

B20 EXTERIOR CLOSURE — B2023 — Exterior Windows

B2023 160 — Aluminum Shutter

	System Description	Freq. (Years)	Crew	Unit	Labor Hours	2019 Bare Costs				Total In-House	Total w/O&P
						Material	Labor	Equipment	Total		
3040	**Replace aluminum shutter - 3rd floor**	60	1 CARP	Ea.							
	Set up and secure scaffold				1.333		69		69	89.50	111
	Remove shutter				.260		13.50		13.50	17.50	22
	Install new shutter				.520	100	26.75		126.75	145	168
	Remove scaffold				1.333		69		69	89.50	111
	Total				3.447	100	178.25		278.25	341.50	412

B2023 162 — Steel Shutter

	System Description	Freq. (Years)	Crew	Unit	Labor Hours	2019 Bare Costs				Total In-House	Total w/O&P
						Material	Labor	Equipment	Total		
1030	**Refinish steel shutter - 1st floor**	5	1 PORD	Ea.							
	Set up and secure scaffold				.444		19.20		19.20	25	30.50
	Prepare shutter for painting				.434		18.72		18.72	24	30
	Paint shutter surface (to 16 S.F.), brushwork, primer + 1 coat				.800	2.27	34.50		36.77	47	58.50
	Remove scaffold				.444		19.20		19.20	25	30.50
	Total				2.123	2.27	91.62		93.89	121	149.50
1040	**Replace steel shutter - 1st floor**	40	1 CARP	Ea.							
	Set up and secure scaffold				.444		23		23	30	37
	Remove old shutter				.308		17.40	.90	18.30	24.50	30
	Install new steel shutter				.617	37.80	34.80	1.86	74.46	91	107
	Remove scaffold				.444		23		23	30	37
	Total				1.814	37.80	98.20	2.76	138.76	175.50	211
2030	**Refinish steel shutter - 2nd floor**	5	1 PORD	Ea.							
	Set up and secure scaffold				.889		38.50		38.50	49.50	61.50
	Prepare shutter for painting				.434		18.72		18.72	24	30
	Paint shutter surface (to 16 S.F.), brushwork, primer + 1 coat				.800	2.27	34.50		36.77	47	58.50
	Remove scaffold				.889		38.50		38.50	49.50	61.50
	Total				3.012	2.27	130.22		132.49	170	211.50

For customer support on your Facilities Maintenance & Repair Costs with RSMeans data, call 800.448.8182.

61

B2023 162 Steel Shutter

	System Description	Freq. (Years)	Crew	Unit	Labor Hours	2019 Bare Costs				Total In-House	Total w/O&P
						Material	Labor	Equipment	Total		
2040	**Replace steel shutter - 2nd floor**	40	1 CARP	Ea.							
	Set up and secure scaffold				.889		46		46	60	74
	Remove old shutter				.308		17.40	.90	18.30	24.50	30
	Install new steel shutter				.617	37.80	34.80	1.86	74.46	91	107
	Remove scaffold				.889		46		46	60	74
	Total				2.703	37.80	144.20	2.76	184.76	235.50	285
3030	**Refinish steel shutter - 3rd floor**	5	1 PORD	Ea.							
	Set up and secure scaffold				1.333		57.50		57.50	74.50	92
	Prepare shutter for painting				.434		18.72		18.72	24	30
	Paint shutter surface (to 16 S.F.), brushwork, primer + 1 coat				.800	2.27	34.50		36.7?	47	58.50
	Remove scaffold				1.333		57.50		57.50	74.50	92
	Total				3.901	2.27	168.22		170.4?	220	272.50
3040	**Replace steel shutter - 3rd floor**	40	1 CARP	Ea.							
	Set up and secure scaffold				1.333		69		69	89.50	111
	Remove old shutter				.308		17.40	.90	18.30	24.50	30
	Install new steel shutter				.617	37.80	34.80	1.86	74.46	91	107
	Remove scaffold				1.333		69		69	89.50	111
	Total				3.592	37.80	190.20	2.76	230.76	294.50	359

B2023 164 Wood Shutter

	System Description	Freq. (Years)	Crew	Unit	Labor Hours	2019 Bare Costs				Total In-House	Total w/O&P
						Material	Labor	Equipment	Total		
1010	**Repair wood shutter - 1st floor**	6	1 CARP	Ea.							
	Remove shutter				.260		13.50		13.50	17.50	22
	Remove damaged panel				.157		8.10		8.10	10.55	13.10
	Install new panel				.314	4.74	16.20		20.94	26.50	32
	Install shutter				.520		26.75		26.75	35	43.50
	Total				1.251	4.74	64.55		69.29	89.55	110.60

For customer support on your Facilities Maintenance & Repair Costs with RSMeans data, call 800.448.8182.

B2023 164 | **Wood Shutter**

	System Description	Freq. (Years)	Crew	Unit	Labor Hours	2019 Bare Costs				Total In-House	Total w/O&P
						Material	Labor	Equipment	Total		
1030	**Refinish wood shutter - 1st floor**	5	1 PORD	Ea.							
	Set up and secure scaffold				.444		19.20		19.20	25	30.50
	Prepare shutter for painting				.434		18.72		18.72	24	30
	Paint shutter surface (to 16 S.F.), brushwork, primer + 1 coat				.800	2.27	34.50		36.77	47	58.50
	Remove scaffold				.444		19.20		19.20	25	30.50
	Total				2.123	2.27	91.62		93.89	121	149.50
1040	**Replace wood shutter - 1st floor**	40	1 CARP	Ea.							
	Set up and secure scaffold				.444		23		23	30	37
	Remove shutter				.260		13.50		13.50	17.50	22
	Install new wood shutter				.520	147	26.75		173.75	197	227
	Remove scaffold				.444		23		23	30	37
	Total				1.669	147	86.25		233.25	274.50	323
2010	**Repair wood shutter - 2nd floor**	6	1 CARP	Ea.							
	Set up and secure scaffold				.889		46		46	60	74
	Remove shutter				.260		13.50		13.50	17.50	22
	Remove damaged panel				.157		8.10		8.10	10.55	13.10
	Install new panel				.314	4.74	16.20		20.94	26.50	32
	Install shutter				.520		26.75		26.75	35	43.50
	Remove scaffold				.889		46		46	60	74
	Total				3.029	4.74	156.55		161.29	209.55	258.60
2030	**Refinish wood shutter - 2nd floor**	5	1 PORD	Ea.							
	Set up and secure scaffold				.889		38.50		38.50	49.50	61.50
	Prepare shutter for painting				.434		18.72		18.72	24	30
	Paint shutter surface (to 16 S.F.), brushwork, primer + 1 coat				.800	2.27	34.50		36.77	47	58.50
	Remove scaffold				.889		38.50		38.50	49.50	61.50
	Total				3.012	2.27	130.22		132.49	170	211.50
2040	**Replace wood shutter - 2nd floor**	40	1 CARP	Ea.							
	Set up and secure scaffold				.889		46		46	60	74
	Remove shutter				.260		13.50		13.50	17.50	22
	Install new wood shutter				.520	147	26.75		173.75	197	227
	Remove scaffold				.889		46		46	60	74
	Total				2.558	147	132.25		279.25	334.50	397

B20	EXTERIOR CLOSURE	B2023	Exterior Windows

B2023 164 — Wood Shutter

	System Description	Freq. (Years)	Crew	Unit	Labor Hours	2019 Bare Costs				Total In-House	Total w/O&P
						Material	Labor	Equipment	Total		
3010	**Repair wood shutter - 3rd floor**	6	1 CARP	Ea.							
	Set up and secure scaffold				1.333		69		69	89.50	111
	Remove shutter				.260		13.50		13.50	17.50	22
	Remove damaged panel				.157		8.10		8.10	10.55	13.10
	Install new panel				.314	4.74	16.20		20.34	26.50	32
	Install shutter				.520		26.75		26.75	35	43.50
	Remove scaffold				1.333		69		69	89.50	111
	Total				3.917	4.74	202.55		207.29	268.55	332.60
3030	**Refinish wood shutter - 3rd floor**	5	1 PORD	Ea.							
	Set up and secure scaffold				1.333		57.50		57.50	74.50	92
	Prepare shutter for painting				.434		18.72		18.72	24	30
	Paint shutter surface (to 16 S.F.), brushwork, primer + 1 coat				.800	2.27	34.50		36.77	47	58.50
	Remove scaffold				1.333		57.50		57.50	74.50	92
	Total				3.901	2.27	168.22		170.49	220	272.50
3040	**Replace wood shutter - 3rd floor**	40	1 CARP	Ea.							
	Set up and secure scaffold				1.333		69		69	89.50	111
	Remove shutter				.260		13.50		13.50	17.50	22
	Install new wood shutter				.520	147	26.75		173.75	197	227
	Remove scaffold				1.333		69		69	89.50	111
	Total				3.447	147	178.25		325.25	393.50	471

B2023 166 — Metal Window Grating

	System Description	Freq. (Years)	Crew	Unit	Labor Hours	2019 Bare Costs				Total In-House	Total w/O&P
						Material	Labor	Equipment	Total		
1020	**Refinish 3'-0" x 4'-0" metal window grating**	5	1 PORD	Ea.							
	Prepare window grating surface				.308		13.30		13.30	17.15	21.50
	Paint grated surface, brushwork, primer + 1 coat				.167	1.68	7.20		8.88	11.30	13.60
	Total				.474	1.68	20.50		22.18	28.45	35.10
1030	**Replace 3'-0" x 4'-0" metal window grating - 1st floor**	30	1 SSWK	Ea.							
	Remove old grating				.800		33		33	43	53
	Fabricate and install new window grating				6.275	256	352		608	760	910
	Total				7.075	256	385		641	803	963

For customer support on your Facilities Maintenance & Repair Costs with RSMeans data, call 800.448.8182.

B20 EXTERIOR CLOSURE — B2023 — Exterior Windows

B2023 166 — Metal Window Grating

	System Description	Freq. (Years)	Crew	Unit	Labor Hours	2019 Bare Costs Material	Labor	Equipment	Total	Total In-House	Total w/O&P
2030	**Replace 3'-0" x 4'-0" metal window grating - 2nd floor**	30	1 SSWK	Ea.							
	Set up and secure scaffold				.889		50		50	67.50	83
	Remove old grating				.800		33		33	43	53
	Fabricate and install new window grating				6.275	256	352		608	760	910
	Remove scaffold				.889		50		50	67.50	83
	Total				8.852	256	485		741	938	1,129
3030	**Replace 3'-0" x 4'-0" metal window grating - 3rd floor**	30	1 SSWK	Ea.							
	Set up and secure scaffold				1.333		74.50		74.50	101	124
	Remove old window grating				.800		33		33	43	53
	Fabricate and install new window grating				6.275	256	352		608	760	910
	Remove scaffold				1.333		74.50		74.50	101	124
	Total				9.741	256	534		790	1,005	1,211

B2023 168 — Metal Wire Mesh Cover

	System Description	Freq. (Years)	Crew	Unit	Labor Hours	2019 Bare Costs Material	Labor	Equipment	Total	Total In-House	Total w/O&P
1020	**Refinish 3'-0" x 4'-0" metal wire mesh window cover**	5	1 PORD	Ea.							
	Prepare metal mesh surface				.103		4.44		4.44	5.70	7.10
	Paint metal mesh surface, brushwork, primer + 1 coat				.167	1.68	7.20		8.88	11.30	13.60
	Total				.269	1.68	11.64		13.32	17	20.70
1030	**Repl. 3'-0" x 4'-0" metal wire mesh window cover - 1st floor**	30	1 SSWK	Ea.							
	Set up and secure scaffold				.444		25		25	33.50	41.50
	Remove old mesh				.800		33		33	43	53
	Install new wire mesh & frame (to 12 S.F.)				.452	62	23.25		85.25	98.50	115
	Remove scaffold				.444		25		25	33.50	41.50
	Total				2.141	62	106.25		168.25	208.50	251
2030	**Repl. 3'-0" x 4'-0" metal wire mesh window cover - 2nd floor**	30	1 SSWK	Ea.							
	Set up and secure scaffold				.889		50		50	67.50	83
	Remove old mesh				.800		33		33	43	53
	Install new wire mesh & frame (to 12 S.F.)				.452	62	23.25		85.25	98.50	115
	Remove scaffold				.889		50		50	67.50	83
	Total				3.030	62	156.25		218.25	276.50	334

For customer support on your Facilities Maintenance & Repair Costs with RSMeans data, call 800.448.8182.

65

B2023 168 | Metal Wire Mesh Cover

	System Description	Freq. (Years)	Crew	Unit	Labor Hours	2019 Bare Costs Material	Labor	Equipment	Total	Total In-House	Total w/O&P
3030	Repl. 3'-0" x 4'-0" metal wire mesh window cover - 3rd floor	30	1 SSWK	Ea.							
	Set up and secure scaffold				1.333		74.50		74.50	101	124
	Remove old mesh				.800		33		33	43	53
	Install new wire mesh & frame (to 12 S.F.)				.452	62	23.25		85.25	98.50	115
	Remove scaffold				1.333		74.50		74.50	101	124
	Total				3.919	62	205.25		267.25	343.50	416

B2023 170 | Aluminum Frame Window Screen

	System Description	Freq. (Years)	Crew	Unit	Labor Hours	2019 Bare Costs Material	Labor	Equipment	Total	Total In-House	Total w/O&P
1010	Repair aluminum frame window screen	20	1 CARP	S.F.							
	Remove screen from mountings				.052		2.91		2.91	3.95	4.85
	Repair damaged screen				.017	3.35	.97		4.32	5	5.80
	Remount screen				.104		5.80		5.80	7.90	9.70
	Total				.173	3.35	9.68		13.00	16.85	20.35
1020	Refinish 3'-0" x 4'-0" aluminum frame window screen	5	1 PORD	Ea.							
	Prepare screen frame surface				.473		20.50		20.50	26.50	32.50
	Refinish screen frame surface, brushwork, primer + 1 coat				.615	1.08	26.50		27.58	35.50	44
	Total				1.088	1.08	47		48.08	62	76.50
1030	Replace 3'-0" x 4'-0" aluminum frame window screen - 1st floor	50	1 CARP	Ea.							
	Set up and secure scaffold				.444		23		23	30	37
	Remove screen from mountings				.624		34.92		34.92	47.50	58
	Install new screen				1.248	276	69.60		345.60	400	460
	Remove scaffold				.444		23		23	30	37
	Total				2.761	276	150.52		426.52	507.50	592
2030	Replace 3'-0" x 4'-0" aluminum frame window screen - 2nd floor	50	1 CARP	Ea.							
	Set up and secure scaffold				.889		46		46	60	74
	Remove screen from mountings				.624		34.92		34.92	47.50	58
	Install new screen				1.248	276	69.60		345.60	400	460
	Remove scaffold				.889		46		46	60	74
	Total				3.650	276	196.52		472.52	567.50	666

For customer support on your Facilities Maintenance & Repair Costs with RSMeans data, call 800.448.8182.

B2023 170 Aluminum Frame Window Screen

System Description	Freq. (Years)	Crew	Unit	Labor Hours	2019 Bare Costs Material	Labor	Equipment	Total	Total In-House	Total w/O&P
3030 Replace 3'-0" x 4'-0" aluminum frame window screen - 3rd floor	50	1 CARP	Ea.							
Set up and secure scaffold				1.333		69		69	89.50	111
Remove screen from mountings				.624		34.92		34.92	47.50	58
Install new screen				1.248	276	69.60		345.60	400	460
Remove scaffold				1.333		69		69	89.50	111
Total				4.539	276	242.52		518.52	626.50	740

B2023 172 Steel Frame Window Screen

System Description	Freq. (Years)	Crew	Unit	Labor Hours	2019 Bare Costs Material	Labor	Equipment	Total	Total In-House	Total w/O&P
1010 Repair steel frame window screen	10	1 CARP	S.F.							
Remove screen from mountings				.052		2.91		2.91	3.95	4.85
Repair damaged screen				.017	3.35	.97		4.32	5	5.80
Remount screen				.104		5.80		5.80	7.90	9.70
Total				.173	3.35	9.68		13.03	16.85	20.35
1020 Refinish 3'-0" x 4'-0" steel frame window screen	5	1 PORD	Ea.							
Prepare screen frame surface				.473		20.50		20.50	26.50	32.50
Refinish screen frame surface, brushwork, primer + 1 coat				.615	1.08	26.50		27.58	35.50	44
Total				1.088	1.08	47		48.08	62	76.50
1030 Replace 3'-0" x 4'-0" steel frame window screen - 1st floor	40	1 CARP	Ea.							
Set up and secure scaffold				.444		23		23	30	37
Remove screen from mountings				.624		34.92		34.92	47.50	58
Install new screen				1.248	54.48	69.60		124.08	155	185
Remove scaffold				.444		23		23	30	37
Total				2.761	54.48	150.52		205	262.50	317
2030 Replace 3'-0" x 4'-0" steel frame window screen - 2nd floor	40	1 CARP	Ea.							
Set up and secure scaffold				.889		46		46	60	74
Remove screen from mountings				.624		34.92		34.92	47.50	58
Install new screen				1.248	54.48	69.60		124.08	155	185
Remove scaffold				.889		46		46	60	74
Total				3.650	54.48	196.52		251	322.50	391

For customer support on your Facilities Maintenance & Repair Costs with RSMeans data, call 800.448.8182.

67

B2023 172 Steel Frame Window Screen

	System Description	Freq. (Years)	Crew	Unit	Labor Hours	2019 Bare Costs				Total In-House	Total w/O&P
						Material	Labor	Equipment	Total		
3030	**Replace 3'-0" x 4'-0" steel frame window screen - 3rd floor**	40	1 CARP	Ea.							
	Set up and secure scaffold				1.333		69		69	89.50	111
	Remove screen from mountings				.624		34.92		34.92	47.50	58
	Install new screen				1.248	54.48	69.60		124.08	155	185
	Remove scaffold				1.333		69		69	89.50	111
	Total				4.539	54.48	242.52		297	381.50	465

B2023 174 Wood Frame Window Screen

	System Description	Freq. (Years)	Crew	Unit	Labor Hours	2019 Bare Costs				Total In-House	Total w/O&P
						Material	Labor	Equipment	Total		
1010	**Repair wood frame window screen**	10	1 CARP	S.F.							
	Remove screen from mountings				.052		2.91		2.91	3.95	4.85
	Repair damaged screen				.017	3.35	.97		4.32	5	5.80
	Remount screen				.104		5.80		5.80	7.90	9.70
	Total				.173	3.35	9.68		13.03	16.85	20.35
1020	**Refinish 3'-0" x 4'-0" wood frame window screen**	5	1 PORD	Ea.							
	Prepare screen frame surface				.473		20.50		20.50	26.50	32.50
	Refinish screen frame surface, brushwork, primer + 1 coat				.615	1.08	26.50		27.58	35.50	44
	Total				1.088	1.08	47		48.08	62	76.50
1030	**Replace 3'-0" x 4'-0" wood frame window screen - 1st floor**	40	1 CARP	Ea.							
	Set up and secure scaffold				.444		23		23	30	37
	Remove screen from mountings				.624		34.92		34.92	47.50	58
	Install new screen				.666	102.60	34.44		137.04	158	184
	Remove scaffold				.444		23		23	30	37
	Total				2.179	102.60	115.36		217.96	265.50	316
2030	**Replace 3'-0" x 4'-0" wood frame window screen - 2nd floor**	40	1 CARP	Ea.							
	Set up and secure scaffold				.889		46		46	60	74
	Remove screen from mountings				.624		34.92		34.92	47.50	58
	Install new screen				.666	102.60	34.44		137.04	158	184
	Remove scaffold				.889		46		46	60	74
	Total				3.067	102.60	161.36		263.96	325.50	390

For customer support on your Facilities Maintenance & Repair Costs with RSMeans data, call 800.448.8182.

B20 EXTERIOR CLOSURE | B2023 | Exterior Windows

B2023 174 Wood Frame Window Screen

	System Description	Freq. (Years)	Crew	Unit	Labor Hours	2019 Bare Costs				Total In-House	Total w/O&P
						Material	Labor	Equipment	Total		
3030	Replace 3'-0" x 4'-0" wood frame window screen - 3rd floor	40	1 CARP	Ea.							
	Set up and secure scaffold				1.333		69		69	89.50	111
	Remove screen from mountings				.624		34.92		34.92	47.50	58
	Install new screen				.666	102.60	34.44		137.04	158	184
	Remove scaffold				1.333		69		69	89.50	111
	Total				3.956	102.60	207.36		309.96	384.50	464

B2023 324 Plate Glass

	System Description	Freq. (Years)	Crew	Unit	Labor Hours	2019 Bare Costs				Total In-House	Total w/O&P
						Material	Labor	Equipment	Total		
1020	Replace plate glass storefront - 1st floor	50	2 CARP	C.S.F.							
	Remove plate glass and frame				3.467		179		179	233	289
	Replace plate glass and frame				16.000	3,150	795		3,945	4,500	5,225
	Total				19.467	3,150	974		4,124	4,733	5,514
2020	Replace plate glass storefront - 2nd floor	50	2 CARP	C.S.F.							
	Set up and secure scaffold				1.500		77.50		77.50	101	125
	Remove plate glass and frame				3.467		179		179	233	289
	Replace plate glass and frame				16.000	3,150	795		3,945	4,500	5,225
	Remove scaffold				1.500		77.50		77.50	101	125
	Total				22.467	3,150	1,129		4,279	4,935	5,764
3020	Replace plate glass storefront - 3rd floor	50	2 CARP	C.S.F.							
	Set up and secure scaffold				2.250		116.25		116.25	151	188
	Remove plate glass and frame				3.467		179		179	233	289
	Replace plate glass and frame				16.000	3,150	795		3,945	4,500	5,225
	Remove scaffold				2.250		116.25		116.25	151	188
	Total				23.967	3,150	1,206.50		4,356.50	5,035	5,890

B2033 111 Glazed Aluminum

	System Description	Freq. (Years)	Crew	Unit	Labor Hours	2019 Bare Costs				Total In-House	Total w/O&P
						Material	Labor	Equipment	Total		
1020	**Repair aluminum storefront door**	12	1 CARP	Ea.							
	Remove weatherstripping				.015		.76		.76	.99	1.23
	Remove door closer				.167		8.60		8.60	11.20	13.90
	Remove door hinge				.167		8.60		8.60	11.20	13.90
	Install new weatherstrip				3.478	51	180		231	290	355
	Install new door closer				.868	102	44.75		146.75	171	200
	Install new hinge (50% of total hinges / 12 years)				.167		8.60		8.60	11.20	13.90
	Hinge, 5 x 5, brass base					64			64	70.50	80
	Oil / lubricate door closer				.065		3.33		3.33	4.34	5.35
	Oil / lubricate hinges				.065		3.33		3.33	4.34	5.35
	Total				4.990	217	257.97		474.97	**574.77**	**688.63**
1030	**Replace 3'-0" x 7'-0" aluminum storefront doors**	50	1 CARP	Ea.							
	Remove door				.650		26.50		26.50	35	43
	Remove door frame				1.300		67		67	87.50	108
	Install new door, frame, and hardware				10.458	925	585		1,510	1,800	2,125
	Install new glass				3.640	149.10	180.60		329.70	400	475
	Total				16.048	1,074.10	859.10		1,933.20	**2,322.50**	**2,751**
2010	**Replace insulating glass - (3% of glass)**	1	1 CARP	S.F.							
	Remove damaged glass				.052		2.13		2.13	2.78	3.44
	Install new insulating glass				.277	33.50	13.75		47.25	54.50	64
	Total				.329	33.50	15.88		49.38	**57.28**	**67.44**
2020	**Repair aluminum storefront sliding door**	12	1 CARP	Ea.							
	Remove weatherstripping				.015		.76		.76	.99	1.23
	Oil / lubricate door closer				.065		3.33		3.33	4.34	5.35
	Oil / inspect wheel & track				.065		3.33		3.33	4.34	5.35
	Install new door track / glide wheels (25% of total / 12 years)				.440	13.88	22.75		36.63	45	54
	Install new weatherstrip				4.167	55	215		270	340	415
	Total				4.750	68.88	245.17		314.05	**394.67**	**480.93**

For customer support on your Facilities Maintenance & Repair Costs with RSMeans data, call 800.448.8182.

B2033 130 | Glazed Wood

System Description	Freq. (Years)	Crew	Unit	Labor Hours	Material	Labor	Equipment	Total	Total In-House	Total w/O&P
						2019 Bare Costs				
1020 Repair glazed wood doors	12	1 CARP	Ea.							
Remove weatherstripping				.015		.76		.76	.99	1.23
Remove door closer				.167		8.60		8.60	11.20	13.90
Remove door hinge				.167		8.60		8.60	11.20	13.90
Install new weatherstrip				3.478	51	180		231	290	355
Install new door closer				.868	102	44.75		146.75	171	200
Install new hinge (50% of total hinges / 12 years)				.167		8.60		8.60	11.20	13.90
Hinge, 5 x 5, brass base					64			64	70.50	80
Oil / lubricate door closer				.065		3.33		3.33	4.34	5.35
Oil / lubricate hinges				.065		3.33		3.33	4.34	5.35
Total				4.990	217	257.97		474.97	574.77	688.63
1040 Replace 3'-0" x 7'-0" glazed wood door	40	1 CARP	Ea.							
Remove door				.650		26.50		26.50	35	43
Remove door frame				.743		38.50		38.50	50	62
Install new door frame				.943	140.25	48.79		189.04	218	254
Install new oak door sill				.832	86	43		129	151	177
Install exterior door				2.105	870	109		979	1,100	1,275
Total				5.274	1,096.25	265.79		1,362.04	1,554	1,811
1050 Refinish glazed wood doors	4	1 PORD	Ea.							
Paint door, door frame & trim, brushwork, primer + 2 coats				1.760	15.90	76		91.90	116	142
Total				1.760	15.90	76		91.90	116	142
2010 Replace insulating glass - (8% of glass)	1	1 CARP	S.F.							
Remove damaged glass				.052		2.13		2.13	2.78	3.44
Install new insulated glass				.277	33.50	13.75		47.25	54.50	64
Total				.329	33.50	15.88		49.38	57.28	67.44
2020 Repair glazed wood sliding door	14	1 CARP	Ea.							
Remove weatherstripping				.015		.76		.76	.99	1.23
Oil / lubricate door closer				.065		3.33		3.33	4.34	5.35
Oil / inspect wheel & track				.065		3.33		3.33	4.34	5.35
Install new door track / glide wheels (25% of total / 14 years)				.440	13.88	22.75		36.63	45	54
Install new weatherstrip				4.167	55	215		270	340	415
Total				4.750	68.88	245.17		314.05	394.67	480.93

For customer support on your Facilities Maintenance & Repair Costs with RSMeans data, call 800.448.8182.

71

B20 EXTERIOR CLOSURE	B2033	Exterior Doors

B2033 130 Glazed Wood

	System Description	Freq. (Years)	Crew	Unit	Labor Hours	2019 Bare Costs				Total In-House	Total w/O&P
						Material	Labor	Equipment	Total		
2030	**Prepare and refinish glazed wood sliding door**	4	1 PORD	Ea.							
	Prepare surface				.310		13.40		13.40	17.25	21.50
	Paint door surface, brushwork, primer + 2 coats				.310	7.28	13.52		20.80	25.50	30.50
	Total				.621	7.28	26.92		34.20	42.75	52
2040	**Replace 6'-0" x 7'-0" glazed wood sliding door**	40	2 CARP	Ea.							
	Remove sliding door				1.733		89.50		89.50	117	144
	Install new sliding door				5.200	1,500	269		1,769	2,000	2,300
	Total				6.933	1,500	358.50		1,858.50	2,117	2,444
2050	**Refinish glazed wood sliding door & frame**	4	1 PORD	Ea.							
	Prepare surface				.465		20.10		20.10	26	32.50
	Paint door, door frame & trim				1.760	15.90	76		91.90	116	142
	Total				2.225	15.90	96.10		112	142	174.50

B2033 206 Solid Core, Painted

	System Description	Freq. (Years)	Crew	Unit	Labor Hours	2019 Bare Costs				Total In-House	Total w/O&P
						Material	Labor	Equipment	Total		
1010	**Repair solid core door, painted**	12	1 CARP	Ea.							
	Remove weatherstripping				.015		.76		.76	.99	1.23
	Remove door hinge				.250		12.90		12.90	16.85	21
	Install new weatherstrip				3.478	51	180		231	290	355
	Install new hinge (50% of total hinges / 12 years)				.250		12.90		12.90	16.85	21
	Hinge, 5 x 5, brass base					96			96	106	120
	Oil / lubricate hinges				.065		3.33		3.33	4.34	5.35
	Total				4.058	147	209.89		356.89	435.03	523.58
1020	**Prepare and refinish solid core door, painted**	4	1 PORD	Ea.							
	Prepare door surface				.333		14.40		14.40	18.55	23
	Paint door surface, brushwork, primer + 2 coats				.549	12.88	23.92		36.80	45	54.50
	Total				.883	12.88	38.32		51.20	63.55	77.50

For customer support on your Facilities Maintenance & Repair Costs with RSMeans data, call 800.448.8182.

B2033 206 Solid Core, Painted

	System Description	Freq. (Years)	Crew	Unit	Labor Hours	Material	Labor	Equipment	Total	Total In-House	Total w/O&P
							2019 Bare Costs				
1030	**Replace 3'-0" x 7'-0" solid core door, painted**	40	1 CARP	Ea.							
	Remove door				.650		26.50		26.50	35	43
	Remove door frame				.743		38.50		38.50	50	62
	Install new door frame				.943	140.25	48.79		189.04	218	254
	Install new oak door sill				.832	86	43		129	151	177
	Install new solid core door				1.143	227	59		286	325	380
	Install new weatherstrip				4.167	55	215		270	340	415
	Install new lockset				.867	79.50	45		124.50	146	171
	Total				9.345	587.75	475.79		1,063.54	1,265	1,502
2010	**Repair solid core sliding wood door**	14	1 CARP	Ea.							
	Remove weatherstripping				.015		.76		.76	.99	1.23
	Oil / lubricate door closer				.065		3.33		3.33	4.34	5.35
	Oil / inspect wheel & track				.065		3.33		3.33	4.34	5.35
	Install new door track / glide wheels (25% of total / 14 years)				.440	13.88	22.75		36.63	45	54
	Install new weatherstrip				4.167	55	215		270	340	415
	Total				4.750	68.88	245.17		314.05	394.67	480.93
2030	**Replace 3'-0" x 7'-0" solid core sliding wood door**	40	1 CARP	Ea.							
	Remove sliding door				1.733		89.50		89.50	117	144
	Install new sliding door frame				.998	121.50	51.66		173.16	201	235
	Install new door track / glide wheels				1.760	55.50	91		146.50	179	216
	Install new solid core door				1.143	227	59		286	325	380
	Install new weatherstrip				3.478	51	180		231	290	355
	Install new lockset				.867	79.50	45		124.50	146	171
	Total				9.980	534.50	516.16		1,050.66	1,258	1,501
3010	**Replace sliding door glass - (3% of glass)**	1	1 CARP	S.F.							
	Remove damaged glass				.052		2.13		2.13	2.78	3.44
	Install new glass				.173	7.10	8.60		15.70	18.95	22.50
	Total				.225	7.10	10.73		17.83	21.73	25.94

For customer support on your Facilities Maintenance & Repair Costs with RSMeans data, call 800.448.8182.

73

B2033 206 | Solid Core, Painted

	System Description	Freq. (Years)	Crew	Unit	Labor Hours	2019 Bare Costs				Total In-House	Total w/O&P
						Material	Labor	Equipment	Total		
3040	Replace 3'-0" x 7'-0" solid core door, w/ safety glass, painted	40	1 CARP	Ea.							
	Remove door				.650		26.50		26.50	35	43
	Remove door frame				.743		38.50		38.50	50	62
	Install new door frame				.943	140.25	48.79		189.04	218	254
	Install new oak door sill				.832	86	43		129	151	177
	Install new solid core door				1.143	227	59		286	325	380
	Install new glass (3% of total glass/year)				.201	8.24	9.98		18.22	22	26.50
	Install new weatherstrip				4.167	55	215		270	340	415
	Install new lockset				.867	79.50	45		124.50	146	171
	Total				9.546	595.99	485.77		1,081.76	1,287	1,528.50

B2033 210 | Aluminum Doors

	System Description	Freq. (Years)	Crew	Unit	Labor Hours	2019 Bare Costs				Total In-House	Total w/O&P
						Material	Labor	Equipment	Total		
1010	Repair aluminum door	12	1 CARP	Ea.							
	Remove door closer				.167		8.60		8.60	11.20	13.90
	Remove door hinge				.250		12.90		12.90	16.85	21
	Install new door closer				1.735	204	89.50		293.50	340	400
	Install new hinge (50% of total hinges / 12 years)				.250		12.90		12.90	16.85	21
	Hinge, 5 x 5, brass base					96			96	106	120
	Oil / lubricate door closer				.065		3.33		3.33	4.34	5.35
	Oil / lubricate hinges				.065		3.33		3.33	4.34	5.35
	Total				2.531	300	130.56		430.56	499.58	586.60
1020	Replace 3'-0" x 7'-0" flush aluminum door	50	1 CARP	Ea.							
	Remove door				.650		26.50		26.50	35	43
	Remove cylinder lock				.400		20.50		20.50	27	33.50
	Reinstall cylinder lock				1.156		59.50		59.50	78	96.50
	Install new door				1.224	1,575	63		1,638	1,825	2,075
	Total				3.431	1,575	169.50		1,744.50	1,965	2,248
2010	Replace wire glass - (3% of glass)	1	1 CARP	S.F.							
	Remove damaged glass				.052		2.13		2.13	2.78	3.44
	Install new wire glass				.179	31.32	8.87		40.19	46	53.50
	Total				.231	31.32	11		42.32	48.78	56.94

For customer support on your Facilities Maintenance & Repair Costs with RSMeans data, call 800.448.8182.

B2033 210 — Aluminum Doors

	System Description	Freq. (Years)	Crew	Unit	Labor Hours	Material	Labor	Equipment	Total	Total In-House	Total w/O&P
								2019 Bare Costs			
2020	**Repair aluminum frame and door**	12	1 CARP	Ea.							
	Remove weatherstripping				.015		.76		.76	.99	1.23
	Remove door closer				.167		8.60		8.60	11.20	13.90
	Remove door hinge				.250		12.90		12.90	16.85	21
	Install new weatherstrip				3.478	51	180		231	290	355
	Install new door closer				1.735	204	89.50		293.50	340	400
	Install new hinge (50% of total hinges / 12 years)				.250		12.90		12.90	16.85	21
	Hinge, 5 x 5, brass base					96			96	106	120
	Oil / lubricate door closer				.065		3.33		3.33	4.34	5.35
	Oil / lubricate hinges				.065		3.33		3.33	4.34	5.35
	Total				6.024	351	311.32		662.32	790.57	942.83
2030	**Replace 3'-0" x 7'-0" aluminum door, incl. vision lite**	50	1 CARP	Ea.							
	Remove door closer				.167		8.60		8.60	11.20	13.90
	Remove door				.650		26.50		26.50	35	43
	Install new door				1.224	1,575	63		1,638	1,825	2,075
	Reinstall door closer				1.735		89.50		89.50	117	145
	Install new glass				.217	8.88	10.75		19.63	23.50	28.50
	Total				3.993	1,583.88	198.35		1,782.23	2,011.70	2,305.40
3010	**Replace float glass - (3% of glass)**	1	1 CARP	S.F.							
	Remove damaged glass				.052		2.13		2.13	2.78	3.44
	Install new glass				.201	8.24	9.98		18.22	22	26.50
	Total				.253	8.24	12.11		20.35	24.78	29.94
3020	**Repair aluminum sliding door**	12	1 CARP	Ea.							
	Remove weatherstripping				.015		.76		.76	.99	1.23
	Oil / lubricate door closer				.065		3.33		3.33	4.34	5.35
	Install new door closer (25% of total / 12 years)				.434	51	22.38		73.38	85.50	100
	Oil / inspect wheel & track				.065		3.33		3.33	4.34	5.35
	Install new door track / glide wheels (25% of total / 12 years)				.440	13.88	22.75		36.63	45	54
	Install new weatherstrip				4.167	55	215		270	340	415
	Total				5.184	119.88	267.55		387.43	480.17	580.93

For customer support on your Facilities Maintenance & Repair Costs with RSMeans data, call 800.448.8182.

75

B2033 210 Aluminum Doors

	System Description	Freq. (Years)	Crew	Unit	Labor Hours	2019 Bare Costs				Total In-House	Total w/O&P
						Material	Labor	Equipment	Total		
3030	Replace 3'-0" x 7'-0" aluminum sliding door	50	1 CARP	Ea.							
	Remove door				.650		26.50		26.50	35	43
	Install new door track / glide wheels				1.760	55.50	91		146.50	179	216
	Install new aluminum door				1.224	1,575	63		1,638	1,825	2,075
	Total				3.634	1,630.50	180.50		1,811	2,039	2,334
4020	Repair flush aluminum door, wood core	12	1 CARP	Ea.							
	Remove weatherstripping				.015		.76		.76	.99	1.23
	Remove door closer				.167		8.60		8.60	11.20	13.90
	Remove door hinge				.250		12.90		12.90	16.85	21
	Install new weatherstrip				3.478	51	180		231	290	355
	Install new door closer				1.735	204	89.50		293.50	340	400
	Install new hinge (50% of total hinges / 12 years)				.250		12.90		12.90	16.85	21
	Hinge, 5 x 5, brass base					96			96	106	120
	Oil / lubricate door closer				.065		3.33		3.33	4.34	5.35
	Oil / lubricate hinges				.065		3.33		3.33	4.34	5.35
	Total				6.024	351	311.32		662.32	790.57	942.83
4030	Replace 3'-0" x 7'-0" aluminum door, wood core	50	1 CARP	Ea.							
	Remove door				.650		26.50		26.50	35	43
	Remove door closer				.167		8.60		8.60	11.20	13.90
	Install new aluminum door				1.224	1,575	63		1,638	1,825	2,075
	Install door closer				1.735		89.50		89.50	117	145
	Total				3.777	1,575	187.60		1,762.60	1,988.20	2,276.90
5010	Replace tempered glass - (3% of glass)	1	1 CARP	S.F.							
	Remove damaged glass				.052		2.13		2.13	2.78	3.44
	Install new glass				.201	8.24	9.98		18.22	22	26.50
	Total				.253	8.24	12.11		20.35	24.78	29.94

For customer support on your Facilities Maintenance & Repair Costs with RSMeans data, call 800.448.8182.

B2033 210 | Aluminum Doors

System Description	Freq. (Years)	Crew	Unit	Labor Hours	2019 Bare Costs				Total In-House	Total w/O&P
					Material	Labor	Equipment	Total		
5020 Repair aluminum door, insulated	12	1 CARP	Ea.							
Remove weatherstripping				.015		.76		.76	.99	1.23
Remove door closer				.167		8.60		8.60	11.20	13.90
Remove door hinge				.250		12.90		12.90	16.85	21
Install new weatherstrip				3.478	51	180		231	290	355
Install new door closer				1.735	204	89.50		293.50	340	400
Install new hinge (50% of total hinges / 12 years)				.250		12.90		12.90	16.85	21
Hinge, 5 x 5, brass base					96			96	106	120
Oil / lubricate door closer				.065		3.33		3.33	4.34	5.35
Oil / lubricate hinges				.065		3.33		3.33	4.34	5.35
Total				6.024	351	311.32		662.32	790.57	942.83
5030 Replace 3'-0" x 7'-0" aluminum door, insulated	50	1 CARP	Ea.							
Remove door				.650		26.50		26.50	35	43
Remove door closer				.167		8.60		8.60	11.20	13.90
Install new aluminum door				1.224	1,575	63		1,638	1,825	2,075
Install door closer				1.735		89.50		89.50	117	145
Total				3.777	1,575	187.60		1,762.60	1,988.20	2,276.90

B2033 220 | Steel, Painted

System Description	Freq. (Years)	Crew	Unit	Labor Hours	2019 Bare Costs				Total In-House	Total w/O&P
					Material	Labor	Equipment	Total		
1010 Repair steel door, painted	14	1 CARP	Ea.							
Remove weatherstripping				.015		.76		.76	.99	1.23
Remove door closer				.167		8.60		8.60	11.20	13.90
Remove door hinge				.250		12.90		12.90	16.85	21
Install new weatherstrip				3.478	51	180		231	290	355
Install new door closer				1.735	204	89.50		293.50	340	400
Install new hinge (50% of total hinges / 14 years)				.250		12.90		12.90	16.85	21
Hinge, 5 x 5, brass base					96			96	106	120
Oil / lubricate door closer				.065		3.33		3.33	4.34	5.35
Oil / lubricate hinges				.065		3.33		3.33	4.34	5.35
Total				6.024	351	311.32		662.32	790.57	942.83

For customer support on your Facilities Maintenance & Repair Costs with RSMeans data, call 800.448.8182.

77

B2033 220   Steel, Painted

	System Description	Freq. (Years)	Crew	Unit	Labor Hours	Material	2019 Bare Costs Labor	2019 Bare Costs Equipment	2019 Bare Costs Total	Total In-House	Total w/O&P
1020	**Refinish 3'-0" x 7'-0" steel door, painted**	4	1 PORD	Ea.							
	Paint on finished metal surface, brushwork, primer + 1 coat				.800	4.50	34.50		39	49.50	61
	Total				.800	4.50	34.50		39	49.50	61
1030	**Replace 3'-0" x 7'-0" steel door, painted**	45	1 CARP	Ea.							
	Remove door				.650		26.50		26.50	35	43
	Remove cylinder lock				.400		20.50		20.50	27	33.50
	Install new door				1.224	470	63		533	600	690
	Reinstall cylinder lock				1.156		59.50		59.50	78	96.50
	Total				3.431	470	169.50		639.50	740	863
2010	**Replace tempered glass - (3% of glass)**	1	1 CARP	S.F.							
	Remove damaged glass				.052		2.13		2.13	2.78	3.44
	Install new glass				.201	8.24	9.98		18.22	22	26.50
	Total				.253	8.24	12.11		20.35	24.78	29.94
2030	**Prepare and refinish 3'-0" x 7'-0" steel door, painted**	4	1 PORD	Ea.							
	Prepare surface				.333		14.40		14.40	18.55	23
	Refinish door surface, brushwork, primer + 1 coat				1.333	4.24	57.50		61.74	79	97.50
	Total				1.667	4.24	71.90		76.14	97.55	120.50
2040	**Replace 3'-0" x 7'-0" steel door w/ wire glass, painted**	45	1 CARP	Ea.							
	Remove door				.650		26.50		26.50	35	43
	Remove door closer				.167		8.60		8.60	11.20	13.90
	Remove door frame & trim				1.300		67		67	87.50	108
	Install new steel door				1.224	400	63		463	520	600
	Install new steel frame				1.301	205	67		272	315	365
	Install new wire glass				.179	35.38	8.87		44.25	50.50	58.50
	Reinstall door closer				1.735		89.50		89.50	117	145
	Total				6.556	640.38	330.47		970.85	1,136.20	1,333.40

For customer support on your Facilities Maintenance & Repair Costs with RSMeans data, call 800.448.8182.

B2033 220 | **Steel, Painted**

System Description	Freq. (Years)	Crew	Unit	Labor Hours	Material	Labor	Equipment	Total	Total In-House	Total w/O&P
						2019 Bare Costs				
3010 Repair steel sliding door, painted	14	1 CARP	Ea.							
Remove weatherstripping				.015		.76		.76	.99	1.23
Oil / lubricate door closer				.065		3.33		3.33	4.34	5.35
Install new door closer (25% of total / 14 years)				.434	51	22.38		73.38	85.50	100
Oil / inspect wheel & track				.065		3.33		3.33	4.34	5.35
Install new door track / glide wheels (25% of total / 14 years)				.440	13.88	22.75		36.63	45	54
Install new weatherstrip				4.167	55	215		270	340	415
Total				5.184	119.88	267.55		387.43	480.17	580.93
3020 Refinish 3'-0" x 7'-0" steel sliding door, painted	4	1 PORD	Ea.							
Prepare surface				.333		14.40		14.40	18.55	23
Refinish door surface, brushwork, primer + 1 coat				1.333	4.24	57.50		61.74	79	97.50
Total				1.667	4.24	71.90		76.14	97.55	120.50
3030 Replace 3'-0" x 7'-0" steel sliding door, painted	45	1 CARP	Ea.							
Remove door				.650		26.50		26.50	35	43
Remove door closer				.167		8.60		8.60	11.20	13.90
Install new steel door				1.224	400	63		463	520	600
Install door closer				1.735		89.50		89.50	117	145
Total				3.777	400	187.60		587.60	683.20	801.90
4010 Replace steel sliding door glass - (3% of glass)	1	1 CARP	S.F.							
Remove damaged glass				.052		2.13		2.13	2.78	3.44
Install new glass				.201	8.24	9.98		18.22	22	26.50
Total				.253	8.24	12.11		20.35	24.78	29.94
4020 Repair steel door, insulated core, painted	14	1 CARP	Ea.							
Remove weatherstripping				.015		.76		.76	.99	1.23
Remove door closer				.167		8.60		8.60	11.20	13.90
Remove door hinge				.167		8.60		8.60	11.20	13.90
Install new weatherstrip				3.478	51	180		231	290	355
Install new door closer				1.735	204	89.50		293.50	340	400
Install new hinge (50% of total hinges / 14 years)				.167		8.60		8.60	11.20	13.90
Hinge, 5 x 5, brass base					64			64	70.50	80
Oil / lubricate door closer				.065		3.33		3.33	4.34	5.35
Oil / lubricate hinges				.065		3.33		3.33	4.34	5.35
Total				5.857	319	302.72		621.72	743.77	888.63

B2033 220   Steel, Painted

	System Description	Freq. (Years)	Crew	Unit	Labor Hours	2019 Bare Costs				Total In-House	Total w/O&P
						Material	Labor	Equipment	Total		
4030	**Refinish 3'-0" x 7'-0" steel door, insulated core, painted**	4	1 PORD	Ea.							
	Prepare surface				.333		14.40		14.40	18.55	23
	Refinish door surface, brushwork, primer + 1 coat				1.333	4.24	57.50		61.74	79	97.50
	Total				1.667	4.24	71.90		76.14	**97.55**	**120.50**
4040	**Replace 3'-0" x 7'-0" steel door, insulated core, painted**	45	1 CARP	Ea.							
	Remove door				.650		26.50		26.50	35	43
	Remove door closer				.167		8.60		8.60	11.20	13.90
	Remove door frame & trim				1.300		67		67	87.50	108
	Install new steel door				1.224	400	63		463	520	600
	Install new steel frame				1.301	205	67		272	315	365
	Install new insulated glass				.322	38.86	15.95		54.81	63.50	74
	Install door closer				1.735		89.50		89.50	117	145
	Total				6.699	643.86	337.55		981.41	**1,149.20**	**1,348.90**

B2033 221   Steel, Unpainted

	System Description	Freq. (Years)	Crew	Unit	Labor Hours	2019 Bare Costs				Total In-House	Total w/O&P
						Material	Labor	Equipment	Total		
1010	**Repair steel door, unpainted**	14	1 CARP	Ea.							
	Remove weatherstripping				.015		.76		.76	.99	1.23
	Remove door closer				.167		8.60		8.60	11.20	13.90
	Remove door hinge				.167		8.60		8.60	11.20	13.90
	Install new weatherstrip				3.478	51	180		231	290	355
	Install new door closer				1.735	204	89.50		293.50	340	400
	Install new hinge (50% of total hinges / 14 years)				.167		8.60		8.60	11.20	13.90
	Hinge, 5 x 5, brass base					64			64	70.50	80
	Oil / lubricate door closer				.065		3.33		3.33	4.34	5.35
	Oil / lubricate hinges				.065		3.33		3.33	4.34	5.35
	Total				5.857	319	302.72		621.72	**743.77**	**888.63**
1020	**Replace 3'-0" x 7'-0" steel door, unpainted**	45	1 CARP	Ea.							
	Remove door				.650		26.50		26.50	35	43
	Remove door frame & trim				1.300		67		67	87.50	108
	Install new steel door				1.224	595	63		658	735	845
	Install new steel frame				1.301	205	67		272	315	365
	Total				4.475	800	223.50		1,023.50	**1,172.50**	**1,361**

For customer support on your Facilities Maintenance & Repair Costs with RSMeans data, call 800.448.8182.

B2033 221 | **Steel, Unpainted**

	System Description	Freq. (Years)	Crew	Unit	Labor Hours	2019 Bare Costs				Total In-House	Total w/O&P
						Material	Labor	Equipment	Total		
2010	**Replace wire glass - (3% of glass)**	1	1 CARP	S.F.							
	Remove damaged glass				.052		2.13		2.13	2.78	3.44
	Install new wire glass				.179	31.32	8.87		40.19	46	53.50
	Total				.231	31.32	11		42.32	48.78	56.94
2020	**Repair steel door with safety glass, unpainted**	14	1 CARP	Ea.							
	Remove weatherstripping				.015		.76		.76	.99	1.23
	Remove door closer				.167		8.60		8.60	11.20	13.90
	Remove door hinge				.167		8.60		8.60	11.20	13.90
	Install new weatherstrip				3.478	51	180		231	290	355
	Install new door closer				1.735	204	89.50		293.50	340	400
	Install new hinge (50% of total hinges / 14 years)				.167		8.60		8.60	11.20	13.90
	Hinge, 5 x 5, brass base					64			64	70.50	80
	Oil / lubricate door closer				.065		3.33		3.33	4.34	5.35
	Oil / lubricate hinges				.065		3.33		3.33	4.34	5.35
	Total				5.857	319	302.72		621.72	743.77	888.63
2030	**Repl. 3'-0" x 7'-0" steel door with safety glass, unpainted**	45	1 CARP	Ea.							
	Remove door				.650		26.50		26.50	35	43
	Remove door closer				.167		8.60		8.60	11.20	13.90
	Remove door frame & trim				1.300		67		67	87.50	108
	Install new steel door				1.301	610	67		677	760	870
	Install new steel frame				1.301	205	67		272	315	365
	Install new wire glass				.308	61	15.30		76.30	87	101
	Install door closer				1.735		89.50		89.50	117	145
	Total				6.762	876	340.90		1,216.90	1,412.70	1,645.90
3010	**Repair steel sliding door, unpainted**	14	1 CARP	Ea.							
	Remove weatherstripping				.015		.76		.76	.99	1.23
	Oil / lubricate door closer				.065		3.33		3.33	4.34	5.35
	Install new door closer (25% of total / 14 years)				.434	51	22.38		73.38	85.50	100
	Oil / inspect wheel & track				.065		3.33		3.33	4.34	5.35
	New door track / glide wheels (25% of total / 14 years)				.440	13.88	22.75		36.63	45	54
	Total				1.018	64.88	52.55		117.43	140.17	165.93

System Description	Freq. (Years)	Crew	Unit	Labor Hours	2019 Bare Costs				Total In-House	Total w/O&P
					Material	Labor	Equipment	Total		
3020 Replace 3'-0" x 7'-0" steel sliding door, unpainted	45	1 CARP	Ea.							
Remove door				.650		26.50		26.50	35	43
Remove door closer				.167		8.60		8.60	11.20	13.90
Install new steel door				1.224	595	63		658	735	845
Reinstall door closer				1.735		89.50		89.50	117	145
Total				3.776	595	187.60		782.60	898.20	1,046.90
4010 Replace tempered glass - (3% of glass)	1	1 CARP	S.F.							
Remove damaged glass				.052		2.13		2.13	2.78	3.44
Install new glass				.201	8.24	9.98		18.22	22	26.50
Total				.253	8.24	12.11		20.35	24.78	29.94
4020 Repair steel door, insulated core, unpainted	14	1 CARP	Ea.							
Remove weatherstripping				.015		.76		.76	.99	1.23
Remove door closer				.167		8.60		8.60	11.20	13.90
Remove door hinge				.167		8.60		8.60	11.20	13.90
Install new weatherstrip				3.478	51	180		231	290	355
Install new door closer (50% of total / 14 years)				.868	102	44.75		146.75	171	200
Install new hinge (50% of total hinges / 14 years)				.167		8.60		8.60	11.20	13.90
Hinge, 5 x 5, brass base					64			64	70.50	80
Oil / lubricate door closer				.065		3.33		3.33	4.34	5.35
Oil / lubricate hinges				.065		3.33		3.33	4.34	5.35
Total				4.990	217	257.97		474.97	574.77	688.63
4030 Repl. 3'-0" x 7'-0" steel door, insulated core, unpainted	45	1 CARP	Ea.							
Remove door				.650		26.50		26.50	35	43
Remove door closer				.167		8.60		8.60	11.20	13.90
Remove door frame & trim				1.300		67		67	87.50	108
Install new insulated steel door				1.301	695	67		762	850	975
Install new steel frame				1.301	205	67		272	315	365
Install new insulated glass				.322	38.86	15.95		54.81	63.50	74
Reinstall door closer				1.735		89.50		89.50	117	145
Total				6.776	938.86	341.55		1,280.41	1,479.20	1,723.90

For customer support on your Facilities Maintenance & Repair Costs with RSMeans data, call 800.448.8182.

B2033 410 Steel Roll-Up, Single

	System Description	Freq. (Years)	Crew	Unit	Labor Hours	2019 Bare Costs				Total In-House	Total w/O&P
						Material	Labor	Equipment	Total		
1010	**Repair 12' x 12' steel single roll-up door**	10	1 CARP	Ea.							
	Remove overhead type door section				1.303		67.25		67.25	87.50	109
	Install new roll-up door section				4.348	293.75	243.75		537.50	655	775
	Total				5.651	293.75	311		604.75	742.50	884
1020	**Refinish 12' x 12' steel single roll-up door**	5	1 PORD	Ea.							
	Set up, secure and take down ladder				.258		11.15		11.15	14.35	17.85
	Prepare door surface				2.000		86.40		86.40	111	138
	Paint overhead door, brushwork, primer + 1 coat, 1 side				1.493	32.76	64.26		97.02	120	144
	Total				3.751	32.76	161.81		194.57	245.35	299.85
1030	**Replace 12' x 12' steel single roll-up door**	35	2 CARP	Ea.							
	Remove overhead type door				5.212		269		269	350	435
	Install new roll-up door				17.391	1,175	975		2,150	2,625	3,100
	Total				22.603	1,175	1,244		2,419	2,975	3,535

B2033 412 Steel Roll-Up, Double

	System Description	Freq. (Years)	Crew	Unit	Labor Hours	2019 Bare Costs				Total In-House	Total w/O&P
						Material	Labor	Equipment	Total		
1020	**Refinish 12' x 24' steel double roll-up door**	5	1 PORD	Ea.							
	Set up, secure and take down ladder				.533		23		23	29.50	37
	Prepare door surface				4.000		172.80		172.80	223	276
	Paint overhead door, brushwork, primer + 1 coat, 1 side				2.986	65.52	128.52		194.04	239	289
	Total				7.519	65.52	324.32		389.84	491.50	602
1040	**Replace 12' x 24' steel double roll-up door**	35	2 CARP	Ea.							
	Remove overhead type door				10.423		538		538	700	870
	Install new door				34.783	2,350	1,950		4,300	5,225	6,200
	Total				45.206	2,350	2,488		4,838	5,925	7,070

For customer support on your Facilities Maintenance & Repair Costs with RSMeans data, call 800.448.8182.

83

B2033 413 | Steel Garage Door, Single Leaf, Spring

	System Description	Freq. (Years)	Crew	Unit	Labor Hours	2019 Bare Costs				Total In-House	Total w/O&P
						Material	Labor	Equipment	Total		
1010	**Repair 9' x 7' steel, single leaf, garage door**	10	1 CARP	Ea.							
	Repair damaged track				.068		3.53		3.53	4.60	5.70
	Remove cremone lockset				.400		20.50		20.50	27	33.50
	Install new cremone lockset				1.040	151	53.50		204.50	236	275
	Retension door coil spring				1.143		59		59	77	95
	Total				2.651	151	136.53		287.53	344.60	409.20
1020	**Refinish 9' x 7' steel, single leaf, garage door**	5	1 PORD	Ea.							
	Set up, secure and take down ladder				.258		11.15		11.15	14.35	17.85
	Prepare surface				2.000		86.40		86.40	111	138
	Paint garage door, brushwork, primer + 1 coat, 1 side				1.493	32.76	64.26		97.02	120	144
	Total				3.751	32.76	161.81		194.57	245.35	299.85
1030	**Replace 9' x 7' steel, single leaf, garage door**	35	2 CARP	Ea.							
	Remove door				2.602		134		134	175	217
	Install new door				2.602	1,100	134		1,234	1,375	1,600
	Total				5.203	1,100	268		1,368	1,550	1,817

B2033 420 | Aluminum Roll-Up, Single

	System Description	Freq. (Years)	Crew	Unit	Labor Hours	2019 Bare Costs				Total In-House	Total w/O&P
						Material	Labor	Equipment	Total		
1010	**Repair 12' x 12' aluminum single roll-up door**	10	1 CARP	Ea.							
	Remove door section				1.303		67.25		67.25	87.50	109
	Install new door section				3.478	850	180		1,030	1,175	1,350
	Total				4.781	850	247.25		1,097.25	1,262.50	1,459
1020	**Refinish 12' x 12' aluminum single roll-up door**	5	1 PORD	Ea.							
	Set up, secure and take down ladder				.258		11.15		11.15	14.35	17.85
	Prepare surface				2.000		86.40		86.40	111	138
	Paint overhead type door				1.493	32.76	64.26		97.02	120	144
	Total				3.751	32.76	161.81		194.57	245.35	299.85
1030	**Replace 12' x 12' aluminum single roll-up door**	35	2 CARP	Ea.							
	Paint overhead door, brushwork, primer + 1 coat, 1 side				1.706	37.44	73.44		110.88	137	165
	Install new door				13.913	3,400	720		4,120	4,675	5,400
	Total				15.619	3,437.44	793.44		4,230.88	4,812	5,565

For customer support on your Facilities Maintenance & Repair Costs with RSMeans data, call 800.448.8182.

B2033 422 Aluminum Roll-Up, Double

	System Description	Freq. (Years)	Crew	Unit	Labor Hours	2019 Bare Costs				Total In-House	Total w/O&P
						Material	Labor	Equipment	Total		
1010	**Repair 12' x 24' aluminum double roll-up door**	10	1 CARP	Ea.							
	Remove door section				2.606		134.50		134.50	175	218
	Install new door section				6.957	1,700	360		2,060	2,350	2,700
	Total				9.562	1,700	494.50		2,194.50	2,525	2,918
1020	**Refinish 12' x 24' aluminum double roll-up door**	5	1 PORD	Ea.							
	Set up, secure and take down ladder				.533		23		23	29.50	37
	Prepare surface				4.000		172.80		172.80	223	276
	Paint overhead door, brushwork, primer + 1 coat, 1 side				2.986	65.52	128.52		194.04	239	289
	Total				7.519	65.52	324.32		389.84	491.50	602
1030	**Replace 12' x 24' aluminum double roll-up door**	35	2 CARP	Ea.							
	Remove overhead type door				10.423		538		538	700	870
	Install new door				27.826	6,800	1,440		8,240	9,350	10,800
	Total				38.250	6,800	1,978		8,778	10,050	11,670

B2033 423 Aluminum Garage Door, Single Leaf, Spring

	System Description	Freq. (Years)	Crew	Unit	Labor Hours	2019 Bare Costs				Total In-House	Total w/O&P
						Material	Labor	Equipment	Total		
1010	**Repair 9' x 7' aluminum, single leaf, garage door**	10	1 CARP	Ea.							
	Repair damaged track				.684		35.30		35.30	46	57
	Remove cremone lockset				.400		20.50		20.50	27	33.50
	Install new cremone lockset				1.040	151	53.50		204.50	236	275
	Retension door coil spring				1.143		59		59	77	95
	Total				3.267	151	168.30		319.30	386	460.50
1020	**Refinish 9' x 7' aluminum, single leaf, garage door**	5	1 PORD	Ea.							
	Set up, secure and take down ladder				.258		11.15		11.15	14.35	17.85
	Prepare surface				2.000		86.40		86.40	111	138
	Paint garage door, brushwork, primer + 1 coat, 1 side				.747	16.38	32.13		48.51	60	72
	Total				3.005	16.38	129.68		146.06	185.35	227.85
1030	**Replace 9' x 7' aluminum, single leaf, garage door**	35	2 CARP	Ea.							
	Remove door				2.602		134		134	175	217
	Install new door				2.602	1,100	134		1,234	1,375	1,600
	Total				5.203	1,100	268		1,368	1,550	1,817

For customer support on your Facilities Maintenance & Repair Costs with RSMeans data, call 800.448.8182.

85

B2033 430 Wood Roll-Up, Single

System Description	Freq. (Years)	Crew	Unit	Labor Hours	2019 Bare Costs				Total In-House	Total w/O&P
					Material	Labor	Equipment	Total		
1010 Repair 12' x 12' wood single roll-up door	8	1 CARP	Ea.							
Remove door section				1.303		67.25		67.25	87.50	109
Install new door section				3.478	600	180		780	895	1,050
Total				4.781	600	247.25		847.25	982.50	1,159
1020 Refinish 12' x 12' wood single roll-up door	5	1 PORD	Ea.							
Set up, secure and take down ladder				.258		11.15		11.15	14.35	17.85
Prepare door surface				2.000		86.40		86.40	111	138
Paint overhead door, brushwork, primer + 1 coat, 1 side				3.150	13.86	136.08		149.94	190	235
Total				5.408	13.86	233.63		247.49	315.35	390.85
1030 Replace 12' x 12' wood single roll-up door	16	2 CARP	Ea.							
Remove overhead type door				5.212		269		269	350	435
Install new door				13.913	2,400	720		3,120	3,575	4,150
Total				19.125	2,400	989		3,389	3,925	4,585

B2033 432 Wood Roll-Up, Double

System Description	Freq. (Years)	Crew	Unit	Labor Hours	2019 Bare Costs				Total In-House	Total w/O&P
					Material	Labor	Equipment	Total		
1010 Repair 12' x 24' wood double roll-up door	8	1 CARP	Ea.							
Remove door section				2.606		134.50		134.50	175	218
Install new door section				6.957	1,200	360		1,560	1,800	2,075
Total				9.562	1,200	494.50		1,694.50	1,975	2,293
1020 Refinish 12' x 24' wood double roll-up door	5	1 PORD	Ea.							
Set up, secure and take down ladder				.533		23		23	29.50	37
Prepare surface				4.000		172.80		172.80	223	276
Paint overhead door, brushwork, primer + 1 coat, 1 side				6.300	27.72	272.16		299.88	380	470
Total				10.833	27.72	467.96		495.68	632.50	783
1030 Replace 12' x 24' wood double roll-up door	16	2 CARP	Ea.							
Remove overhead type door				10.423		538		538	700	870
Install new door				27.826	4,800	1,440		6,240	7,150	8,300
Total				38.250	4,800	1,978		6,778	7,850	9,170

For customer support on your Facilities Maintenance & Repair Costs with RSMeans data, call 800.448.8182.

B2033 433 Wood Garage Door, Single Leaf, Spring

System Description	Freq. (Years)	Crew	Unit	2019 Bare Costs					Total In-House	Total w/O&P
				Labor Hours	Material	Labor	Equipment	Total		
1010 Repair 9' x 7' wood, single leaf, garage door	8	1 CARP	Ea.							
Repair damaged track				.068		3.53		3.53	4.60	5.70
Remove cremone lockset				.400		20.50		20.50	27	33.50
Install new cremone lockset				1.040	151	53.50		204.50	236	275
Retension door coil spring				1.143		59		59	77	95
Total				2.651	151	136.53		287.53	344.60	409.20
1020 Refinish 9' x 7' wood, single leaf, garage door	5	1 PORD	Ea.							
Set up, secure and take down ladder				.258		11.15		11.15	14.35	17.85
Prepare surface				2.000		86.40		86.40	111	138
Paint garage door, brushwork, primer + 1 coat, 1 side				3.150	13.86	136.08		149.94	190	235
Total				5.408	13.86	233.63		247.49	315.35	390.85
1030 Replace 9' x 7' wood, single leaf, garage door	16	2 CARP	Ea.							
Remove door				2.602		134		134	175	217
Install new door				2.602	745	134		879	995	1,150
Total				5.203	745	268		1,013	1,170	1,367

B2033 513 Electric Bifolding Hangar Door

System Description	Freq. (Years)	Crew	Unit	2019 Bare Costs					Total In-House	Total w/O&P
				Labor Hours	Material	Labor	Equipment	Total		
0120 Remove and replace electric bi-folding hangar door motor	15	2 SKWK	Ea.							
Hangar door,electric bi-folding door, replace motor				3.200	485	171		656	755	880
Total				3.200	485	171		656	755	880
0140 Remove and replace electric bi-folding hangar door cables	15	2 SKWK	Ea.							
Hangar door, elec bi-folding door, replace cables				8.000	152	425		577	720	875
Total				8.000	152	425		577	720	875
0160 Remove and replace electric bi-folding hangar door	20	2 SKWK	Ea.							
Hangar door demo				77.568		4,336		4,336	5,900	7,225
Hangar door, bi-fldg dr, remove and replace 20' x 80'				116.368	35,200	6,512		41,712	47,600	55,000
Total				193.936	35,200	10,848		46,048	53,500	62,225

For customer support on your Facilities Maintenance & Repair Costs with RSMeans data, call 800.448.8182.

87

B2033 810 Hinges, Brass

	System Description	Freq. (Years)	Crew	Unit	Labor Hours	2019 Bare Costs				Total In-House	Total w/O&P
						Material	Labor	Equipment	Total		
1010	Replace brass door hinge	60	1 CARP	Ea.							
	Remove door				.650		26.50		26.50	35	43
	Remove door hinge				.167		8.60		8.60	11.20	13.90
	Install new hinge				.167		8.60		8.60	11.20	13.90
	Full mortise hinge, 5 x 5, brass base					64			64	70.50	80
	Reinstall exterior door				1.388		71.50		71.50	93.50	116
	Total				2.371	64	115.20		179.20	221.40	266.80

B2033 820 Lockset, Brass

	System Description	Freq. (Years)	Crew	Unit	Labor Hours	2019 Bare Costs				Total In-House	Total w/O&P
						Material	Labor	Equipment	Total		
1010	Replace brass door lockset	30	1 CARP	Ea.							
	Remove mortise lock set				.421		22		22	28.50	35
	Mortise lock				1.301	345	67		412	465	540
	Total				1.722	345	89		434	493.50	575

B2033 830 Door Closer, Brass

	System Description	Freq. (Years)	Crew	Unit	Labor Hours	2019 Bare Costs				Total In-House	Total w/O&P
						Material	Labor	Equipment	Total		
1010	Replace brass door closer	15	1 CARP	Ea.							
	Remove door closer				.167		8.60		8.60	11.20	13.90
	Install new door closer				1.735	204	89.50		293.50	340	400
	Total				1.902	204	98.10		302.10	351.20	413.90

For customer support on your Facilities Maintenance & Repair Costs with RSMeans data, call 800.448.8182.

B2033 840 Door Opener

System Description	Freq. (Years)	Crew	Unit	Labor Hours	2019 Bare Costs				Total In-House	Total w/O&P
					Material	Labor	Equipment	Total		
1010										
Automatic door opener on existing door	7	1 ELEC	Ea.							
Cutout demolition of exterior wall for actuator button				1.739		72		82.20	105	127
Install 1/2" E.M.T. conduit				2.462	32.40	148	10.20	180.40	218	269
Install conductor, type THWN copper, #14				1.040	9.82	62.40		72.22	88	108
Install junction box				.533	11.40	32		43.40	52	64
Install outlet				.381	9.55	23		32.55	38.50	47.50
Install plate				.100	2.74	6		8.74	10.40	12.70
Automatic opener, button operation				20.000	3,850	1,075		4,925	5,625	6,525
Total				26.255	3,915.91	1,418.40	10.20	5,344.51	6,136.90	7,153.20

B2033 850 Deadbolt, Brass

System Description	Freq. (Years)	Crew	Unit	Labor Hours	2019 Bare Costs				Total In-House	Total w/O&P
					Material	Labor	Equipment	Total		
1010										
Replace brass deadbolt	20	1 CARP	Ea.							
Remove deadbolt				.154		7.95		7.95	10.35	12.80
Install new deadbolt				1.156	190	59.50		249.50	287	335
Total				1.309	190	67.45		257.45	297.35	347.80

B2033 860 Weatherstripping, Brass

System Description	Freq. (Years)	Crew	Unit	Labor Hours	2019 Bare Costs				Total In-House	Total w/O&P
					Material	Labor	Equipment	Total		
1010										
Replace brass door weatherstripping	20	1 CARP	Ea.							
Remove weatherstripping				.015		.76		.76	.99	1.23
Install new weatherstrip				3.478	51	180		231	290	355
Total				3.493	51	180.76		231.76	290.99	356.23

B2033 870 Panic Device

System Description	Freq. (Years)	Crew	Unit	Labor Hours	Material	Labor	Equipment	Total	Total In-House	Total w/O&P
							2019 Bare Costs			
1010 Replace door panic device	25	1 CARP	Ea.							
Remove panic bar				.421		22		22	28.50	35
Install new panic bar				2.080	1,100	107		1,207	1,350	1,550
Total				2.501	1,100	129		1,229	1,378.50	1,585

B2033 905 Hollow Core, Painted

System Description	Freq. (Years)	Crew	Unit	Labor Hours	Material	Labor	Equipment	Total	Total In-House	Total w/O&P
							2019 Bare Costs			
1010 Repair hollow core door	12	1 CARP	Ea.							
Remove weatherstripping				.015		.76		.76	.99	1.23
Remove door hinge				.167		8.60		8.60	11.20	13.90
Install new weatherstrip				3.478	51	180		231	290	355
Install new hinge (50% of total hinges / 12 years)				.167		8.60		8.60	11.20	13.90
Hinge, 5 x 5, brass base					64			64	70.50	80
Oil / lubricate hinges				.065		3.33		3.33	4.34	5.35
Total				3.891	115	201.29		316.29	388.23	469.38
1020 Refinish 3'-0" x 7'-0" hollow core door	4	1 PORD	Ea.							
Prepare door surface				.333		14.40		14.40	18.55	23
Paint door surface, brushwork, primer + 1 coat				.549	12.88	23.92		36.80	45	54.50
Total				.883	12.88	38.32		51.20	63.55	77.50
1030 Replace 3'-0" x 7'-0" hollow core door	30	1 CARP	Ea.							
Remove door				.650		26.50		26.50	35	43
Remove door frame				.743		38.50		38.50	50	62
Install new door frame				.943	140.25	48.79		189.04	218	254
Install new oak door sill				.832	86	43		129	151	177
Install new hollow core door				1.000	165	51.50		216.50	249	290
Install new weatherstrip				3.478	51	180		231	290	355
Install new lockset				.867	79.50	45		124.50	146	171
Total				8.514	521.75	433.29		955.04	1,139	1,352

For customer support on your Facilities Maintenance & Repair Costs with RSMeans data, call 800.448.8182.

B2033 905 Hollow Core, Painted

System Description	Freq. (Years)	Crew	Unit	2019 Bare Costs					Total In-House	Total w/O&P
				Labor Hours	Material	Labor	Equipment	Total		
2010 **Repair hollow core sliding wood door**	14	1 CARP	Ea.							
Remove weatherstripping				.015		.76		.76	.99	1.23
Oil / lubricate door closer				.065		3.33		3.33	4.34	5.35
Oil / inspect wheel & track				.065		3.33		3.33	4.34	5.35
Install new door track / glide wheels (25% of total / 14 years)				.440	13.88	22.75		36.63	45	54
Install new weatherstrip				4.167	55	215		270	340	415
Total				4.750	68.88	245.17		314.05	394.67	480.93
2030 **Replace 3'-0" x 7'-0" hollow core sliding wood door**	30	1 CARP	Ea.							
Remove sliding door				1.733		89.50		89.50	117	144
Install new sliding door frame				4.622	435	261	13.80	709.80	850	995
Install new hollow core door				1.000	165	51.50		216.50	249	290
Install new weatherstrip				3.478	51	180		231	290	355
Lockset				.867	79.50	45		124.50	146	171
Total				11.701	730.50	627	13.80	1,371.30	1,652	1,955

B2033 925 Metal Grated

System Description	Freq. (Years)	Crew	Unit	2019 Bare Costs					Total In-House	Total w/O&P
				Labor Hours	Material	Labor	Equipment	Total		
1010 **Repair metal grated door**	15	1 SSWK	Ea.							
Remove damaged metal bar				.836		46.92		46.92	63.50	78
Oil / lubricate door closer				.065		3.61		3.61	4.90	6
Oil / lubricate hinges				.065		3.33		3.33	4.34	5.35
Metal grating repair (3% of door area / 15 years)				.320	1.40	18.60	3.82	23.82	31	37
Replace damaged metal bar				1.600	13.20	91	9.60	113.80	149	179
Total				2.885	14.60	163.46	13.42	191.48	252.74	305.35
1020 **Prepare and refinish metal grated painted door**	4	1 PORD	Ea.							
Prepare grated door surface				.308		13.30		13.30	17.15	21.50
Paint grated door surface, brushwork, 1 coat				.319	3.22	13.80		17.02	21.50	26
Total				.627	3.22	27.10		30.32	38.65	47.50

For customer support on your Facilities Maintenance & Repair Costs with RSMeans data, call 800.448.8182.

91

B2033 925 Metal Grated

	System Description	Freq. (Years)	Crew	Unit	Labor Hours	Material	Labor	Equipment	Total	Total In-House	Total w/O&P
							2019 Bare Costs				
1030	**Replace 3'-0" x 7'-0" metal grated painted door**	30	2 SSWK	Ea.							
	Remove door				.650		26.50		26.50	35	43
	Remove door closer				.167		8.60		8.60	11.20	13.90
	Remove door hinge				.500		25.80		25.80	33.50	41.50
	Install new grated door				15.686	640	880		1,520	1,900	2,275
	Install new hinge (25% of total hinges / 30 years)				.500		25.80		25.80	33.50	41.50
	Hinge, 5 x 5					96			96	106	120
	Total				17.503	736	966.70		1,702.70	2,119.20	2,534.90

B2033 926 Metal Wire Mesh Door

	System Description	Freq. (Years)	Crew	Unit	Labor Hours	Material	Labor	Equipment	Total	Total In-House	Total w/O&P
							2019 Bare Costs				
1010	**Repair metal wire mesh door**	15	1 SSWK	Ea.							
	Remove wire mesh				1.750		98.18		98.18	133	163
	Wire mesh					9.77			9.77	10.75	12.20
	Install wire mesh				7.000		391.65		391.65	530	650
	Oil / lubricate door closer				.065		3.61		3.61	4.90	6
	Oil / lubricate hinges				.065		3.33		3.33	4.34	5.35
	Total				8.879	9.77	496.77		506.54	682.99	836.55
1020	**Prepare and refinish metal wire mesh door**	4	1 PORD	Ea.							
	Prepare metal mesh surface				.312		13.50		13.50	17.40	21.50
	Paint metal mesh door surface, brushwork, 1 coat				.319	3.22	13.80		17.02	21.50	26
	Total				.631	3.22	27.30		30.52	38.90	47.50
1030	**Replace metal wire mesh door**	30	2 SSWK	Ea.							
	Remove wire mesh door				.640		36		36	48.50	59.50
	Remove wire mesh door frame				2.602		146		146	197	243
	Install new woven wire swinging door				3.471	292	179		471	555	655
	Install new metal door frame				3.200	238	181	9.55	428.55	515	610
	Install new hinge (25% of total hinges / 30 years)				.500		25.80		25.80	33.50	41.50
	Hinge, 5 x 5					96			96	106	120
	Install new deadlock night latch				1.039	51.50	53.50		105	127	151
	Total				11.451	677.50	621.30	9.55	1,308.35	1,582	1,880

For customer support on your Facilities Maintenance & Repair Costs with RSMeans data, call 800.448.8182.

B2033 927 Louvered Door

System Description	Freq. (Years)	Crew	Unit	Labor Hours	2019 Bare Costs Material	2019 Bare Costs Labor	2019 Bare Costs Equipment	2019 Bare Costs Total	Total In-House	Total w/O&P
1010										
Repair aluminum louvered door	12	1 CARP	Ea.							
Remove door				.650		26.50		26.50	35	43
Remove weatherstripping				.015		.76		.76	.99	1.23
Remove damaged louver				.137		7.05		7.05	9.20	11.40
Install new louver				.274	16.50	14.15		30.65	36.50	43.50
Oil / lubricate door closer				.065		3.33		3.33	4.34	5.35
Install new weatherstrip				3.478	51	180		231	290	355
Reinstall exterior door				1.388		71.50		71.50	93.50	116
Total				6.006	67.50	303.29		370.79	469.53	575.48
1020										
Prepare and refinish aluminum louvered door	4	1 PORD	Ea.							
Prepare door surface				.333		14.40		14.40	18.55	23
Paint door surface, brushwork, primer + 1 coat				.549	12.88	23.92		36.80	45	54.50
Total				.883	12.88	38.32		51.20	63.55	77.50
1040										
Replace 3'-0" x 7'-0" aluminum louvered door	50	1 CARP	Ea.							
Remove door				.650		26.50		26.50	35	43
Remove door closer				.167		8.60		8.60	11.20	13.90
Install new aluminum door				1.224	1,575	63		1,638	1,825	2,075
Install new door closer				1.735	204	89.50		293.50	340	400
Total				3.777	1,779	187.60		1,966.60	2,211.20	2,531.90
2010										
Repair steel louvered door	14	1 CARP	Ea.							
Remove door				.650		26.50		26.50	35	43
Remove weatherstripping				.015		.76		.76	.99	1.23
Remove damaged louver				.137		7.05		7.05	9.20	11.40
Install new louver				.274	16.50	14.15		30.65	36.50	43.50
Oil / lubricate door closer				.065		3.33		3.33	4.34	5.35
Install new weatherstrip				3.478	51	180		231	290	355
Reinstall exterior door				1.388		71.50		71.50	93.50	116
Total				6.006	67.50	303.29		370.79	469.53	575.48
2020										
Prepare and refinish steel louvered door	4	1 PORD	Ea.							
Prepare door surfaces				.333		14.40		14.40	18.55	23
Paint door surface, brushwork, primer + 1 coat				.549	12.88	23.92		36.80	45	54.50
Total				.883	12.88	38.32		51.20	63.55	77.50

For customer support on your Facilities Maintenance & Repair Costs with RSMeans data, call 800.448.8182.

93

	System Description	Freq. (Years)	Crew	Unit	Labor Hours	2019 Bare Costs				Total In-House	Total w/O&P
						Material	Labor	Equipment	Total		
2040	**Replace 3'-0" x 7'-0" steel louvered door**	45	1 CARP	Ea.							
	Remove door				.650		26.50		26.50	35	43
	Remove door closer				.167		8.60		8.60	11.20	13.90
	Install new steel door, incl. louver				1.224	595	63		658	735	845
	Install new door closer				1.735	204	89.50		293.50	340	400
	Total				3.776	799	187.60		986.60	1,121.20	1,301.90
3010	**Repair wood louvered door**	7	1 CARP	Ea.							
	Remove door				.650		26.50		26.50	35	43
	Remove weatherstripping				.015		.76		.76	.99	1.23
	Remove damaged louver				.137		7.05		7.05	9.20	11.40
	Install new louver				.274	16.50	14.15		30.65	36.50	43.50
	Oil / lubricate door closer				.065		3.33		3.33	4.34	5.35
	Install new weatherstrip				3.478	51	180		231	290	355
	Reinstall exterior door				1.388		71.50		71.50	93.50	116
	Total				6.006	67.50	303.29		370.79	469.53	575.48
3020	**Prepare and refinish wood louvered door**	4	1 PORD	Ea.							
	Prepare door surfaces				.333		14.40		14.40	18.55	23
	Paint door surface, brushwork, primer + 1 coat				.549	12.88	23.92		36.80	45	54.50
	Total				.883	12.88	38.32		51.20	63.55	77.50
3040	**Replace 3'-0" x 7'-0" wood louvered door**	40	1 CARP	Ea.							
	Remove door				.650		26.50		26.50	35	43
	Remove door closer				.167		8.60		8.60	11.20	13.90
	Install new solid core door				1.143	227	59		286	325	380
	Add for installed wood louver					226			226	249	283
	Install new door closer				1.735	204	89.50		293.50	340	400
	Total				3.695	657	183.60		840.60	960.20	1,119.90

For customer support on your Facilities Maintenance & Repair Costs with RSMeans data, call 800.448.8182.

B2033 935 Aluminum Storm/Screen Door

	System Description	Freq. (Years)	Crew	Unit	Labor Hours	2019 Bare Costs				Total In-House	Total w/O&P
						Material	Labor	Equipment	Total		
1010	**Repair aluminum storm/screen door**	5	1 CARP	Ea.							
	Remove weatherstripping				.015		.76		.76	.99	1.23
	Remove screen / storm door threshold				.519		26.80		26.80	35	43
	Remove door closer				.167		8.60		8.60	11.20	13.90
	Install new screen / storm door closer				1.301	29	67		96	119	144
	Install new weatherstrip				.870	12.75	45		57.75	72.50	88.50
	Install new plain aluminum threshold				.520	9.95	27		36.95	46	56
	Total				3.391	51.70	175.16		226.86	284.69	346.63
1020	**Replace aluminum storm/screen door**	20	1 CARP	Ea.							
	Remove door				.650		26.50		26.50	35	43
	Install new anodized aluminum screen / storm door				1.487	191	77		268	310	365
	Total				2.137	191	103.50		294.50	345	408

B2033 960 Aluminized Steel Pedestrian Gate

	System Description	Freq. (Years)	Crew	Unit	Labor Hours	2019 Bare Costs				Total In-House	Total w/O&P
						Material	Labor	Equipment	Total		
1010	**Repair aluminized steel pedestrian gate**	10	1 CARP	Ea.							
	Remove gate from hinges				.286		14.75		14.75	19.25	24
	Oil / lubricate hinges				.065		3.33		3.33	4.34	5.35
	Install new 6' aluminized steel fencing				.360	73.50	15.30	3.03	91.83	104	120
	Reinstall gate				.615		32		32	41.50	51.50
	Total				1.326	73.50	65.38	3.03	141.91	169.09	200.85
1030	**Prepare and refinish pedestrian gate**	5	1 PORD	Ea.							
	Lay drop cloth				.090		3.90		3.90	5	6.20
	Prepare / brush surface				.042		1.80		1.80	2.34	2.88
	Paint gate, brushwork, 1 coat				.150	1.26	6.48		7.74	9.70	12
	Remove drop cloth				.090		3.90		3.90	5	6.20
	Total				.371	1.26	16.08		17.34	22.04	27.28
1040	**Replace pedestrian gate**	40	1 CARP	Ea.							
	Remove old gate from hinges				.286		14.75		14.75	19.25	24
	Install new 6' x 3' aluminized steel gate				3.117	210	133	26.50	369.50	430	505
	Total				3.403	210	147.75	26.50	384.25	449.25	529

For customer support on your Facilities Maintenance & Repair Costs with RSMeans data, call 800.448.8182.

95

B2033 961 Pedestrian Gate, Galvanized Steel

System Description	Freq. (Years)	Crew	Unit	Labor Hours	2019 Bare Costs				Total In-House	Total w/O&P
					Material	Labor	Equipment	Total		
1010 Repair steel gate	10	1 CARP	Ea.							
Remove gate from hinges				.286		14.75		14.75	19.25	24
Oil / lubricate hinges				.065		3.33		3.33	4.34	5.35
Install new 6' galvanized steel fencing				.360	61.50	15.30	3.03	79.83	91	105
Reinstall gate				.615		32		32	41.50	51.50
Total				1.326	61.50	65.38	3.03	129.91	156.09	185.85
1030 Prepare and refinish steel gate	5	1 PORD	Ea.							
Lay drop cloth				.090		3.90		3.90	5	6.20
Prepare / brush surface				.042		1.80		1.80	2.34	2.88
Paint gate, brushwork, 1 coat				.150	1.26	6.48		7.74	9.70	12
Remove drop cloth				.090		3.90		3.90	5	6.20
Total				.371	1.26	16.08		17.34	22.04	27.28
1040 Replace steel gate	40	1 CARP	Ea.							
Remove old gate from hinges				.286		14.75		14.75	19.25	24
Install new 6' x 3' galvanized steel gate				3.117	207	133	26.50	366.50	430	500
Total				3.403	207	147.75	26.50	381.25	449.25	524

B2033 962 Pedestrian Gate, Wood

System Description	Freq. (Years)	Crew	Unit	Labor Hours	2019 Bare Costs				Total In-House	Total w/O&P
					Material	Labor	Equipment	Total		
1010 Repair wood gate	7	1 CARP	Ea.							
Remove gate from hinges				.286		14.75		14.75	19.25	24
Oil / lubricate hinges				.065		3.33		3.33	4.34	5.35
Replace damaged fencing				1.171	120	49.80	9.90	179.70	208	241
Reinstall gate				.615		32		32	41.50	51.50
Total				2.136	120	99.88	9.90	229.78	273.09	321.85
1030 Prepare and refinish wood gate	4	1 PORD	Ea.							
Lay drop cloth				.090		3.90		3.90	5	6.20
Prepare / brush surface				.042		1.80		1.80	2.34	2.88
Paint gate, brushwork, primer + 1 coat				.138	1.44	5.94		7.38	9.35	11.35
Remove drop cloth				.090		3.90		3.90	5	6.20
Total				.359	1.44	15.54		16.98	21.69	26.63

For customer support on your Facilities Maintenance & Repair Costs with RSMeans data, call 800.448.8182.

B2033 962 Pedestrian Gate, Wood

	System Description	Freq. (Years)	Crew	Unit	Labor Hours	2019 Bare Costs				Total In-House	Total w/O&P
						Material	Labor	Equipment	Total		
1040	**Replace wood gate**	25	1 CARP	Ea.							
	Remove old gate from hinges				.286		14.75		14.75	19.25	24
	New 6'-0" x 3'-0" wood gate				3.468	230	148	29.50	407.50	475	560
	Total				3.754	230	162.75	29.50	422.25	494.25	584

B2033 963 Pedestrian Gate, Wrought Iron

	System Description	Freq. (Years)	Crew	Unit	Labor Hours	2019 Bare Costs				Total In-House	Total w/O&P
						Material	Labor	Equipment	Total		
1010	**Repair wrought iron gate**	11	1 CARP	Ea.							
	Remove gate from hinges				.286		14.75		14.75	19.25	24
	Oil / lubricate hinges				.065		3.33		3.33	4.34	5.35
	Repair hand forged wrought iron				2.600	286.50	145.50		432	515	600
	Reinstall gate				.615		32		32	41.50	51.50
	Total				3.566	286.50	195.58		482.08	580.09	680.85
1030	**Prepare and refinish wrought iron gate**	5	1 PORD	Ea.							
	Lay drop cloth				.090		3.90		3.90	5	6.20
	Prepare / brush surface				.125		5.40		5.40	7	8.65
	Paint gate, brushwork, primer + 1 coat				.167	1.44	7.20		8.64	11	13.30
	Remove drop cloth				.090		3.90		3.90	5	6.20
	Total				.471	1.44	20.40		21.84	28	34.35
1040	**Replace wrought iron gate**	45	1 CARP	Ea.							
	Remove old gate from hinges				.286		14.75		14.75	19.25	24
	Install new hand forged wrought iron gate				3.467	382	194		576	685	800
	Total				3.753	382	208.75		590.75	704.25	824

For customer support on your Facilities Maintenance & Repair Costs with RSMeans data, call 800.448.8182.

97

B3013 105 Built-Up Roofing

ID	System Description	Freq. (Years)	Crew	Unit	Labor Hours	2019 Bare Costs				Total In-House	Total w/O&P
						Material	Labor	Equipment	Total		
0100	**Debris removal and visual inspection**	0.50	2 ROFC	M.S.F.							
	Set up, secure and take down ladder				.052		2.34		2.34	3.41	4.14
	Pick up trash / debris & clean-up				.327		14.70		14.70	21.50	26
	Visual inspection				.327		14.70		14.70	21.50	26
	Total				.705		31.74		31.74	46.41	56.14
0200	**Non-destructive moisture inspection**	5	2 ROFC	M.S.F.							
	Set up, secure and take down ladder				.052		2.34		2.34	3.41	4.14
	Infrared inspection of roof membrane				2.133		96		96	140	170
	Total				2.185		98.34		98.34	143.41	174.14
0300	**Minor BUR membrane repairs - (2% of roof area)**	1	G-5	Sq.							
	Set up, secure and take down ladder				1.000		45		45	66	80
	Sweep / spud ballast clean				.640		29		29	42	51
	Cross cut incision through bitumen				.258		11.65		11.65	16.95	20.50
	Install base sheet and 2 plies of glass mopped				3.294	116	139	30	285	365	425
	Reinstall ballast in bitumen				.390		17.60		17.60	25.50	31
	Clean up				.390		17.60		17.60	25.50	31
	Total				5.973	116	259.85	30	405.85	540.95	638.50
0400	**BUR flashing repairs - (2 S.F. per sq. repaired)**	1	2 ROFC	S.F.							
	Set up, secure and take down ladder				.010		.45		.45	.66	.80
	Sweep / spud ballast clean				.009		.41		.41	.59	.72
	Cut out buckled flashing				.020		.90		.90	1.31	1.59
	Install 2 ply membrane flashing				.013	.21	.60		.81	1.10	1.32
	Reinstall ballast in bitumen				.005		.23		.23	.34	.41
	Clean up				.004		.18		.18	.26	.32
	Total				.062	.21	2.77		2.98	4.26	5.16
0500	**Minor BUR membrane replacement - (25% of roof area)**	15	G-5	Sq.							
	Set up, secure and take down ladder				.078		3.52		3.52	5.10	6.20
	Remove 5 ply built-up roof				3.252		135		135	176	217
	Remove roof insulation board				1.333		55		55	72	89
	Install 2" polystyrene insulation				.641	73	29		102	122	142
	Install 4 ply bituminous roofing				3.733	145	157	34	336	425	495
	Clean up				1.000		45		45	65.50	79.50
	Total				10.037	218	424.52	34	676.52	865.60	1,028.70

For customer support on your Facilities Maintenance & Repair Costs with RSMeans data, call 800.448.8182.

B3013 105   Built-Up Roofing

	System Description	Freq. (Years)	Crew	Unit	Labor Hours	2019 Bare Costs				Total In-House	Total w/O&P
						Material	Labor	Equipment	Total		
0600	**Place new BUR membrane over existing**	20	G-5	Sq.							
	Set up, secure and take down ladder				.020		.90		.90	1.31	1.59
	Sweep / spud ballast clean				.500		22.50		22.50	33	40
	Vent existing membrane				.130		5.85		5.85	8.50	10.35
	Cut out buckled flashing				.024		1.08		1.08	1.58	1.92
	Install 2 ply membrane flashing				.027	.42	1.20		1.62	2.21	2.65
	Install 4 ply bituminous roofing				3.733	145	157	34	336	425	495
	Reinstall ballast				.381		17.15		17.15	25	30.50
	Clean up				.390		17.60		17.60	25.50	31
	Total				5.205	145.42	223.28	34	402.70	522.10	613.01
0700	**Total BUR roof replacement**	28	G-1	Sq.							
	Set up, secure and take down ladder				.020		.90		.90	1.31	1.59
	Sweep / spud ballast clean				.500		22.50		22.50	33	40
	Remove built-up roofing				2.500		104		104	135	167
	Remove insulation board				1.026		43		43	55	69
	Remove flashing				.024		1.08		1.08	1.58	1.92
	Install 2" perlite insulation				.879	110	40		150	179	208
	Install 2 ply membrane flashing				.027	.42	1.20		1.62	2.21	2.65
	Install 4 ply bituminous membrane				2.800	145	118	25.50	288.50	360	420
	Clean up				1.000		45		45	65.50	79.50
	Total				8.776	255.42	375.68	25.50	656.60	832.60	989.66

B3013 120   Modified Bituminous / Thermoplastic

	System Description	Freq. (Years)	Crew	Unit	Labor Hours	2019 Bare Costs				Total In-House	Total w/O&P
						Material	Labor	Equipment	Total		
0100	**Debris removal by hand & visual inspection**	1	2 ROFC	M.S.F.							
	Set up, secure and take down ladder				.052		2.34		2.34	3.41	4.14
	Pick up trash / debris & clean up				.327		14.70		14.70	21.50	26
	Visual inspection				.327		14.70		14.70	21.50	26
	Total				.705		31.74		31.74	46.41	56.14
0200	**Non - destructive moisture inspection**	5	2 ROFC	M.S.F.							
	Set up, secure and take down ladder				.052		2.34		2.34	3.41	4.14
	Infrared inspection of roof membrane				2.133		96		96	140	170
	Total				2.185		98.34		98.34	143.41	174.14

B3013 120 Modified Bituminous / Thermoplastic

System Description	Freq. (Years)	Crew	Unit	Labor Hours	2019 Bare Costs				Total In-House	Total w/O&P
					Material	Labor	Equipment	Total		
0300 Minor thermoplastic membrane repairs - (2% of roof area)	1	G-5	Sq.							
Set up, secure and take down ladder				1.000		45		45	66	80
Clean away loose surfacing				.258		11.65		11.65	16.95	20.50
Install 150 mil mod. bit., fully adhered				2.597	84	106	12	202	260	305
Clean up				.390		17.60		17.60	25.50	31
Total				4.245	84	180.25	12	276.25	368.45	436.50
0500 Thermoplastic flashing repairs - (2 S.F. per sq. repaired)	1	2 ROFC	S.F.							
Set up, secure and take down ladder				.010		.45		.45	.66	.80
Clean away loose surfacing				.003		.12		.12	.17	.21
Remove flashing				.020		.90		.90	1.31	1.59
Install 120 mil mod. bit. flashing				.019	.79	.76	.08	1.63	2.07	2.43
Clean up				.004		.18		.18	.26	.32
Total				.055	.79	2.41	.08	3.28	4.47	5.35
0600 Minor thermoplastic membrane replacement - (25% of roof area)	20	G-5	Sq.							
Set up, secure and take down ladder				.078		3.52		3.52	5.10	6.20
Remove existing membrane / insulation				4.571		206		206	300	365
Install 2" perlite insulation				1.140	110	51		161	196	229
Clean adjacent membrane / patch				1.143		51.50		51.50	75	91
Install fully adhered 150 mil membrane				2.597	84	106	12	202	260	305
Clean up				1.000		45		45	65.50	79.50
Total				10.529	194	463.02	12	669.02	901.60	1,075.70
0700 Total thermoplastic roof replacement	25	G-1	Sq.							
Set up, secure and take down ladder				.020		.90		.90	1.31	1.59
Remove existing membrane / insulation				3.501		158		158	230	279
Remove flashing				.026		1.18		1.18	1.70	2.06
Install 2" perlite insulation				.879	110	40		150	179	208
Install flashing				.037	1.58	1.52	.16	3.26	4.14	4.86
Install fully adhered 180 mil membrane				2.000	96	82	9	187	235	275
Clean up				1.000		45		45	65.50	79.50
Total				7.463	207.58	328.60	9.16	545.34	716.65	850.01

For customer support on your Facilities Maintenance & Repair Costs with RSMeans data, call 800.448.8182.

B3013 125 Thermosetting

System Description	Freq. (Years)	Crew	Unit	Labor Hours	2019 Bare Costs				Total In-House	Total w/O&P
					Material	Labor	Equipment	Total		
0100 Debris removal by hand & visual inspection	1	2 ROFC	M.S.F.							
Set up, secure and take down ladder				.052		2.34		2.34	3.41	4.14
Pick up trash / debris & clean up				.327		14.70		14.70	21.50	26
Visual inspection				.327		14.70		14.70	21.50	26
Total				.705		31.74		31.74	46.41	56.14
0200 Non - destructive moisture inspection	5	2 ROFC	M.S.F.							
Set up, secure and take down ladder				.052		2.34		2.34	3.41	4.14
Infrared inspection of roof membrane				2.133		96		96	140	170
Total				2.185		98.34		98.34	143.41	174.14
0300 Minor thermoset membrane repairs - (2% of roof area)	1	G-5	Sq.							
Set up, secure and take down ladder				1.000		45		45	66	80
Broom roof surface clean				.258		11.65		11.65	16.95	20.50
Remove / vent damaged membrane				.258		11.65		11.65	16.95	20.50
Install thermosetting patch				2.597	80	106	12	198	256	300
Clean up				.390		17.60		17.60	25.50	31
Total				4.503	80	191.90	12	283.90	381.40	452
0400 Thermoset flashing repairs - (2 S.F. per sq. repaired)	1	2 ROFC	S.F.							
Set up, secure and take down ladder				.010		.45		.45	.66	.80
Remove flashing				.020		.90		.90	1.31	1.59
Prime surfaces				.003		.14		.14	.20	.24
Install 2 ply membrane flashing				.007	.11	.30		.41	.56	.66
Clean up				.004		.18		.18	.26	.32
Total				.044	.11	1.97		2.08	2.99	3.61
0500 Minor thermoset membrane replacement - (25% of roof area)	10	G-5	Sq.							
Set up, secure and take down ladder				.078		3.52		3.52	5.10	6.20
Broom roof surface clean				.258		11.65		11.65	16.95	20.50
Remove, strip-in & seal joint				.593		26.50		26.50	39	47
Install base sheet and 2 plies of glass mopped				3.294	116	139	30	285	365	425
Clean up				.390		17.60		17.60	25.50	31
Total				4.613	116	198.27	30	344.27	451.55	529.70

B3013 125 Thermosetting

	System Description	Freq. (Years)	Crew	Unit	Labor Hours	2019 Bare Costs				Total In-House	Total w/O&P
						Material	Labor	Equipment	Total		
0600	**Total thermoset roof replacement**	20	G-1	Sq.							
	Set up, secure and take down ladder				.020		.90		.90	1.31	1.59
	Remove existing membrane / insulation				3.501		158		158	230	279
	Remove flashing				.026		1.18		1.18	1.70	2.06
	Install 2" perlite insulation				.879	110	40		150	179	208
	Install new flashing				.020	.42	.90		1.32	1.77	2.13
	Install thermosetting membrane				2.800	178	118	26	322	395	460
	Clean up				1.000		45		45	65.50	79.50
	Total				8.246	288.42	363.98	26	678.40	874.28	1,032.28

B3013 127 EPDM (Ethylene Propylene Diene Monomer)

	System Description	Freq. (Years)	Crew	Unit	Labor Hours	2019 Bare Costs				Total In-House	Total w/O&P
						Material	Labor	Equipment	Total		
0600	**Total EPDM roof replacement**	25	G-5	Sq.							
	Set up, secure and take down ladder				.020		.90		.90	1.31	1.59
	Remove existing membrane / insulation				3.501		158		158	230	279
	Remove flashing				.026		1.18		1.18	1.70	2.06
	Install 2" perlite insulation				.879	110	40		150	179	208
	Install new flashing				.020	.42	.90		1.32	1.77	2.13
	Fully adhered with adhesive				1.538	113	63	6.95	182.95	224	260
	Clean up				1.000		45		45	65.50	79.50
	Total				6.985	223.42	308.98	6.95	539.35	703.28	832.28

B3013 130 Metal Panel Roofing

	System Description	Freq. (Years)	Crew	Unit	Labor Hours	2019 Bare Costs				Total In-House	Total w/O&P
						Material	Labor	Equipment	Total		
0100	**Debris removal by hand & visual inspection**	1	2 ROFC	M.S.F.							
	Set up, secure and take down ladder				.100		4.51		4.51	6.55	7.95
	Visual inspection				.327		14.70		14.70	21.50	26
	Total				.427		19.21		19.21	28.05	33.95

For customer support on your Facilities Maintenance & Repair Costs with RSMeans data, call 800.448.8182.

System Description	Freq. (Years)	Crew	Unit	Labor Hours	2019 Bare Costs				Total In-House	Total w/O&P
					Material	Labor	Equipment	Total		
0300 Minor metal roof finish repairs - (2% of roof area)	5	2 ROFC	S.F.							
Set up, secure and take down ladder				.010		.45		.45	.66	.80
Wire brush surface to remove oxide				.020		.90		.90	1.31	1.59
Prime and paint prepared surface				.016	.71	.72		1.43	1.83	2.16
Clean up				.004		.18		.18	.26	.32
Total				.050	.71	2.25		2.96	4.06	4.87
0400 Metal roof flashing replacement - (2 S.F. per sq. repaired)	1	2 ROFC	S.F.							
Set up, secure and take down ladder				.010		.45		.45	.66	.80
Dismantle seam				.016		.72		.72	1.05	1.27
Remove metal roof panels				.060		2.70		2.70	3.93	4.77
Remove damaged metal flashing				.019		.86		.86	1.25	1.51
Install new aluminum flashing				.072	1.36	3.25		4.61	6.25	7.45
Replace metal panels				.129		5.82		5.82	8.45	10.25
Clean up				.004		.18		.18	.26	.32
Total				.310	1.36	13.98		15.34	21.85	26.37
0500 Minor metal roof panel replacement - (2.5% of roof area)	20	2 ROFC	S.F.							
Set up, secure and take down ladder				.010		.45		.45	.66	.80
Dismantle seam				.016		.72		.72	1.05	1.27
Remove metal roof panels				.020		.90		.90	1.31	1.59
Install new metal panel				.044	3.03	2.24		5.27	6.20	7.35
Re-assemble seam				.016		.72		.72	1.05	1.27
Prime and paint surface				.016	.71	.72		1.43	1.83	2.16
Clean up				.004		.18		.18	.26	.32
Total				.126	3.74	5.93		9.67	12.36	14.76
0600 Total metal roof panel replacement	30	2 ROFC	Sq.							
Set up, secure and take down ladder				.020		.90		.90	1.31	1.59
Remove existing roofing panel				3.600		162		162	236	287
Remove flashing metal				.038		1.72		1.72	2.50	3.02
Install new aluminum flashing				.072	1.36	3.25		4.61	6.25	7.45
Install new metal panel				4.384	303	224		527	620	735
Clean up				.390		17.60		17.60	25.50	31
Total				8.504	304.36	409.47		713.83	891.56	1,065.06

For customer support on your Facilities Maintenance & Repair Costs with RSMeans data, call 800.448.8182.

103

B3013 140 Slate Tile Roofing

System Description	Freq. (Years)	Crew	Unit	Labor Hours	2019 Bare Costs				Total In-House	Total w/O&P
					Material	Labor	Equipment	Total		
0100 Visual inspection	3	2 ROFC	M.S.F.							
Set up, secure and take down ladder				.100		4.51		4.51	6.55	7.95
Visual inspection				.327		14.70		14.70	21.50	26
Total				.427		19.21		19.21	28.05	33.95
0200 Minor slate tile replacement - (2.5% of roof area)	20	2 ROTS	S.F.							
Set up, secure and take down ladder				.010		.45		.45	.66	.80
Remove and replace damaged slates				.800	7.10	36		43.10	60.50	73
Clean up				.013		.59		.59	.85	1.03
Total				.823	7.10	37.04		44.14	62.01	74.83
0300 Slate tile flashing repairs - (2 S.F. per sq. repaired)	20	2 ROTS	S.F.							
Set up, secure and take down ladder				.010		.45		.45	.66	.80
Removal of adjacent slate tiles				.100		4.50		4.50	6.55	7.95
Remove damaged metal flashing				.052		2.34		2.34	3.41	4.14
Install 16 oz. copper flashing				.089	8.10	4		12.10	14.75	17.25
Reinstall tiles in position				.134		6.04		6.04	8.80	10.70
Clean up				.013		.59		.59	.85	1.03
Total				.398	8.10	17.92		26.02	35.02	41.87
0400 Total slate tile roof replacement	70	2 ROTS	Sq.							
Set up, secure and take down ladder				.020		.90		.90	1.31	1.59
Remove slate shingles				1.600		66		66	86	107
Install 16 oz. copper flashing				.178	16.20	8		24.20	29.50	34.50
Install 15# felt				.138	9.95	6.20		16.15	20	23.50
Install slate roofing				4.571	510	207		717	865	1,000
Clean up				1.333		60		60	87.50	106
Total				7.840	536.15	348.10		884.25	1,089.31	1,272.59

B3013 142 Mineral Fiber Steep Roofing

System Description	Freq. (Years)	Crew	Unit	Labor Hours	2019 Bare Costs				Total In-House	Total w/O&P
					Material	Labor	Equipment	Total		
0300 Visual inspection	3	2 ROFC	M.S.F.							
Set up, secure and take down ladder				.100		4.51		4.51	6.55	7.95
Visual inspection				.327		14.70		14.70	21.50	26
Total				.427		19.21		19.21	28.05	33.95

For customer support on your Facilities Maintenance & Repair Costs with RSMeans data, call 800.448.8182.

B3013 144 | **Clay Tile Roofing**

System Description	Freq. (Years)	Crew	Unit	Labor Hours	2019 Bare Costs				Total In-House	Total w/O&P
					Material	Labor	Equipment	Total		
0300 Visual inspection										
Set up, secure and take down ladder		2 ROFC	M.S.F.	.100		4.51		4.51	6.55	7.95
Visual inspection	3			.327		14.70		14.70	21.50	26
Total				.427		19.21		19.21	28.05	33.95
0400 Minor clay tile repairs - (2% of roof area)										
Set up, secure and take down ladder	20	2 ROFC	S.F.	.010		.45		.45	.66	.80
Remove damaged tiles				.020		.90		.90	1.31	1.59
Install new clay tiles				.073	4.75	3.30		8.05	10.05	11.75
Clean up				.013		.59		.59	.85	1.03
Total				.116	4.75	5.24		9.99	12.87	15.17
0500 Clay tile flashing repairs - (2 S.F. per sq. repaired)										
Set up, secure and take down ladder	1	2 ROFC	S.F.	.010		.45		.45	.66	.80
Remove adjacent tiles				.208		9.36		9.36	13.65	16.55
Remove damaged metal flashing				.052		2.34		2.34	3.41	4.14
Install 16 oz copper flashing				.178	16.20	8		24.20	29.50	34.50
Reinstall clay tiles				.073		3.30		3.30	4.80	5.80
Clean up				.013		.59		.59	.85	1.03
Total				.533	16.20	24.04		40.24	52.87	62.82
0600 Total clay tile roof replacement										
Set up, secure and take down ladder	70	2 ROFC	Sq.	.020		.90		.90	1.31	1.59
Remove clay tile roofing				3.200		144		144	210	255
Remove damaged metal flashing				.040		1.80		1.80	2.62	3.18
Install 16 oz copper flashing				.178	16.20	8		24.20	29.50	34.50
Install felt				.138	9.95	6.20		16.15	20	23.50
Install new clay tiles				7.273	475	330		805	1,000	1,175
Clean up				1.333		60		60	87.50	106
Total				12.182	501.15	550.90		1,052.05	1,350.93	1,598.77

System Description	Freq. (Years)	Crew	Unit	Labor Hours	Material	Labor	Equipment	Total	Total In-House	Total w/O&P
						2019 Bare Costs				
0100 Debris removal by hand & visual inspection	1	2 ROFC	M.S.F.							
Set up, secure and take down ladder				.100		4.51		4.51	6.55	7.95
Visual inspection				.327		14.70		14.70	21.50	26
Total				.427		19.21		19.21	28.05	33.95
0300 Minor asphalt shingle repair - (2% of roof area)	1	2 ROFC	S.F.							
Set up, secure and take down ladder				.010		.45		.45	.66	.80
Remove damaged shingles				.024		.98		.98	1.27	1.57
Install asphalt shingles				.036	1.10	1.60		2.70	3.54	4.21
Clean up				.003		.12		.12	.17	.21
Total				.072	1.10	3.15		4.25	5.64	6.79
0500 Asphalt shingle flashing repairs - (2 S.F. per sq. repaired)	1	2 ROFC	S.F.							
Set up, secure and take down ladder				.010		.45		.45	.66	.80
Remove adjoining shingles				.046		1.90		1.90	2.46	3.06
Remove damaged metal flashing				.019		.86		.86	1.25	1.51
Install new aluminum flashing				.072	1.36	3.25		4.61	6.25	7.45
Install asphalt shingles				.071	2.20	3.20		5.40	7.10	8.40
Clean up				.003		.12		.12	.17	.21
Total				.221	3.56	9.78		13.34	17.89	21.43
0600 Install new asphalt shingles over existing	20	2 ROFC	Sq.							
Set up, secure and take down ladder				.020		.90		.90	1.31	1.59
Remove shingles at eaves and ridge				.460		20.80		20.80	30	36.50
Install new aluminum flashing				.144	2.72	6.50		9.22	12.45	14.90
Install asphalt shingles				1.778	110	80		190	238	280
Clean up				.258		11.65		11.65	16.95	20.50
Total				2.660	112.72	119.85		232.57	298.71	353.49
0700 Total asphalt shingle roof replacement	40	2 ROFC	Sq.							
Set up, secure and take down ladder				.020		.90		.90	1.31	1.59
Remove existing shingles				2.353		98		98	127	157
Remove damaged metal flashing				.038		1.72		1.72	2.50	3.02
Install 15# felt				.138	9.95	6.20		16.15	20	23.50
Install new aluminum flashing				.144	2.72	6.20		9.22	12.45	14.90
Install asphalt shingles				1.778	110	80		190	238	280
Clean up				.258		11.65		11.65	16.95	20.50
Total				4.729	122.67	204.97		327.64	418.21	500.51

For customer support on your Facilities Maintenance & Repair Costs with RSMeans data, call 800.448.8182.

B3013 160 Roll Roofing

System Description	Freq. (Years)	Crew	Unit	Labor Hours	2019 Bare Costs Material	Labor	Equipment	Total	Total In-House	Total w/O&P
0100 Debris removal by hand & visual inspection	1	2 ROFC	M.S.F.							
Set up, secure and take down ladder				.100		4.51		4.51	6.55	7.95
Pick up trash / debris & clean up				.327		14.70		14.70	21.50	26
Visual inspection				.327		14.70		14.70	21.50	26
Total				.753		33.91		33.91	49.55	59.95
0300 Minor roofing repairs - (2% of roof area)	1	2 ROFC	S.F.							
Set up, secure and take down ladder				.010		.45		.45	.66	.80
Cut out damaged area				.015		.68		.68	.98	1.19
Install roll roofing with mastic				.027	.77	1.14	.25	2.16	2.77	3.25
Clean up				.003		.12		.12	.17	.21
Total				.055	.77	2.39	.25	3.41	4.58	5.45
0400 Flashing repairs - (2 S.F. per sq. repaired)	1	2 ROFC	S.F.							
Set up, secure and take down ladder				.010		.45		.45	.66	.80
Remove adjacent roofing				.015		.68		.68	.98	1.19
Remove damaged metal flashing				.019		.86		.86	1.25	1.51
Install aluminum flashing				.072	1.36	3.25		4.61	6.25	7.45
Install roll roofing with mastic				.054	1.53	2.28	.50	4.31	5.55	6.50
Clean up				.003		.12		.12	.17	.21
Total				.173	2.89	7.64	.50	11.03	14.86	17.66
0500 Minor replacement - (25% of roof area)	15	2 ROFC	Sq.							
Set up, secure and take down ladder				.078		3.52		3.52	5.10	6.20
Remove strip-in				1.067		48		48	70	85
Renew striping with 2 ply felts				2.917	124	123	27	274	345	400
Clean up				.258		11.65		11.65	16.95	20.50
Total				4.319	124	186.17	27	337.17	437.05	511.70
0600 Total roof replacement	20	2 ROFC	Sq.							
Set up, secure and take down ladder				.020		.90		.90	1.31	1.59
Remove existing shingles				1.333		60		60	87.50	106
Remove flashing				.019		.86		.86	1.25	1.51
Install aluminum flashing				.144	2.72	6.50		9.22	12.45	14.90
Install 4 ply roll roofing				2.917	124	123	27	274	345	400
Clean up				.258		11.65		11.65	16.95	20.50
Total				4.691	126.72	202.91	27	356.63	464.46	544.50

B3013 165   Corrugated Fiberglass Panel Roofing

	System Description	Freq. (Years)	Crew	Unit	Labor Hours	Material	Labor	Equipment	Total	Total In-House	Total w/O&P
								2019 Bare Costs			
0100	**Debris removal & visual inspection**	5	2 ROFC	M.S.F.							
	Set up, secure and take down ladder				.100		4.51		4.51	6.55	7.95
	Visual inspection				.327		14.70		14.70	21.50	26
	Total				.427		19.21		19.21	28.05	33.95
0200	**Fiberglass roof flashing repairs - (2 S.F. per sq. repaired)**	15	2 ROFC	S.F.							
	Set up, secure and take down ladder				.010		.45		.45	.66	.80
	Cut flashing				.010		.45		.45	.66	.80
	Install new flashing				.036	.33	1.64		1.97	2.75	3.30
	Clean up				.004		.18		.18	.26	.32
	Total				.060	.33	2.72		3.05	4.33	5.22
0400	**Minor fiberglass roof panel replacement - (25% of roof area)**	15	2 ROFC	S.F.							
	Set up, secure and take down ladder				.010		.45		.45	.66	.80
	Remove damaged panel				.030		1.35		1.35	1.97	2.39
	Install new fiberglass panel				.064	2.60	3.26		5.86	7.05	8.45
	Install strip-in with joint seal				.029	.02	1.47		1.49	1.96	2.42
	Clean up				.004		.18		.18	.26	.32
	Total				.137	2.62	6.71		9.33	11.90	14.38
0600	**Total fiberglass roof replacement**	20	2 ROFC	Sq.							
	Set up, secure and take down ladder				.020		.90		.90	1.31	1.59
	Remove existing fiberglass panels				2.500		113		113	164	199
	Install new flashing				.073	.66	3.28		3.94	5.50	6.60
	Install new fiberglass panel				6.400	260	326		586	705	845
	Apply joint sealer to seams				.029	.02	1.47		1.49	1.96	2.42
	Clean up				.390		17.60		17.60	25.50	31
	Total				9.412	260.68	462.25		722.93	903.27	1,085.61

For customer support on your Facilities Maintenance & Repair Costs with RSMeans data, call 800.448.8182.

B3013 170 | **Roof Edges**

System Description	Freq. (Years)	Crew	Unit	Labor Hours	2019 Bare Costs				Total In-House	Total w/O&P
					Material	Labor	Equipment	Total		
1010 Replace Roof edges, aluminum, duranodic, .050" thick, 6" face	25	2 ROFC	L.F.							
Set up, secure, and take down ladder				.010		.45		.45	.66	.80
Remove adjacent roofing				.015		.68		.68	.98	1.19
Remove roof edges				.080		3.28		3.28	4.28	5.30
2" x 8" Edge blocking				.025	1.13	1.27		2.40	2.89	3.44
Cants, 4" x 4", treated timber, cut diagonally				.025	1.96	1.11		3.07	3.78	4.41
Aluminum gravel stop with Duranodic finish				.089	12	5.42		17.42	20	23.50
Roof covering, splice sheet				.011	2.41	.47	.05	2.93	3.39	3.90
Total				.254	17.50	12.68	.05	30.23	35.98	42.54
1020 Replace Roof edges, copper, 20 oz., 8" face	35	2 ROFC	L.F.							
Set up, secure, and take down ladder				.010		.45		.45	.66	.80
Remove adjacent roofing				.015		.68		.68	.98	1.19
Remove roof edge				.080		3.28		3.28	4.28	5.30
2" x 8" Edge blocking				.025	1.13	1.27		2.40	2.89	3.44
Cants, 4" x 4", treated timber, cut diagonally				.025	1.96	1.11		3.07	3.78	4.41
Copper gravel stop				.139	16.20	6.26		22.46	27	31.50
Roof covering, splice sheet				.011	2.41	.47	.05	2.93	3.39	3.90
Total				.305	21.70	13.52	.05	35.27	42.98	50.54
1030 Replace Roof edges, galvanized, 20 ga., 6" face	25	2 ROFC	L.F.							
Set up, secure, and take down ladder				.010		.45		.45	.66	.80
Remove adjacent roofing				.015		.68		.68	.98	1.19
Remove roof edge				.080		3.28		3.28	4.28	5.30
2" x 8" edge blocking				.025	1.13	1.27		2.40	2.89	3.44
Cants, 4" x 4", treated timber, cut diagonally				.025	1.96	1.11		3.07	3.78	4.41
Galvanized steel gravel stop				.092	1.80	4.16		5.96	8.05	9.60
Roof covering, splice sheet				.011	2.41	.47	.05	2.93	3.39	3.90
Total				.258	7.30	11.42	.05	18.77	24.03	28.64

B3013 620 Gutters and Downspouts

System Description	Freq. (Years)	Crew	Unit	Labor Hours	2019 Bare Costs				Total In-House	Total w/O&P
					Material	Labor	Equipment	Total		
1010 Replace aluminum gutter, enameled, 5" K type, .027" thick	40	2 SHEE	L.F.							
Set up, secure and take down ladder				.010		.45		.45	.66	.80
Remove gutter				.033		1.37		1.37	1.78	2.21
Install new gutter				.064	2.92	3.90		6.82	8.10	9.75
Clean up				.003		.12		.12	.17	.21
Total				.110	2.92	5.84		8.76	10.71	12.97
1020 Replace aluminum gutter, enameled, 5" K type, .032" thick	40	2 SHEE	L.F.							
Set up, secure and take down ladder				.010		.45		.45	.66	.80
Remove gutter				.033		1.37		1.37	1.78	2.21
Install new gutter				.064	3.59	3.90		7.49	8.85	10.60
Clean up				.003		.12		.12	.17	.21
Total				.110	3.59	5.84		9.43	11.46	13.82
1060 Replace aluminum downspout, 2" x 3", .024" thick	25	2 SHEE	L.F.							
Set up, secure and take down ladder				.010		.45		.45	.66	.80
Remove downspout				.023		.94		.94	1.22	1.51
Install new downspout				.044	2.14	2.71		4.85	5.75	6.95
Clean up				.003		.12		.12	.17	.21
Total				.080	2.14	4.22		6.36	7.80	9.47
1070 Replace aluminum downspout, 3" x 4", .024" thick	25	2 SHEE	L.F.							
Set up, secure and take down ladder				.010		.45		.45	.66	.80
Remove downspout				.023		.94		.94	1.22	1.51
Install new downspout				.057	2.14	3.48		5.62	6.75	8.15
Clean up				.003		.12		.12	.17	.21
Total				.093	2.14	4.99		7.13	8.80	10.67
2010 Replace galvanized steel gutter, 5" box type, 28 ga.	40	2 SHEE	L.F.							
Set up, secure and take down ladder				.010		.45		.45	.66	.80
Remove gutter				.033		1.37		1.37	1.78	2.21
Install new gutter				.064	2.16	3.90		6.06	7.30	8.80
Clean up				.003		.12		.12	.17	.21
Total				.110	2.16	5.84		8	9.91	12.02

For customer support on your Facilities Maintenance & Repair Costs with RSMeans data, call 800.448.8182.

B3013 620 Gutters and Downspouts

System Description	Freq. (Years)	Crew	Unit	Labor Hours	2019 Bare Costs Material	Labor	Equipment	Total	Total In-House	Total w/O&P
2020 Replace galvanized steel gutter, 6" box type, 26 ga.	40	2 SHEE	L.F.							
Set up, secure and take down ladder				.010		.45		.45	.66	.80
Remove gutter				.033		1.37		1.37	1.78	2.21
Install new gutter				.064	2.67	3.90		6.57	7.85	9.45
Clean up				.003		.12		.12	.17	.21
Total				.110	2.67	5.84		8.51	10.46	12.67
2060 Replace round corrugated galvanized downspout, 3" diameter	25	2 SHEE	L.F.							
Set up, secure and take down ladder				.010		.45		.45	.66	.80
Remove downspout				.023		.94		.94	1.22	1.51
Install new downspout				.042	2.11	2.57		4.68	5.55	6.65
Clean up				.003		.12		.12	.17	.21
Total				.078	2.11	4.08		6.19	7.60	9.17
2070 Replace round corrugated galvanized downspout, 4" diameter	25	2 SHEE	L.F.							
Set up, secure and take down ladder				.010		.45		.45	.66	.80
Remove downspout				.023		.94		.94	1.22	1.51
Install new downspout				.055	2.12	3.36		5.48	6.55	7.90
Clean up				.003		.12		.12	.17	.21
Total				.091	2.12	4.87		6.99	8.60	10.42
3020 Replace vinyl box gutter, 5" wide	20	1 CARP	L.F.	.119						
Set up, secure and take down ladder				.010		.45		.45	.66	.80
Remove gutter				.033		1.37		1.37	1.78	2.21
Install new gutter				.070	1.64	3.59		5.23	6.50	7.85
Clean up				.003		.12		.12	.17	.21
Total				.116	1.64	5.53		7.17	9.11	11.07
3060 Replace vinyl downspout, 2" x 3"	15	1 CARP	L.F.	.074						
Set up, secure and take down ladder				.010		.45		.45	.66	.80
Remove downspout				.023		.94		.94	1.22	1.51
Install new downspout				.038	1.94	2.32		4.26	5.05	6.05
Clean up				.003		.12		.12	.17	.21
Total				.074	1.94	3.83		5.77	7.10	8.57

For customer support on your Facilities Maintenance & Repair Costs with RSMeans data, call 800.448.8182.

111

B3013 620 **Gutters and Downspouts**

System Description	Freq. (Years)	Crew	Unit	Labor Hours	2019 Bare Costs				Total In-House	Total w/O&P
					Material	Labor	Equipment	Total		
4020 **Replace 6" copper box gutters, 16 oz.**	50	1 SHEE	L.F.							
Set up, secure and take down ladder				.010		.45		.45	.66	.80
Remove gutter				.033		1.37		1.37	1.78	2.21
Install new gutter				.064	8.80	3.90		12.70	14.60	17.10
Clean up				.003		.12		.12	.17	.21
Total				.110	8.80	5.84		14.64	17.21	20.32
4060 **Replace copper downspouts, 2" x 3", 16 oz.**	40	1 SHEE	L.F.							
Set up, secure and take down ladder				.010		.45		.45	.66	.80
Remove gutter				.023		.94		.94	1.22	1.51
Install new gutter				.042	8.20	2.57		10.77	12.25	14.30
Clean up				.003		.12		.12	.17	.21
Total				.078	8.20	4.08		12.28	14.30	16.82

For customer support on your Facilities Maintenance & Repair Costs with RSMeans data, call 800.448.8182.

B3023 110 | Single Unit Glass Skylight

	System Description	Freq. (Years)	Crew	Unit	Labor Hours	2019 Bare Costs Material	2019 Bare Costs Labor	2019 Bare Costs Equipment	2019 Bare Costs Total	Total In-House	Total w/O&P
1010	**Repair glass skylight glazing**	6	1 CARP	S.F.							
	Set up, secure and take down ladder				.130		6.70		6.70	8.75	10.85
	Remove glazing				.069		2.85		2.85	3.71	4.59
	Install new glazing				.291	27	14.45		41.45	48.50	57
	Total				.490	27	24		51	60.96	72.44
1020	**Replace continuous skylight and structure**	40	2 CARP	C.S.F.							
	Set up and secure scaffold				.750		38.75		38.75	50.50	62.50
	Remove unit skylight				5.202		269		269	350	435
	Inspect / remove flashing				3.152	205	142		347	435	505
	Replace flashing				3.152	205	142		347	435	505
	Install new unit				12.960	2,200	660		2,860	3,275	3,800
	Remove scaffold				.750		38.75		38.75	50.50	62.50
	Total				25.966	2,610	1,290.50		3,900.50	4,596	5,370
1040	**Replace skylight and structure, double glazed, 10 to 20 S.F.**	40	G-3	S.F.							
	Set up, secure and take down ladder				.010		.45		.45	.66	.80
	Remove skylight, plstc domes,flush/curb mtd				.041		2.09		2.09	2.73	3.37
	10 S.F. to 20 S.F., double				.102	29	5.20		34.20	38.50	44.50
	Total				.152	29	7.74		36.74	41.89	48.67
1080	**Replace skylight and structure, double glazed, 30 to 65 S.F.**	40	G-3	S.F.							
	Set up, secure and take down ladder				.010		.45		.45	.66	.80
	Remove skylight, plstc domes,flush/curb mtd				.041		2.09		2.09	2.73	3.37
	30 S.F. to 65 S.F., double				.069	29	3.51		32.51	36.50	42
	Total				.119	29	6.05		35.05	39.89	46.17

B3023 120 | Hatches

	System Description	Freq. (Years)	Crew	Unit	Labor Hours	2019 Bare Costs Material	2019 Bare Costs Labor	2019 Bare Costs Equipment	2019 Bare Costs Total	Total In-House	Total w/O&P
1020	**Replace aluminum roof hatch & structure**	40	2 CARP	Ea.							
	Set up, secure and take down ladder				.130		6.70		6.70	8.75	10.85
	Remove roof hatch				1.000		41		41	53.50	66
	Install 2'-6" x 3'-0" aluminum hatch				3.200	880	163		1,043	1,175	1,350
	Total				4.330	880	210.70		1,090.70	1,237.25	1,426.85

For customer support on your Facilities Maintenance & Repair Costs with RSMeans data, call 800.448.8182.

113

B3023 120 | Hatches

System Description	Freq. (Years)	Crew	Unit	Labor Hours	2019 Bare Costs				Total In-House	Total w/O&P
					Material	Labor	Equipment	Total		
1120 Replace galvanized roof hatch & structure	40	2 CARP	Ea.							
Set up, secure and take down ladder				.130		6.70		6.70	8.75	10.85
Selective demo, therm & moist protect, roof acc, roof hatch, to 10 S.F.				1.000		41		41	53.50	66
Install 2'-6" x 3'-0" galvanized roof hatch				3.200	850	163		1,013	1,150	1,325
Total				4.330	850	210.70		1,060.70	1,212.25	1,401.85

B3023 130 | Smoke Hatches

System Description	Freq. (Years)	Crew	Unit	Labor Hours	2019 Bare Costs				Total In-House	Total w/O&P
					Material	Labor	Equipment	Total		
1010 Replace galvanized smoke hatch single unit 4' x 4'	40	2 CARP	Ea.							
Set up and secure scaffold				.750		38.75		38.75	50.50	62.50
Inspect / remove flashing				.032	2.05	1.42		3.47	4.33	5.05
Replace flashing				.032	2.05	1.42		3.47	4.33	5.05
Galvanized steel cover and frame				2.462	1,650	126		1,776	1,975	2,250
Remove scaffold				.750		38.75		38.75	50.50	62.50
Total				4.025	1,654.10	206.34		1,860.44	2,084.66	2,385.10
1020 Replace aluminum smoke hatch single unit, 4' x 8'	40	2 CARP	Ea.							
Set up and secure scaffold				.750		38.75		38.75	50.50	62.50
Smoke vent 4' x 8'				4.000		164		164	214	265
Inspect / remove flashing				.756	49.20	34.08		83.28	104	122
Replace flashing				.756	49.20	34.08		83.28	104	122
4' x 8' aluminum cover and frame				5.246	3,200	268		3,468	3,850	4,425
Remove scaffold				.750		38.75		38.75	50.50	62.50
Total				12.259	3,298.40	577.66		3,876.06	4,373	5,059

For customer support on your Facilities Maintenance & Repair Costs with RSMeans data, call 800.448.8182.

C1013 110 — Concrete Block, Painted

	System Description	Freq. (Years)	Crew	Unit	Labor Hours	2019 Bare Costs				Total In-House	Total w/O&P
						Material	Labor	Equipment	Total		
0010	Repair 8″ concrete block wall - (2% of walls)	25	1 BRIC	C.S.F.							
	Remove block				6.420		266	38	304	390	470
	Replace block				11.111	279	520		799	990	1,200
	Total				17.531	279	786	38	1,103	1,380	1,670
0020	Refinish concrete block wall	4	1 PORD	C.S.F.							
	Prepare surface				.250		11		11	14	17
	Fill & paint block, brushwork				1.376	22	60		82	101	123
	Place and remove mask and drops				.370		16		16	21	26
	Total				1.996	22	87		109	136	166
0040	Replace 8″ concrete block wall	75	2 BRIC	C.S.F.							
	Set up and secure scaffold				.750		38.75		38.75	50.50	62.50
	Remove block				6.420		266	38	304	390	470
	Replace block				11.111	279	520		799	990	1,200
	Remove scaffold				.750		38.75		38.75	50.50	62.50
	Total				19.031	279	863.50	38	1,180.50	1,481	1,795

C1013 115 — Glazed CMU Interior Wall Finish

	System Description	Freq. (Years)	Crew	Unit	Labor Hours	2019 Bare Costs				Total In-House	Total w/O&P
						Material	Labor	Equipment	Total		
0010	Repair 4″ glazed C.M.U. wall - (2% of walls)	25	1 BRIC	C.S.F.							
	Remove damaged blocks				5.200		216	31	247	315	380
	Install new blocks				15.072	1,025	705		1,730	2,050	2,425
	Total				20.272	1,025	921	31	1,977	2,365	2,805
0020	Replace 4″ glazed C.M.U. wall	75	2 BRIC	C.S.F.							
	Set up and secure scaffold				.750		38.75		38.75	50.50	62.50
	Remove blocks				5.200		216	31	247	315	380
	Install new blocks				15.072	1,025	705		1,730	2,050	2,425
	Remove scaffold				.750		38.75		38.75	50.50	62.50
	Total				21.772	1,025	998.50	31	2,054.50	2,466	2,930

C1013 120 Glass Block

System Description	Freq. (Years)	Crew	Unit	Labor Hours	2019 Bare Costs Material	Labor	Equipment	Total	Total In-House	Total w/O&P
0010 **Repair glass block wall - (2% of walls)**	25	1 BRIC	C.S.F.							
Remove damaged glass block				17.829		745		745	970	1,200
Repair / replace blocks				45.217	2,350	2,100		4,450	5,375	6,400
Total				63.046	2,350	2,845		5,195	6,345	7,600
0020 **Replace glass block wall**	75	2 BRIC	C.S.F.							
Set up and secure scaffold				.750		38.75		38.75	50.50	62.50
Remove damaged glass block				17.829		745		745	970	1,200
Repair / replace blocks				45.217	2,350	2,100		4,450	5,375	6,400
Remove scaffold				.750		38.75		38.75	50.50	62.50
Total				64.546	2,350	2,922.50		5,272.50	6,446	7,725

C1013 125 Stone Veneer

System Description	Freq. (Years)	Crew	Unit	Labor Hours	2019 Bare Costs Material	Labor	Equipment	Total	Total In-House	Total w/O&P
0010 **Repair 4" stone veneer wall - (2% of walls)**	25	1 BRIC	C.S.F.							
Remove damaged stone				7.429	56	379		435	560	685
Replace stone				41.600	1,875	1,945		3,820	4,625	5,500
Total				49.029	1,931	2,324		4,255	5,185	6,185
0020 **Replace 4" stone veneer wall**	75	2 BRIC	C.S.F.							
Set up and secure scaffold				.750		38.75		38.75	50.50	62.50
Remove damaged stone				7.429	56	379		435	560	685
Replace stone				41.600	1,875	1,945		3,820	4,625	5,500
Remove scaffold				.750		38.75		38.75	50.50	62.50
Total				50.529	1,931	2,401.50		4,332.50	5,286	6,310

For customer support on your Facilities Maintenance & Repair Costs with RSMeans data, call 800.448.8182.

C1013 230 | Demountable Partitions

	System Description	Freq. (Years)	Crew	Unit	Labor Hours	2019 Bare Costs				Total In-House	Total w/O&P
						Material	Labor	Equipment	Total		
1010	**Remove and reinstall demountable partitions**	5	2 CARP	C.L.F.							
	Remove demountable partitions for reuse				16.000		825		825	1,075	1,325
	Vinly clad gypsum wall system, material only to replace damaged					1,210			1,210	1,325	1,525
	Install new and reused vinyl clad gypsum in new configuration				34.286		1,770		1,770	2,300	2,850
	18 ga. prefinished 3'-0" x 6'-8" to 4-7/8" deep frame				3.000	567	154.50		721.50	825	960
	Reinstall doors				3.200		165		165	215	267
	Cleanup				1.000		52		52	67	83
	Total				57.486	1,777	2,966.50		4,743.50	5,807	7,010

C1013 544 | Channel Frame Wire Mesh Wall

	System Description	Freq. (Years)	Crew	Unit	Labor Hours	2019 Bare Costs				Total In-House	Total w/O&P
						Material	Labor	Equipment	Total		
0010	**Repair channel frame wire mesh wall - (2% of walls)**	20	1 SSWK	C.S.F.							
	Remove damaged 1-1/2" diamond mesh panel				12.000		500		500	650	805
	Replace with new 1-1/2" diamond mesh panel				2.857	492.50	147.50		640	735	855
	Total				14.857	492.50	647.50		1,140	1,385	1,660
0020	**Refinish channel frame wire mesh wall**	5	1 PORD	C.S.F.							
	Prepare surface				.370		16		16	21	26
	Scrape surface				1.250		54		54	70	86
	Refinish surface, brushwork, primer + 1 coat				1.538	45	67		112	136	162
	Total				3.158	45	137		182	227	274

C1013 730 | Plate Glass Wall - Interior

	System Description	Freq. (Years)	Crew	Unit	Labor Hours	2019 Bare Costs				Total In-House	Total w/O&P
						Material	Labor	Equipment	Total		
0010	**Repair plate glass interior wall - (2% of total)**	25	1 CARP	C.S.F.							
	Remove plate glass & frame				6.933		285		285	370	460
	Repair damaged glass				23.111	7,750	1,220		8,970	10,100	11,700
	Total				30.044	7,750	1,505		9,255	10,470	12,160

For customer support on your Facilities Maintenance & Repair Costs with RSMeans data, call 800.448.8182.

117

C1013 730 | Plate Glass Wall - Interior

	System Description	Freq. (Years)	Crew	Unit	Labor Hours	2019 Bare Costs				Total In-House	Total w/O&P
						Material	Labor	Equipment	Total		
0020	**Replace plate glass interior wall**	50	2 CARP	C.S.F.							
	Set up and secure scaffold				.750		38.75		38.75	50.50	62.50
	Remove plate glass & frame				6.933		285		285	370	460
	Repair damaged glass				23.111	7,750	1,220		8,970	10,100	11,700
	Remove scaffold				.750		38.75		38.75	50.50	62.50
	Total				31.544	7,750	1,582.50		9,332.50	10,571	12,285

For customer support on your Facilities Maintenance & Repair Costs with RSMeans data, call 800.448.8182.

C10 INTERIOR CONSTRUCTION   C1023   Interior Doors

C1023 108   Fully Glazed Wooden Doors

	System Description	Freq. (Years)	Crew	Unit	Labor Hours	2019 Bare Costs				Total In-House	Total w/O&P
						Material	Labor	Equipment	Total		
1010	**Replace insulating glass (3% of glass)**	1	1 CARP	S.F.							
	Remove damaged glass				.052		2.13		2.13	2.78	3.44
	Install new insulating glass				.277	32.50	13.75		46.25	53.50	62.50
	Total				.329	32.50	15.88		48.38	56.28	65.94
1020	**Repair fully glazed wood door**	10	1 CARP	Ea.							
	Remove lockset				.520		27		27	35	43.50
	Replace lockset				1.040	151	53.50		204.50	236	275
	Oil hinges				.065		3.33		3.33	4.34	5.35
	Oil door closer				.065		3.33		3.33	4.34	5.35
	Total				1.689	151	87.16		238.16	279.68	329.20
1030	**Refinish 3'-0" x 7'-0" fully glazed wood door**	4	1 PORD	Ea.							
	Prepare interior door for painting				.433		18.75		18.75	24	30
	Paint door, roller + brush, 1 coat				1.333	2.51	57.50		60.01	77	95
	Total				1.767	2.51	76.25		78.76	101	125
1040	**Replace 3'-0" x 7'-0" fully glazed wood door**	40	1 CARP	Ea.							
	Remove interior door				.520		21.50		21.50	28	34.50
	Remove interior door frame				1.300		67		67	87.50	108
	Remove door closer				.167		8.60		8.60	11.20	13.90
	Install new wood door and frame				2.600	750	134.50		884.50	1,000	1,150
	Reinstall door closer				1.735		89.50		89.50	117	145
	Total				6.322	750	321.10		1,071.10	1,243.70	1,451.40

C1023 110   Steel Interior Door, Painted

	System Description	Freq. (Years)	Crew	Unit	Labor Hours	2019 Bare Costs				Total In-House	Total w/O&P
						Material	Labor	Equipment	Total		
1010	**Repair steel painted door**	14	1 CARP	Ea.							
	Remove lockset				.520		27		27	35	43.50
	Replace lockset				1.040	151	53.50		204.50	236	275
	Oil hinges				.065		3.33		3.33	4.34	5.35
	Oil door closer				.065		3.33		3.33	4.34	5.35
	Total				1.689	151	87.16		238.16	279.68	329.20

C1023 110   Steel Interior Door, Painted

	System Description	Freq. (Years)	Crew	Unit	Labor Hours	2019 Bare Costs Material	2019 Bare Costs Labor	2019 Bare Costs Equipment	2019 Bare Costs Total	Total In-House	Total w/O&P
1020	**Refinish 3'-0" x 7'-0" steel door, painted**	4	1 PORD	Ea.							
	Prepare interior door for painting				.433		18.75		18.75	24	30
	Paint door, roller + brush, 1 coat				.800	4.50	34.50		39	49.50	61
	Total				1.233	4.50	53.25		57.75	**73.50**	**91**
1030	**Replace 3'-0" x 7'-0" steel door, painted**	60	1 CARP	Ea.							
	Remove interior door				.520		21.50		21.50	28	34.50
	Remove interior door frame				1.300		67		67	87.50	108
	Remove door closer				.167		8.60		8.60	11.20	13.90
	Install new steel door				1.224	595	63		658	735	845
	Install new steel door frame				1.000	205	51.50		256.50	293	340
	Reinstall door closer				1.735		89.50		89.50	117	145
	Total				5.946	800	301.10		1,101.10	**1,271.70**	**1,486.40**
2010	**Safety glass replace, (3% of glass)**	1	1 CARP	S.F.							
	Remove damaged glass				.052		2.13		2.13	2.78	3.44
	Install new safety glass				.173	7.10	8.60		15.70	18.95	22.50
	Total				.225	7.10	10.73		17.83	**21.73**	**25.94**
2030	**Refinish, 3'-0" x 7'-0" steel door w/ safety glass**	4	1 PORD	Ea.							
	Prepare interior door for painting				.433		18.75		18.75	24	30
	Paint door, roller + brush, 1 coat				.800	4.50	34.50		39	49.50	61
	Total				1.233	4.50	53.25		57.75	**73.50**	**91**
2040	**Replace 3'-0" x 7'-0" steel door & frame w/ vision lite**	60	1 CARP	Ea.							
	Remove interior door				.520		21.50		21.50	28	34.50
	Remove interior door frame				1.300		67		67	87.50	108
	Remove door closer				.167		8.60		8.60	11.20	13.90
	Install new steel door				1.224	595	63		658	735	845
	Install new steel door frame				1.301	205	67		272	315	365
	Reinstall door closer				1.735		89.50		89.50	117	145
	Vision lite					108			108	119	135
	Total				6.246	908	316.60		1,224.60	**1,412.70**	**1,646.40**

For customer support on your Facilities Maintenance & Repair Costs with RSMeans data, call 800.448.8182.

C1023 110 | **Steel Interior Door, Painted**

	System Description	Freq. (Years)	Crew	Unit	Labor Hours	2019 Bare Costs				Total In-House	Total w/O&P
						Material	Labor	Equipment	Total		
3010	**Repair 3'-0" x 7'-0" steel sliding door**	14	1 CARP	Ea.							
	Oil door closer				.065		3.33		3.33	4.34	5.35
	Oil / inspect track / glide wheels				.065		3.33		3.33	4.34	5.35
	Remove door track / glide wheels (25% of total / 14 years)				.220		11.38		11.38	14.80	18.25
	Install new door track / glide wheels (25% of total / 14 years)				.440	13.88	22.75		36.63	45	54
	Total				.789	13.88	40.79		54.67	68.48	82.95
3020	**Refinish 3'-0" x 7'-0" steel sliding door**	4	1 PORD	Ea.							
	Prepare interior door for painting				.433		18.75		18.75	24	30
	Paint door, roller + brush, 1 coat				.800	4.50	34.50		39	49.50	61
	Total				1.233	4.50	53.25		57.75	73.50	91
3030	**Replace 3'-0" x 7'-0" steel sliding door & frame**	60	1 CARP	Ea.							
	Remove interior door				.520		21.50		21.50	28	34.50
	Remove interior door frame				1.300		67		67	87.50	108
	Remove door closer				.167		8.60		8.60	11.20	13.90
	Install new steel door				1.224	595	63		658	735	845
	Install new steel sliding door frame				3.200	238	181	9.55	428.55	515	610
	Install new door track / glide wheels				1.760	55.50	91		146.50	179	216
	Install door closer				1.735		89.50		89.50	117	145
	Total				9.905	888.50	521.60	9.55	1,419.65	1,672.70	1,972.40
4010	**Replace wire glass (3% of glass)**	1	1 CARP	S.F.							
	Remove damaged glass				.052		2.13		2.13	2.78	3.44
	Install new wire glass				.179	31.32	8.87		40.19	46	53.50
	Total				.231	31.32	11		42.32	48.78	56.94
4030	**Refinish 3'-0" x 7'-0" steel half glass door**	4	1 PORD	Ea.							
	Prepare interior door for painting				.433		18.75		18.75	24	30
	Paint door, roller + brush, 1 coat				.800	4.50	34.50		39	49.50	61
	Total				1.233	4.50	53.25		57.75	73.50	91

For customer support on your Facilities Maintenance & Repair Costs with RSMeans data, call 800.448.8182.

121

C1023 110 | Steel Interior Door, Painted

	System Description	Freq. (Years)	Crew	Unit	Labor Hours	Material	Labor	Equipment	Total	Total In-House	Total w/O&P
							2019 Bare Costs				
4040	**Replace 3'-0" x 7'-0" steel half glass door & frame**	60	1 CARP	Ea.							
	Remove interior door				.520		21.50		21.50	28	34.50
	Remove door closer				.167		8.60		8.60	11.20	13.90
	Remove interior door frame				1.300		67		67	87.50	108
	Install new steel door frame				1.301	205	67		272	315	365
	Install new half glass interior door				1.224	400	63		463	520	600
	Install new safety glass				1.040	42.60	51.60		94.20	114	136
	Reinstall door closer				1.735		89.50		89.50	117	145
	Total				7.287	647.60	368.20		1,015.80	1,192.70	1,402.40

C1023 111 | Steel Interior Door, Unpainted

	System Description	Freq. (Years)	Crew	Unit	Labor Hours	Material	Labor	Equipment	Total	Total In-House	Total w/O&P
							2019 Bare Costs				
1010	**Repair steel door, unpainted**	14	1 CARP	Ea.							
	Remove lockset				.520		27		27	35	43.50
	Replace lockset				1.040	151	53.50		204.50	236	275
	Oil hinges				.065		3.33		3.33	4.34	5.35
	Oil door closer				.065		3.33		3.33	4.34	5.35
	Total				1.689	151	87.16		238.16	279.68	329.20
1020	**Replace 3'-0" x 7'-0" steel door & frame, unpainted**	60	1 CARP	Ea.							
	Remove interior door				.520		21.50		21.50	28	34.50
	Remove interior door frame				1.300		67		67	87.50	108
	Remove door closer				.167		8.60		8.60	11.20	13.90
	Install new steel door				1.224	595	63		658	735	845
	Install new steel door frame				1.000	205	51.50		256.50	293	340
	Reinstall door closer				1.735		89.50		89.50	117	145
	Total				5.946	800	301.10		1,101.10	1,271.70	1,486.40
2010	**Replace safety glass (3% of glass)**	1	1 CARP	S.F.							
	Remove damaged glass				.052		2.13		2.13	2.78	3.44
	Install new safety glass				.173	7.10	8.60		15.70	18.95	22.50
	Total				.225	7.10	10.73		17.83	21.73	25.94

For customer support on your Facilities Maintenance & Repair Costs with RSMeans data, call 800.448.8182.

C1023 111   Steel Interior Door, Unpainted

	System Description	Freq. (Years)	Crew	Unit	Labor Hours	2019 Bare Costs				Total In-House	Total w/O&P
						Material	Labor	Equipment	Total		
2030	**Replace 3'-0" x 7'-0" steel half glass door & frame, unpainted**	60	1 CARP	Ea.							
	Remove interior door				.520		21.50		21.50	28	34.50
	Remove interior door frame				1.300		67		67	87.50	108
	Remove door closer				.167		8.60		8.60	11.20	13.90
	Install new half glass interior door				1.224	400	63		463	520	600
	Install new safety glass				1.560	63.90	77.40		141.30	171	204
	Install new steel door frame				1.301	205	67		272	315	365
	Reinstall door closer				1.735		89.50		89.50	117	145
	Total				7.807	668.90	394		1,062.90	1,249.70	1,470.40
3010	**Repair 3'-0" x 7'-0" steel sliding door, unpainted**	14	1 CARP	Ea.							
	Oil door closer				.065		3.33		3.33	4.34	5.35
	Oil / inspect track / glide wheels				.065		3.33		3.33	4.34	5.35
	Remove door track / glide wheels (25% of total / 14 years)				.220		11.38		11.38	14.80	18.25
	Install new door track / glide wheels (25% of total / 14 years)				.440	13.88	22.75		36.63	45	54
	Total				.789	13.88	40.79		54.67	68.48	82.95
3020	**Replace 3'-0" x 7'-0" steel sliding door, unpainted**	60	1 CARP	Ea.							
	Remove interior door				.520		21.50		21.50	28	34.50
	Remove interior door frame				1.300		67		67	87.50	108
	Remove door closer				.167		8.60		8.60	11.20	13.90
	Install new steel door				1.224	595	63		658	735	845
	Install new steel sliding door frame				3.200	238	181	9.55	428.55	515	610
	Install new door track / glide wheels				1.760	55.50	91		146.50	179	216
	Reinstall door closer				1.735		89.50		89.50	117	145
	Total				9.905	888.50	521.60	9.55	1,419.65	1,672.70	1,972.40
4010	**Replace wire glass (3% of glass)**	1	1 CARP	S.F.							
	Remove damaged glass				.052		2.13		2.13	2.78	3.44
	Install new wire glass				.179	31.32	8.87		40.19	46	53.50
	Total				.231	31.32	11		42.32	48.78	56.94

C1023 112   Bi-Fold Interior Doors

	System Description	Freq. (Years)	Crew	Unit	Labor Hours	2019 Bare Costs				Total In-House	Total w/O&P
						Material	Labor	Equipment	Total		
2010	**Repair 2'-6" x 6'-8" bi-fold louvered door, wood**	15	1 CARP	Ea.							
	Remove interior door				.520		21.50		21.50	28	34.50
	Remove damaged stile				.078		4.05		4.05	5.30	6.55
	Remove damaged slat				.039		2		2	2.60	3.22
	Install new slat				.077	.82	4		4.82	6.10	7.45
	Install new stile				.157	2.37	8.10		10.47	13.15	16.05
	Reinstall door				.667		34.50		34.50	45	55.50
	Oil / lubricate door				.065		3.33		3.33	4.34	5.35
	Total				1.602	3.19	77.48		80.67	104.49	128.62
2020	**Refinish 2'-6" x 6'-8" bi-fold louvered door, wood**	8	1 PORD	Ea.							
	Prepare door for finish				.967		42		42	54	67
	Paint door, roller + brush, 1 coat				1.143	5.60	49.50		55.10	70	86
	Total				2.110	5.60	91.50		97.10	124	153
2040	**Replace 2'-6" x 6'-8" wood louver bi-fold door & frame**	24	1 CARP	Ea.							
	Remove doors				.520		21.50		21.50	28	34.50
	Install new door				1.600	300	82.50		382.50	440	510
	Install new interior door frame				.998	85.50	51.66		137.16	161	190
	Total				3.118	385.50	155.66		541.16	629	734.50

C1023 114   Aluminum Interior Doors

	System Description	Freq. (Years)	Crew	Unit	Labor Hours	2019 Bare Costs				Total In-House	Total w/O&P
						Material	Labor	Equipment	Total		
1010	**Repair aluminum door**	12	1 CARP	Ea.							
	Remove lockset				.520		27		27	35	43.50
	Replace lockset				1.040	151	53.50		204.50	236	275
	Oil hinges				.065		3.33		3.33	4.34	5.35
	Oil door closer				.065		3.33		3.33	4.34	5.35
	Total				1.689	151	87.16		238.16	279.68	329.20

For customer support on your Facilities Maintenance & Repair Costs with RSMeans data, call 800.448.8182.

C1023 114   Aluminum Interior Doors

System Description	Freq. (Years)	Crew	Unit	Labor Hours	2019 Bare Costs				Total In-House	Total w/O&P
					Material	Labor	Equipment	Total		
1020 Replace 3'-0" x 7'-0" aluminum door & frame	50	1 CARP	Ea.							
Remove interior door				.520		21.50		21.50	28	34.50
Remove interior door frame				1.300		67		67	87.50	108
Remove door closer				.167		8.60		8.60	11.20	13.90
Install new aluminum door				1.224	1,575	63		1,638	1,825	2,075
Install new aluminum door frame				2.971	735	166		901	1,025	1,200
Reinstall door closer				1.735		89.50		89.50	117	145
Total				7.917	2,310	415.60		2,725.60	3,093.70	3,576.40
2010 Replace safety glass (3% of glass)	1	1 CARP	S.F.							
Remove damaged glass				.052		2.13		2.13	2.78	3.44
Install new safety glass				.173	7.10	8.60		15.70	18.95	22.50
Total				.225	7.10	10.73		17.83	21.73	25.94
2030 Replace 3'-0" x 7'-0" aluminum door & frame w/ vision lite	50	1 CARP	Ea.							
Remove interior door				.520		21.50		21.50	28	34.50
Remove interior door frame				1.300		67		67	87.50	108
Remove door closer				.167		8.60		8.60	11.20	13.90
Install new aluminum door				1.224	1,575	63		1,638	1,825	2,075
Install new aluminum door frame				2.971	735	166		901	1,025	1,200
Reinstall door closer				1.735		89.50		89.50	117	145
Vision lite					108			108	119	135
Total				7.917	2,418	415.60		2,833.60	3,212.70	3,711.40
3010 Replace insulating glass (3% of glass)	1	1 CARP	S.F.							
Remove damaged glass				.052		2.13		2.13	2.78	3.44
Install new insulating glass				.277	32.50	13.75		46.25	53.50	62.50
Total				.329	32.50	15.88		48.38	56.28	65.94
3020 Repair 3'-0" x 7'-0" aluminum sliding door	14	1 CARP	Ea.							
Oil door closer				.065		3.33		3.33	4.34	5.35
Oil / inspect track / glide wheels				.065		3.33		3.33	4.34	5.35
Remove door track / glide wheels (25% of total / 14 years)				.220		11.38		11.38	14.80	18.25
Install new door track / glide wheels (25% of total / 14 years)				.440	13.88	22.75		36.63	45	54
Total				.789	13.88	40.79		54.67	68.48	82.95

C1023 114 | Aluminum Interior Doors

System Description	Freq. (Years)	Crew	Unit	Labor Hours	2019 Bare Costs Material	Labor	Equipment	Total	Total In-House	Total w/O&P
3030										
Replace 3'-0" x 7'-0" aluminum sliding door	50	1 CARP	Ea.							
Remove interior door				.520		21.50		21.50	28	34.50
Remove interior door frame				1.300		67		67	87.50	108
Remove door closer				.167		8.60		8.60	11.20	13.90
Install new aluminum door				1.224	1,575	63		1,638	1,825	2,075
Install new aluminum door frame				2.971	735	166		901	1,025	1,200
Install new door track / glide wheels				1.760	55.50	91		146.50	179	216
Install door closer				1.735		89.50		89.50	117	145
Total				9.677	2,365.50	506.60		2,872.10	3,272.70	3,792.40

C1023 116 | Fully Glazed Aluminum Doors

System Description	Freq. (Years)	Crew	Unit	Labor Hours	2019 Bare Costs Material	Labor	Equipment	Total	Total In-House	Total w/O&P
1010										
Replace insulating glass (3% of glass)	1	1 CARP	S.F.							
Remove damaged glass				.052		2.13		2.13	2.78	3.44
Install new insulating glass				.277	32.50	13.75		46.25	53.50	62.50
Total				.329	32.50	15.88		48.38	56.28	65.94
1020										
Repair 3'-0" x 7'-0" fully glazed aluminum door	12	1 CARP	Ea.							
Remove lockset				.520		27		27	35	43.50
Replace lockset				1.040	151	53.50		204.50	236	275
Oil hinges				.065		3.33		3.33	4.34	5.35
Oil door closer				.065		3.33		3.33	4.34	5.35
Total				1.689	151	87.16		238.16	279.68	329.20
1030										
Replace 3'-0" x 7'-0" fully glazed aluminum door	50	1 CARP	Ea.							
Remove interior door				.520		21.50		21.50	28	34.50
Remove interior door frame				1.300		67		67	87.50	108
Remove door closer				.167		8.60		8.60	11.20	13.90
Install new aluminum door and frame				10.458	925	585		1,510	1,800	2,125
Install new safety glass				3.640	149.10	180.60		329.70	400	475
Reinstall door closer				1.735		89.50		89.50	117	145
Total				17.819	1,074.10	952.20		2,026.30	2,443.70	2,901.40

For customer support on your Facilities Maintenance & Repair Costs with RSMeans data, call 800.448.8182.

C1023 120 — Hollow Core Interior Doors

	System Description	Freq. (Years)	Crew	Unit	Labor Hours	2019 Bare Costs				Total In-House	Total w/O&P
						Material	Labor	Equipment	Total		
1010	**Repair hollow core wood door**	7	1 CARP	Ea.							
	Remove lockset				.520		27		27	35	43.50
	Replace lockset				1.040	151	53.50		204.50	236	275
	Oil hinges				.065		3.33		3.33	4.34	5.35
	Oil door closer				.065		3.33		3.33	4.34	5.35
	Total				1.689	151	87.16		238.16	279.68	329.20
1020	**Refinish 3'-0" x 7'-0" hollow core wood door**	4	1 PORD	Ea.							
	Prepare interior door for painting				.433		18.75		18.75	24	30
	Paint door, roller + brush, 1 coat				.533	5.60	23		28.60	36	44
	Total				.967	5.60	41.75		47.35	60	74
1030	**Replace 3'-0" x 7'-0" hollow core wood door**	30	1 CARP	Ea.							
	Remove interior door				.520		21.50		21.50	28	34.50
	Remove interior door frame				1.300		67		67	87.50	108
	Install new interior wood door and frame				.941	165	48.50		213.50	245	285
	Total				2.761	165	137		302	360.50	427.50

C1023 121 — Solid Core Interior Doors

	System Description	Freq. (Years)	Crew	Unit	Labor Hours	2019 Bare Costs				Total In-House	Total w/O&P
						Material	Labor	Equipment	Total		
1010	**Repair solid core wood door**	11	1 CARP	Ea.							
	Remove lockset				.520		27		27	35	43.50
	Replace lockset				1.040	151	53.50		204.50	236	275
	Oil hinges				.065		3.33		3.33	4.34	5.35
	Oil door closer				.065		3.33		3.33	4.34	5.35
	Total				1.689	151	87.16		238.16	279.68	329.20
1020	**Refinish 3'-0" x 7'-0" solid core wood door**	4	1 PORD	Ea.							
	Prepare interior door for painting				.433		18.75		18.75	24	30
	Paint door, roller + brush, 1 coat				.533	5.60	23		28.60	36	44
	Total				.967	5.60	41.75		47.35	60	74

C1023 121 — Solid Core Interior Doors

	System Description	Freq. (Years)	Crew	Unit	Labor Hours	2019 Bare Costs				Total In-House	Total w/O&P
						Material	Labor	Equipment	Total		
1030	**Replace 3'-0" x 7'-0" solid core wood door**	40	1 CARP	Ea.							
	Remove interior door				.520		21.50		21.50	28	34.50
	Remove interior door frame				.650		33.80		33.80	44	54
	Install new interior wood door & frame				.941	365	48.50		413.50	465	535
	Total				2.111	365	103.80		468.80	537	623.50
2010	**Repair solid core sliding wood door**	14	1 CARP	Ea.							
	Oil door closer				.065		3.33		3.33	4.34	5.35
	Oil / inspect track / glide wheels				.065		3.33		3.33	4.34	5.35
	Remove door track / glide wheels (25% of total / 14 years)				.220		11.38		11.38	14.80	18.25
	Install new door track / glide wheels (25% of total / 14 years)				.440	13.88	22.75		36.63	45	54
	Total				.789	13.88	40.79		54.67	68.48	82.95
2030	**Replace 3'-0" x 7'-0" solid core sliding wood door**	40	1 CARP	Ea.							
	Remove interior door				.520		21.50		21.50	28	34.50
	Remove interior door frame				.650		33.80		33.80	44	54
	Remove door closer				.167		8.60		8.60	11.20	13.90
	Install new interior wood door and frame				.941	365	48.50		413.50	465	535
	Install new door track / glide wheels				1.760	55.50	91		146.50	179	216
	Reinstall door closer				1.735		89.50		89.50	117	145
	Total				5.773	420.50	292.90		713.40	844.20	998.40
3040	**Replace 3'-0" x 7'-0" solid core wood door, w/ safety glass**	40	1 CARP	Ea.							
	Remove interior door				.520		21.50		21.50	28	34.50
	Remove interior door frame				.650		33.80		33.80	44	54
	Install new interior wood door and frame				.941	365	48.50		413.50	465	535
	Install new safety glass lite					86			86	94.50	108
	Total				2.111	451	103.80		554.80	631.50	731.50

For customer support on your Facilities Maintenance & Repair Costs with RSMeans data, call 800.448.8182.

C10 INTERIOR CONSTRUCTION C1023 Interior Doors

C1023 129 Interior Gates

	System Description	Freq. (Years)	Crew	Unit	Labor Hours	2019 Bare Costs				Total In-House	Total w/O&P
						Material	Labor	Equipment	Total		
1030	Prepare and refinish interior metal gate	5	1 PORD	Ea.							
	Lay drop cloth				.090		3.90		3.90	5	6.20
	Prepare / brush surface				.125		5.40		5.40	7	8.65
	Paint gate, brushwork, primer + 1 coat				.167	1.44	7.20		8.64	11	13.30
	Remove drop cloth				.090		3.90		3.90	5	6.20
	Total				.471	1.44	20.40		21.84	28	34.35

C1023 324 Hinges, Brass

	System Description	Freq. (Years)	Crew	Unit	Labor Hours	2019 Bare Costs				Total In-House	Total w/O&P
						Material	Labor	Equipment	Total		
0010	Replace 1 1/2 pair brass hinges	60	1 CARP	Ea.							
	Remove door				.650		26.50		26.50	35	43
	Remove hinges				.500		25.80		25.80	33.50	41.50
	Replace hinges				.500		25.80		25.80	33.50	41.50
	Hinges, 1 1/2 pair					192			192	211	240
	Install door				1.224		63		63	82.50	102
	Total				2.875	192	141.10		333.10	395.50	468

C1023 325 Lockset, Brass

	System Description	Freq. (Years)	Crew	Unit	Labor Hours	2019 Bare Costs				Total In-House	Total w/O&P
						Material	Labor	Equipment	Total		
0010	Replace brass lockset	30	1 CARP	Ea.							
	Remove lockset				.400		20.50		20.50	27	33.50
	Install lockset				1.156	234	59.50		293.50	335	390
	Total				1.556	234	80		314	362	423.50

For customer support on your Facilities Maintenance & Repair Costs with RSMeans data, call 800.448.8182.

129

C1023 326 Door Closer, Brass

	System Description	Freq. (Years)	Crew	Unit	Labor Hours	2019 Bare Costs				Total In-House	Total w/O&P
						Material	Labor	Equipment	Total		
0010	Replace brass closer	15	1 CARP	Ea.							
	Remove door closer				.167		8.60		8.60	11.20	13.90
	Install new door closer				1.735	204	89.50		293.50	340	400
	Total				1.902	204	98.10		302.10	351.20	413.90

C1023 327 Deadbolt, Brass

	System Description	Freq. (Years)	Crew	Unit	Labor Hours	2019 Bare Costs				Total In-House	Total w/O&P
						Material	Labor	Equipment	Total		
0010	Replace brass deadbolt	20	1 CARP	Ea.							
	Remove deadbolt				.578		30		30	39	48
	Replace deadbolt				1.156	190	59.50		249.50	287	335
	Total				1.733	190	89.50		279.50	326	383

C1023 328 Weatherstripping, Brass

	System Description	Freq. (Years)	Crew	Unit	Labor Hours	2019 Bare Costs				Total In-House	Total w/O&P
						Material	Labor	Equipment	Total		
0010	Replace brass weatherstripping	20	1 CARP	Ea.							
	Remove weatherstripping				.015		.76		.76	.99	1.23
	Install new weatherstripping				3.478	51	180		231	290	355
	Total				3.493	51	180.76		231.76	290.99	356.23

C1023 329 Panic Bar

	System Description	Freq. (Years)	Crew	Unit	Labor Hours	2019 Bare Costs				Total In-House	Total w/O&P
						Material	Labor	Equipment	Total		
0010	Replace panic bar	25	1 CARP	Ea.							
	Remove panic bar				1.040		53.50		53.50	70	86.50
	Install panic bar				2.080	1,100	107		1,207	1,350	1,550
	Total				3.120	1,100	160.50		1,260.50	1,420	1,636.50

C1033 110 Toilet Partitions

System Description	Freq. (Years)	Crew	Unit	Labor Hours	2019 Bare Costs Material	2019 Bare Costs Labor	2019 Bare Costs Equipment	2019 Bare Costs Total	Total In-House	Total w/O&P
1010 Replace toilet partitions, laminate clad-overhead braced, per stall	20	2 CARP	Ea.							
Remove overhead braced laminate clad partitions				1.333		69		69	89.50	111
Install overhead braced laminate clad partitions				3.478	835	180		1,015	1,150	1,325
Cleanup				.500		26		26	33.50	41.50
Total				5.312	835	275		1,110	1,273	1,477.50
1020 Replace toilet partitions, painted metal-overhead braced, per stall	20	2 CARP	Ea.							
Remove overhead braced painted metal partitions				1.333		69		69	89.50	111
Install overhead braced painted metal partitions				3.478	395	180		575	670	785
Cleanup				.500		26		26	33.50	41.50
Total				5.312	395	275		670	793	937.50
1030 Replace toilet partitions, porcelain enamel-overhead braced, per stall	20	2 CARP	Ea.							
Remove overhead braced partitions				1.333		69		69	89.50	111
Install overhead braced porcelain enamel partitions				3.478	565	180		745	855	995
Cleanup				.500		26		26	33.50	41.50
Total				5.312	565	275		840	978	1,147.50
1040 Replace toilet partitions, phenolic-overhead braced, per stall	20	2 CARP	Ea.							
Remove overhead braced partitions				1.333		69		69	89.50	111
Install phenolic overhead braced partition				3.478	750	180		930	1,050	1,225
Cleanup				.500		26		26	33.50	41.50
Total				5.312	750	275		1,025	1,173	1,377.50
1050 Replace toilet partitions, stainless steel-overhead braced, per stall	30	2 CARP	Ea.							
Remove overhead braced partitions				1.333		69		69	89.50	111
Install stainless steel overhead braced partition				2.667	1,075	138		1,213	1,350	1,575
Cleanup				.500		26		26	33.50	41.50
Total				4.500	1,075	233		1,308	1,473	1,727.50
1060 Replace toilet partitions, stainless steel, ceiling hung, per stall	30	2 CARP	Ea.							
Remove ceiling hung partitions				1.333		69		69	89.50	111
Install stainless steel ceiling hung partition				4.000	1,150	207		1,357	1,525	1,775
Cleanup				.500		26		26	33.50	41.50
Total				5.833	1,150	302		1,452	1,648	1,927.50

For customer support on your Facilities Maintenance & Repair Costs with RSMeans data, call 800.448.8182.

131

C1033 110 | Toilet Partitions

System Description	Freq. (Years)	Crew	Unit	Labor Hours	2019 Bare Costs				Total In-House	Total w/O&P
					Material	Labor	Equipment	Total		
2020 Replace urinal screen, stainless steel	30	2 CARP	Ea.							
Remove stainless steel urinal screen				1.000		41		41	53.50	66
Install stainless steel urinal screen				2.000	535	103		638	725	835
Cleanup				.500		26		26	33.50	41.50
Total				3.500	535	170		705	812	942.50
3010 Replace shower stall, fiberglass, per stall	30	2 SHEE	Ea.							
Remove shower base, partition and door				2.000		82		82	107	132
Install shower base, partition and door				3.556	760	217		977	1,100	1,300
Cleanup				.500		26		26	33.50	41.50
Total				6.056	760	325		1,085	1,240.50	1,473.50

C1033 210 | Metal Lockers

System Description	Freq. (Years)	Crew	Unit	Labor Hours	2019 Bare Costs				Total In-House	Total w/O&P
					Material	Labor	Equipment	Total		
1010 Replace metal lockers, single tier	20	1 SHEE	Ea.							
Remove lockers				.533		22		22	28.50	35.50
Install 12" x 18" x 72" single enameled locker				.400	207	24.50		231.50	258	297
Cleanup				.500		26		26	33.50	41.50
Total				1.433	207	72.50		279.50	320	374

For customer support on your Facilities Maintenance & Repair Costs with RSMeans data, call 800.448.8182.

C2013 110 Interior Concrete Steps

	System Description	Freq. (Years)	Crew	Unit	Labor Hours	2019 Bare Costs				Total In-House	Total w/O&P
						Material	Labor	Equipment	Total		
0010	**Repair interior concrete steps**	15	1 CEFI	S.F.							
	Repair spalls in steps				.104	13.10	5.10		18.20	21	24.50
	Total				.104	13.10	5.10		18.20	**21**	**24.50**
0020	**Replace interior concrete steps**	100	2 CEFI	S.F.							
	Remove concrete				.520		21.50	3.05	24.55	31.50	38.50
	Install new concrete steps				.520	5.30	26	.29	31.59	40	49
	Total				1.040	5.30	47.50	3.34	56.14	**71.50**	**87.50**

C2013 115 Interior Masonry Steps, Painted

	System Description	Freq. (Years)	Crew	Unit	Labor Hours	2019 Bare Costs				Total In-House	Total w/O&P
						Material	Labor	Equipment	Total		
0010	**Repair interior masonry steps, painted**	20	1 BRIC	S.F.							
	Remove damaged masonry				.083		3.42		3.42	4.45	5.50
	Replace with new masonry				.232	4.25	10.60		14.85	18.65	22.50
	Refinish surface				.014	.22	.60		.82	1.01	1.23
	Total				.329	4.47	14.62		19.09	**24.11**	**29.23**
0040	**Replace interior masonry steps, painted**	50	2 BRIC	S.F.							
	Remove masonry steps				4.000		164		164	214	265
	Clean work area				.015		.63	.14	.77	.98	1.18
	Install new masonry steps				.891	92.50	42		134.50	156	183
	Total				4.907	92.50	206.63	.14	299.27	**370.98**	**449.18**

C2013 130 Interior Metal Steps

	System Description	Freq. (Years)	Crew	Unit	Labor Hours	2019 Bare Costs				Total In-House	Total w/O&P
						Material	Labor	Equipment	Total		
0010	**Repair interior metal steps**	15	1 SSWK	S.F.							
	Remove damaged tread				.347		19.40		19.40	26.50	32.50
	Replace with cast iron tread, 1/2" thick				.120	55.44	6.77		62.21	70	80.50
	Total				.467	55.44	26.17		81.61	**96.50**	**113**

For customer support on your Facilities Maintenance & Repair Costs with RSMeans data, call 800.448.8182.

133

C2013 130 Interior Metal Steps

System Description	Freq. (Years)	Crew	Unit	Labor Hours	2019 Bare Costs Material	Labor	Equipment	Total	Total In-House	Total w/O&P
0030 Refinish interior metal steps	9	1 PORD	S.F.							
Prepare steps for painting				.026		1.12		1.12	1.45	1.80
Wipe surface				.001		.05		.05	.07	.08
Refinish surface, roller + brush, 1 coat				.025	.38	1.08		1.46	1.81	2.21
Total				.052	.38	2.25		2.63	3.33	4.09
0040 Replace interior metal steps	50	2 SSWK	S.F.							
Remove steps				.264		10.89		10.89	14.10	17.50
Install pre-erected steel steps				.163	127	9.20	.49	136.69	153	175
Prepare steps for painting				.026		1.12		1.12	1.45	1.80
Wipe surface				.001		.05		.05	.07	.08
Refinish surface				.025	.38	1.08		1.46	1.81	2.21
Total				.479	127.38	22.34	.49	150.21	170.43	196.59

C2013 435 Interior Metal Stair Railing

System Description	Freq. (Years)	Crew	Unit	Labor Hours	2019 Bare Costs Material	Labor	Equipment	Total	Total In-House	Total w/O&P
0030 Refinish interior metal stair railing	7	1 PORD	S.F.							
Prepare surface				.019		.83		.83	1.06	1.32
Refinish surface, roller + brush, 1 coat				.015	.11	.67		.78	.98	1.20
Total				.035	.11	1.50		1.61	2.04	2.52
0040 Replace interior metal stair railing	45	2 SSWK	L.F.							
Remove railing				.098		5.50	.29	5.79	7.80	9.50
Install new railing				.195	25	11.05	.58	36.63	43	50.50
Total				.293	25	16.55	.87	42.42	50.80	60

C2013 437 Interior Iron Stair Railing

System Description	Freq. (Years)	Crew	Unit	Labor Hours	2019 Bare Costs Material	Labor	Equipment	Total	Total In-House	Total w/O&P
0030 Refinish interior wrought iron stair railing	7	1 PORD	L.F.							
Prepare surface				.019		.83		.83	1.06	1.32
Refinish surface, roller + brush, 1 coat				.015	.11	.67		.78	.98	1.20
Total				.035	.11	1.50		1.61	2.04	2.52

For customer support on your Facilities Maintenance & Repair Costs with RSMeans data, call 800.448.8182.

C2013 437 Interior Iron Stair Railing

	System Description	Freq. (Years)	Crew	Unit	Labor Hours	2019 Bare Costs Material	Labor	Equipment	Total	Total In-House	Total w/O&P
0040	Replace interior wrought iron stair railing	45	2 SSWK	L.F.							
	Remove railing				.087		3.56		3.56	4.64	5.75
	Install new railing				1.300	33	73		106	135	162
	Total				1.387	33	76.56		109.56	139.64	167.75

C2013 440 Interior Wood Steps

	System Description	Freq. (Years)	Crew	Unit	Labor Hours	2019 Bare Costs Material	Labor	Equipment	Total	Total In-House	Total w/O&P
0030	Refinish interior wood steps	3	1 PORD	S.F.							
	Prepare surface				.026		1.12		1.12	1.45	1.80
	Refinish surface, roller + brush, 1 coat				.026	.07	1.12		1.19	1.53	1.89
	Total				.052	.07	2.24		2.31	2.98	3.69
0040	Replace interior wood steps	40	1 CARP	S.F.							
	Remove steps				.172		7.10		7.10	9.20	11.40
	Replace steps				.325	97	16.80		113.80	129	148
	Total				.497	97	23.90		120.90	138.20	159.40

C2013 445 Interior Wood Stair Railing

	System Description	Freq. (Years)	Crew	Unit	Labor Hours	2019 Bare Costs Material	Labor	Equipment	Total	Total In-House	Total w/O&P
0030	Refinish interior wood stair railing	7	1 PORD	L.F.							
	Prepare railings for refinishing				.026		1.12		1.12	1.45	1.80
	Refinish surface, roller + brush, 1 coat				.026	.07	1.12		1.19	1.53	1.89
	Total				.052	.07	2.24		2.31	2.98	3.69
0040	Replace interior wood stair railing	40	1 CARP	L.F.							
	Remove railing				.087		3.56		3.56	4.64	5.75
	Replace railing				.130	2.13	6.70		8.83	11.10	13.50
	Total				.217	2.13	10.26		12.39	15.74	19.25

C2023 110 Carpeted Steps

	System Description	Freq. (Years)	Crew	Unit	Labor Hours	Material	Labor	Equipment	Total	Total In-House	Total w/O&P
							2019 Bare Costs				
0020	Replace carpeted steps	8	2 TILF	S.F.							
	Remove old carpet				.010		.43		.43	.56	.69
	Install new carpet				.069	15.18	3.28		18.46	21	24
	Additional labor				.229	30.36	10.96		41.32	47	55
	Total				.308	45.54	14.67		60.21	68.56	79.69

C2023 120 Rubber Steps

	System Description	Freq. (Years)	Crew	Unit	Labor Hours	Material	Labor	Equipment	Total	Total In-House	Total w/O&P
							2019 Bare Costs				
0020	Replace rubber steps	18	2 TILF	L.F.							
	Remove rubber covering				.015		.61		.61	.80	.98
	Prepare floor for installation				.063		3.04		3.04	3.84	4.78
	Install new rubber tread				.090	14.55	4.33		18.88	21.50	25
	Install new rubber riser				.042	8.60	1.99		10.59	12	13.90
	Total				.210	23.15	9.97		33.12	38.14	44.66

C2023 130 Terrazzo Steps

	System Description	Freq. (Years)	Crew	Unit	Labor Hours	Material	Labor	Equipment	Total	Total In-House	Total w/O&P
							2019 Bare Costs				
0020	Replace terrazzo steps	50	2 MSTZ	S.F.							
	Remove old terrazzo tile				.052		2.13		2.13	2.78	3.44
	Clean area				.020		1.03		1.03	1.34	1.66
	Replace with new tread				.743	3.15	32.75	14.75	50.65	61	71.50
	Total				.815	3.15	35.91	14.75	53.81	65.12	76.60

For customer support on your Facilities Maintenance & Repair Costs with RSMeans data, call 800.448.8182.

C30 INTERIOR FINISHES — C3013 — Wall Finishes

C3013 106 — Vinyl Wall Covering

System Description	Freq. (Years)	Crew	Unit	Labor Hours	2019 Bare Costs				Total In-House	Total w/O&P
					Material	Labor	Equipment	Total		
0010 Repair med. wt. vinyl wall covering - (2% of walls)	1	1 PORD	C.S.F.							
Remove damaged vinyl wall covering				1.156		50		50	65	80
Install vinyl wall covering				2.391	99	104		203	243	290
Total				3.547	99	154		253	**308**	**370**
0020 Replace medium weight vinyl wall covering	15	1 PORD	C.S.F.							
Set up and secure scaffold				.750		38.75		38.75	50.50	62.50
Remove vinyl wall covering				1.156		50		50	65	80
Install vinyl wall covering				2.391	99	104		203	243	290
Remove scaffold				.750		38.75		38.75	50.50	62.50
Total				5.047	99	231.50		330.50	**409**	**495**

C3013 202 — Wallpaper

System Description	Freq. (Years)	Crew	Unit	Labor Hours	2019 Bare Costs				Total In-House	Total w/O&P
					Material	Labor	Equipment	Total		
0010 Repair wallpaper - (2% of walls)	8	1 PORD	S.Y.							
Remove damaged wallpaper				.104		4.50		4.50	5.85	7.20
Repair / replace wallpaper				.175	10.98	7.65		18.63	22	26
Total				.279	10.98	12.15		23.13	**27.85**	**33.20**
0020 Replace wallpaper	20	1 PORD	S.Y.							
Set up and secure scaffold				.130		6.70		6.70	8.75	10.85
Remove damaged wallpaper				.104		4.50		4.50	5.85	7.20
Replace wallpaper				.175	10.98	7.65		18.63	22	26
Remove scaffold				.130		6.70		6.70	8.75	10.85
Total				.539	10.98	25.55		36.53	**45.35**	**54.90**

For customer support on your Facilities Maintenance & Repair Costs with RSMeans data, call 800.448.8182.

137

C3013 206 Fabric Interior Wall Finish

System Description	Freq. (Years)	Crew	Unit	Labor Hours	2019 Bare Costs Material	Labor	Equipment	Total	Total In-House	Total w/O&P
0010 **Repair fabric wall finish**	9	1 PORD	S.Y.							
Set up, secure and take down ladder				.130		5.65		5.65	7.30	9.05
Remove fabric wall covering				.104		4.50		4.50	5.85	7.20
Repair / replace fabric				.234	12.51	10.17		22.68	27	32
Total				.468	12.51	20.32		32.83	40.15	48.25
0020 **Replace fabric wall finish**	50	1 PORD	S.Y.							
Set up and secure scaffold				1.350		69.75		69.75	91	113
Remove fabric wall covering				.104		4.50		4.50	5.85	7.20
Replace fabric				.234	12.51	10.17		22.68	27	32
Remove scaffold				1.350		69.75		69.75	91	113
Total				3.038	12.51	154.17		166.68	214.85	265.20

C3013 208 Cork Tile

System Description	Freq. (Years)	Crew	Unit	Labor Hours	2019 Bare Costs Material	Labor	Equipment	Total	Total In-House	Total w/O&P
0010 **Repair 12" x 12" x 3/16" cork tile wall - (2% of walls)**	8	1 CARP	C.S.F.							
Remove old tiles				.620		25		25	33	41
Replace with new tile				4.426	425	193		618	715	840
Total				5.046	425	218		643	748	881
0020 **Replace 12" x 12" x 3/16" cork tile wall**	16	2 CARP	C.S.F.							
Set up and secure scaffold				.750		38.75		38.75	50.50	62.50
Remove old tiles				.620		25		25	33	41
Replace with new tile				4.426	425	193		618	715	840
Remove scaffold				.750		38.75		38.75	50.50	62.50
Total				6.546	425	295.50		720.50	849	1,006

For customer support on your Facilities Maintenance & Repair Costs with RSMeans data, call 800.448.8182.

C3013 210 Acoustical Tile

	System Description	Freq. (Years)	Crew	Unit	Labor Hours	2019 Bare Costs				Total In-House	Total w/O&P
						Material	Labor	Equipment	Total		
0010	**Repair acoustical tile - (2% of walls)**	25	1 CARP	C.S.F.							
	Remove damaged tile				1.067		44		44	57	71
	Install new tile				2.737	390	141		531	615	715
	Total				**3.804**	**390**	**185**		**575**	**672**	**786**
0030	**Refinish acoustical tile**	10	1 PORD	C.S.F.							
	Wipe surface				.894	7	39		46	58	71
	Prepare surface				.894	7	39		46	58	71
	Refinish surface				1.529	23	66		89	110	135
	Total				**3.317**	**37**	**144**		**181**	**226**	**277**
0040	**Replace acoustical tile**	60	2 CARP	C.S.F.							
	Set up and secure scaffold				.750		38.75		38.75	50.50	62.50
	Remove old tiles				1.067		44		44	57	71
	Install new tile				2.737	390	141		531	615	715
	Remove scaffold				.750		38.75		38.75	50.50	62.50
	Total				**5.304**	**390**	**262.50**		**652.50**	**773**	**911**

C3013 212 Stucco

	System Description	Freq. (Years)	Crew	Unit	Labor Hours	2019 Bare Costs				Total In-House	Total w/O&P
						Material	Labor	Equipment	Total		
0010	**Repair stucco wall - (2% of walls)**	20	1 BRIC	S.Y.							
	Prepare surface				.023		.99		.99	1.26	1.53
	Repair stucco				.532	3.78	24.03	1.80	29.61	37	45
	Place and remove mask and drops				.067		2.88		2.88	3.78	4.68
	Total				**.621**	**3.78**	**27.90**	**1.80**	**33.48**	**42.04**	**51.21**
0030	**Refinish stucco wall**	4	1 PORD	S.Y.							
	Wash surface				.023		.99		.99	1.26	1.53
	Prepare surface				.033		1.44		1.44	1.89	2.34
	Refinish surface, brushwork, 2 coats				.138	2.07	5.94		8.01	9.90	12.15
	Place and remove mask and drops				.033		1.44		1.44	1.89	2.34
	Total				**.227**	**2.07**	**9.81**		**11.88**	**14.94**	**18.36**

For customer support on your Facilities Maintenance & Repair Costs with RSMeans data, call 800.448.8182.

139

C3013 212 Stucco

	System Description	Freq. (Years)	Crew	Unit	Labor Hours	2019 Bare Costs Material	Labor	Equipment	Total	Total In-House	Total w/O&P
0040	**Replace stucco wall**	75	2 BRIC	S.Y.							
	Set up and secure scaffold				.130		6.70		6.70	8.75	10.85
	Remove stucco				.279		14.40		14.40	18.80	23.50
	Replace stucco				1.200	6.85	55	3.38	65.23	82	100
	Remove scaffold				.130		6.70		6.70	8.75	10.85
	Total				1.739	6.85	82.80	3.38	93.03	118.30	145.20
4120	**Spray refinish stucco wall**	5	1 PORD	S.Y.							
	Wash surface				.023		.99		.99	1.26	1.53
	Prepare surface				.033		1.44		1.44	1.89	2.34
	Refinish surface, spray, 2 coats				.049	2.25	2.16		4.41	5.20	6.25
	Place and remove mask and drops				.033		1.44		1.44	1.89	2.34
	Total				.138	2.25	6.03		8.28	10.24	12.46

C3013 213 Plaster

	System Description	Freq. (Years)	Crew	Unit	Labor Hours	2019 Bare Costs Material	Labor	Equipment	Total	Total In-House	Total w/O&P
0010	**Repair plaster wall - (2% of walls)**	13	1 PLAS	S.Y.							
	Remove damage				.468		19.22		19.22	25	31
	Replace two coat plaster finish, incl. lath				.752	6.95	34.50	2.12	43.57	54.50	66
	Total				1.220	6.95	53.72	2.12	62.79	79.50	97
0030	**Refinish plaster wall**	4	1 PORD	S.Y.							
	Prepare surface				.033		1.44		1.44	1.89	2.34
	Paint / seal surface, brushwork, 1 coat				.063	.54	2.70		3.24	4.14	5
	Place and remove mask and drops				.033		1.44		1.44	1.89	2.34
	Total				.129	.54	5.58		6.12	7.92	9.68
0040	**Replace plaster wall**	75	2 PLAS	S.Y.							
	Set up, secure and take down ladder				.130		6.70		6.70	8.75	10.85
	Remove material				.468		19.22		19.22	25	31
	Replace two coat plaster wall, incl. lath				.752	6.95	34.50	2.12	43.57	54.50	66
	Total				1.350	6.95	60.42	2.12	69.49	88.25	107.85

For customer support on your Facilities Maintenance & Repair Costs with RSMeans data, call 800.448.8182.

C3013 214 | **Drywall**

	System Description	Freq. (Years)	Crew	Unit	Labor Hours	2019 Bare Costs				Total In-House	Total w/O&P
						Material	Labor	Equipment	Total		
0010	**Repair 5/8" drywall - (2% of walls)**	20	1 CARP	S.F.							
	Remove damage				.008		.33		.33	.43	.53
	Replace 5/8" drywall, taped and finished				.022	.35	1.11		1.46	1.84	2.24
	Total				.030	.35	1.44		1.79	2.27	**2.77**
0030	**Refinish drywall**	4	1 PORD	S.F.							
	Place and remove mask and drops				.004		.16		.16	.21	.26
	Prepare surface				.004		.16		.16	.21	.26
	Paint surface, brushwork, 1 coat				.007	.06	.30		.36	.46	.56
	Total				.014	.06	.62		.68	.88	**1.08**
0040	**Replace 5/8" drywall**	75	2 CARP	S.F.							
	Set up, secure and take down ladder				.014		.74		.74	.97	1.20
	Remove drywall				.008		.33		.33	.43	.53
	Replace 5/8" drywall, taped and finished				.022	.35	1.11		1.46	1.84	2.24
	Total				.044	.35	2.18		2.53	3.24	**3.97**
0050	**Office painting, 10' x 12', 10' high walls**	5	1 PORD	Ea.							
	Spread drop cloths				.004		.16		.16	.21	.26
	Prepare drywall partitions				1.628		70.40		70.40	92.50	114
	Clean drywall partitions				.220	.75	9.02		9.77	12.60	15.45
	Paint drywall partitions, roller + brush, 1 coat				3.062	26.40	132		158.40	202	244
	Remove drop cloths				.004		.16		.16	.21	.26
	Total				4.918	27.15	211.74		238.89	307.52	**373.97**
0060	**Office painting, 10' x 15', 10' high walls**	5	1 PORD	Ea.							
	Spread drop cloths				.005		.20		.20	.26	.33
	Prepare drywall partitions				1.850		80		80	105	130
	Clean drywall partitions				.250	.85	10.25		11.10	14.30	17.55
	Paint drywall partitions, roller + brush, 1 coat				3.480	30	150		180	230	278
	Remove drop cloths				.005		.20		.20	.26	.33
	Total				5.589	30.85	240.65		271.50	349.82	**426.21**

For customer support on your Facilities Maintenance & Repair Costs with RSMeans data, call 800.448.8182.

141

C3013 214 | Drywall

	System Description	Freq. (Years)	Crew	Unit	Labor Hours	2019 Bare Costs Material	Labor	Equipment	Total	Total In-House	Total w/O&P
2030	**Refinish drywall, 12' to 24' high**	5	2 PORD	S.F.							
	Set up and secure scaffold				.015		.78		.78	1.01	1.25
	Place and remove mask and drops				.004		.16		.16	.21	.26
	Prepare surface				.004		.16		.16	.21	.26
	Paint surface, roller + brushwork, 1 coat				.007	.06	.30		.36	.46	.56
	Remove scaffold				.015		.78		.78	1.01	1.25
	Total				.044	.06	2.18		2.24	2.90	3.58
3030	**Refinish drywall, over 24' high**	5	2 PORD	S.F.							
	Set up and secure scaffold				.023		1.16		1.16	1.51	1.88
	Place and remove mask and drops				.004		.16		.16	.21	.26
	Prepare surface				.004		.16		.16	.21	.26
	Paint surface, roller + brushwork, 1 coat				.007	.06	.30		.36	.46	.56
	Remove scaffold				.023		1.16		1.16	1.51	1.88
	Total				.059	.06	2.94		3	3.90	4.84

C3013 215 | Fiberglass Panels, Rigid

	System Description	Freq. (Years)	Crew	Unit	Labor Hours	2019 Bare Costs Material	Labor	Equipment	Total	Total In-House	Total w/O&P
0010	**Repair glass cloth fiberglass panels - (2% of walls)**	9	1 CARP	C.S.F.							
	Remove damaged fiberglass panels				2.600		107		107	139	172
	Install new fiberglass panel				6.710	920	347		1,267	1,475	1,700
	Total				9.310	920	454		1,374	1,614	1,872
0040	**Replace glass cloth fiberglass panels**	35	2 CARP	C.S.F.							
	Set up and secure scaffold				.750		38.75		38.75	50.50	62.50
	Remove old panels				2.600		107		107	139	172
	Remove old furring				.516		21		21	28	34
	Install new furring				2.101	48	109		157	194	235
	Install new glass cloth fiberglass panels				6.710	920	347		1,267	1,475	1,700
	Remove scaffold				.750		38.75		38.75	50.50	62.50
	Total				13.427	968	661.50		1,629.50	1,937	2,266

For customer support on your Facilities Maintenance & Repair Costs with RSMeans data, call 800.448.8182.

C3013 220 | Tile

	System Description	Freq. (Years)	Crew	Unit	Labor Hours	2019 Bare Costs				Total In-House	Total w/O&P
						Material	Labor	Equipment	Total		
0010	**Repair 4″ x 4″ thin set ceramic tile - (2% of walls)**	10	1 TILF	C.S.F.							
	Remove damaged tiles				2.600		107		107	139	172
	Replace tile / grout				10.947	242	469		711	860	1,050
	Total				13.547	242	576		818	**999**	**1,222**
0020	**Replace 4″ x 4″ thin set ceramic tile**	75	1 TILF	C.S.F.							
	Set up and secure scaffold				.750		38.75		38.75	50.50	62.50
	Remove tiles				2.600		107		107	139	172
	Replace tile / grout				10.947	242	469		711	860	1,050
	Remove scaffold				.750		38.75		38.75	50.50	62.50
	Total				15.047	242	653.50		895.50	**1,100**	**1,347**

C3013 230 | Plywood Paneling

	System Description	Freq. (Years)	Crew	Unit	Labor Hours	2019 Bare Costs				Total In-House	Total w/O&P
						Material	Labor	Equipment	Total		
0010	**Repair plywood paneling - (2% of walls)**	10	1 CARP	C.S.F.							
	Remove damaged paneling				1.156		47		47	62	77
	Replace paneling				4.952	121	256		377	465	565
	Total				6.108	121	303		424	**527**	**642**
0020	**Refinish plywood paneling**	10	1 PORD	C.S.F.							
	Place mask and drops				.500		22		22	28	35
	Sand & paint, brushwork, 1 coat				.990	7	43		50	63	77
	Remove mask and drops				.500		22		22	28	35
	Total				1.990	7	87		94	**119**	**147**
0040	**Replace plywood paneling**	30	2 CARP	C.S.F.							
	Set up and secure scaffold				.750		38.75		38.75	50.50	62.50
	Remove damaged paneling				1.156		47		47	62	77
	Replace paneling				4.952	121	256		377	465	565
	Remove scaffold				.750		38.75		38.75	50.50	62.50
	Total				7.608	121	380.50		501.50	**628**	**767**

For customer support on your Facilities Maintenance & Repair Costs with RSMeans data, call 800.448.8182.

143

C3013 234 Wainscot

	System Description	Freq. (Years)	Crew	Unit	Labor Hours	2019 Bare Costs				Total In-House	Total w/O&P
						Material	Labor	Equipment	Total		
0010	**Repair wainscot - (2% of walls)**	10	1 CARP	C.S.F.							
	Remove damaged wainscot				1.733		71		71	93	115
	Replace wainscot				16.000	1,630	825		2,455	2,875	3,375
	Total				17.733	1,630	896		2,526	2,968	**3,490**
0020	**Refinish wainscot**	4	1 PORD	C.S.F.							
	Place mask and drops				.500		22		22	28	35
	Sand & paint, brushwork, 1 coat				.990	7	43		50	63	77
	Remove mask and drops				.500		22		22	28	35
	Total				1.990	7	87		94	119	**147**
0040	**Replace wainscot**	40	2 CARP	C.S.F.							
	Remove wainscot				1.733		71		71	93	115
	Replace wainscot				16.000	1,630	825		2,455	2,875	3,375
	Total				17.733	1,630	896		2,526	2,968	**3,490**

C3013 240 Stainless Steel Interior Finish

	System Description	Freq. (Years)	Crew	Unit	Labor Hours	2019 Bare Costs				Total In-House	Total w/O&P
						Material	Labor	Equipment	Total		
0020	**Replace stainless steel wall**	75	2 SSWK	C.S.F.							
	Set up, secure and take down ladder				1.500		77.50		77.50	101	125
	Remove existing stainless steel sheets				2.342		96		96	125	155
	Replace stainless steel sheets				13.419	470	605		1,075	1,400	1,650
	Total				17.261	470	778.50		1,248.50	1,626	**1,930**

C3013 410 Fireplace Mantles, Wood

	System Description	Freq. (Years)	Crew	Unit	Labor Hours	2019 Bare Costs				Total In-House	Total w/O&P
						Material	Labor	Equipment	Total		
0030	**Refinish wood surface - fireplaces**	7	1 PORD	S.F.							
	Prepare surface				.003		.15		.15	.19	.23
	Varnish, 3 coats, brushwork, sanding included				.025	.28	1.06		1.34	1.68	2.05
	Total				.028	.28	1.21		1.49	1.87	**2.28**

For customer support on your Facilities Maintenance & Repair Costs with RSMeans data, call 800.448.8182.

C3023 112 Concrete, Finished

	System Description	Freq. (Years)	Crew	Unit	Labor Hours	2019 Bare Costs				Total In-House	Total w/O&P
						Material	Labor	Equipment	Total		
0020	Refinish concrete floor	25	2 CEFI	C.S.F.	6.933	69	306	43	418	515	620
	Add topping to existing floor 1″										
	Total				6.933	69	306	43	418	**515**	**620**

C3023 405 Epoxy Flooring

	System Description	Freq. (Years)	Crew	Unit	Labor Hours	2019 Bare Costs				Total In-House	Total w/O&P
						Material	Labor	Equipment	Total		
0020	Replace epoxy flooring	15	2 CEFI	C.S.F.							
	Strip existing flooring				3.200		165		165	215	267
	Repair and seal floor				12.735	459	545	14	1,018	1,225	1,475
	Total				15.935	459	710	14	1,183	**1,440**	**1,742**

C3023 410 Vinyl Tile

	System Description	Freq. (Years)	Crew	Unit	Labor Hours	2019 Bare Costs				Total In-House	Total w/O&P
						Material	Labor	Equipment	Total		
0020	Replace vinyl tile flooring	18	1 TILF	S.Y.							
	Remove damaged floor tile				.234		9.63		9.63	12.50	15.50
	Prepare surface				.645		27.63		27.63	35	43.50
	Install new tiles				.187	10.98	9		19.98	23.50	28
	Total				1.067	10.98	46.26		57.24	**71**	**87**

C3023 412 Vinyl Sheet

	System Description	Freq. (Years)	Crew	Unit	Labor Hours	2019 Bare Costs				Total In-House	Total w/O&P
						Material	Labor	Equipment	Total		
0020	Replace vinyl sheet flooring	18	1 TILF	S.Y.							
	Remove damaged floor tile				.234		9.63		9.63	12.50	15.50
	Prepare surface				.645		27.63		27.63	35	43.50
	Install new vinyl sheet				.407	37.80	19.44		57.24	66	78
	Total				1.286	37.80	56.70		94.50	**113.50**	**137**

For customer support on your Facilities Maintenance & Repair Costs with RSMeans data, call 800.448.8182.

145

C3023 414 Rubber Tile

System Description	Freq. (Years)	Crew	Unit	Labor Hours	2019 Bare Costs				Total In-House	Total w/O&P
					Material	Labor	Equipment	Total		
0020 Replace rubber tile floor	18	1 TILF	S.Y.							
Remove damaged floor tile				.234		9.63		9.63	12.50	15.50
Prepare surface				.645		27.63		27.63	35	43.50
Install new tiles				1.040	107.55	49.95		157.50	181	213
Total				1.920	107.55	87.21		194.76	228.50	272

C3023 418 Rubber / Vinyl Trim

System Description	Freq. (Years)	Crew	Unit	Labor Hours	2019 Bare Costs				Total In-House	Total w/O&P
					Material	Labor	Equipment	Total		
0010 Replace rubber cove base	9	1 TILF	L.F.							
Remove damaged base				.017		.71		.71	.93	1.15
Clean up debris				.001		.05		.05	.06	.08
Install new base				.025	1.30	1.22		2.52	2.97	3.54
Total				.044	1.30	1.98		3.28	3.96	4.77

C3023 420 Ceramic Tile

System Description	Freq. (Years)	Crew	Unit	Labor Hours	2019 Bare Costs				Total In-House	Total w/O&P
					Material	Labor	Equipment	Total		
0010 Ceramic tile floor repairs - (2% of floors)	15	1 TILF	C.S.F.							
Regrout ceramic tile floors				16.640	5	710		715	905	1,125
Total				16.640	5	710		715	905	1,125
0020 Replace 2" x 2" thin set ceramic tile floor	50	1 TILF	C.S.F.							
Remove damaged floor tile				2.600		107		107	139	172
Prepare surface				7.172		307		307	390	485
Install new tiles				10.947	635	469		1,104	1,300	1,525
Total				20.719	635	883		1,518	1,829	2,182

For customer support on your Facilities Maintenance & Repair Costs with RSMeans data, call 800.448.8182.

C3023 428 Ceramic Trim

	System Description	Freq. (Years)	Crew	Unit	Labor Hours	2019 Bare Costs				Total In-House	Total w/O&P
						Material	Labor	Equipment	Total		
0020	**Replace ceramic trim**	50	2 TILF	L.F.							
	Remove ceramic tile trim				.015		.61		.61	.80	.99
	Clean up debris				.010		.51		.51	.67	.83
	Install new ceramic tile trim				.229	4.35	9.80		14.15	17.15	21
	Total				.253	4.35	10.92		15.27	**18.62**	**22.82**

C3023 430 Terrazzo

	System Description	Freq. (Years)	Crew	Unit	Labor Hours	2019 Bare Costs				Total In-House	Total w/O&P
						Material	Labor	Equipment	Total		
0100	**Terrazzo floor repairs - (2% of floors)**	15	1 MSTZ	S.F.							
	Chip existing terrazzo				.089		3.66		3.66	4.77	5.90
	Place terrazzo				.166	3.47	7.35	3.30	14.12	16.75	19.50
	Remove debris				.004		.18		.18	.23	.29
	Total				.260	3.47	11.19	3.30	17.96	**21.75**	**25.69**
0200	**Replace terrazzo floor**	75	2 MSTZ	C.S.F.							
	Break-up existing terrazzo				5.943		244		244	320	395
	Remove debris				.433		17.75		17.75	23	29
	Load into truck				.650		28	11.50	39.50	49	58
	Place terrazzo				16.640	347	735	330	1,412	1,675	1,950
	Total				23.666	347	1,024.75	341.50	1,713.25	**2,067**	**2,432**

C3023 438 Terrazzo Trim

	System Description	Freq. (Years)	Crew	Unit	Labor Hours	2019 Bare Costs				Total In-House	Total w/O&P
						Material	Labor	Equipment	Total		
0020	**Replace precast terrazzo trim**	75	2 MSTZ	L.F.							
	Remove old terrazzo trim				.023		.95		.95	1.24	1.53
	Clean area				.010		.51		.51	.67	.83
	Install new precast terrazzo trim				.416	24.50	19.90		44.40	52	62
	Total				.449	24.50	21.36		45.86	**53.91**	**64.36**

For customer support on your Facilities Maintenance & Repair Costs with RSMeans data, call 800.448.8182.

147

C3023 440 Quarry Tile

System Description	Freq. (Years)	Crew	Unit	Labor Hours	Material	Labor	Equipment	Total	Total In-House	Total w/O&P
						2019 Bare Costs				
0010 Quarry tile floor repairs - (2% of floors)	15	1 TILF	S.F.							
Regrout quarry tile floor				.166	.05	7.10		7.15	9.05	11.25
Total				**.166**	**.05**	**7.10**		**7.15**	**9.05**	**11.25**
0020 Replace quarry tile floor	50	2 TILF	S.F.							
Remove damaged floor tile				.026		1.07		1.07	1.39	1.72
Prepare surface				.072		3.07		3.07	3.88	4.83
Install new tiles				.109	6.35	4.69		11.04	12.90	15.30
Total				**.207**	**6.35**	**8.83**		**15.18**	**18.17**	**21.85**

C3023 450 Brick

System Description	Freq. (Years)	Crew	Unit	Labor Hours	Material	Labor	Equipment	Total	Total In-House	Total w/O&P
						2019 Bare Costs				
0020 Replace thickset brick floor	60	2 BRIC	S.F.							
Remove brick				.035		1.42		1.42	1.85	2.30
Prepare 2" thick mortar bed				.260	1.15	11.90		13.05	16.95	21
Install new brick				.219	9.20	9.35		18.55	22	26.50
Total				**.514**	**10.35**	**22.67**		**33.02**	**40.80**	**49.80**

C3023 460 Marble

System Description	Freq. (Years)	Crew	Unit	Labor Hours	Material	Labor	Equipment	Total	Total In-House	Total w/O&P
						2019 Bare Costs				
0020 Replace thinset marble floor	50	2 TILF	S.F.							
Remove marble				.035		1.42		1.42	1.85	2.30
Prepare floor				.260	1.15	11.90		13.05	16.95	21
Replace marble flooring				.347	11.45	14.85		26.30	31.50	38
Total				**.641**	**12.60**	**28.17**		**40.77**	**50.30**	**61.30**

C3023 470 Plywood

System Description	Freq. (Years)	Crew	Unit	Labor Hours	2019 Bare Costs				Total In-House	Total w/O&P
					Material	Labor	Equipment	Total		
0020 Replace plywood floor	40	2 CARP	C.S.F.							
Remove existing flooring				1.733		90		90	117	144
Install new plywood flooring 3/4"				1.600	147	83		230	270	315
Total				3.333	147	173		320	387	459

C3023 472 Wood Parquet

System Description	Freq. (Years)	Crew	Unit	Labor Hours	2019 Bare Costs				Total In-House	Total w/O&P
					Material	Labor	Equipment	Total		
0020 Sand and refinish parquet floor	10	1 CARP	S.F.							
Sand floor				.035	.22	1.45		1.67	2.13	2.61
Refinish existing floor, brushwork, 2 coats				.026	.22	1.07		1.29	1.63	2
Total				.061	.44	2.52		2.96	3.76	4.61
0030 Replace 5/16" oak parquet floor	40	2 CARP	S.F.							
Remove flooring				.023		1.19		1.19	1.56	1.93
Repair underlayment				.014	1.27	.72		1.99	2.33	2.75
Install new parquet flooring				.065	5.45	3.36		8.81	10.35	12.20
Total				.102	6.72	5.27		11.99	14.24	16.88

C3023 474 Maple Strip

System Description	Freq. (Years)	Crew	Unit	Labor Hours	2019 Bare Costs				Total In-House	Total w/O&P
					Material	Labor	Equipment	Total		
0020 Sand and refinish maple strip floor	10	1 CARP	S.F.							
Sand floor				.026	.22	1.07		1.29	1.63	2
Refinish maple strip floor, brushwork, 2 coats				.035	.22	1.45		1.67	2.13	2.61
Total				.061	.44	2.52		2.96	3.76	4.61
0030 Replace maple floor	40	2 CARP	S.F.							
Remove flooring				.023		1.19		1.19	1.56	1.93
Repair underlayment				.014	1.27	.72		1.99	2.33	2.75
Install new maple flooring				.061	5.05	3.16		8.21	9.70	11.40
Total				.098	6.32	5.07		11.39	13.59	16.08

For customer support on your Facilities Maintenance & Repair Costs with RSMeans data, call 800.448.8182.

149

C3023 478 Wood Trim

System Description	Freq. (Years)	Crew	Unit	Labor Hours	2019 Bare Costs				Total In-House	Total w/O&P
					Material	Labor	Equipment	Total		
0010 Repair 1" x 3" wood trim	13	1 CARP	L.F.							
Remove wood				.021		.85		.85	1.11	1.38
Install new trim				.173	5.40	8.96		14.36	17.65	21
Total				.194	5.40	9.81		15.21	18.76	**22.38**
0030 Refinish 1" x 3" wood trim	7	1 PORD	L.F.							
Prepare surface				.003		.15		.15	.19	.23
Varnish, 3 coats, brushwork, sanding included				.025	.28	1.06		1.34	1.68	2.05
Total				.028	.28	1.21		1.49	1.87	**2.28**
0040 Replace 1" x 3" wood trim	75	2 CARP	L.F.							
Remove wood				.021		.85		.85	1.11	1.38
Install new trim				.173	5.40	8.96		14.36	17.65	21
Prepare surface				.003		.15		.15	.19	.23
Varnish, 3 coats, brushwork, sanding included				.025	.28	1.06		1.34	1.68	2.05
Total				.222	5.68	11.02		16.70	20.63	**24.66**

C3023 510 Carpet

System Description	Freq. (Years)	Crew	Unit	Labor Hours	2019 Bare Costs				Total In-House	Total w/O&P
					Material	Labor	Equipment	Total		
0020 Replace carpet	8	2 TILF	S.Y.							
Remove damaged carpet				.094		3.87		3.87	5.05	6.20
Install new carpet				.138	39.50	6.60		46.10	52	60
Total				.232	39.50	10.47		49.97	57.05	**66.20**

For customer support on your Facilities Maintenance & Repair Costs with RSMeans data, call 800.448.8182.

C3033 105 — Plaster

	System Description	Freq. (Years)	Crew	Unit	Labor Hours	2019 Bare Costs				Total In-House	Total w/O&P
						Material	Labor	Equipment	Total		
0010	**Repair plaster ceiling - (2% of ceilings)**	12	1 PLAS	S.Y.							
	Set up, secure and take down ladder				.130		6.70		6.70	8.75	10.85
	Remove damaged ceiling				.267		10.98		10.98	14.30	17.75
	Replace plaster, 2 coats on lath				.752	6.35	34.50	2.12	42.97	53.50	65.50
	Total				1.149	6.35	52.18	2.12	60.65	76.55	94.10
0030	**Refinish plaster ceiling**	10	1 PORD	S.Y.							
	Set up, secure and take down ladder				.130		6.70		6.70	8.75	10.85
	Place mask and drops				.033		1.44		1.44	1.89	2.34
	Sand & paint				.089	.63	3.87		4.50	5.65	6.90
	Total				.252	.63	12.01		12.64	16.29	20.09
0040	**Replace plaster ceiling**	75	2 PLAS	S.Y.							
	Set up and secure scaffold				.130		6.70		6.70	8.75	10.85
	Remove ceiling				.267		10.98		10.98	14.30	17.75
	Replace plaster, 2 coats on lath				.752	6.35	34.50	2.12	42.97	53.50	65.50
	Remove scaffold				.130		6.70		6.70	8.75	10.85
	Total				1.279	6.35	58.88	2.12	67.35	85.30	104.95

C3033 107 — Gypsum Wall Board

	System Description	Freq. (Years)	Crew	Unit	Labor Hours	2019 Bare Costs				Total In-House	Total w/O&P
						Material	Labor	Equipment	Total		
0010	**Repair gypsum board ceiling - (2% of ceilings)**	20	1 CARP	C.S.F.							
	Set up, secure and take down ladder				1.300		67		67	87.50	109
	Remove damaged gypsum board				2.737		112		112	146	181
	Replace 5/8" gypsum board, taped and finished				3.382	42	175		217	274	335
	Total				7.419	42	354		396	507.50	625
0020	**Refinish gypsum board ceiling, up to 12' high**	20	1 PORD	C.S.F.							
	Set up, secure and take down ladder				1.300		67		67	87.50	109
	Wash surface				.250		11		11	14	17
	Sand & paint				.990	7	43		50	63	77
	Place and remove mask and drops				.370		16		16	21	26
	Total				2.910	7	137		144	185.50	229

For customer support on your Facilities Maintenance & Repair Costs with RSMeans data, call 800.448.8182.

151

C-30 INTERIOR FINISHES — C3033 Ceiling Finishes

C3033 107 — Gypsum Wall Board

	System Description	Freq. (Years)	Crew	Unit	Labor Hours	2019 Bare Costs				Total In-House	Total w/O&P
						Material	Labor	Equipment	Total		
0040	**Replace gypsum board ceiling, up to 12' high**	40	2 CARP	C.S.F.							
	Set up and secure scaffold				.750		38.75		38.75	50.50	62.50
	Remove damaged gypsum board				2.737		112		112	146	181
	Replace 5/8" gypsum board ceiling, taped and finished				3.382	42	175		217	274	335
	Remove scaffold				.750		38.75		38.75	50.50	62.50
	Total				7.619	42	364.50		406.50	521	641
2020	**Refinish gypsum board ceiling, 12' to 24' high**	5	2 PORD	C.S.F.							
	Set up and secure scaffold				1.500		77.50		77.50	101	125
	Place and remove mask and drops				.370		16		16	21	26
	Prepare surface				.370		16		16	21	26
	Paint surface, roller + brushwork, 1 coat				.696	6	30		36	46	55.50
	Remove scaffold				1.500		77.50		77.50	101	125
	Total				4.436	6	217		223	290	357.50
3020	**Refinish gypsum board ceiling, over 24' high**	5	2 PORD	C.S.F.							
	Set up and secure scaffold				2.250		116.25		116.25	151	188
	Place and remove mask and drops				.370		16		16	21	26
	Prepare surface				.370		16		16	21	26
	Paint surface, roller + brushwork, 1 coat				.696	6	30		36	46	55.50
	Remove scaffold				2.250		116.25		116.25	151	188
	Total				5.936	6	294.50		300.50	390	483.50

C3033 108 — Acoustic Tile

	System Description	Freq. (Years)	Crew	Unit	Labor Hours	2019 Bare Costs				Total In-House	Total w/O&P
						Material	Labor	Equipment	Total		
0010	**Acoustic tile repairs - (2% of ceilings)**	9	1 CARP	C.S.F.							
	Set up, secure and take down ladder				1.300		67		67	87.50	109
	Remove damaged tile				1.067		44		44	57	71
	Install new tile				2.737	390	141		531	615	715
	Total				5.104	390	252		642	759.50	895

For customer support on your Facilities Maintenance & Repair Costs with RSMeans data, call 800.448.8182.

	System Description	Freq. (Years)	Crew	Unit	Labor Hours	2019 Bare Costs				Total In-House	Total w/O&P
						Material	Labor	Equipment	Total		
0020	**Replace acoustic tile ceiling, non fire-rated**	20	1 CARP	C.S.F.							
	Set up, secure and take down scaffold				.750		38.75		38.75	50.50	62.50
	Remove old ceiling tiles				.500		21		21	27	33
	Remove old ceiling grid				.625		26		26	33	41
	Install new ceiling grid				1.000	80	52		132	155	183
	Install new ceiling tiles				1.280	83	66		149	177	211
	Sweep and clean debris				.520		30		30	38.50	47.50
	Total				4.675	163	233.75		396.75	481	578
0030	**Refinish acoustic tile ceiling/grid (unoccupied area)**	5	2 PORD	C.S.F.							
	Protect exposed surfaces and mask light fixtures				.080		3.45		3.45	4.46	5.55
	Refinish acoustic tiles, roller + brush, 1 coat				.267	.01	11.50		11.51	14.85	18.40
	Total				.347	.01	14.95		14.96	19.31	23.95
0040	**Refinish acoustic tile ceiling/grid (occupied area)**	5	2 PORD	C.S.F.							
	Protect exposed surfaces and mask light fixtures				.320		13.80		13.80	17.80	22
	Refinish acoustic tiles, roller + brush, 1 coat				.320	.01	13.80		13.81	17.85	22
	Total				.640	.01	27.60		27.61	35.65	44
2030	**Refinish acoustic tile ceiling, 12' to 24' high**	5	2 PORD	C.S.F.							
	Set up and secure scaffold				1.500		77.50		77.50	101	125
	Place and remove mask and drops				.100		2		2	3	3
	Paint surface, roller + brushwork, 1 coat				.267	.01	11.50		11.51	14.85	18.40
	Remove scaffold				1.500		77.50		77.50	101	125
	Total				3.367	.01	168.50		168.51	219.85	271.40
3030	**Refinish acoustic tile ceiling, over 24' high**	5	2 PORD	C.S.F.							
	Set up and secure scaffold				2.250		116.25		116.25	151	188
	Place and remove mask and drops				.100		2		2	3	3
	Paint surface, roller + brushwork, 1 coat				.267	.01	11.50		11.51	14.85	18.40
	Remove scaffold				2.250		116.25		116.25	151	188
	Total				4.867	.01	246		246.01	319.85	397.40

For customer support on your Facilities Maintenance & Repair Costs with RSMeans data, call 800.448.8182.

153

C3033 109 Acoustic Tile, Fire-Rated

	System Description	Freq. (Years)	Crew	Unit	Labor Hours	2019 Bare Costs Material	Labor	Equipment	Total	Total In-House	Total w/O&P
0010	**Replace acoustic tile ceiling, fire-rated**	20	1 CARP	C.S.F.							
	Set up, secure and take down scaffold				.750		38.75		38.75	50.50	62.50
	Remove old ceiling tiles				.500		21		21	27	33
	Remove old grid system				.625		26		26	33	41
	Install new grid system				1.200	96	62.40		158.40	186	220
	Install new ceiling tiles				1.185	131	61		192	224	263
	Sweep & clean debris				.520		30		30	38.50	47.50
	Total				4.780	227	239.15		466.15	559	667

C3033 120 Wood

	System Description	Freq. (Years)	Crew	Unit	Labor Hours	2019 Bare Costs Material	Labor	Equipment	Total	Total In-House	Total w/O&P
0010	**Repair wood ceiling - (2% of ceilings)**	10	1 CARP	C.S.F.							
	Set up, secure and take down ladder				1.300		67		67	87.50	109
	Remove damaged ceiling				1.455		60		60	78	96
	Replace wood				5.200	173	269		442	540	650
	Total				7.955	173	396		569	705.50	855
0030	**Refinish wood ceiling**	6	1 PORD	C.S.F.							
	Set up and secure scaffold				.375		19.38		19.38	25	31.50
	Wash surface				.286		12		12	16	20
	Sand & paint, brushwork, 1 coat				.990	7	43		50	63	77
	Place and remove mask and drops				.370		16		16	21	26
	Remove scaffold				.375		19.38		19.38	25	31.50
	Total				2.396	7	109.76		116.76	150	186
0040	**Replace wood ceiling**	50	2 CARP	C.S.F.							
	Set up and secure scaffold				.750		38.75		38.75	50.50	62.50
	Remove damaged ceiling				1.455		60		60	78	96
	Replace wood				5.200	173	269		442	540	650
	Remove scaffold				.750		38.75		38.75	50.50	62.50
	Total				8.155	173	406.50		579.50	719	871

For customer support on your Facilities Maintenance & Repair Costs with RSMeans data, call 800.448.8182.

D2013 110 | Tankless Water Closet

	System Description	Freq. (Years)	Crew	Unit	Labor Hours	2019 Bare Costs				Total In-House	Total w/O&P
						Material	Labor	Equipment	Total		
0010	**Replace flush valve diaphragm**	10	1 PLUM	Ea.							
	Turn valve off and on				.010		.66		.66	.81	1.02
	Remove / replace diaphragm				.300	1.80	18.95		20.75	25.50	32
	Check for leaks				.013		.82		.82	1.02	1.27
	Total				.323	1.80	20.43		22.23	27.33	**34.29**
0015	**Rebuild flush valve**	20	1 PLUM	Ea.							
	Turn valve off and on				.010		.66		.66	.81	1.02
	Remove flush valve				.100		6.30		6.30	7.85	9.80
	Rebuild valve				.350	64	22		86	98	115
	Reinstall valve				1.000		63		63	78.50	98
	Check operation				.024		1.52		1.52	1.88	2.35
	Total				1.484	64	93.48		157.48	187.04	**226.17**
0020	**Unplug clogged line**	5	1 PLUM	Ea.							
	Turn valve off and on				.010		.66		.66	.81	1.02
	Remove water closet				2.078		131		131	163	204
	Unplug line				.200		12.65		12.65	15.65	19.60
	Reinstall water closet				.510		32		32	40	50
	Check operation				.167		10.55		10.55	13.10	16.35
	Total				2.965		186.86		186.86	232.56	**290.97**
0030	**Replace tankless water closet**	35	2 PLUM	Ea.							
	Turn valve off and on				.010		.66		.66	.81	1.02
	Remove flush valve				.100		6.30		6.30	7.85	9.80
	Remove water closet				.460		29		29	36	45
	Install wall-hung water closet				2.759	1,025	157		1,182	1,325	1,525
	Check operation				.150		9.45		9.45	11.75	14.70
	Total				3.479	1,025	202.41		1,227.41	1,381.41	**1,595.52**
0040	**Replace tankless flush valve**	25	1 PLUM	Ea.							
	Turn valve off and on				.010		.66		.66	.81	1.02
	Remove flush valve				.100		6.30		6.30	7.85	9.80
	Install water closet flush valve				1.000	157	63		220	251	294
	Check operation				.024		1.52		1.52	1.88	2.35
	Total				1.134	157	71.48		228.48	261.54	**307.17**

For customer support on your Facilities Maintenance & Repair Costs with RSMeans data, call 800.448.8182.

155

D2013 110 Tankless Water Closet

	System Description	Freq. (Years)	Crew	Unit	Labor Hours	2019 Bare Costs				Total In-House	Total w/O&P
						Material	Labor	Equipment	Total		
0050	**Replace wax ring gasket**	5	1 PLUM	Ea.							
	Turn valve off and on				.010		.66		.66	.81	1.02
	Remove floor mounted tankless bowl				.667		42		42	52	65.50
	Install new wax gasket				.083	1.74	5.25		6.99	8.45	10.35
	Install floor mounted tankless bowl				1.000		63		63	78.50	98
	Check operation				.150		9.45		9.45	11.75	14.70
	Total				1.910	1.74	120.36		122.10	151.51	189.57
0080	**Replace stainless steel detention water closet flush valve actuator**	20	1 SKWK	Ea.							
	Turn valve off and on				.010		.66		.66	.81	1.02
	Remove flush valve actuator				.100		6.30		6.30	7.85	9.80
	Remove flush valve				.100		6.30		6.30	7.85	9.80
	Detention equipment, replace toilet actuator				1.000	274	53.50		327.50	370	430
	Check operation				.024		1.52		1.52	1.88	2.35
	Total				1.234	274	68.28		342.28	388.39	452.97
0090	**Replace stainless steel detention water closet flush valve**	20	1 SKWK	Ea.							
	Turn valve off and on				.010		.66		.66	.81	1.02
	Remove flush valve				.100		6.30		6.30	7.85	9.80
	Detention equipment, replace water closet valve				1.000	320	53.50		373.50	420	485
	Check operation				.024		1.52		1.52	1.88	2.35
	Total				1.134	320	61.98		381.98	430.54	498.17

D2013 130 Flush-Tank Water Closet

	System Description	Freq. (Years)	Crew	Unit	Labor Hours	2019 Bare Costs				Total In-House	Total w/O&P
						Material	Labor	Equipment	Total		
0010	**Unplug clogged line**	5	1 PLUM	Ea.							
	Turn valve off and on				.010		.66		.66	.81	1.02
	Remove water closet				2.078		131		131	163	204
	Unplug line				.200		12.65		12.65	15.65	19.60
	Reinstall water closet				.510		32		32	40	50
	Check operation				.167		10.55		10.55	13.10	16.35
	Total				2.965		186.86		186.86	232.56	290.97

For customer support on your Facilities Maintenance & Repair Costs with RSMeans data, call 800.448.8182.

System Description	Freq. (Years)	Crew	Unit	Labor Hours	2019 Bare Costs				Total In-House	Total w/O&P
					Material	Labor	Equipment	Total		
0020 Replace washer / diaphragm in ball cock	5	1 PLUM	Ea.							
Turn valve on and off				.010		.66		.66	.81	1.02
Remove and replace washer / diaphragm				.180	2.50	11.35		13.85	16.85	21
Check operation				.024		1.52		1.52	1.88	2.35
Total				.214	2.50	13.53		16.03	**19.54**	**24.37**
0030 Replace valve and ball cock assembly	15	1 PLUM	Ea.							
Turn valve off and on				.010		.66		.66	.81	1.02
Remove flush valve and ball cock assembly				.300		18.95		18.95	23.50	29.50
Install flush valve and ball cock assembly				.610	11.65	38.50		50.15	60.50	74.50
Check operation				.167		10.55		10.55	13.10	16.35
Total				1.087	11.65	68.66		80.31	**97.91**	**121.37**
0040 Install gasket between tank and bowl	20	1 PLUM	Ea.							
Turn valve off and on				.010		.66		.66	.81	1.02
Remove tank and gasket				.200		12.65		12.65	15.65	19.60
Install new gasket and reinstall tank				.300	2.90	18.95		21.85	26.50	33
Inspect connection				.009		.57		.57	.71	.89
Total				.520	2.90	32.83		35.73	**43.67**	**54.51**
0050 Replace two piece water closet	35	2 PLUM	Ea.							
Turn water off and on				.010		.66		.66	.81	1.02
Remove floor-mounted water closet				1.000		63		63	78.50	98
Install new bowl, tank and seat				3.019	203	172		375	435	520
Check for leaks				.013		.82		.82	1.02	1.27
Total				4.042	203	236.48		439.48	**515.33**	**620.29**
0060 Replace one piece water closet	35	2 PLUM	Ea.							
Turn water off and on				.010		.66		.66	.81	1.02
Remove floor-mounted water closet				1.000		63		63	78.50	98
Install floor-mounted water closet with seat				3.019	770	172		942	1,050	1,225
Check for leaks				.013		.82		.82	1.02	1.27
Total				4.042	770	236.48		1,006.48	**1,130.33**	**1,325.29**

For customer support on your Facilities Maintenance & Repair Costs with RSMeans data, call 800.448.8182.

157

D2013 210 Urinal

System Description	Freq. (Years)	Crew	Unit	Labor Hours	2019 Bare Costs				Total In-House	Total w/O&P
					Material	Labor	Equipment	Total		
0010 Replace flush valve diaphragm	7	1 PLUM	Ea.							
Turn valve off and on				.010		.66		.66	.81	1.02
Remove / replace diaphragm				.300	1.80	18.95		20.75	25.50	32
Check for leaks				.013		.82		.82	1.02	1.27
Total				.323	1.80	20.43		22.23	27.33	34.29
0015 Rebuild flush valve	20	1 PLUM	Ea.							
Turn valve off and on				.010		.66		.66	.81	1.02
Remove flush valve				.100		6.30		6.30	7.85	9.80
Rebuild valve				.350	64	22		86	98	115
Reinstall valve				1.000		63		63	78.50	98
Check operation				.024		1.52		1.52	1.88	2.35
Total				1.484	64	93.48		157.48	187.04	226.17
0020 Unplug line	5	1 PLUM	Ea.							
Turn valve off and on				.010		.66		.66	.81	1.02
Remove and reinstall bowl				1.600		101		101	125	157
Unplug line				.200		12.65		12.65	15.65	19.60
Check operation				.167		10.55		10.55	13.10	16.35
Total				1.977		124.86		124.86	154.56	193.97
0030 Replace wall-hung urinal	35	2 PLUM	Ea.							
Turn valve off and on				.010		.66		.66	.81	1.02
Remove urinal				2.667		168		168	209	261
Fit up urinal to wall				.196		12.40		12.40	15.35	19.20
Install flushometer pipe				.096		6.05		6.05	7.50	9.40
Install urinal				5.333	292	305		597	695	835
Check operation				.334		21		21	26	32.50
Total				8.636	292	513.11		805.11	953.66	1,158.12

For customer support on your Facilities Maintenance & Repair Costs with RSMeans data, call 800.448.8182.

D2013 310 Lavatory, Iron, Enamel

System Description	Freq. (Years)	Crew	Unit	Labor Hours	Material	Labor	Equipment	Total	Total In-House	Total w/O&P
0010										
Replace washer in spud connection	7	1 PLUM	Ea.							
Turn valve off and on				.022		1.40		1.40	1.73	2.16
Remove gasket or washer				.017		1.07		1.07	1.32	1.66
Install new gasket / washer set				.026	4.76	1.65		6.41	7.30	8.50
Install spud connection				.026		1.64		1.64	2.03	2.54
Inspect connection				.009		.57		.57	.71	.89
Total				.100	4.76	6.33		11.09	13.09	15.75
0020										
Replace washer in faucet	2	1 PLUM	Ea.							
Turn valves on and off				.022		1.40		1.40	1.73	2.16
Remove valve stem and washer				.091		5.75		5.75	7.10	8.90
Install washer				.026	.48	1.65		2.13	2.57	3.15
Install stem				.017		1.07		1.07	1.32	1.66
Test faucet				.010		.66		.66	.81	1.02
Total				.166	.48	10.53		11.01	13.53	16.89
0040										
Replace faucets	10	1 PLUM	Ea.							
Turn valves on and off				.022		1.40		1.40	1.73	2.16
Remove faucets and tubing				.381		24		24	30	37.50
Install detention sink faucets and tubing				1.040	67	65.50		132.50	155	186
Test faucets				.010		.66		.66	.81	1.02
Total				1.453	67	91.56		158.56	187.54	226.68
0050										
Clean out strainer and P trap	2	1 PLUM	Ea.							
Remove strainer and P trap				.157		9.90		9.90	12.30	15.35
Clean and reinstall strainer and trap				.320		20		20	25	31.50
Total				.477		29.90		29.90	37.30	46.85
0060										
Replace lavatory	40	2 PLUM	Ea.							
Shut off water (hot and cold)				.022		1.40		1.40	1.73	2.16
Disconnect and remove wall-hung lavatory				.800		50.50		50.50	62.50	78.50
Install new wall-hung lavatory with trim, (18" x 15")				2.807	410	177		587	670	790
Total				3.629	410	228.90		638.90	734.23	870.66

For customer support on your Facilities Maintenance & Repair Costs with RSMeans data, call 800.448.8182.

159

	System Description	Freq. (Years)	Crew	Unit	Labor Hours	2019 Bare Costs				Total In-House	Total w/O&P
						Material	Labor	Equipment	Total		
0010	**Replace washer in spud connection**	7	1 PLUM	Ea.							
	Turn valve off and on				.022		1.40		1.40	1.73	2.16
	Loosen locknuts				.011		.69		.69	.86	1.08
	Remove spud connection				.017		1.07		1.07	1.32	1.66
	Remove gasket or washer				.017		1.07		1.07	1.32	1.66
	Clean spud seat				.021		1.31		1.31	1.63	2.04
	Install new gasket / washer set				.026	4.76	1.65		6.41	7.30	8.50
	Install spud connection				.026		1.64		1.64	2.03	2.54
	Tighten locknut				.014		.90		.90	1.12	1.40
	Inspect connection				.009		.57		.57	.71	.89
	Total				.163	4.76	10.30		15.06	18.02	21.93
0020	**Replace washer in faucet**	2	1 PLUM	Ea.							
	Turn valves on and off				.022		1.40		1.40	1.73	2.16
	Remove packing nut and install				.023		1.48		1.48	1.83	2.29
	Remove valve stem and cap				.017		1.07		1.07	1.32	1.66
	Remove seat washer and install				.023		1.48		1.48	1.83	2.29
	Pry out washer				.017		1.07		1.07	1.32	1.66
	Install washer				.026	.48	1.65		2.13	2.57	3.15
	Install stem				.017		1.07		1.07	1.32	1.66
	Tighten main gland valves				.012		.74		.74	.92	1.15
	Test faucet				.010		.66		.66	.81	1.02
	Total				.168	.48	10.62		11.10	13.65	17.04
0040	**Replace faucets**	10	1 PLUM	Ea.							
	Turn valves on and off				.022		1.40		1.40	1.73	2.16
	Remove faucets and tubing				.381		24		24	30	37.50
	Install faucets and tubing				1.040	67	65.50		132.50	155	186
	Test faucets				.010		.66		.66	.81	1.02
	Total				1.453	67	91.56		158.56	187.54	226.68
0050	**Clean out strainer and P trap**	2	1 PLUM	Ea.							
	Remove strainer and P trap				.157		9.90		9.90	12.30	15.35
	Clean and reinstall strainer and trap				.320		20		20	25	31.50
	Total				.477		29.90		29.90	37.30	46.85

For customer support on your Facilities Maintenance & Repair Costs with RSMeans data, call 800.448.8182.

D2013 330 Lavatory, Vitreous China

System Description	Freq. (Years)	Crew	Unit	Labor Hours	2019 Bare Costs				Total In-House	Total w/O&P
					Material	Labor	Equipment	Total		
0060 Replace lavatory	35	2 PLUM	Ea.							
Shut off water (hot and cold)				.022		1.40		1.40	1.73	2.16
Disconnect and remove wall-hung lavatory				1.485		94		94	116	145
Install new lavatory (to 20" x 17")				2.971	153	169		322	380	455
Install faucets and tubing				1.040	67	65.50		132.50	155	186
Total				5.518	220	329.90		549.90	652.73	788.16

D2013 350 Lavatory, Enameled Steel

System Description	Freq. (Years)	Crew	Unit	Labor Hours	2019 Bare Costs				Total In-House	Total w/O&P
					Material	Labor	Equipment	Total		
0010 Replace washer in spud connection	7	1 PLUM	Ea.							
Turn valve off and on				.022		1.40		1.40	1.73	2.16
Loosen locknut				.014		.90		.90	1.12	1.40
Remove spud connection				.017		1.07		1.07	1.32	1.66
Remove gasket or washer				.017		1.07		1.07	1.32	1.66
Clean spud seat				.021		1.31		1.31	1.63	2.04
Install new gasket / washer set				.026	4.76	1.65		6.41	7.30	8.50
Install spud connection				.026		1.64		1.64	2.03	2.54
Tighten locknut				.014		.90		.90	1.12	1.40
Inspect connection				.009		.57		.57	.71	.89
Total				.166	4.76	10.51		15.27	18.28	22.25
0020 Replace washer in faucet	2	1 PLUM	Ea.							
Turn valves on and off				.022		1.40		1.40	1.73	2.16
Remove valve stem and washer				.091		5.75		5.75	7.10	8.90
Install washer				.026	.48	1.65		2.13	2.57	3.15
Install stem				.017		1.07		1.07	1.32	1.66
Test faucet				.010		.66		.66	.81	1.02
Total				.166	.48	10.53		11.01	13.53	16.89
0040 Replace faucets	10	1 PLUM	Ea.							
Turn valves on and off				.022		1.40		1.40	1.73	2.16
Remove faucets and tubing				.381		24		24	30	37.50
Install faucets and tubing				1.040	67	65.50		132.50	155	186
Test faucets				.010		.66		.66	.81	1.02
Total				1.453	67	91.56		158.56	187.54	226.68

For customer support on your Facilities Maintenance & Repair Costs with RSMeans data, call 800.448.8182.

161

D2013 350 Lavatory, Enameled Steel

	System Description	Freq. (Years)	Crew	Unit	Labor Hours	2019 Bare Costs				Total In-House	Total w/O&P
						Material	Labor	Equipment	Total		
0050	**Clean out strainer and P trap**	2	1 PLUM	Ea.							
	Remove strainer and P trap				.157		9.90		9.90	12.30	15.35
	Clean and reinstall strainer and trap				.320		20		20	25	31.50
	Total				.477		29.90		29.90	37.30	46.85
0060	**Replace lavatory**	35	2 PLUM	Ea.							
	Shut off water (hot and cold)				.022		1.40		1.40	1.73	2.16
	Disconnect and remove wall-hung lavatory				1.793		113		113	140	176
	Install new lavatory (to 20" x 17")				3.586	120	204		324	385	465
	Install faucets and tubing				1.040	67	65.50		132.50	155	186
	Total				6.441	187	383.90		570.90	681.73	829.16

D2013 410 Sink, Iron Enamel

	System Description	Freq. (Years)	Crew	Unit	Labor Hours	2019 Bare Costs				Total In-House	Total w/O&P
						Material	Labor	Equipment	Total		
0010	**Replace faucet washer**	2	1 PLUM	Ea.							
	Turn valves on and off				.022		1.40		1.40	1.73	2.16
	Remove valve stem and washer				.091		5.75		5.75	7.10	8.90
	Install washer				.026	.48	1.65		2.13	2.57	3.15
	Install stem				.017		1.07		1.07	1.32	1.66
	Test faucet				.010		.66		.66	.81	1.02
	Total				.166	.48	10.53		11.01	13.53	16.89
0020	**Clean trap**	3	1 PLUM	Ea.							
	Drain trap				.021		1.33		1.33	1.65	2.06
	Remove / install trap				.043		2.73		2.73	3.39	4.24
	Inspect trap				.031		1.97		1.97	2.45	3.06
	Clean trap				.021		1.31		1.31	1.63	2.04
	Total				.116		7.34		7.34	9.12	11.40
0030	**Replace faucets**	10	1 PLUM	Ea.							
	Turn valves on and off				.022		1.40		1.40	1.73	2.16
	Remove faucets and tubing				.381		24		24	30	37.50
	Install faucets and tubing				1.040	67	65.50		132.50	155	186
	Test faucets				.010		.66		.66	.81	1.02
	Total				1.453	67	91.56		158.56	187.54	226.68

For customer support on your Facilities Maintenance & Repair Costs with RSMeans data, call 800.448.8182.

D2013 410 Sink, Iron Enamel

System Description	Freq. (Years)	Crew	Unit	Labor Hours	2019 Bare Costs				Total In-House	Total w/O&P
					Material	Labor	Equipment	Total		
0040 Unstop sink										
Unstop plugged drain	2	1 PLUM	Ea.	.571		36		36	45	56
Total				.571		36		36	**45**	**56**
0060 Replace kitchen sink, P.E.C.I.										
Remove PE sink				1.625		92.50		92.50	115	143
Install countertop porcelain enameled kitchen sink, 24" x 21"	10	1 PLUM	Ea.	2.857	310	162		472	540	640
Check for leaks				.013		.82		.82	1.02	1.27
Total				4.495	310	255.32		565.32	**656.02**	**784.27**
0070 Replace laundry sink, P.E.C.I.										
Remove PE sink				1.625		92.50		92.50	115	143
Install free-standing porcelain enameled laundry sink, 24" x 21"	35	2 PLUM	Ea.	2.667	595	152		747	845	980
Total				4.292	595	244.50		839.50	**960**	**1,123**

D2013 420 Sink, Enameled Steel

System Description	Freq. (Years)	Crew	Unit	Labor Hours	2019 Bare Costs				Total In-House	Total w/O&P
					Material	Labor	Equipment	Total		
0010 Replace faucet washer										
Turn valves on and off				.022		1.40		1.40	1.73	2.16
Remove valve stem and washer				.091		5.75		5.75	7.10	8.90
Install washer	2	1 PLUM	Ea.	.026	.48	1.65		2.13	2.57	3.15
Install stem				.017		1.07		1.07	1.32	1.66
Test faucet				.010		.66		.66	.81	1.02
Total				.166	.48	10.53		11.01	**13.53**	**16.89**
0020 Clean trap										
Drain trap				.021		1.33		1.33	1.65	2.06
Remove / install trap				.043		2.73		2.73	3.39	4.24
Inspect trap	3	1 PLUM	Ea.	.031		1.97		1.97	2.45	3.06
Clean trap				.021		1.31		1.31	1.63	2.04
Total				.116		7.34		7.34	**9.12**	**11.40**

For customer support on your Facilities Maintenance & Repair Costs with RSMeans data, call 800.448.8182.

163

D2013 420 Sink, Enameled Steel

	System Description	Freq. (Years)	Crew	Unit	Labor Hours	2019 Bare Costs Material	Labor	Equipment	Total	Total In-House	Total w/O&P
0030	**Replace faucets**	10	1 PLUM	Ea.							
	Turn valves on and off				.022		1.40		1.40	1.73	2.16
	Remove faucets and tubing				.381		24		24	30	37.50
	Install faucets and tubing				1.040	67	65.50		132.50	155	186
	Test faucets				.010		.66		.66	.81	1.02
	Total				1.453	67	91.56		158.56	187.54	226.68
0040	**Unstop sink**	2	1 PLUM	Ea.							
	Unstop plugged drain				.571		36		36	45	56
	Total				.571		36		36	45	56
0060	**Replace sink, enameled steel**	35	2 PLUM	Ea.							
	Remove sink				1.793		102		102	126	158
	Install countertop enameled steel sink, 24" x 21"				3.714	525	211		736	840	980
	Total				5.507	525	313		838	966	1,138

D2013 430 Sink, Stainless Steel

	System Description	Freq. (Years)	Crew	Unit	Labor Hours	2019 Bare Costs Material	Labor	Equipment	Total	Total In-House	Total w/O&P
0010	**Replace faucet washer**	2	1 PLUM	Ea.							
	Turn valves on and off				.022		1.40		1.40	1.73	2.16
	Remove valve stem and washer				.091		5.75		5.75	7.10	8.90
	Install washer				.026	.48	1.65		2.13	2.57	3.15
	Install stem				.017		1.07		1.07	1.32	1.66
	Test faucet				.010		.66		.66	.81	1.02
	Total				.166	.48	10.53		11.01	13.53	16.89
0020	**Clean trap**	3	1 PLUM	Ea.							
	Drain trap				.021		1.33		1.33	1.65	2.06
	Remove / install trap				.043		2.73		2.73	3.39	4.24
	Inspect trap				.031		1.97		.97	2.45	3.06
	Clean trap				.021		1.31		.31	1.63	2.04
	Total				.116		7.34		7.34	9.12	11.40

For customer support on your Facilities Maintenance & Repair Costs with RSMeans data, call 800.448.8182.

D20 PLUMBING   D2013   Plumbing Fixtures

D2013 430   Sink, Stainless Steel

	System Description	Freq. (Years)	Crew	Unit	Labor Hours	2019 Bare Costs				Total In-House	Total w/O&P
						Material	Labor	Equipment	Total		
0030	**Replace faucets**	10	1 PLUM	Ea.							
	Turn valves on and off				.022		1.40		1.40	1.73	2.16
	Remove faucets and tubing				.381		24		24	30	37.50
	Install faucets and tubing				1.040	67	65.50		132.50	155	186
	Test faucets				.010		.66		.66	.81	1.02
	Total				1.453	67	91.56		158.56	187.54	226.68
0040	**Unstop sink**	2	1 PLUM	Ea.							
	Unstop plugged drain				.571		36		36	45	56
	Total				.571		36		36	45	56
0060	**Replace sink, stainless steel**	40	2 PLUM	Ea.							
	Remove sink				1.625		92.50		92.50	115	143
	Install countertop stainless steel sink, 25" x 22"				3.714	680	211		891	1,000	1,175
	Total				5.339	680	303.50		983.50	1,115	1,318
0070	**Replace stainless steel detention sink faucets**	15	1 SKWK	Ea.							
	Turn valves on and off				.022		1.40		1.40	1.73	2.16
	Remove faucets and tubing				.333		21		21	26	32.50
	Detention equipment, replace fuacet valve and tubing				2.000	425	126		551	625	725
	Check for leaks				.013		.82		.82	1.02	1.27
	Total				2.368	425	149.22		574.22	653.75	760.93

D2013 440   Sink, Plastic

	System Description	Freq. (Years)	Crew	Unit	Labor Hours	2019 Bare Costs				Total In-House	Total w/O&P
						Material	Labor	Equipment	Total		
0010	**Replace faucet washer**	2	1 PLUM	Ea.							
	Turn valves on and off				.022		1.40		1.40	1.73	2.16
	Remove valve stem and washer				.091		5.75		5.75	7.10	8.90
	Install washer				.026	.48	1.65		2.13	2.57	3.15
	Install stem				.017		1.07		1.07	1.32	1.66
	Test faucet				.010		.66		.66	.81	1.02
	Total				.166	.48	10.53		11.01	13.53	16.89

D2013 440 — Sink, Plastic

System Description	Freq. (Years)	Crew	Unit	Labor Hours	2019 Bare Costs Material	Labor	Equipment	Total	Total In-House	Total w/O&P
0020 Clean trap	3	1 PLUM	Ea.							
Drain trap				.021		1.33		1.33	1.65	2.06
Remove / install trap				.043		2.73		2.73	3.39	4.24
Inspect trap				.031		1.97		1.97	2.45	3.06
Clean trap				.021		1.31		1.31	1.63	2.04
Total				.116		7.34		7.34	**9.12**	**11.40**
0030 Replace faucets	10	1 PLUM	Ea.							
Turn valves on and off				.022		1.40		1.40	1.73	2.16
Remove faucets and tubing				.381		24		24	30	37.50
Install faucets and tubing				1.040	67	65.50		132.50	155	186
Test faucets				.010		.66		.66	.81	1.02
Total				1.453	67	91.56		158.56	**187.54**	**226.68**
0040 Unstop, sink	2	1 PLUM	Ea.							
Unstop plugged drain				.571		36		36	45	56
Total				.571		36		36	**45**	**56**
0060 Replace sink and fittings, polyethylene	15	2 PLUM	Ea.							
Remove sink				3.467		197		197	244	305
Install bench mounted polyethylene sink, 18-1/2" x 18-1/2"				4.000	370	227		597	690	820
Total				7.467	370	424		794	**934**	**1,125**
0080 Replace lab sink and fittings, polyethylene	15	2 PLUM	Ea.							
Remove sink				3.467		197		197	244	305
Install bench mounted PE lab sink, 23-1/2" x 20-1/2"				5.333	1,300	305		1,505	1,800	2,100
Total				8.800	1,300	502		1,302	**2,044**	**2,405**

For customer support on your Facilities Maintenance & Repair Costs with RSMeans data, call 800.448.8182.

D2013 444 Laboratory Sink

	System Description	Freq. (Years)	Crew	Unit	Labor Hours	2019 Bare Costs				Total In-House	Total w/O&P
						Material	Labor	Equipment	Total		
0010	**Replace faucet washer**	2	1 PLUM	Ea.							
	Turn valves on and off				.022		1.40		1.40	1.73	2.16
	Remove valve stem and washer				.091		5.75		5.75	7.10	8.90
	Install washer				.026	.48	1.65		2.13	2.57	3.15
	Install stem				.017		1.07		1.07	1.32	1.66
	Test faucet				.010		.66		.66	.81	1.02
	Total				.166	.48	10.53		11.01	13.53	16.89
0030	**Replace faucets**	10	1 PLUM	Ea.							
	Turn valves on and off				.022		1.40		1.40	1.73	2.16
	Remove faucets and tubing				.381		24		24	30	37.50
	Install faucets and tubing				1.040	67	65.50		132.50	155	186
	Test faucets				.010		.66		.66	.81	1.02
	Total				1.453	67	91.56		158.56	187.54	226.68
0060	**Replace sink and fittings, stainless steel, lab.**	15	2 PLUM	Ea.							
	Remove lab. sink				3.467		197		197	244	305
	Install bench mounted stainless steel lab. sink, 47" x 24"				5.333	1,300	305		1,605	1,800	2,100
	Total				8.800	1,300	502		1,802	2,044	2,405

D2013 450 Laundry Sink, Plastic

	System Description	Freq. (Years)	Crew	Unit	Labor Hours	2019 Bare Costs				Total In-House	Total w/O&P
						Material	Labor	Equipment	Total		
0020	**Replace washer in faucet**	2	1 PLUM	Ea.							
	Turn valves on and off				.022		1.40		1.40	1.73	2.16
	Remove valve stem and washer				.091		5.75		5.75	7.10	8.90
	Install washer				.026	.48	1.65		2.13	2.57	3.15
	Install stem				.017		1.07		1.07	1.32	1.66
	Test faucet				.010		.66		.66	.81	1.02
	Total				.166	.48	10.53		11.01	13.53	16.89

For customer support on your Facilities Maintenance & Repair Costs with RSMeans data, call 800.448.8182.

167

D2013 450 Laundry Sink, Plastic

System Description	Freq. (Years)	Crew	Unit	Labor Hours	Material	Labor	Equipment	Total	Total In-House	Total w/O&P
							2019 Bare Costs			
0040 Replace faucets	10	1 PLUM	Ea.							
Turn valves on and off				.022		1.40		1.40	1.73	2.16
Remove faucets and tubing				.381		24		24	30	37.50
Install faucets and tubing				1.040	67	65.50		132.50	155	186
Test faucets				.010		.66		.66	.81	1.02
Total				1.453	67	91.56		158.56	187.54	226.68
0050 Clean out strainer and P trap	2	1 PLUM	Ea.							
Remove strainer and P trap				.157		9.90		9.90	12.30	15.35
Clean and reinstall strainer and trap				.320		20		20	25	31.50
Total				.477		29.90		29.90	37.30	46.85
0060 Replace laundry sink	20	2 PLUM	Ea.							
Shut off water (hot and cold)				.022		1.40		1.40	1.73	2.16
Disconnect and remove laundry sink				1.600		91		91	113	141
Install sink (23" x 18", single)				3.200	146	182		328	385	465
Install faucets and tubing				1.040	67	65.50		132.50	155	186
Total				5.862	213	339.90		552.90	654.73	794.16

D2013 455 Laundry Sink, Stone

System Description	Freq. (Years)	Crew	Unit	Labor Hours	Material	Labor	Equipment	Total	Total In-House	Total w/O&P
							2019 Bare Costs			
0060 Replace laundry sink, molded stone	40	2 PLUM	Ea.							
Remove molded stone sink				1.625		92.50		92.50	115	143
Install laundry sink, w/trim, molded stone, 22"x23" ,sgl compt				2.667	164	152		316	370	440
Total				4.292	164	244.50		408.50	485	583

For customer support on your Facilities Maintenance & Repair Costs with RSMeans data, call 800.448.8182.

System Description	Freq. (Years)	Crew	Unit	Labor Hours	2019 Bare Costs Material	2019 Bare Costs Labor	2019 Bare Costs Equipment	2019 Bare Costs Total	Total In-House	Total w/O&P
0010 Replace faucet washer	2	1 PLUM	Ea.							
Turn valves on and off				.022		1.40		1.40	1.73	2.16
Remove valve stem and washer				.091		5.75		5.75	7.10	8.90
Install washer				.026	.48	1.65		2.13	2.57	3.15
Install stem				.017		1.07		1.07	1.32	1.66
Test faucet				.010		.66		.66	.81	1.02
Total				.166	.48	10.53		11.01	13.53	16.89
0020 Clean trap	3	1 PLUM	Ea.							
Drain trap				.021		1.33		1.33	1.65	2.06
Remove / install trap				.043		2.73		2.73	3.39	4.24
Inspect trap				.031		1.97		1.97	2.45	3.06
Clean trap				.021		1.31		1.31	1.63	2.04
Total				.116		7.34		7.34	9.12	11.40
0030 Replace faucets	10	1 PLUM	Ea.							
Turn valves on and off				.022		1.40		1.40	1.73	2.16
Remove faucets and tubing				.381		24		24	30	37.50
Install faucets and tubing				1.040	67	65.50		132.50	155	186
Test faucets				.010		.66		.66	.81	1.02
Total				1.453	67	91.56		158.56	187.54	226.68
0040 Unstop sink	2	1 PLUM	Ea.							
Unstop plugged drain				.571		36		36	45	56
Total				.571		36		36	45	56
0060 Replace service sink, P.E.C.I.	35	1 PLUM	Ea.							
Remove sink				1.625		92.50		92.50	115	143
Install service/utility sink, 22" x 18"				5.212	825	296		1,121	1,275	1,500
Total				6.837	825	388.50		1,213.50	1,390	1,643

For customer support on your Facilities Maintenance & Repair Costs with RSMeans data, call 800.448.8182.

169

D2013 470 Group Washfountain

	System Description	Freq. (Years)	Crew	Unit	Labor Hours	Material	Labor	Equipment	Total	Total In-House	Total w/O&P
							2019 Bare Costs				
0060	**Replace group wash fountain, 54" diameter**	20	Q-2	Ea.							
	Shut off water (hot and cold)				.022		1.40		1.40	1.73	2.16
	Disconnect and remove group wash fountain, 54" dia.				4.800		283		283	350	440
	Install group wash fountain, terrazzo, 54" diameter				9.600	9,500	565		0,065	11,200	12,800
	Install faucets and tubing				1.040	67	65.50		132.50	155	186
	Total				15.462	9,567	914.90		0,481.90	11,706.73	13,428.16

D2013 510 Bathtub, Cast Iron Enamel

	System Description	Freq. (Years)	Crew	Unit	Labor Hours	Material	Labor	Equipment	Total	Total In-House	Total w/O&P
							2019 Bare Costs				
0010	**Inspect / clean shower head**	3	1 PLUM	Ea.							
	Turn water off and on				.010		.66		.66	.81	1.02
	Remove shower head				.217		13.70		13.70	17	21
	Check shower head				.009		.57		.57	.71	.89
	Install shower head				.433		27.50		27.50	34	42.50
	Total				.669		42.43		42.43	52.52	65.41
0020	**Replace mixing valve barrel**	2	1 PLUM	Ea.							
	Turn valves on and off				.022		1.40		1.40	1.73	2.16
	Remove mixing valve cover / barrel				.650		41		41	51	63.50
	Install new mixing valve barrel / replace cover				.650	163	41		204	230	267
	Test valve				.010		.66		.66	.81	1.02
	Total				1.332	163	84.06		247.06	283.54	333.68
0030	**Replace mixing valve**	10	1 PLUM	Ea.							
	Turn valves off and on				.022		1.40		1.40	1.73	2.16
	Remove mixing valve				.867		54.50		54.50	68	85
	Install new mixing valve				1.733	141	109		250	291	345
	Test valve				.010		.66		.66	.81	1.02
	Total				2.633	141	165.56		306.56	361.54	433.18
0070	**Replace tub**	40	2 PLUM	Ea.							
	Remove cast iron tub				1.891		107		107	133	167
	Install cast iron enameled bathtub				3.782	1,200	215		1,415	1,575	1,825
	Total				5.673	1,200	322		1,522	1,708	1,992

For customer support on your Facilities Maintenance & Repair Costs with RSMeans data, call 800.448.8182.

D2013 530 Bathtub, Enameled Steel

	System Description	Freq. (Years)	Crew	Unit	Labor Hours	2019 Bare Costs Material	Labor	Equipment	Total	Total In-House	Total w/O&P
0010	**Inspect / clean shower head**	3	1 PLUM	Ea.							
	Turn water off and on				.010		.66		.66	.81	1.02
	Remove shower head				.217		13.70		13.70	17	21
	Check shower head				.009		.57		.57	.71	.89
	Install shower head				.433		27.50		27.50	34	42.50
	Total				.669		42.43		42.43	52.52	65.41
0020	**Replace mixing valve barrel**	2	1 PLUM	Ea.							
	Turn valves on and off				.022		1.40		1.40	1.73	2.16
	Remove mixing valve cover / barrel				.650		41		41	51	63.50
	Install new mixing valve barrel / replace cover				.650	163	41		204	230	267
	Test valve				.010		.66		.66	.81	1.02
	Total				1.332	163	84.06		247.06	283.54	333.68
0030	**Replace mixing valve**	10	1 PLUM	Ea.							
	Turn valves off and on				.022		1.40		1.40	1.73	2.16
	Remove mixing valve				.867		54.50		54.50	68	85
	Install new mixing valve				1.733	141	109		250	291	345
	Test valve				.010		.66		.66	.81	1.02
	Total				2.633	141	165.56		306.56	361.54	433.18
0070	**Replace tub**	35	2 PLUM	Ea.							
	Remove tub				1.891		107		107	133	167
	Install enameled steel bathtub				3.782	505	215		720	820	965
	Total				5.673	505	322		827	953	1,132

D2013 550 Bathtub, Fiberglass

	System Description	Freq. (Years)	Crew	Unit	Labor Hours	2019 Bare Costs Material	Labor	Equipment	Total	Total In-House	Total w/O&P
0010	**Inspect / clean shower head**	3	1 PLUM	Ea.							
	Turn water off and on				.010		.66		.66	.81	1.02
	Remove shower head				.217		13.70		13.70	17	21
	Check shower head				.009		.57		.57	.71	.89
	Install shower head				.433		27.50		27.50	34	42.50
	Total				.669		42.43		42.43	52.52	65.41

For customer support on your Facilities Maintenance & Repair Costs with RSMeans data, call 800.448.8182.

171

D2013 550 Bathtub, Fiberglass

	System Description	Freq. (Years)	Crew	Unit	Labor Hours	2019 Bare Costs Material	Labor	Equipment	Total	Total In-House	Total w/O&P
0020	**Replace mixing valve barrel**	2	1 PLUM	Ea.							
	Turn valves on and off				.022		1.40		1.40	1.73	2.16
	Remove mixing valve cover / barrel				.650		41		41	51	63.50
	Install new mixing valve barrel / replace cover				.650	163	41		204	230	267
	Test valve				.010		.66		.66	.81	1.02
	Total				1.332	163	84.06		247.06	283.54	333.68
0030	**Replace mixing valve**	10	1 PLUM	Ea.							
	Turn valves off and on				.022		1.40		1.40	1.73	2.16
	Remove mixing valve				.867		54.50		54.50	68	85
	Install new mixing valve				1.733	141	109		250	291	345
	Test valve				.010		.66		.66	.81	1.02
	Total				2.633	141	165.56		306.56	361.54	433.18
0070	**Replace tub**	20	2 PLUM	Ea.							
	Remove bathtub				2.600		148		148	183	229
	Install fiberglass bathtub with shower wall surround				5.200	770	296		1,066	1,225	1,425
	Total				7.800	770	444		1,214	1,408	1,654

D2013 710 Shower, Terrazzo

	System Description	Freq. (Years)	Crew	Unit	Labor Hours	2019 Bare Costs Material	Labor	Equipment	Total	Total In-House	Total w/O&P
0010	**Inspect / clean shower head**	3	1 PLUM	Ea.							
	Turn water off and on				.010		.66		.66	.81	1.02
	Remove shower head				.217		13.70		13.70	17	21
	Check shower head				.009		.57		.57	.71	.89
	Install shower head				.433		27.50		27.50	34	42.50
	Total				.669		42.43		42.43	52.52	65.41
0020	**Replace mixing valve barrel**	2	1 PLUM	Ea.							
	Turn valves on and off				.022		1.40		1.40	1.73	2.16
	Remove mixing valve cover / barrel				.650		41		41	51	63.50
	Install new mixing valve barrel / replace cover				.650	163	41		204	230	267
	Test valve				.010		.66		.66	.81	1.02
	Total				1.332	163	84.06		247.06	283.54	333.68

For customer support on your Facilities Maintenance & Repair Costs with RSMeans data, call 800.448.8182.

D2013 710 Shower, Terrazzo

	System Description	Freq. (Years)	Crew	Unit	Labor Hours	2019 Bare Costs				Total In-House	Total w/O&P
						Material	Labor	Equipment	Total		
0030	**Replace mixing valve**	10	1 PLUM	Ea.							
	Turn valves off and on				.022		1.40		1.40	1.73	2.16
	Remove mixing valve				.867		54.50		54.50	68	85
	Install new mixing valve				1.733	141	109		250	291	345
	Test valve				.010		.66		.66	.81	1.02
	Total				2.633	141	165.56		306.56	361.54	433.18
0060	**Replace terrazzo shower surface**	30	2 PLUM	Ea.							
	Remove old terrazzo surface				3.004		123.50		123.50	161	199
	Install precast terrazzo tiles				9.014	334.75	432.25		767	915	1,100
	Install curb				2.080	139.50	99		238.50	279	330
	Total				14.098	474.25	654.75		1,129	1,355	1,629

D2013 730 Shower, Enameled Steel

	System Description	Freq. (Years)	Crew	Unit	Labor Hours	2019 Bare Costs				Total In-House	Total w/O&P
						Material	Labor	Equipment	Total		
0010	**Inspect / clean shower head**	3	1 PLUM	Ea.							
	Turn water off and on				.010		.66		.66	.81	1.02
	Remove shower head				.217		13.70		13.70	17	21
	Check shower head				.009		.57		.57	.71	.89
	Install shower head				.433		27.50		27.50	34	42.50
	Total				.669		42.43		42.43	52.52	65.41
0020	**Replace mixing valve barrel**	2	1 PLUM	Ea.							
	Turn valves on and off				.022		1.40		1.40	1.73	2.16
	Remove mixing valve cover / barrel				.650		41		41	51	63.50
	Install new mixing valve barrel / replace cover				.650	163	41		204	230	267
	Test valve				.010		.66		.66	.81	1.02
	Total				1.332	163	84.06		247.06	283.54	333.68
0030	**Replace mixing valve**	10	1 PLUM	Ea.							
	Turn valves off and on				.022		1.40		1.40	1.73	2.16
	Remove mixing valve				.867		54.50		54.50	68	85
	Install new mixing valve				1.733	141	109		250	291	345
	Test valve				.010		.66		.66	.81	1.02
	Total				2.633	141	165.56		306.56	361.54	433.18

D2013 730 Shower, Enameled Steel

	System Description	Freq. (Years)	Crew	Unit	Labor Hours	Material	Labor	Equipment	Total	Total In-House	Total w/O&P
								2019 Bare Costs			
0060	**Replace shower**	35	2 PLUM	Ea.							
	Remove shower				5.200		296		296	365	460
	Install shower				4.522	1,150	257		1,407	1,575	1,850
	Total				9.722	1,150	553		1,703	1,940	2,310

D2013 750 Shower, Fiberglass

	System Description	Freq. (Years)	Crew	Unit	Labor Hours	Material	Labor	Equipment	Total	Total In-House	Total w/O&P
								2019 Bare Costs			
0010	**Inspect / clean shower head**	3	1 PLUM	Ea.							
	Turn water off and on				.010		.66		.66	.81	1.02
	Remove shower head				.217		13.70		13.70	17	21
	Check shower head				.009		.57		.57	.71	.89
	Install shower head				.433		27.50		27.50	34	42.50
	Total				.669		42.43		42.43	52.52	65.41
0020	**Replace mixing valve barrel**	2	1 PLUM	Ea.							
	Turn valves on and off				.022		1.40		1.40	1.73	2.16
	Remove mixing valve cover / barrel				.650		41		41	51	63.50
	Install new mixing valve barrel / replace cover				.650	163	41		204	230	267
	Test valve				.010		.66		.66	.81	1.02
	Total				1.332	163	84.06		247.06	283.54	333.68
0030	**Replace mixing valve**	10	1 PLUM	Ea.							
	Turn valves off and on				.022		1.40		1.40	1.73	2.16
	Remove mixing valve				.867		54.50		54.50	68	85
	Install new mixing valve				1.733	141	109		250	291	345
	Test valve				.010		.66		.66	.81	1.02
	Total				2.633	141	165.56		306.56	361.54	433.18
0060	**Replace shower and fittings**	20	2 PLUM	Ea.							
	Remove shower				4.334		246		246	305	380
	Install shower, fiberglass				3.302	400	188		588	675	790
	Total				7.635	400	434		834	980	1,170

For customer support on your Facilities Maintenance & Repair Costs with RSMeans data, call 800.448.8182.

System Description	Freq. (Years)	Crew	Unit	Labor Hours	2019 Bare Costs				Total In-House	Total w/O&P
					Material	Labor	Equipment	Total		
0010	**Inspect / clean shower head**	3	1 PLUM	Ea.							
	Turn water off and on				.010		.66		.66	.81	1.02
	Remove shower head				.217		13.70		13.70	17	21
	Check shower head				.009		.57		.57	.71	.89
	Install shower head				.433		27.50		27.50	34	42.50
	Total				.669		42.43		42.43	**52.52**	**65.41**
0020	**Replace mixing valve barrel**	2	1 PLUM	Ea.							
	Turn valves on and off				.022		1.40		1.40	1.73	2.16
	Remove mixing valve cover / barrel				.650		41		41	51	63.50
	Install new mixing valve barrel / replace cover				.650	163	41		204	230	267
	Test valve				.010		.66		.66	.81	1.02
	Total				1.332	163	84.06		247.06	**283.54**	**333.68**
0030	**Replace mixing valve**	10	1 PLUM	Ea.							
	Turn valves off and on				.022		1.40		1.40	1.73	2.16
	Remove mixing valve				.867		54.50		54.50	68	85
	Install new mixing valve				1.733	141	109		250	291	345
	Test valve				.010		.66		.66	.81	1.02
	Total				2.633	141	165.56		306.56	**361.54**	**433.18**
0060	**Replace shower and fittings, aluminum**	25	2 PLUM	Ea.							
	Remove shower				4.334		246		246	305	380
	Install shower				3.302	400	188		588	675	790
	Total				7.635	400	434		834	**980**	**1,170**
0070	**Replace shower head with water conserving head**	10	1 PLUM	Ea.							
	Remove shower head				.217		13.70		13.70	17	21
	Install new water conserving head				.400	88.50	25.50		114	129	150
	Turn water on and off				.010		.66		.66	.81	1.02
	Total				.627	88.50	39.86		128.36	**146.81**	**172.02**
0120	**Replace shower, C.M.U.**	20	D-8	Ea.							
	Remove C.M.U. shower enclosure				3.595		148.96	21.28	170.24	217	263
	Install C.M.U. shower enclosure				6.772	163.52	316.40		479.92	595	720
	Install built-in head, arm and valve				2.000	91.50	126		217.50	257	310
	Total				12.367	255.02	591.36	21.28	867.66	**1,069**	**1,293**

For customer support on your Facilities Maintenance & Repair Costs with RSMeans data, call 800.448.8182.

175

D2013 770 Shower, Misc.

System Description	Freq. (Years)	Crew	Unit	Labor Hours	2019 Bare Costs Material	Labor	Equipment	Total	Total In-House	Total w/O&P
0200 Replace shower, glazed C.M.U.	25	D-8	Ea.							
Remove C.M.U. shower enclosure				4.697		195	27.50	222.50	284	345
Install glazed C.M.U. shower enclosure				9.393	652.40	439.60		1,092	1,300	1,525
Install built-in head, arm and valve				2.000	91.50	126		217.50	257	310
Total				16.090	743.90	760.60	27.50	1,532	1,841	2,180
0280 Replace shower surface, ceramic tile	30	D-7	Ea.							
Remove ceramic tile shower enclosure surface				3.066		126		126	164	203
Install ceramic tile in shower enclosure				6.130	135.52	262.64		398.16	480	580
Install built-in head, arm and valve				2.000	91.50	126		217.50	257	310
Total				11.197	227.02	514.64		741.66	901	1,093

D2013 810 Drinking Fountain

System Description	Freq. (Years)	Crew	Unit	Labor Hours	2019 Bare Costs Material	Labor	Equipment	Total	Total In-House	Total w/O&P
0010 Check / minor repairs	1	1 PLUM	Ea.							
Check unit, make minor adjustments				.667		42		42	52	65.50
Total				.667		42		42	52	65.50
0020 Repair internal leaks	4	1 PLUM	Ea.							
Open unit, repair leaks, close				.615		39		39	48	60.50
Total				.615		39		39	48	60.50
0030 Correct water pressure	2	1 PLUM	Ea.							
Adjust water pressure, close				.571		36		36	45	56
Total				.571		36		36	45	56
0040 Replace refrigerant	2	1 PLUM	Ea.							
Add freon to unit				.133	23	8.55		31.55	36	42
Total				.133	23	8.55		31.55	36	42
0050 Repair drain leak	4	1 PLUM	Ea.							
Remove old pipe				.052		3.28		3.28	4.07	5.10
Install new pipe				.186	12.70	11.75		24.45	28.50	34
Total				.238	12.70	15.03		27.73	32.57	39.10

For customer support on your Facilities Maintenance & Repair Costs with RSMeans data, call 800.448.8182.

D20 PLUMBING | D2013 | Plumbing Fixtures

D2013 810 Drinking Fountain

System Description	Freq. (Years)	Crew	Unit	Labor Hours	2019 Bare Costs				Total In-House	Total w/O&P
					Material	Labor	Equipment	Total		
0070 Replace fountain	10	2 PLUM	Ea.							
Remove old water cooler				2.600		148		148	183	229
Install new wall mounted water cooler				5.200	695	296		991	1,125	1,325
Total				7.800	695	444		1,139	1,308	1,554

D2013 910 Emergency Shower Station

System Description	Freq. (Years)	Crew	Unit	Labor Hours	2019 Bare Costs				Total In-House	Total w/O&P
					Material	Labor	Equipment	Total		
0020 Inspect and clean shower head	3	1 PLUM	Ea.							
Turn water off and on				.010		.66		.66	.81	1.02
Remove shower head				.217		13.70		13.70	17	21
Check shower head				.009		.57		.57	.71	.89
Install shower head				.433		27.50		27.50	34	42.50
Total				.669		42.43		42.43	52.52	65.41
0030 Replace shower	25	2 PLUM	Ea.							
Remove emergency shower				2.600		148		148	183	229
Install emergency shower				5.200	395	296		691	800	955
Total				7.800	395	444		839	983	1,184

D2013 920 Emergency Eye Wash

System Description	Freq. (Years)	Crew	Unit	Labor Hours	2019 Bare Costs				Total In-House	Total w/O&P
					Material	Labor	Equipment	Total		
0020 Inspect and clean spray heads	3	1 PLUM	Ea.							
Turn water off and on				.010		.66		.66	.81	1.02
Remove spray head				.217		13.70		13.70	17	21
Check shower head				.009		.57		.57	.71	.89
Install spray head				.433		27.50		27.50	34	42.50
Total				.669		42.43		42.43	52.52	65.41
0030 Replace eye wash station	25	2 PLUM	Ea.							
Remove old eye wash station				2.600		148		148	183	229
Install eye wash station				5.200	475	296		771	890	1,050
Total				7.800	475	444		919	1,073	1,279

For customer support on your Facilities Maintenance & Repair Costs with RSMeans data, call 800.448.8182.

177

D2023 110 | **Pipe & Fittings, Copper**

System Description	Freq. (Years)	Crew	Unit	Labor Hours	2019 Bare Costs				Total In-House	Total w/O&P
					Material	Labor	Equipment	Total		
0010 Resolder joint	10	1 PLUM	Ea.							
Measure, cut & ream both ends				.053		3.37		3.37	4.18	5.20
Solder fitting (3/4" coupling)				.495	2.49	31.50		33.99	41.50	51.50
Total				.549	2.49	34.87		37.36	45.68	**56.70**
0020 Replace 3/4" copper pipe and fittings	20	2 PLUM	L.F.							
Remove old pipe				.071		4.48		4.48	5.55	6.95
Install 3/4" copper tube with couplings and hangers				.141	8.55	8.90		17.45	20.50	24.50
Total				.212	8.55	13.38		21.93	26.05	**31.45**
0030 Replace 1" copper pipe and fittings	25	2 PLUM	L.F.							
Remove old pipe				.073		4.58		4.58	5.70	7.10
Install 1" copper tube with couplings and hangers				.158	12.70	9.95		22.65	26.50	31.50
Total				.230	12.70	14.53		27.23	32.20	**38.60**
0050 Replace 1-1/2" copper pipe and fittings	25	2 PLUM	L.F.							
Remove old pipe				.104		6.55		6.55	8.15	10.20
Install 1 1/2" copper tubing, couplings, hangers				.208	18.15	13.15		31.30	36.50	43
Total				.312	18.15	19.70		37.85	44.65	**53.20**
0060 Replace 2" copper pipe and fittings	25	2 PLUM	L.F.							
Remove old pipe				.130		8.20		8.20	10.20	12.75
Install 2" copper tube with couplings and hangers				.260	27	16.40		43.40	50	59.50
Total				.390	27	24.60		51.60	60.20	**72.25**
0070 Replace 4" copper pipe and fittings	25	2 PLUM	L.F.							
Remove old pipe				.274		15.55		15.55	19.30	24
Install 4" copper tube with couplings and hangers				.547	96	31		127	144	168
Total				.821	96	46.55		142.55	163.30	**192**
0080 Replace 8" copper pipe and fittings	25	2 PLUM	L.F.							
Remove old pipe				.694		41		41	50.50	63.50
Install 8" copper tube with couplings and hangers				.918	435	54		489	545	630
Total				1.612	435	95		530	595.50	**693.50**

For customer support on your Facilities Maintenance & Repair Costs with RSMeans data, call 800.448.8182.

D2023 120 | Pipe & Fittings, Steel/Iron, Threaded

	System Description	Freq. (Years)	Crew	Unit	Labor Hours	2019 Bare Costs				Total In-House	Total w/O&P
						Material	Labor	Equipment	Total		
0030	**Replace 3/4" threaded steel pipe and fittings**	75	2 PLUM	L.F.							
	Remove old pipe				.086		5.45		5.45	6.75	8.40
	Install 3/4" diameter steel pipe, couplings, hangers				.131	4.25	8.30		12.55	14.95	18.15
	Total				.217	4.25	13.75		18	21.70	26.55
0040	**Replace 1-1/2" threaded steel pipe and fittings**	75	2 PLUM	L.F.							
	Remove old pipe				.130		7.40		7.40	9.15	11.45
	Install 1-1/2" diameter steel pipe, couplings, hangers				.260	5.70	14.80		20.50	24.50	30
	Total				.390	5.70	22.20		27.90	33.65	41.45
0050	**Replace 2" threaded steel pipe and fittings**	75	2 PLUM	L.F.							
	Remove old pipe				.163		9.25		9.25	11.50	14.35
	Install 2" diameter steel pipe, couplings, hangers				.325	11.55	18.45		30	35.50	43
	Total				.488	11.55	27.70		39.25	47	57.35
0060	**Replace 4" threaded steel pipe and fittings**	75	2 PLUM	L.F.							
	Remove old pipe				.289		16.40		16.40	20.50	25.50
	Install 4" diameter steel pipe, couplings, hangers				.578	21.50	33		54.50	64.50	78
	Total				.867	21.50	49.40		70.90	85	103.50

D2023 130 | Solar Piping: Pipe & Fittings, PVC

	System Description	Freq. (Years)	Crew	Unit	Labor Hours	2019 Bare Costs				Total In-House	Total w/O&P
						Material	Labor	Equipment	Total		
0010	**Reglue joint, install 1" tee**	10	1 PLUM	Ea.							
	Cut exist pipe, install tee, 1"				.867	1.38	54.50		55.88	69.50	86.50
	Inspect joints				.126		7.95		7.95	9.85	12.35
	Total				.993	1.38	62.45		63.83	79.35	98.85
0110	**Reglue joint, install 1-1/4" tee**	10	1 PLUM	Ea.							
	Cut exist pipe, install tee, 1-1/4"				.945	2.15	59.50		61.65	76.50	95
	Inspect joints				.126		7.95		7.95	9.85	12.35
	Total				1.071	2.15	67.45		69.60	86.35	107.35

For customer support on your Facilities Maintenance & Repair Costs with RSMeans data, call 800.448.8182.

179

D2023 130 | **Solar Piping: Pipe & Fittings, PVC**

System Description	Freq. (Years)	Crew	Unit	Labor Hours	2019 Bare Costs				Total In-House	Total w/O&P
					Material	Labor	Equipment	Total		
0210 Reglue joint, install 1-1/2" tee	10	1 PLUM	Ea.							
Cut exist pipe, install tee, 1-1/2"				1.040	2.61	65.50		68.11	84.50	105
Inspect joints				.126		7.95		7.95	9.85	12.35
Total				1.166	2.61	73.45		76.06	94.35	117.35
0310 Reglue joint, install 2" tee	10	Q-1	Ea.							
Cut exist pipe, install tee, 2"				1.224	3.80	69.50		73.30	90.50	113
Inspect joints				.126		7.95		7.95	9.85	12.35
Total				1.350	3.80	77.45		81.25	100.35	125.35
1020 Install 10' section PVC 1" diameter	20	1 PLUM	Ea.							
Remove broken pipe				1.131		71.50		71.50	88.50	111
Install 10' new PVC pipe 1" diameter				2.939	117	185.90		302.90	360	430
Inspect joints				.084		5.30		5.30	6.60	8.25
Total				4.154	117	262.70		379.70	455.10	549.25
1120 Install 10' PVC 1-1/4" diameter	20	1 PLUM	Ea.							
Remove broken pipe				1.239		78		78	97	122
Install 10' new PVC pipe 1-1/4" diameter				3.219	127.40	203.45		330.85	390	480
Inspect joints				.084		5.30		5.30	6.60	8.25
Total				4.542	127.40	286.75		414.15	493.60	610.25
1220 Install 10' PVC 1-1/2" diameter	20	1 PLUM	Ea.							
Remove broken pipe				1.445		91		91	113	142
Install 10' new PVC pipe 1-1/2" diameter				3.756	128.05	237.25		365.30	435	530
Inspect joints				.084		5.30		5.30	6.60	8.25
Total				5.284	128.05	333.55		461.60	554.60	680.25
1320 Install 10' section PVC 2" diameter	20	Q-1	Ea.							
Remove broken pipe				1.763		100		100	124	156
Install 10' new PVC pipe 2" diameter				4.583	147.55	260		407.55	485	585
Inspect joints				.084		5.30		5.30	6.60	8.25
Total				6.430	147.55	365.30		512.85	615.60	749.25

For customer support on your Facilities Maintenance & Repair Costs with RSMeans data, call 800.448.8182.

D2023 130 | Solar Piping: Pipe & Fittings, PVC

	System Description	Freq. (Years)	Crew	Unit	Labor Hours	2019 Bare Costs				Total In-House	Total w/O&P
						Material	Labor	Equipment	Total		
2030	**Replace 1000' PVC pipe 1" diameter**	30	1 PLUM	M.L.F.							
	Remove broken pipe				113.080		7,150		7,150	8,850	11,100
	Install 1000' new PVC pipe 1" diameter				226.080	9,000	14,300		23,300	27,600	33,300
	Inspect joints				8.399		530		530	660	825
	Total				347.559	9,000	21,980		30,980	37,110	45,225
2130	**Replace 1000' PVC pipe 1-1/4" diameter**	30	1 PLUM	M.L.F.							
	Remove broken pipe				123.850		7,800		7,800	9,700	12,200
	Install 1000' new PVC pipe 1-1/4" diameter				247.620	9,800	15,650		25,450	30,200	36,800
	Inspect joints				8.399		530		530	660	825
	Total				379.869	9,800	23,980		33,780	40,560	49,825
2230	**Replace 1000' PVC pipe 1-1/2" diameter**	30	1 PLUM	M.L.F.							
	Remove broken pipe				144.490		9,100		9,100	11,300	14,200
	Install 1000' new PVC pipe 1-1/2" diameter				288.890	9,850	18,250		28,100	33,500	40,800
	Inspect joints				8.399		530		530	660	825
	Total				441.779	9,850	27,880		37,730	45,460	55,825
2330	**Replace 1000' PVC pipe 2" diameter**	30	Q-1	M.L.F.							
	Remove broken pipe				176.320		10,000		10,000	12,400	15,600
	Install 1000' new PVC pipe 2" diameter				352.540	11,350	20,000		31,350	37,300	45,200
	Inspect joints				8.399		530		530	660	825
	Total				537.259	11,350	30,530		41,880	50,360	61,625

D2023 150 | Valve, Non-Drain, Less Than 1-1/2"

	System Description	Freq. (Years)	Crew	Unit	Labor Hours	2019 Bare Costs				Total In-House	Total w/O&P
						Material	Labor	Equipment	Total		
0020	**Replace old valve**	10	1 PLUM	Ea.							
	Remove old valve				.400		25.50		25.50	31.50	39
	Install new valve				.800	330	50.50		380.50	425	490
	Total				1.200	330	76		406	456.50	529

For customer support on your Facilities Maintenance & Repair Costs with RSMeans data, call 800.448.8182.

181

D2023 152 Valve, Non-Drain, 2"

System Description	Freq. (Years)	Crew	Unit	Labor Hours	2019 Bare Costs				Total In-House	Total w/O&P
					Material	Labor	Equipment	Total		
0020 Replace old valve	10	1 PLUM	Ea.							
Remove old valve (2")				.800		45.50		45.50	56.50	70.50
Install new valve (to 2")				.727	530	46		576	640	735
Total				1.527	530	91.50		621.50	696.50	805.50

D2023 154 Valve, Non-Drain, 3"

System Description	Freq. (Years)	Crew	Unit	Labor Hours	2019 Bare Costs				Total In-House	Total w/O&P
					Material	Labor	Equipment	Total		
0020 Replace old valve	10	1 PLUM	Ea.							
Remove old valve				.800		45.50		45.50	56.50	70.50
Install new valve				1.600	1,750	91		1,841	2,050	2,325
Total				2.400	1,750	136.50		1,886.50	2,106.50	2,395.50

D2023 156 Valve, Non-Drain, 4" and Larger

System Description	Freq. (Years)	Crew	Unit	Labor Hours	2019 Bare Costs				Total In-House	Total w/O&P
					Material	Labor	Equipment	Total		
0020 Replace old valve, 4"	10	2 PLUM	Ea.							
Remove old flanged gate valve, 4"				5.200		305		305	380	475
Install new flanged gate valve, 4"				7.038	990	415		1,405	1,600	1,875
Total				12.239	990	720		1,710	1,980	2,350
0030 Replace old valve, 6"	10	2 PLUM	Ea.							
Remove old flanged gate valve, 6"				4.615		272		272	335	420
Install new flanged gate valve, 6"				10.399	1,675	615		2,290	2,600	3,050
Total				15.014	1,675	887		2,562	2,935	3,470
0040 Replace old valve, 8"	10	2 PLUM	Ea.							
Remove old flanged gate valve, 8"				3.871		228		228	283	355
Install new flanged gate valve, 8"				12.371	2,900	730		3,630	4,100	4,750
Total				16.242	2,900	958		3,858	4,383	5,105

D20 PLUMBING | D2023 | Domestic Water Distribution

D2023 156 | Valve, Non-Drain, 4" and Larger

System Description	Freq. (Years)	Crew	Unit	2019 Bare Costs					Total In-House	Total w/O&P
				Labor Hours	Material	Labor	Equipment	Total		
0050 Replace old valve, 10"										
Remove old flanged gate valve, 10"	10	3 PLUM	Ea.	3.380		199		199	247	310
Install new flanged gate valve, 10"				14.201	5,100	835		5,935	6,650	7,675
Total				17.581	5,100	1,034		6,134	**6,897**	**7,985**

D2023 160 | Insulation, Pipe

System Description	Freq. (Years)	Crew	Unit	2019 Bare Costs					Total In-House	Total w/O&P
				Labor Hours	Material	Labor	Equipment	Total		
0020 Remove old insulation & replace with new (1/2" pipe size, 1" wall)										
Remove pipe insulation	15	1 PLUM	L.F.	.025		1.56		1.56	1.94	2.43
Install fiberglass pipe insulation (1" wall, 1/2" pipe size)				.087	.88	4.47		5.35	6.70	8.25
Total				.111	.88	6.03		6.91	**8.64**	**10.68**
0030 Remove old insulation & replace with new (3/4" pipe size, 1" wall)										
Remove pipe insulation	15	1 PLUM	L.F.	.025		1.56		1.56	1.94	2.43
Install fiberglass pipe insulation (1" wall, 3/4" pipe size)				.090	.96	4.67		5.63	7.05	8.65
Total				.115	.96	6.23		7.19	**8.99**	**11.08**
0040 Remove old insulation & replace with new (1-1/2" pipe size, 1" wall)										
Remove pipe insulation	15	1 PLUM	L.F.	.025		1.56		1.56	1.94	2.43
Install fiberglass pipe insulation (1" wall, 1-1/2" pipe size)				.099	1.20	5.10		6.30	7.90	9.65
Total				.124	1.20	6.66		7.86	**9.84**	**12.08**
0050 Remove old insulation & replace with new (1/2" pipe size, 3/4" wall)										
Remove pipe insulation	15	1 PLUM	L.F.	.025		1.56		1.56	1.94	2.43
Install flexible rubber tubing (3/4" wall, 1/2" pipe size)				.117	1.11	6.70		7.81	9.80	12.10
Total				.142	1.11	8.26		9.37	**11.74**	**14.53**
0060 Remove old insulation & replace with new (3/4" pipe size, 3/4" wall)										
Remove pipe insulation	15	1 PLUM	L.F.	.025		1.56		1.56	1.94	2.43
Install flexible rubber tubing (3/4" wall, 3/4" pipe size)				.117	1.81	6.70		8.51	10.60	12.95
Total				.142	1.81	8.26		10.07	**12.54**	**15.38**

For customer support on your Facilities Maintenance & Repair Costs with RSMeans data, call 800.448.8182.

183

D2023 160 | Insulation, Pipe

	System Description	Freq. (Years)	Crew	Unit	Labor Hours	2019 Bare Costs				Total In-House	Total w/O&P
						Material	Labor	Equipment	Total		
0070	**Remove old insulation & replace w/new (1-1/2" pipe size, 3/4" wall)**	15	1 PLUM	L.F.							
	Remove pipe insulation				.025		1.56		1.56	1.94	2.43
	Install flexible rubber tubing (3/4" wall, 1-1/2" pipe size)				.120	3.04	6.85		9.89	12.15	14.75
	Total				.144	3.04	8.41		11.45	14.09	17.18

D2023 210 | Water Heater, Gas / Oil, 30 Gallon

	System Description	Freq. (Years)	Crew	Unit	Labor Hours	2019 Bare Costs				Total In-House	Total w/O&P
						Material	Labor	Equipment	Total		
0010	**Overhaul heater**	5	1 PLUM	Ea.							
	Turn valves off and on				.022		1.40		1.40	1.73	2.16
	Remove thermostat, check, and insulate				.182		11.50		11.50	14.25	17.80
	Remove burner, clean and install				.229		14.45		14.45	17.90	22.50
	Clean pilot and thermocouple				.296		18.70		18.70	23	29
	Remove burner door and install				.033		2.05		2.05	2.55	3.19
	Check after maintenance				.216		13.65		13.65	16.95	21
	Drain and fill tank				.615		39		39	48	60.50
	Total				1.593		100.75		100.75	124.38	156.15
0020	**Clean and service**	1	1 PLUM	Ea.							
	Check / adjust fuel burner				2.667		168		168	209	261
	Total				2.667		168		168	209	261
0030	**Replace water heater**	10	2 PLUM	Ea.							
	Remove old heater				2.600		164		164	204	255
	Install 30 gallon gas water heater				5.202	1,775	330		2,105	2,350	2,725
	Check after installation				.216		13.65		13.65	16.95	21
	Total				8.018	1,775	507.65		2,232.65	2,570.95	3,001

For customer support on your Facilities Maintenance & Repair Costs with RSMeans data, call 800.448.8182.

D2023 212 | Water Heater, Gas / Oil, 70 Gallon

	System Description	Freq. (Years)	Crew	Unit	Labor Hours	2019 Bare Costs Material	2019 Bare Costs Labor	2019 Bare Costs Equipment	2019 Bare Costs Total	Total In-House	Total w/O&P
0010	**Overhaul heater**	5	1 PLUM	Ea.							
	Remove hot water heater and insulate				.667		42		42	52	65.50
	Remove thermostat, check and install				.182		11.50		11.50	14.25	17.80
	Remove burner, clean and install				.229		14.45		14.45	17.90	22.50
	Clean pilot and thermocouple				.296		18.70		18.70	23	29
	Remove burner door and install				.033		2.05		2.05	2.55	3.19
	Drain and refill tank				.615		39		39	48	60.50
	Turn valve off and on				.211		13.30		13.30	16.50	20.50
	Total				2.232		141		141	174.20	218.99
0020	**Clean & service**	1	1 PLUM	Ea.							
	Check / adjust fuel burner				2.667		168		168	209	261
	Total				2.667		168		168	209	261
0030	**Replace hot water heater**	12	2 PLUM	Ea.							
	Remove old heater				2.600		164		164	204	255
	Install water heater, oil fired, 70 gallon				6.932	2,275	440		2,715	3,050	3,525
	Total				9.532	2,275	604		2,879	3,254	3,780

D2023 214 | Water Heater, Gas / Oil, 1150 GPH

	System Description	Freq. (Years)	Crew	Unit	Labor Hours	2019 Bare Costs Material	2019 Bare Costs Labor	2019 Bare Costs Equipment	2019 Bare Costs Total	Total In-House	Total w/O&P
0010	**Minor repairs, adjustments**	2	2 PLUM	Ea.							
	Reset, adjust controls				1.333		84		84	104	131
	Total				1.333		84		84	104	131
0020	**Clean & service**	2	2 PLUM	Ea.							
	Drain tank, clean, and service				8.000		505		505	625	785
	Total				8.000		505		505	625	785
0030	**Replace hot water heater**	20	2 PLUM	Ea.							
	Remove water heater				26.016		1,475		1,475	1,825	2,300
	Install water heater, gas fired, 1150 GPH				51.948	32,300	2,950		35,250	39,200	45,000
	Total				77.964	32,300	4,425		36,725	41,025	47,300

For customer support on your Facilities Maintenance & Repair Costs with RSMeans data, call 800.448.8182.

185

D2023 220 Water Heater, Electric, 120 Gallon

	System Description	Freq. (Years)	Crew	Unit	Labor Hours	2019 Bare Costs				Total In-House	Total w/O&P
						Material	Labor	Equipment	Total		
0010	**Drain and flush**	7	1 PLUM	Ea.							
	Drain, clean and readjust				4.000		253		253	315	390
	Total				4.000		253		253	315	390
0020	**Check operation**	3	1 PLUM	Ea.							
	Check for relief valve				.010		.66		.66	.81	1.02
	Check for proper water temperature				.003		.16		.16	.20	.25
	Fill out maintenance report				.022		1.40		1.40	1.73	2.16
	Total				.035		2.22		2.22	2.74	3.43
0030	**Replace water heater**	15	2 PLUM	Ea.							
	Remove old heater				4.334		274		274	340	425
	Install water heater				8.667	13,900	545		14445	16,000	18,200
	Total				13.001	13,900	819		14719	16,340	18,625

D2023 222 Water Heater, Electric, 300 Gallon

	System Description	Freq. (Years)	Crew	Unit	Labor Hours	2019 Bare Costs				Total In-House	Total w/O&P
						Material	Labor	Equipment	Total		
0010	**Drain and flush**	7	1 PLUM	Ea.							
	Drain, clean and adjust				4.000		253		253	315	390
	Total				4.000		253		253	315	390
0020	**Check operation**	3	1 PLUM	Ea.							
	Check relief valve				.010		.66		.66	.81	1.02
	Check for proper water temperature				.003		.16		.16	.20	.25
	Fill out maintenance report				.022		1.40		1.40	1.73	2.16
	Total				.035		2.22		2.22	2.74	3.43
0030	**Replace water heater**	15	2 PLUM	Ea.							
	Remove old heater				8.000		455		455	565	705
	Install water heater				16.000	57,000	910		57,910	64,000	72,500
	Total				24.000	57,000	1,365		58,365	64,565	73,205

For customer support on your Facilities Maintenance & Repair Costs with RSMeans data, call 800.448.8182.

D2023 | Domestic Water Distribution

D2023 224 | Water Heater, Electric, 1000 Gallon

System Description	Freq. (Years)	Crew	Unit	Labor Hours	2019 Bare Costs				Total In-House	Total w/O&P
					Material	Labor	Equipment	Total		
0010 Drain and flush										
Drain, clean and adjust	7	1 PLUM	Ea.	4.000		253		253	315	390
Total				4.000		253		253	**315**	**390**
0020 Check operation										
Check relief valve	3	1 PLUM	Ea.	.010		.66		.66	.81	1.02
Check for proper water temperature				.003		.16		.16	.20	.25
Fill out maintenance report				.022		1.40		1.40	1.73	2.16
Total				.035		2.22		2.22	**2.74**	**3.43**
0030 Replace water heater										
Remove heater	15	2 PLUM	Ea.	22.284		1,325		1,325	1,625	2,025
Install water heater				44.610	133,000	2,625		135,625	149,500	170,500
Total				66.894	133,000	3,950		136,950	**151,125**	**172,525**

D2023 226 | Water Heater, Electric, 2000 Gallon

System Description	Freq. (Years)	Crew	Unit	Labor Hours	2019 Bare Costs				Total In-House	Total w/O&P
					Material	Labor	Equipment	Total		
0010 Drain and flush										
Drain, clean and adjust	7	1 PLUM	Ea.	4.000		253		253	315	390
Total				4.000		253		253	**315**	**390**
0020 Check operation										
Check relief valve	3	1 PLUM	Ea.	.010		.66		.66	.81	1.02
Check for proper water temperature				.003		.16		.16	.20	.25
Fill out maintenance report				.022		1.40		1.40	1.73	2.16
Total				.035		2.22		2.22	**2.74**	**3.43**
0030 Replace water heater										
Remove heater	15	2 PLUM	Ea.	31.209		1,850		1,850	2,275	2,850
Install water heater				62.338	165,500	3,675		169,175	186,500	212,500
Total				93.547	165,500	5,525		171,025	**188,775**	**215,350**

For customer support on your Facilities Maintenance & Repair Costs with RSMeans data, call 800.448.8182.

187

D20 PLUMBING | D2023 | Domestic Water Distribution

D2023 230 | Steam Converter, Domestic Hot Water

System Description	Freq. (Years)	Crew	Unit	Labor Hours	2019 Bare Costs				Total In-House	Total w/O&P
					Material	Labor	Equipment	Total		
0020 **Inspect for leaks**	1	1 PLUM	Ea.							
Check relief valve operation				.038		2.38		2.38	2.96	3.70
Check for proper water temperature				.054		3.44		3.44	4.26	5.35
Total				.092		5.82		5.82	7.22	9.05
0030 **Replace steam converter**	20	2 PLUM	Ea.							
Remove heat exchanger				1.733		109		109	136	170
Install new heat exchanger				3.467	2,525	200		2,725	3,025	3,475
Total				5.200	2,525	309		2,834	3,161	3,645

D2023 240 | Storage Tank, Domestic Hot Water

System Description	Freq. (Years)	Crew	Unit	Labor Hours	2019 Bare Costs				Total In-House	Total w/O&P
					Material	Labor	Equipment	Total		
0020 **Replace storage tank, glass lined, P.E., 80 gal.**	50	2 PLUM	Ea.							
Remove hot water storage tank				.578		36.50		36.50	45.50	56.50
Install new glass lined, P.E., 80 Gal. hot water storage tank				1.156	4,225	73		4,298	4,750	5,400
Total				1.733	4,225	109.50		4,334.50	4,795.50	5,456.50

D2023 245 | Solar Storage Tank, 1000 Gallon

System Description	Freq. (Years)	Crew	Unit	Labor Hours	2019 Bare Costs				Total In-House	Total w/O&P
					Material	Labor	Equipment	Total		
0010 **Replace 1000 gallon solar storage tank**	20	Q-9	Ea.							
Remove tank				8.000		440		440	550	690
Install 1000 gallon solar storage tank				16.000	7,425	880		8,305	9,275	10,700
Total				24.000	7,425	1,320		8,745	9,825	11,390

For customer support on your Facilities Maintenance & Repair Costs with RSMeans data, call 800.448.8182.

D2023 250 Expansion Chamber

	System Description	Freq. (Years)	Crew	Unit	Labor Hours	2019 Bare Costs				Total In-House	Total w/O&P
						Material	Labor	Equipment	Total		
0010	**Refill**										
	Refill air chamber	5	1 PLUM	Ea.	.039		2.46		2.46	3.06	3.82
	Total				.039		2.46		2.46	**3.06**	**3.82**
0020	**Remove old chamber, install new**										
	Remove / install expansion device, 15 gal.	10	1 PLUM	Ea.	.941	790	54		844	935	1,075
	Total				.941	790	54		844	**935**	**1,075**

D2023 260 Circulation Pump, 1/12 HP

	System Description	Freq. (Years)	Crew	Unit	Labor Hours	2019 Bare Costs				Total In-House	Total w/O&P
						Material	Labor	Equipment	Total		
0020	**Pump maintenance, 1/12 H.P.**	0.50	1 PLUM	Ea.							
	Check for leaks				.010		.66		.66	.81	1.02
	Check pump operation				.012		.74		.74	.92	1.15
	Lubricate pump and motor				.047		2.95		2.95	3.67	4.58
	Check suction or discharge pressure				.004		.25		.25	.31	.38
	Check packing glands, tighten				.014		.90		.90	1.12	1.40
	Fill out inspection report				.022		1.40		1.40	1.73	2.16
	Total				.109		6.90		6.90	**8.56**	**10.69**
0030	**Replace pump / motor assembly**	10	2 PLUM	Ea.							
	Remove flanged connection pump				1.733		98.50		98.50	122	153
	Install new flanged connection pump (1/12 H.P.)				3.467	755	197		952	1,075	1,250
	Total				5.200	755	295.50		1,050.50	**1,197**	**1,403**

For customer support on your Facilities Maintenance & Repair Costs with RSMeans data, call 800.448.8182.

189

D20 PLUMBING | D2023 | Domestic Water Distribution

D2023 261 Circulation Pump, 1/8 HP

					2019 Bare Costs				Total In-House	Total w/O&P
System Description	Freq. (Years)	Crew	Unit	Labor Hours	Material	Labor	Equipment	Total		
0020 Pump maintenance, 1/8 H.P.	1	1 PLUM	Ea.							
Check for leaks				.010		.66		.66	.81	1.02
Check pump operation				.012		.74		.74	.92	1.15
Lubricate pump and motor				.047		2.95		2.95	3.67	4.58
Check suction or discharge pressure				.004		.25		.25	.31	.38
Check packing glands, tighten				.014		.90		.90	1.12	1.40
Fill out inspection report				.022		1.40		1.40	1.73	2.16
Total				.109		6.90		6.90	8.56	10.69
0030 Replace pump / motor assembly	20	2 PLUM	Ea.							
Remove flanged connection pump				1.733		98.50		98.50	122	153
Install new flanged connection pump (1/8 H.P.)				3.467	1,325	197		1,522	1,700	1,950
Total				5.200	1,325	295.50		1,820.50	1,822	2,103

D2023 262 Circulation Pump, 1/6 HP

					2019 Bare Costs				Total In-House	Total w/O&P
System Description	Freq. (Years)	Crew	Unit	Labor Hours	Material	Labor	Equipment	Total		
0020 Pump maintenance, 1/6 H.P.	1	1 PLUM	Ea.							
Check for leaks				.010		.66		.66	.81	1.02
Check pump operation				.012		.74		.74	.92	1.15
Lubricate pump and motor				.047		2.95		2.95	3.67	4.58
Check suction or discharge pressure				.004		.25		.25	.31	.38
Check packing glands, tighten				.014		.90		.90	1.12	1.40
Fill out inspection report				.022		1.40		1.40	1.73	2.16
Total				.109		6.90		6.90	8.56	10.69
0030 Replace pump / motor assembly	20	2 PLUM	Ea.							
Remove flanged connection pump				2.080		131		131	163	204
Install new flanged connection pump (1/6 H.P.)				4.160	1,650	236		1,886	2,100	2,425
Total				6.240	1,650	367		2,017	2,263	2,629

D2023 264   Circulation Pump, 1/2 HP

	System Description	Freq. (Years)	Crew	Unit	Labor Hours	2019 Bare Costs Material	2019 Bare Costs Labor	2019 Bare Costs Equipment	2019 Bare Costs Total	Total In-House	Total w/O&P
0020	**Pump maintenance, 1/2 H.P.**	1	1 PLUM	Ea.							
	Check for leaks				.010		.66		.66	.81	1.02
	Check pump operation				.012		.74		.74	.92	1.15
	Check alignment				.053		3.37		3.37	4.18	5.20
	Lubricate pump and motor				.047		2.95		2.95	3.67	4.58
	Check suction or discharge pressure				.004		.25		.25	.31	.38
	Check packing glands, tighten				.014		.90		.90	1.12	1.40
	Fill out inspection report				.022		1.40		1.40	1.73	2.16
	Total				.163		10.27		10.27	12.74	15.89
0030	**Replace pump / motor assembly, partial**	20	2 PLUM	Ea.							
	Remove flanged connection pump				2.600		148		148	183	229
	Install new flanged connection pump (1/2 H.P.)				5.200	3,025	296		3,321	3,700	4,250
	Total				7.800	3,025	444		3,469	3,883	4,479

D2023 266   Circulation Pump, Bronze 1 HP

	System Description	Freq. (Years)	Crew	Unit	Labor Hours	2019 Bare Costs Material	2019 Bare Costs Labor	2019 Bare Costs Equipment	2019 Bare Costs Total	Total In-House	Total w/O&P
0020	**Pump maintenance, 1 H.P.**	0.50	1 PLUM	Ea.							
	Check for leaks				.010		.66		.66	.81	1.02
	Check pump operation				.012		.74		.74	.92	1.15
	Lubricate pump and motor				.047		2.95		2.95	3.67	4.58
	Check suction or discharge pressure				.004		.25		.25	.31	.38
	Check packing glands, tighten				.014		.90		.90	1.12	1.40
	Fill out inspection report				.022		1.40		1.40	1.73	2.16
	Total				.109		6.90		6.90	8.56	10.69
0040	**Replace pump / motor assembly**	20	2 PLUM	Ea.							
	Remove pump / motor				2.600		148		148	183	229
	Install new flanged connection pump, bronze (1 H.P.)				5.200	4,800	296		5,096	5,650	6,450
	Total				7.800	4,800	444		5,244	5,833	6,679

D2023 267   Circulation Pump, C.I. 1-1/2 HP

	System Description	Freq. (Years)	Crew	Unit	Labor Hours	2019 Bare Costs				Total In-House	Total w/O&P
						Material	Labor	Equipment	Total		
0020	**Pump maintenance, 1-1/2 H.P.**	0.50	1 PLUM	Ea.							
	Check for leaks				.010		.66		.66	.81	1.02
	Check pump operation				.012		.74		.74	.92	1.15
	Check alignment				.053		3.37		3.37	4.18	5.20
	Lubricate pump and motor				.047		2.95		2.95	3.67	4.58
	Check suction / discharge pressure				.004		.25		.25	.31	.38
	Check packing glands, tighten				.014		.90		.90	1.12	1.40
	Fill out inspection report				.022		1.40		1.40	1.73	2.16
	Total				.163		10.27		10.27	12.74	15.89
0040	**Replace pump / motor assembly**	20	2 PLUM	Ea.							
	Remove flanged connection pump				2.600		148		148	183	229
	Install new flanged connection pump, C.I. (1-1/2 H.P.)				5.200	2,250	296		2,546	2,850	3,275
	Total				7.800	2,250	444		2,694	3,033	3,504

D2023 268   Circulation Pump, 3 HP

	System Description	Freq. (Years)	Crew	Unit	Labor Hours	2019 Bare Costs				Total In-House	Total w/O&P
						Material	Labor	Equipment	Total		
0020	**Pump maintenance, 3 H.P.**	0.50	Q-1	Ea.							
	Check for leaks				.012		.73		.73	.90	1.13
	Check pump operation				.012		.74		.74	.92	1.15
	Check alignment				.059		3.74		3.74	4.64	5.80
	Lubricate pump and motor				.052		3.28		3.28	4.08	5.10
	Check suction / discharge pressure				.004		.25		.25	.31	.38
	Check packing glands, tighten				.016		1		1	1.24	1.56
	Fill out inspection report				.022		1.40		1.40	1.73	2.16
	Total				.176		11.14		11.14	13.82	17.28
0040	**Replace pump / motor assembly**	20	Q-1	Ea.							
	Remove flanged connection pump				4.400		249.70		249.70	310	390
	Install new flanged connection pump, 3 HP				8.889	9,125	505		9,630	10,700	12,200
	Total				13.289	9,125	754.70		9,879.70	11,010	12,590

D20 PLUMBING | D2023 | Domestic Water Distribution

D2023 310 | Hose Bibb

	System Description	Freq. (Years)	Crew	Unit	Labor Hours	2019 Bare Costs Material	Labor	Equipment	Total	Total In-House	Total w/O&P
0020	**Replace old valve with new**	10	1 PLUM	Ea.							
	Remove hose bibb				.217		13.70		13.70	17	21
	Install new hose bibb				.433	10.60	27.50		38.10	45.50	56
	Total				.650	10.60	41.20		51.80	62.50	77

D2023 320 | Water Meter

	System Description	Freq. (Years)	Crew	Unit	Labor Hours	2019 Bare Costs Material	Labor	Equipment	Total	Total In-House	Total w/O&P
0010	**Overhaul**	13	1 PLUM	Ea.							
	Open meter, replace worn parts				.300	1.80	18.95		20.75	25.50	32
	Total				.300	1.80	18.95		20.75	25.50	32
0020	**Remove old 5/8" meter, install new**	25	1 PLUM	Ea.							
	Remove 5/8" water meter				.325		20.50		20.50	25.50	32
	Install 5/8" threaded water meter				.650	51.50	41		92.50	108	128
	Total				.975	51.50	61.50		113	133.50	160
0025	**Remove old 3/4" meter, install new**	25	1 PLUM	Ea.							
	Remove 3/4" water meter				.371		23.50		23.50	29	36.50
	Install 3/4" threaded water meter				.743	94	47		141	162	191
	Total				1.114	94	70.50		164.50	191	227.50
0030	**Remove old 1" meter, install new**	25	1 PLUM	Ea.							
	Remove 1" water meter				.433		27.50		27.50	34	42.50
	Install 1" threaded water meter				.867	143	54.50		197.50	225	264
	Total				1.300	143	82		225	259	306.50
0035	**Remove old 1-1/2" meter, install new**	25	1 PLUM	Ea.							
	Remove 1-1/2" water meter				.650		41		41	51	63.50
	Install 1-1/2" threaded water meter				1.300	350	82		432	485	565
	Total				1.950	350	123		473	536	628.50
0040	**Remove old 2" meter, install new**	25	1 PLUM	Ea.							
	Remove 2" water meter				.867		54.50		54.50	68	85
	Install 2" threaded water meter				1.733	475	109		584	660	765
	Total				2.600	475	163.50		638.50	728	850

For customer support on your Facilities Maintenance & Repair Costs with RSMeans data, call 800.448.8182.

193

D20 PLUMBING | D2023 | Domestic Water Distribution

D2023 320 | Water Meter

System Description	Freq. (Years)	Crew	Unit	Labor Hours	2019 Bare Costs				Total In-House	Total w/O&P
					Material	Labor	Equipment	Total		
0045										
Remove old 3" meter, install new	25	Q-1	Ea.							
Remove 3" water meter				3.467		197		197	244	305
Install 3" flanged water meter				6.932	2,425	395		2,820	3,150	3,650
Total				10.399	2,425	592		3,017	3,394	**3,955**
0050										
Remove old 4" meter, install new	25	Q-1	Ea.							
Remove 4" water meter				6.932		395		395	490	610
Install 4" flanged water meter				13.865	3,875	790		4,665	5,250	6,075
Total				20.797	3,875	1,185		5,060	5,740	**6,685**
0055										
Remove old 6" meter, install new	25	Q-1	Ea.							
Remove 6" water meter				10.403		590		590	735	915
Install 6" flanged water meter				20.806	6,275	1,175		7,450	8,375	9,675
Total				31.209	6,275	1,765		8,040	9,110	**10,590**
0060										
Remove old 8" meter, install new	25	Q-1	Ea.							
Remove 8" water meter				12.998		740		740	915	1,150
Install 8" flanged water meter				26.016	9,800	1,475		11,275	12,600	14,600
Total				39.014	9,800	2,215		12,015	13,515	**15,750**

D2023 370 | Water Softener

System Description	Freq. (Years)	Crew	Unit	Labor Hours	2019 Bare Costs				Total In-House	Total w/O&P
					Material	Labor	Equipment	Total		
0030										
Replace softener	15	2 PLUM	Ea.							
Remove old unit				2.600		164		164	204	255
Install new water softener				5.200	855	330		1,185	1,350	1,575
Total				7.800	855	494		1,349	1,554	**1,830**

For customer support on your Facilities Maintenance & Repair Costs with RSMeans data, call 800.448.8182.

D20 PLUMBING

D2033 110   D2033   Sanitary Waste

Pipe & Fittings, Cast Iron

	System Description	Freq. (Years)	Crew	Unit	Labor Hours	2019 Bare Costs				Total In-House	Total w/O&P
						Material	Labor	Equipment	Total		
0020	**Unclog main drain**	10	1 PLUM	Ea.							
	Unclog main drain using auger				.630		40		40	49.50	62
	Total				.630		40		40	**49.50**	**62**
0030	**Replace cast iron pipe, 4"**	40	2 PLUM	L.F.							
	Remove old pipe				.189		10.75		10.75	13.30	16.65
	Install new 4" cast iron pipe				.378	24.50	21.50		46	53.50	64
	Total				.567	24.50	32.25		56.75	**66.80**	**80.65**

D2033 130   Pipe & Fittings, PVC

	System Description	Freq. (Years)	Crew	Unit	Labor Hours	2019 Bare Costs				Total In-House	Total w/O&P
						Material	Labor	Equipment	Total		
0010	**Unclog floor drain**	20	1 PLUM	Ea.							
	Unclog plugged floor drain				.650		41		41	51	63.50
	Total				.650		41		41	**51**	**63.50**
0020	**Unclog 4" - 12" diameter PVC main drain per L.F.**	10	1 PLUM	L.F.							
	Unclog main drain using auger				.052		3.26		3.26	4.04	5.05
	Total				.052		3.26		3.26	**4.04**	**5.05**
0040	**Repair joint**	10	1 PLUM	Ea.							
	Remove plastic fitting (2 connections)				.650		37		37	46	57.50
	Connect fitting (2 connections)				1.300	9.20	74		83.20	102	127
	Inspect joints				.163		10.30		10.30	12.80	16
	Total				2.113	9.20	121.30		130.50	**160.80**	**200.50**
0060	**Replace pipe, 1-1/2"**	30	2 PLUM	L.F.							
	Remove broken pipe				.145		8.55		8.55	10.60	13.25
	Install new pipe, couplings, hangers				.289	9.85	18.25		28.10	33.50	41
	Inspect joints				.327		20.60		20.60	25.50	32
	Total				.760	9.85	47.40		57.25	**69.60**	**86.25**

D2033 130 Pipe & Fittings, PVC

	System Description	Freq. (Years)	Crew	Unit	Labor Hours	2019 Bare Costs				Total In-House	Total w/O&P
						Material	Labor	Equipment	Total		
0080	**Replace pipe, 2"**	30	2 PLUM	L.F.							
	Remove broken pipe				.176		10		10	12.45	15.55
	Install new pipe, couplings, hangers				.353	11.35	20		31.35	37.50	45
	Inspect joints				.327		20.60		20.60	25.50	32
	Total				.855	11.35	50.60		61.95	75.45	92.55
0100	**Replace pipe, 4"**	30	2 PLUM	L.F.							
	Remove broken pipe				.217		12.30		12.30	15.30	19.10
	Install new pipe, couplings, hangers				.433	29.50	24.50		54	63	75
	Inspect joints				.327		20.60		20.60	25.50	32
	Total				.977	29.50	57.40		86.90	103.80	126.10
0120	**Replace pipe, 6"**	30	2 PLUM	L.F.							
	Remove broken pipe				.267		15.15		15.15	18.80	23.50
	New pipe, couplings, hangers				.533	38.50	33.50		72	84	100
	Inspect joints				.327		20.60		20.60	25.50	32
	Total				1.126	38.50	69.25		107.75	128.30	155.50
0140	**Replace pipe, 8"**	30	2 PLUM	L.F.							
	Remove broken pipe				.325		19.15		19.15	24	29.50
	Install new pipe, couplings, hangers				.650	35.50	38.50		74	86.50	104
	Inspect joints				.327		20.60		20.60	25.50	32
	Total				1.302	35.50	78.25		113.75	136	165.50

D2033 305 Pipe & Fittings

	System Description	Freq. (Years)	Crew	Unit	Labor Hours	2019 Bare Costs				Total In-House	Total w/O&P
						Material	Labor	Equipment	Total		
3010	**Unclog floor drain**	10	1 PLUM	Ea.							
	Unclog plugged floor drain				4.000		256		256	315	395
	Solvent					34			34	37.50	42.50
	Total				4.000	34	256		290	352.50	437.50
3020	**Unclog 4" - 12" diameter main drain per L.F.**	10	2 PLUM	L.F.							
	Unclog main drain using auger				.052		3.26		3.26	4.04	5.05
	Total				.052		3.26		3.26	4.04	5.05

D2033 310 Floor Drain W/O Bucket

	System Description	Freq. (Years)	Crew	Unit	Labor Hours	2019 Bare Costs				Total In-House	Total w/O&P
						Material	Labor	Equipment	Total		
0010	**Clean drain**	4	1 PLUM	Ea.							
	Remove debris from drain				1.590		100		100	125	156
	Total				1.590		100		100	**125**	**156**
0030	**Replace floor drain**										
	Remove floor drain				1.040		65.50		65.50	81.50	102
	Install floor drain	40	1 PLUM	Ea.	2.080	1,150	118		1,268	1,400	1,625
	Total				3.120	1,150	183.50		1,333.50	**1,481.50**	**1,727**

D2033 330 Floor Drain With Bucket

	System Description	Freq. (Years)	Crew	Unit	Labor Hours	2019 Bare Costs				Total In-House	Total w/O&P
						Material	Labor	Equipment	Total		
0010	**Clean out bucket**	5	1 PLUM	Ea.							
	Remove debris from bucket				4.000		253		253	315	390
	Total				4.000		253		253	**315**	**390**
0030	**Replace floor drain**										
	Remove floor drain				1.040		65.50		65.50	81.50	102
	Install floor drain	40	1 PLUM	Ea.	2.080	1,150	118		1,268	1,400	1,625
	Total				3.120	1,150	183.50		1,333.50	**1,481.50**	**1,727**

D2043 110 — Distribution: Gutters, Pipe

	System Description	Freq. (Years)	Crew	Unit	Labor Hours	2019 Bare Costs				Total In-House	Total w/O&P
						Material	Labor	Equipment	Total		
1010	General maintenance & repair	1	1 PLUM	M.L.F.							
	Inspect rainwater pipes				.104		6.55		6.55	8.15	10.20
	Clean out pipes				4.000		253		253	315	390
	Total				4.104		259.55		259.55	323.15	400.20
1020	Replace pipe or gutter	20	1 PLUM	L.F.							
	Remove 4" PVC roof drain system				.217		12.30		12.30	15.30	19.10
	Install 4" roof drain system				.333	10.45	18.95		29.40	35	42.50
	Total				.550	10.45	31.25		41.70	50.30	61.60

D2043 210 — Drain: Roof, Scupper, Area

	System Description	Freq. (Years)	Crew	Unit	Labor Hours	2019 Bare Costs				Total In-House	Total w/O&P
						Material	Labor	Equipment	Total		
1010	General maintenance & repair	1	1 PLUM	Ea.							
	Inspect drain				.104		6.55		6.55	8.15	10.20
	Clean drain				.400		25.50		25.50	31.50	39
	Total				.504		32.05		32.05	39.65	49.20
1020	Replace drain	40	1 PLUM	Ea.							
	Remove drain				.800		45.50		45.50	56.50	70.50
	Install roof drain				1.600	470	91		561	630	730
	Total				2.400	470	136.50		606.50	686.50	800.50

For customer support on your Facilities Maintenance & Repair Costs with RSMeans data, call 800.448.8182.

D2043 310 Rainwater Sump Pump

System Description	Freq. (Years)	Crew	Unit	Labor Hours	2019 Bare Costs Material	Labor	Equipment	Total	Total In-House	Total w/O&P
1020 Pump maintenance, 87 GPM	1	1 PLUM	Ea.							
Check for leaks				.010		.66		.66	.81	1.02
Check pump operation				.012		.74		.74	.92	1.15
Check alignment				.053		3.37		3.37	4.18	5.20
Lubricate pump and motor				.047		2.95		2.95	3.67	4.58
Check suction or discharge pressure				.004		.25		.25	.31	.38
Check packing glands, tighten				.014		.90		.90	1.12	1.40
Fill out inspection report				.022		1.40		1.40	1.73	2.16
Total				.163		10.27		10.27	12.74	15.89
1030 Replace sump pump / motor assembly	20	2 PLUM	Ea.							
Remove sump pump assembly				1.300		82		82	102	127
Install new cast iron sump pump (87 GPM)				2.078	305	131		436	500	585
Total				3.378	305	213		518	602	712

D2093 910 Pipe & Fittings, Industrial Gas

	System Description	Freq. (Years)	Crew	Unit	Labor Hours	2019 Bare Costs				Total In-House	Total w/O&P
						Material	Labor	Equipment	Total		
1010	**General maintenance**	2	1 PLUM	M.L.F.							
	General maintenance				.500		31.50		31.50	39	49
	Total				.500		31.50		31.50	**39**	**49**
1030	**Replace pipe and fittings**	75	2 PLUM	L.F.							
	Shut off valve				1.067		67.50		67.50	83.50	104
	Remove old pipe				.162		9.20		9.20	11.40	14.30
	Install 2" black steel pipe, couplings, and hangers				.325	11.55	18.45		30	35.50	43
	Check for leaks				.889		56		56	69.50	87
	Total				2.443	11.55	151.15		162.70	**199.90**	**248.30**

D2093 920 Pipe & Fittings, Anesthesia

	System Description	Freq. (Years)	Crew	Unit	Labor Hours	2019 Bare Costs				Total In-House	Total w/O&P
						Material	Labor	Equipment	Total		
1010	**Resolder joint**	12	1 PLUM	Ea.							
	Measure, cut & clean both ends				.053		3.37		3.37	4.18	5.20
	Attach fitting				.578	4.94	36.50		41.44	50.50	62.50
	Total				.631	4.94	39.87		44.81	**54.68**	**67.70**
1030	**Replace pipe and fittings**	25	2 PLUM	L.F.							
	Remove old pipe				.069		4.38		4.38	5.45	6.80
	Install new 3/4" copper tubing, couplings, hangers				.141	8.55	8.90		17.45	20.50	24.50
	Total				.210	8.55	13.28		21.83	**25.95**	**31.30**

D2093 930 Pipe & Fittings, Oxygen

	System Description	Freq. (Years)	Crew	Unit	Labor Hours	2019 Bare Costs				Total In-House	Total w/O&P
						Material	Labor	Equipment	Total		
1010	**Resolder joint**	12	1 PLUM	Ea.							
	Measure, cut & clean both ends				.289		18.25		18.25	22.50	28.50
	Attach fitting				.578	4.94	36.50		41.44	50.50	62.50
	Total				.867	4.94	54.75		59.69	**73**	**91**

For customer support on your Facilities Maintenance & Repair Costs with RSMeans data, call 800.448.8182.

D20 PLUMBING | D2093 | Other Plumbing Systems

D2093 930 | Pipe & Fittings, Oxygen

System Description	Freq. (Years)	Crew	Unit	Labor Hours	2019 Bare Costs				Total In-House	Total w/O&P
					Material	Labor	Equipment	Total		
1030 **Replace pipe and fittings**	25	2 PLUM	L.F.							
Remove old pipe				.069		4.38		4.38	5.45	6.80
Install new 3/4" copper tubing, couplings, hangers				.141	8.55	8.90		17.45	20.50	24.50
Total				.210	8.55	13.28		21.83	25.95	31.30

D2093 940 | Pipe & Fittings, Compressed Air

System Description	Freq. (Years)	Crew	Unit	Labor Hours	2019 Bare Costs				Total In-House	Total w/O&P
					Material	Labor	Equipment	Total		
1010 **General maintenance**	2	1 PLUM	M.L.F.							
General maintenance				.500		31.50		31.50	39	49
Total				.500		31.50		31.50	39	49
1030 **Replace pipe and fittings**	75	2 PLUM	L.F.							
Shut off valve				1.067		67.50		67.50	83.50	104
Remove old pipe				.052		3.28		3.28	4.07	5.10
Install 2" black steel pipe, couplings, and hangers				.325	11.55	18.45		30	35.50	43
Check for leaks				.889		56		56	69.50	87
Total				2.333	11.55	145.23		156.78	192.57	239.10

For customer support on your Facilities Maintenance & Repair Costs with RSMeans data, call 800.448.8182.

201

D2093 946 Compressed Air Systems, Compressors

System Description	Freq. (Years)	Crew	Unit	Labor Hours	2019 Bare Costs				Total In-House	Total w/O&P
					Material	Labor	Equipment	Total		
1004										
Check and adjust 3/4 H.P. compressor	1	1 SPRI	Ea.							
Check compressor oil level				.022		1.40		1.40	1.73	2.16
Check V-belt tension				.038		2.38		2.38	2.96	3.70
Drain moisture from air tank				.033		2.05		2.05	2.55	3.19
Clean air intake filter on compressor				.178		11.25		11.25	13.95	17.40
Clean reusable oil filter				.182		11.50		11.50	14.25	17.80
Check pressure relief valve				.108		6.85		6.85	8.45	10.60
Clean cooling fins				.105		6.65		6.65	8.25	10.30
Check motor				.060		3.77		3.77	4.68	5.85
Check control circuits				.182		11.50		11.50	14.25	17.80
Operation checks				.222		14.05		14.05	17.40	22
Fill out maintenance report				.022		1.40		1.40	1.73	2.16
Total				1.151		72.80		72.80	90.20	112.96
1006										
Replace 3/4 H.P. compressor	25	1 SPRI	Ea.							
Remove air compressor				4.457		266		266	330	410
Install new 3/4 H.P. compressor				6.250	1,175	385		1,560	1,775	2,075
Total				10.707	1,175	651		1,826	2,105	2,485
1010										
Check and adjust 2 H.P. compressor	1	1 PLUM	Ea.							
Check compressor oil level				.022		1.40		1.40	1.73	2.16
Check V-belt tension				.038		2.38		2.38	2.96	3.70
Drain moisture from air tank				.033		2.05		2.05	2.55	3.19
Clean air intake filter on compressor				.178		11.25		11.25	13.95	17.40
Clean reusable oil filter				.182		11.50		11.50	14.25	17.80
Check pressure relief valve				.108		6.85		6.85	8.45	10.60
Clean cooling fins				.105		6.65		6.65	8.25	10.30
Check motor				.060		3.77		3.77	4.68	5.85
Check control circuits				.182		11.50		11.50	14.25	17.80
Operation checks				.222		14.05		14.05	17.40	22
Fill out maintenance report				.022		1.40		1.40	1.73	2.16
Total				1.151		72.80		72.80	90.20	112.96
1030										
Replace 2 H.P. compressor	25	2 PLUM	Ea.							
Remove air compressor				4.457		266		266	330	410
Install new 2 H.P. compressor				8.915	3,150	530		3,680	4,125	4,775
Total				13.372	3,150	796		3,946	4,455	5,185

System Description	Freq. (Years)	Crew	Unit	Labor Hours	2019 Bare Costs Material	Labor	Equipment	Total	Total In-House	Total w/O&P
3010										
Check and adjust 10 H.P. compressor	1	1 PLUM	Ea.							
Check compressor oil level				.022		1.40		1.40	1.73	2.16
Check tension on V-belt				.038		2.38		2.38	2.96	3.70
Drain moisture from air tank				.033		2.05		2.05	2.55	3.19
Clean air intake filter on compressor				.178		11.25		11.25	13.95	17.40
Clean reusable oil filter				.182		11.50		11.50	14.25	17.80
Check operation of pressure valve				.108		6.85		6.85	8.45	10.60
Clean cooling fins				.105		6.65		6.65	8.25	10.30
Check motor				.060		3.77		3.77	4.68	5.85
Check compressor circuits				.182		11.50		11.50	14.25	17.80
Perform operation				.222		14.05		14.05	17.40	22
Fill out maintenance report				.022		1.40		1.40	1.73	2.16
Total				1.151		72.80		72.80	90.20	112.96
3030										
Replace 10 H.P. compressor	25	2 PLUM	Ea.							
Remove compressor				17.335		1,000		1,000	1,250	1,550
Install 2-stage compressor 10 H.P.				34.632	6,225	2,000		8,225	9,325	10,900
Total				51.967	6,225	3,000		9,225	10,575	12,450
4010										
Check and adjust 25 H.P. compressor	1	1 PLUM	Ea.							
Check compressor oil level				.022		1.40		1.40	1.73	2.16
Check tension on V-belt				.038		2.38		2.38	2.96	3.70
Drain moisture from air tank				.033		2.05		2.05	2.55	3.19
Clean air intake filter on compressor				.178		11.25		11.25	13.95	17.40
Clean reusable oil filter				.182		11.50		11.50	14.25	17.80
Check operation of pressure valve				.108		6.85		6.85	8.45	10.60
Clean cooling fins				.105		6.65		6.65	8.25	10.30
Check motor				.060		3.77		3.77	4.68	5.85
Check compressor circuits				.182		11.50		11.50	14.25	17.80
Perform operation				.222		14.05		14.05	17.40	22
Fill out maintenance report				.022		1.40		1.40	1.73	2.16
Total				1.151		72.80		72.80	90.20	112.96
4030										
Replace 25 H.P. compressor	25	2 PLUM	Ea.							
Remove compressor				26.002		1,550		1,550	1,925	2,400
Install 2-stage compressor 25 H.P.				51.948	14,000	3,100		17,100	19,200	22,300
Total				77.950	14,000	4,650		18,650	21,125	24,700

D2093 946 Compressed Air Systems, Compressors

| | System Description | Freq. (Years) | Crew | Unit | Labor Hours | 2019 Bare Costs | | | | Total In-House | Total w/O&P |
						Material	Labor	Equipment	Total		
5030	**Check operation**	1	1 STPI	Ea.							
	Check operation of unit				.320		20.50		20.50	25.50	31.50
	Total				.320		20.50		20.50	25.50	31.50

For customer support on your Facilities Maintenance & Repair Costs with RSMeans data, call 800.448.8182.

D30 HVAC — D3013 — Energy Supply

D3013 110 — Fuel Oil Storage Tank, 275 Gallon

	System Description	Freq. (Years)	Crew	Unit	Labor Hours	Material	Labor	Equipment	Total	Total In-House	Total w/O&P
0010	Replace 275 gallon fuel storage tank	30	Q-5	Ea.							
	Remove 275 gallon tank				2.081		120		120	149	186
	Install 275 gallon tank				4.156	520	239		759	870	1,025
	Total				6.236	520	359		879	1,019	1,211

D3013 150 — Fuel Level Meter

	System Description	Freq. (Years)	Crew	Unit	Labor Hours	Material	Labor	Equipment	Total	Total In-House	Total w/O&P
0010	Preventive maintenance	5	1 STPI	Ea.							
	Check fuel level meter calibration				.500		32		32	39.50	49.50
	Total				.500		32		32	39.50	49.50
0020	Replace remote tank fuel gauge	20	1 STPI	Ea.							
	Remove remote read gauge				2.081		133		133	165	206
	Replace remote read gauge 5' pointer travel				4.167	4,100	266		4,366	4,850	5,550
	Total				6.247	4,100	399		4,499	5,015	5,756

D3013 160 — Oil Filter

	System Description	Freq. (Years)	Crew	Unit	Labor Hours	Material	Labor	Equipment	Total	Total In-House	Total w/O&P
0010	Preventive maintenance	1	1 STPI	Ea.							
	Replace filter element				.052	4.05	3.35		7.40	8.60	10.20
	Total				.052	4.05	3.35		7.40	8.60	10.20
0020	Replace filter housing	30	1 STPI	Ea.							
	Replace filter housing				.520	40.50	33.50		74	86	102
	Total				.520	40.50	33.50		74	86	102

For customer support on your Facilities Maintenance & Repair Costs with RSMeans data, call 800.448.8182.

205

D3013 170   Fuel Oil Storage: Pipe & Fittings, Copper

	System Description	Freq. (Years)	Crew	Unit	Labor Hours	2019 Bare Costs				Total In-House	Total w/O&P
						Material	Labor	Equipment	Total		
0010	**Remake flare type joint**	10	1 STPI	M.L.F.							
	Remake flare type joint				.286		18.25		18.25	22.50	28.50
	Total				.286		18.25		18.25	**22.50**	**28.50**
0020	**Install 10' sect. 3/8" type L copper per M.L.F.**	20	1 PLUM	Ea.							
	Remove section 3/8" diameter copper				.619		39.10		39.10	48.50	60.50
	Install 10' section new 3/8" diameter type L				1.238	31.60	78		109.60	132	161
	Total				1.857	31.60	117.10		148.70	**180.50**	**221.50**
0030	**Install 10' sect. 1/2" type L copper per M.L.F.**	20	1 PLUM	Ea.							
	Remove section 1/2" diameter copper				.642		40.60		40.60	50.50	63
	Install 10' section new 1/2" diameter type L				1.284	33.80	81		114.80	138	168
	Total				1.926	33.80	121.60		155.40	**188.50**	**231**
0040	**Install 10' sect. 5/8" type L copper per M.L.F.**	20	1 PLUM	Ea.							
	Remove section 5/8" diameter copper				.658		41.60		41.60	51.50	64.50
	Install 10' section new 5/8" diameter type L				1.316	55.50	83		138.50	164	198
	Total				1.975	55.50	124.60		180.10	**215.50**	**262.50**
0050	**Install 10' sect. 3/4" type L copper per M.L.F.**	20	1 PLUM	Ea.							
	Remove section 3/4" diameter copper				.684		43.20		43.20	53.50	67
	Install 10' section new 3/4" diameter type L				1.369	43.90	86.50		130.40	156	189
	Total				2.053	43.90	129.70		173.60	**209.50**	**256**
0060	**Install 10' section 1" type L copper per M.L.F.**	20	1 PLUM	Ea.							
	Remove section 1" diameter copper				.765		48.30		48.30	60	75
	Install 10' section new 1" diameter type L				1.529	70	96.50		166.50	197	238
	Total				2.294	70	144.80		214.80	**257**	**313**
0130	**Replace 1000' type L 3/8" copper**	25	1 PLUM	M.L.F.							
	Remove section 3/8" diameter copper				61.920		3,910		3,910	4,850	6,050
	Install 1000' section new 3/8" diameter type L				123.800	3,160	7,800		10,960	13,200	16,100
	Total				185.720	3,160	11,710		14,870	**18,050**	**22,150**
0140	**Replace 1000' type L 1/2" copper**	25	1 PLUM	M.L.F.							
	Remove section 1/2" diameter copper				64.220		4,060		4,060	5,025	6,300
	Install 1000' section new 1/2" diameter type L				128.390	3,380	8,100		11,480	13,800	16,800
	Total				192.610	3,380	12,160		15,540	**18,825**	**23,100**

For customer support on your Facilities Maintenance & Repair Costs with RSMeans data, call 800.448.8182.

D3013 170 Fuel Oil Storage: Pipe & Fittings, Copper

	System Description	Freq. (Years)	Crew	Unit	Labor Hours	2019 Bare Costs Material	Labor	Equipment	Total	Total In-House	Total w/O&P
0150	**Replace 1000' type L 5/8" copper**	25	1 PLUM	M.L.F.							
	Remove section 5/8" diameter copper				65.840		4,160		4,160	5,150	6,450
	Install 1000' section new 5/8" diameter type L				131.640	5,550	8,300		13,850	16,400	19,800
	Total				197.480	5,550	12,460		18,010	21,550	26,250
0160	**Replace 1000' type L 3/4" copper**	25	1 PLUM	M.L.F.							
	Remove section 3/4" diameter copper				68.440		4,320		4,320	5,350	6,700
	Install 1000' section new 3/4" diameter type L				136.850	4,390	8,650		13,040	15,600	18,900
	Total				205.290	4,390	12,970		17,360	20,950	25,600
0170	**Replace 1000' type L 1" copper**	25	1 PLUM	M.L.F.							
	Remove section 1" diameter copper				76.480		4,830		4,830	6,000	7,500
	Install 1000' section new 1" diameter type L				152.930	7,000	9,650		16,650	19,700	23,800
	Total				229.410	7,000	14,480		21,480	25,700	31,300

D3013 210 Natural Gas: Pipe & Fittings, Steel/Iron

	System Description	Freq. (Years)	Crew	Unit	Labor Hours	2019 Bare Costs Material	Labor	Equipment	Total	Total In-House	Total w/O&P
0010	**Install new 2" gasket, 1 per M.L.F.**	30	1 STPI	Ea.							
	Disconnect joint, remove gasket & clean faces				.803		51.43		51.43	63.50	79.50
	Install gasket and make up bolts				.800	8.50	51		59.50	73	90
	Total				1.603	8.50	102.43		110.93	136.50	169.50
0020	**Install new 3" gasket, 1 per M.L.F.**	30	1 STPI	Ea.							
	Disconnect joint, remove gasket & clean faces				.945		60.50		60.50	75	93.50
	Install gasket and make up bolts				.945	8.60	60.50		69.10	84.50	104
	Total				1.890	8.60	121		129.60	159.50	197.50
0030	**Install new 4" gasket, 1 per M.L.F.**	30	1 STPI	Ea.							
	Disconnect joint, remove gasket & clean faces				1.300		83		83	103	129
	Install gasket and make up bolts				1.300	16.30	82		98.30	120	147
	Total				2.600	16.30	165		181.30	223	276

For customer support on your Facilities Maintenance & Repair Costs with RSMeans data, call 800.448.8182.

207

System Description	Freq. (Years)	Crew	Unit	Labor Hours	2019 Bare Costs				Total In-House	Total w/O&P
					Material	Labor	Equipment	Total		
0040 Install new 6" gasket, 1 per M.L.F.										
Disconnect joint, remove gasket & clean faces	30	1 STPI	Ea.	1.733		111		111	138	172
Install gasket and make up bolts				1.733	27	111		138	167	206
Total				3.466	27	222		249	305	378
0110 Replace 10' of buried 2" diam. st. pipe/M.L.F.	12	Q-4	Ea.							
Check break or leak				.016		1.02		1.02	1.27	1.59
Turn valve on and off				.010		.64		.64	.79	.99
Excavate by hand				5.771		237.75		237.75	310	380
Remove broken 2" diameter section				.743		44.60	1.30	45.90	57	70.50
Install 10' of pipe coated & wrapped				1.143	81.50	68.50	2	152	177	211
Check for leaks, additional section				.900		58		58	71	89
Backfill by hand				2.101		86.66		86.66	112	139
Total				10.684	81.50	497.17	3.30	581.97	729.06	892.08
0120 Replace 10' of buried 3" diam. st. pipe/M.L.F.	12	Q-4	Ea.							
Check break or leak				.016		1.02		1.02	1.27	1.59
Turn valve on and off				.010		.64		.64	.79	.99
Excavate by hand				5.771		237.75		237.75	310	380
Remove broken 3" diameter section				.800		48.10	1.40	49.50	61	76
Install 10' of pipe coated & wrapped				1.231	134.50	74	2.20	210.70	242	285
Check for leaks, additional section				.900		58		58	71	89
Backfill by hand				2.101		86.66		86.66	112	139
Total				10.829	134.50	506.17	3.60	644.27	798.06	971.58
0130 Replace 10' of buried 4" diam. st. pipe/M.L.F.	12	B-35	Ea.							
Check break or leak				.016		1.02		1.02	1.27	1.59
Turn valve on and off				.010		.64		.64	.79	.99
Excavate by hand				5.771		237.75		237.75	310	380
Remove broken 4" diameter section				1.224		62.50	18.80	81.30	101	120
Install 10' of pipe coated & wrapped				1.882	175	96.50	28.90	300.40	345	405
Check for leaks, additional section				.900		58		58	71	89
Backfill by hand				2.101		86.66		86.66	112	139
Total				11.905	175	543.07	47.70	765.77	941.06	1,135.58

For customer support on your Facilities Maintenance & Repair Costs with RSMeans data, call 800.448.8182.

D3013 210 Natural Gas: Pipe & Fittings, Steel/Iron

System Description	Freq. (Years)	Crew	Unit	Labor Hours	2019 Bare Costs				Total In-House	Total w/O&P
					Material	Labor	Equipment	Total		
0140 Replace 10' of buried 6" diam. st. pipe/M.L.F.	12	B-35	Ea.							
Check break or leak				.016		1.02		1.02	1.27	1.59
Turn valve on and off				.010		.64		.64	.79	.99
Excavate by hand				5.771		237.75		237.75	310	380
Remove broken 6" diameter section				1.734		89	26.60	115.60	142	170
Install 10' of pipe coated & wrapped				2.667	310	136.50	41	487.50	560	650
Check for leaks, additional section				.900		58		58	71	89
Backfill by hand				2.101		86.66		86.66	112	139
Total				13.199	310	609.57	67.60	987.17	1,197.06	1,430.58
0210 Replace 1000 L.F. of buried 2" diam. st. pipe	75	Q-4	M.L.F.							
Check break or leak				1.600		102		102	127	159
Turn valve on and off				1.000		64		64	79.50	99
Excavate with 1/2 C.Y. backhoe				23.111		1,111.11	560	1,671.11	2,050	2,375
Remove broken 2" diameter section				74.250		4,460	130	4,590	5,675	7,050
Install 1000' of pipe coated & wrapped				114.290	8,150	6,850	200	15,200	17,700	21,100
Check for leaks, additional section				9.000		580		580	710	890
Backfill with 1 C.Y. bucket				8.667		437.78	215.56	653.34	790	925
Total				231.918	8,150	13,604.89	1,105.56	22,860.45	27,131.50	32,598
0220 Replace 1000 L.F. of buried 3" diam. st. pipe	75	Q-4	M.L.F.							
Check break or leak				1.600		102		102	127	159
Turn valve on and off				1.000		64		64	79.50	99
Excavate with 1/2 C.Y. backhoe				23.111		1,111.11	560	1,671.11	2,050	2,375
Remove broken 3" diameter section				80.000		4,810	140	4,950	6,125	7,600
Install 1000' of pipe coated & wrapped				123.080	13,450	7,400	220	21,070	24,200	28,500
Check for leaks, additional section				9.000		580		580	710	890
Backfill with 1 C.Y. bucket				8.667		437.78	215.56	653.34	790	925
Total				246.458	13,450	14,504.89	1,135.56	29,090.45	34,081.50	40,548
0230 Replace 1000 L.F. of buried 4" diam. st. pipe	75	B-35	M.L.F.							
Check break or leak				1.600		102		102	127	159
Turn valve on and off				1.000		64		64	79.50	99
Excavate with 1/2 C.Y. backhoe				23.111		1,111.11	560	1,671.11	2,050	2,375
Remove broken 4" diameter section				122.390		6,250	1,880	8,130	10,100	12,000
Install 1000' of pipe coated & wrapped				188.240	17,500	9,650	2,890	30,040	34,700	40,300
Check for leaks, additional section				9.000		580		580	710	890
Backfill with 1 C.Y. bucket				8.667		437.78	215.56	653.34	790	925
Total				354.008	17,500	18,194.89	5,545.56	41,240.45	48,556.50	56,748

For customer support on your Facilities Maintenance & Repair Costs with RSMeans data, call 800.448.8182.

209

D30 HVAC   D3013   Energy Supply

D3013 210   Natural Gas: Pipe & Fittings, Steel/Iron

System Description	Freq. (Years)	Crew	Unit	Labor Hours	2019 Bare Costs				Total In-House	Total w/O&P
					Material	Labor	Equipment	Total		
0240 Replace 1000 L.F. of buried 6″ diam. st. pipe	75	B-35	M.L.F.							
Check break or leak				1.600		102		102	127	159
Turn valve on and off				1.000		64		64	79.50	99
Excavate with 1/2 C.Y. backhoe				23.111		1,111.11	560	1,671.11	2,050	2,375
Remove broken 6″ diameter section				173.390		8,900	2,660	11,560	14,200	17,000
Install 1000′ of pipe coated & wrapped				266.670	31,000	13,650	4,100	48,750	56,000	65,000
Check for leaks, additional section				9.000		580		580	710	890
Backfill with 1 C.Y. bucket				8.667		437.78	215.56	653.34	790	925
Total				483.438	31,000	24,844.89	7,535.56	63,380.45	73,956.50	86,448
0410 Replace 10′ of hung 2″ diam. st. pipe/M.L.F.	12	Q-1	Ea.							
Check break or leak				.016		1.02		1.02	1.27	1.59
Turn valve on and off				1.000		64		64	79	99
Remove broken 2″ diameter section				1.626		92.50		92.50	115	144
Install 10′ of black 2″ diameter steel pipe				3.250	115.50	184.50		300	355	430
Check for leaks, additional section				.900		58		58	71	89
Total				6.792	115.50	400.02		515.52	621.27	763.59
0420 Replace 10′ of hung 3″ diam. st. pipe/M.L.F.	12	Q-15	Ea.							
Check break or leak				.016		1.02		1.02	1.27	1.59
Turn valve on and off				.010		.64		.64	.79	.99
Remove broken 3″ diameter section				2.420		137.50	8.50	146	180	224
Install 10′ of black 3″ diameter steel pipe				4.837	158	275	17	450	535	640
Check for leaks, additional section				.900		58		58	71	89
Total				8.182	158	472.16	25.50	655.66	788.06	955.58
0430 Replace 10′ of hung 4″ diam. st. pipe/M.L.F.	12	Q-15	Ea.							
Check break or leak				.016		1.02		1.02	1.27	1.59
Turn valve on and off				.010		.64		.64	.79	.99
Remove broken 4″ diameter section				2.812		160	9.90	169.90	209	261
Install 10′ of black 4″ diameter steel pipe				5.622	170	320	19.70	509.70	605	730
Check for leaks, additional section				.900		58		58	71	89
Total				9.359	170	539.66	29.60	739.26	887.06	1,082.58

For customer support on your Facilities Maintenance & Repair Costs with RSMeans data, call 800.448.8182.

D3013 210   Natural Gas: Pipe & Fittings, Steel/Iron

System Description	Freq. (Years)	Crew	Unit	Labor Hours	2019 Bare Costs				Total In-House	Total w/O&P
					Material	Labor	Equipment	Total		
0440 Replace 10' of hung 6" diam. st. pipe/M.L.F.	12	Q-16	Ea.							
Check break or leak				.016		1.02		1.02	1.27	1.59
Turn valve on and off				.010		.64		.64	.79	.99
Remove broken 6" diameter section				4.335		255	10.10	265.10	330	405
Install 10' of black 6" diameter steel pipe				8.667	435	510	20.30	965.30	1,125	1,350
Check for leaks, additional section				.900		58		58	71	89
Total				13.928	435	824.66	30.40	1,290.06	1,528.06	1,846.58
0510 Replace 1000 L.F. of hung 2" diam. steel pipe	75	Q-1	M.L.F.							
Check break or leak				1.600		102		102	127	159
Turn valve on and off				100.000		6,400		6,400	7,925	9,900
Remove broken 2" diameter section				162.550		9,250		9,250	11,500	14,400
Install 1000' of black 2" diameter steel pipe				325.000	11,550	18,450		30,000	35,600	42,900
Check for leaks, additional section				9.000		580		580	710	890
Total				598.150	11,550	34,782		46,332	55,862	68,249
0520 Replace 1000 L.F. of hung 3" diam. steel pipe	75	Q-15	M.L.F.							
Check break or leak				1.600		102		102	127	159
Turn valve on and off				1.000		64		64	79.50	99
Remove broken 3" diameter section				241.950		13,750	850	14,600	18,000	22,400
Install 1000' of black 3" diameter steel pipe				483.680	15,800	27,500	1,700	45,000	53,500	64,000
Check for leaks, additional section				9.000		580		580	710	890
Total				737.230	15,800	41,996	2,550	60,346	72,416.50	87,548
0530 Replace 1000 L.F. of hung 4" diam. steel pipe	75	Q-15	M.L.F.							
Check break or leak				1.600		102		102	127	159
Turn valve on and off				1.000		64		64	79.50	99
Remove broken 4" diameter section				281.150		16,000	990	16,990	20,900	26,100
Install 1000' of black 4" diameter steel pipe				562.190	17,000	32,000	1,970	50,970	60,500	73,000
Check for leaks, additional section				9.000		580		580	710	890
Total				854.940	17,000	48,746	2,960	68,706	82,316.50	100,248

D30 HVAC — D3013 — Energy Supply

D3013 210 Natural Gas: Pipe & Fittings, Steel/Iron

System Description	Freq. (Years)	Crew	Unit	Labor Hours	2019 Bare Costs				Total In-House	Total w/O&P
					Material	Labor	Equipment	Total		
0540 Replace 1000 L.F. of hung 6″ diam. steel pipe	75	Q-16	M.L.F.							
Check break or leak				1.600		102		102	127	159
Turn valve on and off				1.000		64		64	79.50	99
Remove broken 6″ diameter section				433.450		25,500	1,010	26,510	32,800	40,600
Install 1000′ of black 6″ diameter steel pipe				866.740	43,500	51,000	2,030	96,530	113,500	135,500
Check for leaks, additional section				9.000		580		580	710	890
Total				1311.790	43,500	77,246	3,040	123,786	147,216.50	177,248

D3013 240 Natural Gas: Pressure Reducing Valve

System Description	Freq. (Years)	Crew	Unit	Labor Hours	2019 Bare Costs				Total In-House	Total w/O&P
					Material	Labor	Equipment	Total		
0010 Check gas pressure	5	1 STPI	Ea.							
Check pressure				.130		8.30		8.30	10.30	12.90
Total				.130		8.30		8.30	10.30	12.90
0110 Replace pressure regulator 1/2″ diam. pipe	14	1 STPI	Ea.							
Turn valve on and off				.010		.64		.64	.79	.99
Remove 1/2″ diameter regulator				.217		13.85		13.85	17.20	21.50
Replace 1/2″ diameter pipe size regulator				.433	60	27.50		87.50	100	118
Total				.660	60	41.99		101.99	117.99	140.49
0120 Replace pressure regulator 1″ diam. pipe	14	1 STPI	Ea.							
Turn valve on and off				.010		.64		.64	.79	.99
Remove 1″ diameter regulator				.274		17.50		17.50	21.50	27
Replace 1″ diameter pipe size regulator				.547	115	35		150	170	198
Total				.831	115	53.14		168.14	192.29	225.99
0130 Replace pressure regulator 1-1/2″ diam. pipe	14	1 STPI	Ea.							
Turn valve on and off				.010		.64		.64	.79	.99
Remove 1-1/2″ diameter regulator				.400		25.50		25.50	31.50	39.50
Replace 1-1/2″ diameter pipe size regulator				.800	730	51		781	865	990
Total				1.210	730	77.14		807.14	897.29	1,030.49

For customer support on your Facilities Maintenance & Repair Costs with RSMeans data, call 800.448.8182.

D3013 240 | Natural Gas: Pressure Reducing Valve

System Description	Freq. (Years)	Crew	Unit	Labor Hours	2019 Bare Costs				Total In-House	Total w/O&P
					Material	Labor	Equipment	Total		
0140 Replace pressure regulator 2″ diam. pipe	14	1 STPI	Ea.							
Turn valve on and off				.010		.64		.64	.79	.99
Remove 2″ diameter regulator				.473		30		30	37.50	47
Replace 2″ diameter pipe size regulator				.946	730	60.50		790.50	880	1,000
Total				1.428	730	91.14		821.14	918.29	1,047.99

D3013 260 | LPG Distribution: Pipe & Fittings, Steel/Iron

System Description	Freq. (Years)	Crew	Unit	Labor Hours	2019 Bare Costs				Total In-House	Total w/O&P
					Material	Labor	Equipment	Total		
0120 Replace 10′ st. pipe 1/2″ diam. per M.L.F.	12	1 PLUM	Ea.							
Check break or leak				.016		1.02		1.02	1.27	1.59
Turn valve on and off				.010		.64		.64	.79	.99
Remove broken 1/2″ diameter section				.826		52		52	64.50	81
Install section of pipe 1/2″ diameter				1.651	39.10	104		143.10	172	210
Check for leaks, additional section				.900		58		58	71	89
Total				3.402	39.10	215.66		254.76	309.56	382.58
0220 Replace 10′ st. pipe 3/4″ diam. per M.L.F.	12	1 PLUM	Ea.							
Check break or leak				.016		1.02		1.02	1.27	1.59
Turn valve on and off				.010		.64		.64	.79	.99
Remove broken 3/4″ diameter section				.853		54		54	67	83.50
Install section of pipe 3/4″ diameter				1.312	42.50	83		125.50	150	182
Check for leaks, additional section				.900		58		58	71	89
Total				3.090	42.50	196.66		239.16	290.06	357.08
0320 Replace 10′ st. pipe 1″ diam. per M.L.F.	12	1 PLUM	Ea.							
Check break or leak				.016		1.02		1.02	1.27	1.59
Turn valve on and off				.010		.64		.64	.79	.99
Remove broken 1″ diameter section				.982		62		62	77	96
Install section of pipe 1″ diameter				1.962	44.20	124		168.20	202	247
Check for leaks, additional section				.900		58		58	71	89
Total				3.870	44.20	245.66		289.86	352.06	434.58

For customer support on your Facilities Maintenance & Repair Costs with RSMeans data, call 800.448.8182.

213

D3013 260 | LPG Distribution: Pipe & Fittings, Steel/Iron

System Description	Freq. (Years)	Crew	Unit	Labor Hours	2019 Bare Costs				Total In-House	Total w/O&P
					Material	Labor	Equipment	Total		
0420 Replace 10' st. pipe 1-1/4" diam. M.L.F.	12	Q-1	Ea.							
Check break or leak				.016		1.02		1.02	1.27	1.59
Turn valve on and off				.010		.64		.64	.79	.99
Remove broken 1-1/4" diameter section				1.169		66.50		66.50	82.50	103
Install section of pipe 1-1/4" diameter				2.337	52	133		185	222	270
Check for leaks, additional section				.900		58		58	71	89
Total				4.432	52	259.16		311.16	377.56	464.58
0520 Replace 10' st. pipe 1-1/2" diam. M.L.F.	12	Q-1	Ea.							
Check break or leak				.016		1.02		1.02	1.27	1.59
Turn valve on and off				.010		.64		.64	.79	.99
Remove broken 1-1/2" diameter section				1.300		74		74	91.50	115
Install section of pipe 1-1/2" diameter				2.600	57	148		205	246	300
Check for leaks, additional section				.900		58		58	71	89
Total				4.826	57	281.66		338.66	410.56	506.58
0620 Replace 10' section st. pipe 2" diam. M.L.F.	12	Q-1	Ea.							
Check break or leak				.016		1.02		1.02	1.27	1.59
Turn valve on and off				.010		.64		.64	.79	.99
Remove broken 2" diameter section				1.626		92.50		92.50	115	144
Install section of pipe 2" diameter				3.250	115.50	184.50		300	355	430
Check for leaks, additional section				.900		58		58	71	89
Total				5.802	115.50	336.66		452.16	543.06	665.58
1130 Replace 1000' of 1/2" diameter steel pipe	75	Q-1	M.L.F.							
Check break or leak				1.600		102		102	127	159
Turn valve on and off				1.000		64		64	79.50	99
Remove broken 1/2" diameter section				82.560		5,200		5,200	6,475	8,100
Install section of pipe 1/2" diameter				165.080	3,910	10,400		14,310	17,200	21,000
Check for leaks, additional section				9.000		580		580	710	890
Total				259.240	3,910	16,346		20,256	24,591.50	30,248

For customer support on your Facilities Maintenance & Repair Costs with RSMeans data, call 800.448.8182.

D3013 260 | LPG Distribution: Pipe & Fittings, Steel/Iron

System Description	Freq. (Years)	Crew	Unit	Labor Hours	2019 Bare Costs				Total In-House	Total w/O&P
					Material	Labor	Equipment	Total		
1230 Replace 1000' of 3/4" diameter steel pipe	75	Q-1	M.L.F.							
Check break or leak				1.600		102		102	127	159
Turn valve on and off				1.000		64		64	79.50	99
Remove broken 3/4" diameter section				85.270		5,400		5,400	6,675	8,350
Install section of pipe 3/4" diameter				131.150	4,250	8,300		12,550	15,000	18,200
Check for leaks, additional section				9.000		580		580	710	890
Total				228.020	4,250	14,446		18,696	22,591.50	27,698
1330 Replace 1000' of 1" diameter steel pipe	75	Q-1	M.L.F.							
Check break or leak				1.600		102		102	127	159
Turn valve on and off				1.000		64		64	79.50	99
Remove broken 1" diameter section				98.150		6,200		6,200	7,700	9,600
Install section of pipe 1" diameter				196.220	4,420	12,400		16,820	20,200	24,700
Check for leaks, additional section				9.000		580		580	710	890
Total				305.970	4,420	19,346		23,766	28,816.50	35,448
1430 Replace 1000' of 1-1/4" diameter steel pipe	75	Q-1	M.L.F.							
Check break or leak				1.600		102		102	127	159
Turn valve on and off				1.000		64		64	79.50	99
Remove broken 1-1/4" diameter section				116.890		6,650		6,650	8,250	10,300
Install section of pipe 1-1/4" diameter				233.710	5,200	13,300		18,500	22,200	27,000
Check for leaks, additional section				9.000		580		580	710	890
Total				362.200	5,200	20,696		25,896	31,366.50	38,448
1530 Replace 1000' of 1-1/2" diameter steel pipe	75	Q-1	M.L.F.							
Check break or leak				1.600		102		102	127	159
Turn valve on and off				1.000		64		64	79.50	99
Remove broken 1-1/2" diameter section				130.040		7,400		7,400	9,175	11,500
Install section of pipe 1-1/2" diameter				260.000	5,700	14,800		20,500	24,600	30,100
Check for leaks, additional section				9.000		580		580	710	890
Total				401.640	5,700	22,946		28,646	34,691.50	42,748

For customer support on your Facilities Maintenance & Repair Costs with RSMeans data, call 800.448.8182.

215

D3013 260 LPG Distribution: Pipe & Fittings, Steel/Iron

	System Description	Freq. (Years)	Crew	Unit	2019 Bare Costs					Total In-House	Total w/O&P
					Labor Hours	Material	Labor	Equipment	Total		
1630	Replace 1000' of 2" diameter steel pipe	75	Q-1	M.L.F.							
	Check break or leak				1.600		102		102	127	159
	Turn valve on and off				1.000		64		64	79.50	99
	Remove broken 2" diameter section				162.550		9,250		9,250	11,500	14,400
	Install section of pipe 2" diameter				325.000	11,550	18,450		30,000	35,600	42,900
	Check for leaks, additional section				9.000		580		580	710	890
	Total				499.150	11,550	28,446		39,996	48,016.50	58,448

D3013 601 Solar Panel, 3' x 8'

	System Description	Freq. (Years)	Crew	Unit	2019 Bare Costs					Total In-House	Total w/O&P
					Labor Hours	Material	Labor	Equipment	Total		
0010	Replace solar panel 3' x 8'	15	Q-1	Ea.							
	Remove panel				1.891		107		107	133	167
	Install 3' x 8' solar panel aluminum frame				3.783	1,075	215		1,290	1,450	1,675
	Total				5.674	1,075	322		1,397	1,583	1,842

For customer support on your Facilities Maintenance & Repair Costs with RSMeans data, call 800.448.8182.

D3023 180 | **Boiler, Gas**

	System Description	Freq. (Years)	Crew	Unit	Labor Hours	2019 Bare Costs Material	Labor	Equipment	Total	Total In-House	Total w/O&P
1010	**Repair boiler, gas, 250 MBH**	7	1 STPI	Ea.							
	Remove / replace burner blower				.600		38.50		38.50	47.50	59.50
	Remove burner blower bearing				1.000		64		64	79.50	99
	Replace burner blower bearing				2.000	65.50	128		193.50	231	280
	Remove burner blower motor				.976		62.50		62.50	77.50	96.50
	Replace burner blower motor				1.951	270	125		395	450	530
	Remove burner fireye				.195		12.50		12.50	15.50	19.35
	Replace burner fireye				.300	249	19.20		268.20	298	340
	Remove burner gas regulator				.274		17.50		17.50	21.50	27
	Replace burner gas regulator				.547	115	35		150	170	198
	Remove burner auto gas valve				.274		17.50		17.50	21.50	27
	Replace burner auto gas valve				.421	181	27		208	233	268
	Remove burner solenoid valve				.279		17.30		17.30	21.50	27
	Replace burner solenoid valve				.557	730	34.50		764.50	845	965
	Repair controls				.500		32		32	39.50	49.50
	Total				9.874	1,610.50	630.50		2,241	2,551	2,985.85
1060	**Replace boiler, gas, 250 MBH**	30	Q-7	Ea.							
	Remove boiler				21.888		1,325		1,325	1,650	2,075
	Replace boiler, 250 MBH				43.776	3,825	2,675		6,500	7,525	8,900
	Total				65.663	3,825	4,000		7,825	9,175	10,975
2010	**Repair boiler, gas, 2000 MBH**	7	Q-5	Ea.							
	Remove / replace burner blower				.600		38.50		38.50	47.50	59.50
	Remove burner blower bearing				1.000		64		64	79.50	99
	Replace burner blower bearing				2.000	65.50	128		193.50	231	280
	Remove burner blower motor				.976		62.50		62.50	77.50	96.50
	Replace burner blower motor				1.951	240	117		357	410	480
	Remove burner fireye				.195		12.50		12.50	15.50	19.35
	Replace burner fireye				.300	249	19.20		268.20	298	340
	Remove burner gas regulator				.727		42		42	52	65
	Replace burner gas regulator				1.455	1,425	83.50		1,508.50	1,675	1,900
	Remove burner auto gas valve				.800		45.50		45.50	56.50	70.50
	Replace burner auto gas valve				1.231	385	70		455	510	590
	Remove burner solenoid valve				2.600		162		162	200	250
	Replace burner solenoid valve				5.199	1,175	325		1,500	1,700	1,975
	Repair controls				.500		32		32	39.50	49.50
	Total				19.534	3,539.50	1,201.70		4,741.20	5,392	6,274.35

D3023 180 | **Boiler, Gas**

System Description	Freq. (Years)	Crew	Unit	Labor Hours	2019 Bare Costs				Total In-House	Total w/O&P
					Material	Labor	Equipment	Total		
2070 Replace boiler, gas, 2000 MBH	30	Q-7	Ea.							
Remove boiler				57.762		3,525		3,525	4,375	5,450
Replace boiler, 2000 MBH				110.000	20,600	6,675		27,275	30,900	36,100
Total				167.762	20,600	10,200		30,800	35,275	41,550
3010 Repair boiler, gas, 10,000 MBH	7	Q-5	Ea.							
Remove / replace burner blower				1.000		64		64	79.50	99
Remove burner blower bearing				1.000		64		64	79.50	99
Replace burner blower bearing				2.000	65.50	128		193.50	231	280
Remove burner motor				1.000		64		64	79.50	99
Replace burner motor				1.951	206	117		323	370	440
Remove burner fireye				.195		12.50		12.50	15.50	19.35
Replace burner fireye				.300	249	19.20		268.20	298	340
Remove burner gas regulator				1.300		75		75	93	116
Replace burner gas regulator				2.600	2,525	150		2,675	2,975	3,400
Remove burner auto gas valve				5.195		305		305	380	475
Replace burner auto gas valve				8.000	6,175	470		6,645	7,375	8,450
Remove burner solenoid valve				2.600		162		162	200	250
Replace burner solenoid valve				10.399	3,225	645		3,870	4,350	5,025
Repair controls				.500		32		32	39.50	49.50
Total				38.040	12,445.50	2,307.70		14,753.20	16,565.50	19,141.85
3070 Replace boiler, gas, 10,000 MBH	30	Q-7	Ea.							
Remove boiler				364.000		22,120		22,120	27,500	34,400
Replace boiler, gas, 10,000 MBH				722.400	174,300	43,960		218,250	246,500	286,000
Total				1086.400	174,300	66,080		240,330	274,000	320,400

For customer support on your Facilities Maintenance & Repair Costs with RSMeans data, call 800.448.8182.

D3023 182 — Boiler, Coal

System Description	Freq. (Years)	Crew	Unit	Labor Hours	2019 Bare Costs Material	2019 Bare Costs Labor	2019 Bare Costs Equipment	2019 Bare Costs Total	Total In-House	Total w/O&P
1010 Repair boiler, coal, 4600 MBH	20	4 STPI	Ea.							
Remove coal feeder				2.703		171.60		171.60	215	268
Replace coal feeder				4.054	1,705.60	260		1,965.60	2,200	2,525
Remove vibrating grates				33.592		2,152.80		2,152.80	2,675	3,325
Replace vibrating grates				62.816	21,736	4,014.40		25,750.40	28,900	33,400
Remove motor				.832		53.04		53.04	66	82.50
Replace motor				.450	314.60	27.04		341.64	380	435
Repair controls				5.200		332.80		332.80	415	515
Total				109.647	23,756.20	7,011.68		30,767.88	34,851	40,550.50
1050 Replace boiler, coal, 4600 MBH	30	5 STPI	Ea.							
Remove boiler				122.304		7,800		7,800	9,700	12,100
Replace boiler, coal, 4600 MBH				244.712	111,384	15,652		127,036	142,000	163,500
Total				367.016	111,384	23,452		134,836	151,700	175,600

D3023 184 — Boiler, Oil

System Description	Freq. (Years)	Crew	Unit	Labor Hours	2019 Bare Costs Material	2019 Bare Costs Labor	2019 Bare Costs Equipment	2019 Bare Costs Total	Total In-House	Total w/O&P
1010 Repair boiler, oil, 250 MBH	7	Q-5	Ea.							
Remove / replace burner blower				.600		38.50		38.50	47.50	59.50
Remove burner blower bearing				1.000		64		64	79.50	99
Replace burner blower bearing				2.000	65.50	128		193.50	231	280
Remove burner blower motor				.976		62.50		62.50	77.50	96.50
Replace burner blower motor				1.951	270	125		395	450	530
Remove burner fireye				.195		12.50		12.50	15.50	19.35
Replace burner fireye				.300	249	19.20		268.20	298	340
Remove burner ignition transformer				.250		16		16	19.85	25
Replace burner ignition transformer				.500	99	32		131	149	173
Remove burner ignition electrode				.200		12.80		12.80	15.85	19.85
Replace burner ignition electrode				.400	14.10	25.50		39.60	47.50	57
Remove burner oil pump				.350		22.50		22.50	28	34.50
Replace burner oil pump				.700	128	45		173	196	230
Remove burner nozzle				.167		10.65		10.65	13.25	16.55
Replace burner nozzle				.300	6.35	19.20		25.55	31	38
Repair controls				.600		38.50		38.50	47.50	59.50
Total				10.489	831.95	671.85		1,503.80	1,746.95	2,077.75

For customer support on your Facilities Maintenance & Repair Costs with RSMeans data, call 800.448.8182.

219

	System Description	Freq. (Years)	Crew	Unit	Labor Hours	2019 Bare Costs				Total In-House	Total w/O&P
						Material	Labor	Equipment	Total		
1060	**Replace boiler, oil, 250 MBH**	30	Q-7	Ea.							
	Remove boiler				24.465		1,500		1,500	1,850	2,300
	Replace boiler, oil, 250 MBH				48.930	3,625	2,975		6,600	7,675	9,150
	Total				73.394	3,625	4,475		8,100	9,525	11,450
1066	**Replace boiler, oil, 300 MBH**	30	Q-7	Ea.							
	Remove boiler				29.091		1,775		1,775	2,200	2,750
	Replace boiler, oil, 300 MBH				59.480	4,525	3,625		8,150	9,475	11,300
	Total				88.570	4,525	5,400		9,925	11,675	14,050
1070	**Replace boiler, oil, 420 MBH**	30	Q-7	Ea.							
	Remove boiler				35.955		2,200		2,200	2,725	3,400
	Replace boiler, oil, 420 MBH				71.749	5,100	4,375		9,475	11,000	13,200
	Total				107.704	5,100	6,575		11,675	13,725	16,600
1080	**Replace boiler, oil, 530 MBH**	30	Q-7	Ea.							
	Remove boiler				42.440		2,575		2,575	3,200	4,000
	Replace boiler, oil, 530 MBH				84.881	6,450	5,175		11,625	13,500	16,100
	Total				127.321	6,450	7,750		14,200	16,700	20,100
2010	**Repair boiler, oil, 2000 MBH**	7	1 STPI	Ea.							
	Remove / replace burner blower				.600		38.50		38.50	47.50	59.50
	Remove burner blower bearing				1.000		64		64	79.50	99
	Replace burner blower bearing				2.000	65.50	128		193.50	231	280
	Remove burner blower motor				.976		58.50		58.50	72	90.50
	Replace burner blower motor				1.951	240	117		357	410	480
	Remove burner fireye				.195		12.50		12.50	15.50	19.35
	Replace burner fireye				.300	249	19.20		268.20	298	340
	Remove burner ignition transformer				.286		18.25		18.25	22.50	28.50
	Replace burner ignition transformer				.571	99	36.50		135.50	154	180
	Remove burner ignition electrode				.200		12.80		12.80	15.85	19.85
	Replace burner ignition electrode				.400	14.10	25.50		39.60	47.50	57
	Remove burner oil pump				.400		25.50		25.50	31.50	39.50
	Replace burner oil pump				.800	150	51		201	228	267
	Remove burner nozzle				.167		10.65		10.65	13.25	16.55
	Replace burner nozzle				.333	9.10	21.50		30.60	36.50	44.50
	Repair controls				.600		38.50		38.50	47.50	59.50
	Total				10.779	826.70	677.90		1,504.60	1,750.10	2,080.75

For customer support on your Facilities Maintenance & Repair Costs with RSMeans data, call 800.448.8182.

D30 HVAC   D3023   Heat Generating Systems

D3023 184   Boiler, Oil

	System Description	Freq. (Years)	Crew	Unit	Labor Hours	2019 Bare Costs				Total In-House	Total w/O&P
						Material	Labor	Equipment	Total		
2060	**Replace boiler, oil, 2000 MBH**	30	Q-7	Ea.							
	Remove boiler				57.762		3,525		3,525	4,375	5,450
	Replace boiler, oil, 2000 MBH				149.000	17,500	9,050		26,550	30,500	35,900
	Total				206.762	17,500	12,575		30,075	34,875	41,350
2070	**Replace boiler, oil, 3000 MBH**	30	Q-7	Ea.							
	Remove boiler				110.000		6,725		6,725	8,325	10,400
	Replace boiler, oil, 3000 MBH				221.000	23,000	13,400		36,400	42,000	49,600
	Total				331.000	23,000	20,125		43,125	50,325	60,000
3010	**Repair boiler, oil, 10,000 MBH**	7	1 STPI	Ea.							
	Remove / replace burner blower				.600		38.50		38.50	47.50	59.50
	Remove burner blower bearing				1.000		64		64	79.50	99
	Replace burner blower bearing				2.000	65.50	128		193.50	231	280
	Remove burner blower motor				.800		51		51	63.50	79.50
	Replace burner blower motor				1.951	206	117		323	370	440
	Remove burner fireye				.195		12.50		12.50	15.50	19.35
	Replace burner fireye				.300	249	19.20		268.20	298	340
	Remove burner ignition transformer				.333		21.50		21.50	26.50	33
	Replace burner ignition transformer				.667	99	42.50		141.50	162	190
	Remove burner ignition electrode				.200		12.80		12.80	15.85	19.85
	Replace burner ignition electrode				.444	14.15	28.50		42.65	51	61.50
	Remove burner oil pump				.500		32		32	39.50	49.50
	Replace burner oil pump				1.000	128	64		192	220	259
	Remove burner nozzle				.167		10.65		10.65	13.25	16.55
	Replace burner nozzle				.400	9.10	25.50		34.60	42	51
	Repair controls				.600		38.50		38.50	47.50	59.50
	Total				11.158	770.75	706.15		1,476.90	1,722.60	2,057.25
3060	**Replace boiler, oil, 10,000 MBH**	30	Q-7	Ea.							
	Remove boiler				373.100		22,673		22,673	28,200	35,300
	Replace boiler, oil, 10,000 MBH				522.340	142,065	31,713.50		173,778.50	195,500	227,000
	Total				895.440	142,065	54,386.50		196,451.50	223,700	262,300

D30 HVAC | D3023 | Heat Generating Systems

D3023 186 | Boiler, Gas/Oil

	System Description	Freq. (Years)	Crew	Unit	Labor Hours	2019 Bare Costs Material	Labor	Equipment	Total	Total In-House	Total w/O&P
1010	**Repair boiler, gas/oil, 2000 MBH**	7	Q-5	Ea.							
	Remove/replace burner blower				.600		38.50		38.50	47.50	59.50
	Remove burner blower bearing				1.000		64		64	79.50	99
	Replace burner blower bearing				2.000	65.50	128		193.50	231	280
	Remove burner blower motor				.976		62.50		62.50	77.50	96.50
	Replace burner blower motor				1.951	240	117		357	410	480
	Remove burner fireye				.195		12.50		12.50	15.50	19.35
	Replace burner fireye				.300	249	19.20		268.20	298	340
	Remove burner ignition transformer				.333		21.50		21.50	26.50	33
	Replace burner ignition transformer				.571	99	36.50		135.50	154	180
	Remove burner ignition electrode				.200		12.80		12.80	15.85	19.85
	Replace burner ignition electrode				.400	14.10	25.50		39.60	47.50	57
	Remove burner oil pump				.500		32		32	39.50	49.50
	Replace burner oil pump				.800	150	51		201	228	267
	Remove burner nozzle				.167		10.65		10.65	13.25	16.55
	Replace burner nozzle				.333	9.10	21.50		30.60	36.50	44.50
	Remove burner gas regulator				.727		42		42	52	65
	Replace burner gas regulator				1.455	1,425	83.50		1,508.50	1,675	1,900
	Remove burner auto gas valve				.800		45.50		45.50	56.50	70.50
	Replace burner auto gas valve				1.231	385	70		455	510	590
	Remove burner solenoid valve				2.600		162		162	200	250
	Replace burner solenoid valve				5.199	1,175	325		1,500	1,700	1,975
	Repair controls				.600		38.50		38.50	47.50	59.50
	Total				22.939	3,811.70	1,419.65		5,231.35	5,961.10	6,951.75
1050	**Replace boiler, gas/oil, 2000 MBH**	30	Q-7	Ea.							
	Remove boiler				57.762		3,525		3,525	4,375	5,450
	Replace boiler, gas/oil, 2000 MBH				116.000	40,800	7,025		47,825	53,500	62,000
	Total				173.762	40,800	10,550		51,350	57,875	67,450

For customer support on your Facilities Maintenance & Repair Costs with RSMeans data, call 800.448.8182.

D3023 186 Boiler, Gas/Oil

System Description	Freq. (Years)	Crew	Unit	Labor Hours	2019 Bare Costs				Total In-House	Total w/O&P
					Material	Labor	Equipment	Total		
2010 Repair boiler, gas/oil, 20,000 MBH	7	Q-5	Ea.							
Remove/replace burner blower				.600		38.50		38.50	47.50	59.50
Remove burner blower bearing				1.000		64		64	79.50	99
Replace burner blower bearing				2.000	65.50	128		193.50	231	280
Remove burner blower motor				.976		58.50		58.50	72	90.50
Replace burner blower motor				1.951	285	117		402	460	535
Remove burner fireye				.195		12.50		12.50	15.50	19.35
Replace burner fireye				.300	249	19.20		268.20	298	340
Remove burner ignition transformer				.333		21.50		21.50	26.50	33
Replace burner ignition transformer				.667	99	42.50		141.50	162	190
Remove burner ignition electrode				.267		17.05		17.05	21	26.50
Replace burner ignition electrode				.444	14.15	28.50		42.65	51	61.50
Remove burner oil pump				.500		32		32	39.50	49.50
Replace burner oil pump				2.602	195	150		345	400	475
Remove burner nozzle				.167		10.65		10.65	13.25	16.55
Replace burner nozzle				.571	9.10	36.50		45.60	55.50	68
Remove burner gas regulator				1.300		75		75	93	116
Replace burner gas regulator				2.600	2,525	150		2,675	2,975	3,400
Remove burner auto gas valve				5.195		305		305	380	475
Replace burner auto gas valve				8.000	6,175	470		6,645	7,375	8,450
Remove burner solenoid valve				2.600		162		162	200	250
Replace burner solenoid valve				10.399	3,225	645		3,870	4,350	5,025
Repair controls				.600		38.50		38.50	47.50	59.50
Total				43.266	12,841.75	2,621.90		15,463.65	17,392.75	20,118.90
2050 Replace boiler, gas/oil, 20,000 MBH	30	Q-7	Ea.							
Remove boiler				1527.360		93,240		93,240	115,500	144,500
Replace boiler, gas/oil, 20,000 MBH				3156.840	278,240	192,400		470,640	544,500	646,000
Total				4684.200	278,240	285,640		563,880	660,000	790,500

D3023 198 Blowoff System

System Description	Freq. (Years)	Crew	Unit	Labor Hours	2019 Bare Costs				Total In-House	Total w/O&P
					Material	Labor	Equipment	Total		
1010 Repair boiler blowoff system	10	1 STPI	Ea.							
Repair leak				1.000		64		64	79.50	99
Total				1.000		64		64	79.50	99

For customer support on your Facilities Maintenance & Repair Costs with RSMeans data, call 800.448.8182.

223

D3023 198 Blowoff System

System Description	Freq. (Years)	Crew	Unit	Labor Hours	2019 Bare Costs				Total In-House	Total w/O&P
					Material	Labor	Equipment	Total		
1020 Replace boiler blowoff system	15	Q-5	Ea.							
Remove boiler blowoff				2.773		160		160	198	247
Replace boiler blowoff				5.556	6,550	320		6,870	7,600	8,675
Total				8.329	6,550	480		7,030	7,798	8,922

D3023 292 Chemical Feed System

System Description	Freq. (Years)	Crew	Unit	Labor Hours	2019 Bare Costs				Total In-House	Total w/O&P
					Material	Labor	Equipment	Total		
1010 Repair chemical feed	15	1 STPI	Ea.							
Repair controls				.040		2.56		2.56	3.17	3.97
Remove / replace agitator motor				1.951	270	125		395	450	530
Remove / replace pump seals / bearings				2.078	34	131		165	200	247
Remove / replace pump motor				1.951	240	117		357	410	480
Total				6.020	544	375.56		919.56	1,063.17	1,260.97
1030 Replace chemical feed	15	2 STPI	Ea.							
Remove / replace feeder				2.500	605	158		763	860	1,000
Total				2.500	605	158		763	860	1,000

D3023 294 Feed Water Supply

System Description	Freq. (Years)	Crew	Unit	Labor Hours	2019 Bare Costs				Total In-House	Total w/O&P
					Material	Labor	Equipment	Total		
1010 Repair feed water supply pump	15	1 STPI	Ea.							
Repair controls				.600		38.50		38.50	47.50	59.50
Remove / replace pump seals / bearings				1.600		102		102	127	159
Remove pump motor				2.078		133		133	165	206
Replace pump motor				4.167	3,600	266		3,866	4,300	4,925
Remove / replace pump coupling				1.000		64		64	79.50	99
Total				9.445	3,600	603.50		4,203.50	4,719	5,448.50
1030 Replace feed water pump	15	Q-2	Ea.							
Remove pump				11.111		655		655	810	1,025
Replace pump				22.222	23,400	1,300		24,700	27,400	31,300
Total				33.333	23,400	1,955		25,355	28,210	32,325

D30 HVAC | D3023 | Heat Generating Systems

D3023 296 Deaerator

	System Description	Freq. (Years)	Crew	Unit	Labor Hours	2019 Bare Costs Material	Labor	Equipment	Total	Total In-House	Total w/O&P
1010	**Repair deaerator**										
	Repair controls	10	1 STPI	Ea.	1.000		64		64	79.50	99
	Total				1.000		64		64	**79.50**	**99**
1030	**Replace deaerator**										
	Remove unit	20	4 STPI	Ea.	7.500		450		450	555	695
	Replace unit				180.000	32,800	11,500		44,300	50,500	59,000
	Total				187.500	32,800	11,950		44,750	**51,055**	**59,695**

D3023 298 Separators for 9000 Ton Chilled Water System

	System Description	Freq. (Years)	Crew	Unit	Labor Hours	2019 Bare Costs Material	Labor	Equipment	Total	Total In-House	Total w/O&P
1010	**Clean separator strainer**										
	Clean separator strainer	10	Q-6	Ea.	10.000		640		640	795	990
	Total				10.000		640		640	**795**	**990**
1030	**Replace separator**										
	Remove unit	20	Q-6	Ea.	22.500		1,350		1,350	1,675	2,075
	Replace separator				42.857	64,200	2,565		66,765	74,000	84,000
	Total				65.357	64,200	3,915		68,115	**75,675**	**86,075**

D3023 310 Metal Flue / Chimney

	System Description	Freq. (Years)	Crew	Unit	Labor Hours	2019 Bare Costs Material	Labor	Equipment	Total	Total In-House	Total w/O&P
0010	**Replace metal flue, all fuel SS, 6" diameter**										
	Replace flue, all fuel stainless steel, 6" diameter	15	Q-9	L.F.	.387	80.48	21.24		101.72	115	134
	Total				.387	80.48	21.24		101.72	**115**	**134**
0020	**Replace metal flue, all fuel SS, 10" diameter**										
	Replace flue, all fuel stainless steel, 10" diameter	15	Q-9	L.F.	.483	102.95	26.54		129.49	147	170
	Total				.483	102.95	26.54		129.49	**147**	**170**

D3023 310 Metal Flue / Chimney

System Description	Freq. (Years)	Crew	Unit	Labor Hours	2019 Bare Costs				Total In-House	Total w/O&P
					Material	Labor	Equipment	Total		
0030 Replace metal flue, all fuel SS, 20" diameter										
Replace flue, all fuel stainless steel, 20" diameter	15	Q-10	L.F.	.967	192.85	55.10		247.95	281	325
Total				.967	192.85	55.10		247.95	**281**	**325**
0040 Replace metal flue, all fuel SS, 32" diameter										
Replace flue, all fuel stainless steel, 32" diameter	15	Q-10	L.F.	1.289	320.45	73.23		393.68	445	515
Total				1.289	320.45	73.23		393.68	**445**	**515**
0050 Replace metal flue, all fuel SS, 48" diameter										
Replace flue, all fuel stainless steel, 48" diameter	15	Q-10	L.F.	1.832	471.25	104.40		575.65	650	755
Total				1.832	471.25	104.40		575.65	**650**	**755**
2010 Replace metal flue, gas vent, galvanized, 48" diameter										
Replace flue, gas vent, galvanized, 48" diameter	10	Q-10	L.F.	1.895	1,537.50	108		1,645.50	1,825	2,100
Total				1.895	1,537.50	108		1,645.50	**1,825**	**2,100**

D3023 388 Pneumatic Coal Spreader

System Description	Freq. (Years)	Crew	Unit	Labor Hours	2019 Bare Costs				Total In-House	Total w/O&P
					Material	Labor	Equipment	Total		
1010 Repair spreader, pneumatic coal	10	2 STPI	Ea.							
Repair controls				1.000		64		64	79.50	99
Remove conveyor bearing				1.111		71		71	88	110
Replace conveyor bearing				2.222	73.50	142		215.50	257	310
Remove conveyor motor				2.597		166		166	206	258
Replace conveyor motor				5.195	3,950	310		4,260	4,725	5,425
Remove / replace blower				1.200		76.50		76.50	95	119
Remove blower bearings				1.000		64		64	79.50	99
Replace blower bearings				2.000	65.50	128		193.50	231	280
Remove blower motor				2.000		128		128	159	198
Replace blower motor				4.000	2,125	240		2,365	2,625	3,025
Total				22.326	6,214	1,389.50		7,603.50	**8,545**	**9,923**
1060 Replace coal spreader	12	4 STPI	Ea.							
Remove spreader				65.041		4,150		4,150	5,150	6,450
Replace spreader				130.000	7,600	8,325		15,925	18,700	22,400
Total				195.041	7,600	12,475		20,075	**23,850**	**28,850**

For customer support on your Facilities Maintenance & Repair Costs with RSMeans data, call 800.448.8182.

D30 HVAC | D3023 | Heat Generating Systems

D3023 390 | Fuel Oil Equipment

System Description	Freq. (Years)	Crew	Unit	Labor Hours	2019 Bare Costs Material	Labor	Equipment	Total	Total In-House	Total w/O&P
1010 Repair fuel oil equipment, pump	10	1 STPI	Ea.							
Remove pump seals / bearings				.600		38.50		38.50	47.50	59.50
Replace pump seals / bearings				1.500	31.50	96		127.50	154	188
Remove / replace pump coupling				.900		57.50		57.50	71.50	89
Remove / replace impeller / shaft				.900		57.50		57.50	71.50	89
Remove / replace motor				.650		41.50		41.50	51.50	64.50
Remove strainers				.289		18.45		18.45	23	28.50
Replace strainers				.578	20	37		57	68	82.50
Total				5.417	51.50	346.45		397.95	487	601
1030 Replace fuel oil 25 GPH pump / motor set	15	Q-5	Ea.							
Remove equipment				1.733		100		100	124	155
Replace equipment				3.468	1,475	200		1,675	1,875	2,150
Total				5.201	1,475	300		1,775	1,999	2,305
1040 Replace fuel oil 45 GPH pump / motor set	15	Q-5	Ea.							
Remove equipment				1.733		100		100	124	155
Replace equipment				3.468	1,475	200		1,675	1,875	2,150
Total				5.201	1,475	300		1,775	1,999	2,305
1050 Replace fuel oil 90 GPH pump / motor set	15	Q-5	Ea.							
Remove equipment				2.081		120		120	149	186
Replace equipment				4.161	1,475	239		1,714	1,925	2,225
Total				6.242	1,475	359		1,834	2,074	2,411
1060 Replace fuel oil 160 GPH pump / motor set	15	Q-5	Ea.							
Remove equipment				2.597		149		149	185	232
Replace equipment				5.195	1,550	299		1,849	2,075	2,400
Total				7.792	1,550	448		1,998	2,260	2,632

D3033 115   **Cooling Tower**

System Description	Freq. (Years)	Crew	Unit	Labor Hours	2019 Bare Costs				Total In-House	Total w/O&P
					Material	Labor	Equipment	Total		
1010										
Repair cooling tower, 50 ton	10	2 STPI	Ea.							
Repair controls				1.000		64		64	79.50	99
Remove bearings				1.200		76.50		76.50	95	119
Replace bearings				2.000	65.50	128		193.50	231	280
Remove fan motor				1.155		74		74	91.50	115
Replace fan motor				2.312	760	139		899	1,000	1,175
Remove / replace float valve				.500	16.50	29		45.50	54	65
Total				8.167	842	510.50		1,352.50	1,551	1,853
1030										
Replace cooling tower, 50 ton	15	Q-6	Ea.							
Remove cooling tower				10.397		621		621	770	965
Replace cooling tower, 50 ton				20.779	11,880	1,230		13,110	14,600	16,800
Total				31.177	11,880	1,851		13,731	15,370	17,765
2010										
Repair cooling tower, 100 ton	10	2 STPI	Ea.							
Repair controls				1.000		64		64	79.50	99
Remove bearings				4.800		306		306	380	475
Replace bearings				8.000	262	512		774	925	1,125
Remove fan motors				4.621		296		296	365	460
Replace fan motors				9.249	3,040	556		3,596	4,025	4,650
Remove / replace float valve				.575	18.98	33.35		52.33	62	75
Total				28.245	3,320.98	1,767.35		5,088.33	5,836.50	6,884
2030										
Replace cooling tower, 100 ton	15	Q-6	Ea.							
Remove cooling tower				14.286		855		855	1,050	1,325
Replace cooling tower, 100 ton				28.571	16,600	1,705		18,305	20,400	23,400
Total				42.857	16,600	2,560		19,160	21,450	24,725
3010										
Repair cooling tower, 300 ton	10	2 STPI	Ea.							
Repair controls				1.000		64		64	79.50	99
Remove bearings				7.200		459		459	570	715
Replace bearings				13.333	441	852		1,293	1,550	1,875
Remove fan motors				6.931		444		444	550	690
Replace fan motors				13.865	3,660	834		4,494	5,050	5,850
Remove / replace float valve				1.250	41.25	72.50		113.75	135	163
Total				43.580	4,142.25	2,725.50		6,867.75	7,934.50	9,392

For customer support on your Facilities Maintenance & Repair Costs with RSMeans data, call 800.448.8182.

D3033 115 Cooling Tower

	System Description	Freq. (Years)	Crew	Unit	Labor Hours	2019 Bare Costs				Total In-House	Total w/O&P
						Material	Labor	Equipment	Total		
3030	**Replace cooling tower, 300 ton**	15	Q-6	Ea.							
	Remove cooling tower				36.273		2,160		2,160	2,675	3,350
	Replace cooling tower, 300 ton				72.582	28,950	4,335		33,285	37,200	42,900
	Total				108.855	28,950	6,495		35,445	39,875	46,250
4010	**Repair cooling tower, 1000 ton**	10	2 STPI	Ea.							
	Repair controls				1.000		64		64	79.50	99
	Remove bearings				12.000		765		765	950	1,200
	Replace bearings				23.529	1,890	1,500		3,390	3,950	4,700
	Remove fan motors				11.552		740		740	915	1,150
	Replace fan motors				25.974	13,500	1,560		15,060	16,800	19,300
	Remove / replace float valve				2.350	77.55	136.30		213.85	253	305
	Total				76.406	15,467.55	4,765.30		20,232.85	22,947.50	26,754
4030	**Replace cooling tower, 1000 ton**	15	Q-6	Ea.							
	Remove cooling tower				104.350		6,250		6,250	7,725	9,650
	Replace cooling tower, 1000 ton				208.700	73,000	12,450		85,450	96,000	110,500
	Total				313.050	73,000	18,700		91,700	103,725	120,150

D3033 130 Chiller, Water Cooled, Reciprocating

	System Description	Freq. (Years)	Crew	Unit	Labor Hours	2019 Bare Costs				Total In-House	Total w/O&P
						Material	Labor	Equipment	Total		
1010	**Repair water cooled chiller, 20 ton**	10	Q-6	Ea.							
	Repair controls				1.000		64		64	79.50	99
	Remove compressor				21.463		1,276		1,276	1,600	1,975
	Replace compressor				42.857	23,500	2,550		26,050	29,000	33,400
	Remove / replace evaporator tube				1.200	300	71.50		371.50	420	485
	Remove / replace condenser tube				1.200	190	71.50		261.50	298	350
	Replace refrigerant				2.667	460	171		631	720	840
	Total				70.387	24,450	4,204		28,654	32,117.50	37,149
1030	**Replace chiller, water cooled, 20 ton, scroll**	20	Q-7	Ea.							
	Remove chiller				50.633		3,075		3,075	3,825	4,775
	Replace chiller, water cooled, 20 ton				88.889	21,200	5,400		26,600	30,000	34,900
	Total				139.522	21,200	8,475		29,675	33,825	39,675

D3033 130 Chiller, Water Cooled, Reciprocating

System Description	Freq. (Years)	Crew	Unit	Labor Hours	2019 Bare Costs Material	Labor	Equipment	Total	Total In-House	Total w/O&P
2010	**Repair water cooled chiller, 50 ton**	10	Q-6	Ea.							
Repair controls				1.000		64		64	79.50	99
Remove compressor				35.610		2,117		2,117	2,625	3,300
Replace compressor				71.006	32,700	4,250		36,950	41,200	47,500
Remove / replace evaporator tube				1.200	300	71.50		371.50	420	485
Remove / replace condenser tube				1.200	190	71.50		261.50	298	350
Replace refrigerant				4.000	690	256.50		946.50	1,075	1,250
Total				114.016	33,880	6,830.50		40,710.50	45,697.50	52,984
2030	**Replace chiller, water cooled 50 ton, scroll**	20	Q-7	Ea.							
Remove chiller				74.074		4,500		4,500	5,600	7,000
Replace chiller, water cooled, 50 ton				114.000	30,900	6,950		37,850	42,600	49,400
Total				188.074	30,900	11,450		42,350	48,200	56,400
3010	**Repair water cooled chiller, 100 ton**	10	Q-6	Ea.							
Repair controls				1.000		64		64	79.50	99
Remove compressor				77.922		4,650		4,650	5,775	7,200
Replace compressor				154.000	37,600	9,375		46,975	53,000	61,500
Remove / replace evaporator tube				4.798	845	292		1,137	1,300	1,500
Remove / replace condenser tube				4.798	845	292		1,137	1,300	1,500
Replace refrigerant				8.000	1,380	513		1,893	2,150	2,525
Total				250.517	40,670	15,186		55,856	63,604.50	74,324
3030	**Replace chiller, water cooled, 100 ton, scroll**	20	Q-7	Ea.							
Remove chiller				117.000		7,100		7,100	8,825	11,000
Replace chiller, water cooled, 100 ton				178.000	61,500	10,800		72,300	81,000	93,500
Total				295.000	61,500	17,900		79,400	89,825	104,500
5010	**Repair water cooled chiller, 200 ton**	10	Q-7	Ea.							
Repair controls				1.000		64		64	79.50	99
Remove compressor				152.099		9,278.50		9,278.50	11,500	14,400
Replace compressor				304.920	66,066	18,480		84,546	95,500	111,000
Remove / replace evaporator tube				9.610	1,700	585		2,285	2,600	3,025
Remove / replace condenser tube				9.610	1,700	585		2,285	2,600	3,025
Replace refrigerant				13.333	2,300	855		3,155	3,600	4,200
Total				490.571	71,766	29,847.50		101,613.50	115,879.50	135,749

D30 HVAC | D3033 Cooling Generating Systems

D3033 130 Chiller, Water Cooled, Reciprocating

System Description	Freq. (Years)	Crew	Unit	Labor Hours	2019 Bare Costs — Material	Labor	Equipment	Total	Total In-House	Total w/O&P
5030 Replace chiller, water cooled, 200 ton	20	Q-7	Ea.							
Remove chiller				163.000		9,950		9,950	12,300	15,400
Replace chiller, water cooled, 200 ton				327.000	96,500	19,900		116,400	131,000	151,500
Total				490.000	96,500	29,850		126,350	143,300	**166,900**

D3033 135 Chiller, Air Cooled, Reciprocating

System Description	Freq. (Years)	Crew	Unit	Labor Hours	2019 Bare Costs — Material	Labor	Equipment	Total	Total In-House	Total w/O&P
1010 Repair recip. chiller, air cooled, 20 ton	10	2 STPI	Ea.							
Repair controls				.800		51		51	63.50	79.50
Remove fan bearing				3.600		229.50		229.50	286	355
Replace fan bearing				6.000	196.50	384		580.50	690	840
Remove fan motor				2.927		187.50		187.50	232	290
Replace fan motor				5.854	855	351		1,206	1,375	1,600
Remove compressor				24.490		1,450		1,450	1,825	2,275
Replace compressor				48.780	28,400	2,900		31,300	34,900	40,000
Replace refrigerant				2.667	460	171		631	720	840
Total				95.117	29,911.50	5,724		35,635.50	40,091.50	**46,279.50**
1030 Replace chiller, 20 ton	20	Q-7	Ea.							
Remove chiller				42.857		2,550		2,550	3,175	3,975
Replace chiller, air cooled, reciprocating, 20 ton				119.000	22,200	7,250		29,450	33,400	39,000
Total				161.857	22,200	9,800		32,000	36,575	**42,975**
2010 Repair recip. chiller, air cooled, 50 ton	10	Q-6	Ea.							
Repair controls				1.000		64		64	79.50	99
Remove fan bearing				3.600		229.50		229.50	286	355
Replace fan bearing				6.667	220.50	426		646.50	770	935
Remove fan motor				3.096		186		186	229	287
Replace fan motor				6.936	2,280	417		2,697	3,025	3,500
Remove compressor				60.976		3,625		3,625	4,525	5,625
Replace compressor				156.000	34,000	9,300		43,300	48,900	57,000
Remove / replace compressor motor				7.201		432		432	535	670
Replace refrigerant				4.000	690	256.50		946.50	1,075	1,250
Total				249.475	37,190.50	14,936		52,126.50	59,424.50	**69,721**

D3033 135 | Chiller, Air Cooled, Reciprocating

System Description	Freq. (Years)	Crew	Unit	Labor Hours	2019 Bare Costs Material	Labor	Equipment	Total	Total In-House	Total w/O&P
2030 Replace chiller, 50 ton	20	Q-7	Ea.							
Remove chiller				73.395		4,475		4,475	5,550	6,925
Replace chiller, air cooled, reciprocating, 50 ton				147.000	39,000	8,925		47,925	54,000	62,500
Total				220.395	39,000	13,400		52,400	59,550	69,425
3010 Repair chiller, air cooled, 100 ton	10	Q-6	Ea.							
Repair controls				1.000		64		64	79.50	99
Remove fan bearing				4.200		268.50		268.50	335	415
Replace fan bearing				6.667	220.50	426		646.50	770	935
Remove fan motor				3.900		234		234	289	360
Replace fan motor				7.792	4,050	468		4,518	5,025	5,775
Remove compressor				97.403		5,812.50		5,812.50	7,225	9,000
Replace compressor				192.500	47,000	11,718.75		58,718.75	66,000	77,000
Remove / replace compressor motor				9.359		562		562	695	870
Replace refrigerant				6.667	1,150	427.50		1,577.50	1,800	2,100
Total				329.487	52,420.50	19,981.25		72,401.75	82,218.50	96,554
3030 Replace chiller, 100 ton	20	Q-7	Ea.							
Remove chiller				84.211		5,125		5,125	6,350	7,950
Replace chiller, air cooled, reciprocating, 100 ton				128.000	69,000	7,800		76,300	85,500	98,500
Total				212.211	69,000	12,925		81,325	91,850	106,450

D3033 137 | Chiller, Water Cooled, Scroll

System Description	Freq. (Years)	Crew	Unit	Labor Hours	2019 Bare Costs Material	Labor	Equipment	Total	Total In-House	Total w/O&P
4010 Repair water cooled chiller, 5 ton	10	1 STPI	Ea.							
Repair controls				.500		32		32	39.50	49.50
Remove fan bearing				1.200		76.50		76.50	95	119
Replace fan bearing				2.000	65.50	128		193.50	231	280
Remove compressor				2.477		158		158	197	246
Replace compressor				4.954	520	315		835	965	1,150
Replace refrigerant				.667	115	42.75		157.75	179	210
Total				11.797	700.50	752.25		1,452.75	1,706.50	2,054.50

For customer support on your Facilities Maintenance & Repair Costs with RSMeans data, call 800.448.8182.

D30 HVAC | D3033 | Cooling Generating Systems

D3033 137 | Chiller, Water Cooled, Scroll

System Description	Freq. (Years)	Crew	Unit	Labor Hours	2019 Bare Costs				Total In-House	Total w/O&P
					Material	Labor	Equipment	Total		
4030 Replace chiller, water cooled, 5 ton	20	Q-5	Ea.							
Remove chiller				18.182		1,050		1,050	1,300	1,625
Replace chiller, water cooled, 5 ton				36.364	3,550	2,100		5,650	6,500	7,700
Total				54.545	3,550	3,150		6,700	7,800	9,325
5010 Repair water cooled chiller, 10 ton	10	2 STPI	Ea.							
Repair controls				1.000		64		64	79.50	99
Remove fan bearing				3.600		229.50		229.50	286	355
Replace fan bearing				6.000	196.50	384		580.50	690	840
Remove compressor				4.954		316		316	395	490
Replace compressor				9.907	1,040	630		1,670	1,925	2,275
Replace refrigerant				1.333	230	85.50		315.50	360	420
Total				26.794	1,466.50	1,709		3,175.50	3,735.50	4,479
5030 Replace chiller, water cooled, 10 ton	20	Q-6	Ea.							
Remove chiller				43.478		2,600		2,600	3,225	4,025
Replace chiller, water cooled, 10 ton				86.957	6,725	5,200		11,925	13,800	16,500
Total				130.435	6,725	7,800		14,525	17,025	20,525
6010 Repair water cooled chiller, 15 ton	10	Q-5	Ea.							
Repair controls				1.000		64		64	79.50	99
Remove fan bearing				24.000		1,275		1,275	1,650	2,050
Replace fan bearing				6.000	196.50	384		580.50	690	840
Remove compressor				6.932		398		398	495	620
Replace compressor				13.865	2,350	800		3,150	3,575	4,175
Replace refrigerant				2.000	345	128.25		473.25	540	630
Total				53.797	2,891.50	3,049.25		5,940.75	7,029.50	8,414
6030 Replace chiller, water cooled, 15 ton, scroll	20	Q-6	Ea.							
Remove chiller				42.857		2,550		2,550	3,175	3,975
Replace chiller, water cooled, 15 ton				66.667	20,500	3,975		24,475	27,500	31,800
Total				109.524	20,500	6,525		27,025	30,675	35,775

For customer support on your Facilities Maintenance & Repair Costs with RSMeans data, call 800.448.8182.

233

D3033 140 Chiller, Hermetic Centrifugal

	System Description	Freq. (Years)	Crew	Unit	Labor Hours	2019 Bare Costs				Total In-House	Total w/O&P
						Material	Labor	Equipment	Total		
1010	**Repair centrifugal chiller, 100 ton**	10	Q-7	Ea.							
	Repair controls				1.000		64		64	79.50	99
	Remove compressor				48.780		2,900		2,900	3,600	4,500
	Replace compressor				109.600	28,240	6,680		34,920	39,400	45,700
	Remove / replace condenser tube				4.800	760	286		1,046	1,200	1,400
	Remove / replace evaporator tube				4.800	1,200	286		1,486	1,675	1,950
	Remove / replace refrigerant filter				1.600	197	97		294	340	395
	Remove refrigerant				.440		28.10		28.10	35	43.50
	Replace refrigerant				8.000	1,380	513		1,893	2,150	2,525
	Total				179.020	31,777	10,854.10		42,631.10	48,479.50	56,612.50
1030	**Replace centrifugal chiller, 100 ton**	20	Q-7	Ea.							
	Remove chiller				112.000		6,800		6,800	8,450	10,600
	Replace chiller, centrifugal, 100 ton				224.000	92,500	13,600		106,100	118,500	136,500
	Total				336.000	92,500	20,400		112,900	126,950	147,100
2010	**Repair centrifugal chiller, 300 ton**	10	Q-7	Ea.							
	Repair controls				1.000		64		64	79.50	99
	Remove compressor				94.390		5,611.50		5,611.50	7,000	8,700
	Replace compressor				187.000	38,900	11,400		50,300	57,000	66,500
	Remove / replace condenser tube				12.000	1,900	715		2,615	2,975	3,475
	Remove / replace evaporator tube				12.000	3,000	715		3,715	4,200	4,850
	Remove / replace refrigerant filter				1.200	810	73	2.10	885.10	985	1,125
	Remove refrigerant				1.319		84		84	105	131
	Replace refrigerant				21.333	3,680	1,368		5,048	5,750	6,700
	Total				330.242	48,290	20,030.50	2.10	68,322.60	78,094.50	91,580
2030	**Replace centrifugal chiller, 300 ton**	20	Q-7	Ea.							
	Remove chiller				150.000		9,150		9,150	11,300	14,200
	Replace chiller, 300 ton				302.000	151,500	18,400		169,900	189,500	218,000
	Total				452.000	151,500	27,550		179,050	200,800	232,200

For customer support on your Facilities Maintenance & Repair Costs with RSMeans data, call 800.448.8182.

D3033 140 Chiller, Hermetic Centrifugal

System Description	Freq. (Years)	Crew	Unit	Labor Hours	2019 Bare Costs Material	Labor	Equipment	Total	Total In-House	Total w/O&P
3010										
Repair centrifugal chiller, 1000 ton	10	Q-7	Ea.							
Repair controls				1.000		64		64	79.50	99
Remove compressor				153.659		9,135		9,135	11,400	14,200
Replace compressor				296.000	77,500	18,000		95,500	107,500	125,000
Remove / replace condenser tube				16.800	2,660	1,071		3,731	4,250	5,000
Remove / replace evaporator tube				12.600	4,200	805		5,005	5,625	6,500
Remove / replace refrigerant filter				1.600	810	102		912	1,025	1,175
Remove refrigerant				3.516		224		224	279	350
Replace refrigerant				29.333	5,060	1,881		6,941	7,900	9,225
Total				514.508	90,230	31,282		121,512	138,058.50	161,549
3030										
Replace centrifugal chiller, 1000 ton	20	Q-7	Ea.							
Remove chiller				242.000		14,800		14,800	18,300	22,900
Replace chiller, 1000 ton				485.000	375,000	29,500		404,500	449,000	514,500
Total				727.000	375,000	44,300		419,300	467,300	537,400
6010										
Repair centrifugal chiller, 9000 ton	10	Q-7	Ea.							
Repair controls				1.000		64		64	79.50	99
Remove compressor				1083.600		64,680		64,680	80,500	100,000
Replace compressor				2112.000	531,000	128,400		659,400	743,500	863,000
Remove / replace condenser tube				151.204	23,940	9,639		33,579	38,300	44,900
Remove / replace evaporator tube				113.399	37,800	7,245		45,045	50,500	58,500
Remove / replace refrigerant filter				14.400	7,290	918		8,208	9,150	10,500
Remove refrigerant				31.648		2,016		2,016	2,500	3,125
Replace refrigerant				263.993	45,540	16,929		62,469	71,000	83,000
Total				3771.244	645,570	229,891		875,461	995,529.50	1,163,124
6030										
Replace centrifugal chiller, 9000 ton	20	Q-7	Ea.							
Remove chiller				1452.000		88,800		88,800	110,000	137,500
Replace chiller, 9000 ton				3312.000	2,997,000	201,600		3,198,600	3,546,500	4,058,500
Total				4764.000	2,997,000	290,400		3,287,400	3,656,500	4,196,000

D3033 142   Chiller, Open Centrifugal

	System Description	Freq. (Years)	Crew	Unit	Labor Hours	Material	Labor	Equipment	Total	Total In-House	Total w/O&P
						2019 Bare Costs					
1010	**Repair open centrifugal chiller, 300 ton**	10	Q-7	Ea.							
	Repair controls				1.000		64		64	79.50	99
	Remove compressor				104.878		6,235		6,235	7,775	9,675
	Replace compressor				205.700	42,790	12,540		55,330	62,500	73,000
	Remove / replace condenser tube				12.000	1,900	715		2,615	2,975	3,475
	Remove / replace evaporator tube				12.000	3,000	715		3,715	4,200	4,850
	Remove / replace refrigerant filter				1.200	810	73	2.10	885.10	985	1,125
	Remove refrigerant				20.000		1,062.50		1,062.50	1,375	1,725
	Replace refrigerant				21.333	3,680	1,368		5,048	5,750	6,700
	Total				378.111	52,180	22,772.50	2.10	71,954.60	85,639.50	100,649
1030	**Replace open centrifugal chiller, 300 ton**	20	Q-7	Ea.							
	Remove chiller				128.000		7,805		7,805	9,675	12,100
	Replace chiller, 300 ton				257.400	166,650	15,620		182,270	202,500	232,500
	Total				385.400	166,650	23,425		190,075	212,175	244,600
2010	**Repair open centrifugal chiller, 1000 ton**	10	Q-7	Ea.							
	Repair controls				1.000		64		64	79.50	99
	Remove compressor				165.854		9,860		9,860	12,300	15,300
	Replace compressor				325.600	85,250	19,800		105,050	118,500	137,500
	Remove / replace condenser tube				16.800	2,660	1,001		3,661	4,175	4,875
	Remove / replace evaporator tube				16.800	4,200	1,001		5,201	5,875	6,800
	Remove / replace refrigerant filter				1.600	1,475	97.50	2.80	1,575.30	1,750	2,000
	Remove refrigerant				35.204		2,248		2,248	2,800	3,475
	Replace refrigerant				34.666	5,980	2,223		8,203	9,325	10,900
	Total				597.523	99,565	36,294.50	2.80	135,862.30	154,804.50	180,949
2030	**Replace open centrifugal chiller, 1000 ton**	20	Q-7	Ea.							
	Remove chiller				219.800		13,370		13,370	16,600	20,700
	Replace chiller, 1000 ton				409.200	412,500	24,970		437,470	484,500	554,000
	Total				629.000	412,500	38,340		450,840	501,100	574,700

For customer support on your Facilities Maintenance & Repair Costs with RSMeans data, call 800.448.8182.

	System Description	Freq. (Years)	Crew	Unit	Labor Hours	2019 Bare Costs				Total In-House	Total w/O&P
						Material	Labor	Equipment	Total		
1010	**Repair chiller, absorption, 100 ton**	10	2 STPI	Ea.							
	Repair controls				1.000		64		64	79.50	99
	Remove / replace concentrator tube				.900	300	52		352	395	455
	Remove / replace condenser tube				6.000	1,500	357.50		1,857.50	2,100	2,425
	Remove / replace evaporator tube				6.000	950	357.50		1,307.50	1,500	1,750
	Remove / replace absorber tube				.900	300	52		352	395	455
	Remove / replace gang pump				1.600	7,750	92		7,842	8,650	9,825
	Remove gang pump motor				3.704		222		222	274	345
	Replace gang pump motor				7.407	5,975	445		6,420	7,125	8,150
	Add refrigerant				8.000	409	510		919	1,075	1,300
	Total				35.511	17,184	2,152		19,336	21,593.50	24,804
1030	**Replace chiller, absorption, 100 ton**	20	Q-7	Ea.							
	Remove chiller				155.000		9,450		9,450	11,700	14,700
	Replace chiller				311.000	132,500	18,900		151,400	169,000	195,000
	Total				466.000	132,500	28,350		160,850	180,700	209,700
2010	**Repair chiller, absorption, 350 ton**	10	2 STPI	Ea.							
	Repair controls				1.000		64		64	79.50	99
	Remove / replace concentrator tube				1.800	600	104		704	790	910
	Remove / replace condenser tube				7.200	1,800	429		2,229	2,525	2,925
	Remove / replace evaporator tube				7.200	1,140	429		1,569	1,775	2,100
	Remove / replace absorber tube				1.800	600	104		704	790	910
	Remove / replace gang pump				1.600	7,750	92		7,842	8,650	9,825
	Remove gang pump motor				3.704		222		222	274	345
	Replace gang pump motor				7.407	5,975	445		6,420	7,125	8,150
	Add refrigerant				8.000	409	510		919	1,075	1,300
	Total				39.711	18,274	2,399		20,673	23,083.50	26,564
2030	**Replace chiller, absorption, 350 ton**	20	Q-7	Ea.							
	Remove chiller				204.000		12,400		12,400	15,400	19,200
	Replace chiller				400.000	368,000	24,400		392,400	435,000	498,000
	Total				604.000	368,000	36,800		404,800	450,400	517,200

For customer support on your Facilities Maintenance & Repair Costs with RSMeans data, call 800.448.8182.

237

D3033 145 Chiller, Absorption

| | | | | 2019 Bare Costs | | | | | Total | Total |
System Description	Freq. (Years)	Crew	Unit	Labor Hours	Material	Labor	Equipment	Total	In-House	w/O&P
3010 Repair chiller, absorption, 950 ton	10	4 STPI	Ea.							
Repair controls				1.000		64		64	79.50	99
Remove / replace concentrator tube				3.600	1,200	208		1,408	1,575	1,825
Remove / replace condenser tube				12.000	3,000	715		3,715	4,200	4,850
Remove / replace evaporator tube				12.000	1,900	715		2,615	2,975	3,475
Remove / replace absorber tube				3.600	1,200	208		1,408	1,575	1,825
Remove / replace gang pump				1.600	7,750	92		7,842	8,650	9,825
Remove gang pump motor				3.704		222		222	274	345
Replace gang pump motor				8.696	7,750	520		8,270	9,175	10,500
Add refrigerant				8.000	409	510		919	1,075	1,300
Total				54.198	23,209	3,254		26,463	29,578.50	34,044
3030 Replace chiller, absorption, 950 ton	20	Q-7	Ea.							
Remove chiller				267.000		16,200		15,200	20,100	25,200
Replace chiller				533.000	752,500	32,500		785,000	868,000	991,000
Total				800.000	752,500	48,700		801,200	888,100	1,016,200

D3033 210 Air Cooled Condenser

| | | | | 2019 Bare Costs | | | | | Total | Total |
System Description	Freq. (Years)	Crew	Unit	Labor Hours	Material	Labor	Equipment	Total	In-House	w/O&P
1010 Repair condenser, air cooled, 5 ton	10	1 STPI	Ea.							
Repair controls				.500		32		32	39.50	49.50
Remove fan motor				1.800		115		115	143	178
Replace fan motor				2.311	305	139		444	505	595
Total				4.611	305	286		591	687.50	822.50
1030 Replace condenser, air cooled, 5 ton	15	Q-5	Ea.							
Remove condenser				5.195		299		299	370	465
Replace condenser, air cooled, 5 ton				10.390	5,225	600		5,825	6,500	7,450
Total				15.584	5,225	899		6,124	6,870	7,915
2010 Repair condenser, air cooled, 20 ton	10	2 STPI	Ea.							
Repair controls				1.000		64		64	79.50	99
Remove fan motor				2.310		148		148	183	230
Replace fan motor				4.622	740	278		1,018	1,150	1,350
Total				7.932	740	490		1,230	1,412.50	1,679

For customer support on your Facilities Maintenance & Repair Costs with RSMeans data, call 800.448.8182.

D3033 210 | **Air Cooled Condenser**

System Description	Freq. (Years)	Crew	Unit	Labor Hours	2019 Bare Costs				Total In-House	Total w/O&P
					Material	Labor	Equipment	Total		
2030 Replace condenser, air cooled, 20 ton	15	Q-5	Ea.							
Remove condenser				10.390		600		600	740	925
Replace condenser, air cooled, 20 ton				20.779	11,500	1,200		12,700	14,100	16,200
Total				31.169	11,500	1,800		13,300	14,840	17,125
3010 Repair condenser, air cooled, 50 ton	10	2 STPI	Ea.							
Repair controls				1.000		64		64	79.50	99
Remove fan motor				3.466		222		222	275	345
Replace fan motor				6.936	2,280	417		2,697	3,025	3,500
Total				11.402	2,280	703		2,983	3,379.50	3,944
3030 Replace condenser, air cooled, 50 ton	15	Q-6	Ea.							
Remove condenser				38.961		2,325		2,325	2,875	3,600
Replace condenser, air cooled, 50 ton				77.922	25,600	4,650		30,250	33,900	39,200
Total				116.883	25,600	6,975		32,575	36,775	42,800
4010 Repair condenser, air cooled, 100 ton	10	2 STPI	Ea.							
Repair controls				1.000		64		64	79.50	99
Remove fan motor				6.931		444		444	550	690
Replace fan motor				13.873	4,560	834		5,394	6,050	6,975
Total				21.804	4,560	1,342		5,902	6,679.50	7,764
4030 Replace condenser, air cooled, 100 ton	15	Q-7	Ea.							
Remove condenser				69.264		4,225		4,225	5,225	6,550
Replace condenser, air cooled, 100 ton				139.000	52,000	8,425		60,425	67,500	78,000
Total				208.264	52,000	12,650		64,650	72,725	84,550

For customer support on your Facilities Maintenance & Repair Costs with RSMeans data, call 800.448.8182.

239

D3033 260   **Evaporative Condenser**

	System Description	Freq. (Years)	Crew	Unit	Labor Hours	2019 Bare Costs Material	Labor	Equipment	Total	Total In-House	Total w/O&P
1010	**Repair evaporative condenser, 20 ton**	10	2 STPI	Ea.							
	Repair controls				1.000		64		64	79.50	99
	Remove fan motors				2.310		148		148	183	230
	Replace fan motors				4.622	740	278		1,018	1,150	1,350
	Remove pump and motor				20.779		1,175		1,175	1,475	1,825
	Replace pump and motor				41.558	4,425	2,350		6,775	7,800	9,200
	Remove / replace pump nozzles				5.000	46.90	290		336.90	410	505
	Remove / replace float valve				.500	16.50	29		45.50	54	65
	Total				75.770	5,228.40	4,334		9,562.40	11,151.50	13,274
1030	**Replace evaporative condenser, 20 ton**	15	Q-5	Ea.							
	Remove condenser				22.130		1,275		1,275	1,575	1,975
	Replace evaporative condenser, 20 ton				44.199	12,800	2,550		15,350	17,200	20,000
	Total				66.329	12,800	3,825		16,625	18,775	21,975
2010	**Repair evaporative condenser, 100 ton**	10	2 STPI	Ea.							
	Repair controls				1.000		64		64	79.50	99
	Remove fan motors				6.931		444		444	550	690
	Replace fan motors				13.873	4,560	834		5,394	6,050	6,975
	Remove pump and motor				20.779		1,175		1,175	1,475	1,825
	Replace pump and motor				41.558	4,425	2,350		6,775	7,800	9,200
	Remove / replace nozzles				25.000	234.50	1,450		1,684.50	2,050	2,525
	Remove / replace float valve				.500	16.50	29		45.50	54	65
	Total				109.642	9,236	6,346		15,582	18,058.50	21,379
2030	**Replace evaporative condenser, 100 ton**	15	Q-7	Ea.							
	Remove condenser				57.762		3,525		3,525	4,375	5,450
	Replace evaporative condenser, 100 ton				116.000	18,400	7,025		25,425	29,000	33,900
	Total				173.762	18,400	10,550		28,950	33,375	39,350

For customer support on your Facilities Maintenance & Repair Costs with RSMeans data, call 800.448.8182.

D3033 260 Evaporative Condenser

System Description	Freq. (Years)	Crew	Unit	Labor Hours	2019 Bare Costs				Total In-House	Total w/O&P
					Material	Labor	Equipment	Total		
3010 **Repair evaporative condenser, 300 ton**	10	2 STPI	Ea.							
Repair controls				1.000		64		64	79.50	99
Remove fan motors				11.552		740		740	915	1,150
Replace fan motors				23.121	7,600	1,390		8,990	10,100	11,600
Remove pump and motor				20.779		1,175		1,175	1,475	1,825
Replace pump and motor				24.000	5,025	1,425		6,450	7,275	8,475
Remove / replace nozzles				35.000	328.30	2,030		2,358.30	2,850	3,525
Remove / replace float valve				1.000	33	58		91	108	130
Total				116.453	12,986.30	6,882		19,868.30	22,802.50	26,804
3030 **Replace evaporative condenser, 300 ton**	15	Q-7	Ea.							
Remove condenser				152.471		9,274.50		9,274.50	11,500	14,400
Replace evaporative condenser, 300 ton				304.560	60,588	18,630		79,218	89,500	104,500
Total				457.031	60,588	27,904.50		88,492.50	101,000	118,900

For customer support on your Facilities Maintenance & Repair Costs with RSMeans data, call 800.448.8182.

241

D3043 120 | **Fan Coil**

System Description	Freq. (Years)	Crew	Unit	Labor Hours	2019 Bare Costs				Total In-House	Total w/O&P
					Material	Labor	Equipment	Total		
1010 Repair fan coil unit, 1 ton	10	1 STPI	Ea.							
Remove fan motor				1.200		76.50		76.50	95	119
Replace fan motor				1.951	240	117		357	410	480
Total				3.151	240	193.50		433.50	505	599
1030 Replace fan coil unit, 1 ton	15	Q-5	Ea.							
Remove fan coil unit				1.733		100		100	124	155
Replace fan coil unit, 1 ton				3.467	855	200		1,055	1,200	1,375
Total				5.200	855	300		1,155	1,324	1,530
2010 Repair fan coil unit, 3 ton	10	1 STPI	Ea.							
Remove fan motor				1.200		76.50		76.50	95	119
Replace fan motor				1.951	285	117		402	460	535
Total				3.151	285	193.50		478.50	555	654
2030 Replace fan coil unit, 3 ton	15	Q-5	Ea.							
Remove fan coil unit				2.597		149		149	185	232
Replace fan coil unit, 3 ton				5.195	1,975	299		2,274	2,550	2,925
Total				7.792	1,975	448		2,423	2,735	3,157
3010 Repair fan coil unit, 5 ton	10	1 STPI	Ea.							
Remove fan motor				1.200		76.50		76.50	95	119
Replace fan motor				2.311	305	139		444	505	595
Total				3.511	305	215.50		520.50	600	714
3030 Replace fan coil unit, 5 ton	15	Q-5	Ea.							
Remove fan coil unit				2.680		154		154	191	239
Replace fan coil unit, 5 ton				5.371	1,786	310.20		2,096.20	2,350	2,700
Total				8.052	1,786	464.20		2,250.20	2,541	2,939
4010 Repair fan coil unit, 10 ton	10	1 STPI	Ea.							
Remove fan motor				1.400		89.50		89.50	111	139
Replace fan motor				2.312	760	139		899	1,000	1,175
Total				3.712	760	228.50		983.50	1,111	1,314

D30 HVAC | D3043 | Distribution Systems

D3043 120 | Fan Coil

System Description	Freq. (Years)	Crew	Unit	Labor Hours	2019 Bare Costs				Total In-House	Total w/O&P
					Material	Labor	Equipment	Total		
4030 Replace fan coil unit, 10 ton	15	Q-6	Ea.							
Remove fan coil unit				11.538		690		690	855	1,075
Replace fan coil unit, 10 ton				23.077	2,325	1,375		3,700	4,275	5,025
Total				34.615	2,325	2,065		4,390	5,130	**6,100**
5010 Repair fan coil unit, 20 ton	10	1 STPI	Ea.							
Remove fan motor				1.600		102		102	127	159
Replace fan motor				2.311	610	139		749	840	975
Total				3.911	610	241		851	967	**1,134**
5030 Replace fan coil unit, 20 ton	15	Q-6	Ea.							
Remove fan coil unit				19.512		1,175		1,175	1,450	1,800
Replace fan coil unit, 20 ton				39.024	3,625	2,325		5,950	6,875	8,125
Total				58.537	3,625	3,500		7,125	8,325	**9,925**
6010 Repair fan coil unit, 30 ton	10	1 STPI	Ea.							
Remove fan motor				1.800		115		115	143	178
Replace fan motor				2.597	1,350	156		1,506	1,675	1,925
Total				4.398	1,350	271		1,621	1,818	**2,103**
6030 Replace fan coil unit, 30 ton	15	Q-6	Ea.							
Remove fan coil unit				25.974		1,550		1,550	1,925	2,400
Replace fan coil unit, 30 ton				51.948	6,475	3,100		9,575	11,000	12,900
Total				77.922	6,475	4,650		11,125	12,925	**15,300**

	System Description	Freq. (Years)	Crew	Unit	Labor Hours	2019 Bare Costs				Total In-House	Total w/O&P
						Material	Labor	Equipment	Total		
1010	**Repair fan coil, DX 1-1/2 ton, cooling only**	10	1 STPI	Ea.							
	Repair controls				.300		19.20		19.20	24	30
	Remove / replace supply fan				.650		41.50		41.50	51.50	64.50
	Remove supply fan bearing				1.000		64		64	79.50	99
	Replace supply fan bearing				2.000	65.50	128		193.50	231	280
	Remove supply fan motor				1.156		69.50		69.50	85.50	107
	Replace supply fan motor				1.951	270	125		395	450	530
	Remove compressor				2.477		158		158	197	246
	Replace compressor				3.714	276	238		514	600	715
	Replace refrigerant				.267	46	17.10		63.10	72	84
	Total				13.515	657.50	860.30		1,517.80	1,790.50	2,155.50
1040	**Replace fan coil, DX 1-1/2 ton, no heat**	15	Q-5	Ea.							
	Remove fan coil unit				2.080		120		120	149	186
	Replace fan coil unit, 1-1/2 ton, no heat				4.160	650	239		889	1,000	1,175
	Total				6.240	650	359		1,009	1,149	1,361
2010	**Repair fan coil, DX 2 ton, cooling only**	10	1 STPI	Ea.							
	Repair controls				.300		19.20		19.20	24	30
	Remove / replace supply fan				.650		41.50		41.50	51.50	64.50
	Remove supply fan bearing				1.000		64		64	79.50	99
	Replace supply fan bearing				2.000	65.50	128		193.50	231	280
	Remove supply fan motor				1.156		69.50		69.50	85.50	107
	Replace supply fan motor				1.951	240	117		357	410	480
	Remove compressor				2.477		158		158	197	246
	Replace compressor				4.000	285	256		541	630	750
	Replace refrigerant				.400	69	25.65		94.65	108	126
	Total				13.934	659.50	878.85		1,538.35	1,816.50	2,182.50
2040	**Replace fan coil, DX 2 ton, no heat**	15	Q-5	Ea.							
	Remove fan coil unit				2.167		125		125	155	193
	Replace fan coil unit, DX 2 ton, no heat				4.334	700	249		949	1,075	1,250
	Total				6.500	700	374		1,074	1,230	1,443

For customer support on your Facilities Maintenance & Repair Costs with RSMeans data, call 800.448.8182.

		System Description	Freq. (Years)	Crew	Unit	Labor Hours	2019 Bare Costs				Total In-House	Total w/O&P
							Material	Labor	Equipment	Total		
3010		**Repair fan coil, DX 2-1/2 ton**	10	1 STPI	Ea.							
		Repair controls				.300		19.20		19.20	24	30
		Remove / replace supply fan				.650		41.50		41.50	51.50	64.50
		Remove supply fan bearing				1.000		64		64	79.50	99
		Replace supply fan bearing				2.000	65.50	128		193.50	231	280
		Remove supply fan motor				1.156		69.50		69.50	85.50	107
		Replace supply fan motor				1.951	206	117		323	370	440
		Remove compressor				2.477		158		158	197	246
		Replace compressor				4.334	335	277		612	710	850
		Replace refrigerant				.533	92	34.20		126.20	144	168
	Total					14.401	698.50	908.40		1,606.90	1,892.50	2,284.50
3040		**Replace fan coil, DX 2-1/2 ton, no heat**	15	Q-5	Ea.							
		Remove fan coil unit				2.363		136		136	169	211
		Replace fan coil, DX 2-1/2 ton, no heat				4.727	735	272		1,007	1,150	1,350
	Total					7.090	735	408		1,143	1,319	1,561
4010		**Repair fan coil, DX 3 ton, cooling only**	10	1 STPI	Ea.							
		Repair controls				.300		19.20		19.20	24	30
		Remove / replace supply fan				.650		41.50		41.50	51.50	64.50
		Remove supply fan bearing				1.000		64		64	79.50	99
		Replace supply fan bearing				2.000	65.50	128		193.50	231	280
		Remove supply fan motor				1.156		69.50		69.50	85.50	107
		Replace supply fan motor				1.951	206	117		323	370	440
		Remove compressor				2.477		158		158	197	246
		Replace compressor				3.844	314.50	245.65		560.15	650	775
		Replace refrigerant				.533	92	34.20		126.20	144	168
	Total					13.911	678	877.05		1,555.05	1,832.50	2,209.50
4040		**Replace fan coil, DX 3 ton, no heat**	15	Q-5	Ea.							
		Remove fan coil unit				2.737		158		158	195	244
		Replace fan coil, DX 3 ton, no heat				5.474	890	315		1,205	1,375	1,600
	Total					8.211	890	473		1,363	1,570	1,844

D3043 122 Fan Coil, DX Air Conditioner, Cooling Only

System Description	Freq. (Years)	Crew	Unit	2019 Bare Costs					Total In-House	Total w/O&P
				Labor Hours	Material	Labor	Equipment	Total		
5050 **Repair fan coil, DX 5 ton, cooling only**	10	1 STPI	Ea.							
Repair controls				.300		19.20		19.20	24	30
Remove / replace supply fan				.650		41.50		41.50	51.50	64.50
Remove supply fan bearing				1.000		64		64	79.50	99
Replace supply fan bearing				2.000	65.50	128		193.50	231	280
Remove supply fan motor				1.156		69.50		69.50	85.50	107
Replace supply fan motor				1.951	206	117		323	370	440
Remove compressor				2.477		158		158	197	246
Replace compressor				4.954	520	315		835	965	1,150
Replace refrigerant				.533	92	34.20		126.20	144	168
Total				15.021	883.50	946.40		1,829.90	2,147.50	2,584.50
6060 **Replace fan coil, DX 5 ton, no heat**	15	Q-5	Ea.							
Remove fan coil unit				3.466		199		199	247	310
Replace fan coil, DX 5 ton, no heat				6.932	1,125	400		1,525	1,725	2,025
Total				10.399	1,125	599		1,724	1,972	2,335
7070 **Repair fan coil, DX 10 ton, cooling only**	10	Q-6	Ea.							
Repair controls				.300		19.20		19.20	24	30
Remove / replace supply fan				.650		41.50		41.50	51.50	64.50
Remove supply fan bearing				1.000		64		64	79.50	99
Replace supply fan bearing				2.000	65.50	128		193.50	231	280
Remove supply fan motor				1.156		69.50		69.50	85.50	107
Replace supply fan motor				2.311	370	139		509	580	675
Remove compressor				14.513		871		871	1,075	1,350
Replace compressor				28.714	15,745	1,708.50		17,453.50	19,400	22,300
Replace refrigerant				.800	138	51.30		189.30	215	252
Total				51.444	16,318.50	3,092		19,410.50	21,741.50	25,157.50
8080 **Replace fan coil, DX 10 ton, no heat**	15	Q-6	Ea.							
Remove fan coil unit				6.000		360		360	445	555
Replace fan coil, DX 10 ton, no heat				12.000	2,125	715		2,340	3,225	3,750
Total				18.000	2,125	1,075		3,200	3,670	4,305

For customer support on your Facilities Maintenance & Repair Costs with RSMeans data, call 800.448.8182.

D3043 122 Fan Coil, DX Air Conditioner, Cooling Only

	System Description	Freq. (Years)	Crew	Unit	Labor Hours	2019 Bare Costs				Total In-House	Total w/O&P
						Material	Labor	Equipment	Total		
9090	**Repair fan coil, DX 20 ton, cooling only**	10	Q-6	Ea.							
	Repair controls				.300		19.20		19.20	24	30
	Remove / replace supply fan				.650		41.50		41.50	51.50	64.50
	Remove supply fan bearing				1.000		64		64	79.50	99
	Replace supply fan bearing				2.000	65.50	128		193.50	231	280
	Remove supply fan motor				1.156		69.50		69.50	85.50	107
	Replace supply fan motor				2.476	1,100	149		1,249	1,400	1,600
	Remove compressor				21.929		1,305		1,305	1,625	2,025
	Replace compressor				43.902	25,560	2,610		28,170	31,400	36,000
	Replace refrigerant				1.333	230	85.50		315.50	360	420
	Total				74.747	26,955.50	4,471.70		31,427.20	35,256.50	40,625.50
9590	**Replace fan coil, DX 20 ton, no heat**	15	Q-6	Ea.							
	Remove fan coil unit				22.284		1,325		1,325	1,650	2,050
	Replace fan coil, DX 20 ton, no heat				44.610	4,075	2,650		6,725	7,775	9,225
	Total				66.894	4,075	3,975		8,050	9,425	11,275
9600	**Repair fan coil, DX 30 ton, cooling only**	10	Q-6	Ea.							
	Repair controls				.300		19.20		19.20	24	30
	Remove / replace supply fan				.650		41.50		41.50	51.50	64.50
	Remove supply fan bearing				1.000		64		64	79.50	99
	Replace supply fan bearing				2.000	65.50	128		193.50	231	280
	Remove supply fan motor				1.156		69.50		69.50	85.50	107
	Replace supply fan motor				2.476	1,100	149		1,249	1,400	1,600
	Remove compressor				24.490		1,450		1,450	1,825	2,275
	Replace compressor				48.780	28,400	2,900		31,300	34,900	40,000
	Replace refrigerant				1.733	299	111.15		410.15	465	545
	Total				82.586	29,864.50	4,932.35		34,796.85	39,061.50	45,000.50
9610	**Replace fan coil, DX 30 ton, no heat**	15	Q-6	Ea.							
	Remove fan coil unit				25.974		1,550		1,550	1,925	2,400
	Replace fan coil, DX 20 ton, no heat				48.000	5,300	2,875		8,175	9,375	11,100
	Total				73.974	5,300	4,425		9,725	11,300	13,500

For customer support on your Facilities Maintenance & Repair Costs with RSMeans data, call 800.448.8182.

247

D3043 124   Fan Coil, DX Air Conditioner W/ Heat

System Description	Freq. (Years)	Crew	Unit	Labor Hours	2019 Bare Costs Material	Labor	Equipment	Total	Total In-House	Total w/O&P
1010 **Replace fan coil, DX 1-1/2 ton, with heat**	15	Q-5	Ea.							
Remove fan coil unit				2.312		133		133	165	206
Replace fan coil unit, 1-1/2 ton, with heat				6.474	910	372.40		1,282.40	1,475	1,725
Total				8.786	910	505.40		1,415.40	**1,640**	**1,931**
2010 **Replace fan coil, DX 2 ton, with heat**	15	Q-5	Ea.							
Remove fan coil unit				2.407		139		139	172	215
Replace fan coil unit, DX 2 ton, with heat				6.741	980	387.80		1,367.80	1,550	1,825
Total				9.148	980	526.80		1,506.80	**1,722**	**2,040**
3010 **Replace fan coil, DX 2-1/2 ton, with heat**	15	Q-5	Ea.							
Remove fan coil unit				2.626		151		151	187	234
Replace fan coil, DX 2-1/2 ton, with heat				7.354	1,029	420		1,449	1,650	1,950
Total				9.979	1,029	571		1,600	**1,837**	**2,184**
4010 **Replace fan coil, DX 3 ton, with heat**	15	Q-5	Ea.							
Remove fan coil unit				3.042		175		175	217	271
Replace fan coil, DX 3 ton, with heat				8.514	1,246	490		1,736	1,975	2,325
Total				11.556	1,246	665		1,911	**2,192**	**2,596**
5010 **Replace fan coil, DX 5 ton, with heat**	10	Q-5	Ea.							
Remove fan coil unit				3.852		222		222	275	345
Replace fan coil, DX 5 ton, with heat				10.785	1,575	623		2,198	2,500	2,925
Total				14.637	1,575	845		2,420	**2,775**	**3,270**
6010 **Replace fan coil, DX 10 ton, with heat**	10	Q-6	Ea.							
Remove fan coil unit				6.667		400		400	495	615
Replace fan coil, DX 10 ton, with heat				18.667	2,975	1,113		4,088	4,650	5,425
Total				25.333	2,975	1,513		4,488	**5,145**	**6,040**
7010 **Replace fan coil, DX 20 ton, with heat**	10	Q-6	Ea.							
Remove fan coil unit				24.793		1,475		1,475	1,825	2,300
Replace fan coil, DX 20 ton, with heat				69.421	5,705	4,130		9,835	11,400	13,600
Total				94.215	5,705	5,605		11,310	**13,225**	**15,900**

For customer support on your Facilities Maintenance & Repair Costs with RSMeans data, call 800.448.8182.

D3043 128   Unit Ventilator

System Description	Freq. (Years)	Crew	Unit	Labor Hours	2019 Bare Costs Material	Labor	Equipment	Total	Total In-House	Total w/O&P
1010 Repair unit ventilator, 750 CFM, 2 ton	10	1 STPI	Ea.							
Repair controls				.200		12.80		12.80	15.85	19.85
Remove / replace fan				.650		41.50		41.50	51.50	64.50
Remove fan motor				1.156		69.50		69.50	85.50	107
Replace fan motor				1.951	240	117		357	410	480
Total				3.957	240	240.80		480.80	562.85	671.35
1030 Replace unit vent., 750 CFM, heat/cool coils	15	Q-6	Ea.							
Remove unit ventilator				7.792		465		465	575	720
Replace unit, 750 CFM, with heat / cool coils				15.605	4,575	930		5,505	6,200	7,175
Total				23.397	4,575	1,395		5,970	6,775	7,895
2010 Repair unit ventilator, 1250 CFM, 3 ton	10	1 STPI	Ea.							
Repair controls				.200		12.80		12.80	15.85	19.85
Remove / replace fan				.650		41.50		41.50	51.50	64.50
Remove fan motor				1.156		69.50		69.50	85.50	107
Replace fan motor				1.951	285	117		402	460	535
Total				3.957	285	240.80		525.80	612.85	726.35
2030 Replace unit vent., 1250 CFM, heat/cool coils	15	Q-6	Ea.							
Remove unit ventilator				11.142		665		665	825	1,025
Replace unit, 1250 CFM, with heat / cool coils				22.284	5,575	1,325		6,900	7,775	9,025
Total				33.426	5,575	1,990		7,565	8,600	10,050
2040 Repair unit ventilator, 2000 CFM, 5 ton	10	1 STPI	Ea.							
Repair controls				.300		19.20		19.20	24	30
Remove / replace fan				.650		41.50		41.50	51.50	64.50
Remove fan motor				1.156		69.50		69.50	85.50	107
Replace fan motor				2.311	305	139		444	505	595
Total				4.417	305	269.20		574.20	666	796.50
2050 Replace unit vent., 2000 CFM, heat/cool coils	15	Q-6	Ea.							
Remove unit ventilator				31.209		1,875		1,875	2,300	2,900
Replace unit, 2000 CFM, with heat / cool coils				62.338	7,000	3,725		10,725	12,300	14,500
Total				93.547	7,000	5,600		12,600	14,600	17,400

D3043 140 Duct Heater

System Description	Freq. (Years)	Crew	Unit	Labor Hours	2019 Bare Costs				Total In-House	Total w/O&P
					Material	Labor	Equipment	Total		
0020 Maintenance and inspection	0.50	1 ELEC	Ea.							
Inspect and clean duct heater				1.143		68.50		68.50	84.50	106
Total				1.143		68.50		68.50	**84.50**	**106**
0030 Replace duct heater	15	1 ELEC	Ea.							
Remove duct heater				.667		40		40	49.50	61.50
Duct heater 12 KW				2.000	1,900	120		2,020	2,250	2,550
Total				2.667	1,900	160		2,060	**2,299.50**	**2,611.50**

D3043 210 Draft Fan

System Description	Freq. (Years)	Crew	Unit	Labor Hours	2019 Bare Costs				Total In-House	Total w/O&P
					Material	Labor	Equipment	Total		
1010 Repair fan, induced draft, 2000 CFM	10	1 STPI	Ea.							
Remove bearings				1.000		64		64	79.50	99
Replace bearings				2.000	65.50	128		193.50	231	280
Total				3.000	65.50	192		257.50	**310.50**	**379**
1030 Replace fan, induced draft, 2000 CFM	20	Q-9	Ea.							
Remove fan				3.151		173		173	218	271
Replace fan, induced draft, 2000 CFM				6.304	2,625	345		2,970	3,325	3,825
Total				9.456	2,625	518		3,143	**3,543**	**4,096**
2010 Repair fan, induced draft, 6700 CFM	10	1 STPI	Ea.							
Remove bearings				1.000		64		64	79.50	99
Replace bearings				2.000	65.50	128		193.50	231	280
Total				3.000	65.50	192		257.50	**310.50**	**379**
2030 Replace fan, induced draft, 6700 CFM	20	Q-9	Ea.							
Remove fan				4.522		248		248	310	390
Replace fan, induced draft, 6700 CFM				9.045	3,225	495		3,720	4,175	4,800
Total				13.567	3,225	743		3,968	**4,485**	**5,190**
3010 Repair fan, induced draft, 17,700 CFM	10	1 STPI	Ea.							
Remove bearings				1.000		64		64	79.50	99
Replace bearings				2.222	73.50	142		215.50	257	310
Total				3.222	73.50	206		279.50	**336.50**	**409**

For customer support on your Facilities Maintenance & Repair Costs with RSMeans data, call 800.448.8182.

D3043 210 Draft Fan

System Description	Freq. (Years)	Crew	Unit	Labor Hours	2019 Bare Costs				Total In-House	Total w/O&P
					Material	Labor	Equipment	Total		
3030 Replace fan, induced draft, 17,700 CFM	20	Q-9	Ea.							
Remove fan				12.998		715		715	895	1,125
Replace fan, induced draft, 17,700 CFM				26.016	8,400	1,425		9,825	11,000	12,800
Total				39.014	8,400	2,140		10,540	11,895	13,925

D3043 220 Exhaust Fan

System Description	Freq. (Years)	Crew	Unit	Labor Hours	2019 Bare Costs				Total In-House	Total w/O&P
					Material	Labor	Equipment	Total		
1010 Replace fan & motor, propeller exh., 375 CFM	15	Q-20	Ea.							
Remove fan and motor				1.300		72.50		72.50	91	114
Replace fan & motor, propeller exhaust 375 cfm				2.600	209	145		354	410	490
Total				3.900	209	217.50		426.50	501	604
1030 Replace fan & motor, propeller exh., 1000 CFM	15	Q-20	Ea.							
Remove fan and motor				1.625		91		91	114	142
Replace fan & motor, propeller exhaust 1,323 CFM				3.250	350	182		532	615	720
Total				4.875	350	273		623	729	862
1040 Replace fan & motor, propeller exh., 4700 CFM	15	Q-20	Ea.							
Remove fan and motor				2.600		145		145	182	227
Replace fan & motor, propeller exhaust 4,700 CFM				5.200	1,325	291		1,616	1,825	2,100
Total				7.800	1,325	436		1,761	2,007	2,327
2030 Replace roof mounted exhaust fan, 800 CFM	20	Q-20	Ea.							
Remove fan & motor				2.600		145		145	182	227
Replace fan & motor, roof mounted exhaust, 800 CFM				5.200	1,025	291		1,316	1,500	1,725
Total				7.800	1,025	436		1,461	1,682	1,952
2040 Replace roof mounted exhaust fan, 2000 CFM	20	Q-20	Ea.							
Remove fan & motor				3.250		182		182	228	284
Replace fan & motor, roof mounted exhaust, 2000 CFM				6.500	1,825	365		2,190	2,475	2,850
Total				9.750	1,825	547		2,372	2,703	3,134

	System Description	Freq. (Years)	Crew	Unit	Labor Hours	2019 Bare Costs				Total In-House	Total w/O&P
						Material	Labor	Equipment	Total		
2050	**Replace roof mounted exhaust fan, 8500 CFM**	20	Q-20	Ea.							
	Remove fan & motor				4.333		242		242	305	380
	Replace fan & motor, roof mounted exhaust, 8500 CFM				8.666	3,200	485		3,685	4,125	4,750
	Total				12.998	3,200	727		3,927	4,430	5,130
2060	**Replace roof mounted exhaust fan, 20,300 CFM**	20	Q-20	Ea.							
	Remove fan & motor				13.004		725		725	910	1,125
	Replace fan & motor, roof mounted exhaust, 20,300 CFM				26.008	8,825	1,450		10,275	11,500	13,300
	Total				39.012	8,825	2,175		11,000	12,410	14,425
3010	**Replace utility set, belt drive, 800 CFM**	10	Q-20	Ea.							
	Remove fan and motor				2.167		121		121	152	189
	Replace utility set, belt drive, 800 CFM				4.334	1,025	242		1,267	1,425	1,650
	Total				6.500	1,025	363		1,388	1,577	1,839
3020	**Replace utility set, belt drive, 3600 CFM**	10	Q-20	Ea.							
	Remove fan and motor				3.250		182		182	228	284
	Replace utility set, belt drive, 3600 CFM				6.500	2,100	365		2,465	2,775	3,200
	Total				9.750	2,100	547		2,647	3,003	3,484
3030	**Replace utility set, belt drive, 11,000 CFM**	10	Q-20	Ea.							
	Remove fan and motor				6.500		365		365	455	570
	Replace utility set, belt drive, 11,000 CFM				13.004	5,550	725		6,275	7,025	8,075
	Total				19.504	5,550	1,090		6,640	7,480	8,645
3040	**Replace utility set, belt drive, 20,000 CFM**	10	Q-20	Ea.							
	Remove fan and motor				16.247		910		910	1,150	1,425
	Replace utility set, belt drive, 20,000 CFM				32.520	7,450	1,825		9,275	10,500	12,200
	Total				48.767	7,450	2,735		10,185	11,650	13,625
4010	**Replace axial flow fan, 3900 CFM**	10	Q-20	Ea.							
	Remove fan and motor				3.823		214		214	268	335
	Replace axial flow fan, 3900 CFM				7.648	1,425	425		1,850	2,100	2,450
	Total				11.472	1,425	639		2,064	2,368	2,785

For customer support on your Facilities Maintenance & Repair Costs with RSMeans data, call 800.448.8182.

D30 HVAC | D3043 | Distribution Systems

D3043 220 Exhaust Fan

	System Description	Freq. (Years)	Crew	Unit	Labor Hours	Material	Labor	Equipment	Total	Total In-House	Total w/O&P
4020	**Replace axial flow fan, 6400 CFM**	10	Q-20	Ea.							
	Remove fan and motor				4.643		259		259	325	405
	Replace axial flow fan, 6400 CFM				9.285	2,000	520		2,520	2,850	3,300
	Total				13.928	2,000	779		2,779	3,175	3,705
4030	**Replace axial flow fan, 16,900 CFM**	10	Q-20	Ea.							
	Remove fan and motor				6.502		365		365	455	570
	Replace axial flow fan, 16,900 CFM				13.004	2,775	725		3,500	3,975	4,600
	Total				19.506	2,775	1,090		3,865	4,430	5,170
4040	**Replace axial flow fan, 29,000 CFM**	10	Q-20	Ea.							
	Remove fan and motor				9.001		505		505	630	785
	Replace axial flow fan, 29,000 CFM				17.986	4,225	1,000		5,225	5,900	6,850
	Total				26.987	4,225	1,505		5,730	6,530	7,635

D3043 250 Fireplaces, Clay Flue

	System Description	Freq. (Years)	Crew	Unit	Labor Hours	Material	Labor	Equipment	Total	Total In-House	Total w/O&P
0010	**Replace baked clay flue, architectural**	75	2 BRIC	L.F.							
	Set up and secure scaffold				.022		1.13		1.13	1.49	1.84
	Masonry demolition				.083		3.42		3.42	4.45	5.50
	Remove damaged flue				.002		.11		.11	.15	.18
	Install new flue lining				.224	11.90	10.25		22.15	26.50	31.50
	Install new chimney				1.518	49	69.50		118.50	145	174
	Remove scaffold				.022		1.13		1.13	1.49	1.84
	Total				1.872	60.90	85.54		146.44	179.08	214.86

D3043 252 Fireplaces, Metal Flue

	System Description	Freq. (Years)	Crew	Unit	Labor Hours	Material	Labor	Equipment	Total	Total In-House	Total w/O&P
0020	**Replace metal pipe flue, architectural**	50	2 SSWK	L.F.							
	Remove damaged pipe				.008		.47		.47	.64	.78
	Install new vent chimney, prefabricated metal				.306	8.35	16.80		25.15	30.50	37
	Total				.314	8.35	17.27		25.62	31.14	37.78

D3043 310 Steam Converter, Commercial

	System Description	Freq. (Years)	Crew	Unit	Labor Hours	2019 Bare Costs Material	Labor	Equipment	Total	Total In-House	Total w/O&P
0010	**Repair steam converter**	5	1 STPI	Ea.	5.944		380		380	470	590
	Repair heat exchanger										
	Total				5.944		380		380	470	590
0020	**Inspect for leaks**	2	1 STPI	Ea.							
	Check relief valve operation				.038		2.41		2.41	2.99	3.74
	Check for proper water temperature				.055		3.49		3.49	4.33	5.40
	Total				.092		5.90		5.90	7.32	9.14
0030	**Replace steam converter**	30	Q-5	Ea.							
	Remove exchanger				2.081		120		120	149	186
	Install new exchanger 10 GPM of 40F to 180F				4.160	3,800	239		4,039	4,475	5,125
	Total				6.241	3,800	359		4,159	4,624	5,311

D3043 320 Flash Tank, 24 Gallon

	System Description	Freq. (Years)	Crew	Unit	Labor Hours	2019 Bare Costs Material	Labor	Equipment	Total	Total In-House	Total w/O&P
0010	**Repair flash tank**	5	1 STPI	Ea.	5.944		380		380	470	590
	Repair flash tank										
	Total				5.944		380		380	470	590
0030	**Replace flash tank 24 gal**	15	Q-5	Ea.							
	Remove old tank				.743		43		43	53	66.50
	Install new tank 24 gallon				1.486	790	85.50		875.50	975	1,125
	Total				2.229	790	128.50		918.50	1,028	1,191.50

D3043 330 Steam Regulator Valve

	System Description	Freq. (Years)	Crew	Unit	Labor Hours	2019 Bare Costs Material	Labor	Equipment	Total	Total In-House	Total w/O&P
0010	**Replace steam regulator valve 1-1/2" diameter**	6	1 STPI	Ea.							
	Remove steam valve				.400		25.50		25.50	31.50	39.50
	Install steam regulator valve 1-1/2" diameter				.800	3,800	51		3,851	4,250	4,825
	Total				1.200	3,800	76.50		3,876.50	4,281.50	4,864.50

D3043 330 Steam Regulator Valve

	System Description	Freq. (Years)	Crew	Unit	Labor Hours	2019 Bare Costs				Total In-House	Total w/O&P
						Material	Labor	Equipment	Total		
0110	**Replace steam regulator valve 2″ diameter**	6	1 STPI	Ea.							
	Remove steam valve				.473		30		30	37.50	47
	Install steam regulator valve 2″ diameter				.945	4,650	60.50		4,710.50	5,200	5,900
	Total				1.418	4,650	90.50		4,740.50	**5,237.50**	**5,947**
0210	**Replace steam regulator valve 2-1/2″ diameter**	6	Q-5	Ea.							
	Remove steam valve				.867		50		50	62	77.50
	Install steam regulator valve 2-1/2″ diameter				1.733	5,775	100		5,875	6,475	7,375
	Total				2.600	5,775	150		5,925	**6,537**	**7,452.50**
0310	**Replace steam regulator valve 3″ diameter**	6	Q-5	Ea.							
	Remove steam valve				.946		54.50		54.50	67.50	84.50
	Install steam regulator valve 3″ diameter flanged				1.891	7,250	109		7,359	8,100	9,225
	Total				2.836	7,250	163.50		7,413.50	**8,167.50**	**9,309.50**

D3043 340 Condensate Meter

	System Description	Freq. (Years)	Crew	Unit	Labor Hours	2019 Bare Costs				Total In-House	Total w/O&P
						Material	Labor	Equipment	Total		
0010	**Repair meter**	15	1 STPI	Ea.							
	Replace drum				1.538	830	97		927	1,025	1,200
	Replace bearings				2.597	136	164		300	355	425
	Total				4.136	966	261		1,227	**1,380**	**1,625**
0030	**Replace condensate meter 500 lb./hr.**	30	1 STPI	Ea.							
	Remove meter				.372		24		24	29.50	37
	Replace meter 500#/hr				.743	3,725	47.50		3,772.50	4,150	4,725
	Total				1.114	3,725	71.50		3,796.50	**4,179.50**	**4,762**
0130	**Replace condensate meter 1500 lb./hr.**	30	1 STPI	Ea.							
	Remove meter				.743		47.50		47.50	59	73.50
	Replace meter 1500#/hr				1.486	4,350	95		4,445	4,900	5,575
	Total				2.229	4,350	142.50		4,492.50	**4,959**	**5,648.50**

For customer support on your Facilities Maintenance & Repair Costs with RSMeans data, call 800.448.8182.

255

D3043 350   Steam Traps

System Description	Freq. (Years)	Crew	Unit	Labor Hours	2019 Bare Costs Material	Labor	Equipment	Total	Total In-House	Total w/O&P
1030 Replace steam trap, 15 PSIG, 3/4" threaded	7	1 STPI	Ea.							
Remove old CI body float & thermostatic trap				.325		21		21	26	32
Install new CI body float & thermostatic trap				.650	175	41.50		216.50	244	283
Total				.975	175	62.50		237.50	270	315
1040 Replace steam trap, 15 PSIG, 1" threaded	7	1 STPI	Ea.							
Remove old CI body float & thermostatic trap				.347		22		22	27.50	34.50
Install new CI body float & thermostatic trap				.693	183	44.50		227.50	256	298
Total				1.040	183	66.50		249.50	283.50	332.50
1050 Replace steam trap, 15 PSIG, 1-1/4" threaded	7	1 STPI	Ea.							
Remove old CI body float & thermostatic trap				.400		25.50		25.50	31.50	39.50
Install new CI body float & thermostatic trap				.800	231	51		282	320	370
Total				1.200	231	76.50		307.50	351.50	409.50
1060 Replace steam trap, 15 PSIG, 1-1/2" threaded	7	1 STPI	Ea.							
Remove old CI body float & thermostatic trap				.578		37		37	46	57.50
Install new CI body float & thermostatic trap				1.156	310	74		384	435	505
Total				1.733	310	111		421	481	562.50
1070 Replace steam trap, 15 PSIG, 2" threaded	7	1 STPI	Ea.							
Remove old CI body float & thermostatic trap				.867		55.50		55.50	69	86
Install new CI body float & thermostatic trap				1.733	965	111		1076	1,200	1,375
Total				2.600	965	166.50		1131.50	1,269	1,461

D3043 410   Radiator Valve

System Description	Freq. (Years)	Crew	Unit	Labor Hours	2019 Bare Costs Material	Labor	Equipment	Total	Total In-House	Total w/O&P
0010 Replace radiator valve 1/2" angle union	50	1 STPI	Ea.							
Remove radiator valve				.217		13.85		13.85	17.20	21.50
Install radiator valve 1/2" angle union				.433	61	27.50		38.50	101	119
Total				.650	61	41.35		1D2.35	118.20	140.50

D30 HVAC | D3043 | Distribution Systems

D3043 410 | Radiator Valve

System Description	Freq. (Years)	Crew	Unit	Labor Hours	2019 Bare Costs				Total In-House	Total w/O&P
					Material	Labor	Equipment	Total		
0020 Replace radiator valve 3/4" angle union	50	1 STPI	Ea.							
Remove radiator valve				.260		16.65		16.65	20.50	26
Install radiator valve 3/4" angle union				.520	68.50	33.50		102	117	137
Total				.780	68.50	50.15		118.65	**137.50**	**163**
0030 Replace radiator valve 1" angle union	50	1 STPI	Ea.							
Remove radiator valve				.274		17.50		17.50	21.50	27
Install radiator valve 1" angle union				.547	77	35		112	128	151
Total				.821	77	52.50		129.50	**149.50**	**178**
0040 Replace radiator valve 1-1/4" angle union	50	1 STPI	Ea.							
Remove radiator valve				.347		22		22	27.50	34.50
Install radiator valve 1-1/4" angle union				.693	101	44.50		145.50	166	195
Total				1.040	101	66.50		167.50	**193.50**	**229.50**

D3043 420 | Cast Iron Radiator, 10' Section

System Description	Freq. (Years)	Crew	Unit	Labor Hours	2019 Bare Costs				Total In-House	Total w/O&P
					Material	Labor	Equipment	Total		
0010 Replace C.I. radiator 4 tube 25"H 10' section	50	Q-5	Section							
Remove radiator				1.818		104.50		104.50	130	162
Install C.I. radiator 4 tube 25" high 10' section				2.167	495	124.50		619.50	700	810
Total				3.985	495	229		724	**830**	**972**

D3043 430 | Baseboard Radiation, 10' Section

System Description	Freq. (Years)	Crew	Unit	Labor Hours	2019 Bare Costs				Total In-House	Total w/O&P
					Material	Labor	Equipment	Total		
0010 Replace radiator, baseboard 10'	20	Q-5	Ea.							
Remove radiator				2.262		130		130	162	200
Install 10' baseboard radiator				4.522	460	260		720	830	980
Total				6.783	460	390		850	**992**	**1,180**

D3043 440 Finned Radiator, Wall, 10' Section

System Description	Freq. (Years)	Crew	Unit	Labor Hours	2019 Bare Costs				Total In-House	Total w/O&P
					Material	Labor	Equipment	Total		
0010 Replace finned radiator	20	Q-5	Ea.							
Replace radiator				3.468		199.50		199.50	248	310
Install 10' finned tube, 2 tier, 1-1/4" cop tube				6.933	580	400		980	1,125	1,350
Total				10.401	580	599.50		1,179.50	1,373	1,660

D3043 450 Duct Coil, 1-Row, Hot Water

System Description	Freq. (Years)	Crew	Unit	Labor Hours	2019 Bare Costs				Total In-House	Total w/O&P
					Material	Labor	Equipment	Total		
1020 Replace coil, hot water boost, 12" x 24"	25	Q-5	Ea.							
Remove coil				.858		49.50		49.50	61.50	76.50
Replace coil, hot water duct, 12" x 24"				1.111	810	64		874	970	1,100
Total				1.969	810	113.50		923.50	1,031.50	1,176.50
1030 Replace coil, hot water boost, 24" x 24"	25	Q-5	Ea.							
Remove coil				1.717		99		99	123	153
Replace coil, hot water duct, 24" x 24"				3.433	1,100	198		1,298	1,450	1,675
Total				5.150	1,100	297		1,397	1,573	1,828
1040 Replace coil, hot water boost, 24" x 36"	25	Q-5	Ea.							
Remove coil				2.581		149		149	184	230
Replace coil, hot water duct, 24" x 36"				5.161	1,175	297		1,472	1,650	1,925
Total				7.742	1,175	446		1,621	1,834	2,155
1050 Replace coil, hot water boost, 36" x 36"	25	Q-5	Ea.							
Remove coil				3.867		223		223	276	345
Replace coil, hot water duct, 36" x 36"				7.729	1,325	445		1,770	2,000	2,350
Total				11.596	1,325	668		1,993	2,276	2,695

For customer support on your Facilities Maintenance & Repair Costs with RSMeans data, call 800.448.8182.

D30 HVAC | D3043 | Distribution Systems

D3043 510 | Pipe & Fittings, Steel/Iron, Flanged

System Description	Freq. (Years)	Crew	Unit	Labor Hours	2019 Bare Costs Material	Labor	Equipment	Total	Total In-House	Total w/O&P
0010 Install new gasket, 4" pipe size	25	1 PLUM	Ea.							
Disconnect joint / remove flange gasket				1.301		82		82	102	127
Install new gasket				1.300	16.30	82		98.30	120	147
Total				2.601	16.30	164		180.30	**222**	**274**
0030 Replace 2" flanged steel pipe and fittings	75	2 PLUM	L.F.							
Remove old pipe				.231		13.15		13.15	16.30	20.50
Install flanged 2" steel pipe with hangers				.356	14.85	20	1.25	36.10	43	51.50
Total				.587	14.85	33.15	1.25	49.25	**59.30**	**72**
0040 Replace 4" flanged steel pipe and fittings	75	2 PLUM	L.F.							
Remove old pipe				.400		22.50		22.50	28	35.50
Install flanged 4" steel pipe with hangers				.800	30.50	45.50	2.81	78.81	93	112
Total				1.200	30.50	68	2.81	101.31	**121**	**147.50**
0050 Replace 6" flanged steel pipe and fittings	75	2 PLUM	L.F.							
Remove old pipe				.624		35.50		35.50	44	55
Install flanged 6" steel pipe with hangers				1.248	60	73.50	2.92	136.42	160	192
Total				1.872	60	109	2.92	171.92	**204**	**247**
0060 Replace 8" flanged steel pipe and fittings	75	2 PLUM	L.F.							
Remove old pipe				.821		48.50		48.50	60	75
Install flanged 8" steel pipe with hangers				1.642	98.50	97	3.84	199.34	233	277
Total				2.463	98.50	145.50	3.84	247.84	**293**	**352**

D3043 520 | Valves

System Description	Freq. (Years)	Crew	Unit	Labor Hours	2019 Bare Costs Material	Labor	Equipment	Total	Total In-House	Total w/O&P
1010 Repack gate valve gland, 3/8" - 1-1/2"	10	1 STPI	Ea.							
Remove / replace gate valve packing				.219	11.40	12.60		24	28	34
Total				.219	11.40	12.60		24	**28**	**34**

For customer support on your Facilities Maintenance & Repair Costs with RSMeans data, call 800.448.8182.

259

D3043 520   Valves

	System Description	Freq. (Years)	Crew	Unit	Labor Hours	2019 Bare Costs				Total In-House	Total w/O&P
						Material	Labor	Equipment	Total		
1020	**Replace gate valve, partial, 3/8" - 1-1/2"**	20	1 STPI	Ea.							
	Turn valve off and on				.017		1.09		1.09	1.35	1.69
	Remove old				.530		34		34	42	52.50
	Install new, 3/8" - 1-1/2" size				.800	216	51		267	300	350
	Total				1.347	216	86.09		302.09	343.35	404.19
2010	**Repack gate valve gland, 2" - 3"**	10	1 STPI	Ea.							
	Remove / replace gate valve packing				.291	15.16	16.76		31.92	37.50	45
	Total				.291	15.16	16.76		31.92	37.50	45
2020	**Replace gate valve, partial, 2" - 3"**	20	Q-1	Ea.							
	Turn valve off and on				.017		1.09		1.09	1.35	1.69
	Remove valve parts				.440		28		28	35	43.50
	Replace valve parts, 2" - 3"				1.600	845	91		936	1,050	1,200
	Total				2.057	845	120.09		965.09	1,086.35	1,245.19
2060	**Repack gate valve gland, 8" - 12"**	5	1 PLUM	Ea.							
	Replace valve packing, 8" to 12" diameter				1.600	45.50	101		146.50	175	214
	Total				1.600	45.50	101		146.50	175	214
3010	**Repack drain valve gland, 3/4"**	3	1 STPI	Ea.							
	Remove / replace drain valve packing				.219	11.40	12.60		24	28	34
	Total				.219	11.40	12.60		24	28	34
3020	**Replace drain valve stem assembly, 3/4"**	8	1 STPI	Ea.							
	Turn valve off and on				.017		1.09		1.09	1.35	1.69
	Remove valve stem assembly				.267		17.05		17.05	21	26.50
	Replace valve stem assembly				.267	13.40	17.05		30.45	36	43.50
	Total				.550	13.40	35.19		48.59	58.35	71.69
3030	**Replace drain valve, 3/4"**	20	1 STPI	Ea.							
	Turn valve off and on				.017		1.09		1.09	1.35	1.69
	Remove old valve				.308		19.70		19.70	24.50	30.50
	Install new valve, 3/4"				.308	13.40	19.70		33.10	39	47.50
	Total				.632	13.40	40.49		53.89	64.85	79.69

For customer support on your Facilities Maintenance & Repair Costs with RSMeans data, call 800.448.8182.

	System Description	Freq. (Years)	Crew	Unit	Labor Hours	2019 Bare Costs				Total In-House	Total w/O&P
						Material	Labor	Equipment	Total		
1010	**Repair circulator pump, 1/12 - 3/4 H.P.**	5	1 STPI	Ea.							
	Loosen coupling from pump shaft				.009		.58		.58	.71	.89
	Remove 4 stud bolts in motor				.035		2.24		2.24	2.78	3.47
	Remove motor				.052		3.33		3.33	4.13	5.15
	Remove impeller nut from shaft				.033		2.11		2.11	2.62	3.27
	Remove seal assembly				.019		1.22		1.22	1.51	1.88
	Inspect bearing brackets face				.005		.32		.32	.40	.50
	Clean bearing				.016		1.02		1.02	1.27	1.59
	Inspect shaft				.040	42.50	2.30		44.80	49.50	56.50
	Clean shaft				.017		1.09		1.09	1.35	1.69
	Install new seal and rubber				.013		.83		.83	1.03	1.29
	Install seal spring and washers				.006		.38		.38	.48	.59
	Install impeller nut on shaft				.040	10.60	2.30		12.90	14.50	16.80
	Clean pump body gasket area				.037		2.37		2.37	2.94	3.67
	Install new gasket (ready made)				.010		.64		.64	.79	.99
	Install impeller				.009		.58		.58	.71	.89
	Install 4 stud bolts				.035		2.24		2.24	2.78	3.47
	Assemble motor to bearing				.024		1.53		1.53	1.90	2.38
	Install 4 stud bolts				.027		1.73		1.73	2.14	2.68
	Tighten coupling				.009		.58		.58	.71	.89
	Final tightening of bolts				.027		1.73		1.73	2.14	2.68
	Test run				.042		2.69		2.69	3.33	4.16
	Total				.505	53.10	31.81		84.91	97.72	115.43
1030	**Replace circulator pump, 1/12 - 3/4 H.P.**	15	Q-1	Ea.							
	Remove pump and motor				2.597		148		148	183	229
	Install new pump and motor, 1/12-3/4 H.P.				5.195	2,975	295		3,270	3,650	4,175
	Total				7.792	2,975	443		3,418	3,833	4,404

For customer support on your Facilities Maintenance & Repair Costs with RSMeans data, call 800.448.8182.

261

D3043 530 Circulator Pump

	System Description	Freq. (Years)	Crew	Unit	Labor Hours	2019 Bare Costs Material	Labor	Equipment	Total	Total In-House	Total w/O&P
2010	**Repair circulator pump, 1 H.P.**	5	1 STPI	Ea.							
	Loosen coupling from pump shaft				.009		.58		.58	.71	.89
	Remove 4 stud bolts in motor				.035		2.24		2.24	2.78	3.47
	Remove motor				.052		3.33		3.33	4.13	5.15
	Remove impeller nut from shaft				.033		2.11		2.11	2.62	3.27
	Remove seal assembly				.019		1.22		1.22	1.51	1.88
	Inspect bearing brackets face				.005		.32		.32	.40	.50
	Clean bearing				.016		1.02		1.02	1.27	1.59
	Inspect shaft				.040	42.50	2.30		44.80	49.50	56.50
	Clean shaft				.017		1.09		1.09	1.35	1.69
	Install new seal and rubber				.013		.83		.83	1.03	1.29
	Install seal spring and washers				.006		.38		.38	.48	.59
	Install impeller nut on shaft				.040	10.60	2.30		12.90	14.50	16.80
	Clean pump body gasket area				.037		2.37		2.37	2.94	3.67
	Install new gasket (ready made)				.013		.83		.83	1.03	1.29
	Install impeller				.009		.58		.58	.71	.89
	Install 4 stud bolts				.035		2.24		2.24	2.78	3.47
	Assemble motor to bearing				.024		1.53		1.53	1.90	2.38
	Install 4 stud bolts				.027		1.73		1.73	2.14	2.68
	Tighten coupling				.009		.58		.58	.71	.89
	Final tightening of bolts				.027		1.73		1.73	2.14	2.68
	Test run				.042		2.69		2.69	3.33	4.16
	Total				.508	53.10	32		85.10	97.96	115.73
2030	**Replace circulator. pump, 1 H.P.**	15	Q-1	Ea.							
	Remove pump				2.597		148		148	183	229
	Install new pump, 1 H.P.				5.200	4,800	296		5,096	5,650	6,450
	Total				7.797	4,800	444		5,244	5,833	6,679

D3043 540 Expansion Tank

	System Description	Freq. (Years)	Crew	Unit	Labor Hours	2019 Bare Costs Material	Labor	Equipment	Total	Total In-House	Total w/O&P
0010	**Refill expansion tank**	5	1 STPI	Ea.							
	Refill air chamber				.200		12.80		12.80	15.85	19.85
	Total				.200		12.80		12.80	15.85	19.85

For customer support on your Facilities Maintenance & Repair Costs with RSMeans data, call 800.448.8182.

D30 HVAC | D3043 | Distribution Systems

D3043 540 — Expansion Tank

System Description	Freq. (Years)	Crew	Unit	Labor Hours	2019 Bare Costs				Total In-House	Total w/O&P
					Material	Labor	Equipment	Total		
0020 Replace expansion tank, 24 gal capacity	50	Q-5	Ea.							
Remove expansion tank				.743		43		43	53	66.50
Install expansion tank, 24 gallon capacity				1.486	1,475	85.50		1,560.50	1,725	1,975
Total				2.228	1,475	128.50		1,603.50	**1,778**	**2,041.50**
0120 Replace expansion tank, 60 gal capacity	50	Q-5	Ea.							
Remove expansion tank				1.300		75		75	93	116
Install expansion tank, 60 gallon capacity				2.602	2,075	150		2,225	2,475	2,825
Total				3.901	2,075	225		2,300	**2,568**	**2,941**
0220 Replace expansion tank, 175 gal capacity	50	Q-5	Ea.							
Remove expansion tank				2.597		149		149	185	232
Install expansion tank, 175 gallon capacity				5.195	5,150	299		5,449	6,025	6,900
Total				7.792	5,150	448		5,598	**6,210**	**7,132**
0320 Replace expansion tank, 400 gal capacity	50	Q-5	Ea.							
Remove expansion tank				3.721		214		214	266	330
Install expansion tank, 400 gallon capacity				7.442	12,900	430		13,330	14,700	16,800
Total				11.163	12,900	644		13,544	**14,966**	**17,130**

D3043 550 — Pipe Insulation

System Description	Freq. (Years)	Crew	Unit	Labor Hours	2019 Bare Costs				Total In-House	Total w/O&P
					Material	Labor	Equipment	Total		
1010 Repair damaged pipe insulation, fiberglass 1/2"	5	Q-14	Ea.							
Remove 2' length old insulation				.086		4.46		4.46	5.70	7.10
Install insulation, fiberglass, 1" wall - 1/2" diam.				.173	1.76	8.94		10.70	13.40	16.50
Total				.260	1.76	13.40		15.16	**19.10**	**23.60**
1110 Repair damaged pipe insulation, fiberglass 3/4"	5	Q-14	Ea.							
Remove 2' length old insulation				.090		4.66		4.66	6	7.45
Install insulation, fiberglass, 1" wall - 3/4" diam.				.181	1.92	9.34		11.26	14.10	17.30
Total				.271	1.92	14		15.92	**20.10**	**24.75**

For customer support on your Facilities Maintenance & Repair Costs with RSMeans data, call 800.448.8182.

263

D3043 550   Pipe Insulation

System Description	Freq. (Years)	Crew	Unit	Labor Hours	2019 Bare Costs				Total In-House	Total w/O&P
					Material	Labor	Equipment	Total		
1120 Repair damaged pipe insulation, fiberglass 1"	5	Q-14	Ea.							
Remove 2' length old insulation				.095		4.88		4.88	6.30	7.80
Install insulation, fiberglass, 1" wall - 1" diam.				.189	2.06	9.78		11.84	14.80	18.20
Total				.284	2.06	14.66		16.72	**21.10**	**26**
1130 Repair damaged pipe insulation, fbgs 1-1/4"	5	Q-14	Ea.							
Remove 2' length old insulation				.099		5.10		5.10	6.55	8.10
Install insulation, fiberglass, 1" wall - 1-1/4" diam.				.198	2.22	10.20		12.42	15.50	19.10
Total				.296	2.22	15.30		17.52	**22.05**	**27.20**
1140 Repair damaged pipe insulation, fbgs 1-1/2"	5	Q-14	Ea.							
Remove 2' length old insulation				.099		5.10		5.10	6.55	8.10
Install insulation, fiberglass, 1" wall - 1-1/2" diam.				.198	2.40	10.20		12.60	15.75	19.30
Total				.297	2.40	15.30		17.70	**22.30**	**27.40**
1150 Repair damaged pipe insulation, fiberglass 2"	5	Q-14	Ea.							
Remove 2' length old insulation				.104		5.36		5.36	6.90	8.55
Install insulation, fiberglass, 1" wall - 2" diam.				.208	3.22	10.70		13.92	17.30	21
Total				.312	3.22	16.06		19.28	**24.20**	**29.55**
1160 Repair damaged pipe insulation, fiberglass 3"	5	Q-14	Ea.							
Remove 2' length old insulation				.116		5.98		5.98	7.70	9.55
Install insulation, fiberglass, 1" wall - 3" diam.				.232	3.56	12		15.56	19.30	23.50
Total				.348	3.56	17.98		21.54	**27**	**33.05**
1170 Repair damaged pipe insulation, fiberglass 4"	5	Q-14	Ea.							
Remove 2' length old insulation				.139		7.18		7.18	9.20	11.40
Install insulation, fiberglass, 1" wall - 4" diam.				.278	4.72	14.40		19.12	23.50	29
Total				.417	4.72	21.58		26.30	**32.70**	**40.40**
1180 Repair damaged pipe insulation, fiberglass 6"	5	Q-14	Ea.							
Remove 2' length old insulation				.173		8.94		8.94	11.45	14.20
Install insulation, fiberglass, 1" wall - 6" diam.				.347	5.64	17.90		23.54	29	35.50
Total				.520	5.64	26.84		32.48	**40.45**	**49.70**

For customer support on your Facilities Maintenance & Repair Costs with RSMeans data, call 800.448.8182.

D3043 550   Pipe Insulation

System Description	Freq. (Years)	Crew	Unit	Labor Hours	2019 Bare Costs				Total In-House	Total w/O&P
					Material	Labor	Equipment	Total		
1220 Replace pipe insulation, fiberglass 1/2"	5	Q-14	M.L.F.							
Remove 1000' length old insulation				33.330		1,720		1,720	2,200	2,750
Install insulation, fiberglass, 1" wall - 1/2" diam.				66.670	880	3,440		4,320	5,400	6,600
Total				100.000	880	5,160		6,040	7,600	9,350
1230 Replace pipe insulation, fiberglass 3/4"	5	Q-14	M.L.F.							
Remove 1000' length old insulation				34.780		1,800		1,800	2,300	2,850
Install insulation, fiberglass, 1" wall - 3/4" diam.				69.570	960	3,590		4,550	5,675	6,900
Total				104.350	960	5,390		6,350	7,975	9,750
1240 Replace pipe insulation, fiberglass 1"	5	Q-14	M.L.F.							
Remove 1000' length old insulation				36.360		1,880		1,880	2,400	3,000
Install insulation, fiberglass, 1" wall - 1" diam.				72.730	1,030	3,760		4,790	5,950	7,300
Total				109.090	1,030	5,640		6,670	8,350	10,300
1250 Replace pipe insulation, fiberglass 1-1/4"	5	Q-14	M.L.F.							
Remove 1000' length old insulation				38.100		1,970		1,970	2,525	3,125
Install insulation, fiberglass, 1" wall - 1-1/4" diam.				76.190	1,110	3,930		5,040	6,275	7,650
Total				114.290	1,110	5,900		7,010	8,800	10,775
1260 Replace pipe insulation, fiberglass 1-1/2"	5	Q-14	M.L.F.							
Remove 1000' length old insulation				38.100		1,970		1,970	2,525	3,125
Install insulation, fiberglass, 1" wall - 1-1/2" diam.				76.190	1,200	3,930		5,130	6,375	7,750
Total				114.290	1,200	5,900		7,100	8,900	10,875
1270 Replace pipe insulation, fiberglass 2"	5	Q-14	M.L.F.							
Remove 1000' length old insulation				40.000		2,070		2,070	2,650	3,300
Install insulation, fiberglass, 1" wall - 2" diam.				80.000	1,610	4,130		5,740	7,075	8,625
Total				120.000	1,610	6,200		7,810	9,725	11,925
1280 Replace pipe insulation, fiberglass 3"	5	Q-14	M.L.F.							
Remove 1000' length old insulation				44.440		2,290		2,290	2,950	3,650
Install insulation, fiberglass, 1" wall - 3" diam.				88.890	1,780	4,590		6,370	7,850	9,525
Total				133.330	1,780	6,880		8,660	10,800	13,175

D3043 550 Pipe Insulation

	System Description	Freq. (Years)	Crew	Unit	Labor Hours	2019 Bare Costs				Total In-House	Total w/O&P
						Material	Labor	Equipment	Total		
1290	**Replace pipe insulation, fiberglass 4"**	5	Q-14	M.L.F.							
	Remove 1000' length old insulation				53.330		2,750		2,750	3,525	4,400
	Install insulation, fiberglass, 1" wall - 4" diam.				106.670	2,360	5,500		7,860	9,675	11,800
	Total				160.000	2,360	8,250		10,610	13,200	16,200
1300	**Replace pipe insulation, fiberglass 6"**	5	Q-14	M.L.F.							
	Remove 1000' length old insulation				66.670		3,440		3,440	4,425	5,500
	Install insulation, fiberglass, 1" wall - 6" diam.				133.330	2,820	6,900		9,720	11,900	14,500
	Total				200.000	2,820	10,340		13,160	16,325	20,000
1410	**Repair damaged pipe insulation rubber 1/2"**	5	1 ASBE	Ea.							
	Remove 2' length old insulation				.117		6.70		6.70	8.60	10.70
	Install insulation, foam rubber, 3/4" wall - 1/2" diam.				.234	2.22	13.40		15.62	19.65	24
	Total				.350	2.22	20.10		22.32	28.25	34.70
1420	**Repair damaged pipe insulation rubber 3/4"**	5	1 ASBE	Ea.							
	Remove 2' length old insulation				.117		6.70		6.70	8.60	10.70
	Install insulation, foam rubber, 3/4" wall - 3/4" diam.				.234	3.62	13.40		17.02	21	26
	Total				.350	3.62	20.10		23.72	29.60	36.70
1430	**Repair damaged pipe insulation rubber 1"**	5	1 ASBE	Ea.							
	Remove 2' length old insulation				.118		6.78		6.78	8.70	10.80
	Install insulation, foam rubber, 3/4" wall - 1" diam.				.236	4.16	13.60		17.76	22	27
	Total				.355	4.16	20.38		24.54	30.70	37.80
1440	**Repair damaged pipe insulation rubber 1-1/4"**	5	1 ASBE	Ea.							
	Remove 2' length old insulation				.120		6.86		6.86	8.80	10.90
	Install insulation, foam rubber, 3/4" wall - 1-1/4" diam.				.239	4.84	13.70		18.54	23	28
	Total				.359	4.84	20.56		25.40	31.80	38.90
1450	**Repair damaged pipe insulation rubber 1-1/2"**	5	1 ASBE	Ea.							
	Remove 2' length old insulation				.120		6.86		6.86	8.80	10.90
	Install insulation, foam rubber, 3/4" wall - 1-1/2" diam.				.239	6.08	13.70		19.78	24.50	29.50
	Total				.359	6.08	20.56		26.64	33.30	40.40

For customer support on your Facilities Maintenance & Repair Costs with RSMeans data, call 800.448.8182.

D3043 550 | **Pipe Insulation**

System Description	Freq. (Years)	Crew	Unit	Labor Hours	2019 Bare Costs				Total In-House	Total w/O&P
					Material	Labor	Equipment	Total		
1460 Repair damaged pipe insulation rubber 2″	5	1 ASBE	Ea.							
Remove 2′ length old insulation				.121		6.96		6.96	8.90	11.10
Install insulation, foam rubber, 3/4″ wall - 2″ diam.				.242	7.14	13.90		21.04	25.50	31
Total				.364	7.14	20.86		28	34.40	**42.10**
1470 Repair damaged pipe insulation rubber 3″	5	1 ASBE	Ea.							
Remove 2′ length old insulation				.122		7.02		7.02	9	11.20
Install insulation, foam rubber, 3/4″ wall - 3″ diam.				.245	9.64	14		23.64	28.50	34.50
Total				.367	9.64	21.02		30.66	37.50	**45.70**
1480 Repair damaged pipe insulation rubber 4″	5	1 ASBE	Ea.							
Remove 2′ length old insulation				.130		7.46		7.46	9.60	11.90
Install insulation, foam rubber, 3/4″ wall - 4″ diam.				.260	12	14.90		26.90	32.50	39
Total				.390	12	22.36		34.36	42.10	**50.90**
1490 Repair damaged pipe insulation rubber 6″	5	1 ASBE	Ea.							
Remove 2′ length old insulation				.130		7.46		7.46	9.60	11.90
Install insulation, foam rubber, 3/4″ wall - 6″ diam.				.260	17.40	14.90		32.30	38.50	45.50
Total				.390	17.40	22.36		39.76	48.10	**57.40**
1510 Replace pipe insulation foam rubber 1/2″	5	1 ASBE	L.F.							
Remove 1000′ length old insulation				.090		5.16		5.16	6.60	8.20
Install insulation, foam rubber, 3/4″ wall - 1/2″ diam.				.180	2.22	10.30		12.52	15.70	19.20
Total				.270	2.22	15.46		17.68	22.30	**27.40**
1520 Replace pipe insulation foam rubber 3/4″	5	1 ASBE	L.F.							
Remove 1000′ length old insulation				.090		5.16		5.16	6.60	8.20
Install insulation, foam rubber, 3/4″ wall - 3/4″ diam.				.180	3.62	10.30		13.92	17.20	21
Total				.270	3.62	15.46		19.08	23.80	**29.20**
1530 Replace pipe insulation foam rubber 1″	5	1 ASBE	L.F.							
Remove 1000′ length old insulation				.091		5.22		5.22	6.70	8.30
Install insulation, foam rubber, 3/4″ wall - 1″ diam.				.182	4.16	10.40		14.56	17.95	22
Total				.273	4.16	15.62		19.78	24.65	**30.30**

D3043 550 Pipe Insulation

System Description	Freq. (Years)	Crew	Unit	Labor Hours	2019 Bare Costs				Total In-House	Total w/O&P
					Material	Labor	Equipment	Total		
1540										
Replace pipe insulation foam rubber 1-1/4"	5	1 ASBE	L.F.							
Remove 1000' length old insulation				.092		5.28		5.28	6.75	8.40
Install insulation, foam rubber, 3/4" wall - 1-1/4" diam.				.184	4.84	10.50		15.34	18.85	23
Total				.276	4.84	15.78		20.62	**25.60**	**31.40**
1550										
Replace pipe insulation foam rubber 1-1/2"	5	1 ASBE	L.F.							
Remove 1000' length old insulation				.092		5.28		5.28	6.75	8.40
Install insulation, foam rubber, 3/4" wall - 1-1/2" diam.				.184	6.08	10.50		16.58	20	24.50
Total				.276	6.08	15.78		21.86	**26.75**	**32.90**
1560										
Replace pipe insulation foam rubber 2"	5	1 ASBE	L.F.							
Remove 1000' length old insulation				.093		5.34		5.34	6.85	8.50
Install insulation, foam rubber, 3/4" wall - 2" diam.				.186	7.14	10.70		17.84	21.50	26
Total				.279	7.14	16.04		23.18	**28.35**	**34.50**
1570										
Replace pipe insulation foam rubber 3"	5	1 ASBE	L.F.							
Remove 1000' length old insulation				.094		5.40		5.40	6.90	8.60
Install insulation, foam rubber, 3/4" wall - 3" diam.				.188	9.64	10.80		20.44	24.50	29.50
Total				.282	9.64	16.20		25.84	**31.40**	**38.10**
1580										
Replace pipe insulation foam rubber 4"	5	1 ASBE	L.F.							
Remove 1000' length old insulation				.100		5.74		5.74	7.35	9.15
Install insulation, foam rubber, 3/4" wall - 4" diam.				.200	12	11.50		23.50	28	33.50
Total				.300	12	17.24		29.24	**35.35**	**42.65**
1590										
Replace pipe insulation foam rubber 6"	5	1 ASBE	L.F.							
Remove 1000' length old insulation				.100		5.74		5.74	7.35	9.15
Install insulation, foam rubber, 3/4" wall - 6" diam.				.200	17.40	11.50		28.90	34	40
Total				.300	17.40	17.24		34.64	**41.35**	**49.15**

For customer support on your Facilities Maintenance & Repair Costs with RSMeans data, call 800.448.8182.

System Description	Freq. (Years)	Crew	Unit	Labor Hours	2019 Bare Costs				Total In-House	Total w/O&P
					Material	Labor	Equipment	Total		
1010 Repair unit heater, 12 MBH, 2 PSI steam	10	1 STPI	Ea.							
Repair controls				.200		12.80		12.80	15.85	19.85
Remove fan motor				.976		62.50		62.50	77.50	96.50
Replace fan motor				1.951	240	117		357	410	480
Total				3.127	240	192.30		432.30	503.35	596.35
1030 Replace unit heater, 12 MBH, 2 PSI steam	15	Q-5	Ea.							
Remove unit heater				.867		50		50	62	77.50
Replace unit heater, 12 MBH, 2 PSI steam				1.733	400	100		500	565	655
Total				2.600	400	150		550	627	732.50
2010 Repair unit heater, 36 MBH, 2 PSI steam	10	1 STPI	Ea.							
Repair controls				.200		12.80		12.80	15.85	19.85
Remove fan motor				.976		62.50		62.50	77.50	96.50
Replace fan motor				1.951	206	117		323	370	440
Total				3.127	206	192.30		398.30	463.35	556.35
2030 Replace unit heater, 36 MBH, 2 PSI steam	15	Q-5	Ea.							
Remove unit heater				1.300		75		75	93	116
Replace unit heater, 36 MBH, 2 PSI steam				2.600	600	150		750	845	980
Total				3.900	600	225		825	938	1,096
3010 Repair unit heater, 85 MBH, 2 PSI steam	10	1 STPI	Ea.							
Repair controls				.200		12.80		12.80	15.85	19.85
Remove fan motor				.976		62.50		62.50	77.50	96.50
Replace fan motor				1.951	285	117		402	460	535
Total				3.127	285	192.30		477.30	553.35	651.35
3030 Replace unit heater, 85 MBH, 2 PSI steam	15	Q-5	Ea.							
Remove unit heater				1.600		92		92	114	143
Replace unit heater, 85 MBH, 2 PSI steam				3.200	715	184		899	1,025	1,175
Total				4.800	715	276		991	1,139	1,318
4010 Repair unit heater, 250 MBH, 2 PSI steam	10	1 STPI	Ea.							
Repair controls				.200		12.80		12.80	15.85	19.85
Remove fan motor				1.156		69.50		69.50	85.50	107
Replace fan motor				2.311	305	139		444	505	595
Total				3.667	305	221.30		526.30	606.35	721.85

D3053 110 Unit Heater

	System Description	Freq. (Years)	Crew	Unit	Labor Hours	2019 Bare Costs Material	Labor	Equipment	Total	Total In-House	Total w/O&P
4030	**Replace unit heater, 250 MBH, 2 PSI steam**	15	Q-5	Ea.							
	Remove unit heater				4.160		239		239	297	370
	Replace unit heater, 250 MBH, 2 PSI steam				8.320	1,600	480		2,080	2,350	2,750
	Total				12.481	1,600	719		2,319	2,647	3,120
5010	**Repair unit heater, 400 MBH, 2 PSI steam**	10	1 STPI	Ea.							
	Repair controls				.200		12.80		12.80	15.85	19.85
	Remove fan motor				1.156		69.50		69.50	85.50	107
	Replace fan motor				2.311	370	139		509	580	675
	Total				3.667	370	221.30		591.30	681.35	801.85
5020	**Replace unit heater, 400 MBH, 2 PSI steam**	15	Q-5	Ea.							
	Remove unit heater				6.499		375		375	465	580
	Replace unit heater, 400 MBH, 2 PSI steam				13.008	2,350	750		3,100	3,525	4,100
	Total				19.507	2,350	1,125		3,475	3,990	4,680

D3053 112 Infrared Heater Suspended, Commercial

	System Description	Freq. (Years)	Crew	Unit	Labor Hours	2019 Bare Costs Material	Labor	Equipment	Total	Total In-House	Total w/O&P
0010	**Maintenance and repair**	1	1 ELEC	Ea.							
	Repair wiring connections				.615		37		37	45.50	57
	Total				.615		37		37	45.50	57
0020	**Maintenance and inspection**	0.50	1 ELEC	Ea.							
	Inspect and clean infrared heater				1.143		68.50		68.50	84.50	106
	Total				1.143		68.50		68.50	84.50	106
0030	**Replace infrared heater**	15	1 ELEC	Ea.							
	Remove infrared heater				.667		40		40	49.50	61.50
	Infrared heater 240 V, 1500 W				2.105	395	126		521	590	690
	Total				2.772	395	166		561	639.50	751.50

For customer support on your Facilities Maintenance & Repair Costs with RSMeans data, call 800.448.8182.

D3053 114 Standard Suspended Heater

	System Description	Freq. (Years)	Crew	Unit	Labor Hours	2019 Bare Costs				Total In-House	Total w/O&P
						Material	Labor	Equipment	Total		
0010	**Maintenance and repair**	2	1 ELEC	Ea.							
	Remove unit heater component				.151		9.05		9.05	11.15	14
	Unit heater component				.444	78.50	26.50		105	119	139
	Total				.595	78.50	35.55		114.05	130.15	153
0020	**Maintenance and inspection**	0.50	1 ELEC	Ea.							
	Inspect and clean unit heater				1.143		68.50		68.50	84.50	106
	Total				1.143		68.50		68.50	84.50	106
0030	**Replace heater**	15	1 ELEC	Ea.							
	Remove unit heater				.667		40		40	49.50	61.50
	Unit heater, heavy duty, 480 V, 3 KW				1.333	535	80		615	685	790
	Total				2.000	535	120		655	734.50	851.50

D3053 116 Explosionproof Industrial Heater

	System Description	Freq. (Years)	Crew	Unit	Labor Hours	2019 Bare Costs				Total In-House	Total w/O&P
						Material	Labor	Equipment	Total		
0010	**Maintenance and repair**	2	1 ELEC	Ea.							
	Remove explosionproof heater part				.200		12		12	14.80	18.50
	Explosionproof heater part				.615	104	37		141	160	187
	Total				.815	104	49		153	174.80	205.50
0020	**Maintenance and inspection**	0.50	1 ELEC	Ea.							
	Inspect / clean explosionproof heater				1.143		68.50		68.50	84.50	106
	Total				1.143		68.50		68.50	84.50	106
0030	**Replace heater**	15	1 ELEC	Ea.							
	Remove explosionproof heater				.800		48		48	59	74
	Explosionproof heater, 3 KW				3.810	4,100	229		4,329	4,800	5,475
	Total				4.610	4,100	277		4,377	4,859	5,549

For customer support on your Facilities Maintenance & Repair Costs with RSMeans data, call 800.448.8182.

271

D3053 150 Wall Mounted/Recessed Heater, With Fan

	System Description	Freq. (Years)	Crew	Unit	Labor Hours	2019 Bare Costs				Total In-House	Total w/O&P
						Material	Labor	Equipment	Total		
0010	**Maintenance and repair**	5	1 ELEC	Ea.							
	Remove heater fan				.250		15		15	18.50	23
	Heater fan				.727	96	43.50		139.50	159	188
	Total				.977	96	58.50		154.50	177.50	211
0020	**Maintenance and inspection**	1	1 ELEC	Ea.							
	Inspect and clean heater with fan				1.143		68.50		68.50	84.50	106
	Total				1.143		68.50		68.50	84.50	106
0030	**Replace heater**	20	1 ELEC	Ea.							
	Remove heater with fan				.667		40		40	49.50	61.50
	Heater with fan, commercial 2000 W				2.667	202	160		362	420	500
	Total				3.333	202	200		402	469.50	561.50

D3053 160 Convector Suspended, Commercial

	System Description	Freq. (Years)	Crew	Unit	Labor Hours	2019 Bare Costs				Total In-House	Total w/O&P
						Material	Labor	Equipment	Total		
0010	**Maintenance and repair**	2	1 ELEC	Ea.							
	Repair wiring connections				.615		37		37	45.50	57
	Total				.615		37		37	45.50	57
0020	**Maintenance and inspection**	0.50	1 ELEC	Ea.							
	Inspect and clean convector heater				1.143		68.50		68.50	84.50	106
	Total				1.143		68.50		68.50	84.50	106
0030	**Replace heater**	15	1 ELEC	Ea.							
	Remove cabinet convector heater				.667		40		40	49.50	61.50
	Cabinet convector heater, 2' long				2.000	2,300	120		2,420	2,675	3,050
	Total				2.667	2,300	160		2,460	2,724.50	3,111.50

For customer support on your Facilities Maintenance & Repair Costs with RSMeans data, call 800.448.8182.

	System Description	Freq. (Years)	Crew	Unit	Labor Hours	Material	Labor	Equipment	Total	Total In-House	Total w/O&P
							2019 Bare Costs				
1010	**Repair terminal reheat, 12″ x 24″ coil**	10	1 STPI	Ea.							
	Fix leak				1.200		76.50		76.50	95	119
	Total				1.200		76.50		76.50	**95**	**119**
1040	**Replace terminal reheat, 12″ x 24″ coil**	15	Q-5	Ea.							
	Remove terminal reheat unit				.858		49.50		49.50	61.50	76.50
	Replace terminal reheat unit, 12″ x 24″ coil				1.716	1,450	99		1,549	1,725	1,975
	Total				2.574	1,450	148.50		1,598.50	**1,786.50**	**2,051.50**
2010	**Repair terminal reheat, 18″ x 24″ coil**	10	1 STPI	Ea.							
	Fix leak				1.400		89.50		89.50	111	139
	Total				1.400		89.50		89.50	**111**	**139**
2040	**Replace terminal reheat, 18″ x 24″ coil**	15	Q-5	Ea.							
	Remove terminal reheat unit				1.284		74		74	91.50	115
	Replace terminal reheat unit, 18″ x 24″ coil				2.568	1,600	148		1,748	1,950	2,225
	Total				3.852	1,600	222		1,822	**2,041.50**	**2,340**
3010	**Repair terminal reheat, 36″ x 36″ coil**	10	1 STPI	Ea.							
	Fix leak				2.200		141		141	175	218
	Total				2.200		141		141	**175**	**218**
3040	**Replace terminal reheat, 36″ x 36″ coil**	15	Q-5	Ea.							
	Remove terminal reheat unit				3.852		222		222	275	345
	Replace terminal reheat unit, 36″ x 36″ coil				7.703	2,450	445		2,895	3,250	3,750
	Total				11.555	2,450	667		3,117	**3,525**	**4,095**
4010	**Repair terminal reheat, 48″ x 126″ coil**	10	1 STPI	Ea.							
	Fix leak				4.000		256		256	315	395
	Total				4.000		256		256	**315**	**395**
4040	**Replace terminal reheat, 48″ x 126″ coil**	15	Q-5	Ea.							
	Remove terminal reheat unit				17.937		1,025		1,025	1,275	1,600
	Replace terminal reheat unit, 48″ x 126″ coil				35.874	6,425	2,075		8,500	9,625	11,200
	Total				53.812	6,425	3,100		9,525	**10,900**	**12,800**

For customer support on your Facilities Maintenance & Repair Costs with RSMeans data, call 800.448.8182.

273

D3053 245   Heat Pump

	System Description	Freq. (Years)	Crew	Unit	Labor Hours	2019 Bare Costs Material	2019 Bare Costs Labor	2019 Bare Costs Equipment	2019 Bare Costs Total	Total In-House	Total w/O&P
1010	**Repair heat pump, 1.5 ton, air to air split**	10	1 STPI	Ea.							
	Repair controls				.600		38.50		38.50	47.50	59.50
	Remove / replace supply fan				.650		41.50		41.50	51.50	64.50
	Remove supply fan motor				.700		45		45	55.50	69.50
	Replace supply fan motor				1.951	270	125		395	450	530
	Remove compressor				1.860		119		119	148	184
	Replace compressor				3.714	276	238		514	600	715
	Remove / replace condenser fan				1.800		115		115	143	178
	Remove condenser fan motor				1.800		115		115	143	178
	Replace condenser fan motor				1.951	240	117		357	410	480
	Replace refrigerant				.133	23	8.55		31.55	36	42
	Remove / replace heater				.500		32		32	39.50	49.50
	Total				15.661	809	994.55		1,303.55	2,124	2,550
1030	**Replace heat pump, 1.5 ton, air to air split**	20	Q-5	Ea.							
	Remove heat pump				4.336		250		250	310	385
	Replace heat pump, 1.5 ton, air to air split				8.649	1,625	500		2,125	2,400	2,800
	Total				12.985	1,625	750		2,375	2,710	3,185
2010	**Repair heat pump, 5 ton, air to air split**	10	1 STPI	Ea.							
	Repair controls				.600		38.50		38.50	47.50	59.50
	Remove / replace supply fan				.650		41.50		41.50	51.50	64.50
	Remove supply fan motor				1.156		69.50		69.50	85.50	107
	Replace supply fan motor				1.951	240	117		357	410	480
	Remove compressor				2.477		158		158	197	246
	Replace compressor				4.954	520	315		835	965	1,150
	Remove / replace condenser fan				1.800		115		115	143	178
	Remove condenser fan motor				.976		62.50		62.50	77.50	96.50
	Replace condenser fan motor				1.951	206	117		323	370	440
	Replace refrigerant				.667	115	42.75		157.75	179	210
	Remove / replace heater				.500		32		32	39.50	49.50
	Total				17.681	1,081	1,108.75		2,189.75	2,565.50	3,081
2030	**Replace heat pump, 5 ton, air to air split**	20	Q-5	Ea.							
	Remove heat pump				20.806		1,200		1,200	1,475	1,850
	Replace heat pump, 5 ton, air to air split				41.558	2,700	2,400		5,100	5,925	7,075
	Total				62.365	2,700	3,600		6,300	7,400	8,925

For customer support on your Facilities Maintenance & Repair Costs with RSMeans data, call 800.448.8182.

D30 HVAC — D3053 — Terminal and Package Units

D3053 245 — Heat Pump

System Description	Freq. (Years)	Crew	Unit	Labor Hours	Material	Labor	Equipment	Total	Total In-House	Total w/O&P
3010 Repair heat pump, 10 ton, air to air split	10	Q-5	Ea.							
Repair controls				.600		38.50		38.50	47.50	59.50
Remove / replace supply fan				.650		41.50		41.50	51.50	64.50
Remove supply fan motor				.700		45		45	55.50	69.50
Replace supply fan motor				2.311	370	139		509	580	675
Remove compressor				4.000		230		230	286	355
Replace compressor				8.000	1,375	460		1,835	2,075	2,425
Remove / replace condenser fan				1.800		115		115	143	178
Remove condenser fan motor				1.800		115		115	143	178
Replace condenser fan motor				2.311	305	139		444	505	595
Replace refrigerant				1.067	184	68.40		252.40	287	335
Remove / replace heater				.500		32		32	39.50	49.50
Total				23.739	2,234	1,423.40		3,657.40	4,213	4,984
3030 Replace heat pump, 10 ton, air to air split	20	Q-6	Ea.							
Remove heat pump				24.490		1,450		1,450	1,825	2,275
Replace heat pump, 10 ton, air to air split				48.980	6,350	2,925		9,275	10,600	12,500
Total				73.469	6,350	4,375		10,725	12,425	14,775
4010 Repair heat pump, 25 ton, air to air split	10	Q-5	Ea.							
Repair controls				.600		38.50		38.50	47.50	59.50
Remove / replace supply fan				1.000		64		64	79.50	99
Remove supply fan motor				1.156		69.50		69.50	85.50	107
Replace supply fan motor				2.311	610	139		749	840	975
Remove compressor				10.000		575		575	715	890
Replace compressor				20.000	3,437.50	1,150		4,587.50	5,200	6,075
Remove / replace condenser fan				1.800		115		115	143	178
Remove condenser fan motor				1.800		115		115	143	178
Replace condenser fan motor				2.312	760	139		899	1,000	1,175
Replace refrigerant				2.000	345	128.25		473.25	540	630
Remove / replace heater				.500		32		32	39.50	49.50
Total				43.479	5,152.50	2,565.25		7,717.75	8,833	10,416
4030 Replace heat pump, 25 ton, air to air split	20	Q-7	Ea.							
Remove heat pump				61.538		3,750		3,750	4,650	5,800
Replace heat pump, 25 ton, air to air split				123.000	21,500	7,500		29,000	32,900	38,500
Total				184.538	21,500	11,250		32,750	37,550	44,300

D3053 245 Heat Pump

	System Description	Freq. (Years)	Crew	Unit	Labor Hours	2019 Bare Costs				Total In-House	Total w/O&P
						Material	Labor	Equipment	Total		
5010	**Repair heat pump, 50 ton, air to air split**	10	Q-6	Ea.							
	Repair controls				.600		38.50		38.50	47.50	59.50
	Remove / replace supply fan				1.000		64		64	79.50	99
	Remove supply fan motor				1.156		69.50		69.50	85.50	107
	Replace supply fan motor				4.167	2,750	250		3,000	3,325	3,825
	Remove compressor				77.922		4,650		4,650	5,775	7,200
	Replace compressor				156.000	34,000	9,300		43,300	48,900	57,000
	Remove / replace condenser fan				2.400		154		154	190	238
	Remove condenser fan motor				1.800		115		115	143	178
	Replace condenser fan motor				4.000	2,125	240		2,365	2,625	3,025
	Replace refrigerant				4.667	805	299.25		1,104.25	1,250	1,475
	Remove / replace heater				1.200		76.50		76.50	95	119
	Total				254.912	39,680	15,256.75		54,936.75	62,515.50	73,325.50
5030	**Replace heat pump, 50 ton, air to air split**	20	Q-7	Ea.							
	Remove heat pump				96.970		5,900		5,900	7,325	9,150
	Replace heat pump, 50 ton, air to air split				208.000	53,500	12,700		66,200	74,500	86,500
	Total				304.970	53,500	18,600		72,100	81,825	95,650
6010	**Repair heat pump, thru-wall unit, 1.5 ton**	10	1 STPI	Ea.							
	Repair controls				.600		38.50		38.50	47.50	59.50
	Remove / replace supply fan				.650		41.50		41.50	51.50	64.50
	Remove supply fan motor				.700		45		45	55.50	69.50
	Replace supply fan motor				1.951	270	125		395	450	530
	Remove compressor				1.860		119		119	148	184
	Replace compressor				3.714	276	238		514	600	715
	Remove / replace condenser fan				1.800		115		115	143	178
	Remove condenser fan motor				1.800		115		115	143	178
	Replace condenser fan motor				1.951	240	117		357	410	480
	Replace refrigerant				.133	23	8.55		31.55	36	42
	Remove / replace heater				.500		32		32	39.50	49.50
	Total				15.661	809	994.55		1,803.55	2,124	2,550
6030	**Replace heat pump, thru-wall unit, 1.5 ton**	20	Q-5	Ea.							
	Remove heat pump				6.709		385		385	480	600
	Replace heat pump, thru-wall, 1.5 ton				13.423	3,325	770		4,095	4,625	5,350
	Total				20.131	3,325	1,155		4,460	5,105	5,950

For customer support on your Facilities Maintenance & Repair Costs with RSMeans data, call 800.448.8182.

D3053 245 Heat Pump

	System Description	Freq. (Years)	Crew	Unit	Labor Hours	2019 Bare Costs Material	Labor	Equipment	Total	Total In-House	Total w/O&P
7010	**Repair heat pump, thru-wall unit, 5 ton**	10	1 STPI	Ea.							
	Repair controls				.600		38.50		38.50	47.50	59.50
	Remove / replace supply fan				.650		41.50		41.50	51.50	64.50
	Remove supply fan motor				1.156		69.50		69.50	85.50	107
	Replace supply fan motor				1.951	240	117		357	410	480
	Remove compressor				2.477		158		158	197	246
	Replace compressor				4.954	520	315		835	965	1,150
	Remove / replace condenser fan				1.800		115		115	143	178
	Remove condenser fan motor				.976		62.50		62.50	77.50	96.50
	Replace condenser fan motor				1.951	206	117		323	370	440
	Replace refrigerant				.667	115	42.75		157.75	179	210
	Remove / replace heater				.500		32		32	39.50	49.50
	Total				17.681	1,081	1,108.75		2,189.75	2,565.50	3,081
7030	**Replace heat pump, thru-wall unit, 5 ton**	20	Q-5	Ea.							
	Remove heat pump				16.000		920		920	1,150	1,425
	Replace heat pump, trhu-wall, 5 ton				32.000	5,050	1,850		6,900	7,850	9,175
	Total				48.000	5,050	2,770		7,820	9,000	10,600

D3053 265 Air Conditioner, Window, 1 Ton

	System Description	Freq. (Years)	Crew	Unit	Labor Hours	2019 Bare Costs Material	Labor	Equipment	Total	Total In-House	Total w/O&P
1010	**Repair air conditioner, window, 1 ton**	8	1 STPI	Ea.							
	Find / fix leak				1.000		64		64	79.50	99
	Replace refrigerant				.067	11.50	4.28		15.78	17.95	21
	Total				1.067	11.50	68.28		79.78	97.45	120
1030	**Replace air conditioner, window, 1 ton**	10	L-2	Ea.							
	Remove air conditioner				1.300		59		59	77	95.50
	Replace air conditioner, 1 ton				2.000	1,950	90.50		2,040.50	2,275	2,575
	Total				3.300	1,950	149.50		2,099.50	2,352	2,670.50

D3053 266 — Air Conditioner, Thru-The-Wall

	System Description	Freq. (Years)	Crew	Unit	Labor Hours	2019 Bare Costs Material	Labor	Equipment	Total	Total In-House	Total w/O&P
2010	Repair air conditioner, thru-the-wall, 2 ton	8	1 STPI	Ea.							
	Find / fix leak				1.000		64		64	79.50	99
	Replace refrigerant				.133	23	8.55		31.55	36	42
	Total				1.133	23	72.55		95.55	**115.50**	**141**
2030	Replace air conditioner, thru-the-wall, 2 ton	10	L-2	Ea.							
	Remove air conditioner				2.600		118		118	154	191
	Replace air conditioner, 2 ton				5.200	930	235		1,165	1,325	1,550
	Total				7.800	930	353		1,283	1,479	1,741

D3053 272 — Air Conditioner, DX Package

	System Description	Freq. (Years)	Crew	Unit	Labor Hours	2019 Bare Costs Material	Labor	Equipment	Total	Total In-House	Total w/O&P
1010	Repair air conditioner, DX, 5 ton	10	1 STPI	Ea.							
	Repair controls				.600		38.50		38.50	47.50	59.50
	Remove / replace supply fan				.650		41.50		41.50	51.50	64.50
	Remove supply fan bearing				1.000		64		64	79.50	99
	Replace supply fan bearing				2.000	65.50	128		193.50	231	280
	Remove supply fan motor				1.156		69.50		69.50	85.50	107
	Replace supply fan motor				2.311	305	139		444	505	595
	Remove compressor				2.477		158		158	197	246
	Replace compressor				4.954	520	315		835	965	1,150
	Remove / replace condenser fan				1.800		115		115	143	178
	Remove condenser fan bearing				1.200		76.50		76.50	95	119
	Replace condenser fan bearing				2.000	65.50	128		193.50	231	280
	Remove condenser fan motor				.976		62.50		62.50	77.50	96.50
	Replace condenser fan motor				1.951	285	117		402	460	535
	Replace refrigerant				.667	115	42.75		157.75	179	210
	Remove / replace heating coils				3.900	660	232.50		892.50	1,025	1,175
	Total				27.641	2,016	1,727.75		3,743.75	4,372.50	5,194.50
1030	Replace air conditioner, DX, 5 ton	20	Q-6	Ea.							
	Remove air conditioner				13.001		775		775	965	1,200
	Replace air conditioner, 5 ton				26.002	4,400	1,550		5,950	6,775	7,900
	Total				39.003	4,400	2,325		6,725	7,740	9,100

For customer support on your Facilities Maintenance & Repair Costs with RSMeans data, call 800.448.8182.

	System Description	Freq. (Years)	Crew	Unit	Labor Hours	2019 Bare Costs				Total In-House	Total w/O&P
						Material	Labor	Equipment	Total		
2010	**Repair air conditioner, DX, 20 ton**	10	Q-6	Ea.							
	Repair controls				.800		51		51	63.50	79.50
	Remove / replace supply fan				.650		41.50		41.50	51.50	64.50
	Remove supply fan bearing				1.000		64		64	79.50	99
	Replace supply fan bearing				2.000	65.50	128		193.50	231	280
	Remove supply fan motor				1.156		69.50		69.50	85.50	107
	Replace supply fan motor				2.312	760	139		899	1,000	1,175
	Remove compressor				24.390		1,450		1,450	1,800	2,250
	Replace compressor				48.780	28,400	2,900		31,300	34,900	40,000
	Remove / replace condenser fan				1.800		115		115	143	178
	Remove condenser fan bearing				1.200		76.50		76.50	95	119
	Replace condenser fan bearing				2.000	65.50	128		193.50	231	280
	Remove condenser fan motor				.976		62.50		62.50	77.50	96.50
	Replace condenser fan motor				2.311	305	139		444	505	595
	Replace refrigerant				1.333	230	85.50		315.50	360	420
	Remove / replace heating coils				6.936	1,965	423.75		2,388.75	2,675	3,125
	Total				97.645	31,791	5,873.25		37,664.25	42,297.50	48,868.50
2030	**Replace air conditioner, DX, 20 ton**	20	Q-7	Ea.							
	Remove air conditioner				23.105		1,400		1,400	1,750	2,175
	Replace air conditioner, 20 ton				46.243	13,100	2,825		15,925	17,900	20,800
	Total				69.347	13,100	4,225		17,325	19,650	22,975

For customer support on your Facilities Maintenance & Repair Costs with RSMeans data, call 800.448.8182.

279

D3053 272   **Air Conditioner, DX Package**

System Description	Freq. (Years)	Crew	Unit	Labor Hours	2019 Bare Costs				Total In-House	Total w/O&P
					Material	Labor	Equipment	Total		
3010 Repair air conditioner, DX, 50 ton	10	Q-6	Ea.							
Repair controls				.800		51		51	63.50	79.50
Remove / replace supply fan				.650		41.50		41.50	51.50	64.50
Remove supply fan bearing				1.200		76.50		76.50	95	119
Replace supply fan bearing				2.222	73.50	142		215.50	257	310
Remove supply fan motor				1.156		69.50		69.50	85.50	107
Replace supply fan motor				2.597	1,350	156		1,506	1,675	1,925
Remove compressor				77.922		4,650		4,650	5,775	7,200
Replace compressor				156.000	34,000	9,300		43,300	48,900	57,000
Remove / replace condenser fan				1.800		115		115	143	178
Remove condenser fan bearing				1.200		76.50		76.50	95	119
Replace condenser fan bearing				2.222	73.50	142		215.50	257	310
Remove condenser fan motor				.976		62.50		62.50	77.50	96.50
Replace condenser fan motor				2.311	305	139		444	505	595
Replace refrigerant				3.333	575	213.75		788.75	895	1,050
Remove / replace heating coil				12.468	7,575	757.50		8,332.50	9,275	10,600
Total				266.857	43,952	15,992.75		59,944.75	68,150	79,753.50
3030 Replace air conditioner, DX, 50 ton	20	Q-7	Ea.							
Remove air conditioner				41.558		2,525		2,525	3,150	3,925
Replace air conditioner, 50 ton				83.117	50,500	5,050		55,550	62,000	71,000
Total				124.675	50,500	7,575		58,075	65,150	74,925

For customer support on your Facilities Maintenance & Repair Costs with RSMeans data, call 800.448.8182.

D3053 274 Computer Room A/C Units, Air Cooled

	System Description	Freq. (Years)	Crew	Unit	Labor Hours	2019 Bare Costs				Total In-House	Total w/O&P
						Material	Labor	Equipment	Total		
1010	**Repair computer room A/C, air cooled, 5 ton**	10	1 STPI	Ea.							
	Repair controls				.600		38.50		38.50	47.50	59.50
	Remove / replace supply fan				.650		41.50		41.50	51.50	64.50
	Remove supply fan bearing				1.000		64		64	79.50	99
	Replace supply fan bearing				2.000	65.50	128		193.50	231	280
	Remove supply fan motor				1.156		69.50		69.50	85.50	107
	Replace supply fan motor				2.311	305	139		444	505	595
	Replace fan belt for supply fan				.400	16.40	25.50		41.90	50	60
	Remove compressor				2.477		158		158	197	246
	Replace compressor				4.954	520	315		835	965	1,150
	Remove / replace condenser fan				1.800		115		115	143	178
	Remove condenser fan bearing				1.200		76.50		76.50	95	119
	Replace condenser fan bearing				2.000	65.50	128		193.50	231	280
	Remove condenser fan motor				.976		62.50		62.50	77.50	96.50
	Replace condenser fan motor				1.951	285	117		402	460	535
	Replace fan belt for condenser fan				.400	16.40	25.50		41.90	50	60
	Replace refrigerant				.667	115	42.75		157.75	179	210
	Remove / replace heating coils				3.900	660	232.50		892.50	1,025	1,175
	Total				28.441	2,048.80	1,778.75		3,827.55	4,472.50	5,314.50
1015	**Replace computer room A/C, air cooled, 5 ton**	20	Q-6	Ea.							
	Remove air conditioner and condenser				13.001		775		775	965	1,200
	Replace air conditioner and remote condenser, 5 ton				35.556	28,400	2,050		30,450	33,800	38,700
	Total				48.557	28,400	2,825		31,225	34,765	39,900

For customer support on your Facilities Maintenance & Repair Costs with RSMeans data, call 800.448.8182.

281

D3053 274 | **Computer Room A/C Units, Air Cooled**

System Description	Freq. (Years)	Crew	Unit	Labor Hours	2019 Bare Costs				Total In-House	Total w/O&P
					Material	Labor	Equipment	Total		
1020 **Repair computer room A/C, air cooled, 10 ton**	10	1 STPI	Ea.							
Repair controls				.400		25.50		25.50	31.50	39.50
Remove / replace supply fan				.650		41.50		41.50	51.50	64.50
Remove supply fan bearing				1.200		76.50		76.50	95	119
Replace supply fan bearing				2.000	65.50	128		193.50	231	280
Remove supply fan motor				1.156		69.50		69.50	85.50	107
Replace supply fan motor				2.311	370	139		509	580	675
Replace fan belt for supply fan				.400	16.40	25.50		41.90	50	60
Remove compressor				14.513		871		871	1,075	1,350
Replace compressor				28.714	15,745	1,708.50		17,453.50	19,400	22,300
Remove / replace condenser fan				1.800		115		115	143	178
Remove condenser fan bearing				1.200		76.50		76.50	95	119
Replace condenser fan bearing				2.000	65.50	128		193.50	231	280
Remove condenser fan motor				.976		62.50		62.50	77.50	96.50
Replace condenser fan motor				1.951	285	117		402	460	535
Replace fan belt for condenser fan				.400	16.40	25.50		41.90	50	60
Replace refrigerant				1.333	230	85.50		315.50	360	420
Remove / replace heating coils				5.200	880	310		1,190	1,350	1,575
Total				66.204	17,673.80	4,005		21,678.80	24,366	28,258.50
1025 **Replace computer room A/C, air cooled, 10 ton**	20	Q-6	Ea.							
Remove air conditioner and condenser				33.898		2,025		2,025	2,500	3,125
Replace air conditioner and remote condenser, 10 ton				64.000	55,500	3,675		59,175	65,500	75,000
Total				97.898	55,500	5,700		61,200	68,000	78,125

For customer support on your Facilities Maintenance & Repair Costs with RSMeans data, call 800.448.8182.

System Description	Freq. (Years)	Crew	Unit	Labor Hours	2019 Bare Costs				Total In-House	Total w/O&P
					Material	Labor	Equipment	Total		
1030										
Repair computer room A/C, air cooled, 15 ton	10	1 STPI	Ea.							
Repair controls				.500		32		32	39.50	49.50
Remove / replace supply fan				.650		41.50		41.50	51.50	64.50
Remove supply fan bearing				1.200		76.50		76.50	95	119
Replace supply fan bearing				2.000	65.50	128		193.50	231	280
Remove supply fan motor				1.156		69.50		69.50	85.50	107
Replace supply fan motor				2.312	760	139		899	1,000	1,175
Replace fan belt for supply fan				.400	16.40	25.50		41.90	50	60
Remove compressor				21.661		1,300		1,300	1,600	2,000
Replace compressor				42.857	23,500	2,550		26,050	29,000	33,400
Remove / replace condenser fan				1.800		115		115	143	178
Remove condenser fan bearing				1.200		76.50		76.50	95	119
Replace condenser fan bearing				2.000	65.50	128		193.50	231	280
Remove condenser fan motor				.976		62.50		62.50	77.50	96.50
Replace condenser fan motor				1.951	285	117		402	460	535
Replace fan belt for condenser fan				.400	16.40	25.50		41.90	50	60
Replace refrigerant				2.000	345	128.25		473.25	540	630
Remove / replace heating coils				6.501	1,100	387.50		1,487.50	1,700	1,975
Total				89.563	26,153.80	5,402.25		31,556.05	35,449	41,128.50
1035										
Replace computer room A/C, air cooled, 15 ton	20	Q-6	Ea.							
Remove air conditioner and condenser				50.314		3,000		3,000	3,725	4,650
Replace air conditioner and remote condenser, 15 ton				72.727	61,000	4,175		65,175	72,500	83,000
Total				123.042	61,000	7,175		68,175	76,225	87,650

For customer support on your Facilities Maintenance & Repair Costs with RSMeans data, call 800.448.8182.

283

D3053 274 Computer Room A/C Units, Air Cooled

System Description	Freq. (Years)	Crew	Unit	Labor Hours	2019 Bare Costs Material	Labor	Equipment	Total	Total In-House	Total w/O&P
1040 Repair computer room A/C, air cooled, 20 ton	10	1 STPI	Ea.							
Repair controls				.800		51		51	63.50	79.50
Remove / replace supply fan				.650		41.50		41.50	51.50	64.50
Remove supply fan bearing				1.000		64		64	79.50	99
Replace supply fan bearing				2.000	65.50	128		193.50	231	280
Remove supply fan motor				1.156		69.50		69.50	85.50	107
Replace supply fan motor				2.312	760	139		899	1,000	1,175
Replace fan belt for supply fan				.400	16.40	25.50		41.90	50	60
Remove compressor				24.390		1,450		1,450	1,800	2,250
Replace compressor				48.780	28,400	2,900		31,300	34,900	40,000
Remove / replace condenser fan				1.800		115		115	143	178
Remove condenser fan bearing				1.200		76.50		76.50	95	119
Replace condenser fan bearing				2.000	65.50	128		193.50	231	280
Remove condenser fan motor				.976		62.50		62.50	77.50	96.50
Replace condenser fan motor				2.311	305	139		444	505	595
Replace fan belt for condenser fan				.400	16.40	25.50		41.90	50	60
Replace refrigerant				2.667	460	171		631	720	840
Remove / replace heating coils				6.936	1,965	423.75		2,388.75	2,675	3,125
Total				99.778	32,053.80	6,009.75		38,063.55	42,757.50	49,408.50
1045 Replace computer room A/C, air cooled, 20 ton	20	Q-6	Ea.							
Remove air conditioner and condenser				23.105		1,400		1,400	1,750	2,175
Replace air conditioner and remote condenser, 20 ton				100.000	74,000	5,975		79,975	89,000	102,000
Total				123.105	74,000	7,375		81,375	90,750	104,175

For customer support on your Facilities Maintenance & Repair Costs with RSMeans data, call 800.448.8182.

D3053 276 | **Computer Room A/C Units, Chilled Water**

System Description	Freq. (Years)	Crew	Unit	Labor Hours	2019 Bare Costs Material	Labor	Equipment	Total	Total In-House	Total w/O&P
1010 Repair computer room A/C, chilled water, 5 ton	10	1 STPI	Ea.							
Repair controls				.600		38.50		38.50	47.50	59.50
Remove / replace supply fan				.650		41.50		41.50	51.50	64.50
Remove supply fan bearing				1.000		64		64	79.50	99
Replace supply fan bearing				2.000	65.50	128		193.50	231	280
Remove supply fan motor				1.156		69.50		69.50	85.50	107
Replace supply fan motor				2.311	305	139		444	505	595
Replace fan belt for supply fan				.400	16.40	25.50		41.90	50	60
Remove / replace heating coils				26.002	4,400	1,550		5,950	6,775	7,900
Total				34.119	4,786.90	2,056		6,842.90	7,825	9,165
1015 Replace computer room A/C, chilled water, 5 ton	20	Q-6	Ea.							
Remove air conditioner				13.001		775		775	965	1,200
Replace air conditioner, 5 ton				28.070	19,200	1,625		20,825	23,100	26,500
Total				41.071	19,200	2,400		21,600	24,065	27,700
1020 Repair computer room A/C, chilled water, 10 ton	10	1 STPI	Ea.							
Repair controls				.400		25.50		25.50	31.50	39.50
Remove / replace supply fan				.650		41.50		41.50	51.50	64.50
Remove supply fan bearing				1.200		76.50		76.50	95	119
Replace supply fan bearing				2.000	65.50	128		193.50	231	280
Remove supply fan motor				1.156		69.50		69.50	85.50	107
Replace supply fan motor				2.311	370	139		509	580	675
Replace fan belt for supply fan				.400	16.40	25.50		41.90	50	60
Remove / replace heating coils				26.002	4,400	1,550		5,950	6,775	7,900
Total				34.119	4,851.90	2,055.50		6,907.40	7,899.50	9,245
1025 Replace computer room A/C, chilled water, 10 ton	20	Q-6	Ea.							
Remove air conditioner				33.898		2,025		2,025	2,500	3,125
Replace air conditioner, 10 ton				42.105	19,900	2,425		22,325	24,900	28,600
Total				76.004	19,900	4,450		24,350	27,400	31,725

For customer support on your Facilities Maintenance & Repair Costs with RSMeans data, call 800.448.8182.

285

D3053 276 Computer Room A/C Units, Chilled Water

System Description	Freq. (Years)	Crew	Unit	Labor Hours	2019 Bare Costs				Total In-House	Total w/O&P
					Material	Labor	Equipment	Total		
1030 Repair computer room A/C, chilled water, 15 ton	10	1 STPI	Ea.							
Repair controls				.500		32		32	39.50	49.50
Remove / replace supply fan				.650		41.50		41.50	51.50	64.50
Remove supply fan bearing				1.200		76.50		76.50	95	119
Replace supply fan bearing				2.000	65.50	128		193.50	231	280
Remove supply fan motor				1.156		69.50		69.50	85.50	107
Replace supply fan motor				2.312	760	139		899	1,000	1,175
Replace fan belt for supply fan				.400	16.40	25.50		41.90	50	60
Remove / replace heating coils				26.002	4,400	1,550		5,950	6,775	7,900
Total				34.220	5,241.90	2,062		7,303.90	8,327.50	9,755
1035 Replace computer room A/C, chilled water, 15 ton	20	Q-6	Ea.							
Remove air conditioner				50.314		3,000		3,000	3,725	4,650
Replace air conditioner, 15 ton				44.444	21,800	2,550		24,350	27,200	31,200
Total				94.759	21,800	5,550		27,350	30,925	35,850
1040 Repair computer room A/C, chilled water, 20 ton	10	1 STPI	Ea.							
Repair controls				.500		32		32	39.50	49.50
Remove / replace supply fan				.650		41.50		41.50	51.50	64.50
Remove supply fan bearing				1.200		76.50		76.50	95	119
Replace supply fan bearing				2.000	65.50	128		193.50	231	280
Remove supply fan motor				1.156		69.50		69.50	85.50	107
Replace supply fan motor				2.312	760	139		899	1,000	1,175
Replace fan belt for supply fan				.400	16.40	25.50		41.90	50	60
Remove / replace heating coils				46.243	13,100	2,825		15,925	17,900	20,800
Total				54.461	13,941.90	3,337		17,278.90	19,452.50	22,655
1045 Replace computer room A/C, chilled water, 20 ton	20	Q-6	Ea.							
Remove air conditioner				23.105		1,400		1,400	1,750	2,175
Replace air conditioner, 20 ton				47.059	23,200	2,700		25,900	28,900	33,200
Total				70.164	23,200	4,100		27,300	30,650	35,375

For customer support on your Facilities Maintenance & Repair Costs with RSMeans data, call 800.448.8182.

D3053 278   Multi-Zone Air Conditioner

	System Description	Freq. (Years)	Crew	Unit	Labor Hours	2019 Bare Costs				Total In-House	Total w/O&P
						Material	Labor	Equipment	Total		
1010	**Repair multi-zone rooftop unit, 15 ton**	10	Q-6	Ea.							
	Repair controls				1.000		64		64	79.50	99
	Remove fan bearings				2.400		153		153	190	238
	Replace fan bearings				4.000	131	256		387	460	560
	Remove fan motor				1.156		69.50		69.50	85.50	107
	Replace fan motor				2.312	760	139		899	1,000	1,175
	Remove compressor				17.778		1,025		1,025	1,275	1,575
	Replace compressor				42.857	23,500	2,550		26,050	29,000	33,400
	Replace refrigerant				1.333	230	85.50		315.50	360	420
	Remove / replace heater igniter				.500	32	32		64	75	89.50
	Total				73.336	24,653	4,374		29,027	32,525	37,663.50
1040	**Replace multi-zone rooftop unit, 15 ton**	15	Q-7	Ea.							
	Remove multi-zone rooftop unit				94.675		5,775		5,775	7,150	8,925
	Replace multi-zone rooftop, 15 ton				70.022	65,000	4,275		69,275	77,000	88,000
	Total				164.696	65,000	10,050		75,050	84,150	96,925
2010	**Repair multi - zone rooftop unit, 25 ton**	10	Q-6	Ea.							
	Repair controls				1.000		64		64	79.50	99
	Remove fan bearings				2.400		153		153	190	238
	Replace fan bearings				4.000	131	256		387	460	560
	Remove fan motor				1.156		69.50		69.50	85.50	107
	Replace fan motor				2.311	610	139		749	840	975
	Remove compressor				22.222		1,281.25		1,281.25	1,575	1,975
	Replace compressor				60.976	35,500	3,625		39,125	43,600	50,000
	Replace refrigerant				2.000	345	128.25		473.25	540	630
	Remove / replace heater igniter				1.000	64	64		128	150	179
	Total				97.065	36,650	5,780		42,430	47,520	54,763
2040	**Replace multi-zone rooftop unit, 25 ton**	15	Q-7	Ea.							
	Remove multi-zone rooftop unit				116.000		7,025		7,025	8,725	10,900
	Replace multi-zone rooftop, 25 ton				95.808	85,500	5,825		91,325	101,500	116,000
	Total				211.808	85,500	12,850		98,350	110,225	126,900

D3053 278 Multi-Zone Air Conditioner

System Description	Freq. (Years)	Crew	Unit	Labor Hours	2019 Bare Costs				Total In-House	Total w/O&P
					Material	Labor	Equipment	Total		
3010 Repair multi-zone rooftop unit, 40 ton	10	Q-6	Ea.							
Repair controls				1.000		64		64	79.50	99
Remove fan bearings				2.400		153		153	190	238
Replace fan bearings				4.444	147	284		431	515	625
Remove fan motor				2.000		120		120	148	185
Replace fan motor				4.000	2,125	240		2,365	2,625	3,025
Remove compressor				35.451		2,125		2,125	2,625	3,275
Replace compressor				71.006	32,700	4,250		36,950	41,200	47,500
Replace refrigerant				2.667	460	171		631	720	840
Remove / replace heater igniter				1.000	64	64		128	150	179
Total				123.967	35,496	7,471		42,967	48,252.50	55,966
3040 Replace multi-zone rooftop unit, 40 ton	15	Q-7	Ea.							
Remove multi-zone				173.000		10,500		10,500	13,100	16,300
Replace multi-zone rooftop, 40 ton				152.000	123,500	9,225		132,725	147,500	168,500
Total				325.000	123,500	19,725		143,225	160,600	184,800
4010 Repair multi-zone rooftop unit, 70 ton	10	Q-7	Ea.							
Repair controls				1.000		64		64	79.50	99
Remove fan bearings				2.400		153		153	190	238
Replace fan bearings				4.444	147	284		431	515	625
Remove fan motor				2.601		156		156	192	241
Replace fan motor				5.195	3,950	310		4,260	4,725	5,425
Remove compressor				76.923		4,675		4,675	5,800	7,250
Replace compressor				154.000	37,600	9,375		46,975	53,000	61,500
Replace refrigerant				4.667	805	299.25		1,104.25	1,250	1,475
Remove / replace heater igniter				1.000	64	64		128	150	179
Total				252.230	42,566	15,380.25		57,946.25	65,901.50	77,032
4040 Replace multi-zone rooftop unit, 70 ton	15	Q-7	Ea.							
Remove multi-zone rooftop unit				232.000		14,100		14,100	17,500	21,900
Replace multi-zone rooftop, 70 ton				264.000	165,000	16,100		181,100	201,500	231,500
Total				496.000	165,000	30,200		195,200	219,000	253,400

For customer support on your Facilities Maintenance & Repair Costs with RSMeans data, call 800.448.8182.

D30 HVAC — D3053 Terminal and Package Units

D3053 278 Multi-Zone Air Conditioner

System Description	Freq. (Years)	Crew	Unit	Labor Hours	2019 Bare Costs Material	2019 Bare Costs Labor	2019 Bare Costs Equipment	2019 Bare Costs Total	Total In-House	Total w/O&P
5010 Repair multi-zone rooftop unit, 105 ton	10	Q-7	Ea.							
Repair controls				1.000		64		64	79.50	99
Remove fan bearings				4.800		306		306	380	475
Replace fan bearings				9.412	756	600		1,356	1,575	1,875
Remove fan motor				4.334		260		260	320	400
Replace fan motor				8.696	5,050	520		5,570	6,200	7,125
Remove compressor				79.012		4,820		4,820	5,975	7,450
Replace compressor				158.400	34,320	9,600		43,920	49,700	58,000
Replace refrigerant				6.667	1,150	427.50		1,577.50	1,800	2,100
Remove / replace heater igniter				1.500	96	96		192	225	269
Total				273.820	41,372	16,693.50		58,065.50	66,254.50	77,793
5040 Replace multi-zone rooftop unit, 105 ton	15	Q-7	Ea.							
Remove multi-zone rooftop unit				348.000		21,200		21,200	26,300	32,800
Replace multi-zone rooftop, 105 ton				390.000	215,000	23,800		238,800	266,000	305,500
Total				738.000	215,000	45,000		260,000	292,300	338,300

D3053 280 Single Zone Air Conditioner

System Description	Freq. (Years)	Crew	Unit	Labor Hours	2019 Bare Costs Material	2019 Bare Costs Labor	2019 Bare Costs Equipment	2019 Bare Costs Total	Total In-House	Total w/O&P
1001 Repair single zone rooftop unit, 3 ton	10	2 STPI	Ea.							
Repair controls				.300		19.20		19.20	24	30
Remove fan bearings				2.400		153		153	190	238
Replace fan bearings				4.000	131	256		387	460	560
Remove fan motor				1.156		69.50		69.50	85.50	107
Replace fan motor				1.951	206	117		323	370	440
Remove compressor				2.477		158		158	197	246
Replace compressor				4.522	370	289		659	765	915
Replace refrigerant				.533	92	34.20		126.20	144	168
Remove/replace heater igniter				.500	32	32		64	75	89.50
Total				17.840	831	1,127.90		1,958.90	2,310.50	2,793.50
1002 Replace single zone rooftop unit, 3 ton	15	Q-5	Ea.							
Remove single zone				8.000		460		460	570	715
Replace single zone rooftop, 3 ton				30.476	3,025	1,750		4,775	5,500	6,500
Total				38.476	3,025	2,210		5,235	6,070	7,215

For customer support on your Facilities Maintenance & Repair Costs with RSMeans data, call 800.448.8182.

289

D3053 280   Single Zone Air Conditioner

	System Description	Freq. (Years)	Crew	Unit	Labor Hours	2019 Bare Costs				Total In-House	Total w/O&P
						Material	Labor	Equipment	Total		
1003	**Repair single zone rooftop unit, 5 ton**	10	2 STPI	Ea.							
	Repair controls				.300		19.20		19.20	24	30
	Remove fan bearings				2.400		153		153	190	238
	Replace fan bearings				4.000	131	256		387	460	560
	Remove fan motor				1.156		69.50		69.50	85.50	107
	Replace fan motor				1.951	206	117		323	370	440
	Remove compressor				2.477		158		158	197	246
	Replace compressor				4.954	520	315		835	965	1,150
	Replace refrigerant				.533	92	34.20		126.20	144	168
	Remove/replace heater igniter				.500	32	32		64	75	89.50
	Total				18.271	981	1,153.90		2,134.90	2,510.50	3,028.50
1004	**Replace single zone rooftop unit, 5 ton**	15	Q-5	Ea.							
	Remove single zone				18.561		1,075		1,075	1,325	1,650
	Replace single zone rooftop, 5 ton				38.005	4,425	2,175		6,600	7,575	8,925
	Total				56.566	4,425	3,250		7,675	8,900	10,575
1005	**Repair single zone rooftop unit, 7.5 ton**	10	2 STPI	Ea.							
	Repair controls				.400		25.50		25.50	31.50	39.50
	Remove fan bearings				2.400		153		153	190	238
	Replace fan bearings				4.000	131	256		387	460	560
	Remove fan motor				1.156		69.50		69.50	85.50	107
	Replace fan motor				2.311	305	139		444	505	595
	Remove compressor				3.467		200		200	248	310
	Replace compressor				6.932	1,175	400		1,575	1,775	2,100
	Replace refrigerant				.667	115	42.75		157.75	179	210
	Remove/replace heater igniter				.500	32	32		64	75	89.50
	Total				21.833	1,758	1,317.75		3,075.75	3,549	4,249
1006	**Replace single zone rooftop unit, 7.5 ton**	15	Q-5	Ea.							
	Remove single zone				26.016		1,500		1,500	1,850	2,325
	Replace single zone rooftop, 7.5 ton				43.011	5,825	2,475		8,300	9,475	11,100
	Total				69.027	5,825	3,975		9,800	11,325	13,425

	System Description	Freq. (Years)	Crew	Unit	Labor Hours	2019 Bare Costs				Total In-House	Total w/O&P
						Material	Labor	Equipment	Total		
1007	**Repair single zone rooftop unit, 10 ton**	10	2 STPI	Ea.							
	Repair controls				.400		25.50		25.50	31.50	39.50
	Remove fan bearings				2.400		153		153	190	238
	Replace fan bearings				4.000	131	256		387	460	560
	Remove fan motor				1.156		69.50		69.50	85.50	107
	Replace fan motor				2.311	370	139		509	580	675
	Remove compressor				14.513		871		871	1,075	1,350
	Replace compressor				28.714	15,745	1,708.50		17,453.50	19,400	22,300
	Replace refrigerant				.800	138	51.30		189.30	215	252
	Remove/replace heater igniter				.500	32	32		64	75	89.50
	Total				54.794	16,416	3,305.80		19,721.80	22,112	25,611
1008	**Replace single zone rooftop unit, 10 ton**	15	Q-6	Ea.							
	Remove single zone				33.898		2,025		2,025	2,500	3,125
	Replace single zone rooftop, 10 ton				48.000	9,200	2,875		12,075	13,700	16,000
	Total				81.898	9,200	4,900		14,100	16,200	19,125
1010	**Repair single zone rooftop unit, 15 ton**	10	2 STPI	Ea.							
	Repair controls				.500		32		32	39.50	49.50
	Remove fan bearings				2.400		153		153	190	238
	Replace fan bearings				4.000	131	256		387	460	560
	Remove fan motor				1.156		69.50		69.50	85.50	107
	Replace fan motor				2.312	760	139		899	1,000	1,175
	Remove compressor				21.661		1,300		1,300	1,600	2,000
	Replace compressor				42.857	23,500	2,550		26,050	29,000	33,400
	Replace refrigerant				1.333	230	85.50		315.50	360	420
	Remove / replace heater igniter				.500	32	32		64	75	89.50
	Total				76.719	24,653	4,617		29,270	32,810	38,039
1040	**Replace single zone rooftop unit, 15 ton**	15	Q-6	Ea.							
	Remove single zone rooftop unit				50.314		3,000		3,000	3,725	4,650
	Replace single zone rooftop, 15 ton				56.075	13,100	3,350		16,450	18,600	21,600
	Total				106.389	13,100	6,350		19,450	22,325	26,250

For customer support on your Facilities Maintenance & Repair Costs with RSMeans data, call 800.448.8182.

291

D3053 280   Single Zone Air Conditioner

	System Description	Freq. (Years)	Crew	Unit	Labor Hours	2019 Bare Costs				Total In-House	Total w/O&P
						Material	Labor	Equipment	Total		
2010	**Repair single zone rooftop unit, 25 ton**	10	Q-6	Ea.							
	Repair controls				.500		32		32	39.50	49.50
	Remove fan bearings				2.400		153		153	190	238
	Replace fan bearings				4.000	131	256		387	460	560
	Remove fan motor				1.156		69.50		69.50	85.50	107
	Replace fan motor				2.311	610	139		749	840	975
	Remove compressor				22.222		1,281.25		1,281.25	1,575	1,975
	Replace compressor				60.976	35,500	3,625		39,125	43,600	50,000
	Replace refrigerant				2.000	345	128.25		473.25	540	630
	Remove / replace heater igniter				1.000	64	64		128	150	179
	Total				96.565	36,650	5,748		42,398	47,480	54,713.50
2040	**Replace single zone rooftop unit, 25 ton**	15	Q-7	Ea.							
	Remove single zone rooftop unit				76.923		4,675		4,675	5,800	7,250
	Replace single zone rooftop, 25 ton				76.739	32,800	4,675		37,475	41,900	48,300
	Total				153.662	32,800	9,350		42,150	47,700	55,550
3010	**Repair single zone rooftop unit, 60 ton**	10	Q-6	Ea.							
	Repair controls				.500		32		32	39.50	49.50
	Remove fan bearings				2.400		153		153	190	238
	Replace fan bearings				4.444	147	284		431	515	625
	Remove fan motor				2.000		120		120	148	185
	Replace fan motor				4.167	2,750	250		3,000	3,325	3,825
	Remove compressor				93.507		5,580		5,580	6,925	8,650
	Replace compressor				187.200	40,800	11,160		51,960	58,500	68,500
	Replace refrigerant				4.667	805	299.25		1,104.25	1,250	1,475
	Remove / replace heater igniter				1.000	64	64		128	150	179
	Total				299.884	44,566	17,942.25		62,508.25	71,042.50	83,726.50
3040	**Replace single zone rooftop unit, 60 ton**	15	Q-7	Ea.							
	Remove single zone rooftop unit				188.000		11,500		11,500	14,200	17,800
	Replace single zone rooftop, 60 ton				182.000	59,500	11,100		70,600	79,000	91,500
	Total				370.000	59,500	22,600		82,100	93,200	109,300

For customer support on your Facilities Maintenance & Repair Costs with RSMeans data, call 800.448.8182.

D3053 280   Single Zone Air Conditioner

	System Description	Freq. (Years)	Crew	Unit	Labor Hours	Material	Labor	Equipment	Total	Total In-House	Total w/O&P
							2019 Bare Costs				
4010	**Repair single zone rooftop unit, 100 ton**	10	Q-7	Ea.							
	Repair controls				.500		32		32	39.50	49.50
	Remove fan bearings				4.800		306		306	380	475
	Replace fan bearings				9.412	756	600		1,356	1,575	1,875
	Remove fan motor				4.334		260		260	320	400
	Replace fan motor				7.407	4,175	445		4,620	5,150	5,900
	Remove compressor				98.765		6,025		6,025	7,450	9,325
	Replace compressor				198.000	42,900	12,000		54,900	62,000	72,000
	Replace refrigerant				6.667	1,150	427.50		1,577.50	1,800	2,100
	Remove / replace heater igniter				1.500	96	96		192	225	269
	Total				331.385	49,077	20,191.50		69,268.50	78,939.50	92,393.50
4040	**Replace single zone rooftop unit, 100 ton**	15	Q-7	Ea.							
	Remove single zone rooftop unit				296.000		18,000		18,000	22,400	28,000
	Replace single zone rooftop, 100 ton				305.000	125,500	18,600		144,100	161,000	185,500
	Total				601.000	125,500	36,600		162,100	183,400	213,500

D3053 282   Multi-Zone Variable Volume

	System Description	Freq. (Years)	Crew	Unit	Labor Hours	Material	Labor	Equipment	Total	Total In-House	Total w/O&P
							2019 Bare Costs				
1010	**Repair multi-zone variable volume, 50 ton**	10	Q-6	Ea.							
	Repair controls				1.300		83		83	103	129
	Remove fan bearings				2.400		153		153	190	238
	Replace fan bearings				4.444	147	284		431	515	625
	Remove fan motor				2.000		120		120	148	185
	Replace fan motor				4.000	2,125	240		2,365	2,625	3,025
	Remove compressor				77.922		4,650		4,650	5,775	7,200
	Replace compressor				156.000	34,000	9,300		43,300	48,900	57,000
	Replace refrigerant				3.333	575	213.75		788.75	895	1,050
	Remove / replace heater igniter				1.000	64	64		128	150	179
	Total				252.400	36,911	15,107.75		52,018.75	59,301	69,631
1040	**Replace multi-zone variable volume, 50 ton**	15	Q-7	Ea.							
	Remove multi-zone variable volume unit				160.000		9,750		9,750	12,100	15,100
	Replace multi-zone variable volume unit, 50 ton				219.000	115,000	13,300		128,300	143,000	164,500
	Total				379.000	115,000	23,050		138,050	155,100	179,600

	System Description	Freq. (Years)	Crew	Unit	Labor Hours	2019 Bare Costs				Total In-House	Total w/O&P
						Material	Labor	Equipment	Total		
2010	**Repair multi-zone variable volume, 70 ton**	10	Q-7	Ea.							
	Repair controls				1.300		83		83	103	129
	Remove fan bearings				2.400		153		153	190	238
	Replace fan bearings				4.444	147	284		431	515	625
	Remove fan motor				2.601		156		156	192	241
	Replace fan motor				5.195	3,950	310		4,260	4,725	5,425
	Remove compressor				76.923		4,675		4,675	5,800	7,250
	Replace compressor				154.000	37,600	9,375		46,975	53,000	61,500
	Replace refrigerant				4.667	805	299.25		1,104.25	1,250	1,475
	Remove / replace heater igniter				1.000	64	64		128	150	179
	Total				252.530	42,566	15,399.25		57,965.25	65,925	77,062
2040	**Replace multi-zone variable volume, 70 ton**	15	Q-7	Ea.							
	Remove multi-zone variable volume unit				232.000		14,100		14,100	17,500	21,900
	Replace multi-zone variable volume unit, 70 ton				305.000	161,000	18,600		179,600	200,000	230,000
	Total				537.000	161,000	32,700		193,700	217,500	251,900
3010	**Repair multi-zone variable volume, 90 ton**	10	Q-7	Ea.							
	Repair controls				1.300		83		83	103	129
	Remove fan bearings				2.400		153		153	190	238
	Replace fan bearings				4.444	147	284		431	515	625
	Remove fan motor				2.601		156		156	192	241
	Replace fan motor				6.504	4,100	390		4,490	5,000	5,725
	Remove compressor				92.308		5,610		5,610	6,975	8,700
	Replace compressor				184.800	45,120	11,250		56,370	63,500	74,000
	Replace refrigerant				6.667	1,150	427.50		1,577.50	1,800	2,100
	Remove / replace heater igniter				1.500	96	96		192	225	269
	Total				302.523	50,613	18,449.50		69,062.50	78,500	92,027
3040	**Replace multi-zone variable volume, 90 ton**	15	Q-7	Ea.							
	Remove multi-zone variable volume unit				296.000		18,000		18,000	22,400	28,000
	Replace multi-zone variable volume unit, 90 ton				395.000	187,500	24,100		211,600	236,000	271,500
	Total				691.000	187,500	42,100		229,600	258,400	299,500

For customer support on your Facilities Maintenance & Repair Costs with RSMeans data, call 800.448.8182.

System Description	Freq. (Years)	Crew	Unit	Labor Hours	Material	Labor	Equipment	Total	Total In-House	Total w/O&P
4010 Repair multi-zone variable volume, 105 ton	10	Q-7	Ea.							
Repair controls				1.300		83		83	103	129
Remove fan bearings				4.800		306		306	380	475
Replace fan bearings				9.412	756	600		1,356	1,575	1,875
Remove fan motor				4.334		260		260	320	400
Replace fan motor				8.696	5,050	520		5,570	6,200	7,125
Remove compressor				79.012		4,820		4,820	5,975	7,450
Replace compressor				158.400	34,320	9,600		43,920	49,700	58,000
Replace refrigerant				7.333	1,265	470.25		1,735.25	1,975	2,300
Remove / replace heater igniter				1.500	96	96		192	225	269
Total				274.787	41,487	16,755.25		58,242.25	66,453	78,023
4040 Replace multi-zone variable volume, 105 ton	15	Q-7	Ea.							
Remove multi-zone variable volume unit				348.000		21,200		21,200	26,300	32,800
Replace multi-zone variable volume unit, 105 ton				444.000	206,500	27,100		233,600	260,500	300,000
Total				792.000	206,500	48,300		254,800	286,800	332,800
5010 Repair multi-zone variable volume, 140 ton	10	Q-7	Ea.							
Repair controls				1.300		83		83	103	129
Remove fan bearings				7.200		459		459	570	715
Replace fan bearings				14.118	1,134	900		2,034	2,375	2,825
Remove fan motor				4.334		260		260	320	400
Replace fan motor				11.594	6,525	695		7,220	8,025	9,225
Remove compressor				106.370		6,488.93		6,488.93	8,025	10,000
Replace compressor				213.246	46,203.30	12,924		59,127.30	67,000	78,000
Replace refrigerant				8.666	1,495	555.75		2,050.75	2,325	2,725
Remove / replace heater igniter				2.000	128	128		256	300	360
Total				368.828	55,485.30	22,493.68		77,978.98	89,043	104,379
5040 Replace multi-zone variable volume, 140 ton	15	Q-7	Ea.							
Remove multi-zone variable volume unit				516.000		31,400		31,400	39,000	48,700
Replace multi-zone variable volume unit, 140 ton				593.000	272,500	36,100		308,600	344,500	396,500
Total				1109.000	272,500	67,500		340,000	383,500	445,200

2019 Bare Costs

For customer support on your Facilities Maintenance & Repair Costs with RSMeans data, call 800.448.8182.

295

D3053 284 Single Zone Variable Volume

System Description	Freq. (Years)	Crew	Unit	Labor Hours	2019 Bare Costs Material	Labor	Equipment	Total	Total In-House	Total w/O&P
1010 Repair single zone variable volume, 20 ton	10	Q-6	Ea.							
Repair controls				1.000		64		64	79.50	99
Remove fan bearings				2.400		153		153	190	238
Replace fan bearings				4.000	131	256		387	460	560
Remove fan motor				1.156		69.50		69.50	85.50	107
Replace fan motor				2.312	760	139		899	1,000	1,175
Remove compressor				17.778		1,025		1,025	1,275	1,575
Replace compressor				48.780	28,400	2,900		31,300	34,900	40,000
Replace refrigerant				1.333	230	85.50		315.50	360	420
Remove / replace heater igniter				1.000	64	64		128	150	179
Total				79.760	29,585	4,756		34,341	38,500	44,353
1040 Replace single zone variable volume, 20 ton	15	Q-7	Ea.							
Remove single zone variable volume unit				71.545		4,345		4,345	5,400	6,775
Replace single zone variable volume unit, 20 ton				64.945	25,850	3,960		29,810	33,300	38,400
Total				136.489	25,850	8,305		34,155	38,700	45,175
2010 Repair single zone variable volume, 30 ton	10	Q-6	Ea.							
Repair controls				1.000		64		64	79.50	99
Remove fan bearings				2.400		153		153	190	238
Replace fan bearings				4.000	131	256		387	460	560
Remove fan motor				1.156		69.50		69.50	85.50	107
Replace fan motor				2.476	1,100	149		1,249	1,400	1,600
Remove compressor				26.667		1,537.50		1,537.50	1,900	2,375
Replace compressor				73.171	42,600	4,350		46,950	52,500	60,000
Replace refrigerant				2.667	460	171		631	720	840
Remove / replace heater igniter				1.000	64	64		128	150	179
Total				114.536	44,355	6,814		51,169	57,485	65,998
2040 Replace single zone variable volume, 30 ton	15	Q-7	Ea.							
Remove single zone variable volume unit				104.142		6,352.50		6,352.50	7,875	9,825
Replace single zone variable volume unit, 30 ton				92.147	32,670	5,610		38,280	42,900	49,500
Total				196.289	32,670	11,962.50		44,632.50	50,775	59,325

For customer support on your Facilities Maintenance & Repair Costs with RSMeans data, call 800.448.8182.

D30 HVAC | D3053 | Terminal and Package Units

D3053 284 Single Zone Variable Volume

System Description	Freq. (Years)	Crew	Unit	Labor Hours	2019 Bare Costs Material	Labor	Equipment	Total	Total In-House	Total w/O&P
3010 Repair single zone variable volume, 40 ton	10	Q-6	Ea.							
Repair controls				1.000		64		64	79.50	99
Remove fan bearings				2.400		153		153	190	238
Replace fan bearings				4.000	131	256		387	460	560
Remove fan motor				1.156		69.50		69.50	85.50	107
Replace fan motor				4.000	2,125	240		2,365	2,625	3,025
Remove compressor				35.451		2,125		2,125	2,625	3,275
Replace compressor				71.006	32,700	4,250		36,950	41,200	47,500
Replace refrigerant				3.333	575	213.75		788.75	895	1,050
Remove / replace heater igniter				1.000	64	64		128	150	179
Total				123.346	35,595	7,435.25		43,030.25	48,310	56,033
3040 Replace single zone variable volume, 40 ton	15	Q-7	Ea.							
Remove single zone variable volume unit				143.000		8,717.50		8,717.50	10,800	13,500
Replace single zone variable volume unit, 40 ton				123.200	37,510	7,480		44,990	50,500	58,500
Total				266.200	37,510	16,197.50		53,707.50	61,300	72,000
4010 Repair single zone variable volume, 60 ton	10	Q-6	Ea.							
Repair controls				.500		32		32	39.50	49.50
Remove fan bearings				2.400		153		153	190	238
Replace fan bearings				4.444	147	284		431	515	625
Remove fan motor				2.000		120		120	148	185
Replace fan motor				4.167	2,750	250		3,000	3,325	3,825
Remove compressor				93.507		5,580		5,580	6,925	8,650
Replace compressor				187.200	40,800	11,160		51,960	58,500	68,500
Replace refrigerant				4.000	690	256.50		946.50	1,075	1,250
Remove / replace heater igniter				1.000	64	64		128	150	179
Total				299.218	44,451	17,899.50		62,350.50	70,867.50	83,501.50
4040 Replace single zone variable volume, 60 ton	15	Q-7	Ea.							
Remove single zone variable volume unit				190.300		11,550		11,550	14,400	17,900
Replace single zone variable volume unit, 60 ton				184.800	53,570	11,330		64,900	73,000	84,500
Total				375.100	53,570	22,880		76,450	87,400	102,400

For customer support on your Facilities Maintenance & Repair Costs with RSMeans data, call 800.448.8182.

297

D3053 286 | Central Station Air Conditioning Air Handling Unit

System Description	Freq. (Years)	Crew	Unit	Labor Hours	2019 Bare Costs Material	2019 Bare Costs Labor	2019 Bare Costs Equipment	2019 Bare Costs Total	Total In-House	Total w/O&P
1010 Repair central station A.H.U., 1300 CFM	10	1 STPI	Ea.							
Repair controls				.300		19.20		19.20	24	30
Remove blower motor				1.200		76.50		76.50	95	119
Replace blower motor				1.951	285	117		402	460	535
Total				3.451	285	212.70		497.70	**579**	**684**
1040 Replace central station A.H.U., 1300 CFM	15	Q-5	Ea.							
Remove central station A.H.U.				8.667		500		500	620	775
Replace central station A.H.U., 1300 CFM				17.335	7,075	1,000		8,075	9,025	10,400
Total				26.002	7,075	1,500		8,575	**9,645**	**11,175**
2010 Repair central station A.H.U., 1900 CFM	10	1 STPI	Ea.							
Repair controls				.300		19.20		19.20	24	30
Remove blower motor				1.200		76.50		76.50	95	119
Replace blower motor				2.311	305	139		444	505	595
Total				3.811	305	234.70		539.70	**624**	**744**
2040 Replace central station A.H.U., 1900 CFM	15	Q-5	Ea.							
Remove central station A.H.U.				9.456		545		545	675	845
Replace central station A.H.U., 1900 CFM				18.913	10,900	1,100		12,000	13,300	15,300
Total				28.369	10,900	1,645		12,545	**13,975**	**16,145**
3010 Repair central station A.H.U., 5400 CFM	10	1 STPI	Ea.							
Repair controls				.300		19.20		19.20	24	30
Remove blower motor				1.400		89.50		89.50	111	139
Replace blower motor				2.312	760	139		899	1,000	1,175
Total				4.012	760	247.70		1,007.70	**1,135**	**1,344**
3040 Replace central station A.H.U., 5400 CFM	15	Q-6	Ea.							
Remove central station A.H.U.				19.496		1,175		1,175	1,450	1,800
Replace central station A.H.U., 5400 CFM				39.024	17,500	2,325		19,825	22,100	25,500
Total				58.521	17,500	3,500		21,000	**23,550**	**27,300**
4010 Repair central station A.H.U., 8000 CFM	10	1 STPI	Ea.							
Repair controls				.300		19.20		19.20	24	30
Remove blower motor				1.600		102		102	127	159
Replace blower motor				2.311	610	139		749	840	975
Total				4.211	610	260.20		870.20	**991**	**1,164**

For customer support on your Facilities Maintenance & Repair Costs with RSMeans data, call 800.448.8182.

D3053 286 | **Central Station Air Conditioning Air Handling Unit**

System Description	Freq. (Years)	Crew	Unit	Labor Hours	2019 Bare Costs				Total In-House	Total w/O&P
					Material	Labor	Equipment	Total		
4040 Replace central station A.H.U., 8000 CFM	15	Q-6	Ea.							
Remove central station A.H.U.				25.974		1,550		1,550	1,925	2,400
Replace central station A.H.U., 8000 CFM				51.948	28,000	3,100		31,100	34,600	39,800
Total				77.922	28,000	4,650		32,650	36,525	42,200
5010 Repair central station A.H.U., 16,000 CFM	10	1 STPI	Ea.							
Repair controls				.300		19.20		19.20	24	30
Remove blower motor				1.800		115		115	143	178
Replace blower motor				2.597	1,350	156		1,506	1,675	1,925
Total				4.698	1,350	290.20		1,640.20	1,842	2,133
5040 Replace central station A.H.U., 16,000 CFM	15	Q-6	Ea.							
Remove central station A.H.U.				41.026		2,450		2,450	3,050	3,800
Replace central station A.H.U., 16,000 CFM				82.192	52,500	4,900		57,400	64,000	73,000
Total				123.217	52,500	7,350		59,850	67,050	76,800
6010 Repair central station A.H.U., 33,500 CFM	10	1 STPI	Ea.							
Repair controls				.300		19.20		19.20	24	30
Remove blower motor				2.601		156		156	192	241
Replace blower motor				5.195	3,950	310		4,260	4,725	5,425
Total				8.096	3,950	485.20		4,435.20	4,941	5,696
6040 Replace central station A.H.U., 33,500 CFM	15	Q-6	Ea.							
Remove central station A.H.U.				82.192		4,900		4,900	6,075	7,600
Replace central station A.H.U., 33,500 CFM				164.000	112,000	9,800		121,800	135,500	155,000
Total				246.192	112,000	14,700		126,700	141,575	162,600
7010 Repair central station A.H.U., 63,000 CFM	10	1 STPI	Ea.							
Repair controls				.300		19.20		19.20	24	30
Remove blower motor				4.334		260		260	320	400
Replace blower motor				8.696	5,050	520		5,570	6,200	7,125
Total				13.329	5,050	799.20		5,849.20	6,544	7,555
7040 Replace central station A.H.U., 63,000 CFM	15	Q-7	Ea.							
Remove central station A.H.U.				160.000		9,750		9,750	12,100	15,100
Replace central station A.H.U., 63,000 CFM				320.000	211,000	19,500		230,500	256,500	294,000
Total				480.000	211,000	29,250		240,250	268,600	309,100

For customer support on your Facilities Maintenance & Repair Costs with RSMeans data, call 800.448.8182.

299

	System Description	Freq. (Years)	Crew	Unit	Labor Hours	2019 Bare Costs				Total In-House	Total w/O&P
						Material	Labor	Equipment	Total		
1010	**Repair furnace, gas, 25 MBH**	10	Q-1	Ea.							
	Repair controls				.500		32		32	39.50	49.50
	Remove fan motor				.976		62.50		62.50	77.50	96.50
	Replace fan motor				1.951	240	117		357	410	480
	Remove auto vent damper				.274		16.70		16.70	21	26
	Replace auto vent damper				.547	114	33.50		147.50	167	195
	Remove burner				1.000		57		57	70.50	88
	Replace burner				8.320	1,000	475		1,475	1,675	1,975
	Total				13.568	1,354	793.70		2,147.70	2,460.50	2,910
1030	**Replace furnace, gas, 25 MBH**	15	Q-9	Ea.							
	Remove furnace				2.000		110		110	138	172
	Replace furnace				4.000	670	219		889	1,025	1,175
	Total				6.000	670	329		999	1,163	1,347
2010	**Repair furnace, gas, 100 MBH**	10	Q-1	Ea.							
	Repair controls				.500		32		32	39.50	49.50
	Remove fan motor				.976		62.50		62.50	77.50	96.50
	Replace fan motor				1.951	206	117		323	370	440
	Remove auto vent damper				.289		17.60		17.60	22	27.50
	Replace auto vent damper				.578	147	35		182	206	239
	Remove burner				1.200		68		68	84.50	106
	Replace burner				10.403	1,675	590		2,265	2,575	3,000
	Total				15.897	2,028	922.10		2,950.10	3,374.50	3,958.50
2030	**Replace furnace, gas, 100 MBH**	15	Q-9	Ea.							
	Remove furnace				3.250		178		178	224	280
	Replace furnace				6.499	785	355		1,140	1,300	1,550
	Total				9.749	785	533		1,318	1,524	1,830

For customer support on your Facilities Maintenance & Repair Costs with RSMeans data, call 800.448.8182.

D3053 310 Residential Furnace, Gas

System Description	Freq. (Years)	Crew	Unit	Labor Hours	2019 Bare Costs Material	Labor	Equipment	Total	Total In-House	Total w/O&P
3010 Repair furnace, gas, 200 MBH	10	Q-1	Ea.							
Repair controls				.500		32		32	39.50	49.50
Remove fan motor				.976		62.50		62.50	77.50	96.50
Replace fan motor				1.951	285	117		402	460	535
Remove auto vent damper				.325		19.80		19.80	25	31
Replace auto vent damper				.650	163	39.50		202.50	229	266
Remove burner				3.077		175		175	217	271
Replace burner				12.298	5,050	700		5,750	6,425	7,400
Total				19.777	5,498	1,145.80		6,643.80	7,473	8,649
3030 Replace furnace, gas, 200 MBH	15	Q-9	Ea.							
Remove furnace				4.000		219		219	276	345
Replace furnace				8.000	3,400	440		3,840	4,300	4,950
Total				12.000	3,400	659		4,059	4,576	5,295

D3053 320 Residential Furnace, Oil

System Description	Freq. (Years)	Crew	Unit	Labor Hours	2019 Bare Costs Material	Labor	Equipment	Total	Total In-House	Total w/O&P
1010 Repair furnace, oil, 55 MBH	10	Q-1	Ea.							
Repair controls				.500		32		32	39.50	49.50
Remove fan motor				.976		62.50		62.50	77.50	96.50
Replace fan motor				1.951	240	117		357	410	480
Remove auto vent damper				.473		26		26	32.50	40.50
Replace auto vent damper				.945	254	52		306	345	400
Remove burner				4.334		246		246	305	380
Replace burner				8.667	360	495		855	1,000	1,225
Total				17.846	854	1,030.50		1,884.50	2,209.50	2,671.50
1030 Replace furnace, oil, 55 MBH	15	Q-9	Ea.							
Remove furnace				2.889		158		158	199	249
Replace furnace, oil, 55 mbh				5.778	3,125	315		3,440	3,825	4,400
Total				8.667	3,125	473		3,598	4,024	4,649

For customer support on your Facilities Maintenance & Repair Costs with RSMeans data, call 800.448.8182.

301

	System Description	Freq. (Years)	Crew	Unit	Labor Hours	2019 Bare Costs				Total In-House	Total w/O&P
						Material	Labor	Equipment	Total		
2010	**Repair furnace, oil, 100 MBH**	10	Q-1	Ea.							
	Repair controls				.500		32		32	39.50	49.50
	Remove fan motor				.976		62.50		62.50	77.50	96.50
	Replace fan motor				1.951	206	117		323	370	440
	Remove auto vent damper				.495		27		27	34	42.50
	Replace auto vent damper				.990	114	54.50		168.50	194	228
	Remove burner				4.334		246		246	305	380
	Replace burner				8.667	360	495		855	1,000	1,225
	Total				17.914	680	1,034		1,714	2,020	2,461.50
2030	**Replace furnace, oil, 100 MBH**	15	Q-1	Ea.							
	Remove furnace				3.059		168		168	211	263
	Replace furnace, oil, 100 mbh				6.119	2,900	335		3,235	3,600	4,150
	Total				9.177	2,900	503		3,403	3,811	4,413
3010	**Repair furnace, oil, 200 MBH**	10	Q-1	Ea.							
	Repair controls				.500		32		32	39.50	49.50
	Remove fan motor				.976		62.50		62.50	77.50	96.50
	Replace fan motor				1.951	285	117		402	460	535
	Remove auto vent damper				.520		28.50		28.50	36	44.50
	Replace auto vent damper				1.040	147	57		204	233	273
	Remove burner				4.334		246		246	305	380
	Replace burner				8.667	360	495		855	1,000	1,225
	Total				17.988	792	1,038		1,830	2,151	2,603.50
3030	**Replace furnace, oil, 200 MBH**	15	Q-1	Ea.							
	Remove furnace				4.000		219		219	276	345
	Replace furnace, oil, 200 mbh				8.000	3,775	440		4,215	4,700	5,400
	Total				12.000	3,775	659		4,434	4,976	5,745

For customer support on your Facilities Maintenance & Repair Costs with RSMeans data, call 800.448.8182.

System Description	Freq. (Years)	Crew	Unit	Labor Hours	2019 Bare Costs Material	Labor	Equipment	Total	Total In-House	Total w/O&P
1010 **Repair furnace, electric, 25 MBH**	10	Q-20	Ea.							
Repair controls				.500		32		32	39.50	49.50
Remove fan motor				.976		62.50		62.50	77.50	96.50
Replace fan motor				1.951	240	117		357	410	480
Remove heat element				.867		48.50		48.50	60.50	75.50
Replace heat element				1.733	1,375	97		1,472	1,625	1,875
Total				6.027	1,615	357		1,972	2,212.50	2,576.50
1030 **Replace furnace, electric, 25 MBH**	15	Q-20	Ea.							
Remove furnace				2.826		158		158	198	247
Replace furnace, electric, 25 mbh				5.653	750	315		1,065	1,225	1,425
Total				8.479	750	473		1,223	1,423	1,672
2010 **Repair furnace, electric, 50 MBH**	10	Q-20	Ea.							
Repair controls				.500		32		32	39.50	49.50
Remove fan motor				.976		62.50		62.50	77.50	96.50
Replace fan motor				1.951	240	117		357	410	480
Remove heat element				.929		52		52	65	81
Replace heat element				1.857	1,400	104		1,504	1,675	1,900
Total				6.212	1,640	367.50		2,007.50	2,267	2,607
2030 **Replace furnace , electric, 50 MBH**	15	Q-20	Ea.							
Remove furnace				3.095		173		173	217	270
Replace furnace, electric, 50 mbh				6.190	590	345		935	1,075	1,275
Total				9.285	590	518		1,108	1,292	1,545
3010 **Repair furnace, electric, 85 MBH**	10	Q-20	Ea.							
Repair controls				.500		32		32	39.50	49.50
Remove fan motor				.976		62.50		62.50	77.50	96.50
Replace fan motor				1.951	206	117		323	370	440
Remove heat element				1.083		60.50		60.50	76	94.50
Replace heat element				2.167	2,550	121		2,671	2,950	3,375
Total				6.677	2,756	393		3,149	3,513	4,055.50
3030 **Replace furnace, electric, 85 MBH**	15	Q-20	Ea.							
Remove furnace				4.000		219		219	276	345
Replace furnace, electric, 85 mbh				6.842	635	380		1,015	1,175	1,400
Total				10.842	635	599		1,234	1,451	1,745

For customer support on your Facilities Maintenance & Repair Costs with RSMeans data, call 800.448.8182.

303

D3053 410 Electric Baseboard Heating Units

	System Description	Freq. (Years)	Crew	Unit	Labor Hours	2019 Bare Costs				Total In-House	Total w/O&P
						Material	Labor	Equipment	Total		
0010	**Maintenance and repair** Repair wiring connections	2	1 ELEC	Ea.	.615		37		37	45.50	57
	Total				.615		37		37	**45.50**	**57**
0020	**Maintenance and inspection** Inspect and clean baseboard heater	0.50	1 ELEC	Ea.	1.143		68.50		68.50	84.50	106
	Total				1.143		68.50		68.50	**84.50**	**106**
0030	**Replace baseboard heater** Remove baseboard heater Baseboard heater, 5' long	20	1 ELEC	Ea.	.667 1.860	46.50	40 112		40 158.50	49.50 189	61.50 230
	Total				2.527	46.50	152		198.50	**238.50**	**291.50**

D3053 420 Cast Iron Radiator, 10' section, 1 side

	System Description	Freq. (Years)	Crew	Unit	Labor Hours	2019 Bare Costs				Total In-House	Total w/O&P
						Material	Labor	Equipment	Total		
0020	**Refinish C.I. radiator, 1 side** Prepare surface Prime and paint 1 coat, brushwork	5	1 PORD	Ea.	.296 .471	4	12.80 20.40		12.80 24.40	16.40 30.50	20.50 37.50
	Total				.767	4	33.20		37.20	**46.90**	**58**

D3053 710 Fireplaces, Firebrick

	System Description	Freq. (Years)	Crew	Unit	Labor Hours	2019 Bare Costs				Total In-House	Total w/O&P
						Material	Labor	Equipment	Total		
0020	**Replace firebrick** Masonry demolition, fireplace, brick, 30" x 24" Install new firebrick	75	2 BRIC	S.F.	1.155 .208	9.90	47.51 9.45		47.51 19.35	62 23.50	76.50 28
	Total				1.363	9.90	56.96		66.86	**85.50**	**104.50**

System Description	Freq. (Years)	Crew	Unit	Labor Hours	2019 Bare Costs				Total In-House	Total w/O&P
					Material	Labor	Equipment	Total		
3010										
Rebuild 4" diam. reduced pressure backflow preventer	10	1 PLUM	Ea.							
Notify proper authorities prior to closing valves				.250		15.80		15.80	19.60	24.50
Close valves at 4" diameter BFP and open by-pass				.167		10.55		10.55	13.05	16.35
Rebuild 4" diameter reduced pressure BFP				2.000	370	126		496	565	660
Test backflow preventer				.500		31.50		31.50	39	49
Total				2.917	370	183.85		553.85	636.65	749.85
3020										
Rebuild 6" diam. reduced pressure backflow preventer	10	1 PLUM	Ea.							
Notify proper authorities prior to closing valves				.250		15.80		15.80	19.60	24.50
Close valves at 6" diameter BFP and open by-pass				.182		11.50		11.50	14.25	17.80
Rebuild 6" diameter reduced pressure BFP				3.008	390	190		580	665	785
Test backflow preventer				.500		31.50		31.50	39	49
Total				3.939	390	248.80		638.80	737.85	876.30
3030										
Rebuild 8" diam. reduced pressure backflow preventer	10	1 PLUM	Ea.							
Notify proper authorities prior to closing valves				.250		15.80		15.80	19.60	24.50
Close valves at 8" diameter BFP and open by-pass				.200		12.65		12.65	15.65	19.60
Rebuild 8" diameter reduced pressure BFP				4.000	485	253		738	845	995
Test backflow preventer				.500		31.50		31.50	39	49
Total				4.950	485	312.95		797.95	919.25	1,088.10
3040										
Rebuild 10" diam. reduced pressure backflow preventer	10	1 PLUM	Ea.							
Notify proper authorities prior to closing valves				.250		15.80		15.80	19.60	24.50
Close valves at 10" diameter BFP and open by-pass				.222		14.05		14.05	17.40	22
Rebuild 10" diameter reduced pressure BFP				5.000	565	315		880	1,025	1,200
Test backflow preventer				.500		31.50		31.50	39	49
Total				5.972	565	376.35		941.35	1,101	1,295.50

For customer support on your Facilities Maintenance & Repair Costs with RSMeans data, call 800.448.8182.

305

D4013 310 Sprinkler System, Fire Supression

	System Description	Freq. (Years)	Crew	Unit	Labor Hours	2019 Bare Costs				Total In-House	Total w/O&P
						Material	Labor	Equipment	Total		
1020	**Inspect sprinkler system**	1	1 PLUM	Ea.							
	Inspect water pressure gauges				.044		2.80		2.80	3.50	4.30
	Test fire alarm (wet or dry)				.030		1.90		1.90	2.30	2.90
	Motor, exposed, inspect				.333		21		21	26	32.50
	Inspect, visual				.060		3.80		3.80	4.70	5.90
	Total				.467		29.50		29.50	36.50	45.60
1030	**Replace sprinkler head**	20	1 PLUM	Ea.							
	Remove old sprinkler head				.325		20		20	25	31
	Install new head				.650	12.75	40		52.75	64	78
	Total				.975	12.75	60		72.75	89	109
1040	**Rebuild double check 3" backflow preventer**	1	1 PLUM	Ea.							
	Rebuild 3" backflow preventer				3.910	279	247		526	615	735
	Test system				1.300		82		82	102	127
	Total				5.210	279	329		608	717	862
1050	**Rebuild double check 4" backflow preventer**	1	1 PLUM	Ea.							
	Rebuild 4" backflow preventer				4.561	305	288		593	695	825
	Test system				1.300		82		82	102	127
	Total				5.861	305	370		675	797	952
1060	**Rebuild double check 6" backflow preventer**	1	1 PLUM	Ea.							
	Rebuild 6" backflow preventer				5.202	365	330		695	810	965
	Test system				1.300		82		82	102	127
	Total				6.502	365	412		777	912	1,092
1070	**Rebuild reduced pressure backflow preventer**	1	1 PLUM	Ea.							
	Rebuild 3" backflow preventer				3.910	279	247		526	615	735
	Test system				1.300		82		82	102	127
	Total				5.210	279	329		608	717	862

For customer support on your Facilities Maintenance & Repair Costs with RSMeans data, call 800.448.8182.

D40	FIRE PROTECTION		D4013		Sprinklers									

D4013 410   Fire Pump

	System Description	Freq. (Years)	Crew	Unit	Labor Hours	2019 Bare Costs				Total In-House	Total w/O&P
						Material	Labor	Equipment	Total		
1030	**Replace fire pump / electric motor assembly 100 H.P.**	25	Q-13	Ea.							
	Remove 100 H.P. electric-drive fire pump				31.373		1,825		1,825	2,275	2,850
	Install new electric-drive fire pump and controls, 100 H.P.				64.000	24,100	3,750		27,850	31,200	36,000
	Total				95.373	24,100	5,575		29,675	33,475	38,850

D5013 110 Primary Transformer, Liquid Filled

	System Description	Freq. (Years)	Crew	Unit	Labor Hours	2019 Bare Costs				Total In-House	Total w/O&P
						Material	Labor	Equipment	Total		
0010	**Repair 500 kVA transformer**	10	1 ELEC	Ea.							
	Remove transformer component				1.000		60		60	74	92.50
	Transformer component				2.963	2,175	178		2,353	2,600	3,000
	Total				3.963	2,175	238		2,413	2,674	3,092.50
0020	**Maintenance and inspection**	0.50	1 ELEC	Ea.							
	Test energized transformer				.500		30		30	37	46.50
	Total				.500		30		30	37	46.50
0030	**Replace transformer**	30	R-3	Ea.							
	Remove transformer				13.793		825	89.50	914.50	1,125	1,375
	Transformer 500 kVA, 5 KV primary, 277/480 V secondary				50.000	22,200	2,975	325	25,500	28,500	32,700
	Total				63.793	22,200	3,800	414.50	26,414.50	29,625	34,075

D5013 120 Primary Transformer, Dry

	System Description	Freq. (Years)	Crew	Unit	Labor Hours	2019 Bare Costs				Total In-House	Total w/O&P
						Material	Labor	Equipment	Total		
0010	**Repair 15 KV primary transformer**	15	1 ELEC	Ea.							
	Repair transformer terminal				2.000	51.50	120		171.50	205	249
	Total				2.000	51.50	120		171.50	205	249
0020	**Maintenance and inspection**	0.50	1 ELEC	Ea.							
	Test energized transformer				.500		30		30	37	46.50
	Total				.500		30		30	37	46.50
0030	**Replace transformer**	30	R-3	Ea.							
	Remove transformer				13.793		825	89.50	914.50	1,125	1,375
	Transformer 500 kVA, 15 KV primary, 277/480 V secondary				57.143	53,500	3,425	370	57,295	63,500	72,500
	Total				70.936	53,500	4,250	459.50	58,209.50	64,625	73,875

For customer support on your Facilities Maintenance & Repair Costs with RSMeans data, call 800.448.8182.

D50 ELECTRICAL — D5013 — Electrical Service/Distribution

D5013 210 — Switchgear, Mainframe

System Description	Freq. (Years)	Crew	Unit	Labor Hours	2019 Bare Costs Material	Labor	Equipment	Total	Total In-House	Total w/O&P
0010 Repair switchgear 1200 A	5	1 ELEC	Ea.							
Remove relay				1.702		102		102	126	158
Relay				5.000	1,175	300		1,475	1,675	1,925
Total				6.702	1,175	402		1,577	1,801	**2,083**
0020 Maintenance and inspection	1	1 ELEC	Ea.							
Check & retighten switchgear connections				.800		48		48	59	74
Total				.800		48		48	**59**	**74**
0030 Replace switchgear 1200 A	20	3 ELEC	Ea.							
Remove switchgear 1200 A				6.000		360		360	445	555
Switchgear 1200 A				21.818	1,675	1,300		2,975	3,450	4,125
Test switchgear				2.000		120		120	148	185
Total				29.818	1,675	1,780		3,455	4,043	**4,865**

D5013 216 — Fuses

System Description	Freq. (Years)	Crew	Unit	Labor Hours	2019 Bare Costs Material	Labor	Equipment	Total	Total In-House	Total w/O&P
0010 Replace fuse	25	1 ELEC	Ea.							
Remove fuse				.151		9.05		9.05	11.15	14
Fuse 600 A				.400	345	24		369	410	470
Total				.551	345	33.05		378.05	421.15	**484**

D5013 220 — Switchgear, Indoor, Less Than 600 V

System Description	Freq. (Years)	Crew	Unit	Labor Hours	2019 Bare Costs Material	Labor	Equipment	Total	Total In-House	Total w/O&P
0010 Repair switchgear, - (5% of total C.B.)	10	1 ELEC	Ea.							
Remove circuit breaker				.400		24		24	29.50	37
Circuit breaker				1.143	291	68.50		359.50	405	470
Total				1.543	291	92.50		383.50	434.50	**507**
0020 Maintenance and inspection	3	1 ELEC	Ea.							
Inspect and clean panel				.500		30		30	37	46.50
Total				.500		30		30	**37**	**46.50**

D5013 220 Switchgear, Indoor, Less Than 600 V

	System Description	Freq. (Years)	Crew	Unit	Labor Hours	2019 Bare Costs Material	Labor	Equipment	Total	Total In-House	Total w/O&P
0030	**Replace switchgear, 225 A**	30	1 ELEC	Ea.							
	Remove panelboard, 4 wire, 120/208 V, 225 A				4.000		240		240	296	370
	Panelboard, 4 wire, 120/208 V, 225 A main circuit breaker				11.940	3,850	715		4,565	5,125	5,925
	Test indoor switchgear				1.509		90.50		90.50	112	140
	Total				17.450	3,850	1,045.50		4,895.50	5,533	6,435
0200	**Replace switchgear, 400 A**	30	2 ELEC	Ea.							
	Turn power off, later turn on				.200		12		12	14.80	18.50
	Remove panelboard, 4 wire, 120/208 V, 100 A				3.333		200		200	247	310
	Remove panelboard, 4 wire, 120/208 V, 400 A				8.333		500		500	615	770
	Panelboard, 4 wire, 120/208 V, 100 A main circuit breaker				17.021	1,875	1,025		2,900	3,325	3,925
	Panelboard, 4 wire, 120/208 V, 400 A main circuit breaker				33.333	5,000	2,000		7,000	7,975	9,325
	Test circuits				.145		8.74		8.74	10.75	13.50
	Total				62.367	6,875	3,745.74		10,620.74	12,187.55	14,362
0240	**Replace switchgear, 600 A**	30	2 ELEC	Ea.							
	Turn power off, later turn on				.300		18		18	22	28
	Remove panelboard, 4 wire, 120/208 V, 200 A				6.667		400		400	495	615
	Remove switchboard, incoming section, 600 A				4.848		291		291	360	450
	Remove switchboard, distribution section, 600 A				4.000		240		240	296	370
	Panelboard, 4 wire, 120/208 V, 225 A main circuit breaker				28.571	3,850	1,725		5,575	6,350	7,475
	Switchboard, incoming section, 600 A				16.000	3,950	960		4,910	5,525	6,425
	Switchboard, distribution section, 600 A				16.000	1,500	960		2,460	2,825	3,350
	Test circuits				.218		13.11		13.11	16.15	20.50
	Total				76.605	9,300	4,607.11		13,907.11	15,889.15	18,733.50
0280	**Replace switchgear, 800 A**	30	2 ELEC	Ea.							
	Turn power off, later turn on				.400		24		24	29.50	37
	Remove panelboard, 4 wire, 120/208 V, 200 A (2 panelboards)				13.333		800		800	985	1,225
	Remove switchboard, incoming section, 800 A				5.517		330		330	410	510
	Remove switchboard, distribution section, 800 A				4.444		267		267	330	410
	Panelboard, 4 wire, 120/208 V, 225 A main lugs (2 panelboards)				47.059	4,300	2,850		7,150	8,200	9,725
	Switchboard, incoming section, 800 A				18.182	3,950	1,100		5,050	5,700	6,625
	Switchboard, distribution section, 800 A				18.182	1,900	1,100		3,000	3,425	4,050
	Test circuits				.291		17.48		17.48	21.50	27
	Total				107.408	10,150	6,488.48		16,663.48	19,101	22,609

For customer support on your Facilities Maintenance & Repair Costs with RSMeans data, call 800.448.8182.

D5013 220 | Switchgear, Indoor, Less Than 600 V

	System Description	Freq. (Years)	Crew	Unit	Labor Hours	2019 Bare Costs				Total In-House	Total w/O&P
						Material	Labor	Equipment	Total		
0320	**Replace switchgear, 1200 A**	30	3 ELEC	Ea.							
	Turn power off, later turn on				.600		36		36	44.50	55.50
	Remove panelboard, 4 wire, 120/208 V, 200 A (4 panelboards)				26.667		1,600		1,600	1,975	2,450
	Remove switchboard, incoming section, 1200 A				6.667		400		400	495	615
	Remove switchboard, distribution section, 1200 A				5.161		310		310	380	480
	Panelboard, 4 wire, 120/208 V, 225 A main lugs (4 panelboards)				94.118	8,600	5,700		14,300	16,400	19,500
	Switchboard, incoming section, 1200 A				22.222	4,750	1,325		6,075	6,875	8,000
	Switchboard, distribution section, 1200 A				22.222	2,950	1,325		4,275	4,900	5,750
	Test circuits				.436		26.22		26.22	32.50	40.50
	Total				178.093	16,300	10,722.22		27,022.22	31,102	36,891
0360	**Replace switchgear, 1600 A**	30	3 ELEC	Ea.							
	Turn power off, later turn on				.800		48		48	59	74
	Remove panelboard, 4 wire, 120/208 V, 200 A (6 panelboards)				40.000		2,400		2,400	2,950	3,700
	Remove switchboard, incoming section, 1600 A				7.273		435		435	540	675
	Remove switchboard, distribution section, 1600 A				5.517		330		330	410	510
	Panelboard, 4 wire, 120/208 V, 225 A main lugs (6 panelboards)				141.176	12,900	8,550		21,450	24,600	29,200
	Switchboard, incoming section, 1600 A				24.242	4,750	1,450		6,200	7,025	8,200
	Switchboard, distribution section, 1600 A				24.242	3,975	1,450		5,425	6,175	7,225
	Test circuits				.582		34.96		34.96	43	54
	Total				243.833	21,625	14,697.96		36,322.96	41,802	49,638
0400	**Replace switchgear, 2000 A**	30	3 ELEC	Ea.							
	Turn power off, later turn on				1.000		60		60	74	92.50
	Remove panelboard, 4 wire, 120/208 V, 200 A (8 panelboards)				53.333		3,200		3,200	3,950	4,925
	Remove switchboard, incoming section, 2000 A				7.619		460		460	565	705
	Remove switchboard, distribution section, 2000 A				5.926		355		355	440	550
	Panelboard, 4 wire, 120/208 V, 225 A main lugs (8 panelboards)				188.235	17,200	11,400		28,600	32,800	38,900
	Switchboard, incoming section, 2000 A				25.806	5,125	1,550		6,675	7,550	8,800
	Switchboard, distribution section, 2000 A				25.806	4,825	1,550		6,375	7,225	8,425
	Test circuits				.727		43.70		43.70	54	67.50
	Total				308.454	27,150	18,618.70		45,768.70	52,658	62,465

D5013 224 Switchgear, Indoor, 600 V

System Description	Freq. (Years)	Crew	Unit	Labor Hours	2019 Bare Costs				Total In-House	Total w/O&P
					Material	Labor	Equipment	Total		
0010 Maintenance and repair, - (5% of total fuses)										
Remove fuse	10	1 ELEC	Ea.	.151		9.05		9.05	11.15	14
Fuse				.400	345	24		369	410	470
Total				.551	345	33.05		378.05	421.15	**484**
0020 Maintenance and inspection										
Inspect and clean panel	3	1 ELEC	Ea.	.500		30		30	37	46.50
Total				.500		30		30	**37**	**46.50**
0030 Replace switchgear										
Remove indoor switchgear	30	2 ELEC	Ea.	4.000		240		240	296	370
Switchgear				14.035	1,275	845		2,120	2,450	2,900
Test indoor switchgear				2.000		120		120	148	185
Total				20.035	1,275	1,205		2,480	2,894	**3,455**

D5013 230 Meters

System Description	Freq. (Years)	Crew	Unit	Labor Hours	2019 Bare Costs				Total In-House	Total w/O&P
					Material	Labor	Equipment	Total		
0010 Repair switchboard meter										
Repair meter	10	1 ELEC	Ea.	5.000	1,025	300		1,325	1,500	1,750
Total				5.000	1,025	300		1,325	**1,500**	**1,750**
0020 Replace switchboard meter										
Remove meter	20	1 ELEC	Ea.	.500		30		30	37	46.50
Meter				1.600	5,150	96		5,246	5,775	6,575
Total				2.100	5,150	126		5,276	5,812	**6,621.50**

D5013 240 Inverter

System Description	Freq. (Years)	Crew	Unit	Labor Hours	2019 Bare Costs				Total In-House	Total w/O&P
					Material	Labor	Equipment	Total		
0010 Maintenance and repair										
Repair inverter component	1	2 ELEC	Ea.	5.000	360	300		660	765	915
Total				5.000	360	300		660	**765**	**915**

For customer support on your Facilities Maintenance & Repair Costs with RSMeans data, call 800.448.8182.

D5013 240 Inverter

	System Description	Freq. (Years)	Crew	Unit	Labor Hours	2019 Bare Costs				Total In-House	Total w/O&P
						Material	Labor	Equipment	Total		
0020	**Maintenance and inspection**										
	Check connections, controls & relays	0.25	1 ELEC	Ea.	1.600		96		96	118	148
	Total				1.600		96		96	**118**	**148**
0030	**Replace 1 kVA inverter**										
	Remove inverter	20	2 ELEC	Ea.	.500		30		30	37	46.50
	Inverter				1.600	3,575	96		3,671	4,050	4,625
	Total				2.100	3,575	126		3,701	4,087	4,671.50

D5013 250 Rectifier, Up To 600 V

	System Description	Freq. (Years)	Crew	Unit	Labor Hours	2019 Bare Costs				Total In-House	Total w/O&P
						Material	Labor	Equipment	Total		
0010	**Maintenance and repair**										
	Repair rectifier component	2	1 ELEC	Ea.	4.000	320	240		560	650	770
	Total				4.000	320	240		560	**650**	**770**
0020	**Maintenance and inspection**										
	Check connections, controls & relays	0.33	1 ELEC	Ea.	1.600		96		96	118	148
	Total				1.600		96		96	**118**	**148**
0030	**Replace rectifier**										
	Remove rectifier	20	2 ELEC	Ea.	.500		30		30	37	46.50
	Rectifier				1.600	965	96		1,061	1,175	1,350
	Total				2.100	965	126		1,091	1,212	1,396.50

D5013 264 Motor Starter, Up To 600 V

	System Description	Freq. (Years)	Crew	Unit	Labor Hours	2019 Bare Costs				Total In-House	Total w/O&P
						Material	Labor	Equipment	Total		
0010	**Maintenance and repair**										
	Remove motor starter coil	5	1 ELEC	Ea.	.500		30		30	37	46.50
	Motor starter coil				1.600	36.50	96		132.50	159	194
	Total				2.100	36.50	126		162.50	**196**	**240.50**

For customer support on your Facilities Maintenance & Repair Costs with RSMeans data, call 800.448.8182.

313

D5013 264 Motor Starter, Up To 600 V

System Description	Freq. (Years)	Crew	Unit	Labor Hours	2019 Bare Costs				Total In-House	Total w/O&P
					Material	Labor	Equipment	Total		
0020 Maintenance and inspection										
Inspect & clean motor starter	0.50	1 ELEC	Ea.	.667		40		40	49.50	61.50
Total				.667		40		40	**49.50**	**61.50**
0030 Replace starter										
Remove motor starter	18	1 ELEC	Ea.	.899		54		54	66.50	83
Magnetic motor starter 5 H.P., size 0				3.478	290	209		499	575	685
Cut, form, align & connect wires				.151		9.05		9.05	11.15	14
Check operation of motor starter				.018		1.10		1.10	1.35	1.70
Total				4.546	290	273.15		563.15	654	783.70

D5013 266 Motor Starter, 600 V

System Description	Freq. (Years)	Crew	Unit	Labor Hours	2019 Bare Costs				Total In-House	Total w/O&P
					Material	Labor	Equipment	Total		
0010 Maintenance and repair										
Remove motor starter coil	3	1 ELEC	Ea.	1.702		102		102	126	158
Motor starter coil				5.000	114	300		414	495	610
Total				6.702	114	402		516	**621**	**768**
0020 Maintenance and inspection										
Inspect & clean motor starter	0.25	1 ELEC	Ea.	.667		40		40	49.50	61.50
Total				.667		40		40	**49.50**	**61.50**
0030 Replace starter										
Remove motor starter	18	2 ELEC	Ea.	3.333		200		200	247	310
Motor starter FVNR 600 V, 3 P, 100 H.P. motor				13.333	2,950	800		3,750	4,225	4,925
Cut, form, align & connect wires				.170		10.20		10.20	12.60	15.75
Check operation of motor starter				.018		1.10		1.10	1.35	1.70
Total				16.855	2,950	1,011.30		3,961.30	4,485.95	5,252.45

D50 ELECTRICAL | D5013 | Electrical Service/Distribution

D5013 272 Secondary Transformer, Liquid Filled

System Description	Freq. (Years)	Crew	Unit	Labor Hours	2019 Bare Costs				Total In-House	Total w/O&P
					Material	Labor	Equipment	Total		
0010 Maintenance and repair										
Remove transformer terminal	25	1 ELEC	Ea.	.348		21		21	25.50	32
Transformer terminal				1.000	13.05	60		73.05	88.50	109
Total				1.348	13.05	81		94.05	**114**	**141**
0020 Maintenance and inspection										
Check connections & fluid	0.50	1 ELEC	Ea.	.667		40		40	49.50	61.50
Total				.667		40		40	**49.50**	**61.50**

D5013 274 Secondary Transformer, Dry

System Description	Freq. (Years)	Crew	Unit	Labor Hours	2019 Bare Costs				Total In-House	Total w/O&P
					Material	Labor	Equipment	Total		
0010 Maintenance and repair										
Remove transformer terminal	10	1 ELEC	Ea.	.500		30		30	37	46.50
Transformer terminal				1.000	39	60		99	117	141
Total				1.500	39	90		129	**154**	**187.50**
0020 Maintenance and inspection										
Inspect connections & tighten	0.50	1 ELEC	Ea.	1.000		60		60	74	92.50
Total				1.000		60		60	**74**	**92.50**
0030 Replace 15 kVA transformer										
Remove transformer 15 KVA, incl. supp., wire & conduit terminations	30	2 ELEC	Ea.	4.360		262		262	325	405
Transformer 15 KVA, 240/480 V primary, 120/208 V secondary				14.545	1,875	875		2,750	3,150	3,700
Total				18.905	1,875	1,137		3,012	**3,475**	**4,105**
0040 Replace 112.5 kVA transformer										
Remove transformer 112.5 KVA, incl. support, wire, conduit termination	30	R-3	Ea.	6.897		410	45	455	560	685
Transformer 112.5 KVA, 240/480 V primary, 120/208 V secondary				22.222	4,225	1,325	144	5,694	6,450	7,500
Total				29.119	4,225	1,735	189	6,149	**7,010**	**8,185**
0050 Replace 500 kVA transformer										
Remove transformer 500 KVA, incl. support,wire & conduit terminations	30	R-3	Ea.	14.286		855	92.50	947.50	1,150	1,425
Transformer 500 KVA, 240/480 V primary, 120/208 V secondary				44.444	18,800	2,650	288	21,738	24,300	27,900
Total				58.730	18,800	3,505	380.50	22,685.50	**25,450**	**29,325**

For customer support on your Facilities Maintenance & Repair Costs with RSMeans data, call 800.448.8182.

315

D5013 280   Lighting Panel, Indoor

System Description	Freq. (Years)	Crew	Unit	Labor Hours	2019 Bare Costs				Total In-House	Total w/O&P
					Material	Labor	Equipment	Total		
0020 Maintenance and inspection	3	1 ELEC	Ea.							
Inspect and clean panels				.500		30		30	37	46.50
Total				.500		30		30	**37**	**46.50**
0030 Replace load center, 100 A	20	2 ELEC	Ea.							
Turn power off, later turn on				.100		6		6	7.40	9.25
Remove load center, 3 wire, 120/240 V				3.077		185		185	228	285
Load centers, 3 Wire, 120/240V, 100 A				6.667	155	400		555	665	810
Test circuits				.073		4.37		4.37	5.40	6.75
Total				9.916	155	595.37		750.37	**905.80**	**1,111**

For customer support on your Facilities Maintenance & Repair Costs with RSMeans data, call 800.448.8182.

D5023 110 Wireway

	System Description	Freq. (Years)	Crew	Unit	Labor Hours	2019 Bare Costs				Total In-House	Total w/O&P
						Material	Labor	Equipment	Total		
0030	**Replace 8" x 8" wireway**	20	2 ELEC	L.F.							
	Remove and reinstall supply box				.050		3		3	3.70	4.65
	Remove wireway, fittings & support				.100		6		6	7.40	9.25
	Wireway 8" x 8", fittings & support				.400	27.50	24		51.50	60	71.50
	Total				.550	27.50	33		60.50	71.10	85.40

D5023 112 Conduit EMT

	System Description	Freq. (Years)	Crew	Unit	Labor Hours	2019 Bare Costs				Total In-House	Total w/O&P
						Material	Labor	Equipment	Total		
0010	**Replace 1" EMT conduit**	50	2 ELEC	M.L.F.							
	Remove & reinstall cover plate				1.290		77.50		77.50	95.50	120
	Remove conduit w/ fittings & hangers				20.300		1,220		1,220	1,500	1,875
	Remove junction box				1.000		60		60	74	92.50
	1" EMT with couplings				38.650	1,570	2,320		3,890	4,600	5,550
	Outlet box square				4.000	52	240		292	355	435
	Box connector				.889	11.50	53.50		65	78.50	97
	Beam clamp				5.000	404	316		720	835	995
	Conduit hanger				5.926	540	356		896	1,025	1,225
	Total				77.055	2,577.50	4,643		7,220.50	8,563	10,389.50

D5023 120 Cable, Non-Metallic (NM) Sheathed

	System Description	Freq. (Years)	Crew	Unit	Labor Hours	2019 Bare Costs				Total In-House	Total w/O&P
						Material	Labor	Equipment	Total		
0010	**Replace NM cable**	50	2 ELEC	M.L.F.							
	Shut off power, later turn on				.100		6		6	7.40	9.25
	Remove & reinstall cover plate				1.290		77.50		77.50	95.50	120
	Remove junction box				1.000		60		60	74	92.50
	Remove NM cable #12, 2 wire				12.720		760		760	940	1,175
	NM cable #12, 2 wire				32.000	245	1,920		2,165	2,650	3,275
	Staples, for Romex or BX cable				4.000	4	240		244	300	375
	NM #12-2 wire cable connectors				.602	15.30	36.10		51.40	61.50	74.50
	Outlet box square 4" for Romex or BX				4.000	90.50	240		330.50	395	485
	Total				55.712	354.80	3,339.60		3,694.40	4,523.40	5,606.25

For customer support on your Facilities Maintenance & Repair Costs with RSMeans data, call 800.448.8182.

317

D5023 122 | Cable, Service

	System Description	Freq. (Years)	Crew	Unit	Labor Hours	Material	Labor	Equipment	Total	Total In-House	Total w/O&P
						2019 Bare Costs					
0010	**Replace service cable**	50	2 ELEC	M.L.F.							
	Shut off power, later turn on				.100		6		6	7.40	9.25
	Remove & reinstall cover plate				1.290		77.50		77.50	95.50	120
	Remove junction box				1.000		60		60	74	92.50
	Remove wire #6 XLPE				2.000		120		120	148	185
	600 V, copper type XLPE-USE				12.308	570	740		1,310	1,550	1,850
	Staples, for service cable				5.000	8	300		308	380	475
	SER cable connectors				3.333	45.50	200		245.50	297	365
	Outlet box square 4"				4.000	52	240		292	355	435
	Total				29.031	675.50	1,743.50		2,419	2,906.90	3,531.75

D5023 124 | Cable, Armored

	System Description	Freq. (Years)	Crew	Unit	Labor Hours	Material	Labor	Equipment	Total	Total In-House	Total w/O&P
						2019 Bare Costs					
0010	**Replace armored cable**	60	2 ELEC	M.L.F.							
	Shut off power, later turn on				.100		6		6	7.40	9.25
	Remove & reinstall cover plate				1.290		77.50		77.50	95.50	120
	Remove junction box				1.000		60		60	74	92.50
	Remove armored cable				13.220		790		790	980	1,225
	BX #12, 2 wire				34.783	410	2,090		2,500	3,025	3,725
	Staples, for Romex or BX cable				4.000	4	240		244	300	375
	Cable connector				2.000	9.30	120		129.30	158	197
	Outlet box square 4" for Romex or BX				4.000	90.50	240		330.50	395	485
	Total				60.393	513.80	3,623.50		4,137.30	5,034.90	6,228.75

D5023 126 Branch Wiring, With Junction Box

	System Description	Freq. (Years)	Crew	Unit	Labor Hours	2019 Bare Costs Material	2019 Bare Costs Labor	2019 Bare Costs Equipment	2019 Bare Costs Total	Total In-House	Total w/O&P
0010	**Replace branch wiring**	50	2 ELEC	M.L.F.							
	Shut off power, later turn on				.100		6		6	7.40	9.25
	Remove & reinstall cover plate				1.290		77.50		77.50	95.50	120
	Remove junction box				1.000		60		60	74	92.50
	Remove #12 wire				1.455		87.50		87.50	108	135
	Conductor THW-THHN, copper, stranded, #12				7.273	105	435		540	655	805
	Outlet box square 4"				4.000	52	240		292	355	435
	Wire connectors				1.000	2.10	60		62.10	76.50	95.50
	Total				16.117	159.10	966		1,125.10	1,371.40	1,692.25

D5023 128 Branch Wiring, 600 V

	System Description	Freq. (Years)	Crew	Unit	Labor Hours	2019 Bare Costs Material	2019 Bare Costs Labor	2019 Bare Costs Equipment	2019 Bare Costs Total	Total In-House	Total w/O&P
0010	**Replace branch wiring**	50	2 ELEC	M.L.F.							
	Shut off power, later turn on				.100		6		6	7.40	9.25
	Remove #12 wire				1.455		87.50		87.50	108	135
	Conductor THW-THHN, copper, stranded, #12				7.273	105	435		540	655	805
	Cut, form and align 4 wires				.410		24.50		24.50	30.50	38
	Splice, solder, insulate 2 wires				2.667		160		160	197	247
	Total				11.904	105	713		818	997.90	1,234.25

D5023 130 Circuit Breaker

	System Description	Freq. (Years)	Crew	Unit	Labor Hours	2019 Bare Costs Material	2019 Bare Costs Labor	2019 Bare Costs Equipment	2019 Bare Costs Total	Total In-House	Total w/O&P
0010	**Maintenance and repair breaker, molded case, 480 V, 1 pole**	20	1 ELEC	Ea.							
	Tighten connections				.889		53.50		53.50	66	82.50
	Total				.889		53.50		53.50	66	82.50
0020	**Maintenance and inspection C.B., molded case, 480 V, 1 pole**	0.50	1 ELEC	Ea.							
	Inspect, tighten C.B. wiring				.400		24		24	29.50	37
	Total				.400		24		24	29.50	37

For customer support on your Facilities Maintenance & Repair Costs with RSMeans data, call 800.448.8182.

319

System Description	Freq. (Years)	Crew	Unit	Labor Hours	2019 Bare Costs				Total In-House	Total w/O&P
					Material	Labor	Equipment	Total		
0030										
Replace C.B. molded case, 480 V, 1 pole	50	1 ELEC	Ea.							
Remove circuit breaker				.348		21		21	25.50	32
Circuit breaker molded case, 480 V, 70 to 100 A, 1 P				1.143	291	68.50		359.50	405	470
Cut, form, align & connect wires				.151		9.05		9.05	11.15	14
Check operation of breaker				.018		1.10		1.10	1.35	1.70
Total				1.660	291	99.65		390.65	443	517.70
1010										
Maintenance and repair breaker, molded case, 480 V, 2 pole	20	1 ELEC	Ea.							
Tighten connections				.889		53.50		53.50	66	82.50
Total				.889		53.50		53.50	66	82.50
1020										
Maintenance and inspection C.B., molded case, 480 V, 2 pole	0.50	1 ELEC	Ea.							
Inspect, tighten C.B. wiring				.400		24		24	29.50	37
Total				.400		24		24	29.50	37
1030										
Replace C.B. molded case, 480 V, 2 pole	50	1 ELEC	Ea.							
Remove circuit breaker				.500		30		30	37	46.50
Circuit breaker molded case, 480 V, 70 to 100 A, 2 P				1.600	585	96		681	760	880
Cut, form, align & connect wires				.151		9.05		9.05	11.15	14
Check operation of breaker				.018		1.10		1.10	1.35	1.70
Total				2.269	585	136.15		721.15	809.50	942.20
1040										
Maintenance and repair breaker, molded case, 480 V, 3 pole	20	1 ELEC	Ea.							
Tighten connections				.889		53.50		53.50	66	82.50
Total				.889		53.50		53.50	66	82.50
1050										
Maintenance and inspection C.B., molded case, 480 V, 3 pole	0.50	1 ELEC	Ea.							
Inspect & tighten C.B. wiring				.400		24		24	29.50	37
Total				.400		24		24	29.50	37
1060										
Replace C.B. molded case, 480 V, 3 pole	50	1 ELEC	Ea.							
Remove circuit breaker				.500		30		30	37	46.50
Circuit breaker molded case, 480 V, 70 to 100 A, 3 P				2.000	695	120		815	915	1,050
Cut, form, align & connect wires				.151		9.05		9.05	11.15	14
Check operation of breaker				.018		1.10		1.10	1.35	1.70
Total				2.669	695	160.15		855.15	964.50	1,112.20

For customer support on your Facilities Maintenance & Repair Costs with RSMeans data, call 800.448.8182.

D50	ELECTRICAL		D5023	Lighting & Branch Wiring

D5023 130 — Circuit Breaker

System Description	Freq. (Years)	Crew	Unit	Labor Hours	2019 Bare Costs Material	Labor	Equipment	Total	Total In-House	Total w/O&P
2010 Repair failed breaker, molded case, 600 V, 2 pole	10	1 ELEC	Ea.							
Remove trip switch				.500		30		30	37	46.50
Trip switch				1.600	149	96		245	282	335
Total				2.100	149	126		275	319	**381.50**
2020 Maintenance and inspection C.B., molded case, 600 V, 2 pole	0.33	1 ELEC	Ea.							
Inspect & tighten C.B. wiring				.400		24		24	29.50	37
Total				.400		24		24	**29.50**	**37**
2030 Replace C.B. molded case, 600 V, 2 pole	50	1 ELEC	Ea.							
Remove circuit breaker				.500		30		30	37	46.50
Circuit breaker molded case, 600 V, 70 to 100 A, 2 P				1.600	665	96		761	850	980
Cut, form, align & connect wires				.170		10.20		10.20	12.60	15.75
Check operation of breakers				.018		1.10		1.10	1.35	1.70
Total				2.289	665	137.30		802.30	**900.95**	**1,043.95**
3010 Repair failed breaker, molded case, 600 V, 3 pole	10	1 ELEC	Ea.							
Remove trip switch				.500		30		30	37	46.50
Trip switch				1.600	149	96		245	282	335
Total				2.100	149	126		275	319	**381.50**
3020 Maintenance and inspection C.B., molded case, 600 V, 3 pole	0.33	1 ELEC	Ea.							
Inspect & tighten C.B. wiring				.400		24		24	29.50	37
Total				.400		24		24	**29.50**	**37**
3030 Replace C.B. molded case, 600 V, 3 pole	50	1 ELEC	Ea.							
Remove circuit breaker				.889		53.50		53.50	66	82.50
Circuit breaker molded case, 600 V, 125 to 400 A, 3 P				3.478	3,500	209		3,709	4,100	4,700
Cut, form, align & connect wires				.170		10.20		10.20	12.60	15.75
Check operation of breakers				.018		1.10		1.10	1.35	1.70
Total				4.556	3,500	273.80		3,773.80	**4,179.95**	**4,799.95**
3040 Maintenance and repair breaker, enclosed, 240 V, 1 pole	25	1 ELEC	Ea.							
Tighten connections				.889		53.50		53.50	66	82.50
Total				.889		53.50		53.50	**66**	**82.50**

For customer support on your Facilities Maintenance & Repair Costs with RSMeans data, call 800.448.8182.

321

D5023 130 Circuit Breaker

	System Description	Freq. (Years)	Crew	Unit	Labor Hours	2019 Bare Costs				Total In-House	Total w/O&P
						Material	Labor	Equipment	Total		
3050	**Maintenance and inspection C.B., enclosed, 240 V, 1 pole**	1	1 ELEC	Ea.							
	Inspect & tighten C.B. wiring				.400		24		24	29.50	37
	Total				.400		24		24	**29.50**	**37**
3060	**Replace C.B. enclosed, 240 V, 1 pole**	50	1 ELEC	Ea.							
	Remove circuit breaker				.500		30		30	37	46.50
	Circuit breaker enclosed, 240 V, 15 to 60 A, 1 P				2.105	249	126		375	430	505
	Cut, form, align & connect wires				.151		9.05		9.05	11.15	14
	Check operation of breaker				.018		1.10		1.10	1.35	1.70
	Total				2.775	249	166.15		415.15	**479.50**	**567.20**
4010	**Maintenance and repair breaker, enclosed, 240 V, 2 pole**	25	1 ELEC	Ea.							
	Tighten connections				.889		53.50		53.50	66	82.50
	Total				.889		53.50		53.50	**66**	**82.50**
4020	**Maintenance and inspection C.B., enclosed, 240 V, 2 pole**	1	1 ELEC	Ea.							
	Inspect & tighten C.B. wiring				.400		24		24	29.50	37
	Total				.400		24		24	**29.50**	**37**
4030	**Replace C.B. enclosed, 240 V, 2 pole**	50	1 ELEC	Ea.							
	Remove circuit breaker				.500		30		30	37	46.50
	Circuit breaker enclosed, 240 V, 15 to 60 A, 2 P				2.286	500	137		637	720	835
	Cut, form, align & connect wires				.151		9.05		9.05	11.15	14
	Check operation of breaker				.018		1.10		1.10	1.35	1.70
	Total				2.955	500	177.15		677.15	**769.50**	**897.20**
4040	**Maintenance and repair breaker, enclosed, 240 V, 3 pole**	25	1 ELEC	Ea.							
	Tighten connections				.889		53.50		53.50	66	82.50
	Total				.889		53.50		53.50	**66**	**82.50**
4050	**Maintenance and inspection C.B., enclosed, 240 V, 3 pole**	1	1 ELEC	Ea.							
	Inspect & tighten C.B. wiring				.400		24		24	29.50	37
	Total				.400		24		24	**29.50**	**37**

System Description	Freq. (Years)	Crew	Unit	Labor Hours	2019 Bare Costs				Total In-House	Total w/O&P
					Material	Labor	Equipment	Total		
4060 Replace C.B. enclosed, 240 V, 3 pole	50	1 ELEC	Ea.							
Remove circuit breaker				.500		30		30	37	46.50
Circuit breaker enclosed, 240 V, 15 to 60 A, 3 P				2.500	695	150		845	950	1,100
Cut, form, align & connect wires				.151		9.05		9.05	11.15	14
Check operation of breaker				.018		1.10		1.10	1.35	1.70
Total				3.169	695	190.15		885.15	999.50	1,162.20
5010 Repair failed breaker, enclosed, 600 V, 2 pole	4	1 ELEC	Ea.							
Repair circuit breaker component				2.105	470	126		596	675	785
Total				2.105	470	126		596	675	785
5020 Maintenance and inspection C.B., enclosed, 600 V, 2 pole	0.33	1 ELEC	Ea.							
Inspect & tighten C.B. wiring				.400		24		24	29.50	37
Total				.400		24		24	29.50	37
5030 Replace C.B. enclosed, 600 V, 2 pole	50	1 ELEC	Ea.							
Remove circuit breaker				.800		48		48	59	74
Circuit breaker enclosed, 600 V, 100 A, 2 P				3.200	665	192		857	970	1,125
Cut, form, align & connect wires				.170		10.20		10.20	12.60	15.75
Check operation of breaker				.018		1.10		1.10	1.35	1.70
Total				4.189	665	251.30		916.30	1,042.95	1,216.45
6010 Repair failed breaker, enclosed, 600 V, 3 pole	4	1 ELEC	Ea.							
Repair circuit breaker component				2.105	415	126		541	610	715
Total				2.105	415	126		541	610	715
6020 Maintenance and inspection C.B., enclosed, 600 V, 3 pole	0.33	1 ELEC	Ea.							
Inspect & tighten C.B. wiring				.400		24		24	29.50	37
Total				.400		24		24	29.50	37
6030 Replace C.B. enclosed, 600 V, 3 pole	50	1 ELEC	Ea.							
Remove circuit breaker				.889		53.50		53.50	66	82.50
Circuit breaker enclosed, 600 V, 100 A, 3 P				3.478	710	209		919	1,050	1,200
Cut, form, align & connect wires				.170		10.20		10.20	12.60	15.75
Check operation of breaker				.018		1.10		1.10	1.35	1.70
Total				4.556	710	273.80		983.80	1,129.95	1,299.95

For customer support on your Facilities Maintenance & Repair Costs with RSMeans data, call 800.448.8182.

323

D5023 130 | Circuit Breaker

	System Description	Freq. (Years)	Crew	Unit	Labor Hours	2019 Bare Costs				Total In-House	Total w/O&P
						Material	Labor	Equipment	Total		
6830	**Replace C.B. enclosed, 600 V, 3 pole 800 A**	50	2 ELEC	Ea.							
	Remove circuit breaker				4.444		267		267	330	410
	Circuit breaker enclosed, 600 V, 800 A, 3 P				17.021	5,300	1,025		6,325	7,100	8,200
	Cut, form, align & connect wires				.340		20.40		20.40	25	31.50
	Check operation of breaker				.037		2.20		2.20	2.70	3.40
	Total				21.843	5,300	1,314.60		6,614.60	7,457.70	8,644.90
6930	**Replace C.B. enclosed, 600 V, 3 pole 1000 A**	50	2 ELEC	Ea.							
	Remove circuit breaker				5.161		310		310	380	480
	Circuit breaker enclosed, 600 V, 1000 A, 3 P				19.048	6,700	1,150		7,850	8,775	10,200
	Cut, form, align & connect wires				.516		30.91		30.91	38	47.50
	Check operation of breaker				.055		3.33		3.33	4.09	5.15
	Total				24.780	6,700	1,494.24		8,194.24	9,197.09	10,732.65
7130	**Replace C.B. enclosed, 600 V, 3 pole 1600 A**	50	2 ELEC	Ea.							
	Remove circuit breaker				5.714		345		345	425	530
	Circuit breaker enclosed, 600 V, 1600 A, 3 P				22.222	15,800	1,325		17,125	19,000	21,800
	Cut, form, align & connect wires				.681		40.80		40.80	50.50	63
	Check operation of breaker				.073		4.40		4.40	5.40	6.80
	Total				28.691	15,800	1,715.20		17,515.20	19,480.90	22,399.80
7230	**Replace C.B. enclosed, 600 V, 3 pole 2000 A**	50	2 ELEC	Ea.							
	Remove circuit breaker				6.400		385		385	475	595
	Circuit breaker enclosed, 600 V, 2000 A, 3 P				25.000	17,100	1,500		18,600	20,700	23,700
	Cut, form, align & connect wires				1.418		85		85	105	131
	Check operation of breaker				.153		9.17		9.17	11.25	14.15
	Total				32.971	17,100	1,979.17		19,079.17	21,291.25	24,440.15

D5023 132 | Safety Switch, Heavy Duty

	System Description	Freq. (Years)	Crew	Unit	Labor Hours	2019 Bare Costs				Total In-House	Total w/O&P
						Material	Labor	Equipment	Total		
0010	**Maintenance and repair safety switch, H.D., 3 pole**	8	1 ELEC	Ea.							
	Resecure lugs				.500		30		30	37	46.50
	Total				.500		30		30	37	46.50

For customer support on your Facilities Maintenance & Repair Costs with RSMeans data, call 800.448.8182.

D50 ELECTRICAL | D5023 | Lighting & Branch Wiring

D5023 132 | Safety Switch, Heavy Duty

ID	System Description	Freq. (Years)	Crew	Unit	Labor Hours	2019 Bare Costs				Total In-House	Total w/O&P
						Material	Labor	Equipment	Total		
0020	**Maintenance and inspection safety switch, 3 pole**	1	1 ELEC	Ea.							
	Inspect & clean safety switch				.500		30		30	37	46.50
	Total				.500		30		30	**37**	**46.50**
0030	**Replace safety switch, H.D. 30 A**	25	2 ELEC	Ea.							
	Remove safety switch				.650		39		39	48	60
	Fused safety switch, heavy duty, 600 V, 3 P, 30 A				2.500	145	150		295	345	415
	Fuses				.600	66	36		102	117	138
	Check operation of switch				.018		1.10		1.10	1.35	1.70
	Total				3.769	211	226.10		437.10	**511.35**	**614.70**
0130	**Replace safety switch, H.D. 100 A**	25	2 ELEC	Ea.							
	Remove safety switch				1.096		66		66	81	101
	Fused safety switch, heavy duty, 600 V, 3 P, 100 A				4.211	315	253		568	660	785
	Fuses				.667	195	40.05		235.05	264	305
	Check operation of switch				.018		1.10		1.10	1.35	1.70
	Total				5.991	510	360.15		870.15	**1,006.35**	**1,192.70**
0230	**Replace safety switch, H.D. 200 A**	25	2 ELEC	Ea.							
	Remove safety switch				1.600		96		96	118	148
	Fused safety switch, heavy duty, 600 V, 3 P, 200 A				6.154	450	370		820	950	1,125
	Fuses				.800	390	48		438	490	560
	Check operation of switch				.018		1.10		1.10	1.35	1.70
	Total				8.572	840	515.10		1,355.10	**1,559.35**	**1,834.70**
0430	**Replace safety switch, H.D. 400 A**	25	2 ELEC	Ea.							
	Remove safety switch				2.353		141		141	174	218
	Fused safety switch, heavy duty, 600 V, 3 P, 400 A				8.889	1,250	535		1,785	2,025	2,400
	Fuses				1.000	831	60		891	990	1,125
	Check operation of switch				.037		2.20		2.20	2.70	3.40
	Total				12.278	2,081	738.20		2,819.20	**3,191.70**	**3,746.40**
0630	**Replace safety switch, H.D. 600 A**	25	2 ELEC	Ea.							
	Remove safety switch				3.478		209		209	257	320
	Fused safety switch, heavy duty, 600 V, 3 P, 600 A				13.333	1,975	800		2,775	3,150	3,700
	Fuses				1.200	1,080	72		1,152	1,275	1,450
	Check operation of switch				.037		2.20		2.20	2.70	3.40
	Total				18.048	3,055	1,083.20		4,138.20	**4,684.70**	**5,473.40**

For customer support on your Facilities Maintenance & Repair Costs with RSMeans data, call 800.448.8182.

325

D5023 140   Safety Switch, General Duty

ID	System Description	Freq. (Years)	Crew	Unit	Labor Hours	2019 Bare Costs Material	Labor	Equipment	Total	Total In-House	Total w/O&P
2010	**Maintenance and repair safety switch gen., 2 pole**	8	1 ELEC	Ea.							
	Resecure lugs				.500		30		30	37	46.50
	Total				**.500**		**30**		**30**	**37**	**46.50**
2020	**Maintenance and inspection safety switch, 2 pole**	1	1 ELEC	Ea.							
	Inspect & clean safety switch				.500		30		30	37	46.50
	Total				**.500**		**30**		**30**	**37**	**46.50**
2030	**Replace safety switch 240 V, 2 pole**	25	2 ELEC	Ea.							
	Remove safety switch				.650		39		39	48	60
	Fused safety switch, general duty, 240 V, 2 P, 30 A				2.299	65	138		203	242	294
	Fuse, dual element, time delay, 250 V, 30 A				.320	19.50	19.20		38.70	45	54
	Cut, form, align & connect wires				.151		9.05		9.05	11.15	14
	Check operation of switch				.018		1.10		1.10	1.35	1.70
	Total				**3.439**	**84.50**	**206.35**		**290.85**	**347.50**	**423.70**
3010	**Maintenance and repair safety switch gen., 3 pole**	8	1 ELEC	Ea.							
	Resecure lugs				.500		30		30	37	46.50
	Total				**.500**		**30**		**30**	**37**	**46.50**
3020	**Maintenance and inspection safety switch, 3 pole**	1	1 ELEC	Ea.							
	Inspect & clean safety switch				.500		30		30	37	46.50
	Total				**.500**		**30**		**30**	**37**	**46.50**
3030	**Replace safety switch 240 V, 3 pole**	25	2 ELEC	Ea.							
	Remove safety switch				.650		39		39	48	60
	Fused safety switch, general duty, 240 V, 3 P, 30 A				2.500	87.50	150		237.50	281	340
	Fuse, dual element, time delay, 250 V, 30 A				.480	29.25	28.80		58.05	67.50	81
	Cut, form, align & connect wires				.151		9.05		9.05	11.15	14
	Check operation of switch				.018		1.10		1.10	1.35	1.70
	Total				**3.800**	**116.75**	**227.95**		**344.70**	**409**	**496.70**
4010	**Replace low voltage cartridge**	50	1 ELEC	Ea.							
	Remove cartridge fuse cutouts				.100		6		6	7.40	9.25
	Cartridge fuse cutouts 3 P				.296	19	17.80		36.80	43	51.50
	Total				**.396**	**19**	**23.80**		**42.80**	**50.40**	**60.75**

For customer support on your Facilities Maintenance & Repair Costs with RSMeans data, call 800.448.8182.

D5023 140 Safety Switch, General Duty

	System Description	Freq. (Years)	Crew	Unit	Labor Hours	2019 Bare Costs				Total In-House	Total w/O&P
						Material	Labor	Equipment	Total		
5010	**Replace plug fuse**	25	1 ELEC	Ea.							
	Remove plug fuse				.080		4.80		4.80	5.90	7.40
	Fuse, plug, 120 V, 15 to 30 A				.160	4.05	9.60		13.65	16.30	19.85
	Total				**.240**	**4.05**	**14.40**		**18.45**	**22.20**	**27.25**

D5023 150 Receptacles And Plugs

	System Description	Freq. (Years)	Crew	Unit	Labor Hours	2019 Bare Costs				Total In-House	Total w/O&P
						Material	Labor	Equipment	Total		
0010	**Maintenance and repair**	20	1 ELEC	Ea.							
	Repair receptacle component (cover)				.500	2.58	30		32.58	40	49.50
	Total				**.500**	**2.58**	**30**		**32.58**	**40**	**49.50**
0020	**Replace receptacle/plug**	20	1 ELEC	Ea.							
	Turn power off, later turn on				.100		6		6		9.25
	Remove & install cover plates				.170		10.20		10.20		15.75
	Remove receptacle				.040		2.40		2.40		3.70
	Wiring devices, receptacle				.381	9.55	23		32.55		47.50
	Cut, form, align & connect wires				.026		1.56		1.56		2.40
	Test for operation				.015		.88		.88		1.35
	Total				**.732**	**9.55**	**44.04**		**53.59**	**64.46**	**79.95**

D5023 154 4-Pin Receptacle

	System Description	Freq. (Years)	Crew	Unit	Labor Hours	2019 Bare Costs				Total In-House	Total w/O&P
						Material	Labor	Equipment	Total		
0010	**Repair 4-pin receptacle cover**	10	1 ELEC	Ea.							
	Repair receptacle(cover)				.615	6.60	37		43.60	53	65.50
	Total				**.615**	**6.60**	**37**		**43.60**	**53**	**65.50**
0020	**Replace 4 - pin receptacle**	20	1 ELEC	Ea.							
	Remove 4-pin receptacle				.200		12		12	14.80	18.50
	Receptacle locking, 4-pin				.615	81	37		118	135	158
	Total				**.815**	**81**	**49**		**130**	**149.80**	**176.50**

For customer support on your Facilities Maintenance & Repair Costs with RSMeans data, call 800.448.8182.

327

D5023 210 Contactors And Relays

System Description	Freq. (Years)	Crew	Unit	Labor Hours	2019 Bare Costs				Total In-House	Total w/O&P
					Material	Labor	Equipment	Total		
0010 Maintenance and repair	3	1 ELEC	Ea.							
Remove contactor coil				.602		36		36	44.50	55.50
Contactor coil				2.000	35	120		55	187	229
Total				2.602	35	156		91	231.50	284.50
0020 Maintenance and inspection	0.50	1 ELEC	Ea.							
Inspect and clean contacts				.250		15		15	18.50	23
Total				.250		15		15	18.50	23
0030 Replace contactor	18	1 ELEC	Ea.							
Remove magnetic contactor				.800		48		48	59	74
Magnetic contactor 600 V, 3 P, 25 H.P.				2.667	425	160		585	665	780
Cut, form, align & connect wires				.151		9.05		9.05	11.15	14
Check operation of contactor				.018		1.10		1.10	1.35	1.70
Total				3.636	425	218.15		643.15	736.50	869.70

D5023 220 Wiring Devices, Switches

System Description	Freq. (Years)	Crew	Unit	Labor Hours	2019 Bare Costs				Total In-House	Total w/O&P
					Material	Labor	Equipment	Total		
0010 Maintenance and repair	10	1 ELEC	Ea.							
Repair switch component (cover)				.500	2.58	30		32.58	40	49.50
Total				.500	2.58	30		32.58	40	49.50
0020 Replace switch	15	1 ELEC	Ea.							
Turn power off, later turn on				.100		6		6	7.40	9.25
Remove & install cover plates				.170		10.20		10.20	12.60	15.75
Remove switch				.040		2.40		2.40	2.96	3.70
Wiring devices, switch				.381	3.26	23		26.26	32	39.50
Cut, form, align & connect wires				.026		1.56		1.56	1.92	2.40
Test for operation				.015		.88		.88	1.08	1.35
Total				.732	3.26	44.04		47.30	57.96	71.95

For customer support on your Facilities Maintenance & Repair Costs with RSMeans data, call 800.448.8182.

D5023 222 — Switch, Pull Cord

System Description	Freq. (Years)	Crew	Unit	Labor Hours	2019 Bare Costs Material	Labor	Equipment	Total	Total In-House	Total w/O&P
0010 Maintenance and repair										
Repair switch component (cover)	5	1 ELEC	Ea.	.500	2.84	30		32.84	40	50
Total				.500	2.84	30		32.84	**40**	**50**
0020 Replace switch	15	1 ELEC	Ea.							
Turn power off, later turn on				.100		6		6	7.40	9.25
Remove & install cover plates				.170		10.20		10.20	12.60	15.75
Remove switch, pull cord				.040		2.40		2.40	2.96	3.70
Switch, pull cord				.381	12.25	23		35.25	41.50	51
Cut, form, align & connect wires				.026		1.56		1.56	1.92	2.40
Test for operation				.015		.88		.88	1.08	1.35
Total				.732	12.25	44.04		56.29	**67.46**	**83.45**

D5023 230 — Light Dimming Panel

System Description	Freq. (Years)	Crew	Unit	Labor Hours	2019 Bare Costs Material	Labor	Equipment	Total	Total In-House	Total w/O&P
0010 Minor repairs to light dimming panel	5	1 ELEC	Ea.							
Check and test components				.400		24		24	29.50	37
Remove dimming panel component				.222		13.35		13.35	16.45	20.50
Dimming panel component				.667	88.50	40		128.50	147	172
Total				1.289	88.50	77.35		165.85	**192.95**	**229.50**
0020 Maintenance and inspection	1	1 ELEC	Ea.							
Check light dimming panel				.533		32		32	39.50	49.50
Total				.533		32		32	**39.50**	**49.50**
0030 Replace light dimming panel	15	1 ELEC	Ea.							
Turn power off, later turn on				.100		6		6	7.40	9.25
Remove dimming panel				1.600		96		96	118	148
Dimming panel				8.000	710	480		1,190	1,375	1,625
Test circuits				.073		4.37		4.37	5.40	6.75
Total				9.773	710	586.37		1,296.37	**1,505.80**	**1,789**

For customer support on your Facilities Maintenance & Repair Costs with RSMeans data, call 800.448.8182.

329

	System Description	Freq. (Years)	Crew	Unit	Labor Hours	2019 Bare Costs				Total In-House	Total w/O&P
						Material	Labor	Equipment	Total		
0010	**Maintenance and repair**	10	1 ELEC	Ea.							
	Remove glass globe				.100		6		6	7.40	9.25
	Glass globe				.296	34.50	17.80		52.30	60	70.50
	Total				.396	34.50	23.80		58.30	67.40	79.75
0020	**Replace lamp**	5	1 ELEC	Ea.							
	Remove incand. lighting fixture lamp				.022		1.32		1.32	1.63	2.04
	Incand. lighting fixture lamp				.065	27	3.91		30.91	34.50	40
	Total				.087	27	5.23		32.23	36.13	42.04
0030	**Replace lighting fixture**	20	1 ELEC	Ea.							
	Turn circuit off and on for fixture				.018		1.10		1.10	1.35	1.70
	Remove incand. lighting fixture				.258		15.50		15.50	19.10	24
	Incand. lighting fixture, 75 W				.800	52.50	48		100.50	117	140
	Total				1.076	52.50	64.60		117.10	137.45	165.70
0120	**Replace lamp, 200 W**	5	1 ELEC	Ea.							
	Remove incand. lighting fixture lamp				.022		1.33		1.33	1.64	2.06
	Incand. lighting fixture lamp, 200 W				.065	3.21	3.91		7.12	8.35	10
	Total				.087	3.21	5.24		8.45	9.99	12.06
0220	**Replace lamp for explosion proof fixture**	5	1 ELEC	Ea.							
	Remove incand. lighting fixture lamp				.044		2.66		2.66	3.28	4.12
	Incand. lighting fixture lamp for explosion proof fixture				.130	6.42	7.82		14.24	16.70	20
	Total				.175	6.42	10.48		16.90	19.98	24.12
0230	**Replace incan. lighting fixt., explosion proof, ceiling mtd., 200 W**	20	1 ELEC	Ea.							
	Turn branch circuit off and on				.018		1.10		1.10	1.35	1.70
	Remove incandescent lighting fixture, explosion proof				.920		55		55	68	85
	Incan. lighting fixture, explosion proof, ceiling mtd, 200 W				2.000	1,450	120		1,570	1,750	2,000
	Total				2.938	1,450	176.10		1,626.10	1,819.35	2,086.70
1230	**Replace incan. lighting fixt., high hat can type, ceiling mtd., 200 W**	20	1 ELEC	Ea.							
	Turn branch circuit off and on				.018		1.10		1.10	1.35	1.70
	Remove incandescent lighting fixture, can type				.258		15.50		15.50	19.10	24
	Incan. lighting fixture, high hat can type, ceiling mtd, 200 W				1.194	98	71.50		169.50	196	234
	Total				1.470	98	88.10		186.10	216.45	259.70

For customer support on your Facilities Maintenance & Repair Costs with RSMeans data, call 800.448.8182.

D5023 250 Quartz Fixture

System Description	Freq. (Years)	Crew	Unit	Labor Hours	2019 Bare Costs Material	Labor	Equipment	Total	Total In-House	Total w/O&P
0010 Maintenance and repair										
Repair glass lens	10	1 ELEC	Ea.	.400	26	24		50	58	69.50
Total				.400	26	24		50	**58**	**69.50**
0020 Replace 1500 W quartz lamp										
Remove quartz lamp	10	1 ELEC	Ea.	.087		5.20		5.20	6.45	8.05
Quartz lamp				.258	17.55	15.50		33.05	38.50	46
Total				.345	17.55	20.70		38.25	**44.95**	**54.05**
0030 Replace fixture										
Turn branch circuit off and on	20	1 ELEC	Ea.	.018		1.10		1.10	1.35	1.70
Remove exterior fixt., quartz, 1500 W				.593		35.50		35.50	44	55
Exterior fixt. quartz, 1500 W				1.905	100	114		214	251	300
Total				2.516	100	150.60		250.60	**296.35**	**356.70**

D5023 260 Fluorescent Lighting Fixture

System Description	Freq. (Years)	Crew	Unit	Labor Hours	2019 Bare Costs Material	Labor	Equipment	Total	Total In-House	Total w/O&P
0010 Replace fluor. ballast										
Remove indoor fluor., ballast	10	1 ELEC	Ea.	.333		20		20	24.50	31
Fluorescent, electronic ballast				.667	43.50	40		83.50	97	116
Test fixture				.018		1.10		1.10	1.35	1.70
Total				1.018	43.50	61.10		104.60	**122.85**	**148.70**
0020 Replace lamps (2 lamps), 4', 34 W energy saver										
Remove fluor. lamps in fixture	10	1 ELEC	Ea.	.078		4.68		4.68	5.80	7.20
Fluor. lamp, 4' energy saver				.232	15.90	13.90		29.80	34.50	41.50
Total				.310	15.90	18.58		34.48	**40.30**	**48.70**
0030 Replace fixture, lay-in, recess mtd., 2' x 4', two 40 W										
Turn branch circuit off and on	20	1 ELEC	Ea.	.018		1.10		1.10	1.35	1.70
Remove fluor. lighting fixture				.485		29		29	36	45
Fluor. 2' x 4', recess mounted, two 40 W				1.509	54	90.50		144.50	171	208
Total				2.013	54	120.60		174.60	**208.35**	**254.70**

For customer support on your Facilities Maintenance & Repair Costs with RSMeans data, call 800.448.8182.

331

D5023 260 Fluorescent Lighting Fixture

	System Description	Freq. (Years)	Crew	Unit	Labor Hours	2019 Bare Costs				Total In-House	Total w/O&P
						Material	Labor	Equipment	Total		
0040	**Replace fixture, lay-in, recess mtd., 2' x 4', four 32 W**	20	1 ELEC	Ea.							
	Turn branch circuit off and on				.018		1.10		1.10	1.35	1.70
	Remove fluor. lighting fixture				.533		32		32	39.50	49.50
	Fluor. lay-in, recess mtd., 2' x 4', four 32 W				1.702	74.50	102		176.50	208	251
	Total				2.254	74.50	135.10		209.60	248.85	302.20
0120	**Replace lamps (2 lamps), 8', 60 W energy saver**	10	1 ELEC	Ea.							
	Remove fluor. lamps in fixture				.078		4.68		4.68	5.80	7.20
	Fluor. lamps, 8' energy saver				.200	7.90	12		19.90	23.50	28.50
	Total				.278	7.90	16.68		24.58	29.30	35.70
0130	**Replace fixture, strip, surface mtd., 8', two 75 W**	20	2 ELEC	Ea.							
	Turn branch circuit off and on				.018		1.10		1.10	1.35	1.70
	Remove fluor. lighting fixture				.400		24		24	29.50	37
	Fluor. strip, surface mounted, 8', two 75 W				1.290	66.50	77.50		144	169	202
	Total				1.709	66.50	102.60		169.10	199.85	240.70
0140	**Replace fixture, strip, pendent mtd., 8', two 75 W**	20	2 ELEC	Ea.							
	Turn branch circuit off and on				.018		1.10		1.10	1.35	1.70
	Remove fluor. lighting fixture				.593		35.50		35.50	44	55
	Fluor. strip, pendent mtd., 8', two 75 W				1.818	98.50	109		207.50	243	291
	Total				2.429	98.50	145.60		244.10	288.35	347.70

D5023 270 Metal Halide Fixture

	System Description	Freq. (Years)	Crew	Unit	Labor Hours	2019 Bare Costs				Total In-House	Total w/O&P
						Material	Labor	Equipment	Total		
2010	**Replace M.H. ballast, 175 W**	10	1 ELEC	Ea.							
	Remove metal halide fixt. ballast				.333		20		20	24.50	31
	Metal halide fixt. 175 W, ballast				.667	92	40		132	151	177
	Test fixture				.018		1.10		1.10	1.35	1.70
	Total				1.018	92	61.10		153.10	176.85	209.70
2020	**Replace lamp, 175 W**	5	1 ELEC	Ea.							
	Remove metal halide fixt. lamp				.116		6.95		6.95	8.60	10.75
	Metal halide lamp, 175 W				.348	10.50	21		31.50	37.50	45
	Total				.464	10.50	27.95		38.45	46.10	55.75

For customer support on your Facilities Maintenance & Repair Costs with RSMeans data, call 800.448.8182.

D5023 270 | **Metal Halide Fixture**

System Description	Freq. (Years)	Crew	Unit	Labor Hours	2019 Bare Costs				Total In-House	Total w/O&P
					Material	Labor	Equipment	Total		
2030 **Replace fixture, 175 W**	20	1 ELEC	Ea.							
Turn branch circuit off and on				.018		1.10		1.10	1.35	1.70
Remove metal halide fixt.				.800		48		48	59	74
Metal halide, recessed, square 175 W				2.353	440	141		581	660	770
Total				3.171	440	190.10		630.10	720.35	845.70
2210 **Replace M.H. ballast, 400 W**	10	1 ELEC	Ea.							
Remove metal halide fixt. ballast				.333		20		20	24.50	31
Metal halide fixt. 400 W, ballast				.762	150	46		196	221	258
Test fixture				.018		1.10		1.10	1.35	1.70
Total				1.114	150	67.10		217.10	246.85	290.70
2220 **Replace lamp, 400 W**	5	1 ELEC	Ea.							
Remove metal halide fixt. lamp				.118		7.05		7.05	8.70	10.90
Metal halide lamp, 400 W				.348	20	21		41	47.50	57
Total				.465	20	28.05		48.05	56.20	67.90
2230 **Replace fixture, 400 W**	20	2 ELEC	Ea.							
Turn branch circuit off and on				.018		1.10		1.10	1.35	1.70
Remove metal halide fixt.				1.067		64		64	79	99
Metal halide, surface mtd, high bay, 400 W				3.478	410	209		619	710	835
Total				4.563	410	274.10		684.10	790.35	935.70
2310 **Replace M.H. ballast, 1000 W**	10	1 ELEC	Ea.							
Remove metal halide fixt. ballast				.333		20		20	24.50	31
Metal halide fixt. 1000 W, ballast				.889	245	53.50		298.50	335	390
Test fixture				.018		1.10		1.10	1.35	1.70
Total				1.241	245	74.60		319.60	360.85	422.70
2320 **Replace lamp, 1000 W**	5	1 ELEC	Ea.							
Remove metal halide fixt. lamp				.118		7.05		7.05	8.70	10.90
Metal halide lamp, 1000 W				.523	33	31.50		64.50	75	90
Total				.641	33	38.55		71.55	83.70	100.90

For customer support on your Facilities Maintenance & Repair Costs with RSMeans data, call 800.448.8182.

333

D5023 270 Metal Halide Fixture

	System Description	Freq. (Years)	Crew	Unit	Labor Hours	2019 Bare Costs Material	2019 Bare Costs Labor	2019 Bare Costs Equipment	2019 Bare Costs Total	Total In-House	Total w/O&P
2330	**Replace fixture, 1000 W**	20	2 ELEC	Ea.							
	Turn branch circuit off and on				.018		1.10		1.10	1.35	1.70
	Remove metal halide fixt.				1.333		80		80	98.50	123
	Metal halide, surface mtd, high bay, 1000 W				4.000	595	240		835	950	1,125
	Total				5.352	595	321.10		916.10	1,049.85	1,249.70

D5023 280 H.P. Sodium Fixture, 250 W

	System Description	Freq. (Years)	Crew	Unit	Labor Hours	2019 Bare Costs Material	2019 Bare Costs Labor	2019 Bare Costs Equipment	2019 Bare Costs Total	Total In-House	Total w/O&P
1010	**Replace H.P. sodium ballast**	10	2 ELEC	Ea.							
	Remove H.P. Sodium fixt. ballast				.333		20		20	24.50	31
	H.P. Sodium fixt., 250 W, ballast				.667	294	40		334	375	430
	Test fixture				.018		1.10		1.10	1.35	1.70
	Total				1.018	294	61.10		355.10	400.85	462.70
1020	**Replace lamp**	10	2 ELEC	Ea.							
	Remove H.P. Sodium fixt. lamp				.116		6.95		6.95	8.60	10.75
	H.P. Sodium fixt. lamp, 250 W				.348	21.50	21		42.50	49.50	59
	Total				.464	21.50	27.95		49.45	58.10	69.75
1030	**Replace fixture**	20	2 ELEC	Ea.							
	Turn branch circuit off and on				.018		1.10		1.10	1.35	1.70
	Remove H.P. Sodium fixt.				.889		53.50		53.50	66	82.50
	H.P. Sodium fixt., recessed, square 250 W				2.667	870	160		1,030	1,150	1,325
	Total				3.574	870	214.60		1,084.60	1,217.35	1,409.20

For customer support on your Facilities Maintenance & Repair Costs with RSMeans data, call 800.448.8182.

D5033 310 | Telephone Cable

System Description	Freq. (Years)	Crew	Unit	Labor Hours	2019 Bare Costs				Total In-House	Total w/O&P
					Material	Labor	Equipment	Total		
0010 Repair cable, 22 A.W.G., 4 pair	8	1 ELEC	M.L.F.							
Cut telephone cable				.080		4.80		4.80	5.90	7.40
Splice and insulate telephone cable				.471	12.90	28.50		41.40	49	59.50
Total				.551	12.90	33.30		46.20	**54.90**	**66.90**
0020 Replace telephone cable , #22-4 conductor	50	2 ELEC	M.L.F.							
Turn power off, later turn on				.100		6		6	7.40	9.25
Remove & reinstall cover plates				1.290		77.50		77.50	95.50	120
Remove telephone cable				2.581		155		155	191	239
Telephone twisted, #22-4 conductor				10.000	150	600		750	905	1,125
Total				13.971	150	838.50		988.50	**1,198.90**	**1,493.25**
1020 Replace telephone jack	20	1 ELEC	Ea.							
Remove telephone jack				.059		3.56		3.56	4.39	5.50
Telephone jack				.250	3.48	15		18.48	22.50	27.50
Total				.309	3.48	18.56		22.04	**26.89**	**33**

D5033 410 | Master Clock Control

System Description	Freq. (Years)	Crew	Unit	Labor Hours	2019 Bare Costs				Total In-House	Total w/O&P
					Material	Labor	Equipment	Total		
0010 Maintenance and repair	10	1 ELEC	Ea.							
Remove time control clock motor				.296		17.80		17.80	22	27.50
Time control clock motor				1.000	75.50	60		135.50	157	187
Total				1.296	75.50	77.80		153.30	**179**	**214.50**
0020 Check operation	1	1 ELEC	Ea.							
Check time control clock operation				.533		32		32	39.50	49.50
Total				.533		32		32	**39.50**	**49.50**
0030 Replace time control clock	15	1 ELEC	Ea.							
Remove time control clock				.296		17.80		17.80	22	27.50
Time control clock				1.333	89	80		169	197	234
Total				1.630	89	97.80		186.80	**219**	**261.50**

D5033 410 Master Clock Control

System Description	Freq. (Years)	Crew	Unit	Labor Hours	2019 Bare Costs				Total In-House	Total w/O&P
					Material	Labor	Equipment	Total		
1120 Maintenance and inspection Check operation of program bell	1	2 ELEC	Ea.	.320		19.20		19.20	23.50	29.50
Total				.320		19.20		19.20	**23.50**	**29.50**
1130 Replace program bell Remove program bell	15	2 ELEC	Ea.	.667		40		40	49.50	61.50
Program bell				1.000	110	60		170	195	230
Total				1.667	110	100		210	**244.50**	**291.50**

D5033 510 TV Cable Outlet

System Description	Freq. (Years)	Crew	Unit	Labor Hours	2019 Bare Costs				Total In-House	Total w/O&P
					Material	Labor	Equipment	Total		
0010 Maintenance and repair Repair TV cable outlet (cover)	10	1 ELEC	Ea.	.615	2.84	37		39.84	48.50	60.50
Total				.615	2.84	37		39.84	**48.50**	**60.50**
0020 Replace TV cable outlet Remove TV cable outlet	20	1 ELEC	Ea.	.200		12		12	14.80	18.50
TV cable outlet				.615	3.85	37		40.85	50	62
Total				.815	3.85	49		52.85	**64.80**	**80.50**

D5033 610 Door Bell

System Description	Freq. (Years)	Crew	Unit	Labor Hours	2019 Bare Costs				Total In-House	Total w/O&P
					Material	Labor	Equipment	Total		
0010 Maintenance and repair Remove door bell transformer	10	1 ELEC	Ea.	.200		12		12	14.80	18.50
Door bell transformer				.667	22.50	40		62.50	74	89.50
Total				.867	22.50	52		74.50	**88.80**	**108**
0020 Maintenance and inspection Check operation of door bell	1	1 ELEC	Ea.	.320		19.20		19.20	23.50	29.50
Total				.320		19.20		19.20	**23.50**	**29.50**

For customer support on your Facilities Maintenance & Repair Costs with RSMeans data, call 800.448.8182.

D5033 610 Door Bell

	System Description	Freq. (Years)	Crew	Unit	Labor Hours	2019 Bare Costs				Total In-House	Total w/O&P
						Material	Labor	Equipment	Total		
0030	**Replace door bell**	15	1 ELEC	Ea.							
	Remove door bell system				.667		40		40	49.50	61.50
	Door bell system				2.667	141	160		301	350	425
	Total				3.333	141	200		341	**399.50**	**486.50**

D5033 620 Sound System Components

	System Description	Freq. (Years)	Crew	Unit	Labor Hours	2019 Bare Costs				Total In-House	Total w/O&P
						Material	Labor	Equipment	Total		
0030	**Replace speaker**	20	1 ELEC	Ea.							
	Turn power off, later turn on				.100		6		6	7.40	9.25
	Remove speaker				.333		20		20	24.50	31
	Speaker				1.000	144	60		204	232	273
	Test for operation				.015		.88		.88	1.08	1.35
	Total				1.448	144	86.88		230.88	**264.98**	**314.60**
0120	**Inspect monitor panel**	0.50	1 ELEC	Ea.							
	Check monitor panel				.533		32		32	39.50	49.50
	Total				.533		32		32	**39.50**	**49.50**
0130	**Replace monitor panel**	15	1 ELEC	Ea.							
	Turn power off, later turn on				.100		6		6	7.40	9.25
	Remove monitor panel				1.667		100		100	123	155
	Monitor panel				2.000	475	120		595	670	780
	Test circuits				.073		4.37		4.37	5.40	6.75
	Total				3.839	475	230.37		705.37	**805.80**	**951**
0220	**Inspect volume control**	1	1 ELEC	Ea.							
	Check operation of volume control				.320		19.20		19.20	23.50	29.50
	Total				.320		19.20		19.20	**23.50**	**29.50**
0230	**Replace volume control**	15	2 ELEC	Ea.							
	Remove volume control				.333		20		20	24.50	31
	Volume control				1.000	57	60		117	137	164
	Total				1.333	57	80		137	**161.50**	**195**

For customer support on your Facilities Maintenance & Repair Costs with RSMeans data, call 800.448.8182.

337

D5033 620 Sound System Components

System Description	Freq. (Years)	Crew	Unit	Labor Hours	2019 Bare Costs Material	Labor	Equipment	Total	Total In-House	Total w/O&P
0330 Replace amplifier	15	2 ELEC	Ea.							
Turn power off, later turn on				.100		6		6	7.40	9.25
Remove amplifier				1.667		100		100	123	155
Amplifier				8.000	1,225	480		1,705	1,950	2,275
Test for operation				.015		.88		.88	1.08	1.35
Total				9.781	1,225	586.88		1,811.88	2,081.48	2,440.60

D5033 630 Intercom System Components

System Description	Freq. (Years)	Crew	Unit	Labor Hours	2019 Bare Costs Material	Labor	Equipment	Total	Total In-House	Total w/O&P
0120 Inspect intercom master station	0.50	1 ELEC	Ea.							
Check intercom master station				2.133		128		128	158	198
Total				2.133		128		128	158	198
0130 Replace intercom master station	15	2 ELEC	Ea.							
Turn power off, later turn on				.100		6		6	7.40	9.25
Remove intercom master station				1.333		80		80	98.50	123
Intercom master station				8.000	2,325	480		2,805	3,150	3,650
Test circuits				.073		4.37		4.37	5.40	6.75
Total				9.506	2,325	570.37		2,895.37	3,261.30	3,789
0220 Inspect intercom remote station	1	1 ELEC	Ea.							
Check operation of intercom remote station				.320		19.20		19.20	23.50	29.50
Total				.320		19.20		19.20	23.50	29.50
0230 Replace intercom remote station	15	1 ELEC	Ea.							
Remove intercom remote station				.333		20		20	24.50	31
Intercom remote station				1.000	199	60		259	293	340
Total				1.333	199	80		279	317.50	371

D5033 640 Security System Components

	System Description	Freq. (Years)	Crew	Unit	Labor Hours	2019 Bare Costs				Total In-House	Total w/O&P
						Material	Labor	Equipment	Total		
0020	**Inspect camera and monitor**	0.50	1 ELEC	Ea.							
	Check camera and monitor				.808		48.48		48.48	60	75
	Total				.808		48.48		48.48	**60**	**75**
0030	**Replace camera and monitor**	12	2 ELEC	Ea.							
	Turn power off, later turn on				.100		6		6	7.40	9.25
	Remove closed circuit TV, one camera & one monitor				2.051		123		123	152	190
	Closed circuit TV, one camera & one monitor				6.154	720	370		1,090	1,250	1,475
	Test for operation				.015		.88		.88	1.08	1.35
	Total				8.320	720	499.88		1,219.88	**1,410.48**	**1,675.60**
0120	**Inspect camera**	0.50	1 ELEC	Ea.							
	Check camera				.533		32		32	39.50	49.50
	Total				.533		32		32	**39.50**	**49.50**
0130	**Replace camera**	12	1 ELEC	Ea.							
	Turn power off, later turn on				.100		6		6	7.40	9.25
	Remove closed circuit TV, one camera				1.000		60		60	74	92.50
	Closed circuit TV, one camera				2.963	315	178		493	565	670
	Test for operation				.015		.88		.88	1.08	1.35
	Total				4.078	315	244.88		559.88	**647.48**	**773.10**

D5033 710 Smoke Detector

	System Description	Freq. (Years)	Crew	Unit	Labor Hours	2019 Bare Costs				Total In-House	Total w/O&P
						Material	Labor	Equipment	Total		
0010	**Repair smoke detector**	10	1 ELEC	Ea.							
	Remove smoke detector lamp				.151		9.05		9.05	11.15	14
	Smoke detector lamp				.444	6.85	26.50		33.35	40.50	49.50
	Total				.595	6.85	35.55		42.40	**51.65**	**63.50**
0020	**Check operation**	1	1 ELEC	Ea.							
	Check smoke detector operation				.200		12		12	14.80	18.50
	Total				.200		12		12	**14.80**	**18.50**

For customer support on your Facilities Maintenance & Repair Costs with RSMeans data, call 800.448.8182.

339

D5033 710 Smoke Detector

	System Description	Freq. (Years)	Crew	Unit	Labor Hours	2019 Bare Costs				Total In-House	Total w/O&P
						Material	Labor	Equipment	Total		
0030	**Replace smoke detector**	15	1 ELEC	Ea.							
	Turn power off, later turn on				.100		6		6	7.40	9.25
	Remove smoke detector				.348		21		21	25.50	32
	Smoke detector, ceiling type				1.290	123	77.50		200.50	231	273
	Test for operation				.015		.88		.88	1.08	1.35
	Total				1.753	123	105.38		228.38	264.98	315.60

D5033 712 Heat Detector

	System Description	Freq. (Years)	Crew	Unit	Labor Hours	2019 Bare Costs				Total In-House	Total w/O&P
						Material	Labor	Equipment	Total		
0010	**Repair heat detector**	10	1 ELEC	Ea.							
	Remove heat detector element				.151		9.05		9.05	11.15	14
	Heat detector element				.444	9.85	26.50		36.35	43.50	53.50
	Total				.595	9.85	35.55		45.40	54.65	67.50
0020	**Check operation**	1	1 ELEC	Ea.							
	Check heat detector operation				.200		12		12	14.80	18.50
	Total				.200		12		12	14.80	18.50
0030	**Replace heat detector**	15	1 ELEC	Ea.							
	Turn power off, later turn on				.100		6		6	7.40	9.25
	Remove heat detector				.348		21		21	25.50	32
	Heat detector				1.103	45.50	66.50		112	132	159
	Test for operation				.015		.88		.88	1.08	1.35
	Total				1.566	45.50	94.38		139.88	165.98	201.60

D5033 720 Manual Pull Station

	System Description	Freq. (Years)	Crew	Unit	Labor Hours	2019 Bare Costs				Total In-House	Total w/O&P
						Material	Labor	Equipment	Total		
0010	**Check and repair manual pull station**	10	1 ELEC	Ea.							
	Remove pull station component				.200		12		12	14.80	18.50
	Manual pull station component				.667	18.30	40		58.30	69.50	84.50
	Total				.867	18.30	52		70.30	84.30	103

D50 ELECTRICAL — D5033 Communications and Security

D5033 720 — Manual Pull Station

	System Description	Freq. (Years)	Crew	Unit	Labor Hours	2019 Bare Costs				Total In-House	Total w/O&P
						Material	Labor	Equipment	Total		
0020	**Replace manual pull station**	15	1 ELEC	Ea.							
	Turn power off, later turn on				.100		6		6	7.40	9.25
	Remove manual pull station				.348		21		21	25.50	32
	Manual pull station				1.000	87.50	60		147.50	170	202
	Test for operation				.015		.88		.88	1.08	1.35
	Total				1.462	87.50	87.88		175.38	203.98	244.60

D5033 760 — Fire Alarm Control Panel

	System Description	Freq. (Years)	Crew	Unit	Labor Hours	2019 Bare Costs				Total In-House	Total w/O&P
						Material	Labor	Equipment	Total		
0010	**Minor repairs to fire alarm**	5	1 ELEC	Ea.							
	Check and test components				.400		24		24	29.50	37
	Remove relay				.222		13.35		13.35	16.45	20.50
	Relay				.667	46	40		86	100	119
	Total				1.289	46	77.35		123.35	145.95	176.50
0020	**Maintenance and inspection**	0.50	1 ELEC	Ea.							
	Check control panel				.533		32		32	39.50	49.50
	Total				.533		32		32	39.50	49.50
0030	**Replace fire alarm panel**	15	1 ELEC	Ea.							
	Turn power off, later turn on				.100		6		6	7.40	9.25
	Remove fire alarm control panel				2.000		120		120	148	185
	Fire alarm control panel				16.000	960	960		1,920	2,250	2,675
	Test circuits				.073		4.37		4.37	5.40	6.75
	Total				18.173	960	1,090.37		2,050.37	2,410.80	2,876

For customer support on your Facilities Maintenance & Repair Costs with RSMeans data, call 800.448.8182.

341

D5033 766 Annunciation Panel

System Description	Freq. (Years)	Crew	Unit	Labor Hours	2019 Bare Costs Material	Labor	Equipment	Total	Total In-House	Total w/O&P
0010 Minor repairs to annunciation panel	5	1 ELEC	Ea.							
Check and test components				.400		24		24	29.50	37
Remove relay				.222		13.35		13.35	16.45	20.50
Relay				.667	46	40		86	100	119
Total				1.289	46	77.35		123.35	145.95	**176.50**
0020 Maintenance and inspection	0.50	1 ELEC	Ea.							
Check annunciator panel				.533		32		32	39.50	49.50
Total				.533		32		32	**39.50**	**49.50**
0030 Replace annunciation panel	15	1 ELEC	Ea.							
Turn power off, later turn on				.100		6		6	7.40	9.25
Remove annunciation panel				1.600		96		96	118	148
Annunciation panel				6.154	415	370		785	910	1,100
Test circuits				.073		4.37		4.37	5.40	6.75
Total				7.927	415	476.37		891.37	**1,040.80**	**1,264**

D5033 770 Fire Alarm Bell

System Description	Freq. (Years)	Crew	Unit	Labor Hours	2019 Bare Costs Material	Labor	Equipment	Total	Total In-House	Total w/O&P
0010 Replace fire alarm bell, 6"	20	1 ELEC	Ea.							
Turn power off, later turn on				.100		6		6	7.40	9.25
Remove fire alarm bell				.348		21		21	25.50	32
Fire alarm bell				1.000	135	60		195	223	261
Test for operation				.015		.88		.88	1.08	1.35
Total				1.462	135	87.88		222.88	**256.98**	**303.60**

342

For customer support on your Facilities Maintenance & Repair Costs with RSMeans data, call 800.448.8182.

D50 ELECTRICAL | D5093 | Other Electrical Systems

D5093 110 | Electrical Service Ground

System Description	Freq. (Years)	Crew	Unit	Labor Hours	2019 Bare Costs				Total In-House	Total w/O&P
					Material	Labor	Equipment	Total		
0010 Maintenance and repair										
Repair service ground (1 termination)	25	1 ELEC	M.L.F.	1.053	4.96	63		67.96	83.50	104
Total				1.053	4.96	63		67.96	**83.50**	**104**
0020 Replace electrical service ground										
Remove ground wire #6	50	1 ELEC	M.L.F.	2.000		120		120	148	185
Excavate hole for ground rod				2.222		133.50		133.50	164	205
Ground rod 5/8", 10'				17.391	235	1,040		1,275	1,550	1,900
Ground clamp				2.500	49.60	150		199.60	240	292
Backfill over top of ground rod				1.053		63		63	78	97.50
Insulated copper wire #6				12.308	485	740		1,225	1,450	1,750
Total				37.474	769.60	2,246.50		3,016.10	**3,630**	**4,429.50**

D5093 120 | Building Structure Ground

System Description	Freq. (Years)	Crew	Unit	Labor Hours	2019 Bare Costs				Total In-House	Total w/O&P
					Material	Labor	Equipment	Total		
0010 Maintenance and repair										
Repair bldg structure ground (1 termination)	7	1 ELEC	M.L.F.	1.053	4.96	63		67.96	83.50	104
Total				1.053	4.96	63		67.96	**83.50**	**104**
0020 Replace building structure ground										
Remove ground wire #4	50	1 ELEC	M.L.F.	3.077		185		185	228	285
Excavate hole for ground rod				2.222		133.50		133.50	164	205
Ground rod 5/8", 10'				17.391	235	1,040		1,275	1,550	1,900
Ground clamp				2.500	49.60	150		199.60	240	292
Backfill over top of ground rod				1.053		63		63	78	97.50
Insulated copper wire #4				15.094	690	905		1,595	1,875	2,275
Total				41.337	974.60	2,476.50		3,451.10	**4,135**	**5,054.50**

D5093 130 Lightning Protection System

	System Description	Freq. (Years)	Crew	Unit	Labor Hours	2019 Bare Costs				Total In-House	Total w/O&P
						Material	Labor	Equipment	Total		
1010	**Maintenance and repair of general wiring**										
	Repair lightning protection (1 termination)	1	1 ELEC	M.L.F.	1.053	11.70	63		74.70	91	112
	Total				1.053	11.70	63		74.70	**91**	**112**
1020	**Replace lightning protection general wiring**										
	Remove lightning protection cable	25	2 ELEC	M.L.F.	4.000		240		240	296	370
	Excavate hole for ground rod				2.222		133.50		133.50	164	205
	Ground rod 5/8", 10'				17.391	235	1,040		1,275	1,550	1,900
	Ground clamp				2.500	49.60	150		199.60	240	292
	Backfill over top of ground rod				1.053		63		63	78	97.50
	Lightning protection cable				25.000	3,050	1,500		4,550	5,200	6,125
	Air terminals, copper 3/8" x 10"				10.000	240	600		840	1,000	1,225
	Total				62.166	3,574.60	3,726.50		7,301.10	**8,528**	**10,214.50**

D5093 140 Lightning Ground Rod

	System Description	Freq. (Years)	Crew	Unit	Labor Hours	2019 Bare Costs				Total In-House	Total w/O&P
						Material	Labor	Equipment	Total		
1010	**Maintenance and repair**										
	Repair ground rod connector	1	1 ELEC	Ea.	1.053	4.96	63		67.96	83.50	104
	Total				1.053	4.96	63		67.96	**83.50**	**104**
1020	**Replace lightning ground rod**										
	Excavate hole for ground rod	25	2 ELEC	Ea.	.333		20		20	24.50	31
	Ground rod 5/8", 10'				1.739	23.50	104		127.50	155	190
	Ground clamp				.250	4.96	15		9.96	24	29
	Backfill over top of ground rod				.157		9.40		9.40	11.60	14.55
	Total				2.479	28.46	148.40		176.86	**215.10**	**264.55**

D5093 150 Computer Ground System

	System Description	Freq. (Years)	Crew	Unit	Labor Hours	2019 Bare Costs				Total In-House	Total w/O&P
						Material	Labor	Equipment	Total		
0010	**Maintenance and repair**										
	Test bonding each connection	4	1 ELEC	Ea.	.267		16		6	19.75	24.50
	Total				.267		16		16	**19.75**	**24.50**

For customer support on your Facilities Maintenance & Repair Costs with RSMeans data, call 800.448.8182.

D5093 150 Computer Ground System

System Description	Freq. (Years)	Crew	Unit	Labor Hours	2019 Bare Costs				Total In-House	Total w/O&P
					Material	Labor	Equipment	Total		
0020 Replace computer ground system	50	1 ELEC	M.L.F.							
Remove ground wire #6				2.000		120		120	148	185
Insulated ground wire #6				12.308	485	740		1,225	1,450	1,750
Grounding connection				2.500	49.60	150		199.60	240	292
Total				16.808	534.60	1,010		1,544.60	**1,838**	**2,227**

D5093 190 Special Ground System

System Description	Freq. (Years)	Crew	Unit	Labor Hours	2019 Bare Costs				Total In-House	Total w/O&P
					Material	Labor	Equipment	Total		
0010 Maintenance and repair	4	1 ELEC	Ea.							
Test bonding each connection				.267		16		16	19.75	24.50
Total				.267		16		16	**19.75**	**24.50**
0020 Replace special ground system	50	2 ELEC	M.L.F.							
Remove ground wire #6				2.000		120		120	148	185
Insulated ground wire #6				12.308	485	740		1,225	1,450	1,750
Grounding connection				2.500	49.60	150		199.60	240	292
Total				16.808	534.60	1,010		1,544.60	**1,838**	**2,227**

D5093 210 Generator, Gasoline, 175 KW

System Description	Freq. (Years)	Crew	Unit	Labor Hours	2019 Bare Costs				Total In-House	Total w/O&P
					Material	Labor	Equipment	Total		
0010 Maintenance and inspection	0.08	1 ELEC	Ea.							
Run test gasoline generator				.800		48		48	59	74
Total				.800		48		48	**59**	**74**
0020 Replace generator component	25	2 ELEC	Ea.							
Remove gasoline generator				36.364		2,175		2,175	2,700	3,375
Gasoline generator, 175 KW				64.000	84,000	3,850		87,850	97,000	111,000
Total				100.364	84,000	6,025		90,025	**99,700**	**114,375**

For customer support on your Facilities Maintenance & Repair Costs with RSMeans data, call 800.448.8182.

345

D5093 220 Generator, Diesel, 750 KW

					2019 Bare Costs				Total In-House	Total w/O&P
System Description	Freq. (Years)	Crew	Unit	Labor Hours	Material	Labor	Equipment	Total		
0010 Maintenance and inspection	0.08	1 ELEC	Ea.							
Run test diesel generator				.800		48		48	59	74
Total				.800		48		48	**59**	**74**
0020 Replace diesel generator component	25	4 ELEC	Ea.							
Remove diesel generator				57.143		3,425		3,425	4,225	5,300
Diesel generator, 750 KW				160.000	148,000	9,600		157,600	174,500	200,000
Total				217.143	148,000	13,025		161,025	178,725	205,300

D5093 230 Transfer Switch

					2019 Bare Costs				Total In-House	Total w/O&P
System Description	Freq. (Years)	Crew	Unit	Labor Hours	Material	Labor	Equipment	Total		
0010 Maintenance and repair	5	1 ELEC	Ea.							
Remove transfer switch coil				.667		40		40	49.50	61.50
Transfer switch coil				2.000	112	120		232	271	325
Total				2.667	112	160		272	**320.50**	**386.50**
0020 Maintenance and inspection	0.50	1 ELEC	Ea.							
Inspect and clean transfer switch				.500		30		30	37	46.50
Total				.500		30		30	**37**	**46.50**
0030 Replace transfer switch	18	1 ELEC	Ea.							
Remove transfer switch				.667		40		40	49.50	61.50
Transfer switch, 480 V, 3 P, 1200 A				22.857	23,100	1,375		24,475	27,100	31,000
Cut, form, align & connect wires				.151		9.05		9.05	11.15	14
Check operation of switch				.018		1.10		1.10	1.35	1.70
Total				23.693	23,100	1,425.15		24,525.15	27,162	31,077.20

For customer support on your Facilities Maintenance & Repair Costs with RSMeans data, call 800.448.8182.

D5093 240 Emergency Lighting Fixture

System Description	Freq. (Years)	Crew	Unit	Labor Hours	2019 Bare Costs				Total In-House	Total w/O&P
					Material	Labor	Equipment	Total		
0020 Replace lamp	2	1 ELEC	Ea.							
Remove emergency remote lamp				.100		6		6	7.40	9.25
Emergency remote lamp				.296	29.50	17.80		47.30	54.50	64.50
Total				.396	29.50	23.80		53.30	**61.90**	**73.75**
0030 Replace fixture	20	1 ELEC	Ea.							
Turn branch circuit off and on				.018		1.10		1.10	1.35	1.70
Remove emergency light units				.667		40		40	49.50	61.50
Emergency light units, battery operated				2.000	345	120		465	530	615
Total				2.685	345	161.10		506.10	**580.85**	**678.20**

D5093 250 Exit Light

System Description	Freq. (Years)	Crew	Unit	Labor Hours	2019 Bare Costs				Total In-House	Total w/O&P
					Material	Labor	Equipment	Total		
0010 Maintenance and repair	20	1 ELEC	Ea.							
Remove exit light face plate				.100		6		6	7.40	9.25
Exit light face plate				.296	5.10	17.80		22.90	27.50	34
Total				.396	5.10	23.80		28.90	**34.90**	**43.25**
0020 Replace lamp	5	1 ELEC	Ea.							
Remove exit light lamp				.022		1.32		1.32	1.63	2.04
Exit light lamp				.065	6.65	3.91		10.56	12.15	14.30
Total				.087	6.65	5.23		11.88	**13.78**	**16.34**
0030 Replace lighting fixture	20	1 ELEC	Ea.							
Turn circuit off and on for fixture				.018		1.10		1.10	1.35	1.70
Remove exit light				.333		20		20	24.50	31
Exit light				1.000	72.50	60		132.50	154	183
Total				1.352	72.50	81.10		153.60	**179.85**	**215.70**

For customer support on your Facilities Maintenance & Repair Costs with RSMeans data, call 800.448.8182.

347

D5093 255 | Exit Light L.E.D.

System Description	Freq. (Years)	Crew	Unit	Labor Hours	2019 Bare Costs				Total In-House	Total w/O&P
					Material	Labor	Equipment	Total		
0020 Replace lamp with exit light L.E.D. retrofit kits	15	1 ELEC	Ea.							
Remove exit light lamp				.022		1.32		1.32	1.63	2.04
Exit light L.E.D. retrofit kits				.133	50.50	8		58.50	65.50	75.50
Total				.155	50.50	9.32		59.82	**67.13**	**77.54**
0030 Replace lighting fixture with exit light L.E.D. std.	20	1 ELEC	Ea.							
Turn circuit off and on for fixture				.018		1.10		1.10	1.35	1.70
Remove exit light				.333		20		20	24.50	31
Exit light, L.E.D. standard, double face				1.194	53	71.50		124.50	147	177
Total				1.546	53	92.60		145.60	**172.85**	**209.70**
0040 Replace lighting fixture with exit light L.E.D. w/battery unit	20	1 ELEC	Ea.							
Turn circuit off and on for fixture				.018		1.10		1.10	1.35	1.70
Remove exit light				.333		20		20	24.50	31
Exit light, L.E.D. with battery unit, double face				2.000	217	120		337	385	455
Total				2.352	217	141.10		358.10	**410.85**	**487.70**

D5093 260 | Battery, Wet

System Description	Freq. (Years)	Crew	Unit	Labor Hours	2019 Bare Costs				Total In-House	Total w/O&P
					Material	Labor	Equipment	Total		
0010 Maintenance and inspection	0.02	1 ELEC	Ea.							
Clean & maintain wet cell battery				.500		30		30	37	46.50
Total				.500		30		30	**37**	**46.50**
0020 Replace battery	10	1 ELEC	Ea.							
Remove wet cell battery				.100		6		6	7.40	9.25
Wet cell battery				.296	630	17.80		647.80	715	815
Total				.396	630	23.80		653.80	**722.40**	**824.25**

D5093 265 Battery, Dry

	System Description	Freq. (Years)	Crew	Unit	Labor Hours	2019 Bare Costs				Total In-House	Total w/O&P
						Material	Labor	Equipment	Total		
0010	**Maintenance and inspection**	0.08	1 ELEC	Ea.	.500		30		30	37	46.50
	Clean & maintain dry cell battery										
	Total				.500		30		30	**37**	**46.50**
0020	**Replace battery**	5	1 ELEC	Ea.	.100		6		6	7.40	9.25
	Remove dry cell battery										
	Dry cell battery				.296	152	17.80		169.80	189	218
	Total				.396	152	23.80		175.80	**196.40**	**227.25**

D5093 270 Battery Charger

	System Description	Freq. (Years)	Crew	Unit	Labor Hours	2019 Bare Costs				Total In-House	Total w/O&P
						Material	Labor	Equipment	Total		
0010	**Maintenance and repair**	2	1 ELEC	Ea.	3.077	46	185		231	278	345
	Repair battery charger component										
	Total				3.077	46	185		231	**278**	**345**
0020	**Maintenance and inspection**	0.25	1 ELEC	Ea.	1.600		96		96	118	148
	Check battery charger										
	Total				1.600		96		96	**118**	**148**
0030	**Replace charger**	20	1 ELEC	Ea.	.615		37		37	45.50	57
	Remove battery charger										
	Battery charger				1.600	785	96		881	980	1,125
	Total				2.215	785	133		918	**1,025.50**	**1,182**

D5093 280 UPS Battery

	System Description	Freq. (Years)	Crew	Unit	Labor Hours	2019 Bare Costs				Total In-House	Total w/O&P
						Material	Labor	Equipment	Total		
0020	**Maintenance and inspection**	0.17	1 ELEC	Ea.	.800		48		48	59	74
	UPS battery electrical test										
	Total				.800		48		48	**59**	**74**

For customer support on your Facilities Maintenance & Repair Costs with RSMeans data, call 800.448.8182.

349

D50 ELECTRICAL | D5093 | Other Electrical Systems

D5093 280 | UPS Battery

	System Description	Freq. (Years)	Crew	Unit	Labor Hours	2019 Bare Costs				Total In-House	Total w/O&P
						Material	Labor	Equipment	Total		
0030	Replace motor generator UPS battery	15	1 ELEC	Ea.							
	Remove UPS battery				1.333		80		80	98.50	123
	UPS battery				4.000	505	240		745	850	1,000
	Total				5.333	505	320		825	948.50	1,123

D5093 920 | Communications Components

	System Description	Freq. (Years)	Crew	Unit	Labor Hours	2019 Bare Costs				Total In-House	Total w/O&P
						Material	Labor	Equipment	Total		
0020	Maintenance and repair voice/data outlet	10	1 ELEC	Ea.							
	Repair voice/data outlet				.615	2.84	37		39.84	48.50	60.50
	Total				.615	2.84	37		39.84	48.50	60.50
0030	Replace voice/data outlet	20	1 ELEC	Ea.							
	Remove voice/data outlet				.057		3.43		3.43	4.23	5.30
	Voice/data outlet, single opening, excludes voice/data device				.167	6.50	10		16.50	19.50	23.50
	Total				.224	6.50	13.43		19.93	23.73	28.80
0120	Maintenance and inspection patch panel	0.50	1 ELEC	Ea.							
	Check patch panel				1.067		64		64	79	99
	Total				1.067		64		64	79	99
0130	Replace patch panel	15	1 ELEC	Ea.							
	Remove patch panel				1.010		60.23		60.23	75	93
	Patch panel				4.000	315	240		555	645	765
	Test circuits				.996		59.86		59.86	73.50	92.50
	Total				6.006	315	360.09		675.09	793.50	950.50

For customer support on your Facilities Maintenance & Repair Costs with RSMeans data, call 800.448.8182.

E1023 710 | **Laboratory Equipment**

System Description	Freq. (Years)	Crew	Unit	Labor Hours	2019 Bare Costs				Total In-House	Total w/O&P
					Material	Labor	Equipment	Total		
0020										
Replace glove box gloves	5	1 SKWK	Ea.							
Laboratory equipment, remove bacteriological glove box gloves				.500		26.50		26.50	34.50	43
Lab equipment, replace fiberglass, bacteriological glove box gloves				1.000	211	53.50		264.50	300	350
Laboratory equipment, test replacement gloves				1.000		53.50		53.50	69	85.50
Total				2.500	211	133.50		344.50	403.50	478.50
0080										
Replace fume hood sash	20	2 SKWK	Ea.							
Laboratory equipment, remove fume hood sash				1.000		53.50		53.50	69	85.50
Laboratory equipment, replace fume hood sash				2.000	885	107		992	1,100	1,275
Total				3.000	885	160.50		1,045.50	1,169	1,360.50

E1033 110 | Automotive Equipment

	System Description	Freq. (Years)	Crew	Unit	Labor Hours	2019 Bare Costs				Total In-House	Total w/O&P
						Material	Labor	Equipment	Total		
0020	**Remove and replace vehicle lift hydraulic pump**	15	1 SKWK	Ea.	4.000	3,150	214		3,364	3,750	4,275
	Automotive hoist, two posts, adj. frame, remove and replace pump										
	Total				4.000	3,150	214		3,364	3,750	4,275
0100	**Remove and replace compressor, electric, 5 H.P.**	10	1 ELEC	Ea.	2.000	690	120		810	905	1,050
	Automotive compressors, remove and replace 5 H.P. motor										
	Total				2.000	690	120		810	905	1,050

For customer support on your Facilities Maintenance & Repair Costs with RSMeans data, call 800.448.8182.

E10 EQUIPMENT | E1093 | Other Equipment

E1093 310 | Loading Dock Equipment

	System Description	Freq. (Years)	Crew	Unit	Labor Hours	2019 Bare Costs				Total In-House	Total w/O&P
						Material	Labor	Equipment	Total		
0020	**Remove and replace hydraulic dock leveler lift cylinder** Loading dock, dock levelers, 10 ton cap., replace lift cylinder	15	2 SKWK	Ea.	4.000	6,850	214		7,064	7,800	8,900
	Total				4.000	6,850	214		7,064	**7,800**	**8,900**
0040	**Remove and replace hydraulic dock leveler hydraulic pump** Loading dock, dock levelers, 10 ton cap, replace lift cylinder	20	1 SKWK	Ea.	3.077	1,250	164		1,414	1,575	1,825
	Total				3.077	1,250	164		1,414	**1,575**	**1,825**

E1093 315 | Dishwasher

	System Description	Freq. (Years)	Crew	Unit	Labor Hours	2019 Bare Costs				Total In-House	Total w/O&P
						Material	Labor	Equipment	Total		
0020	**Replace commercial dishwasher, 10 to 12 racks per hour** Cleaning and disposal, dishwasher, commercial, rack type Install new dishwasher	10	2 PLUM	Ea.	12.000 6.499	3,325	745 370		745 3,695	925 4,125	1,150 4,725
	Total				18.499	3,325	1,115		4,440	**5,050**	**5,875**
0040	**Remove and replace dishwasher pump** Laboratory equip., glassware washer, remove and replace pump	15	1 PLUM	Ea.	1.778	805	112		917	1,025	1,175
	Total				1.778	805	112		917	**1,025**	**1,175**
0080	**Replace commercial dishwasher, to 375 racks per hour** Commercial dishwasher, to 275 racks per hour, selective demo Dishwasher, automatic, 235 to 275 racks/hr.	20	L-4	Ea.	76.000 24.000 60.000	28,900	1,175 3,725		1,175 32,625	1,525 36,400	1,875 41,900
	Total				84.000	28,900	4,900		33,800	**37,925**	**43,775**

E1093 316 | Waste Disposal, Residential

	System Description	Freq. (Years)	Crew	Unit	Labor Hours	2019 Bare Costs				Total In-House	Total w/O&P
						Material	Labor	Equipment	Total		
0010	**Unstop disposal** Unstop clogged unit	1	1 PLUM	Ea.	.889		56		56	69.50	87
	Total				.889		56		56	**69.50**	**87**

For customer support on your Facilities Maintenance & Repair Costs with RSMeans data, call 800.448.8182.

353

E10 EQUIPMENT | E1093 | Other Equipment

E1093 316 | Waste Disposal, Residential

System Description	Freq. (Years)	Crew	Unit	Labor Hours	2019 Bare Costs				Total In-House	Total w/O&P
					Material	Labor	Equipment	Total		
0020 **Replace waste disposal unit, residential**	8	1 PLUM	Ea.							
Remove unit				1.040		64		64	79	99
Install new disposal				2.080	208	128		336	385	460
Total				3.120	208	192		400	**464**	**559**

E1093 320 | Waste Handling Equipment

System Description	Freq. (Years)	Crew	Unit	Labor Hours	2019 Bare Costs				Total In-House	Total w/O&P
					Material	Labor	Equipment	Total		
0020 **Remove and replace waste compactor hydraulic cylinder**	15	2 SKWK	Ea.							
Replace ram, 5 C.Y. capacity				4.000	5,700	214		5,914	6,550	7,475
Total				4.000	5,700	214		5,914	**6,550**	**7,475**
0040 **Remove and replace waste compactor hydraulic pump**	20	1 SKWK	Ea.							
Replace hydraulic pump, 5 C.Y. capacity				3.200	1,025	171		1,196	1,350	1,550
Total				3.200	1,025	171		1,196	**1,350**	**1,550**

E1093 510 | Darkroom Dryer

System Description	Freq. (Years)	Crew	Unit	Labor Hours	2019 Bare Costs				Total In-House	Total w/O&P
					Material	Labor	Equipment	Total		
0020 **Remove and replace darkroom dryer fan**	10	1 SKWK	Ea.							
48" x 25" x 68" high				1.600	168	85.50		253.50	295	345
Total				1.600	168	85.50		253.50	**295**	**345**
0040 **Remove and replace darkroom dryer heating element**	15	1 SKWK	Ea.							
Darkroom equip., dryers, dehumidifier, replace compressor				1.333	89	71		160	190	225
Total				1.333	89	71		160	**190**	**225**

For customer support on your Facilities Maintenance & Repair Costs with RSMeans data, call 800.448.8182.

E1093 610 Dust Collector

System Description	Freq. (Years)	Crew	Unit	Labor Hours	2019 Bare Costs				Total In-House	Total w/O&P
					Material	Labor	Equipment	Total		
0100 Remove and replace 20″ dia dust collector bag	5	1 SKWK	Ea.							
Dust collector bag, 20″ diameter				1.600	585	97.50		682.50	765	885
Total				1.600	585	97.50		682.50	765	885

E1093 910 Pump Systems

System Description	Freq. (Years)	Crew	Unit	Labor Hours	2019 Bare Costs				Total In-House	Total w/O&P
					Material	Labor	Equipment	Total		
0500 Remove and replace 50 HP pump motor	25	2 ELEC	Ea.							
Electrical dml, motors, 230/460 V, 60 Hz, 50 HP				1.667		100		100	123	154
Install 50 HP motor, 230/460 V, totly encld, 1800 RPM				8.000	4,475	480		4,955	5,525	6,325
Total				9.667	4,475	580		5,055	5,648	6,479

For customer support on your Facilities Maintenance & Repair Costs with RSMeans data, call 800.448.8182.

355

G1023 210 Underground Storage Tank Removal

System Description	Freq. (Years)	Crew	Unit	Labor Hours	2019 Bare Costs				Total In-House	Total w/O&P
					Material	Labor	Equipment	Total		
1010										
Remove 500 gal. underground storage tank(non-leaking)	20	B-34P	Ea.							
Saw cut asphalt				2.700	9.50	131	77	217.50	263	305
Disconnect and remove piping				1.500		94.80		94.80	118	147
Transfer liquids, 10% of volume				.250		16		16	19.50	24.50
Insert solid carbon dioxide, 1.5 lbs./100 gal.				.150	9.68	6.15		15.83	18.70	22
Cut accessway to inside of tank				1.501		61.50		61.50	80.50	99.50
Remove sludge, wash and wipe tank				1.000		63		63	78.50	98
Properly dispose of sludge/water					36.50			36.50	40	45.50
Excavate, pull, & load tank, backfill hole				16.000		770	320	1,090	1,325	1,575
Select structural backfill					160			160	176	200
Haul tank to certified dump, 100 miles round trip				8.000		415	195	610	735	865
Total				31.101	215.68	1,557.45	592	2,365.13	2,854.20	3,381.50
1020										
Remove 3000 gal. underground storage tank(non-leaking)	20	B-34R	Ea.							
Saw cut asphalt				5.400	19	262	154	435	525	610
Disconnect and remove piping				2.500		158		158	196	245
Transfer liquids, 10% of volume				1.500		96		96	117	147
Insert solid carbon dioxide, 1.5 lbs./100 gal.				.900	58.05	36.90		94.95	112	132
Cut accessway to inside of tank				1.501		61.50		61.50	80.50	99.50
Remove sludge, wash and wipe tank				1.199		75.50		75.50	94	117
Properly dispose of sludge/water					328.50			328.50	360	410
Excavate, pull, & load tank, backfill hole				32.000		1,550	775	2,325	2,825	3,300
Select structural backfill					600			600	660	750
Haul tank to certified dump, 100 miles round trip				8.000		415	246	661	790	920
Total				53.000	1,005.55	2,654.90	1,175	4,835.45	5,759.50	6,730.50
1030										
Remove 5000 gal. underground storage tank(non-leaking)	20	B-34R	Ea.							
Saw cut asphalt				5.940	20.90	288.20	169.40	478.50	580	670
Disconnect and remove piping				2.500		158		158	196	245
Transfer liquids, 10% of volume				2.500		160		160	195	245
Insert solid carbon dioxide, 1.5 lbs./100 gal.				1.500	96.75	61.50		158.25	187	220
Cut accessway to inside of tank				1.501		61.50		61.50	80.50	99.50
Remove sludge, wash and wipe tank				1.301		82		82	102	127
Properly dispose of sludge/water					547.50			547.50	600	685
Excavate, pull, & load tank, backfill hole				32.000		1,550	775	2,325	2,825	3,300
Select structural backfill					1,000			1,000	1,100	1,250
Haul tank to certified dump, 100 miles round trip				8.000		415	246	661	790	920
Total				55.242	1,665.15	2,776.20	1,190.40	5,631.75	6,655.50	7,761.50

For customer support on your Facilities Maintenance & Repair Costs with RSMeans data, call 800.448.8182.

System Description	Freq. (Years)	Crew	Unit	Labor Hours	2019 Bare Costs				Total In-House	Total w/O&P
					Material	Labor	Equipment	Total		
1040 Remove 8000 gal. underground storage tank(non-leaking)	20	B-34R	Ea.							
Saw cut asphalt				6.480	22.80	314.40	184.80	522	630	730
Disconnect and remove piping				2.500		158		158	196	245
Transfer liquids, 10% of volume				4.000		256		256	310	390
Insert solid carbon dioxide, 1.5 lbs./100 gal.				2.400	154.80	98.40		253.20	299	350
Cut accessway to inside of tank				1.501		61.50		61.50	80.50	99.50
Remove sludge, wash and wipe tank				1.501		95		95	118	147
Properly dispose of sludge/water					876			876	965	1,100
Excavate, pull, & load tank, backfill hole				32.000		1,575	2,150	3,725	4,375	4,875
Select structural backfill					1,400			1,400	1,550	1,750
Haul tank to certified dump, 100 miles round trip				8.000		375	845	1,220	1,425	1,525
Total				58.382	2,453.60	2,933.30	3,179.80	8,566.70	9,948.50	11,211.50
1050 Remove 10000 gal. ugnd. storage tank(non-leaking)	20	B-34S	Ea.							
Saw cut asphalt				10.800	38	524	308	870	1,050	1,225
Disconnect and remove piping				3.750		237		237	294	370
Transfer liquids, 10% of volume				5.000		320		320	390	490
Insert solid carbon dioxide, 1.5 lbs./100 gal.				3.000	193.50	123		316.50	375	440
Cut accessway to inside of tank				1.501		61.50		61.50	80.50	99.50
Remove sludge, wash and wipe tank				1.751		111		111	137	171
Properly dispose of sludge/water					1,095			1,095	1,200	1,375
Excavate, pull, & load tank, backfill hole				32.000		1,575	2,150	3,725	4,375	4,875
Select structural backfill					1,800			1,800	1,975	2,250
Haul tank to certified dump, 100 miles round trip				8.000		375	845	1,220	1,425	1,525
Total				65.801	3,126.50	3,326.50	3,303	9,756	11,301.50	12,820.50
1060 Remove 12000 gal. ugnd. storage tank(non-leaking)	20	B-34S	Ea.							
Saw cut asphalt				12.150	42.75	589.50	346.50	978.75	1,175	1,375
Disconnect and remove piping				5.000		316		316	390	490
Transfer liquids, 10% of volume				6.000		384		384	470	590
Insert solid carbon dioxide, 1.5 lbs./100 gal.				3.600	232.20	147.60		379.80	450	530
Cut accessway to inside of tank				1.501		61.50		61.50	80.50	99.50
Remove sludge, wash and wipe tank				1.900		120		120	149	186
Properly dispose of sludge/water					1,314			1,314	1,450	1,650
Excavate, pull, & load tank, backfill hole				32.000		1,575	2,150	3,725	4,375	4,875
Select structural backfill					2,000			2,000	2,200	2,500
Haul tank to certified dump, 100 miles round trip				8.000		375	845	1,220	1,425	1,525
Total				70.151	3,588.95	3,568.60	3,341.50	10,499.05	12,164.50	13,820.50

For customer support on your Facilities Maintenance & Repair Costs with RSMeans data, call 800.448.8182.

357

G2023 210 Parking Lot Repairs

System Description	Freq. (Years)	Crew	Unit	Labor Hours	2019 Bare Costs				Total In-House	Total w/O&P
					Material	Labor	Equipment	Total		
1010 Parking lot repair and sealcoating	5	B-1	M.S.F.							
Thoroughly clean surface				.312		13.32	2.22	15.54	21	24.50
Patch holes				.240	23.22	10.13	.95	34.30	40	46.50
Fill cracks				.715	79.50	30.50	9	119	137	158
Install 2 coat petroleum resistant emulsion				3.330	157.62	138.75		296.37	355	420
Restripe lot				.291	19.40	12.24	3.08	34.72	40.50	47.50
Total				4.887	279.74	204.94	15.25	499.93	**593.50**	**696.50**
1020 Parking lot repair and resurface	10	B-25B	M.S.F.							
Thoroughly clean surface				.312		13.32	2.22	15.54	21	24.50
Patch holes				.240	23.22	10.13	.95	34.30	40	46.50
Fill cracks				.715	79.50	30.50	9	119	137	158
Emulsion tack coat, .05 gal. per S.Y.				.710	37.74	36.63	35.52	109.89	128	145
Install 1" thick asphaltic concrete wearing course				1.008	384.06	46.62	29.97	460.65	515	590
Restripe lot				.291	19.40	12.24	3.08	34.72	40.50	47.50
Total				3.276	543.92	149.44	80.74	774.10	**881.50**	**1,011.50**

G2023 310 General

System Description	Freq. (Years)	Crew	Unit	Labor Hours	2019 Bare Costs				Total In-House	Total w/O&P
					Material	Labor	Equipment	Total		
1000 Remove and replace steel guard rail	7	B-80	L.F.							
Remove old metal guide/guard rail and posts				.213		9.65	4.21	13.86	17.10	20
Install metal guide/guard rail including posts 6'3"				.107	26	4.83	2.11	32.94	37	42.50
Total				.320	26	14.48	6.32	46.80	**54.10**	**62.50**
1100 Raise MH or catch basin frame and cover	10	D-1	Ea.							
Demolish pavement around frame				4.800		207	23.50	230.50	294	355
Raise MH or catch basin frame with brick				1.332	8.82	60.94		69.76	90	110
Total				6.132	8.82	267.94	23.50	300.26	**384**	**465**

For customer support on your Facilities Maintenance & Repair Costs with RSMeans data, call 800.448.8182.

G2033 130 | Asphalt Sidewalk & Curb

	System Description	Freq. (Years)	Crew	Unit	Labor Hours	2019 Bare Costs				Total In-House	Total w/O&P
						Material	Labor	Equipment	Total		
0100	**Remove & replace asphalt sidewalk, 4' wide**	15	B-37	L.F.							
	Remove existing asphalt pavement				.080		3.62	1.65	5.27	6.45	7.60
	Bank run gravel, 3" avg.				.002	.63	.11	.19	.93	1.04	1.16
	Vibratory plate compaction				.003		.11	.01	.12	.15	.18
	Fine grade gravel base				.026		1.25	1.07	2.32	2.78	3.17
	Place asphalt sidewalk, 2-1/2" thick				.027	2.40	1.15	.09	3.64	4.23	4.95
	Total				.138	3.03	6.24	3.01	12.28	14.65	17.06
1000	**Remove and replace asphalt curb or berm**	10	B-37	L.F.							
	Demolish existing curb or berm				.096		4.14	.47	4.61	5.90	7.15
	Remove broken curb				.032		1.45	.66	2.11	2.59	3.04
	Asphalt curb, 8" wide, 6" high machine formed				.042	.05	1.73	.42	2.20	2.77	3.30
	Total				.170	.05	7.32	1.55	8.92	11.26	13.49

G2033 140 | Concrete Sidewalk & Curb

	System Description	Freq. (Years)	Crew	Unit	Labor Hours	2019 Bare Costs				Total In-House	Total w/O&P
						Material	Labor	Equipment	Total		
0100	**Remove and replace concrete sidewalk, 4' wide**	25	B-37	L.F.							
	Remove existing concrete pavement				.107		4.82	2.20	7.02	8.65	10.10
	Bank run gravel, 3" avg.				.002	.63	.11	.19	.93	1.04	1.16
	Vibratory plate compaction				.003		.11	.01	.12	.15	.18
	Fine grade gravel				.026		1.25	1.07	2.32	2.78	3.17
	Broom finished concrete sidewalk, 4" thick				.209	8.24	9.84		18.08	22	26
	Total				.346	8.87	16.13	3.47	28.47	34.62	40.61
2000	**Remove and replace concrete curb or berm**	25	B-37	L.F.							
	Demolish existing curb or berm				.240		10.35	1.17	11.52	14.70	17.90
	Remove broken curb				.080		3.62	1.65	5.27	6.45	7.60
	Concrete curb, 6" x 18", steel forms				.091	7.25	4.55		11.80	13.90	16.40
	Total				.411	7.25	18.52	2.82	28.59	35.05	41.90

G2033 150 Patios

	System Description	Freq. (Years)	Crew	Unit	Labor Hours	2019 Bare Costs Material	Labor	Equipment	Total	Total In-House	Total w/O&P
1030	**Refinish Concrete Patio**	3	1 PORD	S.F.							
	Prepare surface				.003		.14	.03	.17	.21	.25
	Paint 1 coat, roller & brush				.008	.08	.36		.44	.55	.67
	Total				.012	.08	.50	.03	.61	**.76**	**.92**
2030	**Refinish Masonry Patio**	3	1 PORD	S.F.							
	Prepare surface				.003		.14	.03	.17	.21	.25
	Apply 1 coat of sealer, roller & brush				.005	.28	.21		.49	.58	.68
	Total				.008	.28	.35	.03	.66	**.79**	**.93**
3030	**Refinish Stone Patio**	3	1 PORD	S.F.							
	Prepare surface				.003		.14	.03	.17	.21	.25
	Apply 1 coat of sealer, roller & brush				.005	.28	.21		.49	.58	.68
	Total				.008	.28	.35	.03	.66	**.79**	**.93**
4030	**Refinish Wood Patio**	3	1 PORD	S.F.							
	Prepare surface				.015		.64		.64	.82	1.02
	Primer + 1 top coat, roller & brush				.022	.14	.96		1.10	1.39	1.72
	Total				.037	.14	1.60		1.74	**2.21**	**2.74**

G2033 250 Handicap Ramp

	System Description	Freq. (Years)	Crew	Unit	Labor Hours	2019 Bare Costs Material	Labor	Equipment	Total	Total In-House	Total w/O&P
1030	**Refinish Metal Handicap Ramp**	3	1 PORD	S.F.							
	Prepare surface				.015		.64		.64	.82	1.02
	Prime & paint 1 coat, roller & brush				.015	.31	.63		.94	1.15	1.40
	Total				.029	.31	1.27		1.58	**1.97**	**2.42**
2030	**Refinish Wood Handicap Ramp**	3	1 PORD	S.F.							
	Prepare surface				.015		.64		.64	.82	1.02
	Prime & paint 1 coat, roller & brush				.022	.14	.96		1.10	1.39	1.72
	Total				.037	.14	1.60		1.74	**2.21**	**2.74**

For customer support on your Facilities Maintenance & Repair Costs with RSMeans data, call 800.448.8182.

G2043 105 Chain Link Fence and Gate Repairs

ID	System Description	Freq. (Years)	Crew	Unit	Labor Hours	2019 Bare Costs				Total In-House	Total w/O&P
						Material	Labor	Equipment	Total		
1010	**Minor chain link fence repairs (per 10 L.F.)**	1	2 CLAB	Ea.							
	Straighten bent 2" line post				.167		6.85		6.85	8.90	11.05
	Re-tie fence fabric				.167	1.79	6.85		8.64	10.90	13.25
	Refasten loose barbed wire arm				.083		3.42		3.42	4.46	5.50
	Refasten loose barbed wire, 3 strands				.083		3.42		3.42	4.46	5.50
	Total				.500	1.79	20.54		22.33	28.72	35.30
1110	**Replace bent 1-5/8" top rail (per 20 L.F.)**	2	2 CLAB	Ea.							
	Remove fabric ties, top rail & couplings				.333		13.70		13.70	17.85	22
	Install new top rail, couplings & fabric ties				.667	39.27	27.39		66.66	79	93
	Total				1.000	39.27	41.09		80.36	96.85	115
1120	**Replace broken barbed wire arm**	2	2 CLAB	Ea.							
	Remove fabric ties, top rail & couplings				.333		13.70		13.70	17.85	22
	Release barbed wires & remove old arm				.067		2.74		2.74	3.57	4.41
	Install new arm & refasten barbed wires				.167	.54	6.87		7.41	9.55	11.75
	Install old top rail, couplings & fabric ties				.667		27.40		27.40	35.50	44
	Total				1.234	.54	50.71		51.25	66.47	82.16
1130	**Replace barbed wire, 3 strands (per 100 L.F.)**	5	2 CLAB	Ea.							
	Remove old barbed wire				3.002		123		123	161	199
	Install new barbed wire				3.000	107.10	123		230.10	278	330
	Total				6.002	107.10	246		353.10	439	529
1140	**Replace fence fabric, 6' high, 9 ga. (per 10 L.F.)**	10	2 CLAB	Ea.							
	Remove ties from top rail, bottom tension wire & posts				.167		6.85		6.85	8.90	11.05
	Unstitch & remove section of fabric				.667		27.50		27.50	35.50	44
	Hang & stretch new fabric, stitch into existing fabric				1.333	71.40	54.80		126.20	150	177
	Re-tie fence fabric				.167	1.79	6.85		8.64	10.90	13.25
	Total				2.333	73.19	96		169.19	205.30	245.30
1150	**Replace double swing gates, 6' high, 20' opng.**	5	2 CLAB	Opng.							
	Remove 2 old gates (6'x10' ea.) & 4 hinges				1.111		45.50		45.50	59.50	73.50
	Supply 4 new gate hinges					57.53			57.53	63.50	72
	Install 2 new gates (6'x10' ea.), 4 hinges, adjust alignment				2.412	528.35	99.10		627.45	710	820
	Remove & replace gate latch				.914	14.16	37.60		51.76	64.50	78
	Remove & replace gate cane bolt				.883	14.16	36.40		50.56	63	76
	Total				5.320	614.20	218.60		832.80	960.50	1,119.50

For customer support on your Facilities Maintenance & Repair Costs with RSMeans data, call 800.448.8182.

361

G2043 105 Chain Link Fence and Gate Repairs

	System Description	Freq. (Years)	Crew	Unit	Labor Hours	2019 Bare Costs				Total In-House	Total w/O&P
						Material	Labor	Equipment	Total		
1160	**Replace 6'x18' cantilever slide gate**	5	2 CLAB	Opng.							
	Remove old gate & 4 rollers				.926		38		38	49.50	61.50
	Supply 4 new cantilever gate rollers				.003	345	.05		345.05	380	430
	Install new 6'x18' cantilever slide gate, 4 rollers, adjust				2.148	1,716.44	88.31		1,804.75	2,000	2,300
	Remove & replace gate latch				.914	14.16	37.60		51.76	64.50	78
	Total				3.991	2,075.60	163.96		2,239.56	2,494	2,869.50
1210	**Replace 2″ line post**	20	B-55	Ea.							
	Remove fabric ties, top rail & couplings				.333		13.70		13.70	17.85	22
	Release barbed wires & remove old arm				.067		2.74		2.74	3.57	4.41
	Unstitch & roll up section of fabric				1.000		42.50	42	84.50	101	115
	Remove 2″ post & concrete base from ground				.750		32	31.50	63.50	76	86
	Install new 2″ line post with concrete in old hole				1.500	62.48	63.88	62.65	189.01	221	250
	Install old top rail, couplings & new fabric ties				.667		27.40		27.40	35.50	44
	Unroll & stretch fabric, stitch into existing fabric				2.000		85	83.50	168.50	203	229
	Re-tie fence fabric				.333	3.57	13.70		17.27	22	26.50
	Total				6.650	66.05	280.92	219.65	566.62	679.92	776.91
1220	**Replace 3″ corner post**	10	B-55	Ea.							
	Remove fabric ties, retainer bars & ring fittings				.500		21.50	21	42.50	50.50	57
	Roll back fabric				.250		10.65	10.45	21.10	25.50	28.50
	Remove top rail & ring fittings				.500		21.50	21	42.50	50.50	57
	Remove barbed wires & ring fittings				.750		32	31.50	63.50	76	86
	Remove diagonal & horizontal braces, ring fittings				.500		21.50	21	42.50	50.50	57
	Remove 3″ post & concrete base from ground				1.000		42.50	42	84.50	101	115
	Install new 3″ post with concrete in old hole				2.000	89.50	85	83.50	258	300	340
	Install old diagonal & horizontal braces, ring fittings				.500		21.50	21	42.50	50.50	57
	Install old barbed wires & ring fittings				.750		32	31.50	63.50	76	86
	Install old top rail & ring fittings				.500		21.50	21	42.50	50.50	57
	Hang & stretch fabric, insert retainer bars & ring fittings				1.500		64	62.50	126.50	152	172
	Re-tie fence fabric				.333	3.57	13.70		17.27	22	26.50
	Total				9.083	93.07	387.35	366.45	846.87	1,005	1,139

For customer support on your Facilities Maintenance & Repair Costs with RSMeans data, call 800.448.8182.

G20 SITE IMPROVEMENTS   G2043   Site Development

G2043 105   Chain Link Fence and Gate Repairs

	System Description	Freq. (Years)	Crew	Unit	Labor Hours	2019 Bare Costs Material	Labor	Equipment	Total	Total In-House	Total w/O&P
1230	**Replace 3" gate post**	5	B-55	Ea.							
	Remove gate & hinges				.833		35.50	35	70.50	84.50	95.50
	Remove fabric ties, retainer bars & ring fittings				.500		21.50	21	42.50	50.50	57
	Roll back fabric				.250		10.65	10.45	21.10	25.50	28.50
	Remove top rail & ring fittings				.500		21.50	21	42.50	50.50	57
	Remove barbed wires & ring fittings				.750		32	31.50	63.50	76	86
	Remove diagonal & horizontal braces, ring fittings				.500		21.50	21	42.50	50.50	57
	Remove 3" post & concrete base from ground				1.000		42.50	42	84.50	101	115
	Install new 3" post with concrete in old hole				2.000	89.50	85	83.50	258	300	340
	Install old diagonal & horizontal braces, ring fittings				.500		21.50	21	42.50	50.50	57
	Install old barbed wires & ring fittings				.750		32	31.50	63.50	76	86
	Install old top rail & ring fittings				.500		21.50	21	42.50	50.50	57
	Hang & stretch fabric, insert retainer bars & ring fittings				1.500		64	62.50	126.50	152	172
	Re-tie fence fabric				.333	3.57	13.70		17.27	22	26.50
	Install old hinges & gate				1.807		77	75.50	152.50	183	207
	Total				11.724	93.07	499.85	476.95	1,069.87	1,272.50	1,441.50

G2043 110   Wood Fence

	System Description	Freq. (Years)	Crew	Unit	Labor Hours	2019 Bare Costs Material	Labor	Equipment	Total	Total In-House	Total w/O&P
1020	**Refinish Wood Fence, 6' High**	3	1 PORD	L.F.							
	Prepare surface				.089		3.84		3.84	4.92	6.10
	Prime & 1 top coat, roller & brush				.046	.48	1.98		2.46	3.12	3.78
	Total				.135	.48	5.82		6.30	8.04	9.88

G2043 710   Bleachers, Exterior

	System Description	Freq. (Years)	Crew	Unit	Labor Hours	2019 Bare Costs Material	Labor	Equipment	Total	Total In-House	Total w/O&P
1020	**Refinish Wood Bleachers**	3	1 PORD	S.F.							
	Prepare surface				.015		.64		.64	.82	1.02
	Prime & 1 top coat, roller & brush				.022	.14	.96		1.10	1.39	1.72
	Total				.037	.14	1.60		1.74	2.21	2.74

G2043 750 Tennis Court Resurfacing

	System Description	Freq. (Years)	Crew	Unit	Labor Hours	2019 Bare Costs				Total In-House	Total w/O&P
						Material	Labor	Equipment	Total		
1010	**Resurface asphalt tennis court**	7	2 SKWK	Ea.							
	Remove/reinstall tennis net				1.000		41		41	53.50	66
	Thoroughly clean surface				7.760		320		320	415	510
	Install 3 coat acrylic emulsion sealcoat				16.000	5,760	656		6,416	7,200	8,250
	Total				24.760	5,760	1,017		6,777	7,668.50	8,826
1020	**Resurface cushioned asphalt tennis court**	7	2 SKWK	Ea.							
	Remove/reinstall tennis net				1.000		41		41	53.50	66
	Thoroughly clean surface				7.760		320		320	415	510
	Two cushion coats				16.000	2,575	655		3,230	3,700	4,275
	Install 3 coat acrylic emulsion sealcoat				16.000	5,760	656		6,416	7,200	8,250
	Total				40.760	8,335	1,672		10,007	11,368.50	13,101
2010	**Clay tennis court preparation**	1	2 SKWK	Ea.							
	Remove/reinstall tennis net				1.000		41		41	53.50	66
	Remove and reinstall tapes				6.015		247		247	320	400
	Remove weeds and deteriorated surface material				2.000		82		82	107	132
	Check court for level				1.000		41		41	53.50	66
	Clay court top dressing, 1.5 tons				7.500	667.50	307.50		975	1,125	1,325
	Clay court, roll after dressing				1.500		75.50	22.50	98	121	145
	Total				19.015	667.50	794	22.50	1,484	1,780	2,134

G2043 810 Flag Pole

	System Description	Freq. (Years)	Crew	Unit	Labor Hours	2019 Bare Costs				Total In-House	Total w/O&P
						Material	Labor	Equipment	Total		
1020	**Refinish 25' Wood Flag Pole**	5	1 PORD	Ea.							
	Set up, secure and take down ladder				.295		11.15		11.15	14.35	17.85
	Prepare surface				.258		16		16	20.50	25.50
	Prime & 1 top coat, roller & brush				.556	3.50	24		27.50	35	43
	Total				1.184	3.50	51.15		54.65	69.85	86.35

For customer support on your Facilities Maintenance & Repair Costs with RSMeans data, call 800.448.8182.

G30 SITE MECH. UTILITIES | G3013 | Water Supply

G3013 400 Ground Level Water Storage Tank

System Description	Freq. (Years)	Crew	Unit	Labor Hours	2019 Bare Costs				Total In-House	Total w/O&P
					Material	Labor	Equipment	Total		
0100 Prep & paint 100k gal. ground level water stor tank, 30' dia x 19' tall	10	E-11	Ea.							
Scaffolding, outer ring				72.000		3,720		3,720	4,850	6,000
Scaffolding planks				35.666		1,840.40		1,840.40	2,400	3,000
Sandblast exterior, near white blast (SSPC-SP10), loose scale, fine rust				177.775	2,500	8,100	1,125	11,725	14,900	17,900
Epoxy primer, spray exterior				13.325	625	600		1,225	1,550	1,800
Epoxy topcoat, 2 coats, spray exterior				28.550	1,400	1,300		2,700	3,350	3,950
Clean up				10.000		450		450	650	800
Total				337.316	4,525	16,010.40	1,125	21,660.40	27,700	33,450
0200 Prep & paint 250k gal. ground level water stor tank, 40' dia x 27' tall	10	E-11	Ea.							
Scaffolding, outer ring				126.000		6,510		6,510	8,475	10,500
Scaffolding planks				63.333		3,268		3,268	4,250	5,325
Sandblast exterior, near white blast (SSPC-SP10), loose scale, fine rust				327.106	4,600	14,904	2,070	21,574	27,500	32,900
Epoxy primer, spray exterior				24.518	1,150	1,104		2,254	2,850	3,325
Epoxy topcoat, 2 coats, spray exterior				52.532	2,576	2,392		4,968	6,175	7,275
Clean up				18.400		828		828	1,200	1,475
Total				611.889	8,326	29,006	2,070	39,402	50,450	60,800
0300 Prep & paint 500k gal. ground level water stor tank, 50' dia x 34' tall	10	E-11	Ea.							
Scaffolding, outer ring				192.000		9,920		9,920	12,900	16,000
Scaffolding planks				96.666		4,988		4,988	6,500	8,125
Sandblast exterior, near white blast (SSPC-SP10), loose scale, fine rust				519.103	7,300	23,652	3,285	34,237	43,600	52,000
Epoxy primer, spray exterior				38.909	1,825	1,752		3,577	4,525	5,275
Epoxy topcoat, 2 coats, spray exterior				83.366	4,088	3,796		7,884	9,775	11,500
Clean up				29.200		1,314		1,314	1,900	2,325
Total				959.244	13,213	45,422	3,285	61,920	79,200	95,225
0400 Prep & paint 750k gal. ground level water stor tank, 60' dia x 36' tall	10	E-11	Ea.							
Scaffolding, outer ring				234.000		12,090		12,090	15,700	19,500
Scaffolding planks				117.332		6,054.40		6,054.40	7,900	9,850
Sandblast exterior, near white blast (SSPC-SP10), loose scale, fine rust				675.545	9,500	30,780	4,275	44,555	56,500	68,000
Epoxy primer, spray exterior				50.635	2,375	2,280		4,655	5,900	6,875
Epoxy topcoat, 2 coats, spray exterior				108.490	5,320	4,940		10,260	12,700	15,000
Clean up				38.000		1,710		1,710	2,475	3,050
Total				1224.002	17,195	57,854.40	4,275	79,324.40	101,175	122,275

For customer support on your Facilities Maintenance & Repair Costs with RSMeans data, call 800.448.8182.

365

G30 SITE MECH. UTILITIES | G3013 | Water Supply

G3013 400 | Ground Level Water Storage Tank

	System Description	Freq. (Years)	Crew	Unit	Labor Hours	2019 Bare Costs				Total In-House	Total w/O&P
						Material	Labor	Equipment	Total		
0500	**Prep & paint 1M gal. ground level water stor tank, 70' dia x 35' tall**	10	E-11	Ea.							
	Scaffolding, outer ring				261.000		13,485		13,485	17,600	21,800
	Scaffolding planks				131.332		6,776.80		6,776.80	8,825	11,000
	Sandblast exterior, near white blast (SSPC-SP10), loose scale, fine rust				817.765	11,500	37,260	5,175	53,935	68,500	82,000
	Epoxy primer, spray exterior				61.295	2,875	2,760		5,635	7,125	8,300
	Epoxy topcoat, 2 coats, spray exterior				131.330	6,440	5,980		12,420	15,400	18,200
	Clean up				46.000		2,070		2,070	3,000	3,675
	Total				1448.722	20,815	68,331.80	5,175	94,321.80	120,450	144,975
0600	**Prep & paint 2M gal. ground level water stor tank, 100' dia x 34' tall**	10	E-11	Ea.							
	Scaffolding, outer ring				354.000		18,290		18,290	23,800	29,500
	Scaffolding planks				176.998		9,133.20		9,133.20	11,900	14,900
	Sandblast exterior, near white blast (SSPC-SP10), loose scale, fine rust				1315.535	18,500	59,940	8,325	86,765	110,500	132,500
	Epoxy primer, spray exterior				98.605	4,625	4,440		9,065	11,500	13,400
	Epoxy topcoat, 2 coats, spray exterior				211.270	10,360	9,620		19,980	24,800	29,200
	Clean up				74.000		3,330		3,330	4,800	5,925
	Total				2230.408	33,485	104,753.20	8,325	146,563.20	187,300	225,425
0700	**Prep & paint 4M gal. ground level water stor tank, 130' dia x 40' tall**	10	E-11	Ea.							
	Scaffolding, outer ring				531.000		27,435		27,435	35,700	44,300
	Scaffolding planks				266.664		13,760		13,760	17,900	22,400
	Sandblast exterior, near white blast (SSPC-SP10), loose scale, fine rust				2133.300	30,000	97,200	13,500	140,700	179,000	214,500
	Epoxy primer, spray exterior				159.900	7,500	7,200		14,700	18,600	21,700
	Epoxy topcoat, 2 coats, spray exterior				342.600	16,800	15,600		32,400	40,200	47,400
	Clean up				120.000		5,400		5,400	7,800	9,600
	Total				3553.464	54,300	166,595	13,500	234,395	299,200	359,900

For customer support on your Facilities Maintenance & Repair Costs with RSMeans data, call 800.448.8182.

	System Description	Freq. (Years)	Crew	Unit	Labor Hours	2019 Bare Costs				Total In-House	Total w/O&P
						Material	Labor	Equipment	Total		
0100	**Prep & paint 100k gal. grnd. level water standpipe, 24' dia x 30' tall**	10	E-11	Ea.							
	Scaffolding, outer ring				96.000		4,960		4,960	6,450	8,000
	Scaffolding planks				47.666		2,459.60		2,459.60	3,200	4,000
	Sandblast exterior, near white blast (SSPC-SP10), loose scale, fine rust				191.997	2,700	8,748	1,215	12,663	16,100	19,300
	Epoxy primer, spray exterior				14.391	675	648		1,323	1,675	1,950
	Epoxy topcoat, 2 coats, spray exterior				30.834	1,512	1,404		2,916	3,625	4,275
	Clean up				10.800		486		486	700	865
	Total				391.688	4,887	18,705.60	1,215	24,807.60	31,750	38,390
0200	**Prep & paint 250k gal. grnd. level water standpipe, 32' dia x 42' tall**	10	E-11	Ea.							
	Scaffolding, outer ring				165.000		8,525		8,525	11,100	13,800
	Scaffolding planks				82.666		4,265.60		4,265.60	5,575	6,950
	Sandblast exterior, near white blast (SSPC-SP10), loose scale, fine rust				355.550	5,000	16,200	2,250	23,450	29,900	35,800
	Epoxy primer, spray exterior				26.650	1,250	1,200		2,450	3,100	3,625
	Epoxy topcoat, 2 coats, spray exterior				57.100	2,800	2,600		5,400	6,700	7,900
	Clean up				20.000		900		900	1,300	1,600
	Total				706.966	9,050	33,690.60	2,250	44,990.60	57,675	69,675
0300	**Prep & paint 500k gal. grnd. level water standpipe, 40' dia x 53' tall**	10	E-11	Ea.							
	Scaffolding, outer ring				252.000		13,020		13,020	17,000	21,000
	Scaffolding planks				125.665		6,484.40		6,484.40	8,450	10,600
	Sandblast exterior, near white blast (SSPC-SP10), loose scale, fine rust				568.880	8,000	25,920	3,600	37,520	47,800	57,000
	Epoxy primer, spray exterior				42.640	2,000	1,920		3,920	4,950	5,775
	Epoxy topcoat, 2 coats, spray exterior				91.360	4,480	4,160		8,640	10,700	12,600
	Clean up				32.000		1,440		1,440	2,075	2,550
	Total				1112.545	14,480	52,944.40	3,600	71,024.40	90,975	109,525
0400	**Prep & paint 750k gal. grnd. level water standpipe, 46' dia x 60' tall**	10	E-11	Ea.							
	Scaffolding, outer ring				318.000		16,430		16,430	21,400	26,500
	Scaffolding planks				159.665		8,238.80		8,238.80	10,700	13,400
	Sandblast exterior, near white blast (SSPC-SP10), loose scale, fine rust				739.544	10,400	33,696	4,680	48,776	62,000	74,500
	Epoxy primer, spray exterior				55.432	2,600	2,496		5,096	6,450	7,525
	Epoxy topcoat, 2 coats, spray exterior				118.768	5,824	5,408		11,232	13,900	16,400
	Clean up				41.600		1,872		1,872	2,700	3,325
	Total				1433.009	18,824	68,140.80	4,680	91,644.80	117,150	141,650

For customer support on your Facilities Maintenance & Repair Costs with RSMeans data, call 800.448.8182.

367

G3013 405 Ground Level Water Storage Standpipe

	System Description	Freq. (Years)	Crew	Unit	Labor Hours	2019 Bare Costs				Total In-House	Total w/O&P
						Material	Labor	Equipment	Total		
0500	**Prep & paint 1M gal. grnd. level water standpipe, 50' dia x 68' tall**	10	E-11	Ea.							
	Scaffolding, outer ring				384.000		19,840		9,840	25,800	32,000
	Scaffolding planks				192.998		9,958.80		9,958.80	13,000	16,200
	Sandblast exterior, near white blast (SSPC-SP10), loose scale, fine rust				903.097	12,700	41,148	5,715	59,563	76,000	91,000
	Epoxy primer, spray exterior				67.691	3,175	3,048		6,223	7,875	9,175
	Epoxy topcoat, 2 coats, spray exterior				145.034	7,112	6,604		13,716	17,000	20,100
	Clean up				50.800		2,286		2,286	3,300	4,075
	Total				1743.620	22,987	82,884.80	5,715	111,586.80	142,975	172,550
0600	**Prep & paint 2M gal. grnd. level water standpipe, 64' dia x 83' tall**	10	E-11	Ea.							
	Scaffolding, outer ring				579.000		29,915		29,915	39,000	48,300
	Scaffolding planks				290.664		14,998.40		14,998.40	19,600	24,400
	Sandblast exterior, near white blast (SSPC-SP10), loose scale, fine rust				1422.200	20,000	64,800	9,000	93,800	119,500	143,000
	Epoxy primer, spray exterior				106.600	5,000	4,800		9,800	12,400	14,500
	Epoxy topcoat, 2 coats, spray exterior				228.400	11,200	10,400		21,600	26,800	31,600
	Clean up				80.000		3,600		3,600	5,200	6,400
	Total				2706.864	36,200	128,513.40	9,000	173,713.40	222,500	268,200

G3013 410 Fire Hydrants

	System Description	Freq. (Years)	Crew	Unit	Labor Hours	2019 Bare Costs				Total In-House	Total w/O&P
						Material	Labor	Equipment	Total		
1010	**Remove and replace fire hydrant**	25	B-21	Ea.							
	Isolate hydrant from system				1.000		41		41	53.50	66
	Excavate to expose base and feeder pipe				.974		46.90	19.46	66.36	81.50	96
	Disconnect and remove hydrant and associated piping				4.000		164		164	214	265
	Install pipe to new hydrant				3.200	430	158		588	675	790
	Install hydrant, including lower barrel and shoe				4.000	2,800	190	18.50	3,008.50	3,350	3,825
	Install thrust blocks				1.200	57.90	58.80	.29	116.99	140	167
	Backfill excavation				5.091		210		210	272	335
	Fine grade, hand				8.000		328		328	430	530
	Total				27.465	3,287.90	1,196.70	38.25	4,522.85	5,216	6,074

G3013 470 Post Indicator Valve

	System Description	Freq. (Years)	Crew	Unit	Labor Hours	2019 Bare Costs				Total In-House	Total w/O&P
						Material	Labor	Equipment	Total		
1010	**Remove and replace post indicator valve**	35	B-21	Ea.							
	Isolate PIV from system				.126		7.95		7.95	9.85	12.35
	Excavate to expose base of valve				.139		6.70	2.78	9.48	11.60	13.70
	Disconnect and remove post indicator and valve				4.000		164		164	214	265
	Install new 4" diameter valve				4.000	2,025	178	53.50	2,256.50	2,525	2,875
	Install new post indicator				3.636	1,400	173	16.85	1,589.85	1,775	2,050
	Backfill excavation				.727		30		30	39	48
	Fine grade, hand				8.000		328		328	430	530
	Total				20.629	3,425	887.65	73.13	4,385.78	5,004.45	5,794.05

G3013 510 Elevated Water Storage Tank

	System Description	Freq. (Years)	Crew	Unit	Labor Hours	2019 Bare Costs				Total In-House	Total w/O&P
						Material	Labor	Equipment	Total		
0100	**Prep & paint 50K gal. stg tank, 30' dia x 10' tall, 50' above grnd.**	10	E-11	Ea.							
	Scaffolding, outer ring				225.000		11,625		11,625	15,100	18,800
	Scaffolding, center core under tank				80.000		4,140		4,140	5,375	6,650
	Scaffolding planks				58.333		3,010		3,010	3,925	4,900
	Sandblast exterior, near white blast (SSPC-SP10), loose scale, fine rust				167.464	2,355	7,630.20	1,059.75	11,044.95	14,100	16,800
	Epoxy primer, spray exterior				12.552	588.75	565.20		1,153.95	1,450	1,700
	Epoxy topcoat, 2 coats, spray exterior				26.894	1,318.80	1,224.60		2,543.40	3,150	3,725
	Clean up				9.420		423.90		423.90	610	755
	Total				579.663	4,262.55	28,618.90	1,059.75	33,941.20	43,710	53,330
0200	**Prep & paint 100K gal. stg tank, 30' dia x 20' tall, 50' above grnd.**	10	E-11	Ea.							
	Scaffolding, outer ring				261.000		13,485		13,485	17,600	21,800
	Scaffolding, center core under tank				80.000		4,140		4,140	5,375	6,650
	Scaffolding planks				153.332		7,912		7,912	10,300	12,900
	Sandblast exterior, near white blast (SSPC-SP10), loose scale, fine rust				227.552	3,200	10,368	1,440	15,008	19,100	22,900
	Epoxy primer, spray exterior				17.056	800	768		1,568	1,975	2,300
	Epoxy topcoat, 2 coats, spray exterior				36.544	1,792	1,664		3,456	4,300	5,050
	Clean up				12.800		576		576	830	1,025
	Total				788.284	5,792	38,913	1,440	46,145	59,480	72,625

System Description	Freq. (Years)	Crew	Unit	Labor Hours	2019 Bare Costs				Total In-House	Total w/O&P
					Material	Labor	Equipment	Total		
0300 **Prep & paint 250K gal. stg tank, 40' dia x 27' tall, 50' above grnd.**	10	E-11	Ea.							
Scaffolding, outer ring				360.000		18,600		18,600	24,200	30,000
Scaffolding, center core under tank				134.000		6,934.50		6,934.50	9,025	11,200
Scaffolding planks				219.998		11,352		11,352	14,800	18,500
Sandblast exterior, near white blast (SSPC-SP10), loose scale, fine rust				419.549	5,900	19,116	2,655	27,671	35,200	42,200
Epoxy primer, spray exterior				31.447	1,475	1,416		2,891	3,650	4,275
Epoxy topcoat, 2 coats, spray exterior				67.378	3,304	3,068		6,372	7,900	9,325
Clean up				23.600		1,062		1,062	1,525	1,900
Total				1255.972	10,679	61,548.50	2,655	74,882.50	96,300	117,400
0400 **Prep & paint 500K gal. stg tank, 50' dia x 34' tall, 50' above grnd.**	10	E-11	Ea.							
Scaffolding, outer ring				480.000		24,800		24,800	32,300	40,000
Scaffolding, center core under tank				202.000		10,453.50		10,453.50	13,600	16,800
Scaffolding planks				296.664		15,308		15,308	20,000	24,900
Sandblast exterior, near white blast (SSPC-SP10), loose scale, fine rust				661.323	9,300	30,132	4,185	43,617	55,500	66,500
Epoxy primer, spray exterior				49.569	2,325	2,232		4,557	5,775	6,725
Epoxy topcoat, 2 coats, spray exterior				106.206	5,208	4,836		10,044	12,500	14,700
Clean up				37.200		1,674		1,674	2,425	2,975
Total				1832.962	16,833	89,435.50	4,185	110,453.50	142,100	172,600
0500 **Prep & paint 750K gal. stg tank, 60' dia x 36' tall, 50' above grnd.**	10	E-11	Ea.							
Scaffolding, outer ring				570.000		29,450		29,450	38,400	47,500
Scaffolding, center core under tank				283.000		14,645.25		14,645.25	19,000	23,600
Scaffolding planks				366.663		18,920		18,920	24,700	30,800
Sandblast exterior, near white blast (SSPC-SP10), loose scale, fine rust				881.764	12,400	40,176	5,580	58,156	74,000	88,500
Epoxy primer, spray exterior				66.092	3,100	2,976		6,076	7,700	8,950
Epoxy topcoat, 2 coats, spray exterior				141.608	6,944	6,448		13,392	16,600	19,600
Clean up				49.600		2,232		2,232	3,225	3,975
Total				2358.727	22,444	114,847.25	5,580	142,371.25	183,625	222,925
0600 **Prep & paint 1M gal. stg tank, 65' dia x 40' tall, 50' above grnd.**	10	E-11	Ea.							
Scaffolding, outer ring				645.000		33,325		33,325	43,400	54,000
Scaffolding, center core under tank				329.000		17,025.75		17,025.75	22,100	27,400
Scaffolding planks				413.329		21,328		21,328	27,800	34,700
Sandblast exterior, near white blast (SSPC-SP10), loose scale, fine rust				1059.539	14,900	48,276	6,705	69,881	89,000	106,500
Epoxy primer, spray exterior				79.417	3,725	3,576		7,301	9,250	10,800
Epoxy topcoat, 2 coats, spray exterior				170.158	8,344	7,748		16,092	20,000	23,500
Clean up				59.600		2,682		2,682	3,875	4,775
Total				2756.043	26,969	133,960.75	6,705	167,634.75	215,425	261,675

For customer support on your Facilities Maintenance & Repair Costs with RSMeans data, call 800.448.8182.

G3063 410 Ground Level Fuel Storage Tank

	System Description	Freq. (Years)	Crew	Unit	Labor Hours	2019 Bare Costs				Total In-House	Total w/O&P
						Material	Labor	Equipment	Total		
0100	**Prep & paint 100k gal. ground level fuel storage tank, 24' dia x 30' tall**	25	E-11	Ea.							
	Scaffolding, outer ring				96.000		4,960		4,960	6,450	8,000
	Scaffolding planks				47.333		2,442.40		2,442.40	3,175	3,975
	Sandblast exterior, near white blast (SSPC-SP10), loose scale, fine rust				191.997	2,700	8,748	1,215	12,663	16,100	19,300
	Epoxy primer, spray exterior				14.391	675	648		1,323	1,675	1,950
	Epoxy topcoat, 2 coats, spray exterior				30.834	1,512	1,404		2,916	3,625	4,275
	Clean up				10.800		486		486	700	865
	Total				391.355	4,887	18,688.40	1,215	24,790.40	31,725	38,365

G4013 210 Overhead Service Cables

System Description	Freq. (Years)	Crew	Unit	Labor Hours	2019 Bare Costs				Total In-House	Total w/O&P
					Material	Labor	Equipment	Total		
0010 **Repair cable splice**										
Repair splice cable, polyethylene jacket	12	1 ELEC	M.L.F.	7.619	13	460		473	580	720
Total				7.619	13	460		473	**580**	**720**
0020 **Cable inspection**										
Check cables for damage	5	1 ELEC	M.L.F.	.667		40		40	49.50	61.50
Total				.667		40		40	**49.50**	**61.50**
0030 **Replace service cable**	30	2 ELEC	M.L.F.							
Remove overhead service cables				1.860		112		112	138	172
Overhead service 3 ACSR cable				16.000	247	960		1,207	1,450	1,775
Install 3 conductors at terminations				.941		56.50		56.50	69.50	87
Install conductors to insulators and clipping				2.500		150		150	185	232
Install and remove sag gauge				1.000		60		60	74	92.50
Make up & install conductor jumper				2.000		120		120	148	185
Total				24.302	247	1,458.50		1,705.50	**2,064.50**	**2,543.50**

For customer support on your Facilities Maintenance & Repair Costs with RSMeans data, call 800.448.8182.

G4023 210 Outdoor Pole Lights

System Description	Freq. (Years)	Crew	Unit	Labor Hours	2019 Bare Costs				Total In-House	Total w/O&P
					Material	Labor	Equipment	Total		
1010 Replace 400W H.P.S. lamp, pole-mounted fixture	10	R-26	Ea.							
Position truck, raise and lower boom bucket				.289		17.35	4.42	21.77	26	32
Remove & install 400W H.P.S. lamp, pole-mounted fixture				.533	32	32		64	74.50	89.50
Total				.822	32	49.35	4.42	85.77	100.50	121.50
1020 Replace 400W H.P.S. ballast, pole-mounted fixture	10	R-26	Ea.							
Position truck, raise and lower boom bucket				.289		17.35	4.42	21.77	26	32
Remove & install 400W H.P.S. multi-tap ballast, pole-mounted fixt.				2.974	260	179	45.50	484.50	555	650
Test pole-mounted H.I.D. fixture				.074		4.45	1.13	5.58	6.75	8.10
Total				3.337	260	200.80	51.05	511.85	587.75	690.10
1030 Replace 400W H.P.S. pole-mounted fixt. w/ lamp & blst.	20	R-26	Ea.							
Turn branch cicuit off and on, pole light				.370		22	5.65	27.65	33.50	41
Turn branch cicuit off and on, pole light				.370		22	5.65	27.65	33.50	41
Position truck, raise and lower boom bucket				.289		17.35	4.42	21.77	26	32
Remove & install pole-mounted 400W H.P.S. fixture w/ lamp & blst.				9.456	325	570	145	1,040	1,225	1,450
Test pole-mounted H.I.D. fixture				.074		4.45	1.13	5.58	6.75	8.10
Total				10.560	325	635.80	161.85	1,122.65	1,324.75	1,572.10
1040 Replace light pole, 2 fixtures (conc. base not incl.)	10	R-3	Ea.							
Turn branch cicuit off and on, pole light				.370		22	5.65	27.65	33.50	41
Remove parking lot light pole with fixtures				2.500		149	16.25	165.25	203	249
Install 2 brackets arms onto aluminum light pole				2.602	564	156		720	815	945
Install two 400W H.P.S. fixtures w/ lamps & ballasts onto light pole				9.467	650	568		1,218	1,425	1,700
Install 30' aluminum light pole (incl. arms & fixtures) onto base				11.111	2,050	665	72	2,787	3,150	3,675
Pull new wires in existing underground conduits (3#10 500')				15.605	219.75	937.50		1,157.25	1,400	1,725
Test pole-mounted H.I.D. fixture				.074		4.45	1.13	5.58	6.75	8.10
Total				41.729	3,483.75	2,501.95	95.03	6,080.73	7,033.25	8,343.10

For customer support on your Facilities Maintenance & Repair Costs with RSMeans data, call 800.448.8182.

373

G4023 210 | **Outdoor Pole Lights**

	Freq. (Years)	Crew	Unit	Labor Hours	2019 Bare Costs				Total In-House	Total w/O&P
System Description					Material	Labor	Equipment	Total		
1050 **Replace concrete base for parking lot light pole**	20	B-18	Ea.							
Sawcut asphalt paving				1.296	4.56	62.88	36.96	104.40	126	146
Remove concrete light pole base				4.000		228	500	728	840	910
Excavate by hand				12.987		535		535	695	860
Install 1-1/4" PVC electrical conduit				.612	10.40	36.70		47.10	56.50	69.50
Install 1-1/4" PVC electrical sweep				.433	9.30	26		35.30	42.50	51.50
Install new concrete light pole base, 24" dia. x 8'				12.030	300	590		890	1,100	1,325
Backfill excavation by hand				4.728		195		195	253	315
Compact backfill				.578		23.70	2.25	25.95	33.50	40.50
Install asphalt patch				.960	82	41.40	3.04	126.44	147	172
Total				37.624	406.26	1,738.68	542.25	2,687.19	3,293.50	3,889.50

For customer support on your Facilities Maintenance & Repair Costs with RSMeans data, call 800.448.8182.

H1043 200 | **Temporary Cranes**

| | System Description | Freq. (Years) | Crew | Unit | Labor Hours | 2019 Bare Costs | | | | Total In-House | Total w/O&P |
						Material	Labor	Equipment	Total		
1010	**Daily use of crane, portal to portal, 12-ton**	1	A-3H	Day							
	12-ton truck-mounted hydraulic crane, daily use for small jobs				8.000		460	710	1,170	1,350	1,500
	Total				8.000		460	710	1,170	**1,350**	**1,500**
1020	**Daily use of crane, portal to portal, 25-ton**	1	A-3I	Day							
	25-ton truck-mounted hydraulic crane, daily use for small jobs				8.000		460	785	1,245	1,450	1,575
	Total				8.000		460	785	1,245	**1,450**	**1,575**
1030	**Daily use of crane, portal to portal, 40-ton**	1	A-3J	Day							
	40-ton truck-mounted hydraulic crane, daily use for small jobs				8.000		460	1,275	1,735	2,000	2,125
	Total				8.000		460	1,275	1,735	**2,000**	**2,125**
1040	**Daily use of crane, portal to portal, 55-ton**	1	A-3K	Day							
	55-ton truck-mounted hydraulic crane, daily use for small jobs				16.000		855	1,450	2,305	2,650	2,925
	Total				16.000		855	1,450	2,305	**2,650**	**2,925**
1050	**Daily use of crane, portal to portal, 80-ton**	1	A-3L	Day							
	80-ton truck-mounted hydraulic crane, daily use for small jobs				16.000		855	2,200	3,055	3,500	3,775
	Total				16.000		855	2,200	3,055	**3,500**	**3,775**
1060	**Daily use of crane, portal to portal, 100-ton**	1	A-3M	Day							
	100-ton truck-mounted hydraulic crane, daily use for small jobs				16.000		855	2,325	3,180	3,625	3,900
	Total				16.000		855	2,325	3,180	**3,625**	**3,900**

Preventive Maintenance

Table of Contents

Table of Contents (cont.)

How to Use the Preventive Maintenance Cost Tables

The following is a detailed explanation of a sample Preventive Maintenance Cost Table. The Preventive Maintenance Tables are separated into two parts: 1) the components and related labor-hours and frequencies of a typical system and 2) the costs for systems on an annual or annualized basis. Next to each bold number that follows is the described item with the appropriate component of the sample entry in parentheses.

D30 HVAC	D3025 130 ❶	Boiler, Hot Water, Oil/Gas/Comb.						
PM Components ❷	**Labor-hrs.** ❸	**W**	**M** ❹	**Q**	**S**	**A**

	PM Components	Labor-hrs.	W	M	Q	S	A
	PM System D3025 130 1950						
	Boiler, hot water; oil, gas or combination fired, up to 120 MBH						
1	Check combustion chamber for air or gas leaks.	.077					✓
2	Inspect and clean oil burner gun and ignition assembly, where applicable.	.658					✓
3	Inspect fuel system for leaks and change fuel filter element, where applicable.	.098					✓
4	Check fuel lines and connections for damage.	.023		✓	✓	✓	✓
5	Check for proper operational response of burner to thermostat controls.	.133			✓	✓	✓
6	Check and lubricate burner and blower motors.	.079			✓	✓	✓
7	Check main flame failure protection and main flame detection scanner on boiler equipped with spark ignition (oil burner).	.124		✓	✓	✓	✓
8	Check electrical wiring to burner controls and blower.	.079					✓
9	Clean firebox (sweep and vacuum).	.577					✓
10	Check operation of mercury control switches (i.e., steam pressure, hot water temperature limit, atomizing or combustion air proving, etc.).	.143		✓	✓	✓	✓
11	Check operation and condition of safety pressure relief valve.	.030		✓	✓	✓	✓
12	Check operation of boiler low water cut off devices.	.056		✓	✓	✓	✓
13	Check hot water pressure gauges.	.073		✓	✓	✓	✓
14	Inspect and clean water column sight glass (or replace).	.127		✓	✓	✓	✓
15	Clean fire side of water jacket boiler.	.433					
16	Check condition of flue pipe, damper and exhaust stack.	.147			✓	✓	✓
17	Check boiler operation through complete cycle, up to 30 minutes.	.650					
18	Check fuel level with gauge pole, add as required.	.046		✓	✓	✓	✓
19	Clean area around boiler.	.066		✓	✓	✓	✓
20	Fill out maintenance checklist and report deficiencies.	.022		✓	✓	✓	✓
	Total labor-hours/period ❺			.710	1.069	1.069	3.641
	Total labor-hours/year			5.678	2.138	1.069	3.641

				Cost Each				
					2019 Bare Costs ❽			**Total** ❾	**Total** ❿
Description ❻		**Labor-hrs.** ❼	**Material**	**Labor**	**Equip.**	**Total**	**In-House**	**w/O&P**
1900	Boiler, hot water, O/G/C, up to 120 MBH, annually	3.641	56.50	233		289.50	351.43	430
1950	Annualized	12.528	66	800		866	1,067.40	1,325

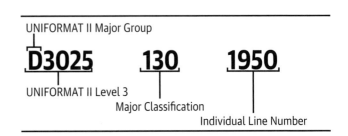

❶ System/Line Numbers (D3025 130 1950)
Each Preventive Maintenance Assembly has been assigned a unique identification number based on the UNIFORMAT II classification system.

UNIFORMAT II Major Group

D3025   **130**   **1950**

UNIFORMAT II Level 3

Major Classification

Individual Line Number

2 PM Components

The individual preventive maintenance operations required to be performed annually are listed separately to show what has been included in the development of the total system price. The cost table below the PM Components listing contains prices for the system on an annual or annualized basis.

3 Labor-Hours

The "Labor-hrs." figure represents the number of labor-hours required to perform the individual operations one time.

4 Frequency

The columns marked with "W" for weekly, "M" for monthly, "Q" for quarterly, "S" for semi-annually, and "A" for annually indicate the recommended frequency for performing a task. For example, a check in the "M" column indicates the recommended frequency of the operation is monthly. Note that, as a result, checks also appear in the "Q," "S," and "A" columns, as they would be part of a monthly frequency.

5 Total Labor-Hour

"Total Labor-hours" are provided on both a per-period and per-year basis. The per-period totals are derived by adding the hours for each operation recommended to be performed (indicated with a check mark) in a given period. In the example given here:

M	$= .023 + .124 + .143 + .030 + .056$ $+ .073 + .127 + .046 + .066 + .022$	$= .710$
Similarly		
Q	$= .023 + .133 + .079 + \ldots + .046$ $+ .066 + .022$	$= 1.069$
S	$= .023 + .133 + .079 + \ldots + .046$ $+ .066 + .022$	$= 1.069$
A	$= .077 + .658 + .098 + .023 + \ldots + .650$ $+ .046 + .066 + .022$	$= 3.641$

The per-year totals are derived by multiplying the per-period total by the net frequency for each time period. The net frequency per period is based on the following table:

Total Labor-hours		Labor-hours		
Annual	=	Annual	X	1
Semi-Annual	=	Semi-Annual	X	1
Quarterly	=	Quarterly	X	2
Monthly	=	Monthly	X	8
Weekly	=	Weekly	X	40

Based on these values, the total labor-hours/year are calculated for the given example as follows:

M	=	(.710)	X	8	=	5.68
Q	=	(1.069)	X	2	=	2.138
S	=	(1.069)	X	1	=	1.069
A	=	(3.641)	X	1	=	3.641
						12.528 hrs.

6 System Description (Boiler, Hot Water, O/G/C, etc.)

The "System Description" corresponds to what's given at the beginning of the PM Components list. A description of the cost basis, annual or annualized, is given as well. Annual costs are based on performance of each operation listed in the PM Components one time in a given year. Annualized costs are based on performance of each operation at the recommended periodic frequencies in a given year.

7 Labor-Hours: Annual (3.641)/ Annualized (12.528)

Annual labor-hours are based on performance of each task once during a given year and are equal to the total labor-hours/period under the "A" column in the PM Components list. Annualized labor-hours are based on the recommended periodic frequencies and are equivalent to the sum of all the values in the total labor-hours/year row at the bottom of the PM Components.

8 Bare Costs

Annualized Material (66.00)

This figure for the system material cost is the "bare" material cost with no overhead and profit allowances included. *Costs shown reflect national average material prices for the current year and generally include delivery to the facility. Small purchases may require an additional delivery charge. No sales taxes are included.*

Labor (800.00)

The labor costs are derived by multiplying bare labor-hour costs by labor-hour units. (The hourly labor costs are listed in the Crew Listings, or for a single trade, the wage rate is found on the inside back cover of the cost data.)

Equip. (Equipment) (0.00)

The unit equipment cost is derived by multiplying the bare equipment hourly cost by the labor-hour units. Equipment costs for each crew are listed in the description of each crew.

Total (866.00)

The total of the bare costs is the arithmetic total of the three previous columns: material, labor, and equipment.

Material	+	Labor	+	Equip.	=	Total
$66.00	+	$800.00	+	$0.00	=	$866.00

9 Total In-House Costs Annualized (1,067.40)

"Total In House Costs" include markups to bare costs for the direct overhead requirements of in-house maintenance staff. The figure in this column is the sum of three components: the bare material cost plus 10% (for purchasing and handling small quantities); the bare labor cost plus workers' compensation and average fixed overhead (per the labor rate table on the inside back cover of the cost data or, if a crew is listed, from the Crew Listings); and the bare equipment cost.

10 Total Costs Including O&P Annualized (1,325)

"Total Costs Including O&P" include suggested markups to bare costs for an outside subcontractor's overhead and profit requirements. The figure in this column is the sum of three components: the bare material cost plus 25% for profit; the bare labor cost plus total overhead and profit (per the labor rate table on the inside back cover of the printed product or the Reference Section of the electronic product, or, if a crew is listed, from the Crew Listings); and the bare equipment cost plus 10% for profit.

PM Components	Labor-hrs.	W	M	Q	S	A
PM System B2035 110 1950						
Door, sliding, electric						
1 Check with operating or area personnel for deficiencies.	.035				√	√
2 Check for proper operation, binding or misalignment; adjust as necessary.	.062				√	√
3 Check and lubricate door guides, pulleys and hinges.	.143				√	√
4 Inspect and lubricate motor gearbox, drive chain (or belt), and motor; adjust as necessary.	.200				√	√
5 Check operation of limit switch; adjust as necessary.	.222				√	√
6 Check electrical operator, wiring, connections and contacts; adjust as necessary.	.471				√	√
7 Clean area around door.	.066				√	√
8 Fill out maintenance checklist and report deficiencies.	.022				√	√
Total labor-hours/period					1.221	1.221
Total labor-hours/year					1.221	1.221

			Cost Each					
			2019 Bare Costs				Total	Total
	Description	Labor-hrs.	Material	Labor	Equip.	Total	In-House	w/O&P
1900	Door, sliding, electric, annually	1.221	22.50	63.50		86	106.41	129
1950	Annualized	2.448	45	126		171	214.42	260

PM Components	Labor-hrs.	W	M	Q	S	A
PM System B2035 110 2950						
Hanger doors, sliding						
1 Check with door operating personnel for deficiencies.	.027			√	√	√
2 Remove debris from door track.	.040			√	√	√
3 Operate door.	.050			√	√	√
4 Check alignment of hanger door, door guides and lubricate.	.650					√
5 Inspect and lubricate wheel drive chain.	.030			√	√	√
6 Adjust brake shoes and check for wear.	.151					√
7 Inspect and lubricate drive wheels, guides, stops and rollers.	.200			√	√	√
8 Fill out maintenance report.	.017			√	√	√
Total labor-hours/period				.364	.364	1.165
Total labor-hours/year				.728	.364	1.165

			Cost Each					
			2019 Bare Costs				Total	Total
	Description	Labor-hrs.	Material	Labor	Equip.	Total	In-House	w/O&P
2900	Hanger doors, sliding, annually	1.165	33.50	60		93.50	115.42	139
2950	Annualized	2.257	78	117		195	238.68	286

PM Components	Labor-hrs.	W	M	Q	S	A
PM System B2035 225 1950						
Door, emergency egress, swinging						
1 Remove obstructions that restrict full movement/swing of door.	.013			√	√	√
2 Check swing of door; door must latch on normal closing.	.013			√	√	√
3 Test operation of panic hardware and local alarm battery.	.013			√	√	√
4 Lubricate hardware.	.013			√	√	√
5 Fill out maintenance checklist and report deficiencies.	.013			√	√	√
Total labor-hours/period				.065	.065	.065
Total labor-hours/year				.130	.065	.065

	Description	Labor-hrs.	Cost Each					
			2019 Bare Costs				Total In-House	Total w/O&P
			Material	Labor	Equip.	Total		
1900	Door, emergency egress, swinging, annually	.065	4.75	3.35		8.10	9.67	11.35
1950	Annualized	.260	14.25	13.45		27.70	33.20	39.50

For customer support on your Facilities Maintenance & Repair Costs with RSMeans data, call 800.448.8182.

PM Components	Labor-hrs.	W	M	Q	S	A
PM System B2035 300 1950						
Door, revolving, manual						
1 Check with operating or area personnel for deficiencies.	.043				✓	✓
2 Check for proper operation, binding or misalignment; adjust as necessary.	.043				✓	✓
3 Check for proper emergency breakaway operation of wings.	.043				✓	✓
4 Check top, bottom, edge and shaft weatherstripping.	.217				✓	✓
5 Check and lubricate pivots and bearings.	.217				✓	✓
6 Clean area around door.	.065				✓	✓
7 Fill out maintenance checklist and report deficiencies.	.022				✓	✓
Total labor-hours/period					.650	.650
Total labor-hours/year					.650	.650

				Cost Each				
			2019 Bare Costs				Total	Total
Description	Labor-hrs.	Material	Labor	Equip.	Total	In-House	w/O&P
1900 Door, revolving, manual, annually	.650	11.15	33.50		44.65	56.13	68
1950 Annualized	1.300	22.50	67.50		90	112.11	136

PM Components	Labor-hrs.	W	M	Q	S	A
PM System B2035 300 2950						
Door, revolving, electric						
1 Check with operating or area personnel for deficiencies.	.043				✓	✓
2 Check for proper operation, binding or misalignment; adjust as necessary.	.043				✓	✓
3 Check for proper emergency breakaway operation of wings.	.043				✓	✓
4 Check top, bottom, edge and shaft weatherstripping.	.217				✓	✓
5 Check elec. operator, safety devices and activators for proper operation.	.433				✓	✓
6 Check and lubricate pivots and bearings and operator gearbox.	.325				✓	✓
7 Clean area around door.	.065				✓	✓
8 Fill out maintenance checklist and report deficiencies.	.022				✓	✓
Total labor-hours/period					1.191	1.191
Total labor-hours/year					1.191	1.191

				Cost Each				
			2019 Bare Costs				Total	Total
Description	Labor-hrs.	Material	Labor	Equip.	Total	In-House	w/O&P
2900 Door, revolving, electric, annually	1.191	11.15	61.50		72.65	92.63	113
2950 Annualized	2.382	22.50	123		145.50	185.11	226

For customer support on your Facilities Maintenance & Repair Costs with RSMeans data, call 800.448.8182.

385

PM Components	Labor-hrs.	W	M	Q	S	A
PM System B2035 400 1950						
Door, overhead, manual, up to 24' high x 25' wide						
1 Check with operating or area personnel for deficiencies.	.035				√	√
2 Check for proper operation, binding or misalignment; adjust as necessary.	.136				√	√
3 Check and lubricate door guides, pulleys and hinges.	.889				√	√
4 Clean area around door.	.026				√	√
5 Fill out maintenance checklist and report deficiencies.	.022				√	√
Total labor-hours/period					1.107	1.107
Total labor-hours/year					1.107	1.107

				Cost Each				
				2019 Bare Costs			**Total**	**Total**
Description		**Labor-hrs.**	**Material**	**Labor**	**Equip.**	**Total**	**In-House**	**w/O&P**
1900	Dr., overhead, manual, up to 24'H x 25'W, annually	1.107	11.15	57.50		68.65	87.23	106
1950	Annualized	2.196	22.50	114		136.50	172.32	211

PM Components	Labor-hrs.	W	M	Q	S	A
PM System B2035 410 1950						
Door, overhead, roll-up, electric, up to 24' high x 25' wide						
1 Check with operating or area personnel for deficiencies.	.035				√	√
2 Check for proper operation, binding or misalignment; adjust as necessary.	.136				√	√
3 Check and lubricate door guides, pulleys and hinges.	.889				√	√
4 Inspect and lubricate motor gearbox, drive chain (or belt), and motor; adjust as necessary.	.200				√	√
5 Check operation of limit switch; adjust as necessary.	.222				√	√
6 Check electrical operator, wiring, connections and contacts; adjust as necessary.	.471				√	√
7 Clean area around door.	.066				√	√
8 Fill out maintenance checklist and report deficiencies.	.022				√	√
Total labor-hours/period					2.040	2.040
Total labor-hours/year					2.040	2.040

				Cost Each				
				2019 Bare Costs			**Total**	**Total**
Description		**Labor-hrs.**	**Material**	**Labor**	**Equip.**	**Total**	**In-House**	**w/O&P**
1900	Dr., overhead, electric, to 24'H x 25'W, annually	2.040	51.50	106		157.50	194.38	234
1950	Annualized	4.070	104	210		314	387.22	470

For customer support on your Facilities Maintenance & Repair Costs with RSMeans data, call 800.448.8182.

PM Components	Labor-hrs.	W	M	Q	S	A
PM System B2035 450 1950						
Shutter, roll-up, electric						
1 Check with operating or area personnel for deficiencies.	.035					√
2 Check for proper operation, binding or misalignment; adjust as necessary.	.136					√
3 Check and lubricate door guides, pulleys and hinges.	.889					√
4 Inspect and lubricate motor gearbox, drive chain (or belt), and motor; adjust as necessary.	.200					√
5 Check operation of limit switch; adjust as necessary.	.222					√
6 Check electrical operator, wiring, connections and contacts; adjust as necessary.	.471					√
7 Clean area around door.	.074					√
8 Fill out maintenance checklist and report deficiencies.	.022					√
Total labor-hours/period						2.048
Total labor-hours/year						2.048

	Description	Labor-hrs.	2019 Bare Costs				Total In-House	Total w/O&P
			Material	Labor	Equip.	Total		
1900	Shutter, roll up, electric, annually	2.048	22.50	106		128.50	162.41	198
1950	Annualized	2.048	22.50	106		128.50	162.41	198

PM Components	Labor-hrs.	W	M	Q	S	A
PM System C1025 110 1950						
Fire Door, Swinging						
1 Remove fusible link hold open devices.	.026			√	√	√
2 Remove obstructions that retard full movement/swing of door.	.013			√	√	√
3 Check swing of door. Door must latch on normal closing.	.013			√	√	√
4 Test operation of panic hardware.	.007			√	√	√
5 Check operation of special devices such as smoke detectors or magnetic door releases.	.013			√	√	√
6 Lubricate hardware.	.013			√	√	√
7 Fill out maintenance checklist and report deficiencies.	.013			√	√	√
Total labor-hours/period				.098	.098	.098
Total labor-hours/year				.196	.098	.098

		Cost Each				Total	Total
		2019 Bare Costs					
Description	Labor-hrs.	Material	Labor	Equip.	Total	In-House	w/O&P
1900 Fire doors, swinging, annually	.098	4.75	5.20		9.95	12.02	14.30
1950 Annualized	.392	14.25	21		35.25	42.74	51.50

PM Components	Labor-hrs.	W	M	Q	S	A
PM System C1025 110 2950						
Fire Door, Sliding						
1 Clean track.	.013			√	√	√
2 Lubricate pulleys.	.013			√	√	√
3 Inspect cable or chain for wear or damage and proper threading through pulleys.	.078			√	√	√
4 Replace fusible links or other heat activated devices that have been painted. Check operation of other heat activated devices.	.130			√	√	√
5 Permanently remove obstructions that prohibit movement of door.	.013			√	√	√
6 Check operation of door by disconnecting or lifting counterweight or other appropriate means.	.129			√	√	√
7 Check fit of door in binders and fit against stay roll.	.013			√	√	√
8 Check door for breaks in covering and examine for dry rot.	.013			√	√	√
9 Fill out maintenance checklist and report deficiencies.	.013			√	√	√
Total labor-hours/period				.415	.415	.415
Total labor-hours/year				.830	.415	.415

		Cost Each				Total	Total
		2019 Bare Costs					
Description	Labor-hrs.	Material	Labor	Equip.	Total	In-House	w/O&P
2900 Fire doors, sliding, annually	.415	18.85	21.50		40.35	48.70	58
2950 Annualized	1.659	56.50	85.50		142	173.93	209

PM Components	Labor-hrs.	W	M	Q	S	A
PM System C1025 110 3950						
Fire Door, Rollup						
1　Check with door operating personnel for deficiencies.	.027			√	√	√
2　Drop test heat activated or fusible link roll up fire door.	.070			√	√	√
3　Adjust rate of descent governor to desired closing speed.	.200					√
4　Inspect, clean and lubricate door gear assembly.	.200			√	√	√
5　Check alignment of overhead door guides, clean and lubricate.	.100			√	√	√
6　Check and adjust door counter balance assembly.	.250					√
7　Check electrical wiring and contacts for wear.	.200					√
8　Check and adjust limit switches.	.170					√
9　Inspect general condition of door for need of paint and repairs.	.005			√	√	√
10　Replace missing or tighten loose nuts and bolts.	.040			√	√	√
11　After maintenance clean around door and ensure access.	.050			√	√	√
12　Fill out maintenance report.	.017			√	√	√
Total labor-hours/period				.509	.509	1.329
Total labor-hours/year				1.018	.509	1.329

			Cost Each					
			2019 Bare Costs				Total	Total
Description	Labor-hrs.	Material	Labor	Equip.	Total	In-House	w/O&P
3900	Fire doors, roll-up, annually	1.329	33.50	68.50		102	126.79	153
3950	Annualized	2.856	79	148		227	279.48	335

For customer support on your Facilities Maintenance & Repair Costs with RSMeans data, call 800.448.8182.

PM Components	Labor-hrs.	W	M	Q	S	A
PM System D1015 100 1950						
Elevator, cable, electric, passenger/freight						
1 Ride car, check for any unusual noise or operation.	.411	√	√	√	√
2 Inspect machine room equipment.	.286	√	√	√	√
3 Motor room:						
A) visually inspect controllers and starters.	.012	√	√	√	√
B) visually inspect selector.	.012	√	√	√	√
C) lubricate and adjust tension of selector.	.113	√	√	√	√
D) inspect sleeve bearings of hoist motor.	.012	√	√	√	√
E) inspect brushes and commutator of hoist motor.	.012	√	√	√	√
F) inspect sleeve bearings of motor generator.	.012	√	√	√	√
G) inspect brushes and commutator of motor generator.	.014	√	√	√	√
H) inspect sleeve bearings of exciter.	.022	√	√	√	√
I) inspect brushes and commutator of exciter.	.022	√	√	√	√
J) inspect sleeve bearings of regulator dampening motors and tach generators.	.044	√	√	√	√
K) inspect brushes and commutator of regulator dampening motors and tach generators.	.047	√	√	√	√
L) inspect and grease sleeve bearings of geared machines.	.044	√	√	√	√
M) inspect oil level in worm and gear of geared machines.	.044	√	√	√	√
N) visually inspect the brakes.	.022	√	√	√	√
O) inspect and grease, if necessary, sleeve bearings of drive deflectors and secondary sheaves.	.186	√	√	√	√
P) visually inspect and adjust contacts of controllers and starters.	.338			√	√	√
Q) inspect and adjust main operating contactor and switches of controllers and starters.	.338			√	√	√
R) inspect selector, adjusting contacts, brushes and cams if needed.	.103			√	√	√
S) inspect selsyn and advancer, motor brushes and commutators of selector.	.022			√	√	√
T) clean hoist motor brush rigging and commutator.	.068			√	√	√
U) clean motor generator and tighten brush rigging and clean commutator.	.068			√	√	√
V) clean exciter brush rigging and commutator.	.066			√	√	√
W) clean brush rigging and commutator of regulator dampening motors and tach generators.	.099			√	√	√
X) inspect machine worm and gear thrust and backlash and spider bolts; replace oil.	.047			√	√	√
Y) clean and lubricate brakes, pins and linkage.	.099			√	√	√
Z) inspect governors; lubricate sleeve bearings.	.099			√	√	√
Aa) inspect controllers and starters; adjust setting of overloads.	.169					√
Ab) tighten connections on controllers and starters; clean fuses and holders.	.169					√
Ac) set controllers and starter timers.	.169					√
Ad) clean controllers and starters.	.169					√
Ae) clean and inspect selector components.	.068					√
Af) inspect group operation controllers; clean and lubricate contacts, timers, steppers of operation controller.	.099					√
Ag) change oil and/or grease roller bearings of the hoist motor.	.094					√
Ah) vacuum and blowout the hoist motor.	.094					√
Ai) tighten all connections on the hoist motor.	.022					√

For customer support on your Facilities Maintenance & Repair Costs with RSMeans data, call 800.448.8182.

PM Components	Labor-hrs.	W	M	Q	S	A
Aj) change oil and/or grease generator's roller bearings.	.069					√
Ak) vacuum and blowout motor generator.	.094					√
Al) inspect all connections on the motor generator and tighten if necessary.	.022					√
Am) inspect roller bearings of the exciter and change oil and/or grease.	.069					√
An) vacuum and blowout exciter.	.094					√
Ao) change oil and/or grease roller bearings of regulator dampening motors and tach generators.	.094					√
Ap) vacuum/blowout regulator dampening motors and tach generators.	.044					√
Aq) inspect all regulator dampening motors and tach generator connections.	.047					√
Ar) change oil and/or grease roller bearings of the geared machines.	.094					√
As) clean and lubricate brake cores.	.099					√
At) inspect drive deflectors and secondary sheaves; change oil and grease roller bearings if necessary.	.170					√
Au) inspect drive deflectors and secondary sheaves' cable grooves for cracks.	.012					√
Av) inspect governors; change oil and grease roller bearings, if necessary.	.094					√
4 Hatch:						
A) lubricate rails of hoistway.	.733		√	√	√	√
B) inspect car top and grease sleeve bearings if needed.	.151		√	√	√	√
C) inspect counterweight and grease sleeve bearings if needed.	.068		√	√	√	√
D) visually inspect cables, chains, hoist, compensating governor for wear and equalization.	.047		√	√	√	√
E) lubricate cables, chain hoist and compensating governor if necessary.	.186		√	√	√	√
F) inspect door operator; clean and lubricate chain and belt tension.	.177			√	√	√
G) inspect door, clean and adjust safety edge, light ray and cables.	.052			√	√	√
H) clean and adjust proximity devices on door.	.103			√	√	√
I) inspect and lubricate governor tape and compensating sheave in pit.	.068			√	√	√
J) clean and lubricate compensating tie down in pit.	.047			√	√	√
K) inspect pit's run-by on counter weight.	.142			√	√	√
L) inspect governor and compensating sheaves or chains for clearance to pit floor.	.022			√	√	√
M) inspect/lubricate overhead hatch switches and cams.	.096					√
N) clean and lubricate rollers, cables, sight guards, chains, motors, closures, limit and zone switches.	.068					√
O) inspect door gibs and fastening.	.022					√
5 Inspect car top; change oil and grease roller bearings, if necessary.	.044					√
6 Check safety switches, indicators, leveling devices, selector tape, switches and hatches.	.044					√
7 Check and lubricate fan motor.	.072					√
8 Inspect counterweight; change oil and grease roller bearings, if necessary.	.094					√
9 Check safety linkage safety and cwt guides.	.098					√
10 Inspect compensation hatch.	.047					√
11 Inspect chains, hoist and compensating governor.	.094					√
12 Inspect traveling cables for tracking and wear.	.070					√

PM Components	Labor-hrs.	W	M	Q	S	A
13 Inspect door operator commutator and brushes.	.012					√
14 Inspect/clean/lubricate door operator clutch, retiring cam, door gib, rollers, tracks, upthrusts and relating cables.	.103					√
15 Under car:						
A) inspect/clean/lubricate safety devices, linkages, car guides, selector tape, switches and hitches.	.020					√
B) inspect cable hitch and loops; load weighing devices.	.027					√
16 Inspect/lubricate pit hatch switches, cams and pit oil buffer.	.083					√
17 In car:						
A) inspect and clean fixtures and signal in operating panel and car position and direction indicator.	.103		√	√	√	√
B) check operation of emergency lights and bells.	.229		√	√	√	√
C) check handrails, ceiling panels and hang on panels for tightness.	.027		√	√	√	√
D) check for tripping hazards.	.044		√	√	√	√
18 Hallway corridor:						
A) inspect hall buttons, signal lamps, lanterns and hall position indicator.	.046		√	√	√	√
B) inspect starter station, key operation and lamps.	.069		√	√	√	√
19 Clean equipment and surrounding area.	.066		√	√	√	√
20 Fill out maintenance checklist.	.022		√	√	√	√
Total labor-hours/period			3.106	5.063	5.063	8.112
Total labor-hours/year			24.848	10.126	5.063	8.112

	Description	Labor-hrs.	Cost Each					
			2019 Bare Costs				Total In-House	Total w/O&P
			Material	Labor	Equip.	Total		
1900	Elevator, cable, electric, passenger / freight, annually	8.112	1,575	680		2,255	2,568.53	3,025
1950	Annualized	47.895	4,125	4,025		8,150	9,477.93	11,300

PM Components	Labor-hrs.	W	M	Q	S	A
PM System D1015 110 1950						
Elevator, hydraulic, passenger/freight						
1 Ride car, checking for any unusual noise or operation.	.052		√	√	√	√
2 In car:						
A) inspect and clean fixtures and signal in operating panel and car position and direction indicator.	.103		√	√	√	√
B) check operation of emergency lights and bell.	.020		√	√	√	√
C) check handrails, ceiling panels and hang on panels for tightness.	.027		√	√	√	√
D) check for tripping hazards.	.012		√	√	√	√
3 Inspect and lubricate rails of hoistway.	.094		√	√	√	√
4 Hallway corridor:						
A) inspect hall buttons, signal lamps, lanterns and hall position indicator.	.046		√	√	√	√
B) inspect starter station, key operation and lamps.	.069		√	√	√	√
5 Motor room:						
A) inspect machine room equipment.	.077		√	√	√	√
B) lockout and log record.	.022		√	√	√	√
C) inspect tank oil level.	.012		√	√	√	√
D) inspect and adjust controller contacts; main operating contactors and switches.	.195			√	√	√
E) inspect pump and valve unit for leaks.	.051			√	√	√
F) inspect and adjust controller overloads; set timers.	.049					√
G) tighten connections and clean controller fuses and holders.	.020					√
H) inspect and lubricate pump motor bearings.	.099					√
6 Hatch:						
A) check hoistway car rails, brackets and fish plates.	.103					√
B) inspect/lubricate overhead hatch switches and cams.	.070					√
C) inspect/clean/lubricate hatch doors locks, rollers, tracks, upthrusts, relating cables, racks, sight guards and closers, motors, gear boxes, limit and zone switches.	.096					√
D) inspect door gibs and fastening.	.022					√
E) inspect/clean/lubricate cab top guides, steading devices, safety switches, inductors, leveling devices, selector tape, switches, hitches and fan motor.	.068					√
F) check traveling cables for wear.	.070					√
G) inspect/clean/lubricate door operator roller tracks, upthrusts, related cables, clutch, retiring cam and door gib.	.078					√
H) inspect door operator; clean and lubricate chain and belt tension.	.052			√	√	√
I) inspect door, clean and adjust safety edge, light ray and cables.	.103			√	√	√
J) clean and adjust proximity devices on door.	.068			√	√	√
7 Inspect pit gland packing.	.010			√	√	√
8 Inspect/clean/lubricate under car guides, selector tape, traveling cable, switches and platen plate assembly.	.068					√
9 Clean equipment and surrounding area.	.074		√	√	√	√
10 Fill out maintenance checklist.	.022		√	√	√	√
Total labor-hours/period			.630	1.109	1.109	1.852
Total labor-hours/year			5.042	2.217	1.109	1.852

	Description	Labor-hrs.	Material	Labor	Equip.	Total	Total In-House	Total w/O&P
			2019 Bare Costs					
1900	Elevator, hydraulic, passenger / freight, annually	1.852	1,400	155		1,555	1,744.08	2,000
1950	Annualized	10.224	1,375	860		2,235	2,586.34	3,050

PM Components	Labor-hrs.	W	M	Q	S	A
PM System D1015 310 1950						
Wheelchair Lift						
1 Check with operating personnel for any deficiencies.	.033				√	√
2 Check voltage and amperage under load.	.091				√	√
3 Check bolts securing drive cabinet and base, tighten accordingly.	.033				√	√
4 Check belt tension.	.033				√	√
5 Check lift nut assembly.	.026				√	√
6 Check cam rollers.	.026				√	√
7 Check wear pads for excessive wear.	.026				√	√
8 Inspect motor and shaft pulleys.	.039				√	√
9 Check Acme screw alignment.	.033				√	√
10 Inspect and lubricate upper and lower bearings.	.039				√	√
11 Check fastening of cable harness.	.026				√	√
12 Check alignment of platform and doors.	.033				√	√
13 Check door interlock switch for proper operation.	.039				√	√
14 Check operation of final limit switch and emergency stop/alarm.	.033				√	√
15 Check call/send controls at each station and on platform.	.039				√	√
16 Lube hinge of flip up ramp.	.033				√	√
17 Fill out maintenance checklist and report deficiencies.	.022				√	√
Total labor-hours/period					.604	.604
Total labor-hours/year					.604	.604

	Description	Labor-hrs.	Cost Each				Total	Total
			2019 Bare Costs				In-House	w/O&P
			Material	Labor	Equip.	Total		
1900	Wheelchair lift, annually	.604	22.50	51		73.50	86.76	106
1950	Annualized	1.208	45	102		147	174.63	212

For customer support on your Facilities Maintenance & Repair Costs with RSMeans data, call 800.448.8182.

PM Components	Labor-hrs.	W	M	Q	S	A
PM System D1025 100 1950						
Escalator						
1 Ride escalator; check operation for smoothness, unusual vibration or noise, condition of handrails, etc.	.104	√	√	√	√	√
2 Deenergize, tag, and lockout the electrical circuit.	.017	√	√	√	√	√
3 Inspect comb plates at both ends of escalator for broken teeth and check for proper clearance between combs and step teeth and check for broken step treads.	.039	√	√	√	√	√
4 Check clearance between steps and skirt panel; look for loose trim, screws or bolts, that could snag or damage clothing or cause injury; check operation of handrail brushes.	.087	√	√	√	√	√
5 Clean escalator machine space.	.113	√	√	√	√	√
6 Lubricate step rollers, step chain, drive gears or chains, handrail drive chains, etc., according to manufacturers' instructions; observe gears and chains for signs of wear, misalignment, etc; adjust as required.	.267	√	√	√	√	√
7 Check motor for signs of overheating.	.004	√	√	√	√	√
8 Inspect controller for loose leads, burned contacts, etc.; repair as required.	.022	√	√	√	√	√
9 Clean handrails as required.	.217	√	√	√	√	√
10 Check escalator lighting; replace bulbs as required.	.099	√	√	√	√	√
11 Operate each emergency stop button and note that the escalator stops; if the escalator has the capabilities of running in both directions, stop buttons should function properly for each direction of travel; observe the stopping distance.	.013	√	√	√	√	√
12 Clean surrounding area.	.066	√	√	√	√	√
13 Fill out maintenance checklist and report deficiencies.	.022	√	√	√	√	√
Total labor-hours/period		1.070	1.070	1.070	1.070	1.070
Total labor-hours/year		40.642	8.556	2.139	1.070	1.070

	Description	Labor-hrs.	Cost Each					Total In-House	Total w/O&P
			2019 Bare Costs						
			Material	Labor	Equip.	Total			
1900	Escalator, annually	1.070	2,425	90		2,515	2,763.11	3,175
1950	Annualized	53.509	3,400	4,500		7,900	9,276.50	11,200

PM Components	Labor-hrs.	W	M	Q	S	A
PM System D1095 100 1950						
Dumbwaiter, electric						
1 Check with operating or area personnel for any obvious deficiencies.	.086					√
2 Operate dumbwaiter through complete cycle and check operation for unusual noises or problems.	.052					√
3 Inspect and clean operating panel, starter station, key operation and lamps.	.072					√
4 Check operation of emergency lights and alarms.	.052					√
5 Inspect, clean and lubricate motor.	.072					√
6 Check operation of indicator lights and buttons.	.052					√
7 Inspect and lubricate rails of guides.	.148					√
8 Clean equipment and surrounding area.	.077					√
9 Fill out maintenance checklist and report deficiencies.	.022					√
Total labor-hours/period						.633
Total labor-hours/year						.633

			Cost Each					
			2019 Bare Costs				Total	Total
	Description	Labor-hrs.	Material	Labor	Equip.	Total	In-House	w/O&P
1900	Dumbwaiter, electric, annually	.633	56.50	53		109.50	127.77	152
1950	Annualized	.633	56.50	53		109.50	127.77	152

PM Components	Labor-hrs.	W	M	Q	S	A
PM System D1095 110 1950						
Dumbwaiter, hydraulic						
1 Check with operating or area personnel for any obvious deficiencies.	.086				√	√
2 Operate dumbwaiter through complete cycle and check operation for unusual noises or problems.	.052				√	√
3 Inspect and clean operating panel, starter station, key operation and lamps.	.072				√	√
4 Check operation of emergency lights and alarms.	.052				√	√
5 Inspect pump and valve unit for leaks.	.072				√	√
6 Inspect tank oil level; add as required.	.039				√	√
7 Inspect and lubricate rails of guides.	.148				√	√
8 Inspect jack seals.	.051				√	√
9 Clean equipment and surrounding area.	.077				√	√
10 Fill out maintenance checklist and report deficiencies.	.022				√	√
Total labor-hours/period						.671	.671
Total labor-hours/year						.671	.671

			Cost Each					
			2019 Bare Costs				Total	Total
	Description	Labor-hrs.	Material	Labor	Equip.	Total	In-House	w/O&P
1900	Dumbwaiter, hydraulic, annually	.671	72	56.50		128.50	148.42	177
1950	Annualized	1.342	144	113		257	297.18	355

For customer support on your Facilities Maintenance & Repair Costs with RSMeans data, call 800.448.8182.

PM Components	Labor-hrs.	W	M	Q	S	A
PM System D1095 200 1950						
Pneumatic Tube System						
1 Check with operating personnel for any deficiencies.	.039			√	√	√
2 Clean air intake screen.	.039			√	√	√
3 Check seal at doors.	.129			√	√	√
4 Inspect check valves	.026			√	√	√
5 Check blower motor and lubricate.	.065			√	√	√
6 Inspect carrier ends, latches and air washers	.026			√	√	√
7 Fill out maintenance checklist and report deficiencies.	.026			√	√	√
Total labor-hours/period				.350	.350	.350
Total labor-hours/year				.700	.350	.350

			Cost Each					
			2019 Bare Costs				Total	Total
Description		Labor-hrs.	Material	Labor	Equip.	Total	In-House	w/O&P
1900	Pneumatic Tube System, annually	.350	23	22.50		45.50	52.97	63.50
1950	Annualized	1.404	68.50	90		158.50	187.30	225

PM Components	Labor-hrs.	W	M	Q	S	A
PM System D2015 100 1950						
Urinals (Time is per fixture)						
1 Urinals - Flush and adjust water flow if required.	.020			√	√	√
2 Inspect for missing or damaged parts/caps and replace.	.020			√	√	√
3 Fill out maintenance checklist and report deficiencies.	.017			√	√	√
Total labor-hours/period				.057	.057	.057
Total labor-hours/year				.114	.057	.057

		Cost Each					
		2019 Bare Costs				Total	Total
Description	Labor-hrs.	Material	Labor	Equip.	Total	In-House	w/O&P
1900 Urinals, annually	.057	5.85	3.05		8.90	10.40	12.20
1950 Annualized	.228	5.85	12.15		18	22.14	27

PM Components	Labor-hrs.	W	M	Q	S	A
PM System D2015 100 2950						
Toilet, vacuum breaker type (Time is per fixture)						
1 Toilet (vacuum breaker type) - Flush and adjust water flow if required.	.020			√	√	√
2 Inspect for missing or damaged parts/caps, seat supports, and replace.	.020			√	√	√
3 Fill out maintenance checklist and report deficiencies.	.017			√	√	√
Total labor-hours/period				.057	.057	.057
Total labor-hours/year				.114	.057	.057

		Cost Each					
		2019 Bare Costs				Total	Total
Description	Labor-hrs.	Material	Labor	Equip.	Total	In-House	w/O&P
2900 Toilet (vacuum breaker type), annually	.057	4.17	3.05		7.22	8.50	10.10
2950 Annualized	.177	8.35	9.45		17.80	21.37	25.50

PM Components	Labor-hrs.	W	M	Q	S	A
PM System D2015 100 3950						
Toilet, tank type (Time is per fixture)						
1 Toilet - Clean flapper seat if leaking, adjust fill level if required.	.050			√	√	√
2 Inspect for missing or damaged parts/caps, seat supports, and replace.	.030			√	√	√
3 Fill out maintenance checklist and report deficiencies.	.017			√	√	√
Total labor-hours/period				.097	.097	.097
Total labor-hours/year				.194	.097	.097

		Cost Each					
		2019 Bare Costs				Total	Total
Description	Labor-hrs.	Material	Labor	Equip.	Total	In-House	w/O&P
3900 Toilet (tank type), annually	.097	7.55	5.20		12.75	14.97	17.75
3950 Annualized	.388	7.55	20.50		28.05	35.04	42.50

PM Components	Labor-hrs.	W	M	Q	S	A
PM System D2015 100 4950						
Lavatories (Time is per fixture)						
1 Lavatories - Operate faucets, replace washers/"O" Rings as necessary.	.040			√	√	√
2 Observe drain flow, clean trap if flow is obstructed.	.030			√	√	√
3 Fill out maintenance checklist and report deficiencies.	.017			√	√	√
Total labor-hours/period				.087	.087	.087
Total labor-hours/year				.174	.087	.087

For customer support on your Facilities Maintenance & Repair Costs with RSMeans data, call 800.448.8182.

Description	Labor-hrs.	2019 Bare Costs Material	Labor	Equip.	Total	Total In-House	Total w/O&P
		Cost Each					
4900 Lavatories, annually	.087	9.25	5.50		14.75	17.03	20
4950 Annualized	.348	9.25	22		31.25	37.75	45.50

PM Components	Labor-hrs.	W	M	Q	S	A
PM System D2015 100 5950						
Showers (Time is per fixture)						
1 Showers - Check for damaged, or missing shower heads/handles and replace if required.	.040			√	√	√
2 Fill out maintenance checklist and report deficiencies.	.017			√	√	√
Total labor-hours/period				.057	.057	.057
Total labor-hours/year				.114	.057	.057

Description	Labor-hrs.	2019 Bare Costs Material	Labor	Equip.	Total	Total In-House	Total w/O&P
		Cost Each					
5900 Showers, annually	.057	12.70	3.60		16.30	18.43	21.50
5950 Annualized	.228	12.70	14.40		27.10	31.85	38.50

PM Components	Labor-hrs.	W	M	Q	S	A
PM System D2015 800 1950						
Drinking fountain						
1 Check unit for proper operation, excessive noise or vibration.	.035					√
2 Clean condenser coils and fan as required.	.222					√
3 Check for water leaks in supply line and drain.	.077					√
4 Check water flow; adjust as necessary.	.039					√
5 Check drinking water temperature to ensure the unit is operating properly.	.039					√
6 Check electrical connections and cord; tighten and repair as necessary.	.120					√
7 Clean area around fountain.	.066					√
8 Fill out maintenance checklist and report deficiencies.	.022					√
Total labor-hours/period						.620
Total labor-hours/year						.620

Description	Labor-hrs.	2019 Bare Costs Material	Labor	Equip.	Total	Total In-House	Total w/O&P
		Cost Each					
1900 Drinking fountain, annually	.620	21.50	33		54.50	66.47	80
1950 Annualized	.620	21.50	33		54.50	66.47	80

For customer support on your Facilities Maintenance & Repair Costs with RSMeans data, call 800.448.8182.

399

PM Components	Labor-hrs.	W	M	Q	S	A
PM System D2025 120 1950						
Valve, butterfly, above 4″						
1 Lubricate valves that have grease fittings.	.022					√
2 Open and close valve using handle, wrench, or hand wheel to check operation.	.049					√
3 Check for leaks.	.007					√
4 Clean valve exterior and area around valve.	.066					√
5 Fill out maintenance report and report deficiencies.	.022					√
Total labor-hours/period						.166
Total labor-hours/year						.166

			Cost Each					
			2019 Bare Costs				Total	Total
Description	Labor-hrs.	Material	Labor	Equip.	Total	In-House	w/O&P
1900 Valve, butterfly, above 4″, annually	.166	11.15	8.85		20	23.69	28
1950 Annualized	.166	11.15	8.85		20	23.69	28

PM Components	Labor-hrs.	W	M	Q	S	A
PM System D2025 120 2950						
Valve, butterfly, auto, above 4″						
1 Lubricate valve actuator stem and valve stem, where possible.	.022					√
2 Check automatic valve for proper operation.	.062					√
3 Check packing gland for leaks; adjust as required.	.160					√
4 Check pneumatic operator and tubing for air leaks.	.047					√
5 Clean valve exterior and area around valve.	.034					√
6 Fill out maintenance report and report deficiencies.	.022					√
Total labor-hours/period						.347
Total labor-hours/year						.347

			Cost Each					
			2019 Bare Costs				Total	Total
Description	Labor-hrs.	Material	Labor	Equip.	Total	In-House	w/O&P
2900 Valve, butterfly, auto, above 4″, annually	.347	11.15	18.55		29.70	36.19	43.50
2950 Annualized	.347	11.15	18.55		29.70	36.19	43.50

PM Components	Labor-hrs.	W	M	Q	S	A
PM System D2025 125 1950						
Valve, check, above 4″						
1 Inspect valve for leaks; repair as necessary.	.009					√
2 Adjust system pressure/flow to verify that valve is opening and closing properly, if applicable.	.160					√
3 Clean valve and area around valve.	.066					√
4 Fill out maintenance report and report deficiencies.	.022					√
Total labor-hours/period						.257
Total labor-hours/year						.257

			Cost Each					
			2019 Bare Costs				Total	Total
Description	Labor-hrs.	Material	Labor	Equip.	Total	In-House	w/O&P
1900 Valve, check, above 4″, annually	.257	11.15	13.75		24.90	29.98	36
1950 Annualized	.257	11.15	13.75		24.90	29.98	36

For customer support on your Facilities Maintenance & Repair Costs with RSMeans data, call 800.448.8182.

PM Components	Labor-hrs.	W	M	Q	S	A
PM System D2025 130 1950						
Valve, ball, above 4"						
1 Lubricate valves that have grease fittings.	.022					√
2 Open and close valve using handle, wrench, or hand wheel to check operation.	.049					√
3 Check for leaks.	.007					√
4 Clean valve and area around valve.	.066					√
5 Fill out maintenance report and report deficiencies.	.022					√
Total labor-hours/period						.166
Total labor-hours/year						.166

			Cost Each					
				2019 Bare Costs			Total	Total
	Description	Labor-hrs.	Material	Labor	Equip.	Total	In-House	w/O&P
1900	Valve, ball, above 4", annually	.166	11.15	8.85		20	23.69	28
1950	Annualized	.166	11.15	8.85		20	23.69	28

PM Components	Labor-hrs.	W	M	Q	S	A
PM System D2025 135 1950						
Valve, diaphragm, above 4"						
1 Lubricate valve stem, close and open valve to check operation.	.022					√
2 Check valve for proper operation and leaks; tighten packing and flange bolts as required.	.044					√
3 Clean valve exterior and area around valve.	.034					√
4 Fill out maintenance report and report deficiencies.	.022					√
Total labor-hours/period						.122
Total labor-hours/year						.122

			Cost Each					
				2019 Bare Costs			Total	Total
	Description	Labor-hrs.	Material	Labor	Equip.	Total	In-House	w/O&P
1900	Valve, diaphragm, above 4", annually	.122	11.15	6.50		17.65	20.65	24.50
1950	Annualized	.122	11.15	6.50		17.65	20.65	24.50

PM Components	Labor-hrs.	W	M	Q	S	A
PM System D2025 140 1950						
Valve, gate, above 4"						
1 Lubricate valve stem; close and open valve to check operation.	.049					√
2 Check packing gland for leaks; tighten packing and flange bolts as req.	.022					√
3 Clean valve exterior and area around valve.	.066					√
4 Fill out maintenance report and report deficiencies.	.022					√
Total labor-hours/period						.159
Total labor-hours/year						.159

			Cost Each					
				2019 Bare Costs			Total	Total
	Description	Labor-hrs.	Material	Labor	Equip.	Total	In-House	w/O&P
1900	Valve, gate, above 4", annually	.159	11.15	8.50		19.65	23.25	27.50
1950	Annualized	.159	11.15	8.50		19.65	23.25	27.50

PM Components	Labor-hrs.	W	M	Q	S	A
PM System D2025 145 1950						
Valve, globe, above 4″						
1 Lubricate stem, close and open valve to check operation.	.049					✓
2 Check packing gland for leaks; tighten packing and flange bolts as req.	.022					✓
3 Clean valve exterior.	.066					✓
4 Fill out maintenance report and report deficiencies.	.022					✓
Total labor-hours/period						.159
Total labor-hours/year						.159

Description	Labor-hrs.	Cost Each				Total In-House	Total w/O&P
		2019 Bare Costs					
		Material	Labor	Equip.	Total		
1900 Valve, globe, above 4″, annually	.159	11.15	8.50		19.65	23.25	27.50
1950 Annualized	.159	11.15	8.50		19.65	23.25	27.50

PM Components	Labor-hrs.	W	M	Q	S	A
PM System D2025 145 2950						
Valve, globe, auto, above 4″						
1 Lubricate valve actuator stem and valve stem, where possible.	.022					✓
2 Check automatic valve for proper operation.	.062					✓
3 Check packing gland for leaks; adjust as required.	.160					✓
4 Check penumatic operator and tubing for air leaks.	.025					✓
5 Clean valve exterior and area around valve.	.034					✓
6 Fill out maintenance report and report deficiencies.	.022					✓
Total labor-hours/period						.325
Total labor-hours/year						.325

Description	Labor-hrs.	Cost Each				Total In-House	Total w/O&P
		2019 Bare Costs					
		Material	Labor	Equip.	Total		
2900 Valve, globe, auto, above 4″, annually	.325	11.15	17.35		28.50	34.67	42
2950 Annualized	.325	11.15	17.35		28.50	34.67	42

PM Components	Labor-hrs.	W	M	Q	S	A
PM System D2025 150 1950						
Valve, motor operated, above 4″						
1 Lubricate valve actuator stem and valve stem, where possible.	.091				✓	✓
2 Check motor and valve for proper operation, including limit switch; adjust as required.	.022				✓	✓
3 Check packing gland for leaks; adjust as required.	.113				✓	✓
4 Check electrical wiring, connections and contacts; repair as necessary.	.119				✓	✓
5 Inspect and lubricate motor gearbox as required.	.099				✓	✓
6 Clean valve exterior and area around valve.	.034				✓	✓
7 Fill out maintenance report and report deficiencies.	.022				✓	✓
Total labor-hours/period					.500	.500
Total labor-hours/year					.500	.500

Description	Labor-hrs.	Cost Each				Total In-House	Total w/O&P
		2019 Bare Costs					
		Material	Labor	Equip.	Total		
1900 Valve, motor operated, above 4″, annually	.500	22.50	26.50		49	58.94	70.50
1950 Annualized	1.002	45	53.50		98.50	118.90	142

For customer support on your Facilities Maintenance & Repair Costs with RSMeans data, call 800.448.8182.

D20 PLUMBING	D2025 155	Valve, OS&Y

PM Components	Labor-hrs.	W	M	Q	S	A
PM System D2025 155 1950						
Valve, OS&Y, above 4″						
1 Lubricate stem; open and close valve to check operation.	.049					√
2 Check packing gland for leaks; tighten packing and flange bolts as required.	.022					√
3 Clean valve exterior and area around valve.	.066					√
4 Fill out maintenance report and report deficiencies.	.022					√
Total labor-hours/period						.159
Total labor-hours/year						.159

	Description	Labor-hrs.	Cost Each				Total In-House	Total w/O&P
			2019 Bare Costs					
			Material	Labor	Equip.	Total		
1900	Valve, OS&Y, above 4″, annually	.159	11.15	8.50		19.65	23.25	27.50
1950	Annualized	.159	11.15	8.50		19.65	23.25	27.50

For customer support on your Facilities Maintenance & Repair Costs with RSMeans data, call 800.448.8182.

PM Components	Labor-hrs.	W	M	Q	S	A
PM System D2025 190 1950						
Solar, closed loop hot water heating system, up to 6 panels						
1 Check with operating or area personnel for deficiencies.	.035				√	√
2 Inspect interior piping and connections for leaks and damaged insulation; tighten connections and repair damaged insulation as necessary.	.125				√	√
3 Check zone and circulating pump motors for excessive overheating; lubricate motor bearings.	.077				√	√
4 Check pressure and air relief valves for proper operation.	.030				√	√
5 Check control panel and differential thermostat for proper operation.	.094				√	√
6 Clean sight glasses, controls, pumps, and flow indicators on tanks.	.127				√	√
7 Check system pressure on closed loop for loss of fluid.	.046				√	√
8 Check fluid level on drain-back systems; add fluid as necessary.	.029				√	√
9 Test glycol strength in closed systems, as applicable; if required, drain system and replace with new fluid mixture.	.222					√
10 Check heat exchanger for exterior leaks.	.077				√	√
11 Clean strainers and traps.	.181					√
12 Check storage and expansion tanks; for leaks and deteriorated insulation.	.077					√
13 Inspect all collector piping for leaks and damaged insulation; tighten connections and repair as required.	.133				√	√
14 Inspect collector glazing for cracks and seals for tightness; tighten or replace seals as necessary.	.124				√	√
15 Wash/clean glazing on collector panels.	.585					√
16 Inspect ferrule around pipe flashing where solar piping runs through roof; repair as necessary.	.086				√	√
17 Check collector mounting brackets and bolts; tighten as required.	.094				√	√
18 Clean area.	.066				√	√
19 Fill out maintenance checklist and report deficiencies.	.022				√	√
Total labor-hours/period					1.165	2.230
Total labor-hours/year					1.165	2.230

	Description	Labor-hrs.	2019 Bare Costs				Total In-House	Total w/O&P
			Material	Labor	Equip.	Total		
1900	Wtr. htng. sys., solar clsd. lp., up to 6 panels, annually	2.230	355	119		474	546.02	635
1950	Annualized	3.395	475	181		656	754.40	885

For customer support on your Facilities Maintenance & Repair Costs with RSMeans data, call 800.448.8182.

PM Components	Labor-hrs.	W	M	Q	S	A
PM System D2025 260 1950						
Water heater, gas, to 120 gallon						
1 Check with operating or area personnel for deficiencies.	.035				√	√
2 Check for water leaks to tank and piping. check for fuel system leaks.	.077				√	√
3 Check gas burner and pilot for proper flame; adjust if required.	.118				√	√
4 Check operation and condition of pressure relief valve.	.010				√	√
5 Check automatic controls for proper operation (temperature	.094				√	√
regulators, thermostatic devices, automatic fuel shut off valve, etc.).						
6 Check draft diverter and clear openings, if clogged.	.027				√	√
7 Check electrical wiring for fraying and loose connections on oil burner.	.072				√	√
8 Check for proper water temperature setting; adjust as required.	.029				√	√
9 Check condition of flue pipe, and chimney.	.148				√	√
10 Drain sediment from tank.	.327					√
11 Clean up area around unit.	.066				√	√
12 Fill out maintenance checklist and report deficiencies.	.022				√	√
Total labor-hours/period					.698	1.024
Total labor-hours/year					.698	1.024

			Cost Each					
			2019 Bare Costs				**Total**	**Total**
Description	**Labor-hrs.**	**Material**	**Labor**	**Equip.**	**Total**		**In-House**	**w/O&P**
1900 Water heater, gas, to 120 gal., annually	1.024	57.50	55		112.50		133.92	159
1950 Annualized	1.721	144	92		236		276.43	325

PM Components	Labor-hrs.	W	M	Q	S	A
PM System D2025 260 2950						
Water heater, oil fired, to 100 gallon						
1 Check with operating or area personnel for deficiencies.	.035				√	√
2 Check for water leaks to tank and piping; check for fuel system leaks.	.077				√	√
3 Check burner flame and pilot on oil burner; adjust if required.	.073				√	√
4 Check operation and condition of pressure relief valve.	.010				√	√
5 Check automatic controls for proper operation (temperature	.094				√	√
regulators, thermostatic devices, automatic fuel shut off valve, etc.).						
6 Check fuel filter element on oil burner.	.068				√	√
7 Check fuel level in tank; check tank, fill pipe and fuel lines	.022				√	√
and connections for damage.						
8 Inspect, clean, and adjust electrodes and nozzles on oil burners;	.254				√	√
inspect fire box and flame detection scanner.						
9 Check electrical wiring for fraying and loose connections on oil burner.	.072				√	√
10 Check for proper water temperature setting; adjust as required.	.029				√	√
11 Clean fire box.	.577					√
12 Check for proper draft adjustment; adjust draft meter if necessary.	.005				√	√
13 Check condition of flue pipe, damper and chimney.	.147				√	√
14 Drain sediment from tank.	.325					√
15 Clean up area around unit.	.066				√	√
16 Fill out maintenance checklist and report deficiencies.	.022				√	√
Total labor-hours/period					.973	1.875
Total labor-hours/year					.973	1.875

			Cost Each					
			2019 Bare Costs				**Total**	**Total**
Description	**Labor-hrs.**	**Material**	**Labor**	**Equip.**	**Total**		**In-House**	**w/O&P**
2900 Water heater, oil fired, to 100 gal., annually	1.875	62.50	100		162.50		198.39	239
2950 Annualized	2.850	154	152		306		365.13	435

PM Components	Labor-hrs.	W	M	Q	S	A
PM System D2025 260 3950						
Water heater, steam coil, to 2500 gallon						
1 Check with operating or area personnel for deficiencies.	.035				√	√
2 Check for water leaks to tank and piping; check steam lines for leaks.	.077				√	√
3 Check operation and condition of pressure relief valve.	.010				√	√
4 Check steam modulating valve and steam condensate trap for proper operation.	.068				√	√
5 Check electrical wiring connections on controls and switches.	.094				√	√
6 Check for proper water temperature setting; adjust as required.	.029				√	√
7 Clean, test and inspect sight gauges, valves and drains.	.040				√	√
8 Check automatic controls for proper operation including temperature regulators and thermostatic devices.	.094				√	√
9 Check insulation on heater; repair as necessary.	.077				√	√
10 Drain sediment from tank.	.325					√
11 Clean up area around unit.	.066				√	√
12 Fill out maintenance checklist and report deficiencies.	.022				√	√
Total labor-hours/period					.612	.937
Total labor-hours/year					.612	.937

			Cost Each					
			2019 Bare Costs				**Total**	**Total**
Description		**Labor-hrs.**	**Material**	**Labor**	**Equip.**	**Total**	**In-House**	**w/O&P**
3900	Water heater, steam coil, to 2500 gal., annually	.937	141	50		191	219.41	257
3950	Annualized	1.549	141	82.50		223.50	262.24	310

For customer support on your Facilities Maintenance & Repair Costs with RSMeans data, call 800.448.8182.

D20 PLUMBING — D2025 262 — Valve, Pressure Relief

PM Components	Labor-hrs.	W	M	Q	S	A
PM System D2025 262 1950						
Valve, pressure relief, above 4″						
1 Inspect valve for leaks, tighten fittings as necessary.	.022					✓
2 Manually operate to check operation.	.038					✓
3 Clean valve and area around valve.	.066					✓
4 Fill out maintenance report and report deficiencies.	.022					✓
Total labor-hours/period						.148
Total labor-hours/year						.148

			Cost Each				
			2019 Bare Costs			**Total**	**Total**
Description	**Labor-hrs.**	**Material**	**Labor**	**Equip.**	**Total**	**In-House**	**w/O&P**
1900 Valve, pressure relief, above 4″, annually	.148	5.60	7.90		13.50	16.35	19.65
1950 Annualized	.148	5.60	7.90		13.50	16.35	19.65

D20 PLUMBING — D2025 265 — Valve, Pressure Regulator

PM Components	Labor-hrs.	W	M	Q	S	A
PM System D2025 265 1950						
Valve, pressure regulator, above 4″						
1 Inspect valve for leaks; repair as necessary.	.160					✓
2 Manually operate to check operation; adjust as required.	.049					✓
3 Check pressure mechanism for proper opening and closing action.	.062					✓
4 Clean valve and area around regulator.	.066					✓
5 Fill out maintenance report and report deficiencies.	.022					✓
Total labor-hours/period						.359
Total labor-hours/year						.359

			Cost Each				
			2019 Bare Costs			**Total**	**Total**
Description	**Labor-hrs.**	**Material**	**Labor**	**Equip.**	**Total**	**In-House**	**w/O&P**
1900 Valve, pressure regular, above 4″, annually	.359	5.60	19.20		24.80	30.93	37.50
1950 Annualized	.359	5.60	19.20		24.80	30.93	37.50

D20 PLUMBING — D2025 270 — Valve, Sediment Strainer

PM Components	Labor-hrs.	W	M	Q	S	A
PM System D2025 270 1950						
Valve, sediment strainer, above 4″						
1 Inspect valve for leaks, tighten fittings as necessary.	.160					✓
2 Open valve drain to remove collected sediment.	.065					✓
3 Clean valve exterior and around strainer.	.066					✓
4 Fill out maintenance checklist and report deficiencies.	.022					✓
Total labor-hours/period						.313
Total labor-hours/year						.313

			Cost Each				
			2019 Bare Costs			**Total**	**Total**
Description	**Labor-hrs.**	**Material**	**Labor**	**Equip.**	**Total**	**In-House**	**w/O&P**
1900 Valve, sediment strainer, above 4″, annually	.313	5.60	16.70		22.30	27.75	34
1950 Annualized	.313	5.60	16.70		22.30	27.75	34

For customer support on your Facilities Maintenance & Repair Costs with RSMeans data, call 800.448.8182.

407

PM Components	Labor-hrs.	W	M	Q	S	A
PM System D2025 310 1950						
Valve, automatic, above 4″						
1 Lubricate valve actuator stem and valve stem, where possible.	.022					√
2 Check automatic valve for proper operation.	.062					√
3 Check packing gland for leaks; adjust as required.	.022					√
4 Check pneumatic operator and tubing for air leaks.	.025					√
5 Clean valve exterior and area around valve.	.034					√
6 Fill out maintenance report and report deficiencies.	.022					√
Total labor-hours/period						.187
Total labor-hours/year						.187

			Cost Each				
			2019 Bare Costs			Total	Total
Description	Labor-hrs.	Material	Labor	Equip.	Total	In-House	w/O&P
1900 Valve, automatic, above 4″, annually	.187	11.15	10		21.15	25.12	30
1950 Annualized	.187	11.15	10		21.15	25.12	30

PM Components	Labor-hrs.	W	M	Q	S	A
PM System D2025 310 2950						
Valve, auto diaphragm, above 4″						
1 Lubricate valve actuator mechanism and valve stem.	.022					√
2 Check valve for proper operation and leaks; tighten packing and flange bolts as required.	.044					√
3 Check pneumatic tubing and operator mechanism for proper alignment and damage; adjust as necessary and soap solution test for leaks after maintenance.	.053					√
4 Clean valve exterior and area around valve.	.034					√
5 Fill out maintenance report and report deficiencies.	.022					√
Total labor-hours/period						.175
Total labor-hours/year						.175

			Cost Each				
			2019 Bare Costs			Total	Total
Description	Labor-hrs.	Material	Labor	Equip.	Total	In-House	w/O&P
2900 Valve, auto diaphragm, above 4″, annually	.175	11.15	9.35		20.50	24.30	29
2950 Annualized	.175	11.15	9.35		20.50	24.30	29

PM Components	Labor-hrs.	W	M	Q	S	A
PM System D2095 905 1950						
Duplex sump pump						
1 Check electrical cords, plugs and connections.	.239				✓	✓
2 Activate float switches and check pumps for proper operation.	.078				✓	✓
3 Lubricate pumps as required.	.094				✓	✓
4 Inspect packing and tighten as required.	.062				✓	✓
5 Check pumps for misalignment and bearings for overheating.	.258				✓	✓
6 Clean out trash from sump bottom.	.072				✓	✓
7 Fill out maintenance checklist and report deficiencies.	.022				✓	✓
Total labor-hours/period					.825	.825
Total labor-hours/year					.825	.825

		Cost Each						
			2019 Bare Costs				Total	Total
Description	Labor-hrs.	Material	Labor	Equip.	Total	In-House	w/O&P
1900	Duplex sump pump, annually	.825	22.50	44		66.50	81.87	98.50
1950	Annualized	1.654	45.50	88.50		134	164.28	199

PM Components	Labor-hrs.	W	M	Q	S	A
PM System D2095 910 1950						
Submersible pump, 1 H.P. and over						
1 Check with operating personnel for any deficiencies.	.035				✓	✓
2 Remove pump from pit.	.585				✓	✓
3 Clean out trash from pump intake.	.338				✓	✓
4 Check electrical plug, cord and connection.	.119				✓	✓
5 Inspect pump body for corrosion; prime and paint as necessary.	.053				✓	✓
6 Check pump and motor operation for excessive vibration, noise and overheating.	.022				✓	✓
7 Lubricate pump and motor, where applicable.	.099				✓	✓
8 Return pump to pit; reset and check float switch for proper operation.	.585				✓	✓
9 Clean area.	.066				✓	✓
10 Fill out maintenance checklist and report deficiencies.	.022				✓	✓
Total labor-hours/period					1.924	1.924
Total labor-hours/year					1.924	1.924

		Cost Each						
			2019 Bare Costs				Total	Total
Description	Labor-hrs.	Material	Labor	Equip.	Total	In-House	w/O&P
1900	Submersible pump, 1 H.P. and over, annually	1.924	19.85	102		121.85	155.01	190
1950	Annualized	3.850	31	205		236	299.39	370

For customer support on your Facilities Maintenance & Repair Costs with RSMeans data, call 800.448.8182.

409

PM Components	Labor-hrs.	W	M	Q	S	A
PM System D2095 930 1950						
Oxygen monitor.						
1 Check alarm set points.	.167			√	√	√
2 Check alarm lights for replacement.	.167			√	√	√
3 Check audible annunciator for proper operation.	.167			√	√	√
4 Check battery backup for proper operation.	.167			√	√	√
5 Check and test oxygen sensor for proper operation.	.276			√	√	√
6 Replace oxygen sensor annually.	1.000					√
7 Clean area.	.034			√	√	√
8 Fill out maintenance checklist and report deficiencies.	.022			√	√	√
Total labor-hours/period				.999	.999	1.999
Total labor-hours/year				1.997	.999	1.999

	Description	Labor-hrs.	Cost Each				Total	Total
			2019 Bare Costs				In-House	w/O&P
			Material	Labor	Equip.	Total		
1900	Oxygen monitor annually	1.999	210	107		317	368.87	435
1950	Annualized	5.000	400	266		666	783.45	925

For customer support on your Facilities Maintenance & Repair Costs with RSMeans data, call 800.448.8182.

PM Components	Labor-hrs.	W	M	Q	S	A
PM System D3025 110 1950						
Boiler, electric, to 1500 gallon						
1 Check with operating or area personnel for deficiencies.	.035			√	√	√
2 Check hot water pressure gauges.	.073			√	√	√
3 Check operation and condition of pressure relief valve.	.030			√	√	√
4 Check for proper operation of primary controls for resistance-type or electrode-type heating elements; check and adjust thermostat.	.133			√	√	√
5 Check electrical wiring to heating elements, blower, motors, overcurrent protective devices, grounding system and other electrical components as required.	.238			√	√	√
6 Check over temperature and over-pressure limit controls for proper operation.	.143			√	√	√
7 Check furnace operation through complete cycle or up to 10 minutes.	.216			√	√	√
8 Clean area around boiler.	.066			√	√	√
9 Fill out maintenance checklist and report deficiencies.	.022			√	√	√
Total labor-hours/period				.956	.956	.956
Total labor-hours/year				1.912	.956	.956

	Description	Labor-hrs.	Cost Each					
			2019 Bare Costs				Total In-House	Total w/O&P
			Material	Labor	Equip.	Total		
1900	Boiler, electric, to 1500 gal., annually	.956	57.50	61		118.50	139.46	166
1950	Annualized	3.828	57.50	245		302.50	367.10	450

For customer support on your Facilities Maintenance & Repair Costs with RSMeans data, call 800.448.8182.

PM Components	Labor-hrs.	W	M	Q	S	A
PM System D3025 130 1950						
Boiler, hot water; oil, gas or combination fired, up to 120 MBH						
1 Check combustion chamber for air or gas leaks.	.077					√
2 Inspect and clean oil burner gun and ignition assembly, where applicable.	.658					√
3 Inspect fuel system for leaks and change fuel filter element, where applicable.	.098					√
4 Check fuel lines and connections for damage.	.023		√	√	√	√
5 Check for proper operational response of burner to thermostat controls.	.133			√	√	√
6 Check and lubricate burner and blower motors.	.079			√	√	√
7 Check main flame failure protection and main flame detection scanner on boiler equipped with spark ignition (oil burner).	.124		√	√	√	√
8 Check electrical wiring to burner controls and blower.	.079					√
9 Clean firebox (sweep and vacuum).	.577					√
10 Check operation of mercury control switches (i.e., steam pressure, hot water temperature limit, atomizing or combustion air proving, etc.).	.143		√	√	√	√
11 Check operation and condition of safety pressure relief valve.	.030		√	√	√	√
12 Check operation of boiler low water cut off devices.	.056		√	√	√	√
13 Check hot water pressure gauges.	.073		√	√	√	√
14 Inspect and clean water column sight glass (or replace).	.127		√	√		√
15 Clean fire side of water jacket boiler.	.433					√
16 Check condition of flue pipe, damper and exhaust stack.	.147			√	√	√
17 Check boiler operation through complete cycle, up to 30 minutes.	.650					√
18 Check fuel level with gauge pole, add as required.	.046		√	√	√	√
19 Clean area around boiler.	.066		√	√	√	√
20 Fill out maintenance checklist and report deficiencies.	.022		√	√	√	√
Total labor-hours/period			.710	1.069	1.069	3.641
Total labor-hours/year			5.678	2.138	1.069	3.641

		Cost Each						
			2019 Bare Costs				Total	Total
Description	Labor-hrs.	Material	Labor	Equip.	Total		In-House	w/O&P
1900 Boiler, hot water, O/G/C, up to 120 MBH, annually	3.641	56.50	233		289.50		351.43	430
1950 Annualized	12.528	66	800		866		1,067.40	1,325

For customer support on your Facilities Maintenance & Repair Costs with RSMeans data, call 800.448.8182.

PM Components	Labor-hrs.	W	M	Q	S	A
PM System D3025 130 2950						
Boiler, hot water; oil, gas or combination fired, 120 to 500 MBH						
1 Check combustion chamber for air or gas leaks.	.077					√
2 Inspect and clean oil burner gun and ignition assembly where applicable.	.835					√
3 Inspect fuel system for leaks and change fuel filter element, where applicable.	.125					√
4 Check fuel lines and connections for damage.	.023		√	√	√	√
5 Check for proper operational response of burner to thermostat controls.	.170			√	√	√
6 Check and lubricate burner and blower motors.	.099			√	√	√
7 Check main flame failure protection and main flame detection scanner on boiler equipped with spark ignition (oil burner).	.155		√	√	√	√
8 Check electrical wiring to burner controls and blower.	.100					√
9 Clean firebox (sweep and vacuum).	.793					√
10 Check operation of mercury control switches (i.e., steam pressure, hot water temperature limit, atomizing or combustion air proving, etc.).	.185		√	√	√	√
11 Check operation and condition of safety pressure relief valve.	.038		√	√	√	√
12 Check operation of boiler low water cut off devices.	.070		√	√	√	√
13 Check hot water pressure gauges.	.073		√	√	√	√
14 Inspect and clean water column sight glass (or replace).	.160		√	√	√	√
15 Clean fire side of water jacket	.433					√
16 Check condition of flue pipe, damper and exhaust stack.	.184			√	√	√
17 Check boiler operation through complete cycle, up to 30 minutes.	.806					√
18 Check fuel level with gauge pole, add as required.	.046		√	√	√	√
19 Clean area around boiler.	.138		√	√	√	√
20 Fill out maintenance checklist and report deficiencies.	.022		√	√	√	√
Total labor-hours/period			.910	1.363	1.363	4.532
Total labor-hours/year			7.278	2.725	1.363	4.532

			Cost Each					
			2019 Bare Costs				Total	Total
Description		Labor-hrs.	Material	Labor	Equip.	Total	In-House	w/O&P
2900	Boiler, hot water, O/G/C, 120 to 500 MBH, annually	4.532	56.50	289		345.50	422.37	520
2950	Annualized	15.881	66	1,025		1,091	1,333.20	1,650

For customer support on your Facilities Maintenance & Repair Costs with RSMeans data, call 800.448.8182.

PM Components	Labor-hrs.	W	M	Q	S	A
PM System D3025 130 3950						
Boiler, hot water; oil, gas or combination fired, 500 to 1000 MBH						
1 Check combustion chamber for air or gas leaks.	.086					√
2 Inspect and clean oil burner gun and ignition assembly where applicable.	.910					√
3 Inspect fuel system for leaks and change fuel filter element, where applicable.	.140					√
4 Check fuel lines and connections for damage.	.026	√	√	√	√	√
5 Check for proper operational response of burner to thermostat controls.	.186			√	√	√
6 Check and lubricate burner and blower motors.	.110			√	√	√
7 Check main flame failure protection and main flame detection scanner on boiler equipped with spark ignition (oil burner).	.169		√	√	√	√
8 Check electrical wiring to burner controls and blower.	.111					√
9 Clean firebox (sweep and vacuum).	.889					√
10 Check operation of mercury control switches (i.e., steam pressure, hot water temperature limit, atomizing or combustion air proving, etc.).	.203		√	√	√	√
11 Check operation and condition of safety pressure relief valve.	.420		√	√	√	√
12 Check operation of boiler low water cut off devices.	.077		√	√	√	√
13 Check hot water pressure gauges.	.081		√	√	√	√
14 Inspect and clean water column sight glass (or replace).	.176		√	√	√	√
15 Clean fire side of water jacket	.432					√
16 Check condition of flue pipe, damper and exhaust stack.	.203			√	√	√
17 Check boiler operation through complete cycle, up to 30 minutes.	.887					√
18 Check fuel level with gauge pole, add as required.	.049		√	√	√	√
19 Clean area around boiler.	.151		√	√	√	√
20 Fill out maintenance checklist and report deficiencies.	.022		√	√	√	√
Total labor-hours/period			1.374	1.872	1.872	5.327
Total labor-hours/year			10.988	3.743	1.872	5.327

		Cost Each						
			2019 Bare Costs				Total	Total
Description	Labor-hrs.	Material	Labor	Equip.	Total		In-House	w/O&P
3900 Boiler, hot water, O/G/C, 500 to 1000 MBH, annually	5.327	56.50	340		396.50		485.39	600
3950 Annualized	17.378	66	1,100		1,166		1,452.45	1,800

For customer support on your Facilities Maintenance & Repair Costs with RSMeans data, call 800.448.8182.

PM Components	Labor-hrs.	W	M	Q	S	A
PM System D3025 130 4950						
Boiler, hot water; oil, gas or combination fired, over 1000 MBH						
1 Check combustion chamber for air or gas leaks.	.117					√
2 Inspect and clean oil burner gun and ignition assembly where applicable.	.987					√
3 Inspect fuel system for leaks and change fuel filter element, where applicable.	.147					√
4 Check fuel lines and connections for damage.	.035		√	√	√	√
5 Check for proper operational response of burner to thermostat controls.	.200			√	√	√
6 Check and lubricate burner and blower motors.	.119			√	√	√
7 Check main flame failure protection and main flame detection scanner on boiler equipped with spark ignition (oil burner).	.186		√	√	√	√
8 Check electrical wiring to burner controls and blower.	.120					√
9 Clean firebox (sweep and vacuum).	.819					√
10 Check operation of mercury control switches (i.e., steam pressure, hot water temperature limit, atomizing or combustion air proving, etc.).	.215		√	√	√	√
11 Check operation and condition of safety pressure relief valve.	.046		√	√	√	√
12 Check operation of boiler low water cut off devices.	.085		√	√	√	√
13 Check hot water pressure gauges.	.109		√	√	√	√
14 Inspect and clean water column sight glass (or replace).	.190		√	√	√	√
15 Clean fire side of water jacket	.433					√
16 Check condition of flue pipe, damper and exhaust stack.	.222			√	√	√
17 Check boiler operation through complete cycle, up to 30 minutes.	.887					√
18 Check fuel level with gauge pole, add as required.	.098		√	√	√	√
19 Clean area around boiler.	.182		√	√	√	√
20 Fill out maintenance checklist and report deficiencies.	.022		√	√	√	√
Total labor-hours/period			1.169	1.710	1.710	5.220
Total labor-hours/year			9.350	3.421	1.710	5.220

			Cost Each					
			2019 Bare Costs				Total	Total
Description		Labor-hrs.	Material	Labor	Equip.	Total	In-House	w/O&P
4900	Boiler, hot water, O/G/C, over 1000 MBH, annually	5.220	56.50	335		391.50	477.05	590
4950	Annualized	19.698	66	1,250		1,316	1,637.20	2,025

For customer support on your Facilities Maintenance & Repair Costs with RSMeans data, call 800.448.8182.

PM Components	Labor-hrs.	W	M	Q	S	A
PM System D3025 140 1950						
Boiler, steam; oil, gas or combination fired, up to 120 MBH						
1 Inspect fuel system for leaks or damage.	.098		√	√	√	√
2 Change fuel filter element and clean strainers; repair leaks, where applicable.	.581					√
3 Check main flame failure protection, positive fuel shutoff and main flame detection scanner on boiler equipped with spark ignition (oil burner)	.124		√	√	√	√
4 Check for proper operational response of burner to thermostat controls.	.133			√	√	√
5 Inspect all gas, steam and water lines, valves, connections for leaks or damage; repair as necessary.	.195			√	√	√
6 Check feedwater system and feedwater makeup control and pump.	.056		√	√	√	√
7 Check and lubricate burner and blower motors as required.	.083			√	√	√
8 Check operation and condition of safety pressure relief valve.	.030		√	√	√	√
9 Check combustion controls, combustion blower and damper modulation control.	.133					√
10 Check all indicator lamps and water/steam pressure gauges.	.073		√	√	√	√
11 Check electrical panels and wiring to burner, blowers and other components.	.079			√	√	√
12 Clean blower air-intake dampers, if required.	.055			√	√	√
13 Check condition of flue pipe, damper and exhaust stack.	.147		√	√	√	√
14 Check boiler operation through complete cycle, up to 30 minutes.	.640		√	√	√	√
15 Check water column sight glass and water level system; clean or replace sight glass, if required.	.127		√	√	√	√
16 Clean firebox (sweep and vacuum).	.577					√
17 Check fuel level with gauge pole for oil burning boilers.	.046		√	√	√	√
18 Inspect and clean oil burner gun and ignition assembly where applicable.	.650					√
19 Clean area around boiler.	.066		√	√	√	√
20 Fill out maintenance checklist and report deficiencies.	.022		√	√	√	√
Total labor-hours/period			.788	1.974	1.974	3.915
Total labor-hours/year			6.304	3.948	1.974	3.915

			Cost Each					
			2019 Bare Costs				Total	Total
Description		Labor-hrs.	Material	Labor	Equip.	Total	In-House	w/O&P
1900	Boiler, steam, O/G/C, up to 120 MBH, annually	3.915	64	250		314	382.19	470
1950	Annualized	16.189	77.50	1,025		1,102.50	1,371.50	1,700

For customer support on your Facilities Maintenance & Repair Costs with RSMeans data, call 800.448.8182.

PM Components	Labor-hrs.	W	M	Q	S	A
PM System D3025 140 2950						
Boiler, steam; oil, gas or combination fired, 120 to 500 MBH						
1 Inspect fuel system for leaks or damage.	.098		√	√	√	√
2 Change fuel filter element and clean strainers; repair leaks, where applicable.	.835					√
3 Check main flame failure protection, positive fuel shutoff and main flame detection scanner on boiler equipped with spark ignition (oil burner).	.125		√	√	√	√
4 Check for proper operational response of burner to thermostat controls.	.168			√	√	√
5 Inspect all gas, steam and water lines, valves, connections for leaks or damage; repair as necessary.	.195			√	√	√
6 Check feedwater system and feedwater makeup control and pump.	.069		√	√	√	√
7 Check and lubricate burner, blowers and motors as required.	.099			√	√	√
8 Check operation and condition of safety pressure relief valve.	.039		√	√	√	√
9 Check combustion controls, combustion blower and damper modulation control.	.169					√
10 Check all indicator lamps and water/steam pressure gauges.	.073		√	√	√	√
11 Check electrical panels and wiring to burner, blowers and other components.	.104			√	√	√
12 Clean blower air-intake dampers, if required.	.069			√	√	√
13 Check condition of flue pipe, damper and exhaust stack.	.147		√	√	√	√
14 Check boiler operation through complete cycle, up to 30 minutes.	.800			√	√	√
15 Check water column sight glass and water level system; clean or replace sight glass, if required.	.127		√	√	√	√
16 Clean firebox (sweep and vacuum).	.793					√
17 Check fuel level with gauge pole for oil burning boilers.	.046		√	√	√	√
18 Inspect and clean oil burner gun and ignition assembly where applicable.	.819					√
19 Clean area around boiler.	.137		√	√	√	√
20 Fill out maintenance checklist and report deficiencies.	.022		√	√	√	√
Total labor-hours/period			.882	2.318	2.318	4.934
Total labor-hours/year			7.059	4.635	2.318	4.934

		Cost Each						
			2019 Bare Costs				Total	Total
Description	Labor-hrs.	Material	Labor	Equip.	Total		In-House	w/O&P
2900 Boiler, steam, O/G/C, 120 to 500 MBH, annually	4.934	64	315		379		462.99	570
2950 Annualized	18.981	77.50	1,225		1,302.50		1,592.90	1,975

For customer support on your Facilities Maintenance & Repair Costs with RSMeans data, call 800.448.8182.

PM Components	Labor-hrs.	W	M	Q	S	A
PM System D3025 140 3950						
Boiler, steam; oil, gas or combination fired, 500 to 1000 MBH						
1 Inspect fuel system for leaks or damage.	.098		√	√	√	√
2 Change fuel filter element and clean strainers; repair leaks, where applicable.	.910					√
3 Check main flame failure protection, positive fuel shutoff and main flame detection scanner on boiler equipped with spark ignition (oil burner).	.140		√	√	√	√
4 Check for proper operational response of burner to thermostat controls.	.186			√	√	√
5 Inspect all gas, steam and water lines, valves, connections for leaks or damage; repair as necessary.	.195			√	√	√
6 Check feedwater system and feedwater makeup control and pump.	.075		√	√	√	√
7 Check and lubricate burner, blowers and motors as required.	.109			√	√	√
8 Check operation and condition of safety pressure relief valve.	.052		√	√	√	√
9 Check combustion controls, combustion blower and damper modulation control.	.185					√
10 Check all indicator lamps and water/steam pressure gauges.	.091		√	√	√	√
11 Check electrical panels and wiring to burner, blowers and other components.	.117			√	√	√
12 Clean blower air-intake dampers, if required.	.069			√	√	√
13 Check condition of flue pipe, damper and exhaust stack.	.147		√	√	√	√
14 Check boiler operation through complete cycle, up to 30 minutes.	.889			√	√	√
15 Check water column sight glass and water level system; clean or replace sight glass, if required.	.127		√	√	√	
16 Clean firebox (sweep and vacuum).	1.053					√
17 Check fuel level with gauge pole for oil burning boilers.	.046		√	√	√	√
18 Inspect and clean oil burner gun and ignition assembly where applicable.	.923					√
19 Clean area around boiler.	.151		√	√	√	√
20 Fill out maintenance checklist and report deficiencies.	.022		√	√	√	√
Total labor-hours/period			.949	2.514	2.514	5.585
Total labor-hours/year			7.593	5.028	2.514	5.585

			Cost Each					
			2019 Bare Costs				Total	Total
Description		Labor-hrs.	Material	Labor	Equip.	Total	In-House	w/O&P
3900	Boiler, steam, O/G/C, 500 to 1000 MBH, annually	5.585	64	355		419	513.86	635
3950	Annualized	20.696	77.50	1,325		1,402.50	1,728.70	2,150

For customer support on your Facilities Maintenance & Repair Costs with RSMeans data, call 800.448.8182.

PM Components	Labor-hrs.	W	M	Q	S	A
PM System D3025 140 4950						
Boiler, steam; oil, gas or combination fired, over 1000 MBH						
1 Inspect fuel system for leaks or damage.	.098		√	√	√	√
2 Change fuel filter element and clean strainers; repair leaks, where applicable.	1.027					√
3 Check main flame failure protection, positive fuel shutoff and main flame detection scanner on boiler equipped with spark ignition (oil burner).	.147		√	√	√	√
4 Check for proper operational response of burner to thermostat controls.	.200			√	√	√
5 Inspect all gas, steam and water lines, valves, connections for leaks abor damage; repair as necessary.	.195			√	√	√
6 Check feedwater system and feedwater makeup control and pump.	.091		√	√	√	√
7 Check and lubricate burner, blowers and motors as required.	.120		√	√	√	√
8 Check operation and condition of safety pressure relief valve.	.069		√	√	√	√
9 Check combustion controls, combustion blower and damper modulation control.	.199					√
10 Check all indicator lamps and water/steam pressure gauges.	.109		√	√	√	√
11 Check electrical panels and wiring to burner, blowers and other components.	.140			√	√	√
12 Clean blower air-intake dampers, if required.	.069			√	√	√
13 Check condition of flue pipe, damper and exhaust stack.	.147		√	√	√	√
14 Check boiler operation through complete cycle, up to 30 minutes.	.889			√	√	√
15 Check water column sight glass and water level system; clean or replace sight glass, if required.	.127		√	√	√	√
16 Clean firebox (sweep and vacuum).	1.144					√
17 Check fuel level with gauge pole for oil burning boilers.	.046		√	√	√	√
18 Inspect and clean oil burner gun and ignition assembly where applicable.	1.196					√
19 Clean area around boiler.	.182		√	√	√	√
20 Fill out maintenance checklist and report deficiencies.	.022		√	√	√	√
Total labor-hours/period			1.037	2.651	2.651	6.217
Total labor-hours/year			8.299	5.302	2.651	6.217

Description	Labor-hrs.	2019 Bare Costs — Material	2019 Bare Costs — Labor	2019 Bare Costs — Equip.	2019 Bare Costs — Total	Total In-House	Total w/O&P
4900 Boiler, steam, O/G/C, over 1000 MBH, annually	6.217	64	400		464	563.84	695
4950 Annualized	22.451	77.50	1,425		1,502.50	1,867.80	2,325

For customer support on your Facilities Maintenance & Repair Costs with RSMeans data, call 800.448.8182.

PM Components	Labor-hrs.	W	M	Q	S	A
PM System D3025 210 1950						
Deaerator tank						
1 Check tank and associated piping for leaks.	.013				√	√
2 Bottom - blow deaerator tank.	.022				√	√
3 Perform sulfite test on deaerator water sample.	.020				√	√
4 Check low and high float levels for proper operation and respective water level alarms.	.340				√	√
5 Clean steam and feedwater strainers.	.143				√	√
6 Check steam pressure regulating valve operation.	.007				√	√
7 Check all indicator lights.	.066				√	√
8 Clean unit and surrounding area.	.022				√	√
9 Fill out maintenance checklist and report deficiencies.	.130				√	√
Total labor-hours/period					.763	.763
Total labor-hours/year					.763	.763

			Cost Each					
			2019 Bare Costs				Total	Total
Description		Labor-hrs.	Material	Labor	Equip.	Total	In-House	w/O&P
1900	Deaerator tank, annually	.763	13.65	49		62.65	75.62	93
1950	Annualized	1.506	23.50	96		119.50	145.15	179

For customer support on your Facilities Maintenance & Repair Costs with RSMeans data, call 800.448.8182.

PM Components	Labor-hrs.	W	M	Q	S	A
PM System D3025 310 1950						
Pump, boiler fuel oil						
1 Check with operating or area personnel for deficiencies.	.035				√	√
2 Check for leaks on discharge piping and seals, etc.	.077				√	√
3 Check pump and motor operation for vibration, noise, overheating, etc.	.022				√	√
4 Check alignment and clearances of shaft and coupler.	.291				√	√
5 Tighten or replace loose, missing, or damaged nuts, bolts, and screws.	.005				√	√
6 Lubricate pump and motor as required.	.099				√	√
7 Clean pump, motor and surrounding area.	.066				√	√
8 Fill out maintenance checklist and report deficiencies.	.022				√	√
Total labor-hours/period					.617	.617
Total labor-hours/year					.617	.617

			Cost Each					
			2019 Bare Costs				Total	Total
	Description	Labor-hrs.	Material	Labor	Equip.	Total	In-House	w/O&P
1900	Pump, boiler fuel oil, annually	.617	67.50	39.50		107	122.90	146
1950	Annualized	1.232	107	78.50		185.50	215.27	256

PM Components	Labor-hrs.	W	M	Q	S	A
PM System D3025 310 2950						
Pump, condensate return, over 1 H.P.						
1 Check for proper operation of pump.	.022				√	√
2 Check for leaks on suction and discharge piping, seals, packing glands, etc.; make minor adjustments as required.	.077				√	√
3 Check pump and motor operation for vibration, noise, overheating, etc.	.022				√	√
4 Check alignment pump and motor; adjust as necessary.	.291				√	√
5 Lubricate pump and motor.	.099				√	√
6 Clean exterior of pump, motor and surrounding area.	.030				√	√
7 Fill out maintenance checklist and report deficiencies.	.022				√	√
Total labor-hours/period					.563	.563
Total labor-hours/year					.563	.563

			Cost Each					
			2019 Bare Costs				Total	Total
	Description	Labor-hrs.	Material	Labor	Equip.	Total	In-House	w/O&P
2900	Pump, steam condensate return, over 1 H.P., annually	.563	79	36		115	131.60	155
2950	Annualized	1.124	67.50	72		139.50	163.23	195

For customer support on your Facilities Maintenance & Repair Costs with RSMeans data, call 800.448.8182.

PM Components	Labor-hrs.	W	M	Q	S	A
PM System D3025 310 3950						
Pump, steam condensate return, duplex						
1 Check with operating or area personnel for deficiencies.	.035				√	√
2 Check for proper operation.	.022				√	√
3 Check for leaks on suction and discharge piping, seals, packing glands, etc.; make minor adjustments as required.	.154				√	√
4 Check pumps and motors operation for excessive vibration, noise and overheating.	.044				√	√
5 Check pump controller for proper operation.	.130				√	√
6 Lubricate pumps and motors.	.099				√	√
7 Clean condensate return unit and surrounding area.	.066				√	√
8 Fill out maintenance checklist and report deficiencies.	.022				√	√
Total labor-hours/period					.572	.572
Total labor-hours/year					.572	.572

	Description	Labor-hrs.	Cost Each				Total In-House	Total w/O&P
			2019 Bare Costs					
			Material	Labor	Equip.	Total		
3900	Pump, steam condensate return, duplex, annually	.572	68	36.50		104.50	119.79	142
3950	Annualized	1.142	79.50	73		152.50	177.98	213

For customer support on your Facilities Maintenance & Repair Costs with RSMeans data, call 800.448.8182.

PM Components	Labor-hrs.	W	M	Q	S	A
PM System D3035 110 1950						
Water cooling tower, forced draft, up to 50 tons						
1　Check with operating or area personnel for deficiencies.	.035				√	√
2　Check operation of unit for water leaks, noise or vibration.	.077				√	√
3　Clean and inspect hot water basin.	.155				√	√
4　Remove access panel.	.048				√	√
5　Check electrical wiring and connections; make appropriate adjustments.	.120				√	√
6　Lubricate all motor and fan bearings.	.047				√	√
7　Check fan blades or blowers for imbalance and tip clearance.	.055				√	√
8　Check belt for wear, tension and alignment; adjust as required.	.029				√	√
9　Drain and flush cold water sump and clean strainer.	.381				√	√
10　Clean inside of water tower using water hose; scrape, brush and wipe as required; heavy deposits of scale should be removed with scale removing compound.	.598				√	√
11　Refill with water, ck. make-up water asm. for leakage, adj. float if nec.	.211				√	√
12　Replace access panel.	.039				√	√
13　Remove, clean and reinstall conductivity and pH electrodes in chemical water treatment system.	.390				√	√
14　Inspect and clean around cooling tower.	.066				√	√
15　Fill out maintenance checklist and report deficiencies.	.022				√	√
Total labor-hours/period					2.273	2.273
Total labor-hours/year					2.273	2.273

			Cost Each				
			2019 Bare Costs			**Total**	**Total**
Description	**Labor-hrs.**	**Material**	**Labor**	**Equip.**	**Total**	**In-House**	**w/O&P**
1900　Cooling tower, forced draft, up to 50 tons, annually	2.273	26	145		171	209.02	259
1950　　　Annualized	4.550	37.50	291		328.50	402.79	495

PM Components	Labor-hrs.	W	M	Q	S	A
PM System D3035 110 2950						
Water cooling tower, forced draft, 51 tons through 500 tons						
1　Check with operating or area personnel for deficiencies.	.074				√	√
2　Check operation of unit for water leaks, noise or vibration.	.163				√	√
3　Clean and inspect hot water basin.	.333				√	√
4　Remove access panel.	.103				√	√
5　Check electrical wiring and connections; make appropriate adjustments.	.254				√	√
6　Lubricate all motor and fan bearings.	.100				√	√
7　Check fan blades or blowers for imbalance and tip clearance.	.246				√	√
8　Check belt for wear, tension and alignment; adjust as required.	.096				√	√
9　Drain and flush cold water sump and clean strainer.	.800				√	√
10　Clean inside of water tower using water hose; scrape, brush and wipe as required; heavy deposits of scale should be removed with scale removing compound.	1.271				√	√
11　Refill with water, ck. make-up water aseb. for leakage, adj. float if nec.	.444				√	√
12　Replace access panel.	.078				√	√
13　Remove, clean and reinstall conductivity and pH electrodes in chemical water treatment system.	.842				√	√
14　Inspect and clean around cooling tower.	.117				√	√
15　Fill out maintenance checklist and report deficiencies.	.022				√	√
Total labor-hours/period					4.944	4.944
Total labor-hours/year					4.944	4.944

			Cost Each				
			2019 Bare Costs			**Total**	**Total**
Description	**Labor-hrs.**	**Material**	**Labor**	**Equip.**	**Total**	**In-House**	**w/O&P**
2900　Cooling tower, forced draft, 50 thru 500 tons, annually	4.944	119	315		434	523.79	640
2950　　　Annualized	9.912	165	635		800	968.04	1,175

PM Components	Labor-hrs.	W	M	Q	S	A
PM System D3035 110 3950						
Water cooling tower, forced draft, 500 tons through 1000 tons						
1 Check with operating or area personnel for deficiencies.	.109				√	√
2 Check operation of unit for water leaks, noise or vibration.	.239				√	√
3 Clean and inspect hot water basin.	.485				√	√
4 Remove access panel.	.151				√	√
5 Check electrical wiring and connections; make appropriate adjustments.	.372				√	√
6 Lubricate all motor and fan bearings.	.147				√	√
7 Check fan blades or blowers for imbalance and tip clearance.	.002				√	√
8 Check belt for wear, tension and alignment; adjust as required.	.219				√	√
9 Drain and flush cold water sump and clean strainer.	1.455				√	√
10 Clean inside of water tower using water hose; scrape, brush and wipe as required; heavy deposits of scale should be removed with scale removing compound.	2.259				√	√
11 Refill with water, check make-up water assembly for leakage, adjust float if necessary.	1.578				√	√
12 Replace access panel.	.117				√	√
13 Remove, clean and reinstall conductivity and ph electrodes in chemical water treatment system.	1.231				√	√
14 Inspect and clean around cooling tower.	.172				√	√
15 Fill out maintenance checklist and report deficiencies.	.022				√	√
Total labor-hours/period					8.857	8.857
Total labor-hours/year					8.857	8.857

	Description	Labor-hrs.	2019 Bare Costs				Total In-House	Total w/O&P
			Material	Labor	Equip.	Total		
3900	Cooling tower, forced draft, 500 to 1000 tons, annually	8.857	119	565		684	832.74	1,025
3950	Annualized	17.728	165	1,125		1,290	1,589.84	1,950

For customer support on your Facilities Maintenance & Repair Costs with RSMeans data, call 800.448.8182.

PM Components	Labor-hrs.	W	M	Q	S	A
PM System D3035 130 1950						
Chiller, reciprocating, air cooled, up to 25 tons						
1 Check unit for proper operation, excessive noise or vibration.	.033		√	√	√	√
2 Run system diagnostics test.	.325		√	√	√	√
3 Check oil level in sight glass of lead compressor only, add oil as necessary.	.042		√	√	√	√
4 Check superheat and subcooling temperatures.	.325					√
5 Check liquid line sight glass, oil and refrigerant pressures.	.036		√	√	√	√
6 Check contactors, sensors and mechanical safety limits.	.094					√
7 Check electrical wiring and connections; tighten loose connections.	.120					√
8 Clean intake side of condenser coils, fans and intake screens.	1.282					√
9 Inspect fan(s) or blower(s) for bent blades of imbalance.	.086					√
10 Lubricate shaft bearings and motor bearings as required.	.291					√
11 Inspect plumbing and valves for leaks, adjust as necessary.	.077		√	√	√	√
12 Check evaporator and condenser for corrosion.	.026		√	√	√	√
13 Clean chiller and surrounding area.	.066		√	√	√	√
14 Fill out maintenance checklist and report deficiencies.	.022		√	√	√	√
Total labor-hours/period			.627	.627	.627	2.825
Total labor-hours/year			5.016	1.254	.627	2.825

			Cost Each					
			2019 Bare Costs				Total	Total
Description	Labor-hrs.	Material	Labor	Equip.	Total	In-House	w/O&P
1900 Chiller, recip., air cooled, up to 25 tons, annually	2.825	29.50	181		210.50	257.49	315
1950 Annualized	9.723	48	620		668	825.75	1,025

PM Components	Labor-hrs.	W	M	Q	S	A
PM System D3035 130 2950						
Chiller, reciprocating, air cooled, over 25 tons						
1 Check unit for proper operation, excessive noise or vibration.	.033		√	√	√	√
2 Run system diagnostics test.	.455		√	√	√	√
3 Check oil level in sight glass of lead compressor only, add oil as necessary.	.042		√	√	√	√
4 Check superheat and subcooling temperatures.	.325					√
5 Check liquid line sight glass, oil and refrigerant pressures.	.036		√	√	√	√
6 Check contactors, sensors and mechanical safety limits.	.094					√
7 Check electrical wiring and connections; tighten loose connections.	.120					√
8 Clean intake side of condenser coils, fans and intake screens.	1.282					√
9 Inspect fan(s) or blower(s) for bent blades of imbalance.	.237					√
10 Lubricate shaft bearings and motor bearings as required.	.341					√
11 Inspect plumbing and valves for leaks, adjust as necessary.	.117		√	√	√	√
12 Check evaporator and condenser for corrosion.	.052		√	√	√	√
13 Clean chiller and surrounding area.	.117		√	√	√	√
14 Fill out maintenance checklist and report deficiencies.	.022		√	√	√	√
Total labor-hours/period			.874	.874	.874	3.273
Total labor-hours/year			6.990	1.747	.874	3.273

			Cost Each					
			2019 Bare Costs				Total	Total
Description	Labor-hrs.	Material	Labor	Equip.	Total	In-House	w/O&P
2900 Chiller, recip., air cooled, over 25 tons, annually	3.273	29.50	210		239.50	292.67	360
2950 Annualized	12.888	39.50	825		864.50	1,068.25	1,325

PM Components	Labor-hrs.	W	M	Q	S	A
PM System D3035 135 1950						
Chiller, reciprocating, water cooled, up to 50 tons						
1 Check unit for proper operation, excessive noise or vibration.	.033		√	√	√	√
2 Run system diagnostics test.	.327		√	√	√	√
3 Check oil level in sight glass of lead compressor only, add oil as necessary.	.042		√	√	√	√
4 Check superheat and subcooling temperatures.	.325					√
5 Check liquid line sight glass, oil and refrigerant pressures.	.036		√	√	√	√
6 Check contactors, sensors and mechanical limits, adjust as necessary.	.094					√
7 Inspect plumbing and valves for leaks, tighten connections as necessary.	.077		√	√	√	√
8 Check condenser and evaporator for corrosion.	.026		√	√	√	√
9 Clean chiller and surrounding area.	.066		√	√	√	√
10 Fill out maintenance checklist and report deficiencies.	.022		√	√	√	√
Total labor-hours/period			.629	.629	.629	1.048
Total labor-hours/year			5.030	1.258	.629	1.048

			Cost Each				
			2019 Bare Costs			**Total**	**Total**
Description	**Labor-hrs.**	**Material**	**Labor**	**Equip.**	**Total**	**In-House**	**w/O&P**
1900 Chiller, recip., water cooled, up to 50 tons, annually	1.048	10.65	67.50		78.15	95.14	117
1950 Annualized	7.944	13.25	505		518.25	646.45	800

PM Components	Labor-hrs.	W	M	Q	S	A
PM System D3035 135 2950						
Chiller, reciprocating, water cooled, over 50 tons						
1 Check unit for proper operation, excessive noise or vibration.	.033		√	√	√	√
2 Run system diagnostics test.	.457		√	√	√	√
3 Check oil level in sight glass of lead compressor only, add oil as necessary.	.042		√	√	√	√
4 Check superheat and subcooling temperatures.	.325					√
5 Check liquid line sight glass, oil and refrigerant pressures.	.036		√	√	√	√
6 Check contactors, sensors and mechanical limits, adjust as necessary.	.094					√
7 Inspect plumbing and valves for leaks, tighten connections as necessary.	.117		√	√	√	√
8 Check condenser and evaporator for corrosion.	.052		√	√	√	√
9 Clean chiller and surrounding area.	.117		√	√	√	√
10 Fill out maintenance checklist and report deficiencies.	.022		√	√	√	√
Total labor-hours/period			.876	.876	.876	1.295
Total labor-hours/year			7.007	1.752	.876	1.295

			Cost Each				
			2019 Bare Costs			**Total**	**Total**
Description	**Labor-hrs.**	**Material**	**Labor**	**Equip.**	**Total**	**In-House**	**w/O&P**
2900 Chiller, recip., water cooled, over 50 tons, annually	1.295	13.25	82.50		95.75	117.69	145
2950 Annualized	10.908	10.65	700		710.65	878.95	1,100

For customer support on your Facilities Maintenance & Repair Costs with RSMeans data, call 800.448.8182.

PM Components	Labor-hrs.	W	M	Q	S	A
PM System D3035 140 1950						
Chiller, centrifugal water cooled, up to 100 tons						
1 Check unit for proper operation.	.035	√	√	√	√	√
2 Check oil level; add oil as necessary.	.022	√	√	√	√	√
3 Check oil temperature.	.038	√	√	√	√	√
4 Check dehydrator or purge system; remove water if observed in sight glass.	.046	√	√	√	√	√
5 Run system control tests.	.327		√	√	√	√
6 Check refrigerant charge/level, add as necessary.	.271		√	√	√	√
7 Check compressor for excessive noise/vibration.	.025		√	√	√	√
8 Check sensor and mechanical safety limits; replace as necessary.	.094				√	√
9 Clean dehydrator float valve.	.195					√
10 Perform spectrochemical analysis of compressor oil; replace oil as necessary.	.039					√
11 Replace oil filters and add oil as necessary.	.081					√
12 Inspect cooler and condenser tubes for leaks; clean screens as necessary.	5.202					√
13 Inspect utility vessel vent piping and safety relief valve; replace as necessary.	.195					√
14 Inspect/clean the economizer (vane) gas line,damper valve and actuator arm.	.650					√
15 Run an insulation test on the centrifugal motor.	1.300					√
16 Clean area around equipment.	.066	√	√	√	√	√
17 Document all maintenance and cleaning procedures.	.022	√	√	√	√	√
Total labor-hours/period		.229	.852	.852	.946	8.607
Total labor-hours/year		8.701	6.814	1.703	.946	8.607

		Cost Each						
			2019 Bare Costs				Total	Total
Description	Labor-hrs.	Material	Labor	Equip.	Total		In-House	w/O&P
1900 Chiller, centrif., water cooled, up to 100 tons, annually	8.607	201	555		756		905.15	1,100
1950 Annualized	26.771	217	1,725		1,942		2,367.60	2,925

For customer support on your Facilities Maintenance & Repair Costs with RSMeans data, call 800.448.8182.

PM Components	Labor-hrs.	W	M	Q	S	A
PM System D3035 140 2950						
Chiller, centrifugal water cooled, over 100 tons						
1 Check unit for proper operation.	.035	√	√	√	√	√
2 Check oil level; add oil as necessary.	.022	√	√	√	√	√
3 Check oil temperature.	.038	√	√	√	√	√
4 Check dehydrator or purge system; remove water if observed in sight glass.	.046	√	√	√	√	√
5 Run system control tests.	.166		√	√	√	√
6 Check refrigerant charge/level, add as necessary.	.381		√	√	√	√
7 Check compressor for excessive noise/vibration.	.039		√	√	√	√
8 Check sensor and mechanical safety limits; replace as necessary.	.133				√	√
9 Clean dehydrator float valve.	.351					√
10 Perform spectrochemical analysis of compressor oil; replace oil as necessary.	.039					√
11 Replace oil filters and add oil as necessary.	.161					√
12 Inspect cooler and condenser tubes for leaks; clean screens as necessary.	5.202					√
13 Inspect utility vessel vent piping and safety relief valve; replace as necessary.	.247					√
14 Inspect/clean the economizer (vane), gas line damper valve and actuator arm.	.650					√
15 Run an insulation test on the centrifugal motor.	1.950					√
16 Clean area around equipment.	.117	√	√	√	√	√
17 Document all maintenance and cleaning procedures.	.022	√	√	√	√	√
Total labor-hours/period		.280	1.155	1.155	1.288	9.887
Total labor-hours/year		10.631	9.238	2.310	1.288	9.887

		Cost Each						
		2019 Bare Costs					Total	Total
Description	Labor-hrs.	Material	Labor	Equip.	Total		In-House	w/O&P
2900 Chiller, centrif., water cooled, over 100 tons, annually	9.887	233	635		868		1,043.28	1,275
2950 Annualized	33.364	210	2,150		2,360		2,886.55	3,575

For customer support on your Facilities Maintenance & Repair Costs with RSMeans data, call 800.448.8182.

PM Components	Labor-hrs.	W	M	Q	S	A
PM System D3035 150 1950						
Chiller, absorption unit, up to 500 tons						
1 Check with operating personnel for deficiencies; check operating log sheets for indications of increased temperature trends.	.035			√	√	√
2 Check unit for proper operation, excessive noise or vibration.	.026			√	√	√
3 Check and clean strainers in all lines as required.	1.300					√
4 Check pulley alignment and belts for condition, proper tension and misalignment on external purge pump system, if applicable; adjust for proper tension and or alignment.	.043			√	√	√
5 Check purge pump vacuum oil level, as required; add/change oil as necessary.	.650			√	√	√
6 Lubricate pump shaft bearings and motor bearings.	.281			√	√	√
7 Check and service system controls, wirings and connections; tighten loose connections.	.119					√
8 Inspect cooling and system water piping circuits for leakage.	.077			√	√	√
9 Clean area around equipment.	.066			√	√	√
10 Fill out maintenance checklist.	.022			√	√	√
Total labor-hours/period				1.200	1.200	2.619
Total labor-hours/year				2.399	1.200	2.619

		Cost Each					
			2019 Bare Costs			Total	Total
Description	Labor-hrs.	Material	Labor	Equip.	Total	In-House	w/O&P
1900 Chiller, absorption unit, up to 500 tons, annually	2.619	41.50	167		208.50	253.78	310
1950 Annualized	6.219	59	400		459	557.30	690

PM Components	Labor-hrs.	W	M	Q	S	A
PM System D3035 150 2950						
Chiller, absorption unit, 500 to 5000 tons						
1 Check with operating personnel for deficiencies; check operating log sheets for indications of increased temperature trends.	.035			√	√	√
2 Check unit for proper operation, excessive noise or vibration.	.026			√	√	√
3 Check and clean strainers in all lines as required.	2.286					√
4 Check pulley alignment and belts for condition, proper tension and misalignment on external purge pump system, if applicable; adjust for proper tension and or alignment.	.086			√	√	√
5 Check purge pump vacuum oil level, as required; add/change oil as necessary.	.975			√	√	√
6 Lubricate pump shaft bearings and motor bearings.	.320			√	√	√
7 Check and service system controls, wirings and connections; tighten loose connections.	.139					√
8 Inspect cooling and system water piping circuits for leakage.	.152			√	√	√
9 Clean area around equipment.	.117			√	√	√
10 Fill out maintenance checklist.	.022			√	√	√
Total labor-hours/period				1.733	1.733	4.158
Total labor-hours/year				3.466	1.733	4.158

		Cost Each					
			2019 Bare Costs			Total	Total
Description	Labor-hrs.	Material	Labor	Equip.	Total	In-House	w/O&P
2900 Chiller, absorption unit, 500 to 5000 tons, annually	4.158	36.50	266		302.50	370.48	455
2950 Annualized	9.361	57.50	600		657.50	806.40	995

PM Components	Labor-hrs.	W	M	Q	S	A
PM System D3035 160 1950						
Chiller, screw, water cooled, up to 100 tons						
1 Check unit for proper operation, excessive noise or vibration.	.033		√	√	√	√
2 Run system diagnostics test.	.327		√	√	√	√
3 Check oil level and oil temperature; add oil as necessary.	.022		√	√	√	√
4 Check refrigerant pressures; add as necessary.	.271		√	√	√	√
5 Replace oil filters and oil, if applicable.	.081				√	√
6 Check contactors, sensors and mechanical safety limits.	.094				√	√
7 Perform spectrochemical analysis of compressor oil.	.039				√	√
8 Check electrical wiring and connections; tighten loose connections.	.119				√	√
9 Inspect cooler and condenser tubes for leaks.	5.202					√
10 Check evaporator and condenser for corrosion.	.026					√
11 Clean chiller and surrounding area.	.066		√	√	√	√
12 Fill out maintenance checklist and report deficiencies.	.022		√	√	√	√
Total labor-hours/period			.741	.741	1.074	6.302
Total labor-hours/year			5.926	1.482	1.074	6.302

			Cost Each					
			2019 Bare Costs				**Total**	**Total**
Description		**Labor-hrs.**	**Material**	**Labor**	**Equip.**	**Total**	**In-House**	**w/O&P**
1900	Chiller, screw, water cooled, up to 100 tons, annually	6.302	96	405		501	607.66	745
1950	Annualized	14.776	104	945		1,049	1,290.55	1,600

PM Components	Labor-hrs.	W	M	Q	S	A
PM System D3035 160 2950						
Chiller, screw, water cooled, over 100 tons						
1 Check unit for proper operation, excessive noise or vibration.	.033		√	√	√	√
2 Run system diagnostics test.	.325		√	√	√	√
3 Check oil level and oil temperature; add oil as necessary.	.022		√	√	√	√
4 Check refrigerant pressures; add as necessary.	.381		√	√	√	√
5 Replace oil filters and oil, if applicable.	.081				√	√
6 Check contactors, sensors and mechanical safety limits.	.094				√	√
7 Perform spectrochemical analysis of compressor oil.	.039				√	√
8 Check electrical wiring and connections; tighten loose connections.	.119				√	√
9 Inspect cooler and condenser tubes for leaks.	5.202					√
10 Check evaporator and condenser for corrosion.	.052					√
11 Clean chiller and surrounding area.	.117		√	√	√	√
12 Fill out maintenance checklist and report deficiencies.	.022		√	√	√	√
Total labor-hours/period			.900	.900	1.233	6.487
Total labor-hours/year			7.200	1.800	1.233	6.487

			Cost Each					
			2019 Bare Costs				**Total**	**Total**
Description		**Labor-hrs.**	**Material**	**Labor**	**Equip.**	**Total**	**In-House**	**w/O&P**
2900	Chiller, screw, water cooled, over 100 tons, annually	6.487	250	420		670	792.31	955
2950	Annualized	16.722	113	1,075		1,188	1,457.15	1,800

For customer support on your Facilities Maintenance & Repair Costs with RSMeans data, call 800.448.8182.

PM Components	Labor-hrs.	W	M	Q	S	A
PM System D3035 170 1950						
Evaporative cooler						
1 Check with operating or area personnel for deficiencies.	.035					✓
2 Check unit for proper operation, noise and vibration.	.016					✓
3 Clean evaporation louver panels.	.083					✓
4 Lubricate fan, bearings and motor.	.052					✓
5 Check belt(s) for excessive wear and deterioration.	.095					✓
6 Visually inspect wiring for damage or loose connections; tighten loose connections.	.119					✓
7 Clean sump.	.039					✓
8 Clean and adjust float.	.258					✓
9 Check interior surfaces and check components for loose paint/lime deposits.	.016					✓
10 Check/clean, adjust nozzles (6).	.039					✓
11 Remove/clean and reinstall evaporative pads.	.356					✓
12 Start unit and check for proper operation.	.055					✓
13 Remove debris from surrounding area.	.066					✓
14 Fill out maintenance checklist.	.022					✓
Total labor-hours/period						1.251
Total labor-hours/year						1.251

				Cost Each				
			2019 Bare Costs				Total	Total
Description	Labor-hrs.	Material	Labor	Equip.	Total	In-House	w/O&P
1900 Evaporative cooler, annually	1.251	42.50	80		122.50	145.61	177
1950 Annualized	1.251	42.50	80		122.50	145.61	177

PM Components	Labor-hrs.	W	M	Q	S	A
PM System D3035 180 1950						
Evaporative cooler, rotating drum						
1 Check with operating or area personnel for deficiencies.	.035					✓
2 Check unit for proper operation, noise and vibration.	.016					✓
3 Clean evaporation louver panels.	.083					✓
4 Lubricate rotary and fan, bearings and motor.	.078					✓
5 Check belt(s) for excessive wear and deterioration.	.095					✓
6 Visually inspect wiring for damage or loose connections; tighten loose connections.	.119					✓
7 Clean sump and pump intake.	.039					✓
8 Clean and adjust float.	.258					✓
9 Check interior surfaces and check components for loose paint/lime deposits.	.055					✓
10 Check/clean, adjust nozzles.	.039					✓
11 Remove/clean and reinstall evaporative pads.	.356					✓
12 Start unit and check for proper operation.	.055					✓
13 Remove debris from surrounding area.	.066					✓
14 Fill out maintenance checklist.	.022					✓
Total labor-hours/period						1.316
Total labor-hours/year						1.316

				Cost Each				
			2019 Bare Costs				Total	Total
Description	Labor-hrs.	Material	Labor	Equip.	Total	In-House	w/O&P
1900 Evaporative cooler, rotating drum, annually	1.316	42.50	84		126.50	150.80	184
1950 Annualized	1.316	42.50	84		126.50	150.80	184

PM Components	Labor-hrs.	W	M	Q	S	A
PM System D3035 210 1950						
Condenser, air cooled, 3 tons to 25 tons						
1 Check with operating or area personnel for deficiencies.	.027			√	√	√
2 Check unit for proper operation, excessive noise or vibration.	.025			√	√	√
3 Pressure wash coils and fans with coil cleaning solution.	.471					√
4 Check electrical wiring and connections; tighten loose connections.	.092					√
5 Lubricate shaft bearings and motor bearings.	.036			√	√	√
6 Inspect fan(s) or blower(s) for bent blades or imbalance; adjust as necessary.	.024			√	√	√
7 Check belt(s) for condition, proper tension, and misalignment; adjust for proper tension and/or alignment, if applicable.	.030			√	√	√
8 Inspect piping and valves for leaks; tighten connections as necessary.	.059			√	√	√
9 Clean area around equipment.	.051			√	√	√
10 Fill out maintenance checklist and report deficiencies.	.017			√	√	√
Total labor-hours/period				.269	.269	.832
Total labor-hours/year				.538	.269	.832

	Description	Labor-hrs.	2019 Bare Costs				Total In-House	Total w/O&P
			Material	Labor	Equip.	Total		
1900	Condenser, air cooled, 3 tons to 25 tons, annually	.832	34	53		87	103.90	125
1950	Annualized	1.708	71.50	109		180.50	214.70	258

PM Components	Labor-hrs.	W	M	Q	S	A
PM System D3035 210 2950						
Condenser, air cooled, 26 tons through 100 tons						
1 Check with operating or area personnel for deficiencies.	.027			√	√	√
2 Check unit for proper operation, excessive noise or vibration.	.025			√	√	√
3 Pressure wash coils and fans with coil cleaning solution.	.571					√
4 Check electrical wiring and connections; tighten loose connections.	.092					√
5 Lubricate shaft bearings and motor bearings.	.042			√	√	√
6 Inspect fan(s) or blower(s) for bent blades or imbalance; adjust as necessary.	.030			√	√	√
7 Check belt(s) for condition, proper tension, and misalignment; adjust for proper tension and/or alignment, if applicable.	.080			√	√	√
8 Inspect piping and valves for leaks; tighten connections as necessary.	.059			√	√	√
9 Clean area around equipment.	.051			√	√	√
10 Fill out maintenance checklist and report deficiencies.	.017			√	√	√
Total labor-hours/period				.331	.331	.994
Total labor-hours/year				.662	.331	.994

	Description	Labor-hrs.	2019 Bare Costs				Total In-House	Total w/O&P
			Material	Labor	Equip.	Total		
2900	Condenser, air cooled, 26 tons to 100 tons, annually	.994	58	63.50		121.50	142.97	171
2950	Annualized	1.697	80.50	109		189.50	223.53	269

For customer support on your Facilities Maintenance & Repair Costs with RSMeans data, call 800.448.8182.

PM Components	Labor-hrs.	W	M	Q	S	A
PM System D3035 210 3950						
Condenser, air cooled, over 100 tons						
1 Check with operating or area personnel for deficiencies.	.027			√	√	√
2 Check unit for proper operation, excessive noise or vibration.	.031			√	√	√
3 Pressure wash coils and fans with coil cleaning solution.	.667					√
4 Check electrical wiring and connections; tighten loose connections.	.092					√
5 Lubricate shaft bearings and motor bearings.	.042			√	√	√
6 Inspect fan(s) or blower(s) for bent blades or imbalance; adjust as necessary.	.030			√	√	√
7 Check belt(s) for condition, proper tension, and misalignment; adjust for proper tension and/or alignment, if applicable.	.100			√	√	√
8 Inspect piping and valves for leaks; tighten connections as necessary.	.059			√	√	√
9 Clean area around equipment.	.051			√	√	√
10 Fill out maintenance checklist and report deficiencies.	.017			√	√	√
Total labor-hours/period				.357	.357	1.116
Total labor-hours/year				.714	.357	1.116

			Cost Each					
			2019 Bare Costs				**Total**	**Total**
Description		**Labor-hrs.**	**Material**	**Labor**	**Equip.**	**Total**	**In-House**	**w/O&P**
3900	Condenser, air cooled, over 100 tons, annually	1.116	90.50	71.50		162	188.45	223
3950	Annualized	2.187	128	140		268	314.55	375

For customer support on your Facilities Maintenance & Repair Costs with RSMeans data, call 800.448.8182.

PM Components	Labor-hrs.	W	M	Q	S	A
PM System D3035 220 1950						
Condensing unit, air cooled, 3 tons to 25 tons						
1 Check with operating or area personnel for deficiencies.	.035			√	√	√
2 Check unit for proper operation, excessive noise or vibration.	.033			√	√	√
3 Pressure wash coils and fans with coil cleaning solution.	.615					√
4 Check electrical wiring and connections; tighten loose connections.	.119					√
5 Lubricate shaft bearings and motor bearings.	.047			√	√	√
6 Inspect fan(s) or blower(s) for bent blades or imbalance; adjust as necessary.	.031			√	√	√
7 Check belt(s) for condition, proper tension, and misalignment; adjust for proper tension and/or alignment, if applicable.	.078			√	√	√
8 Inspect piping and valves for leaks; tighten connections as necessary.	.077			√	√	√
9 Check refrigerant pressure; add refrigerant, if necessary.	.271					√
10 Clean area around equipment.	.066			√	√	√
11 Fill out maintenance checklist and report deficiencies.	.022			√	√	√
Total labor-hours/period				.389	.389	1.395
Total labor-hours/year				.778	.389	1.395

			Cost Each					
			2019 Bare Costs				**Total**	**Total**
Description		**Labor-hrs.**	**Material**	**Labor**	**Equip.**	**Total**	**In-House**	**w/O&P**
1900	Condensing unit, air cooled, 3 to 25 tons, annually	1.395	89	89.50		178.50	208.09	249
1950	Annualized	2.562	111	164		275	325.35	395

PM Components	Labor-hrs.	W	M	Q	S	A
PM System D3035 220 2950						
Condensing unit, air cooled, 26 tons through 100 tons						
1 Check with operating or area personnel for deficiencies.	.035			√	√	√
2 Check unit for proper operation, excessive noise or vibration.	.033			√	√	√
3 Pressure wash coils and fans with coil cleaning solution.	.727					√
4 Check electrical wiring and connections; tighten loose connections.	.119					√
5 Lubricate shaft bearings and motor bearings.	.055			√	√	√
6 Inspect fan(s) or blower(s) for bent blades or imbalance; adjust as necessary.	.039			√	√	√
7 Check belt(s) for condition, proper tension, and misalignment; adjust for proper tension and/or alignment, if applicable.	.104			√	√	√
8 Inspect piping and valves for leaks; tighten connections as necessary.	.077			√	√	√
9 Check refrigerant pressure; add refrigerant, if necessary.	.390					√
10 Clean area around equipment.	.066			√	√	√
11 Fill out maintenance checklist and report deficiencies.	.022			√	√	√
Total labor-hours/period				.431	.431	1.668
Total labor-hours/year				.862	.431	1.668

			Cost Each					
			2019 Bare Costs				**Total**	**Total**
Description		**Labor-hrs.**	**Material**	**Labor**	**Equip.**	**Total**	**In-House**	**w/O&P**
2900	Condensing unit, air cooled, 26 to 100 tons, annually	1.668	136	107		243	282.83	335
2950	Annualized	2.961	195	189		384	449.40	540

For customer support on your Facilities Maintenance & Repair Costs with RSMeans data, call 800.448.8182.

PM Components	Labor-hrs.	W	M	Q	S	A
PM System D3035 220 3950						
Condensing unit, air cooled, over 100 tons						
1 Check with operating or area personnel for deficiencies.	.035			√	√	√
2 Pressure wash coils and fans with coil cleaning solution.	.871					√
3 Clean intake side of condenser coils, fans and intake screens.	.091			√	√	√
4 Check electrical wiring and connections; tighten loose connections.	.120					√
5 Lubricate shaft bearings and motor bearings.	.055			√	√	√
6 Inspect fan(s) or blower(s) for bent blades or imbalance; adjust as necessary.	.039			√	√	√
7 Check belt(s) for condition, proper tension, and misalignment; adjust for proper tension and/or alignment, if applicable.	.130			√	√	√
8 Inspect piping and valves for leaks; tighten connections as necessary.	.077			√	√	√
9 Check refrigerant pressure; add refrigerant, if necessary.	.455					√
10 Clean area around equipment.	.066			√	√	√
11 Fill out maintenance checklist and report deficiencies.	.022			√	√	√
Total labor-hours/period				.515	.515	1.961
Total labor-hours/year				1.030	.515	1.961

			Cost Each					
			2019 Bare Costs				Total	Total
Description		Labor-hrs.	Material	Labor	Equip.	Total	In-House	w/O&P
3900	Condensing unit, air cooled, over 100 tons, annually	1.961	136	125		261	305.46	365
3950	Annualized	3.440	195	220		415	487.74	585

For customer support on your Facilities Maintenance & Repair Costs with RSMeans data, call 800.448.8182.

D30 HVAC | D3035 240 | Condensing Unit, Water Cooled

PM Components	Labor-hrs.	W	M	Q	S	A
PM System D3035 240 1950						
Condensing unit, water cooled, 3 tons to 24 tons						
1 Check with operating or area personnel for deficiencies.	.035			√	√	√
2 Check unit for proper operation, excessive noise or vibration.	.033			√	√	√
3 Check electrical wiring and connections; tighten loose connections.	.119					√
4 Inspect piping and valves for leaks; tighten connections as necessary.	.077			√	√	√
5 Check refrigerant pressure; add refrigerant, if necessary.	.271					√
6 Clean area around equipment.	.066			√	√	√
7 Fill out maintenance checklist and report deficiencies.	.022			√	√	√
Total labor-hours/period				.233	.233	.624
Total labor-hours/year				.466	.233	.624

			Cost Each					
			2019 Bare Costs				**Total**	**Total**
Description	**Labor-hrs.**	**Material**	**Labor**	**Equip.**	**Total**	**In-House**	**w/O&P**	
1900 Condensing unit, water cooled, 3 to 24 tons, annually	.624	31	40		71	83.13	101	
1950 Annualized	1.323	36.50	84.50		121	145	177	

PM Components	Labor-hrs.	W	M	Q	S	A
PM System D3035 240 2950						
Condensing unit, water cooled, 25 tons through 100 tons						
1 Check with operating or area personnel for deficiencies.	.035			√	√	√
2 Check unit for proper operation, excessive noise or vibration.	.033			√	√	√
3 Run system diagnostics test.	.216			√	√	√
4 Check electrical wiring and connections; tighten loose connections.	.119			√	√	√
5 Check oil level in sight glass of lead compressor only; add oil as necessary.	.029			√	√	√
6 Check contactors, sensors and mechanical limits, adjust as necessary.	.157					√
7 Check condenser for corrosion.	.026					√
8 Inspect piping and valves for leaks; tighten connections as necessary.	.077			√	√	√
9 Check refrigerant pressure; add refrigerant, if necessary.	.271					√
10 Clean area around equipment.	.066			√	√	√
11 Fill out maintenance checklist and report deficiencies.	.022			√	√	√
Total labor-hours/period				.598	.598	1.052
Total labor-hours/year				1.195	.598	1.052

			Cost Each					
			2019 Bare Costs				**Total**	**Total**
Description	**Labor-hrs.**	**Material**	**Labor**	**Equip.**	**Total**	**In-House**	**w/O&P**	
2900 Condensing unit, WC, 25 to 100 tons, annually	1.052	41	67.50		108.50	128.49	155	
2950 Annualized	2.846	57	182		239	288.06	355	

For customer support on your Facilities Maintenance & Repair Costs with RSMeans data, call 800.448.8182.

PM Components	Labor-hrs.	W	M	Q	S	A
PM System D3035 240 3950						
Condensing unit, water cooled, over 100 tons						
1 Check with operating or area personnel for deficiencies.	.035			√	√	√
2 Check unit for proper operation, excessive noise or vibration.	.033			√	√	√
3 Run system diagnostics test.	.216			√	√	√
4 Check electrical wiring and connections; tighten loose connections.	.119			√	√	√
5 Check oil level in sight glass of lead compressor only; add oil as necessary.	.029			√	√	√
6 Check contactors, sensors and mechanical limits, adjust as necessary.	.157					√
7 Check condenser for corrosion.	.026					√
8 Inspect piping and valves for leaks; tighten connections as necessary.	.077			√	√	√
9 Check refrigerant pressure; add refrigerant, if necessary.	.271					√
10 Clean area around equipment.	.066			√	√	√
11 Fill out maintenance checklist and report deficiencies.	.022			√	√	√
Total labor-hours/period				.598	.598	1.052
Total labor-hours/year				1.195	.598	1.052

			Cost Each					
			2019 Bare Costs				Total	Total
	Description	Labor-hrs.	Material	Labor	Equip.	Total	In-House	w/O&P
3900	Condensing unit, WC, over 100 tons, annually	1.052	41	67.50		108.50	128.49	155
3950	Annualized	2.846	52	182		234	283.06	345

For customer support on your Facilities Maintenance & Repair Costs with RSMeans data, call 800.448.8182.

PM Components	Labor-hrs.	W	M	Q	S	A
PM System D3035 260 1950						
Compressor, DX Refrigeration, to 25 tons						
1 Remove/replace access panel/cover.	.070			√	√	√
2 Check unit for proper operation, excessive noise or vibration.	.030			√	√	√
3 Run systems diagnostics test, if applicable.	.060			√	√	√
4 Check oil level in compressor, if possible, and add oil as required.	.030			√	√	√
5 Check refrigerant pressures; add refrigerant as necessary.	.200			√	√	√
6 Check contactors, sensors and mechanical limits, adjust as necessary.	.100			√	√	√
7 Inspect refrigerant piping and insulation.	.050			√	√	√
8 Clean compressor and surrounding area.	.050			√	√	√
9 Fill out maintenance checklist and report deficiencies.	.020			√	√	√
Total labor-hours/period				.610	.610	.610
Total labor-hours/year				1.220	.610	.610

	Description	Labor-hrs.	2019 Bare Costs				Total In-House	Total w/O&P
			Material	Labor	Equip.	Total		
1900	Compressor, DX refrigeration, to 25 tons, annually	.610	39	39		78	91.43	109
1950	Annualized	2.440	51.50	156		207.50	249.60	305

PM Components	Labor-hrs.	W	M	Q	S	A
PM System D3035 260 2950						
Compressor, DX refrigeration, 25 tons to 100 tons						
1 Check unit for proper operation, excessive noise or vibration.	.030			√	√	√
2 Run system diagnostics test, if applicable.	.098			√	√	√
3 Check oil level in compressor; add oil as necessary.	.033			√	√	√
4 Check refrigerant pressures; add refrigerant as necessary.	.271			√	√	√
5 Check contactors, sensors and mechanical limits, adjust as necessary.	.216			√	√	√
6 Inspect refrigerant piping and insulation.	.077			√	√	√
7 Clean compressor and surrounding area.	.066			√	√	√
8 Fill out maintenance checklist and report deficiencies.	.022			√	√	√
Total labor-hours/period				.814	.814	.814
Total labor-hours/year				1.627	.814	.814

	Description	Labor-hrs.	2019 Bare Costs				Total In-House	Total w/O&P
			Material	Labor	Equip.	Total		
2900	Compressor, DX refrigeration, 25 to 100 tons, annually	.814	39	52		91	107.37	130
2950	Annualized	3.256	51.50	208		259.50	315	390

For customer support on your Facilities Maintenance & Repair Costs with RSMeans data, call 800.448.8182.

PM Components	Labor-hrs.	W	M	Q	S	A
PM System D3035 290 1950						
Fluid cooler, 2 fans (no compressor)						
1 Check with operating or area personnel for deficiencies.	.035					√
2 Check unit for proper operation, excessive noise or vibration.	.159					√
3 Clean intake side of condenser coils, fans and intake screens.	.473					√
4 Check electrical wiring and connections; tighten loose connections.	.120					√
5 Inspect fan(s) for bent blades or unbalance; adjust as necessary.	.040					√
6 Check belts for condition, proper tension and misalignment; adjust for proper tension and/or alignment, if required.	.029					√
7 Lubricate shaft bearings and motor bearings.	.047					√
8 Inspect piping and valves for leaks; tighten connections as necessary.	.077					√
9 Lubricate and check operation of dampers, if applicable.	.055					√
10 Clean area around fluid cooler.	.066					√
11 Fill out maintenance checklist and report deficiencies.	.022					√
Total labor-hours/period						1.123
Total labor-hours/year						1.123

	Description	Labor-hrs.	Cost Each						
			2019 Bare Costs				**Total In-House**	**Total w/O&P**	
			Material	Labor	Equip.	Total			
1900	Fluid cooler, 2 fans (no compressor), annually	1.123	46	72		118	139.70	169	
1950	Annualized	1.123	46	72		118	139.70	169	

For customer support on your Facilities Maintenance & Repair Costs with RSMeans data, call 800.448.8182.

PM Components	Labor-hrs.	W	M	Q	S	A
PM System D3045 110 1950						
Air handling unit, 3 tons through 24 tons						
1 Check with operating or area personnel for deficiencies.	.035			√	√	√
2 Check controls and unit for proper operation.	.033			√	√	√
3 Check for unusual noise or vibration.	.033			√	√	√
4 Check tension, condition and alignment of belts, adjust as necessary.	.029			√	√	√
5 Clean coils, evaporator drain pan, blower, motor and drain piping, as required.	.381					√
6 Lubricate shaft and motor bearings.	.047			√	√	√
7 Replace air filters.	.078			√	√	√
8 Inspect exterior piping and valves for leaks; tighten connections as required.	.077			√	√	√
9 Clean area around equipment.	.066			√	√	√
10 Fill out maintenance checklist and report deficiencies.	.022			√	√	√
Total labor-hours/period				.421	.421	.802
Total labor-hours/year				.841	.421	.802

		Cost Each						
			2019 Bare Costs				Total	Total
Description	Labor-hrs.	Material	Labor	Equip.	Total	In-House	w/O&P
1900	Air handling unit, 3 thru 24 tons, annually	.802	105	43		148	171.35	200
1950	Annualized	2.061	239	110		349	405.60	475

PM Components	Labor-hrs.	W	M	Q	S	A
PM System D3045 110 2950						
Air handling unit, 25 tons through 50 tons						
1 Check with operating or area personnel for deficiencies.	.035			√	√	√
2 Check controls and unit for proper operation.	.033			√	√	√
3 Check for unusual noise or vibration.	.033			√	√	√
4 Clean coils, evaporator drain pan, blower, motor and condensate drain piping, as required.	.381					√
5 Lubricate shaft and motor bearings.	.047			√	√	√
6 Check belts for wear, proper tension, and alignment; adjust as necessary.	.029			√	√	√
7 Inspect exterior piping and valves for leaks; tighten connections as required.	.077			√	√	√
8 Check operation and clean dampers, louvers and shutters; lubricate all pivot points and linkages.	.078					√
9 Replace air filters.	.078			√	√	√
10 Clean area around equipment.	.066			√	√	√
11 Fill out maintenance checklist and report deficiencies.	.022			√	√	√
Total labor-hours/period				.421	.421	.880
Total labor-hours/year				.841	.421	.880

		Cost Each						
			2019 Bare Costs				Total	Total
Description	Labor-hrs.	Material	Labor	Equip.	Total	In-House	w/O&P
2900	Air handling unit, 25 thru 50 tons, annually	.880	234	47		281	319	370
2950	Annualized	1.940	440	104		544	619.11	715

For customer support on your Facilities Maintenance & Repair Costs with RSMeans data, call 800.448.8182.

PM Components	Labor-hrs.	W	M	Q	S	A
PM System D3045 110 3950						
Air handling unit, over 50 tons						
1 Check with operating or area personnel for deficiencies.	.035			√	√	√
2 Check controls and unit for proper operation.	.033			√	√	√
3 Check for unusual noise or vibration.	.033			√	√	√
4 Clean coils, evaporator drain pan, blower, motor and drain piping, as required.	.516					√
5 Lubricate shaft and motor bearings.	.047			√	√	√
6 Check belts for wear, proper tension, and alignment; adjust as necessary.	.055			√	√	√
7 Inspect exterior piping and valves for leaks; tighten connections as required.	.077			√	√	√
8 Check operation and clean dampers, louvers and shutters; lubricate all pivot points and linkages.	.078					√
9 Clean centrifugal fan.	.065					√
10 Replace air filters.	.286			√	√	√
11 Clean area around equipment.	.066			√	√	√
12 Fill out maintenance checklist and report deficiencies	.022			√	√	√
Total labor-hours/period				.654	.654	1.313
Total labor-hours/year				1.307	.654	1.313

			Cost Each					
			2019 Bare Costs				Total	Total
	Description	Labor-hrs.	Material	Labor	Equip.	Total	In-House	w/O&P
3900	Air handling unit, over 50 tons, annually	1.313	258	70		328	375.48	435
3950	Annualized	3.276	465	175		640	739.73	860

PM Components	Labor-hrs.	W	M	Q	S	A
PM System D3045 112 1950						
Air handling unit, computer room						
1 Check with operating or area personnel for deficiencies.	.035			√	√	√
2 Run microprocessor check, if available, or check controls and unit for proper operation.	.216			√	√	√
3 Check for unusual noise or vibration.	.033			√	√	√
4 Clean coils, evaporator drain pan, blower, motor and drain piping, as required.	.380					√
5 Lubricate shaft and motor bearings.	.047			√	√	√
6 Check belts for wear, proper tension, and alignment; adjust as necessary.	.029			√	√	√
7 Check humidity lamp, replace if necessary.	.157			√	√	√
8 Inspect exterior piping and valves for leaks; tighten connections as required.	.077			√	√	√
9 Replace air filters.	.078			√	√	√
10 Clean area around equipment.	.066			√	√	√
11 Fill out maintenance checklist and report deficiencies.	.022			√	√	√
Total labor-hours/period				.761	.761	1.141
Total labor-hours/year				1.521	.761	1.141

			Cost Each					
			2019 Bare Costs				Total	Total
	Description	Labor-hrs.	Material	Labor	Equip.	Total	In-House	w/O&P
1900	Air handling unit, computer room, annually	1.141	64	61		125	149.42	178
1950	Annualized	3.416	142	183		325	392	470

For customer support on your Facilities Maintenance & Repair Costs with RSMeans data, call 800.448.8182.

PM Components	Labor-hrs.	W	M	Q	S	A
PM System D3045 120 1950						
Fan coil unit						
1 Check with operating or area personnel for deficiencies.	.035			√	√	√
2 Check coil unit while operating.	.120				√	√
3 Remove access panel and vacuum inside of unit and coils.	.471				√	√
4 Check coils and piping for leaks, damage and corrosion; repair as necessary.	.077			√	√	√
5 Lubricate blower shaft and fan motor bearings.	.047				√	√
6 Clean coil, drip pan, and drain line with solvent.	.471				√	√
7 Replace filters as required.	.009			√	√	√
8 Replace access panel.	.023				√	√
9 Check operation after repairs.	.120				√	√
10 Clean area.	.066			√	√	√
11 Fill out maintenance checklist and report deficiencies.	.022			√	√	√
Total labor-hours/period				.209	1.461	1.461
Total labor-hours/year				.418	1.461	1.461

			Cost Each				
			2019 Bare Costs			Total	Total
Description	Labor-hrs.	Material	Labor	Equip.	Total	In-House	w/O&P
1900 Fan coil unit, annually	1.461	47	77.50		124.50	152.99	183
1950 Annualized	3.338	122	178		300	364.17	440

PM Components	Labor-hrs.	W	M	Q	S	A
PM System D3045 150 1950						
Filters, Electrostatic						
1 Review manufacturers instructions.	.091			√	√	√
2 Check indicators for defective tubes or broken ionizing wires.	.130			√	√	√
3 De-energize power supply,tag and lockout disconnect switch.	.130			√	√	√
4 Ground bus trips, top and bottom.	.065			√	√	√
5 Secure filter and fan unit.	.065			√	√	√
6 Wash each manifold until clean, approximately 4 minutes with hot water, 7 minutes with cold water.	.390			√	√	√
7 Clean or replace dry filters as necessary.	.258			√	√	√
8 Inspect for broken on hum supressors and wipe insulators with soft dry cloth.	.258			√	√	√
9 Disassemble unit as required, check it thoroughly, clean and adjust.	.640			√	√	√
10 Restore to service and check for shorts.	.130			√	√	√
11 Fill out maintenance checklist and report deficiencies.	.022			√	√	√
Total labor-hours/period				2.180	2.180	2.180
Total labor-hours/year				4.359	2.180	2.180

			Cost Each				
			2019 Bare Costs			Total	Total
Description	Labor-hrs.	Material	Labor	Equip.	Total	In-House	w/O&P
1900 Air filter, electrostatic, annually	2.180	10.10	116		126.10	160.88	199
1950 Annualized	7.602	10.10	405		415.10	534.95	665

For customer support on your Facilities Maintenance & Repair Costs with RSMeans data, call 800.448.8182.

PM Components	Labor-hrs.	W	M	Q	S	A
PM System D3045 160 1950						
VAV Boxes						
1 Open/close VAV control box access.	.060				√	√
2 Check that pneumatic tubing/electrical connections are in place and tight.	.040				√	√
3 Tighten arm on motor output shaft.	.040				√	√
4 Cycle actuator while watching for proper operation.	.200				√	√
Verify that blades fully open and close.	.040				√	√
5 Lubricate actuator linkage and damper blade pivot points.	.070				√	√
6 Fill out maintenance checklist and report deficiencies.	.017				√	√
Total labor-hours/period					.467	.467
Total labor-hours/year					.467	.467

			Cost Each					
			2019 Bare Costs				Total	Total
	Description	Labor-hrs.	Material	Labor	Equip.	Total	In-House	w/O&P
1900	VAV Boxes, annually	.467	4.75	30		34.75	42.28	52.50
1950	Annualized	.934	9.50	59.50		69	84.25	104

PM Components	Labor-hrs.	W	M	Q	S	A
PM System D3045 170 1950						
Fire dampers						
1 Remove/replace access door.	.800					√
2 Clean out debris/dirt blown against damper.	.200					√
3 Remove fusible link and check that blades operate freely.	.070					√
4 Lubricate pivot points.	.050					√
5 Replace fusible link.	.020					√
6 Fill out maintenance checklist and report deficiencies.	.017					√
Total labor-hours/period						1.157
Total labor-hours/year						1.157

			Cost Each					
			2019 Bare Costs				Total	Total
	Description	Labor-hrs.	Material	Labor	Equip.	Total	In-House	w/O&P
1900	Fire dampers, annually	1.157	12.45	74		86.45	105.50	131
1950	Annualized	1.157	12.45	74		86.45	105.50	131

PM Components	Labor-hrs.	W	M	Q	S	A
PM System D3045 210 1950						
Fan, axial, up to 5,000 CFM						
1 Start and stop fan with local switch.	.012				√	√
2 Check fan for noise and vibration.	.091				√	√
3 Check electrical wiring and connections; tighten loose connections.	.029				√	√
4 Check motor and fan shaft bearings for noise, vibration, overheating; lubricate as required.	.327				√	√
5 Clean area around fan.	.143				√	√
6 Fill out maintenance checklist and report deficiencies.	.022				√	√
Total labor-hours/period					.623	.623
Total labor-hours/year					.623	.623

				Cost Each				
			2019 Bare Costs				Total	Total
Description	Labor-hrs.	Material	Labor	Equip.	Total	In-House	w/O&P
1900 Fan, axial, up to 5,000 CFM, annually	.623	11.15	33.50		44.65	55.45	67.50
1950 Annualized	1.244	22.50	66		88.50	110.93	134

PM Components	Labor-hrs.	W	M	Q	S	A
PM System D3045 210 2950						
Fan, axial, 5,000 to 10,000 CFM						
1 Start and stop fan with local switch.	.012				√	√
2 Check motor and fan shaft bearings for noise, vibration, overheating; lubricate bearings.	.327				√	√
3 Check belts for wear, tension, and alignment, if applicable; adjust as required.	.057				√	√
4 Check fan pitch operator, lubricate; if applicable.	.029				√	√
5 Check electrical wiring and connections; tighten loose connections.	.057				√	√
6 Clean fan and surrounding fan.	.143				√	√
7 Fill out maintenance checklist and report deficiencies.	.022				√	√
Total labor-hours/period					.647	.647
Total labor-hours/year					.647	.647

				Cost Each				
			2019 Bare Costs				Total	Total
Description	Labor-hrs.	Material	Labor	Equip.	Total	In-House	w/O&P
2900 Fan, axial, 5,000 to 10,000 CFM, annually	.647	31	34.50		65.50	78.89	94.50
2950 Annualized	1.290	47	68.50		115.50	141.18	169

For customer support on your Facilities Maintenance & Repair Costs with RSMeans data, call 800.448.8182.

PM Components	Labor-hrs.	W	M	Q	S	A
PM System D3045 210 3950						
Fan, axial, 36″ to 48″ dia (over 10,000 CFM)						
1 Start and stop fan with local switch.	.012				√	√
2 Check motor and fan shaft bearings for noise, vibration, overheating; lubricate bearings.	.327				√	√
3 Check belts for wear, tension, and alignment, if applicable; adjust as required.	.086				√	√
4 Check fan pitch operator, lubricate; if applicable.	.029				√	√
5 Check electrical wiring and connections; tighten loose connections.	.078				√	√
6 Clean fan and surrounding area.	.143				√	√
7 Fill out maintenance checklist and report deficiencies.	.022				√	√
Total labor-hours/period					.696	.696
Total labor-hours/year					.696	.696

			Cost Each					
			2019 Bare Costs				Total	Total
	Description	Labor-hrs.	Material	Labor	Equip.	Total	In-House	w/O&P
3900	Fan, axial, 36″ to 48″ dia (over 10,000 CFM), annually	.696	31	37		68	82.30	98.50
3950	Annualized	1.390	47	74		121	148.08	178

PM Components	Labor-hrs.	W	M	Q	S	A
PM System D3045 220 1950						
Fan, centrifugal, up to 5,000 CFM						
1 Start and stop fan with local switch.	.012				√	√
2 Check motor and fan shaft bearings for noise, vibration, overheating; lubricate bearings.	.327				√	√
3 Check belts for wear, tension, and alignment, if applicable; adjust as required.	.057				√	√
4 Check blower intake dampers, lubricate; if applicable.	.029				√	√
5 Check electrical wiring and connections; tighten loose connections.	.029				√	√
6 Clean fan and surrounding area.	.066				√	√
7 Fill out maintenance checklist and report deficiencies.	.022				√	√
Total labor-hours/period					.542	.542
Total labor-hours/year					.542	.542

			Cost Each					
			2019 Bare Costs				**Total**	**Total**
Description	**Labor-hrs.**	**Material**	**Labor**	**Equip.**	**Total**		**In-House**	**w/O&P**
1900 Fan, centrifugal, up to 5,000 CFM, annually	.542	31	29		60		71.66	85.50
1950 Annualized	1.080	47	57.50		104.50		126.73	151

PM Components	Labor-hrs.	W	M	Q	S	A
PM System D3045 220 2950						
Fan, centrifugal, 5,000 to 10,000 CFM						
1 Start and stop fan with local switch.	.012				√	√
2 Check motor and fan shaft bearings for noise, vibration, overheating; lubricate bearings.	.327				√	√
3 Check belts for wear, tension, and alignment, if applicable; adjust as required.	.057				√	√
4 Check blower intake dampers, lubricate; if applicable.	.029				√	√
5 Check electrical wiring and connections; tighten loose connections.	.057				√	√
6 Clean fan and surrounding area.	.066				√	√
7 Fill out maintenance checklist and report deficiencies.	.022				√	√
Total labor-hours/period					.570	.570
Total labor-hours/year					.570	.570

			Cost Each					
			2019 Bare Costs				**Total**	**Total**
Description	**Labor-hrs.**	**Material**	**Labor**	**Equip.**	**Total**		**In-House**	**w/O&P**
2900 Fan, centrifugal, 5,000 to 10,000 CFM, annually	.570	31	30.50		61.50		73.60	88
2950 Annualized	1.136	47	60.50		107.50		130.58	156

For customer support on your Facilities Maintenance & Repair Costs with RSMeans data, call 800.448.8182.

PM Components	Labor-hrs.	W	M	Q	S	A
PM System D3045 220 3950						
Fan, centrifugal, over 10,000 CFM						
1 Start and stop fan with local switch.	.007				√	√
2 Check motor and fan shaft bearings for noise, vibration, overheating; lubricate bearings.	.327				√	√
3 Check belts for wear, tension, and alignment, if applicable; adjust as required.	.086				√	√
4 Check blower intake dampers, lubricate; if applicable.	.029				√	√
5 Check electrical wiring and connections; tighten loose connections.	.057				√	√
6 Clean fan and surrounding area.	.066				√	√
7 Fill out maintenance checklist and report deficiencies.	.022				√	√
Total labor-hours/period					.594	.594
Total labor-hours/year					.594	.594

		Cost Each					
		2019 Bare Costs				Total	Total
Description	Labor-hrs.	Material	Labor	Equip.	Total	In-House	w/O&P
3900 Fan, centrifugal, over 10,000 CFM, annually	.594	31	31.50		62.50	75.25	90
3950 Annualized	1.184	47	63		110	133.90	160

PM Components	Labor-hrs.	W	M	Q	S	A
PM System D3045 250 1950						
Hood and blower						
1 Check with operating or area personnel for any deficiencies.	.044			√	√	√
2 Check unit for proper operation, including switches, controls and thermostat; calibrate thermostat and repair components as required.	.109			√	√	√
3 Check operation of spray nozzles for spray coverage and drainage, if applicable.	.022					√
4 Inspect soap and spray solution feeder lines.	.007			√	√	√
5 Check components for automatic cleaning system and automatic fire protection system - solenoid valve, line strainer, shut-off valve, detergent tank, etc.	.055			√	√	√
6 Clean spray nozzles.	1.104					√
7 Tighten or replace loose, missing or damaged nuts, bolts or screws.	.005			√	√	√
8 Check operation of exhaust blower on roof; lubricate bearings; adjust tension of fan belts and clean fan blades as required.	.056			√	√	√
9 Fill out maintenance checklist and report deficiencies.	.022			√	√	√
Total labor-hours/period				.298	.298	1.424
Total labor-hours/year				.595	.298	1.424

		Cost Each					
		2019 Bare Costs				Total	Total
Description	Labor-hrs.	Material	Labor	Equip.	Total	In-House	w/O&P
1900 Hood and blower, annually	1.424	64.50	76		140.50	169.17	203
1950 Annualized	2.318	87	124		211	255.68	305

For customer support on your Facilities Maintenance & Repair Costs with RSMeans data, call 800.448.8182.

PM Components	Labor-hrs.	W	M	Q	S	A
PM System D3045 410 1950						
Centrifugal pump over 1 H.P.						
1 Check for proper operation of pump.	.022				√	√
2 Check for leaks on suction and discharge piping, seals, packing glands, etc.; make minor adjustments as required.	.077				√	√
3 Check pump and motor operation for excessive vibration, noise and overheating.	.022				√	√
4 Check alignment of pump and motor; adjust as necessary.	.258				√	√
5 Lubricate pump and motor.	.099				√	√
6 Clean exterior of pump, motor and surrounding area.	.096				√	√
7 Fill out maintenance checklist and report deficiencies.	.022				√	√
Total labor-hours/period					.596	.596
Total labor-hours/year					.596	.596

			Cost Each					
			2019 Bare Costs				Total	Total
	Description	Labor-hrs.	Material	Labor	Equip.	Total	In-House	w/O&P
1900	Centrifugal pump, over 1 H.P., annually	.596	11.15	32		43.15	53.31	65
1950	Annualized	1.196	22.50	64		86.50	107.44	130

PM Components	Labor-hrs.	W	M	Q	S	A
PM System D3045 410 2950						
Centrifugal pump w/reduction gear, over 1 H.P.						
1 Check with operating or area personnel for deficiencies.	.035				√	√
2 Clean pump exterior and check for corrosion on pump exterior and base plate.	.030				√	√
3 Check for leaks on suction and discharge piping, seals, packing glands, etc.	.077				√	√
4 Check pump, gear and motor operation for vibration, noise, overheating, etc.	.022				√	√
5 Check alignment and clearances of shaft reduction gear and coupler.	.258				√	√
6 Tighten or replace loose, missing, or damaged nuts, bolts and screws.	.005				√	√
7 Lubricate pump and motor as required.	.099				√	√
8 When available, check suction or discharge, pressure gauge readings and flow rate.	.022				√	√
9 Clean area around pump.	.066				√	√
10 Fill out maintenance checklist and report deficiencies.	.022				√	√
Total labor-hours/period					.636	.636
Total labor-hours/year					.636	.636

			Cost Each					
			2019 Bare Costs				Total	Total
	Description	Labor-hrs.	Material	Labor	Equip.	Total	In-House	w/O&P
2900	Centrifugal pump, w/ red. gear, over 1H.P., annually	.636	11.15	34		45.15	56.09	68.50
2950	Annualized	1.276	22.50	68.50		91	112.95	137

For customer support on your Facilities Maintenance & Repair Costs with RSMeans data, call 800.448.8182.

PM Components	Labor-hrs.	W	M	Q	S	A
PM System D3045 420 1950						
Pump w/oil reservoir, over 1 H.P.						
1 Check for proper operation of pump.	.022				√	√
2 Check for leaks on suction and discharge piping, seals, packing glands, etc.; make minor adjustments as required.	.077				√	√
3 Check pump and motor operation for excessive vibration, noise and overheating.	.022				√	√
4 Check alignment of pump and motor; adjust as necessary.	.258				√	√
5 Lubricate motor; check oil level in pump reservoir.	.099				√	√
6 Clean exterior of pump and surrounding area.	.096				√	√
7 Fill out maintenance checklist and report deficiencies.	.022				√	√
Total labor-hours/period					.596	.596
Total labor-hours/year					.596	.596

			Cost Each					
			2019 Bare Costs				Total	Total
	Description	Labor-hrs.	Material	Labor	Equip.	Total	In-House	w/O&P
1900	Pump w/ oil reservoir, electric, annually	.596	24	32		56	67.76	81
1950	Annualized	1.196	12.85	64		76.85	96.94	118

For customer support on your Facilities Maintenance & Repair Costs with RSMeans data, call 800.448.8182.

PM Components	Labor-hrs.	W	M	Q	S	A
PM System D3045 600 1950						
Heat exchanger, steam						
1 Check with operating or area personnel for deficiencies.	.035				√	√
2 Check temperature gauges for proper operating temperatures.	.091				√	√
3 Check steam modulating valve and steam condensate trap for proper operation.	.101				√	√
4 Inspect heat exchanger and adjacent piping for torn or deteriorated insulation.	.147				√	√
5 Clean heat exchanger and surrounding area.	.066				√	√
6 Fill out maintenance checklist and report deficiencies.	.022				√	√
Total labor-hours/period					.462	.462
Total labor-hours/year					.462	.462

			Cost Each					
			2019 Bare Costs				Total	Total
Description		Labor-hrs.	Material	Labor	Equip.	Total	In-House	w/O&P
1900	Heat exchanger, steam, annually	.462	20	24.50		44.50	53.84	64.50
1950	Annualized	.924	20	49.50		69.50	86.01	104

For customer support on your Facilities Maintenance & Repair Costs with RSMeans data, call 800.448.8182.

PM Components	Labor-hrs.	W	M	Q	S	A
PM System D3055 110 1950						
Unit heater, gas radiant						
1 Check with operating or area personnel for deficiencies.	.035					√
2 Inspect, clean and adjust control valves and thermo sensing bulbs on gas burners.	.254					√
3 Inspect fuel system for leaks.	.016					√
4 Check for proper operation of burner controls. check and adjust thermostat.	.133					√
5 Check fan and motor for vibration and noise. lubricate bearings.	.056					√
6 Check electrical wiring to blower motor.	.079					√
7 Check condition of flue pipe, damper and stack.	.147					√
8 Check unit heater operation through complete cycle or up to ten minutes.	.133					√
9 Clean area around unit heater.	.133					√
10 Fill out maintenance checklist and report deficiencies.	.022					√
Total labor-hours/period						1.009
Total labor-hours/year						1.009

			Cost Each					
				2019 Bare Costs			Total	Total
	Description	Labor-hrs.	Material	Labor	Equip.	Total	In-House	w/O&P
1900	Unit heater, gas radiant, annually	1.009	2.79	54		56.79	72.63	89.50
1950	Annualized	1.009	2.79	54		56.79	72.63	89.50

PM Components	Labor-hrs.	W	M	Q	S	A
PM System D3055 110 2950						
Unit heater, gas infrared						
1 Check with operating or area personnel for deficiencies.	.035					√
2 Inspect fuel system for leaks around unit.	.016					√
3 Replace primary air intake filter.	.048					√
4 Blow out burner.	.130					√
5 Clean spark electrode and reset gap, replace if necessary.	.254					√
6 Clean pilot and thermo-sensor bulb, replace if necessary.	.133					√
7 Check alignment of thermo-sensor bulb.	.133					√
8 Check wiring and connections; tighten any loose connections.	.079					√
9 Check and clean pilot sight glass.	.091					√
10 Vacuum pump (fan):						
A) check motor bearings for overheating; adjust or repair as required.	.022					√
B) lubricate motor bearings.	.099					√
C) check motor mounting; adjust as required.	.030					√
D) check fan blade clearance; adjust as required.	.042					√
E) check wiring, connections, switches, etc.; tighten any loose connections.	.079					√
11 Operate unit to ensure that it is in proper working condition.	.225					√
12 Clean equipment and surrounding area.	.066					√
13 Fill out maintenance checklist and report deficiencies.	.022					√
Total labor-hours/period						1.505
Total labor-hours/year						1.505

			Cost Each					
				2019 Bare Costs			Total	Total
	Description	Labor-hrs.	Material	Labor	Equip.	Total	In-House	w/O&P
2900	Unit heater, gas infrared, annually	1.505	78.50	80.50		159	189.98	226
2950	Annualized	1.505	78.50	80.50		159	189.98	226

PM Components	Labor-hrs.	W	M	Q	S	A
PM System D3055 110 3950						
Unit heater, steam						
1 Check with operating or area personnel for deficiencies.	.035					√
2 Inspect, clean and adjust control valves and thermostat.	.254					√
3 Inspect coils, connections, trap and steam piping for leaks; repair as necessary.	.195					√
4 Check fan and motor for vibration and noise; lubricate bearings.	.056					√
5 Check electrical wiring to motor.	.079					√
6 Check unit heater operation through complete cycle or up to ten minutes.	.133					√
7 Clean equipment and surrounding area.	.066					√
8 Fill out maintenance checklist and report deficiencies.	.022					√
Total labor-hours/period						.841
Total labor-hours/year						.841

			Cost Each					
			2019 Bare Costs				Total	Total
Description		Labor-hrs.	Material	Labor	Equip.	Total	In-House	w/O&P
3900	Unit heater, steam, annually	.841	47.50	45		92.50	110.19	131
3950	Annualized	.841	47.50	45		92.50	110.19	131

For customer support on your Facilities Maintenance & Repair Costs with RSMeans data, call 800.448.8182.

PM Components	Labor-hrs.	W	M	Q	S	A
PM System D3055 122 1950						
Forced air heater, oil or gas fired, up to 120 MBH						
1 Check with operating or area personnel for deficiencies.	.035			√	√	√
2 Inspect, clean and adjust electrodes and nozzles on oil burners or controls, valves and thermo-sensing bulbs on gas burners; lubricate oil burner motor bearings as applicable.	.254					√
3 Inspect fuel system for leaks.	.016			√	√	√
4 Change fuel filter element on oil burner, where applicable.						
5 Check for proper operation of burner primary controls. check and adjust thermostat.	.133			√	√	√
6 Replace air filters in air handler.	.009			√	√	√
7 Check blower and motor for vibration and noise. lubricate bearings.	.042			√	√	√
8 Check belts for wear and proper tension, tighten if required.	.029			√	√	√
9 Check electrical wiring to burner controls and blower.	.079					√
10 Inspect and clean firebox.	.571					√
11 Clean blower and air plenum.	.296					√
12 Check condition of flue pipe, damper and stack.	.147			√	√	√
13 Check furnace operation through complete cycle or up to 10 minutes.	.640			√	√	√
14 Clean area around furnace.	.066			√	√	√
15 Fill out maintenance checklist and report deficiencies.	.022			√	√	√
Total labor-hours/period				1.139	1.139	2.340
Total labor-hours/year				2.278	1.139	2.340

			Cost Each					
			2019 Bare Costs				Total	Total
Description		Labor-hrs.	Material	Labor	Equip.	Total	In-House	w/O&P
1900	Forced air heater, oil/gas, up to 120 MBH, annually	2.340	72.50	150		222.50	265.39	325
1950	Annualized	5.628	59.50	360		419.50	511.91	635

PM Components	Labor-hrs.	W	M	Q	S	A
PM System D3055 122 2950						
Forced air heater, oil or gas fired, over 120 MBH						
1 Check with operating or area personnel for deficiencies.	.035			√	√	√
2 Inspect, clean and adjust electrodes and nozzles on oil burners or controls valves and thermo-sensing bulbs on gas burners; lubricate oil burner motor bearings as applicable.	.358					√
3 Inspect fuel system for leaks.	.155			√	√	√
4 Change fuel filter element on oil burner, where applicable.	.119					√
5 Check for proper operation of burner primary controls. check and adjust thermostat.	.133			√	√	√
6 Replace air filters in air handler.	.182			√	√	√
7 Check blower and motor for vibration and noise. lubricate bearings.	.047			√	√	√
8 Check belts for wear and proper tension, tighten if required.	.057			√	√	√
9 Check electrical wiring to burner controls and blower.	.079					√
10 Inspect and clean firebox.	.577					√
11 Clean blower and air plenum.	.294					√
12 Check condition of flue pipe, damper and stack.	.147			√	√	√
13 Check furnace operation through complete cycle or up to 10 minutes.	.640			√	√	√
14 Clean area around furnace.	.066			√	√	√
15 Fill out maintenance checklist and report deficiencies.	.022			√	√	√
Total labor-hours/period				1.484	1.484	2.912
Total labor-hours/year				2.968	1.484	2.912

			Cost Each					
			2019 Bare Costs				Total	Total
Description		Labor-hrs.	Material	Labor	Equip.	Total	In-House	w/O&P
2900	Forced air heater, oil/gas, over 120 MBH, annually	2.912	123	187		310	367.32	440
2950	Annualized	7.408	237	475		712	848.60	1,025

For customer support on your Facilities Maintenance & Repair Costs with RSMeans data, call 800.448.8182.

453

PM Components	Labor-hrs.	W	M	Q	S	A
PM System D3055 210 1950						
Package unit, air cooled, 3 tons through 24 tons						
1 Check with operating or area personnel for deficiencies.	.035			√	√	√
2 Check tension, condition, and alignment of belts; adjust as necessary.	.029			√	√	√
3 Lubricate shaft and motor bearings.	.047			√	√	√
4 Replace air filters.	.055			√	√	√
5 Clean electrical wiring and connections; tighten loose connections.	.119					√
6 Clean coils, evaporator drain pan, blowers, fans, motors and drain piping as required.	.381					√
7 Perform operational check of unit; make adjustments on controls and other components as required.	.077			√	√	√
8 During operation of unit, check refrigerant pressure; add refrigerant as necessary.	.134			√	√	√
9 Check compressor oil level; add oil as required.	.033					√
10 Clean area around unit.	.066			√	√	√
11 Fill out maintenance checklist and report deficiencies.	.022			√	√	√
Total labor-hours/period				.465	.465	.999
Total labor-hours/year				.930	.465	.999

	Description	Labor-hrs.	Cost Each				Total In-House	Total w/O&P
			2019 Bare Costs					
			Material	Labor	Equip.	Total		
1900	Package unit, air cooled, 3 thru 24 ton, annually	.999	99.50	64		163.50	188.61	224
1950	Annualized	2.397	177	153		330	385.05	460

PM Components	Labor-hrs.	W	M	Q	S	A
PM System D3055 210 2950						
Package unit, air cooled, 25 tons through 50 tons						
1 Check with operating or area personnel for deficiencies.	.035			√	√	√
2 Check tension, condition, and alignment of belts; adjust as necessary.	.029			√	√	√
3 Lubricate shaft and motor bearings.	.047			√	√	√
4 Replace air filters.	.078			√	√	√
5 Clean electrical wiring and connections; tighten loose connections.	.119					√
6 Clean coils, evaporator drain pan, blowers, fans, motors and drain piping as required.	.381					√
7 Perform operational check of unit; make adjustments on controls and other components as required.	.130			√	√	√
8 During operation of unit, check refrigerant pressure; add refrigerant as necessary.	.271			√	√	√
9 Check compressor oil level; add oil as required.	.033					√
10 Clean area around unit.	.066			√	√	√
11 Fill out maintenance checklist and report deficiencies.	.022			√	√	√
Total labor-hours/period				.679	.679	1.212
Total labor-hours/year				1.358	.679	1.212

	Description	Labor-hrs.	Cost Each				Total In-House	Total w/O&P
			2019 Bare Costs					
			Material	Labor	Equip.	Total		
2900	Package unit, air cooled, 25 thru 50 ton, annually	1.212	99.50	77.50		177	205.31	245
2950	Annualized	3.249	177	208		385	453.55	540

For customer support on your Facilities Maintenance & Repair Costs with RSMeans data, call 800.448.8182.

PM Components	Labor-hrs.	W	M	Q	S	A
PM System D3055 220 1950						
Package unit, water cooled, 3 tons through 24 tons						
1 Check with operating or area personnel for deficiencies.	.035			√	√	√
2 Check tension, condition, and alignment of belts; adjust as necessary.	.029			√	√	√
3 Lubricate shaft and motor bearings.	.047			√	√	√
4 Replace air filters.	.055			√	√	√
5 Clean electrical wiring and connections; tighten loose connections.	.120					√
6 Clean coils, evaporator drain pan, blowers, fans, motors and drain piping as required.	.385					√
7 Perform operational check of unit; make adjustments on controls and other components as required.	.077			√	√	√
8 During operation of unit, check refrigerant pressure; add refrigerant as necessary.	.134			√	√	√
9 Check compressor oil level; add oil as required.	.033					√
10 Clean area around unit.	.066			√	√	√
11 Fill out maintenance checklist and report deficiencies.	.022			√	√	√
Total labor-hours/period				.465	.465	1.003
Total labor-hours/year				.931	.465	1.003

	Description	Labor-hrs.	Cost Each				Total In-House	Total w/O&P
			2019 Bare Costs					
			Material	Labor	Equip.	Total		
1900	Package unit, water cooled, 3 thru 24 ton, annually	1.003	99.50	64		163.50	189.17	224
1950	Annualized	2.402	177	153		330	385.60	460

PM Components	Labor-hrs.	W	M	Q	S	A
PM System D3055 220 2950						
Package unit, water cooled, 25 tons through 50 tons						
1 Check with operating or area personnel for deficiencies.	.035			√	√	√
2 Check tension, condition, and alignment of belts; adjust as necessary.	.029			√	√	√
3 Lubricate shaft and motor bearings.	.047			√	√	√
4 Replace air filters.	.078			√	√	√
5 Clean electrical wiring and connections; tighten loose connections.	.119					√
6 Clean coils, evaporator drain pan, blowers, fans, motors and drain piping as required.	.381					√
7 Perform operational check of unit; make adjustments on controls and other components as required.	.130			√	√	√
8 During operation of unit, check refrigerant pressure; add refrigerant as necessary.	.271			√	√	√
9 Check compressor oil level; add oil as required.	.033					√
10 Clean area around unit.	.066			√	√	√
11 Fill out maintenance checklist and report deficiencies.	.022			√	√	√
Total labor-hours/period				.679	.679	1.212
Total labor-hours/year				1.358	.679	1.212

	Description	Labor-hrs.	Cost Each				Total In-House	Total w/O&P
			2019 Bare Costs					
			Material	Labor	Equip.	Total		
2900	Package unit, water cooled, 25 thru 50 ton, annually	1.212	99.50	77.50		177	205.31	245
2950	Annualized	3.249	177	208		385	453.55	540

PM Components	Labor-hrs.	W	M	Q	S	A
PM System D3055 230 1950						
Package unit, computer room						
1 Check with operating or area personnel for deficiencies.	.035			√	√	√
2 Run microprocessor check, if available, or check controls and unit for proper operation.	.216			√	√	√
3 Check for unusual noise or vibration.	.033			√	√	√
4 Clean coils, evaporator drain pan, humidifier pan, blower, motor and drain piping as required.	.381				√	√
5 Replace air filters.	.078			√	√	√
6 Lubricate shaft and motor bearings.	.047			√	√	√
7 Check belts for wear, proper tension, and alignment; adjust as necessary.	.029			√	√	√
8 Check humidity lamp, replace if necessary.	.155			√	√	√
9 During operation of unit, check refrigerant pressures; add refrigerant as necessary.	.271				√	√
10 Inspect exterior piping and valves for leaks; tighten connections as required.	.077			√	√	√
11 Clean area around unit.	.066			√	√	√
12 Fill out maintenance checklist and report deficiencies.	.022			√	√	√
Total labor-hours/period				.759	1.411	1.411
Total labor-hours/year				1.518	1.411	1.411

			Cost Each					
			2019 Bare Costs				Total	Total
	Description	Labor-hrs.	Material	Labor	Equip.	Total	In-House	w/O&P
1900	Package unit, computer room, annually	1.411	91	90.50		181.50	211.93	254
1950	Annualized	4.336	177	278		455	539.05	650

For customer support on your Facilities Maintenance & Repair Costs with RSMeans data, call 800.448.8182.

PM Components	Labor-hrs.	W	M	Q	S	A
PM System D3055 240 1950						
Package unit, with duct gas heater						
1 Check with operating or area personnel for deficiencies.	.035			√	√	√
2 Check tension, condition and alignment of belts; adjust as necessary.	.029			√	√	√
3 Lubricate shaft and motor bearings.	.047			√	√	√
4 Replace air filters.	.078			√	√	√
5 Check electrical wiring and connections; tighten loose connections.	.119					√
6 Clean coils, evaporator drain pan, blowers, fans, motors and drain piping as required.	.385					√
7 Perform operational check of unit; make adjustments on controls and other components as required.	.077			√	√	√
8 During operation of unit, check refrigerant pressures; add refrigerant as necessary.	.271			√	√	√
9 Check compressor oil level; add oil as required.	.033					√
10 Inspect, clean and adjust control valves and thermo-sensing bulbs on gas burners.	.254					√
11 Inspect fuel system for leaks.	.016			√	√	√
12 Check for proper operation of burner primary controls. Check and adjust thermostat.	.133					√
13 Check electrical wiring to burner controls.	.079					√
14 Inspect and clean firebox.	.571					√
15 Check condition of flue pipe, damper and stack.	.147			√	√	√
16 Check heater operation through complete cycle or up to 10 minutes.	.225					√
17 Clean area around entire unit.	.066			√	√	√
18 Fill out maintenance checklist and report deficiencies.	.022			√	√	√
Total labor-hours/period				.788	.788	2.588
Total labor-hours/year				1.577	.788	2.588

			Cost Each					
			2019 Bare Costs				**Total**	**Total**
Description	Labor-hrs.	Material	Labor	Equip.	Total	In-House	w/O&P
1900	Package unit with duct gas heater, annually	2.588	103	165		268	318.78	385
1950	Annualized	4.956	181	315		496	592.90	715

457

PM Components	Labor-hrs.	W	M	Q	S	A
PM System D3055 250 1950						
Air conditioning split system, DX, air cooled, up to 10 tons						
1 Check with operating or area personnel for deficiencies.	.035			√	√	√
2 Clean intake side of condenser coils, fans and intake screens.	.055			√	√	√
3 Lubricate shaft and motor bearings.	.047			√	√	√
4 Pressure wash condenser coils with coil clean solution, as required.	.611					√
5 Replace air filters.	.078				√	√
6 Clean electrical wiring and connections; tighten loose connections.	.120					√
7 Clean evaporator coils, drain pan, blowers, fans, motors and drain piping as required.	.380					√
8 Perform operational check of unit; make adjustments on controls and other components as required.	.033			√	√	√
9 During operation of unit, check refrigerant pressure; add refrigerant as necessary.	.272			√	√	√
10 Clean area around equipment.	.066			√	√	√
11 Fill out maintenance checklist and report deficiencies.	.022			√	√	√
Total labor-hours/period				.608	.608	1.719
Total labor-hours/year				1.216	.608	1.719

			Cost Each				
			2019 Bare Costs			**Total**	**Total**
Description	Labor-hrs.	Material	Labor	Equip.	Total	In-House	w/O&P
1900 A/C, split sys., DX, air cooled, to 10 tons, annually	1.719	82.50	110		192.50	226.60	273
1950 Annualized	3.543	176	227		403	475.55	570

PM Components	Labor-hrs.	W	M	Q	S	A
PM System D3055 250 2950						
Air conditioning split system, DX, air cooled, over 10 tons						
1 Check with operating or area personnel for deficiencies.	.035			√	√	√
2 Check tension, condition, and alignment of belts; adjust as necessary.	.029			√	√	√
3 Lubricate shaft and motor bearings.	.047			√	√	√
4 Pressure wash condenser coils with coil clean solution, as required.	.611			√	√	√
5 Replace air filters.	.078			√	√	√
6 Clean electrical wiring and connections; tighten loose connections.	.120					√
7 Clean evaporator coils, drain pan, blowers, fans, motors and drain piping as required.	.380					√
8 Perform operational check of unit; make adjustments on controls and other components as required.	.033			√	√	√
9 During operation of unit, check refrigerant pressure; add refrigerant as necessary.	.271			√	√	√
10 Check compressor oil level; add oil as required.	.033			√	√	√
11 Clean area around equipment.	.066			√	√	√
12 Fill out maintenance checklist and report deficiencies.	.022			√	√	√
Total labor-hours/period				.615	.615	1.726
Total labor-hours/year				1.230	.615	1.726

			Cost Each				
			2019 Bare Costs			**Total**	**Total**
Description	Labor-hrs.	Material	Labor	Equip.	Total	In-House	w/O&P
2900 A/C, split sys., DX, air cooled, over 10 T, annually	1.726	99.50	110		209.50	246.18	295
2950 Annualized	3.571	177	228		405	479.05	575

For customer support on your Facilities Maintenance & Repair Costs with RSMeans data, call 800.448.8182.

PM Components	Labor-hrs.	W	M	Q	S	A
PM System D3055 310 1950						
Heat pump, air cooled, up to 5 tons						
1 Check with operating or area personnel for deficiencies.	.035			√	√	√
2 Check unit for proper operation, excessive noise or vibration.	.033			√	√	√
3 Clean intake side of condenser coils, fans and intake screens.	.055			√	√	√
4 Check electrical wiring and connections; tighten loose connections.	.119					√
5 Inspect fan(s) for bent blades or unbalance; adjust and clean as necessary.	.027					√
6 Check belts for condition, proper tension and misalignment; adjust as required.	.029			√	√	√
7 Lubricate shaft bearings and motor bearings.	.047			√	√	√
8 Inspect piping and valves for leaks; tighten connections as necessary.	.077			√	√	√
9 Replace air filters.	.078			√	√	√
10 Check refrigerant pressure; add refrigerant as necessary.	.134			√	√	√
11 Clean evaporative drain pan, and drain piping as required.	.390					√
12 Cycle the reverse cycle valve to insure proper operation.	.091			√	√	√
13 Clean area around equipment.	.066			√	√	√
14 Fill out maintenance checklist and report deficiencies.	.022			√	√	√
Total labor-hours/period				.667	.667	1.204
Total labor-hours/year				1.335	.667	1.204

		Cost Each					
			2019 Bare Costs			Total	Total
Description	Labor-hrs.	Material	Labor	Equip.	Total	In-House	w/O&P
1900 Heat pump, air cooled, up to 5 ton, annually	1.204	81	77		158	184.45	220
1950 Annualized	3.002	189	192		381	446.33	535

PM Components	Labor-hrs.	W	M	Q	S	A
PM System D3055 310 2950						
Heat pump, air cooled, over 5 tons						
1 Check with operating or area personnel for deficiencies.	.035			√	√	√
2 Check unit for proper operation, excessive noise or vibration.	.033			√	√	√
3 Clean intake side of condenser coils, fans and intake screens.	.055			√	√	√
4 Check electrical wiring and connections; tighten loose connections.	.119					√
5 Inspect fan(s) for bent blades or unbalance; adjust and clean as necessary.	.055					√
6 Check belts for condition, proper tension and misalignment; adjust as required.	.057			√	√	√
7 Lubricate shaft bearings and motor bearings.	.047			√	√	√
8 Inspect piping and valves for leaks; tighten connections as necessary.	.077					√
9 Replace air filters.	.155			√	√	√
10 Check refrigerant pressure; add refrigerant as necessary.	.134			√	√	√
11 Lubricate and check operation of dampers, if applicable.	.029					√
12 Check compressor oil level and add oil, if required.	.033			√	√	√
13 Cycle the reverse cycle valve to insure proper operation.	.091			√	√	√
14 Clean evaporative drain pan, and drain piping as required.	.390					√
15 Clean area around equipment.	.066			√	√	√
16 Fill out maintenance checklist and report deficiencies.	.022			√	√	√
Total labor-hours/period				.729	.729	1.400
Total labor-hours/year				1.458	.729	1.400

		Cost Each					
			2019 Bare Costs			Total	Total
Description	Labor-hrs.	Material	Labor	Equip.	Total	In-House	w/O&P
2900 Heat pump, air cooled, over 5 ton, annually	1.400	104	89.50		193.50	225.81	269
2950 Annualized	3.591	184	230		414	488.03	585

PM Components	Labor-hrs.	W	M	Q	S	A
PM System D3055 320 1950						
Heat pump, water cooled, up to 5 tons						
1 Check with operating or area personnel for deficiencies.	.035			√	√	√
2 Check unit for proper operation, excessive noise or vibration.	.033			√	√	√
3 Clean intake side of evaporator coil, fans and intake screens.	.055			√	√	√
4 Check electrical wiring and connections; tighten loose connections.	.119					√
5 Inspect fan(s) for bent blades or unbalance; adjust and clean as necessary.	.027					√
6 Check belts for condition, proper tension and misalignment; adjust as required.	.029			√	√	√
7 Lubricate shaft bearings and motor bearings.	.047			√	√	√
8 Inspect piping and valves for leaks; tighten connections as necessary.	.077			√	√	√
9 Replace air filters.	.078			√	√	√
10 Check refrigerant pressure; add refrigerant as necessary.	.136			√	√	√
11 Lubricate and check operation of dampers, if applicable.	.029					√
12 Clean evaporator drain pan and drain line with solvent.	.381					√
13 Cycle reverse cycle valve to insure proper operation.	.091			√	√	√
14 Backwash condenser coil to remove sediment.	.327					√
15 Clean area around equipment.	.066			√	√	√
16 Fill out maintenance checklist and report deficiencies.	.022			√	√	√
Total labor-hours/period				.668	.668	1.551
Total labor-hours/year				1.337	.668	1.551

			Cost Each					
				2019 Bare Costs			Total	Total
	Description	Labor-hrs.	Material	Labor	Equip.	Total	In-House	w/O&P
1900	Heat pump, water cooled, up to 5 ton, annually	1.551	94	99.50		193.50	226.37	272
1950	Annualized	3.555	188	228		416	489.19	590

PM Components	Labor-hrs.	W	M	Q	S	A
PM System D3055 320 2950						
Heat pump, water cooled, over 5 tons						
1 Check with operating or area personnel for deficiencies.	.035			√	√	√
2 Check unit for proper operation, excessive noise or vibration.	.033			√	√	√
3 Clean intake side of condenser coils, fans and intake screens.	.055			√	√	√
4 Check electrical wiring and connections; tighten loose connections.	.119					√
5 Inspect fan(s) for bent blades or unbalance; adjust and clean as necessary.	.027					√
6 Check belts for condition, proper tension and misalignment; adjust as required.	.029			√	√	√
7 Lubricate shaft bearings and motor bearings.	.047			√	√	√
8 Inspect piping and valves for leaks; tighten connections as necessary.	.077			√	√	√
9 Replace air filters.	.078			√	√	√
10 Check refrigerant pressure; add refrigerant as necessary.	.136			√	√	√
11 Lubricate and check operation of dampers, if applicable.	.029					√
12 Clean evaporator drain pan and drain line with solvent.	.381					√
13 Cycle reverse cycle valve to insure proper operation.	.091			√	√	√
14 Clean area around equipment.	.066			√	√	√
15 Fill out maintenance checklist and report deficiencies.	.022			√	√	√
Total labor-hours/period				.668	.668	1.225
Total labor-hours/year				1.337	.668	1.225

			Cost Each					
				2019 Bare Costs			Total	Total
	Description	Labor-hrs.	Material	Labor	Equip.	Total	In-House	w/O&P
2900	Heat pump, water cooled, over 5 ton, annually	1.225	94	78.50		172.50	200.37	240
2950	Annualized	3.228	188	207		395	463.19	555

For customer support on your Facilities Maintenance & Repair Costs with RSMeans data, call 800.448.8182.

PM Components	Labor-hrs.	W	M	Q	S	A
PM System D3065 100 1950						
Controls, central system, electro/pneumatic						
1 With panel disconnected from power source, clean patrol panel compartment with a vacuum.	.471					√
2 Inspect wiring/components for loose connections; tighten, as required.	.119					√
3 Check set point of controls temperature, humidity or pressure.	.033					√
4 Check unit over its range of control.	.033					√
5 Check for correct pressure differential on all two position controllers.	.195					√
6 Check source of the signal and its amplification on electronic controls.	.471					√
7 Check air systems for leaks; repair as necessary.	.155					√
8 Check relays, pilot valves and pressure regulators for proper operation; repair or replace as necessary.	.327					√
9 Replace air filters in sensors, controllers, and thermostats as necessary.	.029					√
10 Clean area around equipment.	.066					√
11 Fill out maintenance checklist and report deficiencies.	.022					√
Total labor-hours/period						1.921
Total labor-hours/year						1.921

			Cost Each					
			2019 Bare Costs				Total	Total
Description	Labor-hrs.	Material	Labor	Equip.	Total	In-House	w/O&P
1900	Controls, central system, electro/pneumatic, annually	1.921	68.50	123		191.50	228.53	276
1950	Annualized	1.921	68.50	123		191.50	228.53	276

For customer support on your Facilities Maintenance & Repair Costs with RSMeans data, call 800.448.8182.

PM Components	Labor-hrs.	W	M	Q	S	A
PM System D3095 110 1950						
Air compressor, gas engine						
1 Check with operating or area personnel for any obvious deficiencies.	.035		√	√	√	√
2 Check compressor oil level; add oil as required.	.022		√	√	√	√
3 Replace compressor oil.	.222				√	√
4 Clean air intake filter on air compressor(s); replace if necessary.	.178				√	√
5 Clean cylinder cooling fins and air cooler on compressor(s).	.155				√	√
6 Check tension, condition, and alignment of v-belts; adjust as necessary.	.030				√	√
7 Clean oil and water traps.	.178				√	√
8 Drain moisture from air storage tank and check discharge for indication of interior corrosion.	.046				√	√
9 Perform operation check of compressor; check the operation of low pressure cut-in and high pressure cut-out switches.	.221		√	√	√	√
10 Check operation of safety pressure relief valve.	.030				√	√
11 Check and tighten compressor foundation anchor bolts.	.022				√	√
12 Check radiator coolant; add if required.	.012		√	√	√	√
13 Check battery water; add if required.	.242		√	√	√	√
14 Check wiring, connections, switches, etc.; tighten loose connections.	.119				√	√
15 Check engine oil level; add if required.	.014		√	√	√	√
16 Change engine oil.	.225				√	√
17 Change engine oil filter.	.059				√	√
18 Check spark plug and reset cap.	.035				√	√
19 Check condition of engine air filter; replace if necessary.	.039				√	√
20 Test run engine for proper operation.	.410		√	√	√	√
21 Wipe dust and dirt from engine and compressor.	.109		√	√	√	√
22 Check muffler/exhaust system for corrosion.	.020				√	√
23 Clean area around equipment.	.066		√	√	√	√
24 Fill out maintenance checklist and report deficiencies.	.022		√	√	√	√
Total labor-hours/period			1.154	1.154	2.512	2.512
Total labor-hours/year			9.228	2.307	2.512	2.512

			Cost Each					
			2019 Bare Costs				Total	Total
Description		Labor-hrs.	Material	Labor	Equip.	Total	In-House	w/O&P
1900	Air compressor, gas engine powered, annually	2.512	121	160		281	331.92	400
1950	Annualized	16.575	213	1,050		1,263	1,553.17	1,925

For customer support on your Facilities Maintenance & Repair Costs with RSMeans data, call 800.448.8182.

PM Components	Labor-hrs.	W	M	Q	S	A
PM System D3095 114 1950						
Air compressor, centrifugal, to 40 H.P.						
1 Check compressor oil level; add oil as necessary.	.022			√	√	√
2 Perform operational check of compressor system and adjust as required.	.222			√	√	√
3 Check motor for excessive vibration, noise and overheating; lubricate.	.039			√	√	√
4 Check operation of pressure relief valve.	.030			√	√	√
5 Clean cooling fans and air cooler.	.023			√	√	√
6 Check tension, condition, and alignment of V-belts; adjust as necessary.	.030			√	√	√
7 Drain moisture from air storage tank and check low pressure cut-in; while draining, check discharge for indication of interior corrosion.	.046			√	√	√
8 Replace air intake filter, as needed.	.178			√	√	√
9 Clean oil and water trap.	.178			√	√	√
10 Clean compressor and surrounding area.	.066			√	√	√
11 Fill out maintenance checklist and report deficiencies.	.022			√	√	√
Total labor-hours/period				.856	.856	.856
Total labor-hours/year				1.712	.856	.856

			Cost Each					
			2019 Bare Costs				Total	Total
Description	Labor-hrs.	Material	Labor	Equip.	Total	In-House	w/O&P
1900 Air compressor, centrifugal, to 40 H.P., annually	.856	54	54.50		108.50	127.74	153
1950 Annualized	3.412	76.50	219		295.50	354.55	435

For customer support on your Facilities Maintenance & Repair Costs with RSMeans data, call 800.448.8182.

PM Components	Labor-hrs.	W	M	Q	S	A
PM System D3095 114 2950						
Air compressor, centrifugal, over 40 H.P.						
1 Check with operating or area personnel for any obvious deficiencies.	.035		√	√	√	√
2 Perform control system check.	.327				√	√
3 Lubricate main driver coupling if necessary.	.047				√	√
4 Lubricate prelube pump motor and pump coupling, if necessary.	.047				√	√
5 Perform operation check of air compressor, adjust as required.	.222		√	√	√	√
6 Check main driver coupling alignment; realign if necessary.	.327				√	√
7 Check compressor and motor operation for excessive vibration, noise and overheating; lubricate motor.	.039		√	√	√	√
8 Check compressor scrolls, piping, and impeller housing for all leaks or cracks.	.327		√	√	√	√
9 Check intercoolers and aftercoolers for high cooling water temperatures or cooling water leakage.	.195		√	√	√	√
10 Check oil level in compressor oil reservoir; add oil as necessary.	.022		√	√	√	√
11 Replace compressor oil and oil filters.	.282					√
12 Record oil pressure and oil temperature.	.013		√	√	√	√
13 Check compressor for oil leaks.	.026		√	√	√	√
14 Visually inspect oil mist arrestor, clean housing, lines, and replace element if saturated, if applicable.	.177				√	√
15 Check operation of pressure relief valve.	.030		√	√	√	√
16 Visually inspect discharge check valve.	.009				√	√
17 Check oil and water trap.	.022		√	√	√	√
18 Check indicating lamps or gauges for proper operation if appropriate; replace burned out lamps or repair/replace gauges.	.020					√
19 Visually check all air intake filter elements; replace if necessary.	.022		√	√	√	√
20 Replace all air intake filter elements.	.177					√
21 Lubricate motor.	.047				√	√
22 Clean compressor, motor and surrounding area.	.066		√	√	√	√
23 Fill out maintenance check and report deficiencies.	.022		√	√	√	√
Total labor-hours/period			1.041	1.041	2.021	2.500
Total labor-hours/year			8.325	2.081	2.021	2.500

			Cost Each					
			2019 Bare Costs				**Total**	**Total**
Description		**Labor-hrs.**	**Material**	**Labor**	**Equip.**	**Total**	**In-House**	**w/O&P**
2900	Air compressor, centrifugal, over 40 H.P., annually	2.500	98.50	160		258.50	307.25	370
2950	Annualized	14.889	104	950		1,054	1,296.78	1,600

For customer support on your Facilities Maintenance & Repair Costs with RSMeans data, call 800.448.8182.

PM Components	Labor-hrs.	W	M	Q	S	A
PM System D3095 118 1950						
Air compressor, reciprocating, less than 5 H.P.						
1 Replace compressor oil.	.340			√	√	√
2 Perform operation check of compressor system and adjust as required.	.222			√	√	√
3 Check motor operation for excessive vibration, noise and overheating.	.042			√	√	√
4 Lubricate motor.	.047			√	√	√
5 Check operation of pressure relief valve.	.030			√	√	√
6 Check tension, condition, and alignment of V-belts; adjust as necessary.	.030			√	√	√
7 Drain moisture from air storage tank and check low pressure cut-in; while draining, check discharge for indication of interior corrosion.	.046			√	√	√
8 Clean air intake filter on compressor.	.178			√	√	√
9 Clean oil and water trap.	.178			√	√	√
10 Clean exterior of compressor, motor and surrounding area.	.066			√	√	√
11 Fill out maintenance checklist and report deficiencies.	.022			√	√	√
Total labor-hours/period				1.201	1.201	1.201
Total labor-hours/year				2.403	1.201	1.201

			Cost Each				
			2019 Bare Costs			Total	Total
Description	Labor-hrs.	Material	Labor	Equip.	Total	In-House	w/O&P
1900 Air compressor, recip., less than 5 H.P., annually	1.201	48	77		125	148.11	179
1950 Annualized	4.796	76	305		381	464.45	570

PM Components	Labor-hrs.	W	M	Q	S	A
PM System D3095 118 2950						
Air compressor, reciprocating, 5 to 40 H.P.						
1 Replace compressor oil.	.340			√	√	√
2 Perform operation check of compressor system and adjust as required.	.222			√	√	√
3 Check motor operation for excessive vibration, noise and overheating; lubricate motor.	.042			√	√	√
4 Check operation of pressure relief valve.	.043			√	√	√
5 Clean cooling fans and air cooler on compressor.	.023			√	√	√
6 Check tension, condition, and alignment of V-belts; adjust as necessary.	.030			√	√	√
7 Drain moisture from air storage tank and check low pressure cut-in; while draining, check discharge for indication of interior corrosion.	.059			√	√	√
8 Clean air intake filter on compressor.	.178			√	√	√
9 Clean oil and water trap.	.190			√	√	√
10 Clean compressor and surrounding area.	.066			√	√	√
11 Fill out maintenance checklist and report deficiencies.	.022			√	√	√
Total labor-hours/period				1.216	1.216	1.216
Total labor-hours/year				2.432	1.216	1.216

			Cost Each				
			2019 Bare Costs			Total	Total
Description	Labor-hrs.	Material	Labor	Equip.	Total	In-House	w/O&P
2900 Air compressor, reciprocating, 5 to 40 H.P., annually	1.216	48	78		126	149.25	181
2950 Annualized	4.856	115	310		425	511.70	625

PM Components	Labor-hrs.	W	M	Q	S	A
PM System D3095 118 3950						
Air compressor, reciprocating, over 40 H.P.						
1 Check with operating or area personnel for deficiencies.	.035			√	√	√
2 Perform operation check of compressor system and adjust as required.	.222			√	√	√
3 Replace compressor oil.	.340			√	√	√
4 Check motor(s) operation for excessive vibration, noise and overheating; lubricate motor(s).	.042			√	√	√
5 Clean cylinder cooling fins and air cooler on compressor.	.020			√	√	√
6 Check tension, condition, and alignment of V-belts; adjust as necessary.	.056			√	√	√
7 Check operation of pressure relief valve.	.030			√	√	√
8 Check low pressure cut in and high pressure cut out switches.	.120			√	√	√
9 Drain moisture from air storage tank and check low pressure cut-in; while draining, check discharge for indication of interior corrosion.	.072			√	√	√
10 Clean air intake filter on air compressor(s); replace if necessary.	.178			√	√	√
11 Clean oil and water trap.	.205			√	√	√
12 Check indicating lamps or gauges for proper operation if appropriate; replace burned out lamps or repair/replace gauges.	.020			√	√	√
13 Clean area around equipment.	.066			√	√	√
14 Fill out maintenance checklist and report deficiencies.	.022			√	√	√
Total labor-hours/period				1.432	1.432	1.432
Total labor-hours/year				2.864	1.432	1.432

			Cost Each					
			2019 Bare Costs				Total	Total
Description		Labor-hrs.	Material	Labor	Equip.	Total	In-House	w/O&P
3900	Air compressor, reciprocating, over 40 H.P., annually	1.432	69	92		161	189.61	228
3950	Annualized	5.716	137	365		502	605.15	735

For customer support on your Facilities Maintenance & Repair Costs with RSMeans data, call 800.448.8182.

PM Components	Labor-hrs.	W	M	Q	S	A
PM System D3095 210 1950						
Steam Humidification System						
1 Operate humidistat through its throttling range to verify activation and deactivation.	.117				√	√
2 Inspect steam trap for proper operation.	.117				√	√
3 Turn off steam supply.	.065				√	√
4 Secure electrical service before servicing humidification unit.	.065				√	√
5 Clean strainer.	.195				√	√
6 Clean and/or replace water/steam nozzles as necessary.	.457				√	√
7 Inspect pneumatic controller for air leaks.	.039				√	√
8 Inspect steam lines for leaks and corrosion and repair leaks.	.195				√	√
9 Fill out maintenance checklist and report deficiencies	.022				√	√
Total labor-hours/period					1.272	1.272
Total labor-hours/year					1.272	1.272

			Cost Each					
			2019 Bare Costs				Total	Total
Description	Labor-hrs.	Material	Labor	Equip.	Total	In-House	w/O&P
1900	Steam humidification system, annually	1.272	25.50	81		106.50	129.14	158
1950	Annualized	2.540	25.50	162		187.50	229.39	283

PM Components	Labor-hrs.	W	M	Q	S	A
PM System D3095 210 2950						
Evaporative Pan with Heating Coil Humidification System						
1 Operate humidistat through its throttling range to verify activation and deactivation.	.117				√	√
2 Inspect steam trap for proper operation.	.117				√	√
3 Turn off water and steam supply.	.065				√	√
4 Secure electrical service before servicing humidification unit.	.065				√	√
5 Drain and flush water pans, clean drains, etc.	.258				√	√
6 Check condition of heating element/steam coils and clean.	.065				√	√
7 Inspect pneumatic controller for air leaks.	.039				√	√
8 Inspect steam lines for leaks and corrosion and repair leaks.	.195				√	√
9 Fill out maintenance checklist and report deficiencies.	.022				√	√
Total labor-hours/period					.943	.943
Total labor-hours/year					.943	.943

			Cost Each					
			2019 Bare Costs				Total	Total
Description	Labor-hrs.	Material	Labor	Equip.	Total	In-House	w/O&P
2900	Evap. pan with heating coil humidif. system, annually	.943	32	60.50		92.50	109.64	134
2950	Annualized	1.890	32	121		153	185.39	227

For customer support on your Facilities Maintenance & Repair Costs with RSMeans data, call 800.448.8182.

PM Components	Labor-hrs.	W	M	Q	S	A
PM System D3095 220 1950						
Dehumidifier, desiccant wheel						
1 Check with operating or area personnel for deficiencies.	.035		√	√	√	√
2 Check filters for any blockage or fouling, remove and clean as required.	.036		√	√	√	√
3 Check wheel seals for tears or punctures.	.055		√	√	√	√
4 Check the control valve, thermo sensor bulb and burner, if it has heaters; or check for scaling or leaking on steam heating coils, as applicable.	.094				√	√
5 Check that the outlet air temperature is within the proper heat range.	.105		√	√	√	√
6 Clean the desiccant wheel and check for softening of wheel faces.	.031				√	√
7 Check desiccant wheel and motor for vibration and noise, adjust as required.	.109				√	√
8 Check gear reducer oil level; add as required.	.035				√	√
9 Check wheel belt(s) for wear, proper tension and alignment; adjust as required.	.029				√	√
10 Check blower and motor for excessive vibration and noise; adjust as required.	.109				√	√
11 Check blower belt(s) for wear, proper tension and alignment; adjust as required.	.029				√	√
12 Lubricate wheel, blower and motor bearings.	.047				√	√
13 Check electrical wiring and connections; make appropriate adjustments.	.120				√	√
14 Check reactivation ductwork for condensation and air leaks.	.013				√	√
15 Clean the equipment and the surrounding area.	.066		√	√	√	√
16 Fill out maintenance checklist and report deficiencies.	.022		√	√	√	√
Total labor-hours/period			.320	.320	.935	.935
Total labor-hours/year			2.556	.639	.935	.935

			Cost Each					
			2019 Bare Costs				**Total**	**Total**
Description		**Labor-hrs.**	**Material**	**Labor**	**Equip.**	**Total**	**In-House**	**w/O&P**
1900	Dehumidifier, desiccant wheel, annually	.935	76.50	50		126.50	148.51	176
1950	Annualized	5.060	81	270		351	438.46	535

For customer support on your Facilities Maintenance & Repair Costs with RSMeans data, call 800.448.8182.

PM Components	Labor-hrs.	W	M	Q	S	A
PM System D4015 100 1950						
Backflow prevention device, up to 4″						
NOTE: Test frequency may vary depending on local regulations and application.						
1 Test and calibrate check valve operation of backflow prevention device with test set.	.190					√
2 Bleed air from backflow preventer.	.047					√
3 Inspect for leaks under pressure.	.007					√
4 Clean backflow preventer and surrounding area.	.066					√
5 Fill out maintenance checklist and report deficiencies.	.022					√
Total labor-hours/period						.333
Total labor-hours/year						.333

			Cost Each					
			2019 Bare Costs				Total	Total
Description	Labor-hrs.	Material	Labor	Equip.	Total	In-House	w/O&P
1900	Backflow prevention device, up to 4″, annually	.333	13.15	21		34.15	40.51	49
1950	Annualized	.333	13.15	21		34.15	40.51	49

PM Components	Labor-hrs.	W	M	Q	S	A
PM System D4015 100 2950						
Backflow prevention device, over 4″						
NOTE: Test frequency may vary depending on local regulations and application.						
1 Test and calibrate check valve operation of backflow prevention device with test set.	.333					√
2 Bleed air from backflow preventer.	.065					√
3 Inspect for leaks under pressure.	.007					√
4 Clean backflow preventer and surrounding area.	.066					√
5 Fill out maintenance checklist and report deficiencies.	.022					√
Total labor-hours/period						.493
Total labor-hours/year						.493

			Cost Each					
			2019 Bare Costs				Total	Total
Description	Labor-hrs.	Material	Labor	Equip.	Total	In-House	w/O&P
2900	Backflow prevention device, over 4″, annually	.493	13.15	31		44.15	53.02	64.50
2950	Annualized	.493	13.15	31		44.15	53.02	64.50

For customer support on your Facilities Maintenance & Repair Costs with RSMeans data, call 800.448.8182.

PM Components	Labor-hrs.	W	M	Q	S	A
PM System D4015 150 1950						
Extinguishing system, wet pipe						
1 Notify proper authorities prior to testing any alarm systems.	.130		√	√	√	√
2 Open and close post indicator valve (PIV) to check operation; make minor adjustments such as lubricating valve stem, cleaning and/or replacing target windows as required.	.176					√
3 Open and close OS&Y (outside stem and yoke) cut-off valve to check operation; make minor repairs such as lubricating stems and tightening packing glands as required.	.176					√
4 Perform operational test of water flow detectors; make minor adjustments and restore system to proper operating condition.	.148					√
5 Check to ensure that alarm drain is open; clean drain line if necessary.	.081		√	√	√	√
6 Open water motor alarm test valve and ensure that outside alarm operates; lubricate alarm, make adjustments as required.	.176		√	√	√	√
7 Conduct main drain test by opening 2″ test valve; maintain a continuous record of drain tests; make minor adjustments if applicable; restore system to proper operating condition.	.333			√	√	√
8 Check general condition of sprinklers and sprinkler system; make minor adjustments as required.	.229					√
9 Check equipment gaskets, piping, packing glands, and valves for leaks; tighten flange bolts and loose connections to stop all leaks.	.013					√
10 Check condition of fire department connections; replace missing or broken covers as required.	.103					√
11 Trip test wet pipe system using test valve furthest from wet pipe valve (control valve); make minor adjustments as necessary and restore system to proper operating condition.	1.733					√
12 Inspect OS&Y and PIV cut-off valves for open position.	.066		√	√	√	√
13 Clean area around system components.	.432			√	√	√
14 Fill out maintenance checklist and report deficiencies.	.022		√	√	√	√
Total labor-hours/period			.475	1.241	1.241	3.818
Total labor-hours/year			3.798	2.481	1.241	3.818

		Cost Each						
			2019 Bare Costs				Total	Total
Description	Labor-hrs.	Material	Labor	Equip.	Total	In-House	w/O&P
1900 Extinguishing system, wet pipe, annually	3.818	104	241		345	413.54	505
1950 Annualized	11.342	118	715		833	1,017.52	1,250

For customer support on your Facilities Maintenance & Repair Costs with RSMeans data, call 800.448.8182.

PM Components	Labor-hrs.	W	M	Q	S	A
PM System D4015 180 1950						
Extinguishing system, deluge/preaction						
1 Notify proper authorites prior to testing any alarm systems.	.130		√	√	√	√
2 Open and close post indicator valve (PIV) to check operation; make minor repairs such as lubricating valve stem, cleaning and/or replacing target windows as required.	.176					√
3 Open and close outside stem and yoke (OS&Y) cut-off valve to check operation; make minor repairs such as lubricating stems, tightening packing glands as required.	.176					√
4 Perform operational test of supervisory initiating devices and water flow detectors; make minor adjustments and restore system to proper operating condition.	.148					√
5 Check to ensure that alarm drain is open; clean drain if necessary.	.081		√	√	√	√
6 Open water motor alarm test valve and ensure that outside alarm operates; lubricate alarm and adjust as required.	.176		√	√	√	√
7 Visually check water pressure to ensure adequate operating pressure is available; make adjustments as required.	.004		√	√	√	√
8 Conduct main drain test by opening 2″ test valve and observing drop in water pressure on gauge; pressure drop should not exceed 20 PSI; maintain a continuous record of drain tests; make minor adjustments; restore system to proper operating condition.	.333			√	√	√
9 Check general condition of sprinklers and sprinkler system; make minor adjustments as required.	.228					√
10 Check equipment gaskets, piping, packing glands, and valves for leaks; tighten flange bolts and loose connections to stop all leaks.	.025					√
11 Check condition of fire department connections; replace missing or broken covers as required.	.103					√
12 Check and inspect pneumatic system for physical damage and proper operation; make minor adjustments as required.	.320					√
13 Trip test deluge system (control valve closed); make minor adjustments as necessary and restore system to fully operational condition.	1.956					√
14 Inspect OS&Y and PIV cut-off valves for open position.	.066		√	√	√	√
15 Clean area around system components.	.066					√
16 Ensure that system is restored to proper operating condition.	.109		√	√	√	√
17 Fill out maintenance checklist and report deficiencies.	.022		√	√	√	√
Total labor-hours/period			.588	.921	.921	4.119
Total labor-hours/year			4.700	1.842	.921	4.119

			Cost Each					
			2019 Bare Costs				Total	Total
Description	Labor-hrs.	Material	Labor	Equip.	Total	In-House	w/O&P
1900 Extinguishing system, deluge / preaction, annually	4.119	104	260		364	436.47	535
1950 Annualized	11.586	118	730		848	1,035.35	1,275

PM Components	Labor-hrs.	W	M	Q	S	A
PM System D4015 210 1950						
Fire pump, electric motor driven						
1 Check control panel and wiring for loose connections; tighten connections as required.	.109			√	√	√
2 Ensure all valves relating to water system are in correct position.	.008	√	√	√	√	√
3 Open and close OS&Y(outside steam and yoke) cut-off valve to check operation; make minor repairs such as lubricating stems and tightening packing glands as required.	.176			√	√	√
4 Centrifugal pump:						
A) perform 10 minute pump test run; check for proper operation and adjust if required.	.216	√	√	√	√	√
B) check for leaks on suction and discharge piping, seals, packing glands, etc.	.077	√	√	√	√	√
C) check for excessive vibration, noise, overheating, etc.	.022	√	√	√	√	√
D) check alignment, clearances, and rotation of shaft and coupler (includes removing and reinstalling safety cover).	.160	√	√	√	√	√
E) tighten or replace loose, missing or damaged nuts, bolts, or screws.	.005	√	√	√	√	√
F) lubricate pump and motor as required.	.099			√	√	√
G) check suction or discharge pressure gauge readings and flow rate.	.078	√	√	√	√	√
H) check packing glands and tighten or repack as required; note that slight dripping is required for proper lubrication of shaft.	.113			√	√	√
5 Inspect and clean strainers after each use and flow test.	.258	√	√	√	√	√
6 Clean equipment and surrounding area.	.066	√	√	√	√	√
7 Fill out maintenance checklist and report deficiencies.	.022	√	√	√	√	√
Total labor-hours/period		.912	.912	1.408	1.408	1.408
Total labor-hours/year		34.654	7.296	2.816	1.408	1.408

			Cost Each					
			2019 Bare Costs				Total	Total
Description		Labor-hrs.	Material	Labor	Equip.	Total	In-House	w/O&P
1900	Fire pump, electric motor driven, annually	1.408	33.50	89		122.50	146.86	180
1950	Annualized	47.752	101	3,025		3,126	3,853.60	4,825

For customer support on your Facilities Maintenance & Repair Costs with RSMeans data, call 800.448.8182.

PM Components	Labor-hrs.	W	M	Q	S	A
PM System D4015 250 1950						
Fire pump, engine driven						
1 Ensure all valves relating to water system are in correct position.	.008	√	√	√	√	√
2 Open and close OS&Y (outside steam and yoke) cut-off valve to check operation; make minor repairs such as lubricating stems and tightening packing glands as required.	.176			√	√	√
3 Controller panel:						
A) check controller panel for proper operation in accordance with standard procedures for the type of panel installed, replace burned out bulbs and fuses; make other repairs as necessary.	.039			√	√	√
B) check wiring for loose connections and fraying; tighten or replace as required.	.109			√	√	√
4 Diesel engine:						
A) check oil level in crankcase with dipstick; add oil as necessary.	.014	√	√	√	√	√
B) drain and replace engine oil.	.511					√
C) replace engine oil filter.	.059					√
D) check radiator coolant level; add water if low.	.012	√	√	√	√	√
E) drain and replace radiator coolant.	.511					√
F) check battery water level; add water if low.	.012			√	√	√
G) check battery terminals for corrosion; clean if necessary.	.124			√	√	√
H) check belts for tension and wear; adjust or replace if required.	.012			√	√	√
I) check fuel level in tank; refill as required.	.046			√	√	√
J) check electrical wiring, connections, switches, etc.; adjust or tighten as necessary.	.120			√	√	√
K) perform 10 minute pump check for proper operation and adjust as required.	.216	√	√	√	√	√
5 Centrifugal pump:						
A) check for leaks on suction and discharge piping, seals, packing glands, etc.	.077	√	√	√	√	√
B) check pump operation; vibration, noise, overheating, etc.	.022	√	√	√	√	√
C) check alignment, clearances, and rotation of shaft and coupler (includes removing and reinstalling safety cover).	.160	√	√	√	√	√
D) tighten or replace loose, missing or damaged nuts, bolts, or screws.	.005	√	√	√	√	√
E) lubricate pump as required.	.099			√	√	√
F) check suction or discharge pressure gauge readings and flow rate.	.078	√	√	√	√	√
G) check packing glands and tighten or repack as required; note that slight dripping is required for proper lubrication of shaft.	.113			√	√	√
6 Inspect and clean strainers after each use and flow test.	.262	√	√	√	√	√
7 Clean equipment and surrounding area.	.432	√	√	√	√	√
8 Fill out maintenance checklist and report deficiencies.	.022	√	√	√	√	√
Total labor-hours/period		1.309	1.309	2.158	2.158	3.239
Total labor-hours/year		49.723	10.468	4.316	2.158	3.239

			Cost Each					
			2019 Bare Costs				Total	Total
Description		Labor-hrs.	Material	Labor	Equip.	Total	In-House	w/O&P
1900	Fire pump, motor/engine driven, annually	3.239	142	205		347	410.25	495
1950	Annualized	69.866	280	4,425		4,705	5,789.51	7,200

For customer support on your Facilities Maintenance & Repair Costs with RSMeans data, call 800.448.8182.

473

PM Components	Labor-hrs.	W	M	Q	S	A
PM System D4015 310 1950						
Extinguishing system, dry pipe						
1 Notify proper authorities prior to testing any alarm systems.	.130		√	√	√	√
2 Open and close post indicator valve to check operation; make minor repairs such as lubricating valve stem, cleaning and/or replacing target windows as required.	.176					√
3 Open and close OS&Y (outside stem and yoke) cut-off valve to check operation; make minor repairs such as lubricating stems and tightening packing glands as required.	.176					√
4 Perform operational test of water flow detectors; make minor adjustments and restore system to proper operating condition.	.148					√
5 Check to ensure that alarm drain is open; clean drain line if necessary.	.081		√	√	√	√
6 Open water motor alarm test valve and ensure that outside alarm operates; lubricate alarm, make adjustments as required.	.176		√	√	√	√
7 Visually check water pressure to ensure that adequate operating pressure is available; make adjustments as required.	.003		√	√	√	√
8 Conduct main drain test by opening 2″ test valve; maintain a continuous record of drain tests; make minor adjustments if applicable; restore system to proper operating condition.	.333			√	√	√
9 Check general condition of sprinklers and sprinkler system; make minor adjustments as required.	.228					√
10 Check equipment gaskets, piping, packing glands, and valves for leaks; tighten flange bolts and loose connections to stop all leaks.	.013					√
11 Check condition of fire department connections; replace missing or broken covers as required.	.103					√
12 Visually check system air pressure; pump up system and make minor adjustments as required.	.135		√	√	√	√
13 Trip test dry pipe system using test valve furthest from dry pipe valve (control valve partially open); make minor adjustments as necessary and restore system to fully operational condition.	1.733					√
14 Check operation of quick opening device.	.022					√
15 Inspect OS&Y and PIV cut-off valves for open position.	.066		√	√	√	√
16 Clean area around system components.	.433			√	√	√
17 Fill out maintenance checklist and report deficiencies.	.022		√	√	√	√
Total labor-hours/period			.613	1.379	1.379	3.978
Total labor-hours/year			4.902	2.758	1.379	3.978

		Labor-hrs.	Cost Each					
			2019 Bare Costs				Total	Total
	Description		Material	Labor	Equip.	Total	In-House	w/O&P
1900	Extinguishing system, dry pipe, annually	3.978	142	251		393	468	570
1950	Annualized	13.019	142	820		962	1,175.01	1,450

For customer support on your Facilities Maintenance & Repair Costs with RSMeans data, call 800.448.8182.

PM Components	Labor-hrs.	W	M	Q	S	A
PM System D4095 100 1950						
Extinguishing system, CO_2						
1 Check that nozzles, heads and hand hose lines are clear from obstructions, have not been damaged and in proper position.	.098		√	√	√	√
2 Check to ensure that all operating controls are properly set.	.066		√	√	√	√
3 Clean nozzles as required.	.098		√	√	√	√
4 Visually inspect the control panel for obstructions or physical damage; clean dirt and dust from interior and exterior; make sure that all cards are plugged in tightly, tighten loose connections and make other minor adjustments as necessary.	.222		√	√	√	√
5 Check battery voltages where installed, recharge or replace as required.	.012		√	√	√	√
6 Weigh CO_2 cylinders; replace any that show a weight loss greater than 10%.	.640				√	√
7 Blow out entire system to make sure that the system is not plugged or restricted.	.327					√
8 Run out hose reels and check hoses.	.216					√
9 Conduct operational test of actuating devices, both automatic, manual and alarm system. Prior to conducting tests, close valves, disable CO_2dumping control and manually override computer shutdown feature if applicable, all in accordance with the manufacturer's specifications.	.432					√
10 Check the operation and timing of the time delay control and check operation of the abort station.	.108					√
11 Restore system to proper operating condition and notify personnel upon completion of tests.	.079					√
12 Clean up around system.	.066		√	√	√	√
13 Fill out maintenance checklist and report deficiencies.	.022		√	√	√	√
Total labor-hours/period			.585	.585	1.225	2.387
Total labor-hours/year			4.678	1.169	1.225	2.387

		Cost Each					
			2019 Bare Costs			Total	Total
Description	Labor-hrs.	Material	Labor	Equip.	Total	In-House	w/O&P
1900 Extinguishing system, CO_2, annually	2.387	204	127		331	388.74	460
1950 Annualized	9.470	710	505		1,215	1,431.50	1,700

PM Components	Labor-hrs.	W	M	Q	S	A
PM System D4095 200 1950						
Extinguishing system, foam bottle						
1 Check that all nozzles, heads and hand held lines are clear from obstructions, have not been damaged and are in proper position.	.195		√	√	√	√
2 Check to ensure that all operating controls are properly set.	.022		√	√	√	√
3 Observe pressure on system to ensure that proper pressure is being maintained.	.004		√	√	√	√
4 Clean area around system.	.000		√	√	√	√
5 Fill out maintenance checklist and report deficiencies.	.022		√	√	√	√
Total labor-hours/period			.309	.309	.309	.309
Total labor-hours/year			2.473	.618	.309	.309

			Cost Each					
			2019 Bare Costs				Total	Total
Description		Labor-hrs.	Material	Labor	Equip.	Total	In-House	w/O&P
1900	Extinguishing system, foam bottle, annually	.309	4.49	16.50		20.99	26.27	32
1950	Annualized	3.708	17.95	198		215.95	275.21	335

For customer support on your Facilities Maintenance & Repair Costs with RSMeans data, call 800.448.8182.

PM Components	Labor-hrs.	W	M	Q	S	A
PM System D4095 210 1950						
Extinguishing system, foam, electric pump, deluge system						
1 Check foam concentrate level in tank; add concentrate as required to maintain proper level.	.043		√	√	√	√
2 Ensure all valves relating to foam/water system are in correct position.	.008		√	√	√	√
3 Visually check proportioning devices, pumps, and foam nozzles; correct any observed deficiencies.	.386		√	√	√	√
4 Check water supply pressure; adjust as necessary.	.029		√	√	√	√
5 Open and close OS&Y (outside stem and yoke) cut-off valve to check operation; make minor repairs such as lubricating stems and tightening packing glands as required.	.059			√	√	√
6 Centrifugal pump:						
A) perform 10 minute pump test run, check for proper operation and adjust as required.	.217			√	√	√
B) test start pump without foam discharger.	.022		√	√	√	√
C) check for leaks on suction and discharge piping, seals, packing glands, etc.	.077			√	√	√
D) check for excessive vibration, noise, overheating, etc.	.022			√	√	√
E) check alignment, clearances, and rotation of shaft and coupler (includes removing and reinstalling safety cover).	.160			√	√	√
F) tighten or replace loose, missing or damaged nuts, bolts, or screws.	.005			√	√	√
G) lubricate pump and motor as required.	.099			√	√	√
H) check suction or discharge pressure gauge readings and flow rate.	.078			√	√	√
I) check packing glands and tighten or repack as required; note that slight dripping is required for proper lubrication of shaft.	.113			√	√	√
7 Inspect and clean strainers after each use and flow test.	.258		√	√	√	√
8 Clean equipment and surrounding area.	.432		√	√	√	√
9 Fill out maintenance checklist.	.022		√	√	√	√
Total labor-hours/period			1.200	2.029	2.029	2.029
Total labor-hours/year			9.604	4.059	2.029	2.029

			Cost Each					
			2019 Bare Costs				Total	Total
Description	Labor-hrs.	Material	Labor	Equip.	Total	In-House	w/O&P
1900 Extinguishing system, foam, electric pump, annually	2.029	94.50	108		202.50	243.50	292
1950 Annualized	17.721	355	945		1,300	1,607.93	1,975

PM Components	Labor-hrs.	W	M	Q	S	A
PM System D4095 220 1950						
Extinguishing system, foam, diesel pump, deluge system						
1 Check foam concentrate level in tank; add concentrate as required to maintain proper level.	.043		√	√	√	√
2 Ensure all valves relating to foam/water system are in correct position.	.008		√	√	√	√
3 Visually check proportioning devices, pumps, and foam nozzles; correct any observed deficiencies.	.386		√	√	√	√
4 Check water supply pressure, adjust as necessary.	.003		√	√	√	√
5 Open and close OS&Y (outside stem and yoke) cut-off valve to check operation; make minor repairs such as lubricating stems and tightening packing glands as required.	.059			√	√	√
6 Controller panel:						
A) check controller panel for proper operation in accordance with standard procedures for the type of panel installed, replace burned out bulbs and fuses; make other repairs as necessary.	.039			√	√	√
B) check wiring for loose connections and fraying; tighten or replace as required.	.109			√	√	√
7 Diesel engine:						
A) check oil level in crankcase with dipstick; add oil as necessary.	.014		√	√	√	√
B) drain and replace engine oil.	.511					√
C) replace engine oil filter.	.059					√
D) check radiator coolant level; add water if low.	.012		√	√	√	√
E) drain and replace radiator coolant.	.511					√
F) check battery water level; add water if low.	.012			√	√	√
G) check battery terminals for corrosion; clean if necessary.	.124			√	√	√
H) check belts for tension and wear; adjust or replace if required.	.012			√	√	√
I) check fuel level in tank; refill as required.	.046			√	√	√
J) check electrical wiring, connections, switches, etc.; adjust or tighten as necessary.	.120			√	√	√
K) perform 10 minute pump check for proper operation and adjust as required, start pump without foam discharger.	.216		√	√	√	√
8 Centrifugal pump:						
A) check for leaks on suction and discharge piping, seals, packing glands, etc.	.077			√	√	√
B) check pump operation; vibration, noise, overheating, etc.	.022			√	√	√
C) check alignment, clearances, and rotation of shaft and coupler (includes removing and reinstalling safety cover).	.160			√	√	√
D) tighten or replace loose, missing or damaged nuts, bolts, or screws.	.005			√	√	√
E) lubricate pump as required.	.099			√	√	√
F) check suction or discharge pressure gauge readings and flow rate.	.078			√	√	√
G) check packing glands and tighten or repack as as required; note that slight dripping is required for proper lubrication of shaft.	.113			√	√	√
9 Inspect and clean strainers after each use and flow test.	.262		√	√	√	√
10 Clean equipment and surrounding area.	.432		√	√	√	√
11 Fill out maintenance checklist.	.022		√	√	√	√
Total labor-hours/period			1.399	2.473	2.473	3.554
Total labor-hours/year			11.192	4.946	2.473	3.554

	Description	Labor-hrs.	2019 Bare Costs				Total In-House	Total w/O&P
			Material	Labor	Equip.	Total		
1900	Extinguishing system, foam, diesel pump, annually	3.554	330	190		520	607.86	720
1950	Annualized	22.156	910	1,175		2,085	2,528.11	3,050

For customer support on your Facilities Maintenance & Repair Costs with RSMeans data, call 800.448.8182.

PM Components	Labor-hrs.	W	M	Q	S	A
PM System D4095 400 1950						
Extinguishing system, dry chemical						
1 Check nozzles, heads, and fusible links or heat detectors to ensure that they are clean, have adequate clearance from obstructions and have not been damaged.	.098		√	√	√	√
2 Check hand hose lines, if applicable, to ensure that they are clear, in proper position and in good condition.	.098		√	√	√	√
3 Check to ensure that all operating controls are properly set.	.066		√	√	√	√
4 Observe pressure on system to ensure that proper pressure is maintained.	.004		√	√	√	√
5 Check expellant gas cylinders of gas cartridge systems for proper pressure or weight to ensure that sufficient gas is available.	.258				√	√
6 Test all actuating and operating devices; turn valve on dry chemical bottle or cylinder to off position in accordance with manufacturers specifications, and test operation of control head by removing fusible link, activating heat detectors or pulling test control. Restore system to normal operation and make minor adjustments as applicable.	.433					√
7 Open and check dry chemical in cylinder and stored pressure systems to ensure it is free flowing and without lumps.	.327					√
8 When an alarm system is reactivated, ensure that it functions in accordance with manufacturers specifications.	.079					√
9 Clean area around system.	.066		√	√	√	√
10 Fill out maintenance checklist and report deficiencies.	.022		√	√	√	√
Total labor-hours/period			.355	.355	.613	1.451
Total labor-hours/year			2.836	.709	.613	1.451

	Description	Labor-hrs.	Cost Each						
			2019 Bare Costs				**Total In-House**	**Total w/O&P**	
			Material	Labor	Equip.	Total			
1900	Extinguishing system, dry chemical, annually	1.451	51.50	77.50		129	156.87	188	
1950	Annualized	5.606	103	300		403	499.46	610	

For customer support on your Facilities Maintenance & Repair Costs with RSMeans data, call 800.448.8182.

PM Components	Labor-hrs.	W	M	Q	S	A
PM System D4095 450 1950						
Extinguishing system, FM200						
1 Check that nozzles, heads and hand hose lines are clear from obstructions, have not been damaged and in proper position.	.098		√	√	√	√
2 Check to ensure that all operating controls are properly set.	.066		√	√	√	√
3 Clean nozzles as required.	.098		√	√	√	√
4 Visually inspect the control panel for obstructions or physical damage; clean dirt and dust from interior and exterior; make sure that all cards are plugged in tightly, tighten loose connections and make other minor adjustments as necessary.	.222		√	√	√	√
5 Check battery voltages where installed, recharge or replace as required.	.012		√	√	√	√
6 Weigh cylinders; replace any that show a weight loss greater than 5% or a pressure loss greater than 10%.	.640				√	√
7 Blow out entire system to make sure that the system is not plugged or restricted.	.333					√
8 Conduct operational test of actuating devices, both automatic, manual and alarm system; prior to conducting tests, close valves disable CO_2 dumping control and manually override computer shutdown feature if applicable, all in accordance with manufacturers specifications.	.432					√
9 Check the operation and timing of the time delay control and check operation of abort station.	.108					√
10 Restore system to proper operating condition and notify cognizant personnel upon completion of tests.	.079					√
11 Clean out around system.	.066		√	√	√	√
12 Fill out maintenance checklist and report deficiencies.	.022		√	√	√	√
Total labor-hours/period			.585	.585	1.225	2.178
Total labor-hours/year			4.678	1.169	1.225	2.178

		Cost Each						
		2019 Bare Costs				**Total**	**Total**	
Description	**Labor-hrs.**	**Material**	**Labor**	**Equip.**	**Total**	**In-House**	**w/O&P**	
1900 Extinguishing system, FM200, annually	2.178	178	116		294	345.34	410	
1950 Annualized	9.261	660	495		1,155	1,365.10	1,625	

For customer support on your Facilities Maintenance & Repair Costs with RSMeans data, call 800.448.8182.

PM Components	Labor-hrs.	W	M	Q	S	A
PM System D5015 210 1950						
Switchboard, electrical						
1 Check with operating or area personnel for deficiencies.	.044					√
2 Check indicating lamps for proper operation, if appropriate; replace burned out lamps.	.018					√
3 Remove and reinstall cover.	.195					√
4 Check for discolorations, hot spots, odors and charred insulation.	.356					√
5 Clean switchboard exterior and surrounding area.	.066					√
6 Fill out maintenance checklist and report deficiencies.	.022					√
Total labor-hours/period						.701
Total labor-hours/year						.701

			Cost Each					
			2019 Bare Costs				Total	Total
	Description	Labor-hrs.	Material	Labor	Equip.	Total	In-House	w/O&P
1900	Switchboard, electrical, annually	.701	3.92	42		45.92	56.37	70
1950	Annualized	.701	3.92	42		45.92	56.37	70

PM Components	Labor-hrs.	W	M	Q	S	A
PM System D5015 214 1950						
Switchboard w/air circuit breaker						
1 Check with operating or area personnel for deficiencies.	.035					√
2 Check indicating lamps for proper operation, if appropriate; replace burned out lamps.	.020					√
3 Remove switchboard from service, coordinate with appropriate authority.	.038					√
4 Remove and reinstall cover.	.051					√
5 Check for discolorations, hot spots, odors and charred insulation.	.016					√
6 Check the records on the circuit breaker for any full current rated trips which may indicate a potential problem.	.020					√
7 Examine the exterior of the switchboard for any damage or moisture.	.020					√
8 Check operation of circuit breaker tripping mechanism and relays.	.096					√
9 Check the operation of the circuit breaker charging motor.	.020					√
10 Check hardware for tightness.	.120					√
11 Check interlock for proper operation.	.025					√
12 Check that the charger is operating properly.	.022		√	√	√	√
13 Check for any signs of corrosion on the battery terminals or wires.	.020		√	√	√	√
14 Check the electrolyte level in the batteries; add if required.	.624		√	√	√	√
15 Check the specific gravity of the electrolyte in a 10% sample of the batteries.	.069		√	√	√	√
16 Check 25% of terminal-to-cell connection resistance; rehabilitate connections as required; add anti-corrosion grease to battery terminals and connections.	.113					√
17 Measure and record individual cell and string float voltages.	2.808					√
18 Place switchboard back in service, notify appropriate authority.	.057					√
19 Clean switchboard, charger and the surrounding area.	.066		√	√	√	√
20 Fill out maintenance checklist and report deficiencies.	.022		√	√	√	√
Total labor-hours/period			.823	.823	.823	4.262
Total labor-hours/year			6.585	1.646	.823	4.262

			Cost Each					
			2019 Bare Costs				Total	Total
	Description	Labor-hrs.	Material	Labor	Equip.	Total	In-House	w/O&P
1900	Switchboard, with air circuit breaker, annually	4.262	15.85	256		271.85	332.87	415
1950	Annualized	13.318	15.85	800		815.85	1,004.49	1,250

For customer support on your Facilities Maintenance & Repair Costs with RSMeans data, call 800.448.8182.

481

PM Components	Labor-hrs.	W	M	Q	S	A
PM System D5015 217 1950						
Switchboard w/air circuit breaker and tie switch						
1 Check with operating or area personnel for deficiencies.	.035					√
2 Check indicating lamps for proper operation, if appropriate; replace burned out lamps.	.020					√
3 Remove switchboard from service, coordinate with appropriate authority.	.038					√
4 Remove and reinstall cover.	.051					√
5 Check for discolorations, hot spots, odors and charred insulation.	.016					√
6 Check the records on the circuit breaker for any full current rated trips which may indicate a potential problem.	.020					√
7 Examine the exterior of the switchboard for any damage or moisture.	.020					√
8 Check operation of circuit breaker tripping mechanism and relays.	.096					√
9 Check the operation of the circuit breaker charging motor.	.020					√
10 Check hardware for tightness.	.120					√
11 Check interlock for proper operation.	.013					√
12 Operate the switching handle and check the locking mechanism.	.052					√
13 Check that the charger is operating properly.	.022		√	√	√	√
14 Check for any signs of corrosion on the battery terminals or wires.	.020		√	√	√	√
15 Check the electrolyte level in the batteries; add if required.	.624		√	√	√	√
16 Check the specific gravity of the electrolyte in a 10% sample of the batteries.	.069		√	√	√	√
17 Check 25% of terminal-to-cell connection resistance; rehabilitate connections as required; add anti-corrosion grease to battery terminals and connections.	.113					√
18 Measure and record individual cell and string float voltages.	2.808					√
19 Place switchboard back in service, notify appropriate authority.	.057					√
20 Clean switchboard, charger and the surrounding area.	.066		√	√	√	√
21 Fill out maintenance checklist and report deficiencies.	.022		√	√	√	√
Total labor-hours/period			.823	.823	.823	4.302
Total labor-hours/year			6.585	1.646	.823	4.302

	Description	Labor-hrs.	Material	Labor	Equip.	Total	Total In-House	Total w/O&P
			Cost Each					
			2019 Bare Costs					
1900	Switchboard, with air C.B. and tie switch, annually	4.302	19.80	259		278.80	340.15	425
1950	Annualized	13.358	19.80	800		819.80	1,011.77	1,275

For customer support on your Facilities Maintenance & Repair Costs with RSMeans data, call 800.448.8182.

PM Components	Labor-hrs.	W	M	Q	S	A
PM System D5015 220 1950						
Circuit breaker, HV air						
1 Check the records on the circuit breaker for any full current rated trips which may indicate a potential problem.	.066					√
2 Examine the exterior of the circuit breaker for any damage or moisture.	.060					√
3 Remove the circuit breaker from service, coordinate with appropriate authority.	.038					√
4 Check operation of tripping mechanisms and relays.	.096					√
5 Check the operation of the charging motor, if applicable.	.038					√
6 Check hardware for tightness.	.005					√
7 Check interlock for proper operation.	.039					√
8 Clean area around switch.	.066					√
9 Place the circuit breaker back in service, coordinate with appropriate authority.	.040					√
10 Fill out maintenance checklist.	.022					√
Total labor-hours/period						.470
Total labor-hours/year						.470

			Cost Each				
			2019 Bare Costs			Total	Total
Description	Labor-hrs.	Material	Labor	Equip.	Total	In-House	w/O&P
1900 Circuit breaker, high voltage air, annually	.470	14.40	28		42.40	50.85	61.50
1950 Annualized	.470	14.40	28		42.40	50.85	61.50

PM Components	Labor-hrs.	W	M	Q	S	A
PM System D5015 222 1950						
Circuit breaker, HV oil						
1 Check the records on the circuit breaker for any full current rated trips which may indicate a potential problem.	.066					√
2 Examine the exterior of the circuit breaker for any damage, cracks, rust or leaks; check around the gaskets and the drain valves.	.074					√
3 Check the entrance bushings for cracks in the porcelain.	.320					√
4 Check the condition of the ground system.	.170					√
5 Draw oil sample from bottom of circuit breaker; have sample tested; replace removed oil to the proper level.	.042					√
6 Clean the exterior of the circuit breaker and operating mechanism.	.094					√
7 Clean the area around the circuit breaker.	.066					√
8 Fill out maintenance checklist.	.022					√
Total labor-hours/period						.855
Total labor-hours/year						.855

			Cost Each				
			2019 Bare Costs			Total	Total
Description	Labor-hrs.	Material	Labor	Equip.	Total	In-House	w/O&P
1900 Circuit breaker, high voltage oil, annually	.855	33	51.50		84.50	99.46	120
1950 Annualized	.855	33	51.50		84.50	99.46	120

For customer support on your Facilities Maintenance & Repair Costs with RSMeans data, call 800.448.8182.

483

D50 ELECTRICAL | D5015 230 | Switch, Selector, HV Air

PM Components	Labor-hrs.	W	M	Q	S	A
PM System D5015 230 1950						
Switch, selector, HV, air						
1 Examine the exterior of the selector switch for any damage.	.060					√
2 Check that the switching handle is locked.	.044					√
3 Check the condition of the ground system.	.170					√
4 Clean the exterior of the selector switch and operating mechanism.	.046					√
5 Clean the area around the selector switch.	.066					√
6 Fill out maintenance checklist and report deficiencies.	.022					√
Total labor-hours/period						.408
Total labor-hours/year						.408

			Cost Each					
			2019 Bare Costs				**Total**	**Total**
	Description	Labor-hrs.	Material	Labor	Equip.	Total	In-House	w/O&P
1900	Switch, selector, high voltage, air, annually	.408	14.40	24.50		38.90	46.07	56
1950	Annualized	.408	14.40	24.50		38.90	46.07	56

D50 ELECTRICAL | D5015 232 | Switch, Selector, HV Oil

PM Components	Labor-hrs.	W	M	Q	S	A
PM System D5015 232 1950						
Switch, selector, HV, oil						
1 Examine the exterior of the interrupt switch for any damage, cracks, rust or leaks; check around the gaskets and the drain valves.	.060					√
2 Check that the switching handle is locked.	.044					√
3 Check the condition of the ground system.	.170					√
4 Draw oil sample from bottom of selector switch; have sample tested; replace removed oil to the proper level.	.042					√
5 Clean the exterior of the selector switch and operating mechanism.	.046					√
6 Clean the area around the selector switch.	.066					√
7 Fill out maintenance checklist and report deficiencies.	.022					√
Total labor-hours/period						.450
Total labor-hours/year						.450

			Cost Each					
			2019 Bare Costs				**Total**	**Total**
	Description	Labor-hrs.	Material	Labor	Equip.	Total	In-House	w/O&P
1900	Switch, selector, high voltage, oil, annually	.450	33	27		60	69.72	83
1950	Annualized	.450	33	27		60	69.72	83

For customer support on your Facilities Maintenance & Repair Costs with RSMeans data, call 800.448.8182.

PM Components	Labor-hrs.	W	M	Q	S	A
PM System D5015 234 1950						
Switch, automatic transfer						
1 Check with operating or area personnel for deficiencies.	.044		√	√	√	√
2 Inspect wiring, wiring connections and fuse blocks for looseness, charring evidence of short circuiting, overheating and tighten all connections.	.263		√	√	√	√
3 Inspect gen. cond. of transf. switch and clean exter. and surrounding area.	.114		√	√	√	√
4 Fill out maintenance checklist and report deficiencies.	.022		√	√	√	√
Total labor-hours/period			.443	.443	.443	.443
Total labor-hours/year			3.546	.886	.443	.443

				Cost Each				
				2019 Bare Costs			Total	Total
	Description	Labor-hrs.	Material	Labor	Equip.	Total	In-House	w/O&P
1900	Switch, automatic transfer, annually	.443	14.40	26.50		40.90	48.83	59
1950	Annualized	5.316	14.40	320		334.40	409.55	515

PM Components	Labor-hrs.	W	M	Q	S	A
PM System D5015 236 1950						
Switch, interrupt, HV, fused air						
1 Examine the exterior of the interrupt switch for any damage.	.060					√
2 Check the condition of the ground system.	.170					√
3 Clean the exterior of the interrupt switch and operating mechanism.	.046					√
4 Clean the area around the interrupt switch.	.066					√
5 Fill out maintenance checklist and report deficiencies.	.022					√
Total labor-hours/period						.364
Total labor-hours/year						.364

				Cost Each				
				2019 Bare Costs			Total	Total
	Description	Labor-hrs.	Material	Labor	Equip.	Total	In-House	w/O&P
1900	Switch, interrupt, high voltage, fused air, annually	.364	14.40	22		36.40	42.82	51.50
1950	Annualized	.364	14.40	22		36.40	42.82	51.50

PM Components	Labor-hrs.	W	M	Q	S	A
PM System D5015 238 1950						
Switch, interrupt, HV, w/auxiliary fuses, air						
1 Examine the exterior of the interrupt switch for any damage.	.060					√
2 Check the condition of the ground system.	.170					√
3 Clean the exterior of the interrupt switch and operating mechanism.	.046					√
4 Clean the area around the interrupt switch.	.066					√
5 Fill out maintenance checklist and report deficiencies.	.022					√
Total labor-hours/period						.364
Total labor-hours/year						.364

				Cost Each				
				2019 Bare Costs			Total	Total
	Description	Labor-hrs.	Material	Labor	Equip.	Total	In-House	w/O&P
1900	Switch, interrupt, HV, w/ aux fuses, air, annually	.364	14.40	22		36.40	42.82	51.50
1950	Annualized	.364	14.40	22		36.40	42.82	51.50

PM Components	Labor-hrs.	W	M	Q	S	A
PM System D5015 240 1950						
Transformer, dry type 500 KVA and over						
1 Examine the exterior of the transformer for any damage.	.060					✓
2 Remove the covers or open the doors.	.195					✓
3 Check for signs of moisture or overheating.	.059					✓
4 Check for voltage creeping over insulated surfaces, such as evidenced by tracking or carbonization.	.018					✓
5 Check fans, motors and other auxiliary devices for proper operation; where applicable.	.049					✓
6 Check the condition of the ground system.	.170					✓
7 Replace the covers or close the doors.	.195					✓
8 Fill out maintenance checklist and report deficiencies.	.022					✓
Total labor-hours/period						.769
Total labor-hours/year						.769

			Cost Each					
			2019 Bare Costs				Total	Total
Description	**Labor-hrs.**	**Material**	**Labor**	**Equip.**	**Total**	**In-House**	**w/O&P**
1900	Transformer, dry type 500 KVA and over, annually	.769	14.40	46		60.40	72.76	89
1950	Annualized	.769	14.40	46		60.40	72.76	89

PM Components	Labor-hrs.	W	M	Q	S	A
PM System D5015 240 2950						
Transformer, oil, pad mounted						
1 Examine the exterior of the transformer for any damage, cracks, rust or leaks; check around bushings, gaskets and pressure relief device.	.014					✓
2 Check the condition of the ground system.	.170					✓
3 Check and record the oil level, pressure and temperature readings.	.263					✓
4 Check fans, motors and other auxiliary devices for proper operation; where applicable.	.022					✓
5 Draw oil sample from top of transformer and have sample tested for dielectric strength; replace removed oil.	.525					✓
6 Clean transformer exterior and the surrounding area.	.066					✓
7 Fill out maintenance checklist and report deficiencies.	.022					✓
Total labor-hours/period						1.082
Total labor-hours/year						1.082

			Cost Each					
			2019 Bare Costs				Total	Total
Description	**Labor-hrs.**	**Material**	**Labor**	**Equip.**	**Total**	**In-House**	**w/O&P**
2900	Transformer, oil, pad mounted, annually	1.082	22	65		87	104.23	128
2950	Annualized	1.082	22	65		87	104.23	128

For customer support on your Facilities Maintenance & Repair Costs with RSMeans data, call 800.448.8182.

PM Components	Labor-hrs.	W	M	Q	S	A
PM System D5015 240 3950						
Transformer, oil, pad mounted, PCB						
1 Check with operating or area personnel for deficiencies.	.040			√	√	√
2 Examine the exterior of the transformer for any damage, cracks, rust or leaks; check around the gaskets, the pressure relief device and the cooling fins.	.014			√	√	√
3 Check and record the oil levels, pressure and temperature readings: level ____ ; pressure ____ ; temperature ____ .	.263					√
4 Check fans, motors and other auxiliary devices for proper operation; where applicable.	.051					√
5 Check the exterior bushings for leaking oil.	.321			√	√	√
6 Check the condition of the ground system.	.170					√
7 Draw oil sample from top of transformer and have sample tested; replace removed oil.	.525					√
8 Clean transformer exterior and the surrounding area.	.066					√
9 Verify the PCB warning markings are intact and clearly visible.	.022					√
10 Fill out maintenance checklist and report deficiencies. Note: the EPA PCB regulations under 40 CFR 716.30 shall apply to all inspections, samples, disposal, cleanups, decontamination, reporting and record-keeping.	.022			√	√	√
Total labor-hours/period				.397	.397	1.494
Total labor-hours/year				.794	.397	1.494

			Cost Each					
			2019 Bare Costs				Total	Total
Description		Labor-hrs.	Material	Labor	Equip.	Total	In-House	w/O&P
3900	Transformer, oil, pad mounted, PCB, annually	1.494	66	89.50		155.50	183.96	221
3950	Annualized	2.685	66	161		227	271.82	330

PM Components	Labor-hrs.	W	M	Q	S	A
PM System D5015 260 1950						
Panelboard, 225 amps and above						
1 Check with operating or area personnel for deficiencies.	.066					√
2 Check for excessive heat, odors, noise, and vibration.	.239					√
3 Clean and check general condition of panel.	.114					√
4 Fill out maintenance checklist and report deficiencies.	.022					√
Total labor-hours/period						.441
Total labor-hours/year						.441

			Cost Each					
			2019 Bare Costs				Total	Total
Description	Labor-hrs.	Material	Labor	Equip.	Total		In-House	w/O&P
1900 Panelboard, 225 A and above, annually	.441	22.50	26.50		49		57.67	68.50
1950 Annualized	.441	22.50	26.50		49		57.67	68.50

PM Components	Labor-hrs.	W	M	Q	S	A
PM System D5015 280 1950						
Motor control center, over 400 amps						
1 Check with operating or area personnel for deficiencies.	.044					√
2 Check starter lights, replace if required.	.018					√
3 Check for excessive heat, odors, noise, and vibration.	.239					√
4 Clean motor control center exterior and surrounding area.	.066					√
5 Fill out maintenance checklist and report deficiencies.	.022					√
Total labor-hours/period						.389
Total labor-hours/year						.389

			Cost Each					
			2019 Bare Costs				Total	Total
Description	Labor-hrs.	Material	Labor	Equip.	Total		In-House	w/O&P
1900 Motor control center, over 400 A, annually	.389	22.50	23.50		46		53.03	64
1950 Annualized	.389	22.50	23.50		46		53.03	64

For customer support on your Facilities Maintenance & Repair Costs with RSMeans data, call 800.448.8182.

PM Components	Labor-hrs.	W	M	Q	S	A
PM System D5035 610 1950						
Clocks, Central System						
1 Clean dirt and dust from interior and exterior of cabinet.	.260				√	√
2 Adjust relays and check transmission of signal.	.117				√	√
3 Tighten contacts and terminal screws.	.065				√	√
4 Burnish contacts.	.195				√	√
5 Fill out maintenance checklist and report deficiencies.	.022				√	√
Total labor-hours/period					.659	.659
Total labor-hours/year					.659	.659

	Description	Labor-hrs.	Material	Labor	Equip.	Total	Total In-House	Total w/O&P
				2019 Bare Costs				
1900	Central clock systems, annually	.659	5.75	35		40.75	52	63
1950	Annualized	1.318	11.50	71		82.50	103.63	127

For customer support on your Facilities Maintenance & Repair Costs with RSMeans data, call 800.448.8182.

489

PM Components	Labor-hrs.	W	M	Q	S	A
PM System D5035 710 1950						
Fire alarm annunciator system						
1 Visually inspect all alarm equipment for obstructions or physical damage, clean dirt and dust from interior and exterior of panel/pull boxes, tighten loose connections.	.222		√	√	√	√
2 Notify cogniant personnel prior to testing.	.112		√	√	√	√
3 Conduct operational test of initiating and signal transmitting devices in populated buildings by building zone/area; for those circuits which do not operate properly, check detectors, control units, and annunciators for dust on defective components; make minor adjustments as required.	.112		√	√	√	√
4 Check battery voltages where installed; replace as required.	.012		√	√	√	√
5 Conduct operational test of 10% of total number of spot type heat detectors and all smoke detectors; for those circuits which do not operate properly, check to determine if problem relates to circuit, device, or control unit, make minor adjustments as required; if detector is defective but no replacement is immediately available, remove detector, re-establish Initiating circuit and tag location until a replacement detector is installed.	2.172				√	√
6 Restore system to proper operating condition and notify personnel upon completion of tests.	.079		√	√	√	√
7 Fill out maintenance checklist and report deficiencies.	.022		√	√	√	√
Total labor-hours/period			.559	.559	2.731	2.731
Total labor-hours/year			4.474	1.119	2.731	2.731

		Cost Each					
			2019 Bare Costs			Total	Total
Description	Labor-hrs.	Material	Labor	Equip.	Total	In-House	w/O&P
1900 Fire alarm annunciator system, annually	2.731	166	146		312	371.37	440
1950 Annualized	11.051	175	590		765	954.70	1,175

PM Components	Labor-hrs.	W	M	Q	S	A
PM System D5035 810 1950						
Security / intrusion alarm system						
1 Check in and out with area security officer, notify operating/facility personnel, obtain necessary alarm keys, alarm codes and escort when required.	.520			√	√	√
2 Inspect alarm control panel and conduct operational test of initiating and signal transmitting devices; make minor adjustments as required.	.241			√	√	√
3 Check indicating lamps for proper operation, replace if necessary.	.018			√	√	√
4 Check battery voltages where installed, replace as required.	.012			√	√	√
5 Restore system to proper operating condition and notify personnel upon completion of tests.	.079			√	√	√
6 Clean exterior of cabinet and surrounding area.	.066			√	√	√
7 Fill out maintenance checklist and report deficiencies.	.022			√	√	√
Total labor-hours/period				.958	.958	.958
Total labor-hours/year				1.916	.958	.958

		Cost Each					
			2019 Bare Costs			Total	Total
Description	Labor-hrs.	Material	Labor	Equip.	Total	In-House	w/O&P
1900 Security, intrusion alarm system, annually	.958	160	51.50		211.50	242.03	282
1950 Annualized	3.832	160	205		365	439.35	530

For customer support on your Facilities Maintenance & Repair Costs with RSMeans data, call 800.448.8182.

PM Components	Labor-hrs.	W	M	Q	S	A
PM System D5095 210 1950						
Emergency generator up to 15 KVA						
1 Check with the operating or area personnel for any obvious deficiencies.	.044		√	√	√	√
2 Check engine oil level; add as required.	.010		√	√	√	√
3 Change engine oil and oil filter.	.170					√
4 Check battery charge and electrolyte specific gravity, add water as required; check terminals for corrosion, clean as required.	.120		√	√	√	√
5 Check belt tension and wear; adjust as required, if applicable.	.012		√	√	√	√
6 Check engine air filter; change as required.	.042					√
7 Check spark plug or injector nozzle condition; service or replace as required.	.211					√
8 Check wiring, connections, switches, etc., adjust as required.	.036		√	√	√	√
9 Perform 30 minute generator test run; check for proper operation.	.650		√	√	√	√
10 Check fuel level; add as required.	.046		√	√	√	√
11 Wipe dust and dirt from engine and generator.	.056		√	√	√	√
12 Clean area around generator.	.066		√	√	√	√
13 Fill out maintenance checklist and report deficiencies.	.022		√	√	√	√
Total labor-hours/period			1.062	1.062	1.062	1.486
Total labor-hours/year			8.499	2.125	1.062	1.486

			Cost Each					
			2019 Bare Costs				Total	Total
Description	Labor-hrs.	Material	Labor	Equip.	Total	In-House	w/O&P
1900 Emergency diesel generator, up to 15 KVA, annually	1.486	124	79		203	239.57	282
1950 Annualized	13.164	115	700		815	1,037.35	1,275

PM Components	Labor-hrs.	W	M	Q	S	A
PM System D5095 210 2950						
Emergency diesel generator, over 15 KVA						
1 Check with the operating or area personnel for any obvious deficiencies.	.044		√	√	√	√
2 Check engine oil level; add as required.	.010		√	√	√	√
3 Change engine oil and oil filter.	.506					√
4 Check battery charge and electrolyte specific gravity, add water as required; check terminals for corrosion, clean as required.	.241		√	√	√	√
5 Check belts for wear and proper tension; adjust as necessary.	.012		√	√	√	√
6 Check that crank case heater is operating.	.038		√	√	√	√
7 Check engine air filter; change as required.	.042					√
8 Check wiring, connections, switches, etc.; adjust as required.	.036		√	√	√	√
9 Check spark plug or injector nozzle condition; service or replace as required.	.281					√
10 Perform 30 minute generator test run; check for proper operation.	.650		√	√	√	√
11 Check fuel level with gauge pole, add as required.	.046		√	√	√	√
12 Wipe dust and dirt from engine and generator.	.109		√	√	√	√
13 Clean area around generator.	.066		√	√	√	√
14 Fill out maintenance checklist and report deficiencies.	.025		√	√	√	√
Total labor-hours/period			1.277	1.277	1.277	2.106
Total labor-hours/year			10.216	2.554	1.277	2.106

			Cost Each					
			2019 Bare Costs				Total	Total
Description	Labor-hrs.	Material	Labor	Equip.	Total	In-House	w/O&P
2900 Emergency diesel generator, over 15 KVA, annually	2.106	132	112		244	290.71	345
2950 Annualized	16.150	132	860		992	1,260.15	1,550

For customer support on your Facilities Maintenance & Repair Costs with RSMeans data, call 800.448.8182.

PM Components	Labor-hrs.	W	M	Q	S	A
PM System D5095 210 3950						
Emergency diesel generator, turbine						
1 Check with the operating or area personnel for any obvious deficiencies.	.044		√	√	√	√
2 Check turbine oil level; add oil as required.	.036		√	√	√	√
3 Change turbine oil and oil filter; check transmission oil level.	.569					√
4 Check that the crankcase heater is operating properly.	.022		√	√	√	√
5 Replace turbine air filter.	.224					√
6 Check wiring, connections, switches, etc.; adjust as required.	.060					√
7 Check starter for proper operation; lubricate as necessary.	.047		√	√	√	√
8 Check fuel nozzles, fuel regulator and ignition device condition; service or replace as required.	.813					√
9 Perform 30 minute generator test run; check for proper operation.	7.797		√	√	√	√
10 Check and record transmission oil pressure and temperature, and natural gas pressure: oil press:____ oil temp:____ gas press:____.	.021					√
11 Record running time: beginning ____ hours; ending ____ hours.	.010					√
12 Check that the charger is operating properly.	.127		√	√	√	√
13 Check for any signs of corrosion on battery terminals or wires.	.117		√	√	√	√
14 Check the electrolyte level in the batteries; add if required.	.049		√	√	√	√
15 Check the specific gravity of the electrolyte in a 10% sample of the batteries.	.069					√
16 Check 25% of terminal-to-cell connection resistance; rehabilitate connections as required; add anti-corrosion grease to battery terminals and connections.	.038					√
17 Measure and record individual cell and string float voltages.	.140					√
18 Clean area around generator.	.066		√	√	√	√
19 Fill out maintenance checklist and report deficiencies.	.022		√	√	√	√
Total labor-hours/period			8.327	8.327	8.327	10.271
Total labor-hours/year			66.618	16.654	8.327	10.271

		Labor-hrs.	Cost Each						
			2019 Bare Costs				Total In-House	Total w/O&P
	Description		Material	Labor	Equip.	Total		
3900	Emergency diesel generator, turbine, annually	10.271	162	545		707	889.48	1,075
3950	Annualized	16.633	147	885		1,032	1,311.84	1,600

For customer support on your Facilities Maintenance & Repair Costs with RSMeans data, call 800.448.8182.

PM Components	Labor-hrs.	W	M	Q	S	A
PM System D5095 220 1950						
Power stabilizer						
1 Remove the covers or open the doors.	.218					√
2 Check for signs of moisture or overheating.	.022					√
3 Check fans and other auxiliary devices for proper operation.	.027					√
4 Check for voltage creeping over transformer insulated surfaces, such as evidenced by tracking or carbonization.	.018					√
5 Check condition of the ground system.	.170					√
6 Check wiring, connections, relays, etc; tighten and adjust as necessary.	.060					√
7 Replace the covers or close the doors.	.022					√
8 Clean the surrounding area.	.066					√
9 Fill out maintenance checklist and report deficiencies.	.022					√
Total labor-hours/period						.625
Total labor-hours/year						.625

			Cost Each					
			2019 Bare Costs				Total	Total
Description		Labor-hrs.	Material	Labor	Equip.	Total	In-House	w/O&P
1900	Power stablizer, annually	.625	9	33.50		42.50	52.99	65
1950	Annualized	.625	9	33.50		42.50	52.99	65

For customer support on your Facilities Maintenance & Repair Costs with RSMeans data, call 800.448.8182.

PM Components	Labor-hrs.	W	M	Q	S	A
PM System D5095 230 1950						
Uninterruptible power system, up to 200 KVA						
1 Check with operating or area personnel for any obvious deficiencies.	.044		√	√	√	√
2 Check electrolyte level of batteries; add water as required; check terminals for corrosion, clean as required.	.878		√	√	√	√
3 Check 25% of the batteries for charge and electrolyte specific gravity.	.777			√	√	√
4 Check batteries for cracks or leaks.	.059		√	√	√	√
5 Check 25% of the terminal-to-cell connection resistances; rehabilitate connections as required; add anti-corrosion grease to battery terminals and connections.	.143			√	√	√
6 Measure and record individual cell and string float voltages.	.618			√	√	√
7 Check integrity of battery rack.	.176		√	√	√	√
8 Check battery room temperature and ventilation systems.	.010		√	√	√	√
9 Replace air filters on UPS modules.	.036		√	√	√	√
10 Check output voltage and amperages, from control panel.	.007		√	√	√	√
11 Check UPS room temperature and ventilation system.	.062		√	√	√	√
12 Notify personnel and test UPS fault alarm system.	.112			√	√	√
13 Clean around batteries and UPS modules.	.066		√	√	√	√
14 Fill out maintenance checklist and report deficiencies.	.022		√	√	√	√
Total labor-hours/period			1.360	3.010	3.010	3.010
Total labor-hours/year			10.880	6.020	3.010	3.010

		Cost Each					
			2019 Bare Costs			Total	Total
Description	Labor-hrs.	Material	Labor	Equip.	Total	In-House	w/O&P
1900 Uninter. power system, up to 200 KVA, annually	3.010	192	161		353	418.31	500
1950 Annualized	22.924	273	1,225		1,498	1,879.25	2,300

PM Components	Labor-hrs.	W	M	Q	S	A
PM System D5095 230 2950						
Uninterrupted power system, 200 KVA to 800 KVA						
1 Check with operating or area personnel for any obvious deficiencies.	.044		√	√	√	√
2 Check electrolyte level of batteries; add water as required; check terminals for corrosion, clean as required.	3.510		√	√	√	√
3 Check 25% of the batteries for charge and electrolyte specific gravity.	3.200			√	√	√
4 Check batteries for cracks or leaks.	.593		√	√	√	√
5 Check 25% of the terminal-to-cell connection resistances; rehabilitate connections as required; add anti-corrosion grease to battery terminals and connections.	.234			√	√	√
6 Measure and record individual cell and string float voltages.	.941			√	√	√
7 Check integrity of battery rack.	.696		√	√	√	√
8 Check battery room temperature and ventilation systems.	.010		√	√	√	√
9 Replace air filters on UPS modules.	.036		√	√	√	√
10 Check output voltage and amperages, from control panel.	.007		√	√	√	√
11 Check UPS room temperature and ventilation system.	.062		√	√	√	√
12 Notify personnel and test UPS fault alarm system.	.112			√	√	√
13 Clean around batteries and UPS modules.	.066		√	√	√	√
14 Fill out maintenance checklist and report deficiencies.	.022		√	√	√	√
Total labor-hours/period			5.046	9.533	9.533	9.533
Total labor-hours/year			40.365	19.065	9.533	9.533

		Cost Each					
			2019 Bare Costs			Total	Total
Description	Labor-hrs.	Material	Labor	Equip.	Total	In-House	w/O&P
2900 UPS, 200 KVA to 800 KVA, annually	9.533	192	510		702	869.51	1,050
2950 Annualized	78.058	283	4,175		4,458	5,677.25	7,000

For customer support on your Facilities Maintenance & Repair Costs with RSMeans data, call 800.448.8182.

PM Components	Labor-hrs.	W	M	Q	S	A
PM System D5095 240 1950						
Battery system and charger						
1 Check that the charger is operating properly.	.051		√	√	√	√
2 Check for any signs of corrosion on the battery terminals or wires.	.117		√	√	√	√
3 Check the electrolyte level in the batteries; add if required.	.062		√	√	√	√
4 Check the specific gravity of the electrolyte in a 10% sample of the batteries.	.069		√	√	√	√
5 Check 25% of terminal-to-cell connection resistance; rehabilitate connections as required; add anti-corrosion grease to battery terminals and connections.	.741			√	√	√
6 Measure and record individual cell and string float voltages.	.281			√	√	√
7 Clean area around batteries.	.066		√	√	√	√
8 Fill out maintenance checklist and report deficiencies.	.022		√	√	√	√
Total labor-hours/period			.387	1.409	1.409	1.409
Total labor-hours/year			3.097	2.818	1.409	1.409

			Cost Each					
			2019 Bare Costs				Total	Total
Description		Labor-hrs.	Material	Labor	Equip.	Total	In-House	w/O&P
1900	Battery system and charger, annually	1.409	11.95	75		86.95	110.11	136
1950	Annualized	8.732	18.50	465		483.50	622.20	770

For customer support on your Facilities Maintenance & Repair Costs with RSMeans data, call 800.448.8182.

PM Components	Labor-hrs.	W	M	Q	S	A
PM System D5095 250 1950						
Light, emergency, hardwired system						
1 Test for proper operation, replace lamps as required.	.013				✓	✓
2 Clean exterior of cabinet and light heads.	.025				✓	✓
3 Check wiring for obvious defects.	.026				✓	✓
4 Adjust lamp heads for maximum illumination of area.	.039				✓	✓
5 Fill out maintenance checklist and report deficiencies.	.022				✓	✓
Total labor-hours/period					125	125
Total labor-hours/year					.125	.125

		Cost Each					
		2019 Bare Costs				**Total**	**Total**
Description	Labor-hrs.	Material	Labor	Equip.	Total	In-House	w/O&P
1900 Light, emergency, hardwired system, annually	.125	5.85	6.65		12.50	15.02	18
1950 Annualized	.250	7.75	13.35		21.10	25.77	31

PM Components	Labor-hrs.	W	M	Q	S	A
PM System D5095 250 2950						
Light, emergency, dry cell						
1 Test for proper operation, replace lamps as required.	.013				✓	✓
2 Clean exterior of emergency light cabinet and lamp heads.	.025				✓	✓
3 Check condition of batteries and change as required.	.012				✓	✓
4 Clean interior of cabinet, top of battery, and battery terminal.	.009				✓	✓
5 Inspect cabinet, relay, relay contacts, pilot light, wiring, and gen. cond.	.065				✓	✓
6 Adjust lamp heads for maximum illumination of area.	.039				✓	✓
7 Fill out maintenance checklist and report deficiencies.	.022				✓	✓
Total labor-hours/period					.185	.185
Total labor-hours/year					.185	.185

		Cost Each					
		2019 Bare Costs				**Total**	**Total**
Description	Labor-hrs.	Material	Labor	Equip.	Total	In-House	w/O&P
2900 Light, emergency, dry cell, annually	.185	40.50	9.90		50.40	56.77	66.50
2950 Annualized	.356	32.50	19.05		51.55	60.58	71

PM Components	Labor-hrs.	W	M	Q	S	A
PM System D5095 250 3950						
Light, emergency, wet cell						
1 Test for proper operation; correct deficiencies as required.	.013				✓	✓
2 Clean exterior of emergency light cabinet and lamp heads.	.025				✓	✓
3 Clean interior of cabinet, top of battery, and battery terminal.	.009				✓	✓
4 Check battery charge and electrolyte specific gravity; add water if nec.	.012				✓	✓
5 Inspect cabinet, relay, relay contacts, pilot light, wiring, and gen. cond.	.065				✓	✓
6 Apply anti-corrosion coating to battery terminals.	.010				✓	✓
7 Adjust lamp heads for maximum illumination of area.	.039				✓	✓
8 Fill out maintenance checklist and report deficiencies.	.022				✓	✓
Total labor-hours/period					.195	.195
Total labor-hours/year					.195	.195

		Cost Each					
		2019 Bare Costs				**Total**	**Total**
Description	Labor-hrs.	Material	Labor	Equip.	Total	In-House	w/O&P
3900 Light, emergency, wet cell, annually	.195	22.50	10.40		32.90	38.04	45
3950 Annualized	.390	33	21		54	63.17	75

For customer support on your Facilities Maintenance & Repair Costs with RSMeans data, call 800.448.8182.

PM Components	Labor-hrs.	W	M	Q	S	A
PM System E1015 810 1950						
Personal Computers						
1 Check with operating personnel for any deficiencies.	.065				√	√
2 Check all cables for fraying, cracking, or exposed wires; ensure connections are secure.	.065				√	√
3 Remove metal casing and vacuum interior, power supply vents, keyboard, monitor, and surrounding work area.	.130				√	√
4 Check for secure connections of all interior wires and cables. Secure all peripheral cards.	.065				√	√
5 Replace PC cover and clean case, keyboard, monitor screen and housing with an anti-static spray cleaner. Apply cleaner to towel and not the components.	.130				√	√
6 Disassemble the mouse; clean the tracking ball, interior rollers and housing. Blow out the interior of the mouse with clean compressed air.	.065				√	√
7 Turn on PC and check for unusual noises.	.026				√	√
8 Clean floppy drive heads with cleaning diskette and solution.	.039				√	√
9 Defragment the hard drive using a defragmentation utility.	.234				√	√
10 Run diagnostics software to test all the PC's components.	.234				√	√
11 Fill out maintenance checklist and report deficiencies.	.029				√	√
Total labor-hours/period					1.082	1.082
Total labor-hours/year					1.082	1.082

			Cost Each					
			2019 Bare Costs				Total	Total
Description		Labor-hrs.	Material	Labor	Equip.	Total	In-House	w/O&P
1900	Personnel Computers, annually	1.082	3.18	58		61.18	78.10	96.50
1950	Annualized	2.164	6.35	116		122.35	156.69	192

For customer support on your Facilities Maintenance & Repair Costs with RSMeans data, call 800.448.8182.

PM Components	Labor-hrs.	W	M	Q	S	A
PM System E1035 100 1950						
Hydraulic lift						
1 Check for proper operation of pump.	.035				√	√
2 Check for leaks on suction and discharge piping, seals, packing glands, etc.; make minor adjustments as required.	.077				√	√
3 Check pump and motor operation for excessive vibration, noise and overheating.	.022				√	√
4 Check alignment of pump and motor; adjust as necessary.	.258				√	√
5 Lubricate pump and motor.	.000				√	√
6 Inspect hydraulic lift post(s) for wear or leaks.	.065				√	√
7 Inspect, clean and tighten valves.	.034				√	√
8 Inspect and clean motor contactors.	.026				√	√
9 Inspect and test control relays and check wiring terminals.	.034				√	√
10 Clean pump unit and surrounding area.	.099				√	√
11 Fill out maintenance checklist and report deficiencies.	.022				√	√
Total labor-hours/period					.771	.771
Total labor-hours/year					.771	.771

			Cost Each				
			2019 Bare Costs			Total	Total
Description	Labor-hrs.	Material	Labor	Equip.	Total	In-House	w/O&P
1900 Hydraulic lift, annually	.771	247	41		288	325.32	375
1950 Annualized	1.546	495	82.50		577.50	653.06	750

PM Components	Labor-hrs.	W	M	Q	S	A
PM System E1035 310 1950						
Hydraulic lift, loading dock						
1 Check for proper operation of pump.	.022				√	√
2 Check for leaks on suction and discharge piping, seals, packing glands, etc.; make minor adjustments as required.	.077				√	√
3 Check pump and motor operation for excessive vibration, noise and overheating.	.022				√	√
4 Check alignment of pump and motor; adjust as necessary.	.100				√	√
5 Lubricate pump and motor.	.099				√	√
6 Inspect hydraulic lift post(s) for wear or leaks.	.051				√	√
7 Inspect, clean and tighten valves.	.072				√	√
8 Inspect and clean motor contactors.	.148				√	√
9 Inspect and test control relays and check wiring terminals.	.137				√	√
10 Clean pump unit and surrounding area.	.066				√	√
11 Fill out maintenance checklist and report deficiencies.	.022				√	√
Total labor-hours/period					.816	.816
Total labor-hours/year					.816	.816

			Cost Each				
			2019 Bare Costs			Total	Total
Description	Labor-hrs.	Material	Labor	Equip.	Total	In-House	w/O&P
1900 Hydraulic lift, loading dock, annually	.816	40.50	43.50		84	100.95	121
1950 Annualized	1.632	61	87		148	179.49	216

For customer support on your Facilities Maintenance & Repair Costs with RSMeans data, call 800.448.8182.

PM Components	Labor-hrs.	W	M	Q	S	A
PM System E1095 122 1950						
Crane, electric bridge, up to 5 ton						
1 Check with operating or area personnel for deficiencies.	.035					√
2 Inspect crane for damaged structural members, alignment, worn pins, column fasteners, stops and capacity markings.	.235					√
3 Inspect and lubricate power cable tension reel and inspect condition of cable.	.485					√
4 Check hoist operation - travel function, up, down, speed stepping, upper and lower travel limit switches, etc.	.072					√
5 Inspect, clean and lubricate trolley assembly and reduction gear bearings.	.112					√
6 Inspect and clean hook assembly and chain or wire rope sheaves.	.457					√
7 Inspect, clean and lubricate hoist motor.	.094					√
8 Inspect hoist brake mechanism.	.444					√
9 Lubricate rails.	.114					√
10 Weight capacity test hoisting equipment at 125% of rated capacity in accordance with 29 CFR 1910.179.	.780					√
11 Clean work area.	.066					√
12 Fill out maintenance report and report deficiencies.	.022					√
Total labor-hours/period						2.917
Total labor-hours/year						2.917

		Labor-hrs.	Cost Each						
			2019 Bare Costs				Total In-House	Total w/O&P
	Description		Material	Labor	Equip.	Total		
1900	Crane, electric bridge, up to 5 ton, annually	2.917	160	158		318	371.93	445
1950	Annualized	2.917	160	158		318	371.93	445

PM Components	Labor-hrs.	W	M	Q	S	A
PM System E1095 122 2950						
Crane, electric bridge 5 to 15 ton						
1 Check with operating or area personnel for deficiencies.	.035			√	√	√
2 Inspect crane for damaged structural members, alignment, worn pins, column fasteners, stops and capacity markings.	.235			√	√	√
3 Inspect and lubricate power cable tension reel and inspect condition of cable.	.491			√	√	√
4 Check operation of upper and lower travel limit switches.	.072			√	√	√
5 Inspect brake mechanism.	.250					√
6 Inspect, clean and lubricate trolley assembly and reduction gear bearings.	.099			√	√	√
7 Drain and replace reduction gear housing oil.	.170					√
8 Inspect and clean hook assembly and chain or wire rope sheaves.	.099			√	√	√
9 Inspect, clean and lubricate hoist motor.	.094			√	√	√
10 Check pendent control for hoist operation - travel function, up, down, speed stepping, etc.	.088			√	√	√
11 Lubricate rails.	.325			√	√	√
12 Weight capacity test hoisting equipment at 125% of rated capacity in accordance with 29 CFR 1910.179.	1.040					√
13 Clean work area.	.066			√	√	√
14 Fill out maintenance report and report deficiencies.	.022			√	√	√
Total labor-hours/period				1.626	1.626	3.086
Total labor-hours/year				3.252	1.626	3.086

		Labor-hrs.	Cost Each						
			2019 Bare Costs				Total In-House	Total w/O&P
	Description		Material	Labor	Equip.	Total		
2900	Crane, electric bridge, 5 to 15 ton, annually	3.086	185	167		352	411.33	490
2950	Annualized	7.667	415	415		830	971.80	1,175

For customer support on your Facilities Maintenance & Repair Costs with RSMeans data, call 800.448.8182.

PM Components	Labor-hrs.	W	M	Q	S	A
PM System E1095 122 3950						
Crane, electric bridge, over 15 tons						
1 Check with operating or area personnel for deficiencies.	.035			√	√	√
2 Inspect crane for damaged structural members, alignment, worn pins, column fasteners, stops and capacity markings.	.235			√	√	√
3 Inspect and lubricate power cable tension reel and inspect condition of cable.	.491			√	√	√
4 Check operation of upper and lower travel limit switches.	.072			√	√	√
5 Inspect brake mechanism.	.250					√
6 Inspect, clean and lubricate trolley assembly and reduction gear bearings.	.099			√	√	√
7 Drain and replace reduction gear housing oil.	.170					√
8 Inspect and clean hook assembly and chain or wire rope sheaves.	.099			√	√	√
9 Inspect, clean and lubricate hoist motor.	.094			√	√	√
10 Check pendent control for hoist operation - travel function, up, down, speed stepping, etc.	.088			√	√	√
11 Lubricate rails.	.325			√	√	√
12 Inspect control cabinet.	.238					√
13 Weight capacity test hoisting equipment, at 125% of rated capacity in accordance with 29 CFR 1910.179.	1.170					√
14 Clean work area.	.066			√	√	√
15 Fill out maintenance report and report deficiencies.	.022			√	√	√
Total labor-hours/period				1.626	1.626	3.454
Total labor-hours/year				3.252	1.626	3.454

			Cost Each					
			2019 Bare Costs				**Total**	**Total**
Description		**Labor-hrs.**	**Material**	**Labor**	**Equip.**	**Total**	**In-House**	**w/O&P**
3900	Crane, electric bridge, over 15 tons, annually	3.454	94.50	187		281.50	336.43	410
3950	Annualized	8.266	320	450		770	911.38	1,100

For customer support on your Facilities Maintenance & Repair Costs with RSMeans data, call 800.448.8182.

PM Components	Labor-hrs.	W	M	Q	S	A
PM System E1095 123 1950						
Crane, manual bridge, up to 5 tons						
1 Check with operating or area personnel for deficiencies.	.035					√
2 Operate crane and hoist, inspect underhung hoist monorail for damaged structural members, braces, stops and capacity markings.	.075					√
3 Inspect and lubricate trolley wheels and bearings.	.096					√
4 Inspect, clean and lubricate hooks and chain sheaves.	.099					√
5 Check operation of crane and hoist brakes.	.016					√
6 Make minor adjustments as required for operational integrity; tighten bolts, etc.	.134					√
7 Weight capacity test hoisting equipment at 125% of rated capacity in accordance with 29 CFR 1910.179.	.390					√
8 Clean work area.	.066					√
9 Fill out maintenance report and report deficiencies.	.022					√
Total labor-hours/period						.933
Total labor-hours/year						.933

			Cost Each					
			2019 Bare Costs				Total	Total
Description	Labor-hrs.	Material	Labor	Equip.	Total	In-House	w/O&P
1900	Crane, manual bridge, up to 5 ton, annually	.933	91	50.50		141.50	162.96	192
1950	Annualized	.933	91	50.50		141.50	162.96	192

PM Components	Labor-hrs.	W	M	Q	S	A
PM System E1095 123 2950						
Crane, manual bridge, 5 to 15 tons						
1 Check with operating or area personnel for deficiencies.	.035					√
2 Operate crane and hoist, inspect underhung hoist monorail for damaged structural members, braces, stops and capacity markings.	.130					√
3 Inspect and lubricate trolley wheels and bearings.	.096					√
4 Inspect, clean and lubricate hooks and chain sheaves.	.099					√
5 Check operation of crane and hoist brakes.	.016					√
6 Make minor adjustments as required for operational integrity; tighten bolts, etc.	.134					√
7 Weight capacity test hoisting equipment at 125% of rated capacity in accordance with 29 CFR 1910.179.	.772					√
8 Clean work area.	.066					√
9 Fill out maintenance report and report deficiencies.	.022					√
Total labor-hours/period						1.370
Total labor-hours/year						1.370

			Cost Each					
			2019 Bare Costs				Total	Total
Description	Labor-hrs.	Material	Labor	Equip.	Total	In-House	w/O&P
2900	Crane, manual bridge, 5 to 15 ton, annually	1.370	46	74.50		120.50	143.16	173
2950	Annualized	1.370	46	74.50		120.50	143.16	173

PM Components	Labor-hrs.	W	M	Q	S	A
PM System E1095 123 3950						
Crane, manual bridge, over 15 tons						
1 Check with operating or area personnel for deficiencies.	.035					✓
2 Operate crane and hoist, inspect for damaged structural members, alignment, worn pins, fasteners, stops and capacity markings.	.235					✓
3 Inspect and lubricate trolley wheels and bearings.	.096					✓
4 Inspect, clean and lubricate hooks and chain sheaves.	.099					✓
5 Check operation of crane and hoist brakes.	.016					✓
6 Make minor adjustments as required for operational integrity; tighten bolts, etc.	.134					✓
7 Weight capacity test hoisting equipment at 125% of rated capacity in accordance with 29 CFR 1910.179.	1.162					✓
8 Clean work area.	.066					✓
9 Fill out maintenance report and report deficiencies.	.022					✓
Total labor-hours/period						1.865
Total labor-hours/year						1.865

				Cost Each				
				2019 Bare Costs			Total	Total
	Description	Labor-hrs.	Material	Labor	Equip.	Total	In-House	w/O&P
3900	Crane, manual bridge, over 15 tons, annually	1.865	46	101		147	176.21	215
3950	Annualized	1.865	46	101		147	176.21	215

PM Components	Labor-hrs.	W	M	Q	S	A
PM System E1095 125 1950						
Hoist, pneumatic						
1 Check with operating or area personnel for deficiencies.	.035					✓
2 Inspect for damaged structural members, alignment, worn pins, column fasteners, stops and capacity markings.	.235					✓
3 Inspect and lubricate cable tension reel and inspect condition of cable.	.452					✓
4 Check pendent control for hoist operation - travel function, up, down, speed stepping, upper and lower travel limit switches, etc.	.088					✓
5 Inspect, clean and lubricate trolley assembly.	.099					✓
6 Drain and replace reduction gear housing oil.	.170					✓
7 Inspect brake mechanism.	.016					✓
8 Inspect and clean hook assembly and chain or wire rope sheaves.	.099					✓
9 Inspect, clean and lubricate hoist motor.	.029					✓
10 Clean work area.	.066					✓
11 Fill out maintenance report and report deficiencies.	.022					✓
Total labor-hours/period						1.311
Total labor-hours/year						1.311

				Cost Each				
				2019 Bare Costs			Total	Total
	Description	Labor-hrs.	Material	Labor	Equip.	Total	In-House	w/O&P
1900	Hoist, pneumatic, annually	1.311	162	71		233	266.21	315
1950	Annualized	1.311	162	71		233	266.21	315

For customer support on your Facilities Maintenance & Repair Costs with RSMeans data, call 800.448.8182.

PM Components	Labor-hrs.	W	M	Q	S	A
PM System E1095 126 1950						
Hoist/winch, chain fall/cable, electric						
1 Check with operating or area personnel for deficiencies.	.035					√
2 Inspect for damaged structural members, alignment, worn pins, column fasteners, stops and capacity markings.	.235					√
3 Inspect and lubricate power cable tension reel; inspect condition of cable.	.491					√
4 Check hoist operation - travel function, up, down, speed stepping, upper and lower travel limit switches, etc.	.088					√
5 Inspect, clean and lubricate trolley assembly and reduction gear bearings.	.099					√
6 Drain and replace reduction gear housing oil.	.170					√
7 Inspect brake mechanism.	.250					√
8 Inspect and clean hook assembly and chain or wire rope sheaves.	.099					√
9 Inspect, clean and lubricate hoist motor.	.094					√
10 Clean work area.	.066					√
11 Fill out maintenance report and report deficiencies.	.022					√
Total labor-hours/period						1.649
Total labor-hours/year						1.649

		Cost Each						
			2019 Bare Costs				Total	Total
	Description	Labor-hrs.	Material	Labor	Equip.	Total	In-House	w/O&P
1900	Hoist / winch, chain / cable, electric, annually	1.649	149	89		238	274.98	325
1950	Annualized	1.649	149	89		238	274.98	325

PM Components	Labor-hrs.	W	M	Q	S	A
PM System E1095 302 1950						
Beverage dispensing unit						
1 Check with operating or area personnel for deficiencies.	.044			√	√	√
2 Remove access panels and clean coils, fans, fins, and other areas; vacuum, blow down with compressed air or CO_2, and/or wipe off.	.517			√	√	√
3 Lubricate motor bearings.	.008			√	√	√
4 Check electrical wiring and connections; tighten loose connections.	.120			√	√	√
5 Visually check for refrigerant, water, and other liquid leaks.	.153			√	√	√
6 Check door gaskets for proper sealing; adjust door catch as required.	.033			√	√	√
7 Inspect dispensing valves; adjust as necessary.	.108			√	√	√
8 Perform operational check of unit; make adjustments as necessary.	.040			√	√	√
9 Clean area around equipment.	.066			√	√	√
10 Fill out maintenance checklist and report deficiencies.	.022			√	√	√
Total labor-hours/period				1.111	1.111	1.111
Total labor-hours/year				2.223	1.111	1.111

		Cost Each						
			2019 Bare Costs				Total	Total
	Description	Labor-hrs.	Material	Labor	Equip.	Total	In-House	w/O&P
1900	Beverage dispensing unit, annually	1.111	40.50	59		99.50	120.95	146
1950	Annualized	4.444	53	237		290	365	445

For customer support on your Facilities Maintenance & Repair Costs with RSMeans data, call 800.448.8182.

503

PM Components	Labor-hrs.	W	M	Q	S	A
PM System E1095 304 1950						
Braising pan, tilting, gas/electric						
1 Check with operating or area personnel for deficiencies.	.044			√	√	√
2 Check electric transformer, heating elements, insulators, connections and wiring; tighten connections and repair as required.	.239			√	√	√
3 Check surface temperature with meter for hot spots.	.073			√	√	√
4 Check piping and valves on gas operated units for leaks.	.077			√	√	√
5 Check pilot/igniter, burner and combustion chamber for proper operation; adjust as required.	.079			√	√	√
6 Check temperature control system for proper operation; adjust as required.	.458			√	√	√
7 Lubricate tilting gear mechanism and trunnion bearings.	.047			√	√	√
8 Check nuts, bolts and screws for tightness; tighten as required.	.007			√	√	√
9 Check complete operation of unit.	.055			√	√	√
10 Fill out maintenance checklist and report deficiencies.	.022			√	√	√
Total labor-hours/period				1.101	1.101	1.101
Total labor-hours/year				2.202	1.101	1.101

			Cost Each					
			2019 Bare Costs				Total	Total
Description		Labor-hrs.	Material	Labor	Equip.	Total	In-House	w/O&P
1900	Braising pan, tilting, gas / electric, annually	1.101	40.50	59		99.50	120.08	145
1950	Annualized	4.404	53	235		288	361.78	440

PM Components	Labor-hrs.	W	M	Q	S	A
PM System E1095 306 1950						
Bread slicer, electric						
1 Check with operating or area personnel for deficiencies.	.044			√	√	√
2 Check electric motor, controls, wiring connections, insulation and contacts for damage and/or proper operation; adjust as required.	.338			√	√	√
3 Check V-belt for wear, tension and alignment; change/adjust belt as necessary.	.029			√	√	√
4 Lubricate bushings and check oil level in gear drive mechanism; add oil if required.	.065			√	√	√
5 Clean machine of dust, grease and food particles.	.074			√	√	√
6 Check motor and motor bearings for overheating.	.039			√	√	√
7 Check for loose, missing or damaged nuts, bolts, or screws; tighten or replace as necessary.	.005			√	√	√
8 Fill out maintenance checklist and report deficiencies.	.022			√	√	√
Total labor-hours/period				.616	.616	.616
Total labor-hours/year				1.232	.616	.616

			Cost Each					
			2019 Bare Costs				Total	Total
Description		Labor-hrs.	Material	Labor	Equip.	Total	In-House	w/O&P
1900	Bread slicer, electric, annually	.616	25.50	33		58.50	70.68	85
1950	Annualized	2.464	78.50	131		209.50	256.33	310

For customer support on your Facilities Maintenance & Repair Costs with RSMeans data, call 800.448.8182.

PM Components	Labor-hrs.	W	M	Q	S	A
PM System E1095 310 1950						
Broiler, conveyor type, gas						
1 Check with operating or area personnel for deficiencies.	.044			√	√	√
2 Check broiler pilot/igniters and jets for uniform flame; adjust as required.	.166			√	√	√
3 Check air filter; change as required.	.059			√	√	√
4 Check operation of chains and counterweights.	.012			√	√	√
5 Check piping and valves for leaks.	.077			√	√	√
6 Check electrical connections and wiring for defects; tighten loose connections.	.120			√	√	√
7 Check thermocouple and thermostat; calibrate thermostat as required.	.640			√	√	√
8 Tighten and replace any loose nuts, bolts, or screws.	.005			√	√	√
9 Check operation of unit.	.055			√	√	√
10 Fill out maintenance checklist and report deficiencies.	.022			√	√	√
Total labor-hours/period				1.200	1.200	1.200
Total labor-hours/year				2.400	1.200	1.200

		Cost Each					
		2019 Bare Costs				Total	Total
Description	Labor-hrs.	Material	Labor	Equip.	Total	In-House	w/O&P
1900 Broiler, conveyor type, gas, annually	1.200	21	64		85	105.72	128
1950 Annualized	4.824	63.50	258		321.50	402.51	495

PM Components	Labor-hrs.	W	M	Q	S	A
PM System E1095 310 2950						
Broiler, hot dog, electric						
1 Check with operating or area personnel for deficiencies.	.044			√	√	√
2 Check electric motor, controls, contacts, wiring connections for damage and proper operation; adjust as required.	.152			√	√	√
3 Lubricate bushings.	.007			√	√	√
4 Inspect lines and unit for mechanical defects; adjust or repair as required.	.021			√	√	√
5 Clean machine of dust and grease.	.109			√	√	√
6 Check unit for proper operation.	.082			√	√	√
7 Fill out maintenance checklist and report deficiencies.	.022			√	√	√
Total labor-hours/period				.437	.437	.437
Total labor-hours/year				.874	.437	.437

		Cost Each					
		2019 Bare Costs				Total	Total
Description	Labor-hrs.	Material	Labor	Equip.	Total	In-House	w/O&P
2900 Broiler, hot dog type, electric, annually	.437	9	23.50		32.50	40.05	49
2950 Annualized	1.748	27	93.50		120.50	150.10	184

PM Components	Labor-hrs.	W	M	Q	S	A
PM System E1095 310 3950						
Broiler/oven (salamander), gas/electric						
1　Check with operating or area personnel for deficiencies.	.044			√	√	√
2　Check broiler pilot and oven for uniform flame; adjust as required.	.167			√	√	√
3　Check piping and valves for leaks.	.077			√	√	√
4　Check doors, shelves for warping, alignment, and seals; adjust as required.	.148			√	√	√
5　Check electrical connections and wiring for defects; tighten loose connections.	.242			√	√	√
6　Tighten and replace any loose nuts, bolts and/or screws.	.012			√	√	√
7　Check operation of broiler arm, adjust as required.	.007			√	√	√
8　Check thermocouple and thermostat; calibrate thermostat as required.	.571			√	√	√
9　Fill out maintenance checklist and report deficiencies.	.022			√	√	√
Total labor-hours/period				1.291	1.291	1.291
Total labor-hours/year				2.581	1.291	1.291

		Cost Each					
		2019 Bare Costs				Total	Total
Description	Labor-hrs.	Material	Labor	Equip.	Total	In-House	w/O&P
3900　Broiler / oven, (salamander), gas / electric, annually	1.291	4.23	69		73.23	93.73	115
3950　Annualized	5.144	12.70	275		287.70	368.88	455

PM Components	Labor-hrs.	W	M	Q	S	A
PM System E1095 312 1950						
Chopper, electric						
1　Check with operating or area personnel for deficiencies.	.044			√	√	√
2　Check cutting surfaces, clutch assembly and guards for damage and proper alignment; adjust as required.	.065			√	√	√
3　Check nuts, bolts, and screws for tightness; tighten or replace as required.	.005			√	√	√
4　Check switch and controls for proper operation; adjust as required.	.010			√	√	√
5　Clean machine of dust, grease and food particles.	.014			√	√	√
6　Check operation of motor and check motor bearings for overheating; lubricate as required.	.069			√	√	√
7　Check belt tension; replace belt and/or adjust if required.	.314			√	√	√
8　Fill out maintenance checklist and report deficiencies.	.022			√	√	√
Total labor-hours/period				.543	.543	.543
Total labor-hours/year				1.085	.543	.543

		Cost Each					
		2019 Bare Costs				Total	Total
Description	Labor-hrs.	Material	Labor	Equip.	Total	In-House	w/O&P
1900　Chopper, electric, annually	.543	41.50	29		70.50	82.67	98.50
1950　Annualized	2.176	90	116		206	250.11	299

For customer support on your Facilities Maintenance & Repair Costs with RSMeans data, call 800.448.8182.

PM Components	Labor-hrs.	W	M	Q	S	A
PM System E1095 314 1950						
Coffee maker/urn, gas/electric/steam						
1 Check thermostat, switch, and temperature gauge; calibrate if required.	.465			√	√	√
2 Check pilots and flame on gas operated units; adjust as required.	.038			√	√	√
3 Check working pressure on steam operated unit.	.005			√	√	√
4 Check electrical connections and wiring for defects; tighten connections.	.120			√	√	√
5 Check for clogged or defective steam trap.	.100			√	√	√
6 Inspect and clean steam strainer.	.207			√	√	√
7 Examine equipment, valves, and piping for leaks.	.077			√	√	√
8 Inspect for leaks at water gauge glasses and at valves; repack valves if necessary.	.046			√	√	√
9 Tighten or replace loose, missing, or damaged nuts, bolts, or screws.	.005			√	√	√
10 Lubricate water filter valve, check rings, tighten as required.	.033			√	√	√
11 Check operation of lights.	.027			√	√	√
12 Check timer mechanism and operation of unit.	.018			√	√	√
13 Fill out maintenance checklist and report deficiencies.	.022			√	√	√
Total labor-hours/period				1.163	1.163	1.163
Total labor-hours/year				2.326	1.163	1.163

			Cost Each					
			2019 Bare Costs				Total	Total
Description	Labor-hrs.	Material	Labor	Equip.	Total		In-House	w/O&P
1900 Coffee maker / urn, gas / electric / steam, annually	1.163	20.50	62.50		83		102.92	125
1950 Annualized	4.652	62	248		310		388.19	480

507

For customer support on your Facilities Maintenance & Repair Costs with RSMeans data, call 800.448.8182.

PM Components	Labor-hrs.	W	M	Q	S	A
PM System E1095 316 1950						
Cooker, gas						
1 Check with operating or area personnel for any deficiencies.	.044			√	√	√
2 Examine doors and door gaskets; make necessary adjustments.	.021			√	√	√
3 Visually inspect hinges and latches; lubricate hinges.	.091			√	√	√
4 Tighten or replace any loose nuts, bolts, screws; replace broken knobs.	.007			√	√	√
5 Check gas piping and valves for leaks.	.077			√	√	√
6 Check pilot and thermosensor bulb.	.079			√	√	√
7 Check gas burner for uniform flame.	.007			√	√	√
8 Check calibration of thermostats.	.457			√	√	√
9 Operate unit to ensure that it is in proper working condition.	.130			√	√	√
10 Fill out maintenance checklist and report deficiencies.	.022			√	√	√
Total labor-hours/period				.935	.935	.935
Total labor-hours/year				1.870	.935	.935

		Cost Each						
			2019 Bare Costs				Total	Total
Description	Labor-hrs.	Material	Labor	Equip.	Total	In-House	w/O&P
1900 Cooker, gas, annually	.935	9	50		59	74.33	91.50
1950 Annualized	3.744	27	200		227	287.33	355

PM Components	Labor-hrs.	W	M	Q	S	A
PM System E1095 316 2950						
Cooker, vegetable steamer						
1 Check with operating or area personnel for any deficiencies.	.044			√	√	√
2 Examine doors and door gaskets; make necessary adjustments.	.021			√	√	√
3 Visually inspect hinges and latches; lubricate hinges.	.091			√	√	√
4 Tighten or replace any loose nuts, bolts, screws; replace broken knobs.	.005			√	√	√
5 Check working pressure on steam gauge.	.005			√	√	√
6 Inspect piping and valves for leaks; tighten as necessary.	.077			√	√	√
7 Check operation of low water cut-off; adjust as required.	.030			√	√	√
8 Check water level sight glass; adjust as required.	.007			√	√	√
9 Fill out maintenance checklist and report deficiencies.	.022			√	√	√
Total labor-hours/period				.302	.302	.302
Total labor-hours/year				.603	.302	.302

		Cost Each						
			2019 Bare Costs				Total	Total
Description	Labor-hrs.	Material	Labor	Equip.	Total	In-House	w/O&P
2900 Cooker, vegetable steamer, annually	.302	9	16.10		25.10	30.68	37.50
2950 Annualized	1.208	27	64.50		91.50	112.41	137

For customer support on your Facilities Maintenance & Repair Costs with RSMeans data, call 800.448.8182.

PM Components	Labor-hrs.	W	M	Q	S	A
PM System E1095 316 3950						
Cooker, steam						
1 Check with operating or area personnel for any deficiencies.	.044			√	√	√
2 Examine doors and door gaskets; make necessary adjustments.	.021			√	√	√
3 Visually inspect hinges and latches; lubricate hinges.	.091			√	√	√
4 Tighten or replace any loose nuts, bolts, screws; replace broken knobs.	.005			√	√	√
5 Check for clogged or defective steam trap; clean trap.	.098			√	√	√
6 Inspect and clean steam strainer.	.207			√	√	√
7 Check working pressure on steam gauge.	.005			√	√	√
8 Turn unit on, allow pressure to build, inspect piping, valve and doors for leaks; adjust as required.	.077			√	√	
9 Check operation of pressure relief valve; adjust as required.	.010			√	√	√
10 Check operation of low water cut-off; adjust as required.	.030			√	√	√
11 Check water level sight glass; adjust as required.	.007			√	√	√
12 Check timer mechanism and operation; adjust as required.	.018			√	√	√
13 Fill out maintenance checklist and report deficiencies.	.022			√	√	√
Total labor-hours/period				.635	.635	.635
Total labor-hours/year				1.270	.635	.635

			Cost Each					
				2019 Bare Costs			Total	Total
Description	Labor-hrs.	Material	Labor	Equip.	Total	In-House	w/O&P	
3900	Cooker, steam, annually	.635	9	34		43	53.61	66
3950	Annualized	2.540	27	136		163	204.13	251

PM Components	Labor-hrs.	W	M	Q	S	A
PM System E1095 318 1950						
Cookie maker, electric						
1 Check with operating or area personnel for any deficiencies.	.044			√	√	√
2 Check electrical panel switches, connections, timing device for loose, frayed wiring; tighten connections as required.	.120			√	√	√
3 Check operation of electric motor and bearings for overheating; lubricate motor bearings.	.086			√	√	√
4 Check belt; adjust tension or replace belt as required.	.372			√	√	√
5 Check conveyor and rollers by observing movement; lubricate rollers and adjust as required.	.127			√	√	√
6 Tighten or replace loose, missing or damaged nuts, bolts or screws.	.012			√	√	√
7 Wipe off and clean unit.	.096			√	√	√
8 Fill out maintenance checklist and report deficiencies.	.022			√	√	√
Total labor-hours/period				.879	.879	.879
Total labor-hours/year				1.758	.879	.879

			Cost Each					
				2019 Bare Costs			Total	Total
Description	Labor-hrs.	Material	Labor	Equip.	Total	In-House	w/O&P	
1900	Cookie maker, electric, annually	.879	28.50	47		75.50	92.10	111
1950	Annualized	3.504	86	187		273	336.45	410

For customer support on your Facilities Maintenance & Repair Costs with RSMeans data, call 800.448.8182.

509

PM Components	Labor-hrs.	W	M	Q	S	A
PM System E1095 320 1950						
Deep fat fryer, pressurized broaster, gas/electric						
1 Check with operating or area personnel for any deficiencies.	.044			√	√	√
2 Check piping and valves for leaks; tighten fittings as required.	.125			√	√	√
3 Check pilot and flame on gas operated unit; adjust as required.	.009			√	√	√
4 Check elements, switches, controls, contacts, and wiring on electrically heated units; repair or adjust as required.	.025			√	√	√
5 Check thermostat; calibrate if necessary.	.444			√	√	√
6 Check baskot/racks for bends, breaks or defects, straighten bends or repair as necessary.	.013			√	√	√
7 Check cooking oil filter on circulating system; change as required.	.068			√	√	√
8 Check operation of unit.	.163			√	√	√
9 Check operation of motor for excessive noise and overheating.	.022			√	√	√
10 Check nuts, bolts, and screws for tightness; tighten or replace as necessary.	.005			√	√	√
11 Fill out maintenance checklist and report deficiencies.	.022			√	√	√
Total labor-hours/period				.940	.940	.940
Total labor-hours/year				1.881	.940	.940

		Cost Each						
			2019 Bare Costs				Total	Total
Description	Labor-hrs.	Material	Labor	Equip.	Total	In-House	w/O&P
1900 Fryer, pressurized broaster, gas / electric, annually	.940	12.10	50		62.10	77.95	95.50
1950 Annualized	3.744	36	200		236	298.07	365

PM Components	Labor-hrs.	W	M	Q	S	A
PM System E1095 320 2950						
Deep fat fryer, automatic conveyor belt type, gas/electric						
1 Check with operating or area personnel for any deficiencies.	.044			√	√	√
2 Check compartments, valves and piping for leaks; tighten as required.	.077			√	√	√
3 Check pilot and flame on gas operated fryers; adjust as required.	.168			√	√	√
4 Check elements, switches, controls, contacts and wiring on electrically heated units for defects; repair or adjust as necessary.	.051			√	√	√
5 Check thermostats; calibrate as necessary.	.444			√	√	√
6 Check cooking oil filter on circulating system; change as required.	.068			√	√	√
7 Check operation of motor for excessive noise and overheating; lubricate bearings.	.022			√	√	√
8 Check operation of belt hoist; lubricate and adjust as required.	.052			√	√	√
9 Check nuts, bolts and screws for tightness; replace or tighten as necessary.	.017			√	√	√
10 Clean machine of dust, grease and food particles.	.044			√	√	√
11 Clean area around unit.	.066			√	√	√
12 Fill out maintenance checklist and report deficiencies.	.022			√	√	√
Total labor-hours/period				1.076	1.076	1.076
Total labor-hours/year				2.151	1.076	1.076

		Cost Each						
			2019 Bare Costs				Total	Total
Description	Labor-hrs.	Material	Labor	Equip.	Total	In-House	w/O&P
2900 Fryer, conveyor belt type, gas / electric, annually	1.076	21.50	57		78.50	97.76	119
2950 Annualized	4.280	64.50	229		293.50	366.75	445

For customer support on your Facilities Maintenance & Repair Costs with RSMeans data, call 800.448.8182.

PM Components	Labor-hrs.	W	M	Q	S	A
PM System E1095 320 3950						
Deep fat fryer, conventional type, gas/electric						
1 Check with operating or area personnel for any deficiencies.	.044			√	√	√
2 Clean machine thoroughly of dust, grease, and food particles.	.014			√	√	√
3 Check compartments, valves, and piping for leaks; tighten as required.	.077			√	√	√
4 Check pilot and flame on gas operated deep fat fryer; adjust as required.	.005			√	√	√
5 Check thermostat; calibrate if necessary.	.439			√	√	√
6 On electrically heated units check elements, switches, controls, contacts, and wiring for defects; repair or adjust as necessary.	.025			√	√	√
7 Check basket or rack for bends, breaks or defects, straighten bends or repair as necessary.	.013			√	√	√
8 Check nuts, bolts and screws for tightness; tighten or replace as necessary.	.005			√	√	√
9 Fill out maintenance checklist and report deficiencies.	.022			√	√	√
Total labor-hours/period				.644	.644	.644
Total labor-hours/year				1.288	.644	.644

			Cost Each					
			2019 Bare Costs				Total	Total
Description		Labor-hrs.	Material	Labor	Equip.	Total	In-House	w/O&P
3900	Fryer, conventional type, gas / electric, annually	.644	4.23	34.50		38.73	49.27	60.50
3950	Annualized	2.576	12.70	138		150.70	191.28	236

For customer support on your Facilities Maintenance & Repair Costs with RSMeans data, call 800.448.8182.

PM Components	Labor-hrs.	W	M	Q	S	A
PM System E1095 322 1950						
Dishwasher, electric						
1 Check with operating or area personnel for any deficiencies.	.044			√	√	√
2 Check electric insulators, connections, and wiring (includes removing access panels).	.120			√	√	√
3 Check motor and bearings for excessive noise, vibration and overheating.	.039			√	√	√
4 Check dish conveyor by turning on switch and observing conveyor movement, adjust chain if necessary.	.070			√	√	√
5 Check operation of wash and rinse spray mechanism for spray coverage and drainage.	.022			√	√	√
6 Inspect soap and spray solution feeder lines; clean as necessary.	.013			√	√	√
7 Inspect water lines and fittings for leaks; tighten fittings as necessary.	.077			√	√	√
8 Lubricate conveyor drive bushings and chain drive.	.055			√	√	√
9 Check belt tension; adjust if required.	.100			√	√	√
10 Check doors for: operation of chains and counterweights, warping, alignment and water tightness; adjust if necessary.	.022			√	√	√
11 Check packing glands on wash, rinse, and drain valves; add or replace packing as required.	.147			√	√	√
12 Check lubricant in gear case; add oil if required.	.049			√	√	√
13 Inspect splash curtain for tears, clearance and water tightness; adjust if required.	.016			√	√	√
14 Check proper operation of solenoid valve and float in fill tank; adjust as required.	.096			√	√	√
15 Check pumps for leakage and obstructions; adjust as required.	.120			√	√	√
16 Check proper operation of micro-switch.	.010			√	√	√
17 Check water for proper temperature.	.004			√	√	√
18 Check operation of safety stop and clutch on conveyor belt motor; adjust cut-off switch mechanism as required.	.049			√	√	√
19 Check temperature regulator and adjust if necessary.	.017			√	√	√
20 Clean lime off thermostatic probe and heating elements.	.020			√	√	√
21 Clean area around equipment.	.066			√	√	√
22 Fill out maintenance checklist and report deficiencies.	.022			√	√	√
Total labor-hours/period				1.179	1.179	1.179
Total labor-hours/year				2.357	1.179	1.179

		Cost Each						
			2019 Bare Costs				Total	Total
Description	Labor-hrs.	Material	Labor	Equip.	Total		In-House	w/O&P
1900 Dishwasher, electric, annually	1.179	133	63		196		227.48	267
1950 Annualized	4.712	400	252		652		763.10	905

For customer support on your Facilities Maintenance & Repair Costs with RSMeans data, call 800.448.8182.

	PM Components	Labor-hrs.	W	M	Q	S	A
PM System E1095 322 2950								
Dishwasher, steam								
1	Check with operating or area personnel for any deficiencies.	.044			√	√	√
2	Check electric insulators, connections, and wiring (includes removing access panels).	.120			√	√	√
3	Check motor and bearings for excessive noise, vibration and overheating.	.039			√	√	√
4	Check dish conveyor by turning on switch and observing conveyor movement; adjust chain if necessary.	.070			√	√	√
5	Check operation of wash and rinse spray mechanism for spray coverage and drainage.	.022			√	√	√
6	Inspect soap and spray solution feeder lines; clean as necessary.	.013			√	√	√
7	Inspect steam and water lines and fittings for leaks; tighten fittings as necessary.	.077			√	√	√
8	Lubricate conveyor drive bushings and chain drive.	.055			√	√	√
9	Check belt tension; adjust if required.	.100			√	√	√
10	Check doors for: operation of chains and counterweights, warping, alignment and water tightness; adjust if necessary.	.022			√	√	√
11	Tighten loose nuts, bolts and screws.	.005			√	√	√
12	Check packing glands on wash, rinse, and drain valves; add or replace packing as required.	.147			√	√	√
13	Check lubricant in gear case; add oil if required.	.049			√	√	√
14	Inspect splash curtain for tears, clearance and water tightness; adjust if required.	.016			√	√	√
15	Check proper operation of solenoid valve and float in fill tank; adjust as required.	.096			√	√	√
16	Check pumps for leakage and obstructions; adjust as required.	.120			√	√	√
17	Check proper operation of micro-switch.	.010			√	√	√
18	Check water for proper temperature readings.	.004			√	√	√
19	Check operation of safety stop and clutch on conveyor belt motor; adjust cut-off switch mechanism as required.	.049			√	√	√
20	Check temperature regulator and adjust if necessary.	.017			√	√	√
21	Clean lime off thermostatic probe.	.020			√	√	√
22	Clean area around equipment.	.066			√	√	√
23	Fill out maintenance checklist and report deficienceis.	.022			√	√	√
	Total labor-hours/period				1.184	1.184	1.184
	Total labor-hours/year				2.367	1.184	1.184

			Cost Each					
			2019 Bare Costs				Total	Total
	Description	Labor-hrs.	Material	Labor	Equip.	Total	In-House	w/O&P
2900	Dishwasher, steam, annually	1.184	133	63		196	227.82	267
2950	Annualized	4.732	400	253		653	764.48	905

For customer support on your Facilities Maintenance & Repair Costs with RSMeans data, call 800.448.8182.

PM Components	Labor-hrs.	W	M	Q	S	A
PM System E1095 328 1950						
Dispenser, carbonated beverage						
1 Check with operating or area personnel for any deficiencies.	.044			√	√	√
2 Check all lines, connections and valves for leaks.	.077			√	√	√
3 Check pressure gauges; adjust as required.	.018			√	√	√
4 Check water strainer screen on carbonator; clean as required.	.158			√	√	√
5 Check wiring, connectors and switches; repair as required.	.120			√	√	√
6 Check nuts, bolts and screws for tightness; replace or tighten as required.	.005			√	√	√
7 Wipe and clean unit.	.056			√	√	√
8 Fill out maintenance checklist and report deficiencies.	.022			√	√	√
Total labor-hours/period				.501	.501	.501
Total labor-hours/year				1.001	.501	.501

			Cost Each				
			2019 Bare Costs			Total	Total
Description	Labor-hrs.	Material	Labor	Equip.	Total	In-House	w/O&P
1900 Dispenser, carbonated beverage, annually	.501	25.50	26.50		52	62.84	75
1950 Annualized	2.004	77.50	107		184.50	223.96	269

PM Components	Labor-hrs.	W	M	Q	S	A
PM System E1095 332 1950						
Disposal, garbage, electric						
1 Check with operating or area personnel for any deficiencies.	.044			√	√	√
2 Check motor and drive shaft for excessive noise, vibration, overheating, etc.	.044			√	√	√
3 Inspect electrical wiring.	.014			√	√	√
4 Visually examine grinder; check for obstructions; adjust cutters if required.	.074			√	√	√
5 Check drive belt(s) for wear and tension; adjust if required.	.014			√	√	√
6 Check for leaks to supply and drain connections; tighten if required.	.103			√	√	√
7 Check electrical switch for proper operation.	.014			√	√	√
8 Lubricate grinder, drive, and motor.	.022			√	√	√
9 Check water sprayer operation.	.030			√	√	√
10 Fill out maintenance checklist and report deficiencies.	.022			√	√	√
Total labor-hours/period				.381	.381	.381
Total labor-hours/year				.762	.381	.381

			Cost Each				
			2019 Bare Costs			Total	Total
Description	Labor-hrs.	Material	Labor	Equip.	Total	In-House	w/O&P
1900 Disposal, garbage, electric, annually	.381	4.75	20.50		25.25	31.51	38.50
1950 Annualized	1.524	14.25	81.50		95.75	121.18	148

For customer support on your Facilities Maintenance & Repair Costs with RSMeans data, call 800.448.8182.

PM Components	Labor-hrs.	W	M	Q	S	A
PM System E1095 334 1950						
Dough divider						
1 Check with operating or area personnel for any deficiencies.	.044			√	√	√
2 Check motor and drive shaft for excessive noise, vibration, overheating, etc.	.039			√	√	√
3 Inspect electrical wiring.	.120			√	√	√
4 Visually examine plates; check for obstructions; adjust cutters if required.	.074			√	√	√
5 Check electrical switch for proper operation.	.014			√	√	√
6 Lubricate press, drive, and motor.	.219			√	√	√
7 Fill out maintenance checklist and report deficiencies.	.022			√	√	√
Total labor-hours/period				.533	.533	.533
Total labor-hours/year				1.065	.533	.533

		Cost Each						
			2019 Bare Costs				**Total**	**Total**
Description	**Labor-hrs.**	**Material**	**Labor**	**Equip.**	**Total**		**In-House**	**w/O&P**
1900 Dough divider, annually	.533	4.75	28.50		33.25		42.11	51.50
1950 Annualized	2.124	14.25	113		127.25		161.81	200

PM Components	Labor-hrs.	W	M	Q	S	A
PM System E1095 334 2950						
Dough roller						
1 Check with operating or area personnel for any deficiencies.	.044			√	√	√
2 Check electric motor, switches, controls, wiring, connections and insulation for defective materials; repair as necessary.	.241			√	√	√
3 Check belt tension; adjust or replace as required.	.444			√	√	√
4 Check operation of conveyor-roller clearance; adjust as necessary.	.109			√	√	√
5 Lubricate all moving parts, motors, pivot points and chain drives.	.017			√	√	√
6 Clean unit and treat for rust and/or corrosion.	.083			√	√	√
7 Tighten or replace loose, missing or damaged nuts, bolts and screws.	.005			√	√	√
8 Fill out maintenance checklist and report deficiencies.	.022			√	√	√
Total labor-hours/period				.965	.965	.965
Total labor-hours/year				1.930	.965	.965

		Cost Each						
			2019 Bare Costs				**Total**	**Total**
Description	**Labor-hrs.**	**Material**	**Labor**	**Equip.**	**Total**		**In-House**	**w/O&P**
2900 Dough roller, annually	.965	24	51.50		75.50		92.74	113
2950 Annualized	3.868	72	207		279		346.55	420

PM Components	Labor-hrs.	W	M	Q	S	A
PM System E1095 336 1950						
Drink cooler, with external condenser						
1 Check with operating or area personnel for deficiencies.	.035				√	√
2 Clean condenser coils, fans, and intake screens; lubricate motor.	.500				√	√
3 Inspect door gaskets for damage and proper fit; adjust as necessary.	.033				√	√
4 Check starter panels and controls for proper operation, burned or loose contacts, and loose connections.	.094				√	√
5 Clean coils, evaporator drain pan, blowers, fans, motors and drain piping as required, lubricate motor(s).	.473				√	√
6 During operation of unit, check refrigerant pressures and compressor oil level; add refrigerant and/or oil as necessary.	.066				√	√
7 Check operation of low pressure cut-out; adjust or replace as necessary.	.057				√	√
8 Inspect defrost systems for proper operation; adjust as required.	.027				√	√
9 Clean area around equipment.	.066				√	√
10 Fill out maintenance checklist and report deficiencies.	.022				√	√
Total labor-hours/period					1.373	1.373
Total labor-hours/year					1.373	1.373

			Cost Each				
			2019 Bare Costs			Total	Total
Description	Labor-hrs.	Material	Labor	Equip.	Total	In-House	w/O&P
1900 Drink cooler with external condenser, annually	1.373	105	73.50		178.50	210.13	249
1950 Annualized	2.754	148	147		295	352.18	420

PM Components	Labor-hrs.	W	M	Q	S	A
PM System E1095 338 1950						
Fluid cooler, air cooled condenser						
1 Check with operating or area personnel for deficiencies.	.035					√
2 Check unit for proper operation, excessive noise or vibration.	.158					√
3 Clean intake side of condenser coils, fans and intake screens.	.471					√
4 Check electrical wiring and connections; tighten loose connections.	.120					√
5 Inspect fan(s) for bent blades or unbalance; adjust as necessary.	.014					√
6 Check belts for condition, proper tension and misalignment, if required.	.029					√
7 Lubricate shaft bearings and motor bearings.	.047					√
8 Inspect piping and valves for leaks; tighten connections as necessary.	.077					√
9 Lubricate and check operation of dampers, if applicable.	.055					√
10 Clean area around fluid cooler.	.066					√
11 Fill out maintenance checklist and report deficiencies.	.022					√
Total labor-hours/period						1.094
Total labor-hours/year						1.094

			Cost Each				
			2019 Bare Costs			Total	Total
Description	Labor-hrs.	Material	Labor	Equip.	Total	In-House	w/O&P
1900 Fluid cooler, annually	1.094	9.50	58.50		68	86.01	105
1950 Annualized	1.094	9.50	58.50		68	86.01	105

For customer support on your Facilities Maintenance & Repair Costs with RSMeans data, call 800.448.8182.

PM Components	Labor-hrs.	W	M	Q	S	A
PM System E1095 340 1950						
Food saw, electric						
1 Check with operating or area personnel for any deficiencies.	.044			√	√	√
2 Check operation of saw for excessive vibration, blade tension and proper adjustment of safety guards; repair or adjust as required.	.095			√	√	√
3 Check motor and bearings for overheating; lubricate motor bearings.	.039			√	√	√
4 Check switches, motor, controls and wiring for damage and proper operation; repair as required.	.120			√	√	√
5 Clean machine of dust and grease.	.044			√	√	√
6 Check nuts, bolts and screws for tightness; tighten as required.	.005			√	√	√
7 Check belt tension and alignment; adjust as required.	.029			√	√	√
8 Fill out maintenance checklist and report deficiencies.	.022			√	√	√
Total labor-hours/period				.398	.398	.398
Total labor-hours/year				.796	.398	.398

			Cost Each					
			2019 Bare Costs				Total	Total
Description	Labor-hrs.	Material	Labor	Equip.	Total	In-House	w/O&P
1900 Food saw, electric, annually	.398	26.50	21		47.50	56.42	67
1950 Annualized	1.592	79.50	85		164.50	197.23	235

PM Components	Labor-hrs.	W	M	Q	S	A
PM System E1095 340 2950						
Food slicer, electric						
1 Check with operating or area personnel for any deficiencies.	.044			√	√	√
2 Check nuts, bolts and screws for tightness; tighten or replace as required.	.005			√	√	√
3 Lubricate food slicer and add lubricant to gear case if required.	.085			√	√	√
4 Check condition of blade, blade guard, guides, and controls; adjust as required.	.047			√	√	√
5 Check operation of motor and check motor bearings for overheating; lubricate motor bearings.	.039			√	√	√
6 Check sharpening stones; install new stones as required.	.013			√	√	√
7 Clean machine of dust, grease, and food particles.	.014			√	√	√
8 Fill out maintenance checklist and report deficiencies.	.022			√	√	√
Total labor-hours/period				.269	.269	.269
Total labor-hours/year				.538	.269	.269

			Cost Each					
			2019 Bare Costs				Total	Total
Description	Labor-hrs.	Material	Labor	Equip.	Total	In-House	w/O&P
2900 Food slicer, electric, annually	.269	150	14.35		164.35	183.80	211
2950 Annualized	1.076	445	57.50		502.50	565.86	650

For customer support on your Facilities Maintenance & Repair Costs with RSMeans data, call 800.448.8182.

PM Components	Labor-hrs.	W	M	Q	S	A
PM System E1095 346 1950						
Grill, gas/electric						
1 Check with operating or area personnel for any deficiencies.	.044			√	√	√
2 Check nuts, bolts and screws for tightness; tighten or replace as required.	.005			√	√	√
3 On gas operated units, check piping and valves for leaks.	.077			√	√	√
4 On gas operated units, check pilot and gas burners for uniform flame; adjust as required.	.079			√	√	√
5 On electrically operated units, check switches, connections, and wiring for proper operation; adjust as required.	.134			√	√	√
6 Check calibration of thermostats; adjust as required.	.457			√	√	√
7 Fill out maintenance checklist and report deficiencies.	.022			√	√	√
Total labor-hours/period				.819	.819	.819
Total labor-hours/year				1.637	.819	.819

		Cost Each						
			2019 Bare Costs				Total	Total
Description	Labor-hrs.	Material	Labor	Equip.	Total		In-House	w/O&P
1900 Grill, gas / electric, annually	.819	25.50	44		69.50		84.79	102
1950 Annualized	3.280	77.50	175		252.50		311.55	380

PM Components	Labor-hrs.	W	M	Q	S	A
PM System E1095 348 1950						
Ice machine, flake or cube						
1 Check with operating or area personnel for any deficiencies.	.044			√	√	√
2 Remove and install access panel.	.044			√	√	√
3 Lubricate all moving parts, pivot points and fan motor(s).	.023			√	√	√
4 Visually check for refrigerant, oil or water leaks.	.022			√	√	√
5 Open and close water valve.	.007			√	√	√
6 Replace in-line water filter.	.009			√	√	√
7 Check and clear ice machine draining system (drain vent and trap).	.198			√	√	√
8 Clean motor, compressor and condenser coil.	.400			√	√	√
9 Check and tighten any loose screw-type electrical connections.	.005			√	√	√
10 Inspect door(s) hinges, gaskets, handles; lubricate as required.	.053			√	√	√
11 Clean area around equipment.	.066			√	√	√
12 Fill out maintenance checklist and report deficiencies.	.022			√	√	√
Total labor-hours/period				.893	.893	.893
Total labor-hours/year				1.785	.893	.893

		Cost Each						
			2019 Bare Costs				Total	Total
Description	Labor-hrs.	Material	Labor	Equip.	Total		In-House	w/O&P
1900 Ice machine, flake or cube, annually	.893	88	48		136		158.33	186
1950 Annualized	3.584	265	192		457		537.91	635

For customer support on your Facilities Maintenance & Repair Costs with RSMeans data, call 800.448.8182.

PM Components	Labor-hrs.	W	M	Q	S	A
PM System E1095 350 1950						
Kettle, steam, fixed or tilt						
1 Check with operating or area personnel for any deficiencies.	.044			√	√	√
2 Check piping and fittings for leaks; tighten as required.	.085			√	√	√
3 Check operation of electric water valve.	.030			√	√	√
4 Inspect cover, hinges, and seals on units to equipped; lubricate hinge.	.053			√	√	√
5 Lubricate tilting gear mechanism and trunnion bearings, if applicable.	.047			√	√	√
6 Fill out maintenance checklist and report deficiencies.	.022			√	√	√
Total labor-hours/period				.281	.281	.281
Total labor-hours/year				.562	.281	.281

			Cost Each					
			2019 Bare Costs				Total	Total
Description		Labor-hrs.	Material	Labor	Equip.	Total	In-House	w/O&P
1900	Kettle, steam, fixed or tilt, annually	.281	9.50	15		24.50	29.87	36
1950	Annualized	1.124	28.50	60		88.50	108.95	132

PM Components	Labor-hrs.	W	M	Q	S	A
PM System E1095 354 1950						
Mixer, counter, electric						
1 Check with operating or area personnel for any deficiencies.	.044			√	√	√
2 Check operation of mixer at varying speeds for excessive noise and vibrations; align or adjust as required.	.033			√	√	√
3 Check belt for proper tension and alignment; adjust as required.	.068			√	√	√
4 Lubricate mixer gears.	.042			√	√	√
5 Clean machine thoroughly of dust, grease, and food particles.	.014			√	√	√
6 Check motor, switches, controls and motor bearings for overheating; lubricate motor bearings.	.039			√	√	√
7 Check anchor bolts for tightness; tighten as required.	.010			√	√	√
8 Fill out maintenance checklist and report deficiencies.	.022			√	√	√
Total labor-hours/period				.272	.272	.272
Total labor-hours/year				.544	.272	.272

			Cost Each				
			2019 Bare Costs			Total	Total
	Description	Labor-hrs.	Material	Labor	Equip.	Total	In-House	w/O&P

	Description	Labor-hrs.	Material	Labor	Equip.	Total	In-House	w/O&P
1900	Mixer, counter, electric, annually	.272	9.50	14.50		24	29.26	35.50
1950	Annualized	1.088	28.50	58		86.50	106.67	129

PM Components	Labor-hrs.	W	M	Q	S	A
PM System E1095 354 2950						
Mixer, floor, electric						
1 Check with operating or area personnel for any deficiencies.	.044			√	√	√
2 Check operation of mixer at varying speeds for excessive noise and vibrations; align or adjust as required.	.120			√	√	√
3 Tighten loose bolts, nuts, and screws.	.005			√	√	√
4 Check oil level in transmission; add or replace oil if required.	.158			√	√	√
5 Lubricate mixer gears and bowl lift mechanism.	.042			√	√	√
6 Check switches and control for damage and proper operation; adjust as required.	.068			√	√	√
7 Check belt for proper tension and alignment; adjust as required.	.029			√	√	√
8 Clean machine of dust and grease.	.030			√	√	√
9 Check anchor bolts for tightness; tighten as required.	.010			√	√	√
10 Clean area around equipment.	.066			√	√	√
11 Fill out maintenance checklist and report deficiencies.	.022			√	√	√
Total labor-hours/period				.595	.595	.595
Total labor-hours/year				1.189	.595	.595

			Cost Each				
			2019 Bare Costs			Total	Total
	Description	Labor-hrs.	Material	Labor	Equip.	Total	In-House	w/O&P

	Description	Labor-hrs.	Material	Labor	Equip.	Total	In-House	w/O&P
2900	Mixer, floor, electric, annually	.595	17.60	31.50		49.10	60.33	73
2950	Annualized	2.380	53	127		180	222.54	270

For customer support on your Facilities Maintenance & Repair Costs with RSMeans data, call 800.448.8182.

PM Components	Labor-hrs.	W	M	Q	S	A
PM System E1095 356 1950						
Oven, convection, gas/electric						
1 Check with operating or area personnel for any deficiencies.	.044		√	√	√	√
2 Check doors and seals for warping and misalignment; lubricate hinges and repair as necessary.	.027		√	√	√	√
3 Check piping and valves for leaks.	.077		√	√	√	√
4 Check nuts, bolts, and screws for tightness; replace or tighten as required.	.005		√	√	√	√
5 Check pilot and gas burner for uniform flame; adjust as required.	.083		√	√	√	√
6 Check element, switches, controls and wiring on electrically heated units for defects; repair as required.	.120		√	√	√	√
7 Check fan blades and fan motor for proper operation.	.120		√	√	√	√
8 Check operation of thermostat; calibrate as required.	.440		√	√	√	√
9 Fill out maintenance checklist and report deficiencies.	.022		√	√	√	√
Total labor-hours/period			.938	.938	.938	.938
Total labor-hours/year			7.501	1.875	.938	.938

			Cost Each				
			2019 Bare Costs			**Total**	**Total**
Description	**Labor-hrs.**	**Material**	**Labor**	**Equip.**	**Total**	**In-House**	**w/O&P**
1900 Oven, convection, gas / electric, annually	.938	9	50		59	74.69	91.50
1950 Annualized	11.242	36	600		636	817.20	1,000

PM Components	Labor-hrs.	W	M	Q	S	A
PM System E1095 356 2950						
Oven, rotary, electric						
1 Check with operating or area personnel for any deficiencies.	.044		√	√	√	√
2 Remove and reinstall access panel; lubricate main bearing bushings.	.064		√	√	√	√
3 Turn motor on and check operation of chain drive mechanism; lubricate chain.	.064		√	√	√	√
4 Inspect motor and wiring; sensory inspect motor and bearings for overheating; lubricate motor bearings.	.055		√	√	√	√
5 Check V-belt; adjust tension and/or pulley as required.	.029		√	√	√	√
6 Inspect shelves for defects and level if required.	.142		√	√	√	√
7 Tighten or replace any loose nuts, bolts, screws.	.005		√	√	√	√
8 Check thermostat with thermometer; adjust as required.	.444		√	√	√	√
9 Check operation of stop and reverse switch.	.008		√	√	√	√
10 Check timer mechanism and operation.	.018		√	√	√	√
11 Check lubricant in gear case; add as required.	.055		√	√	√	√
12 Check ventilator for proper operation.	.160		√	√	√	√
13 Clean area around equipment.	.066		√	√	√	√
14 Fill out maintenance checklist and report deficiencies.	.022		√	√	√	√
Total labor-hours/period			1.176	1.176	1.176	1.176
Total labor-hours/year			9.410	2.353	1.176	1.176

			Cost Each				
			2019 Bare Costs			**Total**	**Total**
Description	**Labor-hrs.**	**Material**	**Labor**	**Equip.**	**Total**	**In-House**	**w/O&P**
2900 Oven, rotary, electric, annually	1.176	26.50	62.50		89	110.23	134
2950 Annualized	14.050	106	750		856	1,087.20	1,325

For customer support on your Facilities Maintenance & Repair Costs with RSMeans data, call 800.448.8182.

521

PM Components	Labor-hrs.	W	M	Q	S	A
PM System E1095 356 3950						
Oven, rotary, gas						
1 Check with operating or area personnel for any deficiencies.	.044		√	√	√	√
2 Examine doors and door gaskets; make necessary adjustments.	.021		√	√	√	√
3 Visually inspect hinges and latches; lubricate hinges.	.091		√	√	√	√
4 Remove and reinstall access panel; lubricate main bearing bushings.	.025		√	√	√	√
5 Turn motor on and chk. operation of chain drive mechanism; lubricate chain.	.064		√	√	√	√
6 Inspect motor and wiring; sensory inspect motor and bearings for overheating; lubricate motor bearings.	.055		√	√	√	√
7 Check V-belt; adjust tension and/or pulley as required.	.029		√	√	√	√
8 Check pilot and thermosensor bulb.	.042		√	√	√	√
9 Check gas burner for uniform flame.	.042		√	√	√	√
10 Check piping and valves for leaks.	.077		√	√	√	√
11 Inspect shelves for defects and level if required.	.142		√	√	√	√
12 Tighten or replace any loose nuts, bolts, screws.	.005		√	√	√	√
13 Check thermostat with thermometer; calibrate if required.	.444		√	√	√	√
14 Check operation of stop and reverse switch.	.008		√	√	√	√
15 Check timer mechanism and operation.	.018		√	√	√	√
16 Check lubricant in gear case, add as required.	.055		√	√	√	√
17 Check ventilator for proper operation.	.160		√	√	√	√
18 Clean area around equipment.	.066		√	√	√	√
19 Fill out maintenance checklist and report deficiencies.	.022		√	√	√	√
Total labor-hours/period			1.410	1.410	1.410	1.410
Total labor-hours/year			11.283	2.821	1.410	1.410

			Cost Each					
			2019 Bare Costs				**Total**	**Total**
Description		**Labor-hrs.**	**Material**	**Labor**	**Equip.**	**Total**	**In-House**	**w/O&P**
3900	Oven, rotary, gas, annually	1.410	31.50	75		106.50	131.58	160
3950	Annualized	16.858	125	900		1,025	1,302.10	1,600

PM Components	Labor-hrs.	W	M	Q	S	A
PM System E1095 366 1950						
Peeler, vegetable, electric						
1 Check with operating or area personnel for any deficiencies.	.044			√	√	√
2 Check vegetable peeler for proper operation, including switches and controls; lubricate motor bearings.	.156			√	√	√
3 Lubricate vegetable peeler bushings, as required.	.062			√	√	√
4 Check door for loose hinges and latch fittings; adjust if required.	.007			√	√	√
5 Clean peeler of dust, grease, and food particles.	.014			√	√	√
6 Check piping and valves for leaks.	.077			√	√	√
7 Inspect abrasive disk.	.018			√	√	√
8 Tighten or replace loose, missing or damaged nuts, bolts, screws.	.005			√	√	√
9 Check belt for proper tension and alignment; adjust or replace as required.	.029			√	√	√
10 Fill out maintenance checklist and report deficiencies.	.022			√	√	√
Total labor-hours/period				.434	.434	.434
Total labor-hours/year				.868	.434	.434

			Cost Each					
			2019 Bare Costs				**Total**	**Total**
Description		**Labor-hrs.**	**Material**	**Labor**	**Equip.**	**Total**	**In-House**	**w/O&P**
1900	Peeler, vegetable, electric, annually	.434	24	23		47	56.33	67
1950	Annualized	1.736	72	93		165	198.80	238

For customer support on your Facilities Maintenance & Repair Costs with RSMeans data, call 800.448.8182.

E10 EQUIPMENT — E1095 368 | Pie Maker, Electric

PM Components	Labor-hrs.	W	M	Q	S	A
PM System E1095 368 1950						
Pie maker, electric						
1 Check with operating or area personnel for any deficiencies.	.044			√	√	√
2 Check electrical system for loose connections and frayed wiring; tighten connections as required.	.120			√	√	√
3 Check operation of motors and bearings for overheating; lubricate motor bearings.	.168			√	√	√
4 Check V-belt for alignment and proper tension; adjust or replace belt as required.	.348			√	√	√
5 Check drain drive and sprocket for excessive wear; lubricate if applicable.	.087			√	√	√
6 Tighten or replace loose, missing or damaged nuts, bolts or screws.	.005			√	√	√
7 Wipe clean the machine.	.096			√	√	√
8 Fill out maintenance checklist and report deficiencies.	.022			√	√	√
Total labor-hours/period				.891	.891	.891
Total labor-hours/year				1.781	.891	.891

	Description	Labor-hrs.	Material	Labor	Equip.	Total	Total In-House	Total w/O&P
1900	Pie maker, electric, annually	.891	28.50	47.50		76	93.09	112
1950	Annualized	3.552	86	189		275	339.55	415

E10 EQUIPMENT — E1095 370 | Proofer, Automatic

PM Components	Labor-hrs.	W	M	Q	S	A
PM System E1095 370 1950						
Proofer, automatic, electric						
1 Check with operating or area personnel for any deficiencies.	.044			√	√	√
2 Check operation of electric motor, switches, controls and bearings; adjust, repair and lubricate as required.	.281			√	√	√
3 Check belt tension; adjust as required.	.029			√	√	√
4 Check chain drive operation for proper alignment and tension; repair as required.	.129			√	√	√
5 Check action of flour sifter mechanism, connecting linkage and conveyor; lubricate bushings as required.	.291			√	√	√
6 Check roller conveyor and connecting drive mechanism.	.113			√	√	√
7 Check door latch; adjust if needed.	.014			√	√	√
8 Check nuts, bolts and screws for tightness; replace or tighten as required.	.029			√	√	√
9 Fill out maintenance checklist and report deficiencies.	.022			√	√	√
Total labor-hours/period				.951	.951	.951
Total labor-hours/year				1.903	.951	.951

	Description	Labor-hrs.	Material	Labor	Equip.	Total	Total In-House	Total w/O&P
1900	Proofer, automatic, electric, annually	.951	13.75	51		64.75	80.82	98.50
1950	Annualized	3.800	41	203		244	306.56	375

For customer support on your Facilities Maintenance & Repair Costs with RSMeans data, call 800.448.8182.

523

PM Components	Labor-hrs.	W	M	Q	S	A
PM System E1095 380 1950						
Refrigerator, domestic						
1 Check with operating or area personnel for deficiencies.	.035					√
2 Clean coils, fans, fan motors, drip pan and other areas with vacuum, brush or wiping as necessary.	.066					√
3 Inspect door gaskets for damage and proper fit; adjust gaskets as required and lubricate hinges.	.022					√
4 Check door latch and adjust as necessary.	.023					√
5 Clean area around equipment.	.066					√
6 Fill out maintenance checklist and report deficiencies.	.022					√
Total labor-hours/period						.234
Total labor-hours/year						.234

			Cost Each					
				2019 Bare Costs			Total	Total
	Description	Labor-hrs.	Material	Labor	Equip.	Total	In-House	w/O&P
1900	Refrigerator, domestic, annually	.234	4.75	12.50		17.25	21.38	26
1950	Annualized	.234	4.75	12.50		17.25	21.38	26

PM Components	Labor-hrs.	W	M	Q	S	A
PM System E1095 380 2950						
Refrigerator, display case						
1 Check with operating or area personnel for deficiencies.	.035					√
2 Clean coils, fans, fan motors, drip pan and other areas with vacuum, brush or wiping as necessary.	.066					√
3 Inspect door gaskets for damage and proper fit; adjust gaskets as required and lubricate hinges.	.044					√
4 Check door latch and adjust as necessary.	.047					√
5 Clean area around equipment.	.066					√
6 Fill out maintenance checklist and report deficiencies.	.022					√
Total labor-hours/period						.280
Total labor-hours/year						.280

			Cost Each					
				2019 Bare Costs			Total	Total
	Description	Labor-hrs.	Material	Labor	Equip.	Total	In-House	w/O&P
2900	Refrigerator, display case, annually	.280	4.75	14.95		19.70	24.58	30
2950	Annualized	.280	4.75	14.95		19.70	24.58	30

For customer support on your Facilities Maintenance & Repair Costs with RSMeans data, call 800.448.8182.

PM Components	Labor-hrs.	W	M	Q	S	A
PM System E1095 380 3950						
Refrigerator/display, walk-in w/external condenser						
1 Check with operating or area personnel for deficiencies.	.035				√	√
2 Clean condenser coils, fans, and intake screens; lubricate motor.	.066				√	√
3 Inspect door gaskets for damage and proper fit; adjust gaskets as required and lubricate hinges.	.137				√	√
4 Check starter panels and controls for proper operation, burned or loose contacts, and loose connections.	.182				√	√
5 Clean coils, evaporator drain pan, blowers, fans, motors and drain piping as required; lubricate motor(s).	.473				√	√
6 During operation of unit, check refrigerant pressures and compressor oil level; add refrigerant and/or oil as necessary.	.066				√	√
7 Check operation of low pressure cut-out; adjust or replace as required.	.047				√	√
8 Inspect defrost systems for proper operation; adjust as required.	.094				√	√
9 Clean area around equipment.	.066				√	√
10 Fill out maintenance checklist and report deficiencies.	.022				√	√
Total labor-hours/period					1.188	1.188
Total labor-hours/year					1.188	1.188

		Cost Each						
			2019 Bare Costs				Total	Total
Description	Labor-hrs.	Material	Labor	Equip.	Total	In-House	w/O&P
3900 Refrig. display, walk-in w/ ext. condenser, annually	1.188	98.50	63.50		162	189.54	225
3950 Annualized	2.376	198	127		325	380.96	450

For customer support on your Facilities Maintenance & Repair Costs with RSMeans data, call 800.448.8182.

525

PM Components	Labor-hrs.	W	M	Q	S	A
PM System E1095 382 1950						
Refrigerator/freezer, walk-in box w/external condenser						
1 Check with operating or area personnel for deficiencies.	.035				√	√
2 Clean condenser coils, fans, and intake screens; lubricate motor.	.036				√	√
3 Inspect door gaskets for damage and proper fit; adjust gaskets as required and lubricate hinges.	.022				√	√
4 Check starter panels and controls for proper operation, burned or loose contacts, and loose connections.	.182				√	√
5 Clean coils, evaporator drain pan, blowers, fans, motors and drain piping as required; lubricate motor(s).	.114				√	√
6 During operation of unit, check refrigerant pressures and compressor oil level; add refrigerant and/or oil as necessary.	.083				√	√
7 Check operation of low pressure cut-out; adjust or replace as required.	.079				√	√
8 Inspect defrost systems for proper operation; adjust as required.	.094				√	√
9 Clean area around equipment.	.066				√	√
10 Fill out maintenance checklist and report deficiencies.	.022				√	√
Total labor-hours/period					.733	.733
Total labor-hours/year					.733	.733

		Cost Each					
		2019 Bare Costs				Total	Total
Description	Labor-hrs.	Material	Labor	Equip.	Total	In-House	w/O&P
1900 Refrig. freezer, walk-in box w/ext. condenser, annually	.733	98.50	39		137.50	158.39	186
1950 Annualized	1.466	198	78.50		276.50	318.01	375

PM Components	Labor-hrs.	W	M	Q	S	A
PM System E1095 382 2950						
Refrigerator unit/display case/freezer w/external condenser						
1 Check with operating or area personnel for deficiencies.	.035				√	√
2 Clean condenser coils, fans, and intake screens; lubricate motor.	.500				√	√
3 Inspect door gaskets for damage and proper fit; adjust gaskets as required and lubricate hinges.	.033				√	√
4 Check starter panels and controls for proper operation, burned or loose contacts, and loose connections.	.213				√	√
5 Clean coils, evaporator drain pan, blowers, fans, motors and drain piping as required; lubricate motor(s).	.473				√	√
6 During operation of unit, check refrigerant pressures and compressor oil level; add refrigerant and/or oil as necessary.	.066				√	√
7 Check operation of low pressure cut-out; adjust or replace as required.	.057				√	√
8 Inspect defrost systems for proper operation; adjust as required.	.027				√	√
9 Clean area around equipment.	.066				√	√
10 Fill out maintenance checklist and report deficiencies.	.022				√	√
Total labor-hours/period					1.492	1.492
Total labor-hours/year					1.492	1.492

		Cost Each					
		2019 Bare Costs				Total	Total
Description	Labor-hrs.	Material	Labor	Equip.	Total	In-House	w/O&P
2900 Refrig., display case, freezer w/ ext. cond., annually	1.492	98.50	79.50		178	211.10	251
2950 Annualized	2.992	198	160		358	423.18	505

For customer support on your Facilities Maintenance & Repair Costs with RSMeans data, call 800.448.8182.

PM Components	Labor-hrs.	W	M	Q	S	A
PM System E1095 386 1950						
Steam table						
1 Check with operating or area personnel for any deficiencies.	.044			√	√	√
2 Inspect water compartment, steam coil, valves, and piping for leaks.	.117			√	√	√
3 Check steam trap and strainer; clean as required.	.308			√	√	√
4 Check operation of pressure regulating valve and gauge.	.035			√	√	√
5 Check pilots and flame on gas burner units; adjust as required.	.085			√	√	√
6 Check insulators, connections, and wiring, if applicable; tighten connections.	.120			√	√	√
7 Check condition of covers and receptacles; adjust as required.	.025			√	√	√
8 Check thermostat and temperature gauge; calibrate thermostat if necessary.	.152			√	√	√
9 Fill out maintenance checklist and report deficiencies.	.022			√	√	√
Total labor-hours/period				.908	.908	.908
Total labor-hours/year				1.816	.908	.908

			Cost Each					
			2019 Bare Costs				Total	Total
Description	Labor-hrs.	Material	Labor	Equip.	Total		In-House	w/O&P
1900 Steam table, annually	.908		48.50		48.50		62.38	78
1950 Annualized	3.624		194		194		250.25	310

PM Components	Labor-hrs.	W	M	Q	S	A
PM System E1095 388 1950						
Steamer, vegetable, direct connected units						
1 Check with operating or area personnel for any deficiencies.	.044			√	√	√
2 Inspect doors, door hardware and gaskets; lubricate hinges.	.112			√	√	√
3 Tighten or replace loose, missing, or damaged nuts, bolts, screws.	.005			√	√	√
4 Check for clogged or defective steam and strainers; clean as required.	.302			√	√	√
5 Check working pressure on steam gauge and inspect piping, valves, and doors for leaks; make necessary repairs as required.	.082			√	√	√
6 Check operation of pressure relief valve, low water cut-off, timer operation and water level in sight glass.	.065			√	√	√
7 Fill out maintenance checklist and report deficiencies.	.022			√	√	√
Total labor-hours/period				.632	.632	.632
Total labor-hours/year				1.264	.632	.632

			Cost Each					
			2019 Bare Costs				Total	Total
Description	Labor-hrs.	Material	Labor	Equip.	Total		In-House	w/O&P
1900 Steamer, vegetable, annually	.632	9	33.50		42.50		53.62	65.50
1950 Annualized	2.536	27	136		163		204.50	251

For customer support on your Facilities Maintenance & Repair Costs with RSMeans data, call 800.448.8182.

PM Components	Labor-hrs.	W	M	Q	S	A
PM System E1095 390 1950						
Toaster, rotary, gas/electric						
1 Check with operating or area personnel for any deficiencies.	.044			√	√	√
2 Check operation of toaster by toasting sample slice of bread.	.034			√	√	√
3 Clean toaster.	.014			√	√	√
4 Check valves and piping for leaks on gas units.	.077			√	√	√
5 Check pilot light and burner flame on gas units; adjust when required.	.017			√	√	√
6 Check insulators, heating elements, connections, and wiring on electric units; tighten connections and repair as necessary.	.134			√	√	√
7 Inspect alignment of baskets and conveyor chains; lubricate chains and align as necessary.	.044			√	√	√
8 Check motor and motor bearings for overheating; lubricate motor bearings.	.086			√	√	√
9 Tighten or replace loose, missing or damaged nuts, bolts, screws.	.005			√	√	√
10 Fill out maintenance checklist and report deficiencies.	.022			√	√	√
Total labor-hours/period				.477	.477	.477
Total labor-hours/year				.954	.477	.477

	Description	Labor-hrs.	Cost Each				Total In-House	Total w/O&P
			2019 Bare Costs					
			Material	Labor	Equip.	Total		
1900	Toaster, rotary, gas / electric, annually	.477	13.75	25.50		39.25	48.08	58
1950	Annualized	1.908	41	102		143	176.50	215

PM Components	Labor-hrs.	W	M	Q	S	A
PM System E1095 930 1950						
Vacuum						
1 Check with operating or area personnel for deficiencies.	.035				√	√
2 Check for leaks on suction and discharge piping, seals, etc.	.077				√	√
3 Check pump and motor operation for excessive vibration, noise and overheating.	.022				√	√
4 Check alignment and clearances of shaft and coupler.	.258				√	√
5 Tighten or replace loose, missing, or damaged nuts, bolts and screws.	.005				√	√
6 Lubricate pump and motor as required.	.099				√	√
7 Clean pump, motor and surrounding area.	.096				√	√
8 Fill out maintenance checklist and report deficiencies.	.022				√	√
Total labor-hours/period					.613	.613
Total labor-hours/year					.613	.613

	Description	Labor-hrs.	Cost Each				Total In-House	Total w/O&P
			2019 Bare Costs					
			Material	Labor	Equip.	Total		
1900	Vacuum, annually	.613	9	32.50		41.50	52.19	64
1950	Annualized	1.232	17.95	66		83.95	104.74	127

For customer support on your Facilities Maintenance & Repair Costs with RSMeans data, call 800.448.8182.

PM Components	Labor-hrs.	W	M	Q	S	A
PM System F1045 110 1950						
Swimming pool						
Quantities of chemicals used vary significantly by climate and pool usage, therefore the price of these materials are not provided.						
1 Clean strainer basket.	.050	√	√	√	√	√
2 Backwash pool water filter.	.120	√	√	√	√	√
3 Add soda ash to feeder.	.080	√	√	√	√	√
4 Clean/flush soda ash pump head.	.400		√	√	√	√
5 Check chlorine, change bottles and check for leaks.	.120	√	√	√	√	√
6 Clean chlorine pump, flush with acid, and check for signs of deterioration	.200		√	√	√	√
7 Check acid source, add acid, check for leaks.	.100	√	√	√	√	√
8 Clean acid pump head.	.400		√	√	√	√
9 Check circulating pump for leaks and unusual noises. Lubricate as required.	.050	√	√	√	√	√
10 Check diving board, tighten or replace missing hardware. Inspect for structural defects or deterioration.	.080	√	√	√	√	√
11 Wash exposed stainless steel ladders and scum gutters with tap water to remove oil and dirt.	.302	√	√	√	√	√
12 Fill out maintenance report.	.017	√	√	√	√	√
Total labor-hours/period		.919	1.919	1.919	1.919	1.919
Total labor-hours/year		34.940	15.356	3.839	1.919	1.919

	Description	Labor-hrs.	Cost Each					
			2019 Bare Costs				Total In-House	Total w/O&P
			Material	Labor	Equip.	Total		
1900	Swimming pool outdoor, annually	1.919	4.75	103		107.75	137.62	170
1950	Annualized	57.859	24	3,100		3,124	4,007.50	4,975

PM Components	Labor-hrs.	W	M	Q	S	A
PM System G2015 610 1950						
Traffic signal light						
1 Inspect and clean control cabinet.	.030					√
2 Check wiring for obvious defects and tighten control electrical connections.	.090					√
3 Check operation of cabinet exhaust fan and inspect vents.	.020					√
4 Inspect and tighten power line connections to control cabinet.	.034					√
5 Check operation of cabinet heater.	.037					√

6 Perform the following operational checks of traffic controller.
    A) Check controller phase timing with stop watch.
    B) Check controller pedestrian phase timing with stop watch.
    C) Check controller clocks for correct time and log settings.
	.082			√	√	√

7 Walk a circuit of the intersection to ensure all lights are functioning, replace bulbs as required.	.070			√	√	√
8 Check operation of pedestrian push buttons, average two per intersection.	.084			√	√	√
9 Check operation of all informational signs.	.042			√	√	√
10 Check operation of phase detectors (loop amplifiers).	.144			√	√	√
11 Visually inspect road sensors.	.074			√	√	√
12 Check operation of flashers.	.018					√
13 Fill out maintenance checklist and report deficiencies.	.017			√	√	√
Total labor-hours/period				.514	.514	.743
Total labor-hours/year				1.027	.514	.743

			Cost Each					
			2019 Bare Costs				Total	Total
Description		Labor-hrs.	Material	Labor	Equip.	Total	In-House	w/O&P
1900	Traffic signal light, annually	.743	10.05	39.50		49.55	62.24	76
1950	Annualized	2.281	21	122		143	180.01	221

PM Components	Labor-hrs.	W	M	Q	S	A
PM System G2045 150 1950						
Manual swing gate						
1 Check gate and hinge alignment.	.327				√	√
2 Tighten diagonal brace turnbuckle.	.108				√	√
3 Check cane bolt alignment.	.108				√	√
4 Grease hinges.	.216				√	√
5 Oil latch.	.108				√	√
Total labor-hours/period					.867	.867
Total labor-hours/year					.867	.867

			Cost Each				
			2019 Bare Costs			Total	Total
Description	Labor-hrs.	Material	Labor	Equip.	Total	In-House	w/O&P
1900 Manual swing gate, annually	.867	5.15	46.50		51.65	65.40	81
1950 Annualized	1.732	10.35	92		102.35	130.80	161

PM Components	Labor-hrs.	W	M	Q	S	A
PM System G2045 150 2950						
Electric swing gate						
1 Check gate and hinge alignment.	.327				√	√
2 Tighten diagonal brace turnbuckle.	.108				√	√
3 Check cane bolt alignment.	.108				√	√
4 Grease hinges.	.216				√	√
5 Oil latch.	.108				√	√
6 Check electric motor and connections.	.327				√	√
7 Grease gearbox.	.216				√	√
Total labor-hours/period					1.410	1.410
Total labor-hours/year					1.410	1.410

			Cost Each				
			2019 Bare Costs			Total	Total
Description	Labor-hrs.	Material	Labor	Equip.	Total	In-House	w/O&P
2900 Electric swing gate, annually	1.410	10.75	75.50		86.25	108.90	134
2950 Annualized	2.841	21.50	151		172.50	219.80	269

PM Components	Labor-hrs.	W	M	Q	S	A
PM System G2045 150 3950						
Manual slide gate						
1 Check alignment of rollers and gate.	.327				√	√
2 Grease rollers.	.327				√	√
3 Oil latch.	.108				√	√
Total labor-hours/period					.761	.761
Total labor-hours/year					.761	.761

			Cost Each				
			2019 Bare Costs			Total	Total
Description	Labor-hrs.	Material	Labor	Equip.	Total	In-House	w/O&P
3900 Manual slide gate, annually	.761	5.15	40.50		45.65	58.05	72
3950 Annualized	1.516	10.35	80.50		90.85	116	142

PM Components	Labor-hrs.	W	M	Q	S	A
PM System G2045 150 4950						
Electric slide gate						
1 Check alignment of rollers and gate.	.327				√	√
2 Grease rollers.	.327				√	√
3 Oil latch.	.108				√	√
4 Check electric motor and connections.	.327				√	√
5 Clean and grease chain.	.640			√	√	√
Total labor-hours/period				.640	1.728	1.728
Total labor-hours/year				1.280	1.728	1.728

	Description	Labor-hrs.	Cost Each				Total In-House	Total w/O&P
			2019 Bare Costs					
			Material	Labor	Equip.	Total		
4900	Electric slide gate, annually	1.728	7.95	92		99.95	127.55	158
4950	Annualized	4.766	18.70	254		272.70	349	430

For customer support on your Facilities Maintenance & Repair Costs with RSMeans data, call 800.448.8182.

G30 SITE MECH. UTILITIES | G3015 108 | Storage Tank

PM Components	Labor-hrs.	W	M	Q	S	A
PM System G3015 108 1950						
Storage tank, tower						
1 Inspect exterior of tank for leaks or damage, including tank base and base plates.	.151					√
2 Inspect condition of ladders, sway bracing hardware and structural forms.	.624					√
3 Tighten and lubricate altitude valve and inspect valve vault.	.039					√
4 Check operation of storage tank lighting.	.121					√
5 Check cathodic protection system.	.123					√
6 Operate elevated tank heat system, if installed, and inspect insulation and coils.	.107					√
7 Clean up around tank area.	.134					√
8 Fill out maintenance checklist and report deficiencies.	.022					√
Total labor-hours/period						1.321
Total labor-hours/year						1.321

				Cost Each				
			2019 Bare Costs				Total	Total
Description	Labor-hrs.	Material	Labor	Equip.	Total	In-House	w/O&P
1900 Storage tank, tower, annually	1.321	19	70.50		89.50	111.97	137
1950 Annualized	1.321	19	70.50		89.50	111.97	137

PM Components	Labor-hrs.	W	M	Q	S	A
PM System G3015 108 2950						
Storage tank, ground level						
1 Inspect exterior of tank for leaks or damage.	.053					√
2 Check condition of roof and roof hatches including locks; remove trash from roof.	.147					√
3 Inspect condition of ladders and tighten or replace missing hardware.	.088					√
4 Inspect valves, fittings, drains and controls for leakage or damage.	.039					√
5 Check for deterioration of tank foundation.	.077					√
6 Inspect ground drainage slope, and vent screens; where applicable.	.043					√
7 Inspect tank lighting system; replace bulbs as needed.	.121					√
8 Clean up around tank.	.134					√
9 Fill out maintenance checklist and report deficiencies.	.022					√
Total labor-hours/period						.724
Total labor-hours/year						.724

				Cost Each				
			2019 Bare Costs				Total	Total
Description	Labor-hrs.	Material	Labor	Equip.	Total	In-House	w/O&P
2900 Storage tank, ground level, annually	.724	12.60	38.50		51.10	63.78	78
2950 Annualized	.724	12.60	38.50		51.10	63.78	78

For customer support on your Facilities Maintenance & Repair Costs with RSMeans data, call 800.448.8182.

533

PM Components	Labor-hrs.	W	M	Q	S	A
PM System G3015 116 1950						
Water flow meter, turbine						
1 Inspect meter for leaks, corrosion or broken sight glass; replace broken glass as required.	.035				√	√
2 Check meter operation for unusual noise.	.034				√	√
3 Clean meter to remove mineral buildup.	.039				√	√
4 Clean out trash or sludge from bottom of pit.	.165				√	√
5 Fill out maintenance report and report deficiencies.	.022				√	√
Total labor-hours/period					.295	.295
Total labor-hours/year					.295	.295

			Cost Each					
			2019 Bare Costs				**Total**	**Total**
Description		**Labor-hrs.**	**Material**	**Labor**	**Equip.**	**Total**	**In-House**	**w/O&P**
1900	Water flow meter, turbine, annually	.295	12.85	15.75		28.60	34.46	41
1950	Annualized	.590	12.85	31.50		44.35	55.12	66

PM Components	Labor-hrs.	W	M	Q	S	A
PM System G3015 118 1950						
Fresh water distribution reservoir controls						
1 Check with reservoir personnel for any known deficiencies.	.086			√	√	√
2 Visually inspect and clean interior and exterior of controls cabinet.	.061			√	√	√
3 Inspect electrical system for frayed wires or loose connections; repair as required.	.120			√	√	√
4 Inspect water level receiver at reservoir.	.193			√	√	√
5 Inspect and adjust chronoflo transmitter.	.471			√	√	√
6 Perform operational check of relays, safety switches and electrical contactors.	.542			√	√	√
7 Perform operational check of controls system.	.659			√	√	√
8 Clean up controls area and dispose of debris.	.066			√	√	√
9 Fill out maintenance checklist and report deficiencies.	.022			√	√	√
Total labor-hours/period				2.220	2.220	2.220
Total labor-hours/year				4.440	2.220	2.220

			Cost Each					
			2019 Bare Costs				**Total**	**Total**
Description		**Labor-hrs.**	**Material**	**Labor**	**Equip.**	**Total**	**In-House**	**w/O&P**
1900	Reservoir controls, fresh water distribution, annually	2.220	18.90	118		136.90	174.28	214
1950	Annualized	8.860	28	475		503	641.25	790

For customer support on your Facilities Maintenance & Repair Costs with RSMeans data, call 800.448.8182.

PM Components	Labor-hrs.	W	M	Q	S	A
PM System G3015 126 0950						
Pump, air lift, well						
1 Check with pump operating personnel for any known deficiencies.	.035				√	√
2 Check operation of compressor and pump.	.033				√	√
3 Check water sample for air or oil contamination.	.009				√	√
4 Check compressor oil level; add oil as necessary and lubricate pump and compressor shaft bearings as applicable.	.121				√	√
5 Inspect electrical system for frayed wires or loose connections; repair as required.	.120				√	√
6 Check and cycle high and low shut off valves.	.130				√	√
7 Inspect and clean air intake filter.	.094				√	√
8 Clean pump body and tighten or replace loose hardware.	.086				√	√
9 Calibrate and adjust pressure gauge.	.190				√	√
10 Clean surrounding area.	.066				√	√
11 Fill out maintenance checklist and report deficiencies.	.022				√	√
Total labor-hours/period					.907	.907
Total labor-hours/year					.907	.907

		Cost Each						
			2019 Bare Costs				Total	Total
Description	Labor-hrs.	Material	Labor	Equip.	Total		In-House	w/O&P
0900 Pump, air lift, well, annually	.907	11.05	48.50		59.55		74.74	91.50
0950 Annualized	1.814	11.05	97		108.05		137.55	169

PM Components	Labor-hrs.	W	M	Q	S	A
PM System G3015 126 1950						
Pump, centrifugal ejector						
1 Check for proper operation of pump.	.022				√	√
2 Check for leaks on suction and discharge piping, seals, packing glands, etc.; make minor adjustments as required.	.077				√	√
3 Check pump and motor operation for excessive vibration, noise and overheating.	.022				√	√
4 Check alignment of pump and motor; adjust as necessary.	.258				√	√
5 Lubricate pump and motor.	.099				√	√
6 Clean exterior of pump, motor and surrounding area.	.096				√	√
7 Fill out maintenance checklist and report deficiencies.	.022				√	√
Total labor-hours/period					.596	.596
Total labor-hours/year					.596	.596

		Cost Each						
			2019 Bare Costs				Total	Total
Description	Labor-hrs.	Material	Labor	Equip.	Total		In-House	w/O&P
1900 Pump, centrifugal ejector, annually	.596	22.50	32		54.50		66.26	79
1950 Annualized	1.196	22.50	64		86.50		107.44	130

PM Components	Labor-hrs.	W	M	Q	S	A
PM System G3015 126 3950						
Pump, metering (slurry)						
1 Check with pump operating personnel for any obvious deficiencies.	.035				√	√
2 Obtain and put on safety equipment for working with fluoride: goggles, apron, gloves and respirator.	.014				√	√
3 Check oil level in pump with dipstick; add oil as necessary.	.009				√	√
4 Inspect belt condition and tension on pump; adjust as required.	.029				√	√
5 Flush slurry pump with fresh water.	.139				√	√
6 Pressure test pump.	.012				√	√
7 Inspect pump motor.	.039				√	√
8 Operate pump and check for leaks, excessive noise and vibration.	.099				√	√
9 Clean up area.	.066				√	√
10 Fill out maintenance checklist and report deficiencies.	.022				√	√
Total labor-hours/period					.464	.464
Total labor-hours/year					.464	.464

		Cost Each					
			2019 Bare Costs			Total	Total
Description	Labor-hrs.	Material	Labor	Equip.	Total	In-House	w/O&P
3900 Pump, metering (slurry), annually	.464	28	25		53	62.63	74.50
3950 Annualized	.928	28	49.50		77.50	94.64	115

PM Components	Labor-hrs.	W	M	Q	S	A
PM System G3015 126 4950						
Pump, mixed or axial flow velocity						
1 Check with pump operating personnel for any obvious deficiencies.	.035				√	√
2 Clean pump exterior.	.014				√	√
3 Check impeller and shaft for correct alignment and clearances.	.031				√	√
4 Check oil level; add oil as needed.	.009				√	√
5 Inspect packing for leaks and tighten as needed.	.031				√	√
6 Inspect bearings for wear, damage or overheating; lubricate as required.	.053				√	√
7 Inspect float operation and adjust as needed.	.098				√	√
8 Inspect V-belt tension and condition; adjust as required.	.029				√	√
9 Operate pump and check for leaks, excessive noise and vibration.	.099				√	√
10 Tighten loose nuts and bolts and replace missing hardware.	.005				√	√
11 Check electrical system for frayed wires and loose connections.	.120				√	√
12 Clean up area.	.066				√	√
13 Fill out maintenance checklist and report deficiencies.	.022				√	√
Total labor-hours/period					.612	.612
Total labor-hours/year					.612	.612

		Cost Each					
			2019 Bare Costs			Total	Total
Description	Labor-hrs.	Material	Labor	Equip.	Total	In-House	w/O&P
4900 Pump, mixed or axial flow, annually	.612	39	32.50		71.50	85.08	101
4950 Annualized	1.224	50.50	65.50		116	140.22	168

For customer support on your Facilities Maintenance & Repair Costs with RSMeans data, call 800.448.8182.

PM Components	Labor-hrs.	W	M	Q	S	A
PM System G3015 126 5950						
Pump, reciprocating positive displacement						
1 Check for proper operation of pump.	.022				√	√
2 Chcck for leaks on suction and discharge piping, seals, packing glands, etc.; make minor adjustments as required.	.077				√	√
3 Check pump and motor operation for excessive vibration, noise and overheating.	.022				√	√
4 Check operation of pressure controls.	.133				√	√
5 Check alignment of pump and motor; adjust as necessary.	.258				√	√
6 Lubricate pump and motor.	.099				√	√
7 Clean exterior of pump, motor and surrounding area.	.094				√	√
8 Fill out maintenance checklist and report deficiencies.	.022				√	√
Total labor-hours/period					.727	.727
Total labor-hours/year					.727	.727

			Cost Each				
			2019 Bare Costs			Total	Total
Description	Labor-hrs.	Material	Labor	Equip.	Total	In-House	w/O&P
5900 Pump, reciprocating displacement, annually	.727	11.15	39		50.15	62.41	76
5950 Annualized	1.458	22.50	78		100.50	125.49	152

PM Components	Labor-hrs.	W	M	Q	S	A
PM System G3015 126 6950						
Pump, rotary positive displacement						
1 Check with operating personnel for obvious defects.	.035				√	√
2 Clean surface of pump with solvent.	.014				√	√
3 Operate pump and check for leaks at pipes and connections and for excessive noise and vibration.	.099				√	√
4 Lubricate pump.	.047				√	√
5 Check shaft alignment and clearances and pump rotation.	.031				√	√
6 Inspect packing for leaks and tighten.	.031				√	√
7 Tighten loose bolts and nuts; replace loose and missing hardware as needed.	.009				√	√
8 Inspect electrical system for loose connections and frayed wires; repair as necessary.	.120				√	√
9 Operate pump upon completion of maintenance, read and record pressure gauge readings and check discharge pressure.	.221				√	√
10 Clean up area around pump after repairs.	.066				√	√
11 Fill out maintenance checklist and report deficiencies.	.022				√	√
Total labor-hours/period					.695	.695
Total labor-hours/year					.695	.695

			Cost Each				
			2019 Bare Costs			Total	Total
Description	Labor-hrs.	Material	Labor	Equip.	Total	In-House	w/O&P
6900 Pump, rotary displacement, annually	.695	16.75	37		53.75	66.34	80.50
6950 Annualized	1.390	33.50	74		107.50	133.14	161

PM Components	Labor-hrs.	W	M	Q	S	A
PM System G3015 126 7950						
Pump, sump, up to 1 H.P.						
1 Check electrical plug, cord and connections.	.120				√	√
2 Activate float switch and check pump for proper operation.	.039				√	√
3 Lubricate pump as required.	.047				√	√
4 Inspect packing and tighten as required.	.031				√	√
5 Check pump for misalignment and bearings for overheating.	.199				√	√
6 Clean out trash from sump.	.065				√	√
7 Fill out maintenance checklist and report deficiencies.	.022				√	√
Total labor-hours/period					.523	.523
Total labor-hours/year					.523	.523

		Cost Each					
			2019 Bare Costs			Total	Total
Description	Labor-hrs.	Material	Labor	Equip.	Total	In-House	w/O&P
7900 Pump, sump, up to 1 H.P., annually	.523	11.15	28		39.15	48.33	58.50
7950 Annualized	1.046	22.50	56		78.50	97.20	118

PM Components	Labor-hrs.	W	M	Q	S	A
PM System G3015 126 8950						
Pump, turbine, well						
1 Check with pump operating personnel for any known deficiencies.	.035				√	√
2 Remove turbine and pump from deep well.	.585				√	√
3 Inspect pump body, bowls, water passages and impeller for corrosion, wear or pitting; repair as necessary.	.057				√	√
4 Check packing for leaks; tighten or replace as necessary.	.327				√	√
5 Adjust impeller for optimum operation.	.229				√	√
6 Inspect wear ring clearances, thrust and pump bearings for wear; repair as necessary.	.142				√	√
7 Lubricate all pump parts as required.	.047				√	√
8 Clean suction strainer or replace if damaged.	.338				√	√
9 Inspect operation of discharge valve.	.026				√	√
10 Inspect electrical system for frayed wires or loose connections; repair as necessary.	.120				√	√
11 Check operation of pressure and thermal controls.	.052				√	√
12 Paint pump with underwater paint.	.259				√	√
13 Run pump and check for noise and vibration.	.022				√	√
14 Reinstall pump in deep well and test operation.	.585				√	√
15 Fill out maintenance checklist and report deficiencies.	.022				√	√
Total labor-hours/period					2.845	2.845
Total labor-hours/year					2.845	2.845

		Cost Each					
			2019 Bare Costs			Total	Total
Description	Labor-hrs.	Material	Labor	Equip.	Total	In-House	w/O&P
8900 Pump, turbine, well, annually	2.845	145	151		296	356.12	425
8950 Annualized	5.686	164	305		469	571.78	690

For customer support on your Facilities Maintenance & Repair Costs with RSMeans data, call 800.448.8182.

PM Components	Labor-hrs.	W	M	Q	S	A
PM System G3015 126 9950						
Pump, vacuum						
1 Check with pump operating personnel for any known deficiencies.	.035				√	√
2 Check for leaks on suction and discharge piping, seals, etc.	.077				√	√
3 Check pump and motor operation for vibration, noise, overheating, etc.	.022				√	√
4 Check alignment and clearances of shaft and coupler.	.258				√	√
5 Tighten or replace loose, missing, or damaged nuts, bolts, and screws.	.005				√	√
6 Lubricate pump and motor as required.	.099				√	√
7 Clean pump, motor and surrounding area.	.099				√	√
8 Fill out maintenance checklist and report deficiencies.	.022				√	√
Total labor-hours/period					.617	.617
Total labor-hours/year					.617	.617

	Description	Labor-hrs.	Cost Each				Total In-House	Total w/O&P
			2019 Bare Costs					
			Material	Labor	Equip.	Total		
9900	Pump, vacuum, annually	.617	16.75	33		49.75	60.99	73.50
9950	Annualized	1.238	33.50	66.50		100	122.68	148

For customer support on your Facilities Maintenance & Repair Costs with RSMeans data, call 800.448.8182.

PM Components	Labor-hrs.	W	M	Q	S	A
PM System G3015 310 1950						
Turbine well pump						
1 Check with operating personnel for any deficiencies.	.035				√	√
2 Remove turbine pump from deep well.	.585				√	√
3 Inspect pump body, bowls, water passages and impeller for corrosion, wear or pitting; repair as necessary.	.057				√	√
4 Check packing for leaks; tighten or repack as necessary.	.333				√	√
5 Adjust impeller for optimun operation.	.229				√	√
6 Inspect wear ring clearances, thrust and pump bearings for wear; repair as necessary.	.142				√	√
7 Lubricate all pump parts as required.	.047				√	√
8 Clean suction strainer or replace if damaged.	.338				√	√
9 Inspect operation of discharge valve.	.030				√	√
10 Inspect electrical system for frayed wires or loose connections; repair as necessary.	.120				√	√
11 Check operation of pressure and thermal controls.	.124				√	√
12 Paint pump with underwater paint.	.259				√	√
13 Run pump and check for noise and vibration.	.022				√	√
14 Reinstall pump in deep well and test operation.	.585				√	√
15 Fill out maintenance checklist and report deficiencies.	.022				√	√
Total labor-hours/period					2.928	2.928
Total labor-hours/year					2.928	2.928

		Cost Each					
			2019 Bare Costs			**Total**	**Total**
Description	**Labor-hrs.**	**Material**	**Labor**	**Equip.**	**Total**	**In-House**	**w/O&P**
1900 Turbine well pump, annually	2.928	76.50	156		232.50	286.37	345
1950 Annualized	5.848	96.50	310		406.50	508.78	620

PM Components	Labor-hrs.	W	M	Q	S	A
PM System G3015 320 1950						
Vertical lift pump, over 1 H.P.						
1 Check for proper operation of pump.	.130				√	√
2 Check for leaks on suction and discharge piping, seals, packing glands, etc.; make minor adjustments as required.	.077				√	√
3 Check pump and motor operation for excessive vibration, noise and overheating.	.022				√	√
4 Check alignment of pump and motor; adjust as necessary.	.258				√	√
5 Lubricate pump and motor.	.099				√	√
6 When available, check and record suction or discharge gauge pressure and flow rate.	.022				√	√
7 Clean exterior of pump and surrounding area.	.096				√	√
8 Fill out maintenance checklist and report deficiencies.	.022				√	√
Total labor-hours/period					.726	.726
Total labor-hours/year					.726	.726

		Cost Each					
			2019 Bare Costs			**Total**	**Total**
Description	**Labor-hrs.**	**Material**	**Labor**	**Equip.**	**Total**	**In-House**	**w/O&P**
1900 Vertical lift pump, over 1 H.P., annually	.726	30	38.50		68.50	82.86	99.50
1950 Annualized	1.456	47	78		125	152.04	183

PM Components	Labor-hrs.	W	M	Q	S	A
PM System G3015 410 1950						
Fire hydrant						
1 Remove hydrant caps and check condition of gaskets; replace as required.	.242					√
2 Flush hydrant, lubricate cap threads and reinstall cap.	.372					√
3 Fill out maintenance checklist and report deficiencies.	.022					√
Total labor-hours/period						.636
Total labor-hours/year						.636

				Cost Each				
				2019 Bare Costs			Total	Total
Description		Labor-hrs.	Material	Labor	Equip.	Total	In-House	w/O&P
1900	Fire hydrant, annually	.636	17.90	34		51.90	63.52	77
1950	Annualized	.636	17.90	34		51.90	63.52	77

PM Components	Labor-hrs.	W	M	Q	S	A
PM System G3015 420 1950						
Valve, post indicator						
1 Remove set screw, cap and wire seal.	.111				√	√
2 Apply lubricant to threads, open and close valve.	.291				√	√
3 Check operation of and sign and clean glass indicator windows.	.130				√	√
4 Install cap and tighten screw, install handle and wire seal with valve open.	.088				√	√
5 Clean valve exterior and area around valve.	.066				√	√
6 Fill out maintenance checklist and report deficiencies.	.022				√	√
Total labor-hours/period					.708	.708
Total labor-hours/year					.708	.708

				Cost Each				
				2019 Bare Costs			Total	Total
Description		Labor-hrs.	Material	Labor	Equip.	Total	In-House	w/O&P
1900	Valve, post indicator, annually	.708	13.75	38		51.75	64.12	77.50
1950	Annualized	1.420	23	76		99	123.43	150

For customer support on your Facilities Maintenance & Repair Costs with RSMeans data, call 800.448.8182.

PM Components	Labor-hrs.	W	M	Q	S	A
PM System G3015 612 1950						
Filter plant						
1 Check daily operating records for unusual occurances.	.035			√	√	√
2 Inspect plant sand or gravel filters.	.694			√	√	√
3 Inspect clear well for cleanliness.	.018			√	√	√
4 Inspect and lubricate flocculator chain and baffle mechanism and waste-line valves.	.163			√	√	√
5 Inspect and adjust rate-of-flow controller.	.036			√	√	√
6 Calibrate and adjust loss-of-head gauge.	.190			√	√	√
7 Inspect clear well vent screen and manhole.	.143			√	√	√
8 Inspect operation of pumps, water mixers, chemical feeders and flow measuring devices.	1.145			√	√	√
9 Check aerator nozzles for obstructions; clean as required.	.065			√	√	√
10 Drain, clean and inspect mixing chambers.	1.455					√
11 Drain sedimentation basins and clean with fire hose.	2.220					√
12 Drain, inspect and clean flocculator.	2.401					√
13 Clean up filter plant area.	.334			√	√	√
14 Fill out maintenance checklist and report deficiencies.	.022			√	√	√
Total labor-hours/period				2.845	2.845	8.921
Total labor-hours/year				5.691	2.845	8.921

			Cost Each					
			2019 Bare Costs				Total	Total
Description	Labor-hrs.	Material	Labor	Equip.	Total	In-House	w/O&P
1900	Filter plant, annually	8.921	20	475		495	637.55	790
1950	Annualized	17.459	51.50	935		986.50	1,259.61	1,575

PM Components	Labor-hrs.	W	M	Q	S	A
PM System G3015 630 1950						
Water de-ionization station						
1 Check with operating or area personnel for deficiencies.	.035				√	√
2 Check bulk storage tanks, pumps, valves and piping for damage, corrosion or leaks.	.130				√	√
3 Lubricate pumps.	.047				√	√
4 Check venting of bulk storage tanks.	.043				√	√
5 Check operation of sodium hydroxide tank and piping heating system.	.124				√	√
6 Check operation of resin vent, blower and motor; lubricate as required.	.033				√	√
7 Check calibration of meters.	.571				√	√
8 Clean and check probes of metering system.	.040				√	√
9 Check condition of carbon filter bed; replace annually.	.696					√
10 Check water softener system for proper operation.	.195				√	√
11 Clean equpment and surrounding area.	.066				√	√
12 Fill out maintenance checklist and report deficiencies.	.022				√	√
Total labor-hours/period					2.002	2.002
Total labor-hours/year					2.002	2.002

			Cost Each					
			2019 Bare Costs				Total	Total
Description	Labor-hrs.	Material	Labor	Equip.	Total	In-House	w/O&P
1900	Water de-ionization system, annually	2.002	385	107		492	562.72	650
1950	Annualized	4.012	770	214		984	1,126.46	1,300

For customer support on your Facilities Maintenance & Repair Costs with RSMeans data, call 800.448.8182.

PM Components	Labor-hrs.	W	M	Q	S	A
PM System G3015 632 1950						
Reverse osmosis system, 750 gallons/month						
1 Check with operating personnel for report of treatment effectiveness.	.046			√	√	√
2 Inspect unit, piping and connections for leaks.	.077			√	√	√
3 Check operation of pumps and motors; check shafts for alignment and lubricate as required.	.333			√	√	√
4 Inspect electrical system for frayed wires or loose connections; repair as required.	.120			√	√	√
5 Check filters; replace as required.	.036			√	√	√
6 Check water level controls; adjust as required.	.030			√	√	√
7 Check holding tank temperature control valve operation.	.030			√	√	√
8 Replace charcoal filters as indicated by water test.	.143			√	√	√
9 Check to ensure that chemical supply level is properly maintained in the chemical feed system.	.120			√	√	√
Total labor-hours/period				.936	.936	.936
Total labor-hours/year				1.871	.936	.936

			Cost Each				
			2019 Bare Costs			Total	Total
	Description	Labor-hrs.	Material	Labor	Equip.	Total	In-House	w/O&P
1900	Reverse osmosis system, annually	.936	63.50	50		113.50	135.21	159
1950	Annualized	3.728	134	199		333	403.70	490

PM Components	Labor-hrs.	W	M	Q	S	A
PM System G3015 634 1950						
Water ion exchange system						
1 Check with operating personnel for deficiencies.	.035			√	√	√
2 Clean unit and piping and check for leaks.	.077			√	√	√
3 Clean brine pump and check for proper operation.	.120			√	√	√
4 Clean and check strainers pilot valve, vents and orifices.	.593					√
5 Inspect electrical system including pump contactor and contacts; repair as required.	.120			√	√	√
6 Check regeneration cycle; make adjustments as required.	.327			√	√	√
7 Clean surrounding area.	.066			√	√	√
8 Fill out maintenance report and report deficiencies.	.022			√	√	√
Total labor-hours/period				.767	.767	1.360
Total labor-hours/year				1.534	.767	1.360

			Cost Each				
			2019 Bare Costs			Total	Total
	Description	Labor-hrs.	Material	Labor	Equip.	Total	In-House	w/O&P
1900	Water ion exchange system, annually	1.360	11.20	72.50		83.70	106.04	130
1950	Annualized	3.653	22.50	195		217.50	275.90	340

G30 SITE MECH. UTILITIES | G3015 650 | Water Softener

PM Components	Labor-hrs.	W	M	Q	S	A
PM System G3015 650 1950						
Water softener						
1 Check with building personnel for report of water softener effectiveness.	.086			√	√	√
2 Check pressure gauges for proper operation.	.190			√	√	√
3 Check density of brine solution in salt tank.	.124			√	√	√
4 Check operation of float control in brine.	.040			√	√	√
5 Inspect water softener piping, fittings and valves for leaks.	.077			√	√	√
6 Lubricate valves and motors.	.333			√	√	√
7 Make minor adjustments to water softener controls if required.	.014			√	√	√
8 Inspect softener base and brine tank for corrosion and repair as needed.	.258			√	√	√
9 Check operation of automatic fill valve in brine tank.	.030			√	√	√
10 Check softener electrical wiring and phasing.	.120			√	√	√
11 Clean up area around softener.	.066			√	√	√
12 Fill out maintenance report and report deficiencies.	.022			√	√	√
Total labor-hours/period				1.361	1.361	1.361
Total labor-hours/year				2.721	1.361	1.361

			Cost Each					
			2019 Bare Costs				Total	Total
	Description	Labor-hrs.	Material	Labor	Equip.	Total	In-House	w/O&P
1900	Water softener, annually	1.361	17.65	72.50		90.15	113.83	138
1950	Annualized	5.436	53.50	290		343.50	432.91	530

G30 SITE MECH. UTILITIES | G3015 660 | Fluoride Saturator

PM Components	Labor-hrs.	W	M	Q	S	A
PM System G3015 660 1950						
Fluoride saturator						
1 Check with operating personnel for any obvious discrepancies.	.035	√	√	√	√	
2 Obtain and put on safety equipment such as goggles, apron, gloves and respirator prior to working on the fluoride saturator.	.014	√	√	√	√	
3 Check oil level in gearbox, agitator and injector pump; add oil as necessary.	.018	√	√	√	√	
4 Check water level in fluoride mixing tank.	.007	√	√	√	√	
5 Check operation of fluoride flow regulator, float switch and solenoid.	.127	√	√	√	√	
6 Check saturator for leaks.	.077	√	√	√	√	
7 Check operation of unit.	.098	√	√	√	√	
8 Clean fluoride tank with brush and detergent.	.085	√	√	√	√	
9 Clean surrounding area.	.066	√	√	√	√	
10 Fill out maintenance checklist and report deficiencies.	.022	√	√	√	√	
Total labor-hours/period			.549	.549	.549	.549
Total labor-hours/year			4.394	1.098	.549	.549

			Cost Each					
			2019 Bare Costs				Total	Total
	Description	Labor-hrs.	Material	Labor	Equip.	Total	In-House	w/O&P
1900	Fluoride saturator, annually	.549	15.65	29.50		45.15	55.04	66.50
1950	Annualized	6.588	24.50	355		379.50	481.10	595

For customer support on your Facilities Maintenance & Repair Costs with RSMeans data, call 800.448.8182.

PM Components	Labor-hrs.	W	M	Q	S	A
PM System G3015 670 1950						
Chlorinator						
1 Check with operating or area personnel for deficiencies.	.035			√	√	√
2 Put on safety equipment prior to checking equipment.	.014			√	√	√
3 Clean chlorinator and check for leaks.	.057			√	√	√
4 Check water trap for proper level and bleed air.	.017			√	√	√
5 Check and clean water strainers.	.675			√	√	√
6 Grease fittings.	.023			√	√	√
7 Make minor adjustments to chlorinator as needed.	.094			√	√	√
8 Fill out maintenance checklist and report deficiencies.	.022			√	√	√
Total labor-hours/period				.937	.937	.937
Total labor-hours/year				1.874	.937	.937

			Cost Each				
			2019 Bare Costs			Total	Total
Description	Labor-hrs.	Material	Labor	Equip.	Total	In-House	w/O&P
1900 Chlorinator, annually	.937	8.35	50		58.35	73.80	90.50
1950 Annualized	3.748	25	200		225	285.95	350

PM Components	Labor-hrs.	W	M	Q	S	A
PM System G3015 670 2950						
Chlorine detector						
1 Put on safety equipment prior to checking equipment.	.014			√	√	√
2 Test alarm and inspect fan and sensor wiring.	.254			√	√	√
3 Test detector with chlorine test kit.	.044			√	√	√
4 Flush sensor unit with fresh water and drain and fill reservoir.	.068			√	√	√
5 Check drip rate and reservoir level in sight glass.	.013			√	√	√
6 Check "O" rings; replace as required.	.029			√	√	√
7 Inspect sensor mounting grommet, fan operation, detector mounting bolts and drain tube connection.	.147			√	√	√
8 Check charcoal filter cartridge; replace as needed.	.009			√	√	√
9 Check basket for cracked or worn areas.	.059			√	√	√
10 Check detector for leaks under pressure.	.077			√	√	√
11 Clean unit and surrounding area.	.066			√	√	√
12 Fill out maintenance checklist and report deficiencies.	.022			√	√	√
Total labor-hours/period				.801	.801	.801
Total labor-hours/year				1.603	.801	.801

			Cost Each				
			2019 Bare Costs			Total	Total
Description	Labor-hrs.	Material	Labor	Equip.	Total	In-House	w/O&P
2900 Chlorine detector, annually	.801	42	43		85	101.56	121
2950 Annualized	3.200	127	171		298	360.35	435

PM Components	Labor-hrs.	W	M	Q	S	A
PM System G3025 410 1950						
Ejector, pump, sewage						
1 Check for proper operation of pump.	.130				✓	✓
2 Check for leaks on suction and discharge piping, seals, packing glands, etc.; make minor adjustments as required.	.069				✓	✓
3 Check ejector and motor operation for excessive vibration, noise and overheating.	.022				✓	✓
4 Check float or pressure controls for proper operation.	.039				✓	✓
5 Lubricate ejector and motor.	.099				✓	✓
6 Clean pump unit and surrounding area.	.096				✓	✓
7 Fill out maintenance checklist and report deficiencies.	.022				✓	✓
Total labor-hours/period					.477	.477
Total labor-hours/year					.477	.477

			Cost Each					
			2019 Bare Costs				Total	Total
Description		Labor-hrs.	Material	Labor	Equip.	Total	In-House	w/O&P
1900	Ejector pump, sewage, annually	.477	24.50	25.50		50	59.53	71.50
1950	Annualized	.954	49	51		100	119.66	143

PM Components	Labor-hrs.	W	M	Q	S	A
PM System G3025 412 1950						
Ejector, pump						
1 Check for proper operation of pump.	.130				√	√
2 Check for leaks on suction and discharge piping, seals, packing glands, etc.; make minor adjustments as required.	.077				√	√
3 Check ejector and motor operation for excessive vibration, noise and overheating.	.022				√	√
4 Check float or pressure controls for proper operation.	.039				√	√
5 Lubricate ejector and motor.	.099				√	√
6 Clean pump unit and surrounding area.	.096				√	√
7 Fill out maintenance checklist and report deficiencies.	.022				√	√
Total labor-hours/period					.485	.485
Total labor-hours/year					.485	.485

			Cost Each					
				2019 Bare Costs			Total	Total
Description		Labor-hrs.	Material	Labor	Equip.	Total	In-House	w/O&P
1900	Ejector pump, annually	.485	24.50	26		50.50	60.08	72
1950	Annualized	.970	49	52		101	120.66	144

PM Components	Labor-hrs.	W	M	Q	S	A
PM System G3025 412 2950						
Ejector pump, sump type						
1 Operate float switch, if applicable, and check for proper operation of pump.	.130				√	√
2 Check for leaks on suction and discharge piping, seals, packing glands, etc.; make minor adjustments as required.	.077				√	√
3 Check pump and motor operation for excessive vibration, noise and overheating.	.022				√	√
4 Check alignment of pump and motor; adjust as necessary.	.258				√	√
5 Lubricate motor.	.099				√	√
6 Check electrical wiring and connections and repair as applicable.	.120				√	√
7 Clean exterior of pump and surrounding area.	.096				√	√
8 Fill out maintenance checklist and report deficiencies.	.022				√	√
Total labor-hours/period					.824	.824
Total labor-hours/year					.824	.824

			Cost Each					
				2019 Bare Costs			Total	Total
Description		Labor-hrs.	Material	Labor	Equip.	Total	In-House	w/O&P
2900	Ejector pump, sump type, annually	.824	24.50	44		68.50	83.49	101
2950	Annualized	1.652	49	88.50		137.50	167.81	202

PM Components	Labor-hrs.	W	M	Q	S	A
PM System G3025 420 1950						
Sewage lift pump, over 1 H.P.						
1 Check for proper operation of pump.	.130				√	√
2 Check for leaks on suction and discharge piping, seals, packing glands, etc.; make minor adjustments as required.	.077				√	√
3 Check pump and motor operation for excessive vibration, noise and overheating.	.022				√	√
4 Check alignment of pump and motor; adjust as necessary.	.258				√	√
5 Lubricate pump and motor.	.099				√	√
6 When available, check and record suction or discharge gauge pressure and flow rate.	.022				√	√
7 Clean exterior of pump, motor and surrounding area.	.096				√	√
8 Fill out maintenance checklist and report deficiencies.	.022				√	√
Total labor-hours/period					.726	.726
Total labor-hours/year					.726	.726

			Cost Each					
			2019 Bare Costs				Total	Total
Description		Labor-hrs.	Material	Labor	Equip.	Total	In-House	w/O&P
1900	Sewage lift pump, over 1 H.P., annually	.726	24.50	38.50		63	76.81	92.50
1950	Annualized	1.456	35.50	78		113.50	139.94	168

For customer support on your Facilities Maintenance & Repair Costs with RSMeans data, call 800.448.8182.

PM Components	Labor-hrs.	W	M	Q	S	A
PM System G3025 530 1950						
Aerator, floating						
1 Check with operating or area personnel for deficiencies.	.035			√	√	√
2 Remove aerator from pond.	.390			√	√	√
3 Clean exterior of pump and motor.	.066			√	√	√
4 Check pump and motor operation for excessive vibration, noise and overheating.	.022			√	√	√
5 Check alignment of pump and motor; adjust as necessary.	.258			√	√	√
6 Lubricate pump and motor.	.099			√	√	√
7 Position aerator back in pond.	.390			√	√	√
8 Fill out maintenance checklist and report deficiencies.	.022			√	√	√
Total labor-hours/period				1.282	1.282	1.282
Total labor-hours/year				2.565	1.282	1.282

			Cost Each				
			2019 Bare Costs			Total	Total
Description	Labor-hrs.	Material	Labor	Equip.	Total	In-House	w/O&P
1900 Aerator, floating, annually	1.282	16.75	69		85.75	107	131
1950 Annualized	5.136	50.50	275		325.50	410.25	505

PM Components	Labor-hrs.	W	M	Q	S	A
PM System G3025 532 1950						
Barminutor						
1 Check with operating or area personnel for deficiencies.	.044		√	√	√	√
2 Check condition and clearance of cutting knives and inspect base seal.	.217		√	√	√	√
3 Check oil level in gearbox; add oil as necessary.	.035		√	√	√	√
4 Change oil in gearbox.	.681					√
5 Wire brush and lubricate directional flow valve stem.	.060		√	√	√	√
6 Check for rust and corrosion; scrape, wire brush and spot paint as necessary.	.105		√	√	√	√
7 Fill out maintenance checklist and report deficiencies.	.022		√	√	√	√
Total labor-hours/period			.483	.483	.483	1.164
Total labor-hours/year			3.864	.966	.483	1.164

			Cost Each				
			2019 Bare Costs			Total	Total
Description	Labor-hrs.	Material	Labor	Equip.	Total	In-House	w/O&P
1900 Barminutor, annually	1.164	49	62.50		111.50	134.60	161
1950 Annualized	6.477	118	345		463	577.20	705

PM Components	Labor-hrs.	W	M	Q	S	A
PM System G3025 534 1950						
Blower, aerator						
1 Check with operating or area personnel for deficiencies.	.035		√	√	√	√
2 Check blower oil level; add oil as necessary.	.046		√	√	√	√
3 Change blower oil.	.390				√	√
4 Check bolts and tighten as necessary: foundation, cylinder head, belt guard, etc.	.155		√	√	√	√
5 Check tension, condition, and alignment of V-belts on blower; adjust or replace as necessary.	.030		√	√	√	√
6 Check pump for excessive heat or vibration.	.042		√	√	√	√
7 Clean air intake filter on blower; replace as necessary.	.178		√	√	√	√
8 Perform operation check of air blower and adjust as required.	.222		√	√	√	√
9 Clean area around equipment.	.066		√	√	√	√
10 Fill out maintenance checklist and report deficiencies.	.022		√	√	√	√
Total labor-hours/period			.796	.796	1.187	1.187
Total labor-hours/year			6.371	1.593	1.187	1.187

			Cost Each					
			2019 Bare Costs				**Total**	**Total**
Description		Labor-hrs.	Material	Labor	Equip.	Total	In-House	w/O&P
1900	Blower, aerator, annually	1.187	70.50	63.50		134	159.24	190
1950	Annualized	10.308	231	550		781	965.20	1,175

PM Components	Labor-hrs.	W	M	Q	S	A
PM System G3025 536 1950						
Clarifier						
1 Check with operating or area personnel for deficiencies.	.035		√	√	√	√
2 Check chain and sprocket for wear, where applicable.	.033		√	√	√	√
3 Grease drive gear fittings.	.150		√	√	√	√
4 Check gearbox oil level; add if necessary.	.046		√	√	√	√
5 Drain and refill gearbox oil reservoir.	.681					√
6 Check unit for rust and corrosion; scrape, wire brush, and paint as required.	.105					√
7 Clean clarifier drive and surrounding area.	.098		√	√	√	√
8 Fill out maintenance checklist and report deficiencies.	.022		√	√	√	√
Total labor-hours/period			.384	.384	.384	1.170
Total labor-hours/year			3.072	.768	.384	1.170

			Cost Each					
			2019 Bare Costs				**Total**	**Total**
Description		Labor-hrs.	Material	Labor	Equip.	Total	In-House	w/O&P
1900	Clarifier, annually	1.170	54	62.50		116.50	139.81	168
1950	Annualized	5.394	126	288		414	510.25	620

For customer support on your Facilities Maintenance & Repair Costs with RSMeans data, call 800.448.8182.

PM Components	Labor-hrs.	W	M	Q	S	A
PM System G3025 538 1950						
Comminutor						
1 Check with operating or area personnel for deficiencies.	.035			√	√	√
2 Change gearbox oil and lubricate fittings.	.681				√	√
3 Check comminutor and motor for excessive noise, vibration and overheating.	.022			√	√	√
4 Lubricate fittings and check oil in gearbox.	.150			√	√	√
5 Check operation.	.108			√	√	√
6 Fill out maintenance checklist and report deficiencies.	.022			√	√	√
Total labor-hours/period				.337	1.018	1.018
Total labor-hours/year				.674	1.018	1.018

			Cost Each				
			2019 Bare Costs			**Total**	**Total**
Description	**Labor-hrs.**	**Material**	**Labor**	**Equip.**	**Total**	**In-House**	**w/O&P**
1900 Comminutor, annually	1.018	36.50	54.50		91	110.40	133
1950 Annualized	2.710	85	144		229	281.25	335

PM Components	Labor-hrs.	W	M	Q	S	A
PM System G3025 540 1950						
Filter, trickling						
1 Lubricate fittings and check oil level in oil reservoir.	.035				√	√
2 Check distribution arms, support cables, clevis pins, and turnbuckles for deterioration.	.327				√	√
3 Check for leaks.	.077				√	√
4 Check operation.	.108				√	√
5 Fill out maintenance checklist and report deficiencies.	.022				√	√
Total labor-hours/period					.569	.569
Total labor-hours/year					.569	.569

			Cost Each				
			2019 Bare Costs			**Total**	**Total**
Description	**Labor-hrs.**	**Material**	**Labor**	**Equip.**	**Total**	**In-House**	**w/O&P**
1900 Filter, trickling, annually	.569	24	30.50		54.50	65.77	78.50
1950 Annualized	1.134	48	60.50		108.50	131.53	157

PM Components	Labor-hrs.	W	M	Q	S	A
PM System G3025 542 1950						
Grit drive						
1 Check with operating personnel for obvious deficiencies.	.035				√	√
2 Lubricate fittings and check oil level in gearbox.	.047				√	√
3 Change oil in gearbox.	.681					√
4 Check chain linkage and operation.	.033				√	√
5 Clean area around drive mechanism.	.066				√	√
6 Fill out maintenance checklist and report deficiencies.	.022				√	√
Total labor-hours/period					.203	.884
Total labor-hours/year					.203	.884

			Cost Each					
				2019 Bare Costs			Total	Total
	Description	Labor-hrs.	Material	Labor	Equip.	Total	In-House	w/O&P
1900	Grit drive, annually	.884	49.50	47.50		97	115.26	137
1950	Annualized	1.087	73.50	58		131.50	156.01	185

PM Components	Labor-hrs.	W	M	Q	S	A
PM System G3025 544 1950						
Mixer, sewage						
1 Check with operating or area personnel for deficiencies.	.035			√	√	√
2 Clean drive motor exterior and check for corrosion on motor exterior and base plate.	.053			√	√	√
3 Check motor operation for vibration, noise, overheating; etc.	.022			√	√	√
4 Check alignment and clearances of shaft and coupler.	.031			√	√	√
5 Tighten or replace loose, missing, or damaged nuts, bolts and screws.	.005			√	√	√
6 Lubricate motor and shaft bearings as required.	.047			√	√	√
7 Clean area around motor.	.066			√	√	√
8 Fill out maintenance checklist and report deficiencies.	.022			√	√	√
Total labor-hours/period				.281	.281	.281
Total labor-hours/year				.562	.281	.281

			Cost Each					
				2019 Bare Costs			Total	Total
	Description	Labor-hrs.	Material	Labor	Equip.	Total	In-House	w/O&P
1900	Mixer, sewage, annually	.281	11.15	15		26.15	31.64	38
1950	Annualized	1.124	33.50	60		93.50	114.48	138

For customer support on your Facilities Maintenance & Repair Costs with RSMeans data, call 800.448.8182.

PM Components	Labor-hrs.	W	M	Q	S	A
PM System G3065 310 1950						
Fuel oil storage tank, above ground						
1 Inspect tank for corrosion.	.192					√
2 Inspect floating tank roof, lifter or cone and automatic float gauge, if applicable.	.233					√
3 Inspect fuel tank fire protection devices.	.022					√
4 Inspect berms around fuel storage tank for erosion.	.047					√
5 Inspect non-freeze draw valves.	.060					√
6 Inspect storage tank vacuum and pressure vents.	.179					√
7 Inspect vapor recovery line for deterioration.	.082					√
8 Run pump and check operation.	.159					√
9 Calibrate and adjust fuel oil pressure.	.382					√
10 Inspect fuel strainer and clean as needed.	.415					√
11 Inspect tank electrical grounds for breaks or loose fittings.	1.292					√
12 Inspect liquid level gauge for proper operation.	.191					√
13 Fill out maintenance checklist.	.022					√
Total labor-hours/period						3.276
Total labor-hours/year						3.276

			Cost Each					
			2019 Bare Costs				Total	Total
Description		Labor-hrs.	Material	Labor	Equip.	Total	In-House	w/O&P
1900	Fuel oil storage tank, above ground, annually	3.276	66.50	175		241.50	298.57	360
1950	Annualized	3.276	66.50	175		241.50	298.57	360

PM Components	Labor-hrs.	W	M	Q	S	A
PM System G4095 110 1950						
Cathodic protection system						
1 Inspect exterior of rectifier cabinet, all exposed connecting conduit, and fittings for corrosion, deteriorated paint, mechanical integrity and cleanliness, make minor repairs and clean/touch-up paint as required.	.060				√	√
2 Open and inspect interior of rectifier cabinet, inspect wiring for fraying, signs of overheating, deterioration mechanical damage, and loose connections, make adjustments as required.	.138				√	√
3 Check electrical contacts for overheating and pitting; make adjustments.	.075				√	√
4 Clean interior of rectifier cabinet.	.008				√	√
5 Check panel voltmeter and ampmeter with standard test instruments.	.075				√	√
6 Read and record panel meter voltage and current values.	.007				√	√
7 Measure and record rectifier voltage and current with portable instrument.	.059				√	√
8 Check potential at cathode locations; tighten loose connections.	.286				√	√
9 Fill out maintenance checklist and record deficiencies.	.022				√	√
Total labor-hours/period					.729	.729
Total labor-hours/year					.729	.729

	Description	Labor-hrs.	Material	Labor	Equip.	Total	Total In-House	Total w/O&P
1900	Cathodic protection system, annually	.729	8.70	44		52.70	63.39	78.50
1950	Annualized	1.460	13.10	88		101.10	122.93	151

For customer support on your Facilities Maintenance & Repair Costs with RSMeans data, call 800.448.8182.

General Maintenance

Table of Contents

How to Use the General Maintenance Section

The following is a detailed explanation of a sample entry in the General Maintenance Section. Next to each bold number that follows is the described item with the appropriate component of the sample entry in parentheses.

1 Division Number/Title
(01 93/Facility Maintenance)
Use the General Maintenance Section Table of Contents to locate specific items. This section is classified according to the CSI MasterFormat® 2016 system.

2 Line Numbers (01 93 04.40 0030)
Each General Maintenance unit price line item has been assigned a unique 12-digit code based on the CSI MasterFormat classification.

```
                    ┌─── MasterFormat Level 3
                ┌───┤
            ┌───┤   └─── MasterFormat Level 2 (01  93  00)
            │   └─────── MasterFormat Division (01)
   01   93   04.40           0030
                └── MasterFormat Level 4
        └───── Gordian's RSMeans 12-Digit Line Number
```

3 Description (Lawn Renovation, etc.)
Each line item is described in detail. Sub-items and additional sizes are indented beneath the appropriate line items. The first line or two after the main item (in boldface) may contain descriptive information that pertains to all line items beneath this boldface listing.

4 Crew (A-18)
The "Crew" column designates the typical trade or crew used for the task. If the task can be accomplished by one trade and requires no power equipment, that trade and the number of workers are listed (for example, "2 Clam"). If the task requires a composite crew, a crew code designation is listed (for example, "A-18"). You'll find full details on all composite crews in the Crew Listings.
- For a complete list of all trades utilized in this cost data and their abbreviations, see the inside back cover of the cost data.

Crews - Maintenance

Crew No.	Bare Costs		In-House Costs		Incl. Subs O&P		Cost Per Labor-Hour		
Crew A-18	Hr.	Daily	Hr.	Daily	Hr.	Daily	Bare Costs	In House	Incl. O&P
1 Maintenance Laborer	$30.80	$246.40	$40.10	$320.80	$49.65	$397.20	$30.80	$40.10	$49.65
1 Farm Tractor w/ Attachment		319.35		319.35		351.29	39.92	39.92	43.91
8 L.H., Daily Totals		$565.75		$640.15		$748.49	$70.72	$80.02	$93.56

5 Productivity:
Daily Output (750)/Labor-Hours (.011)
The "Daily Output" represents the typical number of units the designated crew will perform in a normal eight-hour day. To find out the number of days the given crew would require to complete your task, divide your quantity by the daily output. See the second table for an example:

1 01 93 Facility Maintenance

01 93 04 – Landscaping Maintenance

2 01 93 04.15 Edging	Crew	Daily Output	Labor-Hours	Unit	Material	2019 Bare Costs Labor	2019 Bare Costs Equipment	Total	Total In-House	Total Incl O&P
0010 **EDGING**										
0020 Hand edging, at walks	1 Clam	16	.500	C.L.F.		15.40		15.40	20.00	25
0030 At planting, mulch or stone beds		7	1.143			35		35	46.00	56.50
0040 Power edging, at walks		88	.091			2.80		2.80	3.65	4.51
0050 At planting, mulch or stone beds		24	.333			10.25		10.25	13.35	16.55
3 01 93 04.40 Lawn Renovation	**4**	**5**		**6**		**7**			**8**	**9**
0010 **LAWN RENOVATION**										
0020 Lawn renovations, aerating, 18" walk behind cultivator	1 Clam	95	.084	M.S.F.		2.59		2.59	3.38	4.18
0030 48" tractor drawn cultivator	A-18	750	.011			.33	.43	.76	.90	1
0040 72" tractor drawn cultivator	"	1100	.007			.22	.29	.51	.61	.68
0050 Fertilizing, dry granular, 4#/M.S.F., drop spreader	1 Clam	24	.333		2.14	10.25		12.39	15.70	19.25
0060 Rotary spreader	"	140	.057		2.14	1.76		3.90	4.64	5.50

Quantity	÷	Daily Output	=	Duration
2000 M.S.F.	÷	750 M.S.F./ Crew Day	=	2.7 Crew Days

The "Labor-Hours" figure represents the number of labor-hours required to perform one unit of work. To find out the number of labor-hours required for your particular task, multiply the quantity of the item and the number of labor-hours shown. For example:

Quantity	X	Productivity Rate	=	Duration
2000 M.S.F.	X	.011 Labor-Hours/ MSF	=	22 Labor-Hours

6 Unit (M.S.F.)

The abbreviated designation indicates the unit of measure upon which the price, production, and crew are based (M.S.F. = Thousand Square Feet). For a complete listing of abbreviations, refer to the Abbreviations Listing in the Reference Section of this cost data.

7 Bare Costs:

The unit material cost is the "bare" material cost with no overhead and profit included. *Costs shown reflect national average material prices for January of the current year and include delivery to the facility. Small purchases may require an additional delivery charge. No sales taxes are included.*

Labor (.33)

The unit labor cost is derived by multiplying bare labor-hour costs for Crew A-18 by labor-hour units. The bare labor-hour cost is found in the Crew Section under A-18. (If a trade is listed, the hourly labor cost—the wage rate—is found on the inside back cover of the cost data.)

Labor-Hour Cost Crew A-18	X	Labor-Hour Units	=	Labor
$30.80	X	.011	=	$0.33

Equip. (Equipment) (.43)

Equipment costs for each crew are listed in the description of each crew. Tools or equipment whose value justifies purchase or ownership by a contractor are considered overhead as shown on the inside back cover of the cost data. The unit equipment cost is derived by multiplying the bare equipment hourly cost and the labor-hour units.

Equipment Cost Crew A-18	X	Labor-Hour Units	=	Equip.
$39.92	X	.011	=	$0.43

Total (.76)

The total of the bare costs is the arithmetic total of the three previous columns: material, labor, and equip.

Material	+	Labor	+	Equip.	=	Total
$0.00	+	$0.33	+	$0.43	=	$0.76

8 Total In-House Costs (.90)

"Total In-House Costs" include suggested markups to bare costs for the direct overhead requirements of in-house maintenance staff. The figure in this column is the sum of three components: the bare material cost plus 10% (for purchasing and handling small quantities); the bare labor cost plus workers' compensation and average fixed overhead (per the labor rate table on the inside back cover of the cost data or, if a crew is listed, from the Crew Listings); and the bare equipment cost.

Material is Bare Material Cost + 10% = $0.00 + $0.00	=	$0.00
Labor for Crew A-18 = Labor-Hour Cost ($40.10) x Labor-Hour Units (.011)	=	$0.44
Equip. is Bare Equip. Cost	=	$0.43
Total	=	$0.90

9 Total Costs Including O&P (1.00)

"Total Costs Including O&P" include suggested markups to bare costs for an outside subcontractor's overhead and profit requirements. The figure in this column is the sum of three components: the bare material cost plus 25% for profit; the bare labor cost plus total overhead and profit (per the labor rate table on the inside back cover of the printed product or the Reference Section of the electronic product, or, if a crew is listed, from the Crew Listings); and the bare equipment cost plus 10% for profit.

Material is Bare Material Cost + 25% = $0.00 + $0.00	=	$0.00
Labor for Crew A-18 = Labor-Hour Cost ($49.65) x Labor-Hour Units (.011)	=	$0.54
Equip. is Bare Equip. Cost + 10% = $0.43 + $0.05	=	$0.48
Total	=	$1.00

01 93 04 – Landscaping Maintenance

01 93 04.15 Edging

		Crew	Daily Output	Labor-Hours	Unit	Material	2019 Bare Costs Labor	2019 Bare Costs Equipment	Total	Total In-House	Total Incl O&P
0010	**EDGING**										
0020	Hand edging, at walks	1 Clam	16	.500	C.L.F.		15.40		15.40	20.00	25
0030	At planting, mulch or stone beds		7	1.143			35		35	46.00	56.50
0040	Power edging, at walks		88	.091			2.80		2.80	3.65	4.51
0050	At planting, mulch or stone beds		24	.333			10.25		10.25	13.35	16.55

01 93 04.20 Flower, Shrub and Tree Care

		Crew	Daily Output	Labor-Hours	Unit	Material	2019 Bare Costs Labor	2019 Bare Costs Equipment	Total	Total In-House	Total Incl O&P
0010	**FLOWER, SHRUB & TREE CARE**										
0020	Flower or shrub beds, bark mulch, 3" deep hand spreader	1 Clam	100	.080	S.Y.	3.55	2.46		6.01	7.10	8.40
0030	Peat moss, 1" deep hand spreader		900	.009		4.88	.27		5.15	5.75	6.55
0040	Wood chips, 2" deep hand spreader		220	.036		1.63	1.12		2.75	3.25	3.85
0050	Cleaning		1	8	M.S.F.		246		246	320.00	395
0060	Fertilizing, dry granular, 3#/M.S.F.		85	.094		1.14	2.90		4.04	5.00	6.10
0070	Weeding, mulched bed		20	.400			12.30		12.30	16.05	19.85
0080	Unmulched bed		8	1			31		31	40.00	49.50
0090	Trees, pruning from ground, 1-1/2" caliper		84	.095	Ea.		2.93		2.93	3.82	4.73
0100	2" caliper		70	.114			3.52		3.52	4.58	5.65
0110	2-1/2" caliper		50	.160			4.93		4.93	6.40	7.95
0120	3" caliper		30	.267			8.20		8.20	10.70	13.25
0130	4" caliper	2 Clam	21	.762			23.50		23.50	30.50	38
0140	6" caliper		12	1.333			41		41	53.50	66
0150	9" caliper		7.50	2.133			65.50		65.50	85.50	106
0160	12" caliper		6.50	2.462			76		76	98.50	122
0170	Fertilize, slow release tablets	1 Clam	100	.080		2.32	2.46		4.78	5.75	6.85
0180	Pest control, spray		24	.333		27	10.25		37.25	43.00	50.50
0190	Systemic		48	.167		27	5.15		32.15	36.50	42
0200	Watering, under 1-1/2" caliper		34	.235			7.25		7.25	9.45	11.70
0210	1-1/2" to 4" caliper		14.50	.552			17		17	22.00	27.50
0220	4" caliper and over		10	.800			24.50		24.50	32.00	39.50
0230	Shrubs, prune, entire bed		7	1.143	M.S.F.		35		35	46.00	56.50
0240	Per shrub, 3' height		190	.042	Ea.		1.30		1.30	1.69	2.09
0250	4' height		90	.089			2.74		2.74	3.56	4.41
0260	6' height		50	.160			4.93		4.93	6.40	7.95
0270	Fertilize, dry granular, 3#/M.S.F.		85	.094	M.S.F.	1.14	2.90		4.04	5.00	6.10
0280	Watering, entire bed		7	1.143	"		35		35	46.00	56.50
0290	Per shrub		32	.250	Ea.		7.70		7.70	10.05	12.40

01 93 04.35 Lawn Care

		Crew	Daily Output	Labor-Hours	Unit	Material	2019 Bare Costs Labor	2019 Bare Costs Equipment	Total	Total In-House	Total Incl O&P
0010	**LAWN CARE**										
0020	Mowing lawns, power mower, 18"-22"	1 Clam	80	.100	M.S.F.		3.08		3.08	4.01	4.97
0030	22"-30"		120	.067			2.05		2.05	2.67	3.31
0040	30"-32"		140	.057			1.76		1.76	2.29	2.84
0050	Self propelled or riding mower, 36"-44"	A-16	300	.027			.82	.33	1.15	1.43	1.68
0060	48"-58"		480	.017			.51	.21	.72	.90	1.06
0070	Tractor, 3 gang reel, 7' cut		930	.009			.26	.11	.37	.46	.55
0080	5 gang reel, 12' cut		1200	.007			.21	.08	.29	.36	.42
0090	Edge trimming with weed whacker	1 Clam	5760	.001	L.F.		.04		.04	.06	.07

01 93 04.40 Lawn Renovation

		Crew	Daily Output	Labor-Hours	Unit	Material	2019 Bare Costs Labor	2019 Bare Costs Equipment	Total	Total In-House	Total Incl O&P
0010	**LAWN RENOVATION**										
0020	Lawn renovations, aerating, 18" walk behind cultivator	1 Clam	95	.084	M.S.F.		2.59		2.59	3.38	4.18
0030	48" tractor drawn cultivator	A-18	750	.011			.33	.43	.76	.90	1
0040	72" tractor drawn cultivator	"	1100	.007			.22	.29	.51	.61	.68
0050	Fertilizing, dry granular, 4#/M.S.F., drop spreader	1 Clam	24	.333		2.14	10.25		12.39	15.70	19.25
0060	Rotary spreader	"	140	.057		2.14	1.76		3.90	4.64	5.50

01 93 Facility Maintenance

01 93 04 – Landscaping Maintenance

01 93 04.40 Lawn Renovation

		Crew	Daily Output	Labor-Hours	Unit	Material	2019 Bare Costs Labor	Equipment	Total	Total In-House	Total Incl O&P
0070	Tractor drawn 8' spreader	A-18	500	.016	M.S.F.	2.14	.49	.64	3.27	3.69	4.17
0080	Tractor drawn 12' spreader	"	800	.010		2.14	.31	.40	2.85	3.19	3.62
0090	Overseeding, utility mix, 7#/M.S.F., drop spreader	1 Clam	10	.800		10.95	24.50		35.45	44.00	53
0100	Tractor drawn spreader	A-18	52	.154		10.95	4.74	6.15	21.84	25.00	28
0110	Watering, 1" of water, applied by hand	1 Clam	21	.381			11.75		11.75	15.30	18.90
0120	Soaker hoses	"	82	.090			3		3	3.91	4.84

01 93 04.65 Raking

		Crew	Daily Output	Labor-Hours	Unit	Material	2019 Bare Costs Labor	Equipment	Total	Total In-House	Total Incl O&P
0010	**RAKING**										
0020	Raking, leaves, by hand	1 Clam	7.50	1.067	M.S.F.		33		33	43.00	53
0030	Power blower		45	.178			5.50		5.50	7.15	8.85
0040	Grass clippings, by hand		7.50	1.067			33		33	43.00	53

01 93 04.85 Walkways and Parking Care

		Crew	Daily Output	Labor-Hours	Unit	Material	2019 Bare Costs Labor	Equipment	Total	Total In-House	Total Incl O&P
0010	**WALKWAYS AND PARKING CARE**										
0020	General cleaning, parking area and walks, sweeping by hand	1 Clam	15	.533	M.S.F.		16.45		16.45	21.50	26.50
0030	Power vacuum		100	.080			2.46		2.46	3.21	3.97
0040	Power blower		100	.080			2.46		2.46	3.21	3.97
0050	Hose down		30	.267			8.20		8.20	10.70	13.25

01 93 04.90 Yard Waste Disposal

		Crew	Daily Output	Labor-Hours	Unit	Material	2019 Bare Costs Labor	Equipment	Total	Total In-House	Total Incl O&P
0010	**YARD WASTE DISPOSAL**										
0030	On site dump or compost heap	2 Clam	24	.667	C.Y.		20.50		20.50	26.50	33
0040	Into 6 C.Y. dump truck, 2 mile haul to compost heap	A-17	125	.064			1.97	1.56	3.53	4.29	4.90
0050	4 mile haul to compost heap, fees not included	"	85	.094			2.90	2.30	5.20	6.30	7.20

01 93 05 – Exterior General Maintenance

01 93 05.95 Window Washing

		Crew	Daily Output	Labor-Hours	Unit	Material	2019 Bare Costs Labor	Equipment	Total	Total In-House	Total Incl O&P
0010	**WINDOW WASHING**										
0030	Minimum productivity	1 Clam	1	8	M.S.F.	3.81	246		249.81	325.00	400
0040	Average productivity		2.50	3.200		3.81	98.50		102.31	133.00	164
0050	Maximum productivity		4	2		3.81	61.50		65.31	84.50	104

01 93 06 – Interior General Maintenance

01 93 06.20 Carpet Care

		Crew	Daily Output	Labor-Hours	Unit	Material	2019 Bare Costs Labor	Equipment	Total	Total In-House	Total Incl O&P
0010	**CARPET CARE**										
0100	Carpet cleaning, portable extractor with floor wand	1 Clam	20	.400	M.S.F.	8.35	12.30		20.65	25.00	30.50
0200	Deep carpet cleaning, self-contained extractor										
0210	24" path @ 100 feet per minute	1 Clam	64	.125	M.S.F.	10.45	3.85		14.30	16.50	19.25
0220	20" path		54	.148		10.45	4.56		15.01	17.45	20.50
0230	16" path		43	.186		10.45	5.75		16.20	18.95	22.50
0240	12" path		32	.250		10.45	7.70		18.15	21.50	25.50
0300	Carpet pile lifting, self-contained extractor										
0310	24" path at 90 feet per minute	1 Clam	60	.133	M.S.F.	11.50	4.11		15.61	18.00	21
0320	20" path		50	.160		11.50	4.93		16.43	19.05	22.50
0330	16" path		40	.200		11.50	6.15		17.65	20.50	24.50
0340	12" path		30	.267		11.50	8.20		19.70	23.50	27.50
0400	Carpet restoration cleaning, self-contained extractor										
0410	24" path at 50 feet per minute	1 Clam	32	.250	M.S.F.	12.55	7.70		20.25	24.00	28
0420	20" path		27	.296		12.55	9.15		21.70	25.50	30.50
0430	16" path		21	.381		12.55	11.75		24.30	29.00	34.50
0440	12" path		16	.500		12.55	15.40		27.95	34.00	40.50

01 93 06.25 Ceiling Care

		Crew	Daily Output	Labor-Hours	Unit	Material	2019 Bare Costs Labor	Equipment	Total	Total In-House	Total Incl O&P
0010	**CEILING CARE**										
0020	Washing hard ceiling from ladder, hand	1 Clam	2.05	3.902	M.S.F.	1.70	120		121.70	158.00	196

01 93 Facility Maintenance

01 93 06 — Interior General Maintenance

01 93 06.25 Ceiling Care

		Crew	Daily Output	Labor-Hours	Unit	Material	2019 Bare Costs Labor	Equipment	Total	Total In-House	Total Incl O&P
0030	Machine	1 Clam	2.52	3.175	M.S.F.	1.70	98		99.70	129.00	160
0035	Acoustic tile cleaning, chemical spray, including masking										
0040	Unobstructed floor	4 Clam	5400	.006	S.F.	.07	.18		.25	.32	.38
0045	Obstructed floor	"	4000	.008	"	.07	.25		.32	.40	.49
0048	Chemical spray cleaning and coating, including masking										
0050	Unobstructed floor	4 Clam	4800	.007	S.F.	.14	.21		.35	.42	.51
0055	Obstructed floor	"	3400	.009	"	.14	.29		.43	.53	.65

01 93 06.40 General Cleaning

		Crew	Daily Output	Labor-Hours	Unit	Material	2019 Bare Costs Labor	Equipment	Total	Total In-House	Total Incl O&P
0010	**GENERAL CLEANING**										
0020	Sweep stairs & landings, damp wipe handrails	1 Clam	80	.100	Floor	.03	3.08		3.11	4.04	5
0030	Dust mop stairs & landings, damp wipe handrails		62.50	.128		.03	3.94		3.97	5.15	6.40
0040	Damp mop stairs & landings, damp wipe handrails		45	.178		.23	5.50		5.73	7.40	9.15
0050	Vacuum stairs & landings, damp wipe handrails, canister		41	.195		.03	6		6.03	7.85	9.75
0060	Vacuum stairs & landings, damp wipe handrails, backpack		58.22	.137		.03	4.23		4.26	5.55	6.85
0070	Vacuum stairs & landings, damp wipe handrails, upright		31.57	.253		.03	7.80		7.83	10.20	12.65
0080	Clean carpeted elevators		48	.167	Ea.	.59	5.15		5.74	7.35	9.05
0090	Clean uncarpeted elevators		32	.250		.43	7.70		8.13	10.50	12.95
0100	Clean escalator		8	1		.34	31		31.34	40.50	50
0110	Restrooms, fixture cleaning, toilets		180	.044		.05	1.37		1.42	1.84	2.27
0120	Urinals		192	.042		.05	1.28		1.33	1.73	2.13
0130	Sinks		206	.039		.01	1.20		1.21	1.57	1.94
0140	Utility sink		175	.046		.02	1.41		1.43	1.85	2.30
0150	Bathtub		85	.094		.06	2.90		2.96	3.84	4.75
0160	Shower		85	.094		1.43	2.90		4.33	5.35	6.45
0170	Accessories cleaning, soap dispenser		1280	.006		.01	.19		.20	.26	.32
0180	Sanitary napkin dispenser		640	.013		.01	.39		.40	.51	.63
0190	Paper towel dispenser		1455	.006		.01	.17		.18	.23	.28
0200	Toilet tissue dispenser		1455	.006		.01	.17		.18	.23	.28
0210	Accessories, refilling/restocking, soap dispenser (1 bag)		640	.013		9.90	.39		10.29	11.40	13
0220	Sanitary napkin/tampon dispenser (25/50)		384	.021		19.70	.64		20.34	22.50	25.50
0230	Paper towel dispenser, folded (2 pkgs of 250)		545	.015		6.55	.45		7	7.80	8.90
0240	Paper towel dispenser, rolled (350' roll)		340	.024		5.10	.72		5.82	6.55	7.55
0250	Toilet tissue dispenser (500 2-ply sheet roll)		1440	.006		1.36	.17		1.53	1.72	1.98
0260	Sanitary seat covers (pkg of 250)		960	.008		5.55	.26		5.81	6.45	7.35
0270	Toilet partition cleaning		96	.083		.13	2.57		2.70	3.48	4.30
0280	Ceramic tile wall cleaning		2400	.003	S.F.	.03	.10		.13	.16	.21
0290	General cleaning, empty to larger portable container, wastebaskets		960	.008	Ea.		.26		.26	.33	.41
0300	Ash trays		2880	.003			.09		.09	.11	.14
0310	Cigarette stands		960	.008			.26		.26	.33	.41
0320	Sand urns		640	.013			.39		.39	.50	.62
0330	Pencil sharpeners		2880	.003			.09		.09	.11	.14
0340	Remove trash to central location		290	.028	Lb.		.85		.85	1.11	1.37
0347	Bench		480	.017	Ea.	.02	.51		.53	.69	.86
0350	Chairs		960	.008		.01	.26		.27	.34	.42
0360	Chalkboard erasers		960	.008			.26		.26	.33	.41
0370	Chalkboards		16	.500	M.S.F.	1.70	15.40		17.10	22.00	27
0380	Desk tops		225	.036	Ea.	.03	1.10		1.13	1.46	1.81
0390	Glass doors		168	.048	"	.16	1.47		1.63	2.09	2.56
0400	Hard surface audience type seating		40	.200	M.S.F.	2.13	6.15		8.28	10.35	12.60
0409	Locker, metal		320	.025	Ea.	.05	.77		.82	1.06	1.30
0410	Microwave ovens		240	.033		.01	1.03		1.04	1.35	1.66
0420	Office partitions, to 48 S.F.		160	.050		.08	1.54		1.62	2.10	2.58

For customer support on your Facilities Maintenance & Repair Costs with RSMeans data, call 800.448.8182.

561

01 93 06.40 General Cleaning

		Crew	Daily Output	Labor-Hours	Unit	Material	2019 Bare Costs Labor	2019 Bare Costs Equipment	Total	Total In-House	Total Incl O&P
0430	Overhead air vents	1 Clam	175	.046	Ea.	.01	1.41		1.42	1.84	2.28
0440	Refrigerators		35	.229		.14	7.05		7.19	9.30	11.55
0450	Revolving doors		35	.229		.95	7.05		8	10.20	12.55
0460	Tables		240	.033		.03	1.03		1.06	1.37	1.69
0470	Telephone		288	.028		.01	.86		.87	1.12	1.39
0480	Telephone booth		190	.042		.49	1.30		1.79	2.23	2.70
0490	Water fountain		480	.017		.01	.51		.52	.68	.84
0500	Wastebaskets		260	.031		.01	.95		.96	1.24	1.54
0510	Dusting, beds		360	.022			.68		.68	.89	1.10
0520	Bedside stand		960	.008			.26		.26	.33	.41
0530	Bookcase		960	.008			.26		.26	.33	.41
0540	Calculator		1920	.004			.13		.13	.17	.21
0550	Chair		1340	.006			.18		.18	.24	.30
0560	Coat rack		400	.020			.62		.62	.80	.99
0570	Couch		320	.025			.77		.77	1.00	1.24
0580	Desk		550	.015			.45		.45	.58	.72
0590	Desk lamp		1200	.007			.21		.21	.27	.33
0600	Desk trays		1920	.004			.13		.13	.17	.21
0610	Dictaphone		1920	.004			.13		.13	.17	.21
0620	Dresser		480	.017			.51		.51	.67	.83
0630	File cabinets		960	.008			.26		.26	.33	.41
0640	Lockers		2400	.003			.10		.10	.13	.17
0650	Piano		425	.019			.58		.58	.75	.93
0660	Radio		960	.008			.26		.26	.33	.41
0665	Shelf, storage, 1' x 20'		480	.017			.51		.51	.67	.83
0670	Table		670	.012			.37		.37	.48	.59
0680	Table lamp		960	.008			.26		.26	.33	.41
0690	Telephones		960	.008			.26		.26	.33	.41
0700	Television		960	.008			.26		.26	.33	.41
0710	Typewriter		960	.008			.26		.26	.33	.41
0720	Vacuuming, armless chair		825	.010			.30		.30	.39	.48
0730	Armchair		480	.017			.51		.51	.67	.83
0740	Couch		320	.025			.77		.77	1.00	1.24
0750	Entry mats		8	1	M.S.F.		31		31	40.00	49.50
0760	Overhead air vents		480	.017	Ea.		.51		.51	.67	.83
0770	Upholstered auditorium seating		5.27	1.518	M.S.F.		47		47	61.00	75.50

01 93 06.45 Hard Floor Care

		Crew	Daily Output	Labor-Hours	Unit	Material	2019 Bare Costs Labor	2019 Bare Costs Equipment	Total	Total In-House	Total Incl O&P
0010	**HARD FLOOR CARE**										
0020	Damp mop w/bucket & wringer, 12 oz. mop head, unobstructed	1 Clam	25	.320	M.S.F.	5.40	9.85		15.25	18.75	22.50
0030	Obstructed		17	.471		5.40	14.50		19.90	25.00	30.50
0040	20 oz. mop head, unobstructed		29	.276		5.40	8.50		13.90	17.00	20.50
0050	Obstructed		20	.400		5.40	12.30		17.70	22.00	26.50
0060	24 oz. mop head, unobstructed		35	.229		5.40	7.05		12.45	15.10	18.10
0070	Obstructed		23	.348		5.40	10.70		16.10	19.90	24
0080	32 oz. mop head, unobstructed		43	.186		5.40	5.75		11.15	13.40	16
0090	Obstructed		29	.276		5.40	8.50		13.90	17.00	20.50
0100	Dust mop, 12" mop, unobstructed		33	.242			7.45		7.45	9.70	12.05
0110	Obstructed		19	.421			12.95		12.95	16.90	21
0120	18" mop, unobstructed		49	.163			5.05		5.05	6.55	8.10
0130	Obstructed		28	.286			8.80		8.80	11.45	14.20
0140	24" mop, unobstructed		61	.131			4.04		4.04	5.25	6.50
0150	Obstructed		34	.235			7.25		7.25	9.45	11.70

01 93 06.45 Hard Floor Care	Crew	Daily Output	Labor-Hours	Unit	Material	2019 Bare Costs Labor	Equipment	Total	Total In-House	Total Incl O&P
0160	30" mop, unobstructed	1 Clam	74	.108	M.S.F.		3.33		3.33	4.34	5.35
0170	Obstructed		42	.190			5.85		5.85	7.65	9.45
0180	36" mop, unobstructed		92	.087			2.68		2.68	3.49	4.32
0190	Obstructed		52	.154			4.74		4.74	6.15	7.65
0200	42" mop, unobstructed		124	.065			1.99		1.99	2.59	3.20
0210	Obstructed		70	.114			3.52		3.52	4.58	5.65
0220	48" mop, unobstructed		185	.043			1.33		1.33	1.73	2.15
0230	Obstructed		104	.077			2.37		2.37	3.08	3.82
0240	60" mop, unobstructed		246	.033			1		1	1.30	1.61
0250	Obstructed		138	.058			1.79		1.79	2.32	2.88
0260	72" mop, unobstructed		369	.022			.67		.67	.87	1.08
0270	Obstructed		208	.038			1.18		1.18	1.54	1.91
0280	Sweeping, 8" broom, light soil, unobstructed		19	.421			12.95		12.95	16.90	21
0290	Obstructed		14	.571			17.60		17.60	23.00	28.50
0300	Heavy soil, unobstructed		11	.727			22.50		22.50	29.00	36
0310	Obstructed		9	.889			27.50		27.50	35.50	44
0320	12" push broom, light soil, unobstructed		32	.250			7.70		7.70	10.05	12.40
0330	Obstructed		24	.333			10.25		10.25	13.35	16.55
0340	Heavy soil, unobstructed		19	.421			12.95		12.95	16.90	21
0350	Obstructed		14	.571			17.60		17.60	23.00	28.50
0360	16" push broom, light soil, unobstructed		40	.200			6.15		6.15	8.00	9.95
0370	Obstructed		30	.267			8.20		8.20	10.70	13.25
0380	Heavy soil, unobstructed		24	.333			10.25		10.25	13.35	16.55
0390	Obstructed		18	.444			13.70		13.70	17.80	22
0400	18" push broom, light soil, unobstructed		44	.182			5.60		5.60	7.30	9.05
0410	Obstructed		33	.242			7.45		7.45	9.70	12.05
0420	Heavy soil, unobstructed		36	.222			6.85		6.85	8.90	11.05
0430	Obstructed		20	.400			12.30		12.30	16.05	19.85
0440	24" push broom, light soil, unobstructed		57	.140			4.32		4.32	5.65	6.95
0450	Obstructed		43	.186			5.75		5.75	7.45	9.25
0460	Heavy soil, unobstructed		34	.235			7.25		7.25	9.45	11.70
0470	Obstructed		26	.308			9.50		9.50	12.35	15.30
0480	30" push broom, light soil, unobstructed		81	.099			3.04		3.04	3.96	4.90
0490	Obstructed		60	.133			4.11		4.11	5.35	6.60
0500	Heavy soil, unobstructed		48	.167			5.15		5.15	6.70	8.30
0510	Obstructed		36	.222			6.85		6.85	8.90	11.05
0520	36" push broom, light soil, unobstructed		100	.080			2.46		2.46	3.21	3.97
0530	Obstructed		75	.107			3.29		3.29	4.28	5.30
0540	Heavy soil, unobstructed		60	.133			4.11		4.11	5.35	6.60
0550	Obstructed		45	.178			5.50		5.50	7.15	8.85
0560	42" push broom, light soil, unobstructed		133	.060			1.85		1.85	2.41	2.99
0570	Obstructed		100	.080			2.46		2.46	3.21	3.97
0580	Heavy soil, unobstructed		80	.100			3.08		3.08	4.01	4.97
0590	Obstructed		60	.133			4.11		4.11	5.35	6.60
0600	48" push broom, light soil, unobstructed		200	.040			1.23		1.23	1.60	1.99
0610	Obstructed		150	.053			1.64		1.64	2.14	2.65
0620	Heavy soil, unobstructed		120	.067			2.05		2.05	2.67	3.31
0630	Obstructed		90	.089			2.74		2.74	3.56	4.41
0640	Wet mop/rinse w/bucket & wringer, 12 oz. mop, unobstructed		11	.727		1.70	22.50		24.20	31.00	38
0650	Obstructed		8	1		1.70	31		32.70	42.00	51.50
0660	20 oz. mop head, unobstructed		14	.571		1.70	17.60		19.30	25.00	30.50
0670	Obstructed		10	.800		1.70	24.50		26.20	34.00	41.50
0680	24 oz. mop head, unobstructed		20	.400		1.70	12.30		14	17.90	22

01 93 06.45 Hard Floor Care	Crew	Daily Output	Labor-Hours	Unit	Material	2019 Bare Costs Labor	Equipment	Total	Total In-House	Total Incl O&P
0690	Obstructed	1 Clam	15	.533	M.S.F.	1.70	16.45		18.15	23.50	28.50
0700	32 oz. mop head, unobstructed		27	.296		1.70	9.15		10.85	13.75	16.85
0710	Obstructed		20	.400		1.70	12.30		14	17.90	22
0720	Dry buffing or polishing, 175 RPM, 12" diameter machine		14.21	.563			17.35		17.35	22.50	28
0730	14" diameter machine		16.51	.485			14.90		14.90	19.45	24
0740	17" diameter machine		19.20	.417			12.05		12.85	16.70	20.50
0750	20" diameter machine		22.85	.350			10.80		10.80	14.05	17.40
0760	24" diameter machine		28.99	.276			8.50		8.50	11.05	13.70
0770	350 RPM, 17" diameter machine		30.76	.260			8		8	10.45	12.90
0780	20" diameter machine		40.60	.197			6.05		6.05	7.90	9.80
0790	24" diameter machine		47.06	.170			5.25		5.25	6.80	8.45
0800	1000 RPM, 17" diameter machine		62.09	.129			3.97		3.97	5.15	6.40
0810	20" diameter machine		67.68	.118			3.64		3.64	4.74	5.85
0820	24" diameter machine		88.79	.090			2.78		2.78	3.61	4.47
0830	27" diameter machine		99.34	.081			2.48		2.48	3.23	4
0840	Spray buffing, 175 RPM, 12" diameter		10.66	.750		.98	23		23.98	31.00	38.50
0850	14" diameter machine		11.95	.669		.98	20.50		21.48	28.00	34
0860	17" diameter machine		13.80	.580		.98	17.85		18.83	24.50	30
0870	20" diameter machine		16	.500		.98	15.40		16.38	21.00	26
0880	24" diameter machine		19.05	.420		.98	12.95		13.93	17.90	22
0890	350 RPM, 17" diameter machine		17.39	.460		.98	14.15		15.13	19.55	24
0900	20" diameter machine		22.13	.362		.98	11.15		12.13	15.60	19.20
0910	24" diameter machine		29.22	.274		.98	8.45		9.43	12.05	14.85
0920	1000 RPM, 17" diameter machine		41.02	.195		.98	6		6.98	8.90	10.95
0930	20" diameter machine		44.14	.181		.98	5.60		6.58	8.35	10.25
0940	24" diameter machine		52.21	.153		.98	4.72		5.70	7.20	8.85
0950	27" diameter machine		57.43	.139		.98	4.29		5.27	6.65	8.15
0960	Burnish, 1000 RPM, 17" diameter machine		60	.133			4.11		4.11	5.35	6.60
0970	2000 RPM, 17" diameter machine		72	.111			3.42		3.42	4.46	5.50
0980	Scrubbing, 12" brush by hand		2	4		5.40	123		128.40	166.00	206
0990	175 RPM, 12" diameter machine		10.05	.796		5.40	24.50		29.90	38.00	46.50
1000	14" diameter machine		11.62	.688		5.40	21		26.40	33.50	41
1010	17" diameter machine		14	.571		5.40	17.60		23	29.00	35.50
1020	20" diameter machine		17.61	.454		5.40	14		19.40	24.00	29.50
1030	24" diameter machine		23.58	.339		5.40	10.45		15.85	19.55	23.50
1040	350 RPM, 17" diameter machine		24.12	.332		5.40	10.20		15.60	19.25	23
1050	20" diameter machine		28.91	.277		5.40	8.50		13.90	17.05	20.50
1060	24" diameter machine		36.92	.217		5.40	6.65		12.05	14.65	17.50
1070	Automatic, 21" machine		60.75	.132		5.40	4.06		9.46	11.20	13.30
1080	24" machine		83.47	.096		5.40	2.95		8.35	9.80	11.50
1090	32" machine		128.68	.062		5.40	1.91		7.31	8.45	9.85
1100	Stripping, separate wet pick-up required, 175 RPM, 17" diameter		6.86	1.166		3.74	36		39.74	51.00	62.50
1110	20" diameter machine		7.68	1.042		3.74	32		35.74	46.00	56
1120	350 RPM, 17" diameter machine		8.80	.909		3.74	28		31.74	40.50	49.50
1130	20" diameter machine		10.50	.762		3.74	23.50		27.24	34.50	42.50
1140	Wet pick-up, 12" opening attachment		18	.444			13.70		13.70	17.80	22
1150	20" opening attachment		24.12	.332			10.20		10.20	13.30	16.45
1160	Finish application, mop		14	.571		15.15	17.60		32.75	39.50	47.50
1170	Gravity fed applicator		21	.381		15.15	11.75		26.90	32.00	38
1180	Lambswool applicator		17	.471		15.15	14.50		29.65	35.50	42.50
1190	Sealer application, mop		14	.571		23.50	17.60		41.10	49.00	58
1200	Gravity fed applicator		21	.381		23.50	11.75		35.25	41.00	48.50
1210	Lambswool applicator		17	.471		23.50	14.50		38	44.50	53

564

01 93 06.80 Vacuuming	Crew	Daily Output	Labor-Hours	Unit	Material	2019 Bare Costs Labor	Equipment	Total	Total In-House	Total Incl O&P
0010	**VACUUMING**										
0020	Vacuum carpet, 12" upright, spot areas	1 Clam	73.68	.109	M.S.F.		3.34		3.34	4.35	5.40
0030	Traffic areas		44.21	.181			5.55		5.55	7.25	9
0040	All areas		21.05	.380			11.70		11.70	15.25	18.85
0050	14" upright, spot areas		79.98	.100			3.08		3.08	4.01	4.97
0060	Traffic areas		47.99	.167			5.15		5.15	6.70	8.30
0070	All areas		22.85	.350			10.80		10.80	14.05	17.40
0080	16" upright, spot areas		87.50	.091			2.82		2.82	3.67	4.54
0090	Traffic areas		52.50	.152			4.69		4.69	6.10	7.55
0100	All areas		25	.320			9.85		9.85	12.85	15.90
0110	18" upright, spot areas		96.53	.083			2.55		2.55	3.32	4.12
0120	Traffic areas		57.92	.138			4.25		4.25	5.55	6.85
0130	All areas		27.58	.290			8.95		8.95	11.65	14.40
0140	20" upright, spot areas		107.66	.074			2.29		2.29	2.98	3.69
0150	Traffic areas		64.60	.124			3.81		3.81	4.97	6.15
0160	All areas		30.76	.260			8		8	10.45	12.90
0170	22" upright, spot areas		121.73	.066			2.02		2.02	2.64	3.26
0180	Traffic areas		73.04	.110			3.37		3.37	4.39	5.45
0190	All areas		34.78	.230			7.10		7.10	9.20	11.40
0200	24" upright, spot areas		140	.057			1.76		1.76	2.29	2.84
0210	Traffic areas		84	.095			2.93		2.93	3.82	4.73
0220	All areas		40	.200			6.15		6.15	8.00	9.95
0230	14" twin motor upright, spot areas		98.81	.081			2.49		2.49	3.25	4.02
0240	Traffic areas		59.28	.135			4.16		4.16	5.40	6.70
0250	All areas		28.23	.283			8.75		8.75	11.35	14.05
0260	18" twin motor upright, spot areas		112	.071			2.20		2.20	2.86	3.55
0270	Traffic areas		67.20	.119			3.67		3.67	4.77	5.90
0280	All areas		32	.250			7.70		7.70	10.05	12.40
0290	Backpack, 12" opening carpet attachment, spot areas		77.77	.103			3.17		3.17	4.13	5.10
0300	Traffic areas		46.66	.171			5.30		5.30	6.90	8.50
0310	All areas		22.22	.360			11.10		11.10	14.45	17.90
0320	14" opening carpet attachment, spot areas		84.84	.094			2.90		2.90	3.78	4.68
0330	Traffic areas		50.90	.157			4.84		4.84	6.30	7.80
0340	All areas		24.24	.330			10.15		10.15	13.25	16.40
0350	16" opening carpet attachment, spot areas		93.35	.086			2.64		2.64	3.44	4.26
0360	Traffic areas		56.01	.143			4.40		4.40	5.75	7.10
0370	All areas		26.67	.300			9.25		9.25	12.05	14.90
0380	18" opening carpet attachment, spot areas		103.67	.077			2.38		2.38	3.09	3.83
0390	Traffic areas		62.20	.129			3.96		3.96	5.15	6.40
0400	All areas		29.62	.270			8.30		8.30	10.85	13.40
0410	20" opening carpet attachment, spot areas		116.66	.069			2.11		2.11	2.75	3.41
0420	Traffic areas		69.99	.114			3.52		3.52	4.58	5.70
0430	All areas		33.33	.240			7.40		7.40	9.60	11.90
0440	22" opening carpet attachment, spot areas		133.32	.060			1.85		1.85	2.41	2.98
0450	Traffic areas		79.99	.100			3.08		3.08	4.01	4.97
0460	All areas		38.09	.210			6.45		6.45	8.40	10.45
0470	24" opening carpet attachment, spot areas		155.54	.051			1.58		1.58	2.06	2.55
0480	Traffic areas		93.32	.086			2.64		2.64	3.44	4.26
0490	All areas		44.44	.180			5.55		5.55	7.20	8.95
0500	Tank/canister, 12" opening carpet attachment, spot areas		70	.114			3.52		3.52	4.58	5.65
0510	Traffic areas		42	.190			5.85		5.85	7.65	9.45
0520	All areas		20	.400			12.30		12.30	16.05	19.85

01 93 06 – Interior General Maintenance

01 93 06.80 Vacuuming

	Crew	Daily Output	Labor-Hours	Unit	Material	2019 Bare Costs Labor	2019 Bare Costs Equipment	Total	Total In-House	Total Incl O&P
0530	14" opening carpet attachment, spot areas	1 Clam	75.67	.106	M.S.F.		3.26		3.26	4.24	5.25
0540	Traffic areas		45.40	.176			5.45		5.45	7.05	8.75
0550	All areas		21.62	.370			11.40		11.40	14.85	18.35
0560	16" opening carpet attachment, spot areas		82.32	.097			2.99		2.99	3.90	4.83
0570	Traffic areas		49.39	.162			4.99		4.99	6.50	8.05
0580	All areas		23.52	.340			10.50		10.50	13.65	16.90
0590	18" opening carpet attachment, spot areas		90.30	.089			2.73		2.73	3.55	4.40
0600	Traffic areas		54.18	.148			4.55		4.55	5.90	7.35
0610	All areas		25.80	.310			9.55		9.55	12.45	15.40
0620	20" opening carpet attachment, spot areas		100	.080			2.46		2.46	3.21	3.97
0630	Traffic areas		60	.133			4.11		4.11	5.35	6.60
0640	All areas		28.57	.280			8.60		8.60	11.25	13.90
0650	22" opening carpet attachment, spot areas		112	.071			2.20		2.20	2.86	3.55
0660	Traffic areas		67.20	.119			3.67		3.67	4.77	5.90
0670	All areas		32	.250			7.70		7.70	10.05	12.40
0680	24" opening carpet attachment, spot areas		127.26	.063			1.94		1.94	2.52	3.12
0690	Traffic areas		76.36	.105			3.23		3.23	4.20	5.20
0700	All areas		36.36	.220			6.80		6.80	8.80	10.90
0710	Large area machine, no obstructions, 26" opening		44.44	.180			5.55		5.55	7.20	8.95
0720	28" opening		64	.125			3.85		3.85	5.00	6.20
0730	30" opening		80	.100			3.08		3.08	4.01	4.97
0740	32" opening		120	.067			2.05		2.05	2.67	3.31

01 93 06.90 Wall Care

	Crew	Daily Output	Labor-Hours	Unit	Material	2019 Bare Costs Labor	2019 Bare Costs Equipment	Total	Total In-House	Total Incl O&P
0010	**WALL CARE**										
0020	Dusting walls, hand duster	1 Clam	87	.092	M.S.F.		2.83		2.83	3.69	4.57
0030	Treated cloth		40	.200			6.15		6.15	8.00	9.95
0040	Hand held duster vacuum		53	.151			4.65		4.65	6.05	7.50
0050	Canister vacuum		32	.250			7.70		7.70	10.05	12.40
0060	Backpack vacuum		45	.178			5.50		5.50	7.15	8.85
0070	Washing walls, hand, painted surfaces		2.50	3.200		1.70	98.50		100.20	130.00	161
0080	Vinyl surfaces		3.25	2.462		1.70	76		77.70	101.00	124
0090	Machine, painted surfaces		4.80	1.667		1.70	51.50		53.20	68.50	85
0100	Washing walls from ladder, hand, painted surfaces		1.65	4.848		1.70	149		150.70	196.00	243
0110	Vinyl surfaces		2.15	3.721		1.70	115		116.70	151.00	187
0120	Machine, painted surfaces		3.17	2.524		1.70	77.50		79.20	103.00	127

01 93 06.95 Window and Accessory Care

	Crew	Daily Output	Labor-Hours	Unit	Material	2019 Bare Costs Labor	2019 Bare Costs Equipment	Total	Total In-House	Total Incl O&P
0010	**WINDOW AND ACCESSORY CARE**										
0025	Wash windows, trigger sprayer & wipe cloth, 12 S.F.	1 Clam	4.50	1.778	M.S.F.	3.81	55		58.81	75.50	93.50
0030	Over 12 S.F.		12.75	.627		3.81	19.35		23.16	29.50	36
0040	With squeegee & bucket, 12 S.F. and under		5.50	1.455		3.81	45		48.81	62.50	77
0050	Over 12 S.F.		15.50	.516		3.81	15.90		19.71	25.00	30.50
0060	Dust window sill		8640	.001	L.F.		.03		.03	.04	.05
0070	Damp wipe, venetian blinds		160	.050	Ea.	.03	1.54		1.57	2.04	2.52
0080	Vacuum, venetian blinds, horizontal		87.27	.092			2.82		2.82	3.68	4.55
0090	Draperies, in place, to 64 S.F.		160	.050			1.54		1.54	2.01	2.48

01 93 08 – Facility Maintenance Equipment

01 93 08.50 Equipment

	Crew	Daily Output	Labor-Hours	Unit	Material	2019 Bare Costs Labor	2019 Bare Costs Equipment	Total	Total In-House	Total Incl O&P
0010	**EQUIPMENT**, Purchase										
0090	Carpet care equipment										
0110	Dual motor vac, 1 HP, 16" brush				Ea.	700			700	770.00	875
0120	Upright vacuum, 12" brush					232			232	255.00	290
0130	14" brush					435			435	480.00	545

01 93 08 – Facility Maintenance Equipment

01 93 08.50 Equipment	Crew	Daily Output	Labor-Hours	Unit	Material	2019 Bare Costs Labor	Equipment	Total	Total In-House	Total Incl O&P
0140	Soil extractor, hot water 5' wand, 12" head				Ea.	3,075			3,075	3,375.00	3,850
0150	Dry foam, 13" brush					2,150			2,150	2,375.00	2,700
0160	24" brush				↓	4,200			4,200	4,625.00	5,250
0240	Floor care equipment										
0260	Polishing, buffing, waxing machine, 175 RPM										
0270	.33 HP, 20" diam. brush				Ea.	820			820	900.00	1,025
0280	1 HP, 16" diam. brush					1,150			1,150	1,275.00	1,450
0290	17" diam. brush					1,175			1,175	1,300.00	1,475
0300	1.5 HP, 20" diam. brush					1,300			1,300	1,425.00	1,625
0310	20" diam. brush, 1500 RPM				↓	1,550			1,550	1,700.00	1,950
0330	Scrubber, automatic, 2 stage 1 HP vacuum motor										
0340	20" diam. brush				Ea.	12,400			12,400	13,600.00	15,500
0350	28" diam. brush				"	17,200			17,200	18,900.00	21,500
1200	Plumbing maintenance equipment										
1220	Kinetic water ram				Ea.	226			226	249.00	283
1240	Cable pipe snake, 104 ft., self feed, electric, .5 HP				"	3,050			3,050	3,350.00	3,825
1500	Specialty equipment										
1550	Litterbuggy, 9 HP gasoline				Ea.	4,900			4,900	5,400.00	6,125
1560	Propane					4,125			4,125	4,550.00	5,150
1570	Pressure cleaner, hot water, 1000 psi					3,175			3,175	3,500.00	3,975
1580	3500 psi				↓	7,425			7,425	8,175.00	9,275
1590	Vacuum cleaners, steel canister										
1600	Dry only, two stage, 1 HP				Ea.	385			385	425.00	480
1610	Wet/dry, two stage, 2 HP					705			705	775.00	880
1620	4 HP					965			965	1,050.00	1,200
1630	Wet/dry, three stage, 1.5 HP					705			705	775.00	880
1670	Squeegee wet only, 2 stage, 1 HP					1,100			1,100	1,200.00	1,375
1680	2 HP					1,575			1,575	1,725.00	1,975
1690	Asbestos and hazardous waste dry vacs, 1 HP					945			945	1,050.00	1,175
1700	2 HP					665			665	730.00	830
1710	Three stage, 1 HP					550			550	605.00	690
1800	Upholstery soil extractor, dry-foam				↓	3,425			3,425	3,775.00	4,275
1900	Snow removal equipment										
1920	Thrower, 3 HP, 20", single stage, gas				Ea.	735			735	810.00	920
1930	Electric start					880			880	970.00	1,100
1940	5 HP, 24", 2 stage, gas					1,150			1,150	1,275.00	1,450
1950	11 HP, 36"				↓	2,375			2,375	2,625.00	2,975
2000	Tube cleaning equip., for heat exchangers, condensers, evap.										
2010	Tubes/pipes 1/4"-1", 115 volt, .5 HP drive				Ea.	1,675			1,675	1,850.00	2,100
2020	Air powered					1,700			1,700	1,875.00	2,125
2040	1" and up, 115 volt, 1 HP drive				↓	2,025			2,025	2,225.00	2,525

Facilities
Audits

Auditing the Facility

Introduction

Capital budget planning begins with compiling a complete inventory of the facility, including equipment, that identifies existing deficiencies. The Facilities Audit is a vehicle for producing such an inventory; it is a system for thoroughly assessing the existing physical condition and functional performance of buildings, grounds, utilities, and equipment. The results of this audit are used to address major and minor, urgent and long-term needs for corrective action, for short- and long-term financial planning. The basic principles presented here can be applied to any scale operation, from a single structure to a facility consisting of multiple building complexes in dispersed locations.

The first step is determining the scope of the audit. Next, the audit team is selected, and a set of forms is designed. A survey team then records field observations on the forms for individual building components. Finally, the information from these forms is summarized, prioritized, presented, and used to:

- compare a building's condition and functional performance to that of other buildings within a facility
- define regular maintenance requirements
- define capital repair and replacement projects in order to eliminate deferred maintenance
- develop cost estimates for capital repair and replacement projects
- restore functionally obsolete facilities to a usable condition
- eliminate conditions that are either potentially damaging to property or present life safety hazards
- identify energy conservation measures

The audit ultimately produces a database on the condition of a facility's buildings, identifying the physical adequacy of construction, material, and equipment, and assessing the functional adequacy of the facility to perform its designated purpose. This comprehensive inventory can be used to guide property management decisions and strategic planning of capital assets, such as determining those expenditures necessary to extend the life of a facility.

The audit procedures and forms included here are for the purpose of identifying the condition of buildings and assessing the cost and priority of their repair. While this is a comprehensive stand-alone process, it is for the limited purposes of assessing costs and establishing priorities for repairs. It represents only a portion of the requirements of a full facilities audit. For further information and guidance on conducting a full audit, contact the RSMeans Engineering staff.

The Rating Forms

The audit approach involves conducting a thorough inspection of a facility, then employing a systematic process of completing individual forms for each component, following the order in which building components are assembled. This is a self-evaluation process designed to fit all types of structures. The audit procedure described is a "generic" approach that can be used as a starting point; the facilities manager should incorporate the special characteristics of an organization and its facilities into an individualized facilities audit.

Audits are best performed by the in-house personnel familiar with the building being audited. However, outside consultants may be required when staff is not available or when special expertise is required to further examine suspected deficiencies.

Four types of forms that can be used to conduct the building audit are described here. A complete set of the blank forms ready for photocopying follows.

1. *Building Data Summary*
   This form can be used to summarize all buildings to be covered by the audit.

2. *Facilities Inventory Form*
   This form is designed to collect the basic data about a specific building's characteristics (name, identification number, location, area), including land data. One form is prepared for each building to be audited as summarized on the *Building Data Summary* form.

3. *Standard Inspection Form*
   The standard form for recording observed deficiencies, it provides a uniform way to comment on, prioritize, and estimate the cost to correct those deficiencies. The *Standard Inspection Form* is designed to make use of cost data from the Maintenance & Repair and Preventive Maintenance sections of this publication. Cost data from publications such as *RSMeans Facilities Construction Cost Data* may be utilized as well.

4. *System Checklists*
   For each major building system from foundations through site work, there is a checklist to guide the audit and record the review process. The checklists are intended as general guides and may not be totally comprehensive. Be sure to make any necessary modifications (additions or deletions) required for the specific building being audited. The forms prompt the user on what to check and provide a place to check off or indicate by "Yes" or "No" the presence of deficiencies requiring further detailed analysis on the *Standard Inspection Form*.

Completing the Inspection Forms

The standard inspection forms are designed to be used in conjunction with the system checklists. A system checklist of typical deficiencies for each building system is completed as part of the inspection process. Then for each building deficiency, a standard inspection report is generated listing the identifying information as required on the form. With the deficiency defined and described, the specific work items required to correct the deficiency can be entered, prioritized, and costed in Part 3 of the form. The priority rating is further described below.

Priority Ratings

Priority rating is the inspector's judgment of the priority of an observed deficiency. The rating provides guidance for facilities managers who must review overall deficiencies and develop maintenance schedules and capital budgets. The following priority rating terms, commonly used in maintenance management, are a suggested guideline for priority values. The scale may be expanded as required to fit specific needs.

1. *Emergency*: This designation indicates work that demands immediate attention to repair an essential operating system or building component, or because life safety or property is endangered. A response should occur within two hours of notification.

2. *Urgent*: Work demanding prompt attention to alleviate temporary emergency repairs or a condition requiring correction within one to five days to prevent a subsequent emergency is classified as urgent.

3. *Routine*: A specific completion date can be requested or required for routine work. This includes work that has no short range effect on life safety, security of personnel, or property. Routine work can be planned in detail and incorporated into a trade backlog for scheduling within twelve months.

4. *Deferred*: Projects that can be deferred into the following year's work planning are classified as deferred.

Cost Estimates for Maintenance

A major characteristic of the audit described here is the application of detailed cost estimates. To prepare cost estimates for work required to correct deficiencies, detailed information about the nature and extent of the facility's condition is required at the beginning of the audit. However, when cost estimates are incorporated, the audit process generates highly useful summaries outlining major maintenance costs and priorities. Including cost estimates allows for comparisons to be made in order to prioritize repair and replacement needs on a project-by-project and building-by-building basis.

Other Considerations

A problem which may have to be settled by the audit leader is what to do with mixed-use buildings, those that have recent additions to the original construction, or buildings that have, under emergency circumstances, been forced to house new activities for which they were not designed. Such problems will have to be dealt with individually by the audit team.

The data gathered in a facilities audit must be kept current, gathered consistently, and regularly updated. An annual cycle of inspections is recommended. Each organization initiating a comprehensive audit and continuing with annual cycles of inspections will benefit from organizing data in a systematic format. Conducting annual audits in a consistent format provides the basis for monitoring the backlog of deferred maintenance and preparing effective long-term plans and budgets for maintenance and repair operations.

The forms are designed for ease of field entry of data and manual preparation of summaries and reports. The forms can also be created using basic word processing and spreadsheet software or database programming. Although the self-evaluation process is readily adaptable to data processing programs, not every organization has the capabilities or resources available. For those without computer capabilities, a manual method of storing and updating must be developed and used.

Summarizing the Audit Data

Information collected on the audit forms is compiled and used to determine the condition of the facility's buildings and what should be done to maintain and/or improve them. Summaries of a building audit can serve several purposes, some of which are listed below.

- *Routine Maintenance Needs:* A current property owner may be interested in planning and estimating a program for the cost of routine operations and maintenance.

- *Major Maintenance:* A current or prospective property owner may want to evaluate major maintenance needs to plan and estimate the cost of a corrective program.
- *Deferred Maintenance:* A backlog of routine and major maintenance can be addressed using a survey of conditions to plan and budget a deferred maintenance program.
- *Renovations and/or Additions:* Prior to developing feasibility studies on alternatives to renovations and/or additions, a survey of existing conditions may be necessary.
- *Capital Budgeting and Planning:* A comprehensive audit of facilities conditions will incorporate major maintenance requirements into overall capital needs.

Conclusion

The information and forms for conducting an audit of the facility are included in *RSMeans Facilities Maintenance & Repair Cost Data* to provide a method for applying the cost data to the actual maintenance and repair needs of a specific facility. The Facility Audit presented here is a subset of a more comprehensive audit program. The more comprehensive approach includes assessment of the building's functional performance, as well as additional forms and procedures to incorporate the cost of maintenance and repair into a comprehensive multi-year facility management plan.

BUILDING DATA SUMMARY

Building Number	Building Name	Building Use*	Location / Address	Ownership** Own/Lea/O-L

* Principal and Special Uses
** Own = Owned
    Lea = Leased
    O-L = Owned/Leased

For customer support on your Facilities Maintenance & Repair Costs with RSMeans data, call 800.448.8182.

FACILITIES INVENTORY FORM

Building Data

1. Organization Name _____

2. Approved by: Name _____ Title _____ Date _____

3. Building Number _____

4. Property Name _____

5. Property Address _____

6. Property Main Use/Special Use _____

7. Gross Area (Sq. Ft.) _____ 8. Net Area (Sq. Ft.) _____

9. Type of Ownership: Owned _____ Leased _____ Owned / Leased _____

10. Book Value $ _____ Year Appraised _____

11. Replacement Value $ _____

12. Year Built _____ Year Acquired _____

13. Year Major Additions _____

Land Data

14. Type of Ownership: Owned _____ Leased _____ Owned / Leased _____

15. Book Value $ _____ Year Appraised _____

16. Year Acquired _____

17. Size in Acres _____

18. Market Value $ _____

Notes

_____
_____
_____

For customer support on your Facilities Maintenance & Repair Costs with RSMeans data, call 800.448.8182.

STANDARD INSPECTION FORM

1. Facility Inspection Data

Facility #: _____

Component #: _____

Inspector: _____

Name: _____

Name: _____

Date: _____

2. Component Description

_____

_____

_____

3. Component Repair Evaluation

Defi-ciency #	Component Description	Line Number	Pri-ority	Unit	Quant.	Unit Cost	Total In-House	Unit Cost	Total W/O&P

For customer support on your Facilities Maintenance & Repair Costs with RSMeans data, call 800.448.8182.

Deficiencies		Causes	
_____	Settlement, alignment changes or cracks	_____	Soils - changes in load bearing capacity due to shrinkage, erosion, or compaction. Adjacent construction undermining foundations. Reduced soil cover resulting in frost exposure.
		_____	Design loads - building equipment loads exceeding design loads. Vibration from heavy equipment requiring isolated foundations.
		_____	Structural or occupancy changes - inadequate bearing capacities. Foundation settling.
		_____	Earthquake resistance non-functioning.
_____	Moisture penetration	_____	Water table changes - inadequate drainage.
		_____	Ineffective drains or sump pump/sump pits.
		_____	Roof drainage - storm sewer connections inadequate or defective. Installation of roof drain restrictors, gutters, and downspouts where required.
		_____	Surface drainage - exterior grades should slope away from building and structures.
		_____	Utilities - broken or improperly functioning utility service lines or drains.
		_____	Leakage - wall cracks, opening of construction joints, inadequate or defective waterproofing.
		_____	Condensation - inadequate ventilation, vapor barrier, and/or dehumidification.
_____	Temperature changes	_____	Insulation - improperly selected for insulating value, fire ratings, and vermin resistance.
_____	Surface material deterioration	_____	Concrete, masonry, or stucco - spalling, corrosion of reinforcing, moisture penetration, or chemical reaction between cement and soil.
		_____	Steel or other ferrous metals - corrosion due to moisture or contact with acid-bearing soils.
		_____	Wood - decay due to moisture or insect infestation.
_____	Openings deterioration	_____	Non-functioning of doors, windows, hatchways, and stairways.
		_____	Utilities penetration due to damage, weather, wear, or other cause.

For customer support on your Facilities Maintenance & Repair Costs with RSMeans data, call 800.448.8182.

Deficiencies		Causes	
_____	Floors, concrete - cracking or arching	_____	Shrinkage, or settlement of subsoil.
		_____	Inadequate drainage.
		_____	Movement in exterior walls, or frost heave.
		_____	Improper compaction of base. Erosion of base.
		_____	Heaving from hydraulic pressure.
_____	Floors, wood - rotting or arching	_____	Excessive dampness or insect infestation.
		_____	Leak in building exterior.
		_____	Lack of ventilation.
_____	Wall deterioration	_____	Concrete or masonry (see *A10 Foundations*)
_____	Crawl space ventilation and maintenance	_____	Inadequate air circulation due to blockage of openings in foundation walls.
		_____	Moisture barrier ineffective.
		_____	Pest control, housekeeping, and proper drainage.

For customer support on your Facilities Maintenance & Repair Costs with RSMeans data, call 800.448.8182.

The primary materials encountered in the superstructure inspection are concrete, steel, and wood. Typical observations of deficiencies will be observed by: failures in the exterior closure systems of exterior walls, openings, and roofs; cracks; movement of materials; moisture penetration; and discoloration. The exterior visual survey will detect failures of surface materials or at openings that will require further inspection to determine whether the cause was the superstructure system.

Deficiencies	Causes

Concrete (columns, walls, and floor and roof slabs)

_____ Overall alignment

        _____ Settlement due to design and construction techniques.
        _____ Under designed for loading conditions (see *A10 Foundations*).

_____ Deflection

        _____ Expansion and/or contraction due to changes in design loads.
        _____ Original design deficient.
        _____ Original materials deficient.

_____ Surface Conditions

    _____ Cracks

        _____ Inadequate design and/or construction due to changes in design loads.
        _____ Stress concentration.
        _____ Extreme temperature changes; secondary effects of freeze-thaw.

    _____ Scaling, spalls, and pop-outs

        _____ Extreme temperature changes.
        _____ Reinforcement corrosion.
        _____ Environmental conditions.
        _____ Mechanical damage.
        _____ Poor materials.

    _____ Stains

        _____ Chemical reaction of reinforcing.
        _____ Reaction of materials in concrete mixture.
        _____ Environmental conditions.

    _____ Exposed reinforcing

        _____ Corrosion of steel.
        _____ Insufficient cover.
        _____ Mechanical damage.

Deficiencies	**Causes**

Steel (structural members, stairs, and connections)

_____ Overall alignment	_____ Settlement due to design and construction techniques; improper fabrication.
_____ Deflection or cracking	_____ Expansion and/or contraction.
	_____ Changes in design loads.
	_____ Fatigue due to vibration or impact.
_____ Corrosion	_____ Electrochemical reaction.
	_____ Failure of protective coating.
	_____ Excessive moisture exposure.
_____ Surface deterioration	_____ Excessive wear.

Wood (structural members and connections)

_____ Overall alignment	_____ Settlement; design and construction techniques.
_____ Deflection or cracking	_____ Expansion and/or contraction.
	_____ Changes in design loads.
	_____ Fatigue due to vibration or impact.
	_____ Failure of compression members.
	_____ Poor construction techniques.
	_____ General material failure.
_____ Rot (Decay)	_____ Direct contact with moisture.
	_____ Condensation.
	_____ Omission or deterioration of vapor barrier.
	_____ Poor construction techniques.
	_____ Damage from rodents or insects.

B20 EXTERIOR CLOSURE INSPECTION CHECKLIST

General Inspection

- Overall appearance
- Displacement
- Paint conditions
- Caulking
- Window & door fit
- Flashing condition

- Material integrity
- Cracks
- Evidence of moisture
- Construction joints
- Hardware condition

Exterior Walls

_____ Wood (Shingles, weatherboard siding, plywood).

   _____ Paint or surface treatment condition
   _____ Rot or decay

   _____ Moisture penetration
   _____ Loose, cracked, warped, or broken boards and shingles.

_____ Concrete, Masonry, and Tile (Concrete, brick, concrete masonry units, structural tile, glazed tile, stucco, stone).

   _____ Settlement
   _____ Construction and expansion joints
   _____ Surface deterioration
   _____ Parapet movement

   _____ Structural frame movement causing cracks
   _____ Condition of caulking and mortar
   _____ Efflorescence and staining
   _____ Tightness of fasteners

_____ Metal (Corrugated iron or steel, aluminum, enamel coated steel, protected metals).

   _____ Settlement
   _____ Condition of bracing
   _____ Tightness of fasteners
   _____ Flashings

   _____ Structural frame movement
   _____ Surface damage due to impact
   _____ Caulking
   _____ Corrosion

_____ Finishes (Mineral products, fiberglass, polyester resins, and plastics).

   _____ Settlement
   _____ Surface damage due to impact
   _____ Stains
   _____ Adhesion to substrate
   _____ Flashings
   _____ Asbestos products

   _____ Structural frame movement
   _____ Cracks
   _____ Fasteners
   _____ Caulking
   _____ Lead paint

_____ Insulation

   _____ Insulation present

   _____ Satisfactory condition

Windows and Doors

   _____ Frame fitting
   _____ Paint or surface finish
   _____ Hardware and operating parts
   _____ Cleanliness
   _____ Rot or corrosion

   _____ Frame and molding condition
   _____ Putty and weatherstripping
   _____ Security
   _____ Material condition (glass, wood, & metal Panels)
   _____ Screens and storm windows

Shading Devices

   _____ Material conditions
   _____ Cleanliness

   _____ Operations

For customer support on your Facilities Maintenance & Repair Costs with RSMeans data, call 800.448.8182.

B30 ROOFING INSPECTION CHECKLIST

General Appearance

Good_____   Fair_____   Poor_____

_____ **Water tightness**

_____	Evidence of leaks on undersurface	_____	Surface weathering
_____	Faulty material	_____	Faulty design
_____	Faulty application	_____	Standing water
_____	Weather damage	_____	Mechanical damage
_____	Fastening failure	_____	Flashing failure

_____ **Roofing Surface**

_____ Built Up (Felt or bitumen surfacing)

_____	Adhesion	_____	Moisture meter readings
_____	Bare areas	_____	Blisters, wrinkles
_____	Cracks, holes, tears	_____	Fish mouths
_____	Alligatoring	_____	Ballast

_____ Single Ply (Thermosetting, thermoplastic, composites)

_____	Adhesion	_____	Moisture meter readings
_____	Bare areas	_____	Blisters, wrinkles
_____	Cracks	_____	Holes, tears
_____	Seam conditions	_____	Protective coating
_____	Ballast		

_____ Metal Roofing (Preformed, formed)

_____	Corrosion (%)	_____	Protective coating
_____	Seams	_____	Cracks or breaks
_____	Holes	_____	Expansion joints

_____ Shingles & Tiles (Metal, clay, mission, concrete, or others)

_____	Disintegration	_____	Broken or cracked (%)
_____	Missing (%)	_____	Fasteners
_____	Underlayment		

_____ Wood shingles

_____	Cracked	_____	Curled
_____	Missing (%)		

For customer support on your Facilities Maintenance & Repair Costs with RSMeans data, call 800.448.8182.

B30 ROOFING INSPECTION CHECKLIST

_____ **Insulation**

    _____ Full coverage (%)   _____ Full thickness (%)
    _____ Satisfactory condition

_____ **Roof Penetrations**

    Weather tightness   _____ Operable

_____ **Flashing**

    _____ Deterioration   _____ Open joints
    _____ Holes or damage   _____ Anchoring
    _____ Protective coating

_____ **Drainage**

    _____ Alignment   _____ Free flowing
    _____ Clamping rings secure   _____ Screens
    _____ Corrosion

General Inspection

Overall Appearance

    Good _____   Fair _____   Poor _____

    _____ Evidence of moisture   _____ Settlement
    _____ Cracks   _____ Surface condition
    _____ Cleanliness   _____ Stains
    _____ Discoloration

_____ **Paint**

    _____ Repainting necessary   _____ Peeling, cracking, flaking
    _____ Reflectivity   _____ Maintainability
    _____ Lead paint

_____ **Coverings & Coatings**

    _____ Replacement necessary   _____ Peeling
    _____ Rips, tears   _____ Holes
    _____ Adhesion   _____ Seams

_____ **Interior glazing**

    _____ Cracks   _____ Seals
    _____ Frame condition   _____ Missing panes
    _____ Shading devices

B30 ROOFING INSPECTION CHECKLIST

For customer support on your Facilities Maintenance & Repair Costs with RSMeans data, call 800.448.8182.

General Inspection

- Strength & stability
- Acoustical quality
- Maintainability

- Physical condition
- Adaptability
- Code compliance

Deficiencies   **Causes**

Wall Material

	Deficiencies		Causes
_____	Cracks	_____	Settlement
_____	Holes	_____	Defective material
_____	Looseness	_____	Operational abuse or vibrations
_____	Missing segments	_____	Environmental attack
_____	Water stains	_____	Moisture
_____	Joints	_____	Structural expansion or contraction
_____	Surface appearance	_____	Wind pressure

Hardware

_____	Overall condition	_____	Appearance
_____	Keying system	_____	Operations
_____	Fit	_____	Locksets
_____	Cylinders	_____	Panic devices
_____	Maintainability	_____	Security operations

For customer support on your Facilities Maintenance & Repair Costs with RSMeans data, call 800.448.8182.

General Inspection

Code compliance _____

Maintainability _____

Means of egress _____

Fire ratings _____

Audible & visual device condition _____

Extinguishing systems (see also *08 Mechanical/Plumbing*): _____

     Type _____

Lighting system (see also *09 Electrical Lighting/Power*): _____

     Type _____

Handicapped accessibility _____

Building user comments _____

_____

Exterior Lighting

Adequacy:

    Good _____   Fair _____   Poor _____

Condition _____

Controls (type & location) _____

Fire Alarm Systems

_____	Panel visible	_____	Operational
_____	Pull station condition	_____	Detector conditions

Stairs and Ramps

_____	Exits marked	_____	Hardware operational
_____	Tripping hazards	_____	Surface conditions
_____	Lighting adequate	_____	Handrails

For customer support on your Facilities Maintenance & Repair Costs with RSMeans data, call 800.448.8182.

General Inspection

Overall Appearance

Good_____   Fair_____   Poor_____

- Evidence of moisture
- Irregular surface
- Handicapped hazards
- Visible settlement
- Tripping hazards
- Replacement necessary

Carpet (Tufted, tile)

_____ Age   _____ Excessive wear
_____ Stains   _____ Discoloration
_____ Holes, tears   _____ Seam conditions

Resilient (Asphalt tile, cork tile, linoleum, rubber, vinyl)

_____ Broken tiles   _____ Loose tiles
_____ Shrinkage   _____ Lifting, cupping
_____ Fading   _____ Cuts, holes
_____ Porosity   _____ Asbestos present

Masonry (Stone, brick)

_____ Cracks   _____ Deterioration
_____ Joints   _____ Stains
_____ Porosity   _____ Sealing

Monolithic Topping (Concrete, granolithic, terrazzo, magnesite)

_____ Cracks   _____ Porosity
_____ Joints   _____ Sealing

Wood (Plank, strips, block, parquet)

_____ Shrinkage   _____ Cupping, warpage
_____ Excessive wear   _____ Unevenness
_____ Decay   _____ Sealing

587

For customer support on your Facilities Maintenance & Repair Costs with RSMeans data, call 800.448.8182.

General Inspection

Overall appearance

Good ____   Fair ____   Poor ____

- Settlement or sagging
- Attachment
- Stains, discoloration
- Suitability
- Code compliance

- Alignment
- Evidence of moisture
- Missing units
- Acoustic quality

Exposed Systems (Unpainted, painted, spray-on, decorative)

_____ Cracks
_____ Surface deterioration
_____ Missing elements
_____ Adhesion

Applied to Structure & Suspended

General Condition:

Good ____   Fair ____   Poor ____

_____ Fasteners
_____ Trim condition

Openings:

_____ Panels
_____ Lighting fixtures
_____ Fire protection

_____ Inserts
_____ Air distribution
_____ Other

For customer support on your Facilities Maintenance & Repair Costs with RSMeans data, call 800.448.8182.

General Inspection (Passenger Conveying)

Overall appearance (interior)

        Good _____   Fair _____   Poor _____

Overall appearance (exterior)

        Good _____   Fair _____   Poor _____

_____ Maintenance history available

    _____ Regular inspection frequency

    _____ Door operations

    _____ Control systems

    _____ Noise

    _____ Code compliance

    _____ Handicapped access

    _____ Major repairs necessary

    _____ Replacement necessary

For customer support on your Facilities Maintenance & Repair Costs with RSMeans data, call 800.448.8182.

D20 MECHANICAL/PLUMBING INSPECTION CHECKLIST

General Inspection

    General appearance

        Good _____   Fair _____   Poor _____

    Leaks, dripping, running faucets and valves _____

    Maintenance history _____

    Supply adequacy _____

    Sanitation hazards _____

    Drain & waste connection _____

    Backflow protection _____

    Cross connections _____

    Fixture quantity _____

    Fixture types & conditions _____

    Handicapped fixtures _____

    Female facilities _____

    Metal pipe & fittings corrosion _____

    Pipe joints & sealing _____

    Pipe insulation _____

    Hanger supports & clamps _____

    Filters _____

    Building user comments _____

_____ **Water System**

_____	Water pressure adequate	_____	Odors, tastes
_____	Main cutoff operable	_____	Water heating temperature setting
_____	Pump condition	_____	Insulation condition

_____ **Sanitary & Storm System**

_____	Flow adequate	_____	Cleanouts access
_____	Floor drains	_____	Chemical resistance
_____	Gradient	_____	On-site disposal system

_____ **Code Requirements**

_____	EPA/local permits	_____	Other

_____ **Fire protection system**

_____	Regular inspections	_____	Sprinkler heads operable
_____	Complies with code	_____	Controls operable
_____	Hose cabinets functional	_____	Water pressure sufficient

For customer support on your Facilities Maintenance & Repair Costs with RSMeans data, call 800.448.8182.

General Inspection

General appearance

Good _____   Fair _____   Poor _____

Lubrication: bearings and moving parts _____

Rust and corrosion _____

Motors, fans, drive assemblies, and pumps _____

Wiring and electrical controls _____

Thermostats and automatic temperature controls _____

Thermal insulation and protective coatings _____

Guards, casings, hangers, supports, platforms, and mounting bolts _____

Piping system identification _____

Solenoid valves _____

Burner assemblies _____

Combustion chambers, smokepipes, and breeching _____

Electrical heating units _____

Guards, casings, hangers, supports, platforms, and mounting bolts _____

Steam and hot water heating equipment _____

Accessible steam, water, and fuel piping _____

Traps _____

Humidifier assemblies and controls _____

Strainers _____

Water sprays, weirs, and similar devices _____

Shell-and tube-type condensers _____

Self-contained evaporative condensers _____

Air cooled condensers _____

Compressors _____

Liquid receivers _____

Refrigerant driers, strainers, valves, oil traps, and accessories _____

Building user comments _____

For customer support on your Facilities Maintenance & Repair Costs with RSMeans data, call 800.448.8182.

Cleaning, maintenance, repair, and replacement

_____	Registers	_____	Grills
_____	Dampers	_____	Draft diverters
_____	Plenum chambers	_____	Supply and return ducts
_____	Louvers	_____	Fire dampers

Air filters

_____	Correct type	_____	Replacement schedule

Heating System Evaluation

_____	Heating capacity	_____	Temperature control
_____	Heating:		
_____	Seasonal	_____	All year
_____	Noise level	_____	Energy consumption
_____	Air circulation & ventilation	_____	Filtration
_____	Humidity control		

Cooling System Evaluation

_____	Cooling capacity	_____	Temperature and humidity control
_____	Cooling all season	_____	Noise level
_____	Energy consumption	_____	Air circulation & ventilation
_____	Filtration	_____	Reliability

Ventilation System Evaluation

_____	Air velocity	_____	Exhaust air systems
_____	Bag collection	_____	Wet collectors
_____	Steam and hot water coils	_____	Electrical heating units
_____	Fire hazards	_____	Fire protective devices

For customer support on your Facilities Maintenance & Repair Costs with RSMeans data, call 800.448.8182.

General Inspection

Safety conditions _____

Service capacity, % used, and age _____

Switchgear capacity, % used, and age _____

Feeder capacity, % used, and age _____

Panel capacity _____

Thermo-scanning _____

Maintenance records available _____

Convenience outlets _____

Building user comments _____

Exterior Service

_____ Line drawing
_____ Feed source:
    _____ Utility/owned   _____ Above/below ground
_____ Transformer:
    _____ Transformer tested   _____ Transformer arcing or burning
    _____ Transformer PCB's   _____ Ownership (facility or utility)

Interior Distribution System

_____ Line drawing   _____ Incoming conduit marked
_____ Main circuit breaker marked   _____ Panel boards, junction boxes covered
_____ All wiring in conduit   _____ Conduit properly secured
_____ Panels marked   _____ Panel schedules
_____ Missing breakers

Emergency Circuits

Emergency generator(s):

_____ Condition and age   _____ Auto start and switchover
_____ Testing schedule   _____ Test records available
_____ Service schedule   _____ Service schedule records available
_____ Circuits appropriate   _____ Cooling & exhaust
_____ Fuel storage (capacity)

Emergency lighting/power systems:

_____ Battery operation   _____ Separate power feed
_____ Exit signs   _____ Stairways/corridors
_____ Elevators   _____ Interior areas
_____ HVAC   _____ Exterior

593

For customer support on your Facilities Maintenance & Repair Costs with RSMeans data, call 800.448.8182.

General Inspection

 Overall appearance

  Good _____ Fair _____ Poor _____

 Maintainability _____

 Repairs/replacements _____

 Code compliance _____

Roads, Walks, and Parking Lots

 Surface conditions _____

 Subsurface conditions _____

 Settling and uplift _____

 Cracks, holes _____

 Drainage and slope _____

Curbing

 Alignment _____

 Erosion _____

 Repairs/replacements _____

Drainage and Erosion Controls

 Surface drainage _____

 Manholes, catch basins _____

 Vegetation _____

 Channels, dikes _____

 Retention, detention _____

 Drains _____

Parking Lot Controls

 Location _____

 Operation _____

 Repairs/replacements _____

For customer support on your Facilities Maintenance & Repair Costs with RSMeans data, call 800.448.8182.

H10 SAFETY INSPECTION CHECKLIST

General Inspection

Code compliance _____

Maintainability _____

Means of egress _____

Fire ratings _____

Audible & visual device condition _____

Extinguishing systems (see also *D20 Mechanical/Plumbing*): _____

      Type _____

Lighting system (see also *D50 Electrical Lighting/Power*): _____

      Type _____

Handicapped accessibility _____

Building user comments _____

Exterior Lighting

Adequacy:

    Good _____   Fair _____   Poor_____

Condition _____

Controls (type & location) _____

Fire Alarm Systems

_____	Panel visible	_____	Operational
_____	Pull station condition	_____	Detector conditions

Stairs and Ramps

_____	Exits marked	_____	Hardware operational
_____	Tripping hazards	_____	Surface conditions
_____	Lighting adequate	_____	Handrails

For customer support on your Facilities Maintenance & Repair Costs with RSMeans data, call 800.448.8182.

Reference Section

All the reference information is in one section, making it easy to find what you need to know ... and easy to use the data set on a daily basis. This section is visually identified by a vertical black bar on the page edges.

In this Reference Section we've included information on Equipment Rental Costs, Crew Listings, Travel Costs, Reference Tables, Life Cycle Costing, and a listing of Abbreviations.

Table of Contents

597

Estimating Tips

- This section contains the average costs to rent and operate hundreds of pieces of construction equipment. This is useful information when one is estimating the time and material requirements of any particular operation in order to establish a unit or total cost. Bare equipment costs shown on a unit cost line include, not only rental, but also operating costs for equipment under normal use.

Rental Costs

- Equipment rental rates are obtained from the following industry sources throughout North America: contractors, suppliers, dealers, manufacturers, and distributors.

- Rental rates vary throughout the country, with larger cities generally having lower rates. Lease plans for new equipment are available for periods in excess of six months, with a percentage of payments applying toward purchase.

- Monthly rental rates vary from 2% to 5% of the purchase price of the equipment depending on the anticipated life of the equipment and its wearing parts.

- Weekly rental rates are about 1/3 of the monthly rates, and daily rental rates are about 1/3 of the weekly rate.

- Rental rates can also be treated as reimbursement costs for contractor-owned equipment. Owned equipment costs include depreciation, loan payments, interest, taxes, insurance, storage, and major repairs.

Operating Costs

- The operating costs include parts and labor for routine servicing, such as the repair and replacement of pumps, filters, and worn lines. Normal operating expendables, such as fuel, lubricants, tires, and electricity (where applicable), are also included.

- Extraordinary operating expendables with highly variable wear patterns, such as diamond bits and blades, are excluded. These costs can be found as material costs in the Unit Price section.

- The hourly operating costs listed do not include the operator's wages.

Equipment Cost/Day

- Any power equipment required by a crew is shown in the Crew Listings with a daily cost.

- This daily cost of equipment needed by a crew includes both the rental cost and the operating cost and is based on dividing the weekly rental rate by 5 (the number of working days in the week), then adding the hourly operating cost multiplied by 8 (the number of hours in a day). This "Equipment Cost/Day" is shown in the far right column of the Equipment Rental section.

- If equipment is needed for only one or two days, it is best to develop your own cost by including components for daily rent and hourly operating costs. This is important when the listed Crew for a task does not contain the equipment needed, such as a crane for lifting mechanical heating/cooling equipment up onto a roof.

- If the quantity of work is less than the crew's Daily Output shown for a Unit Price line item that includes a bare unit equipment cost, the recommendation is to estimate one day's rental cost and operating cost for equipment shown in the Crew Listing for that line item.

- Please note, in some cases the equipment description in the crew is followed by a time period in parenthesis. For example: (daily) or (monthly). In these cases the equipment cost/day is calculated by adding the rental cost per time period to the hourly operating cost multiplied by 8.

Mobilization, Demobilization Costs

- The cost to move construction equipment from an equipment yard or rental company to the job site and back again is not included in equipment rental costs listed in the Reference Section. It is also not included in the bare equipment cost of any Unit Price line item or in any equipment costs shown in the Crew Listings.

- Mobilization (to the site) and demobilization (from the site) costs can be found in the Unit Price section.

- If a piece of equipment is already at the job site, it is not appropriate to utilize mobilization or demobilization costs again in an estimate. ■

Did you know?

RSMeans data is available through our online application:
- Search for costs by keyword
- Leverage the most up-to-date data
- Build and export estimates

Try it free
rsmeans.com/2019freetrial

No part of this cost data may be reproduced, stored in a retrieval system, or transmitted in any form or by any means without prior written permission of Gordian.

01 54 33 | Equipment Rental

		UNIT	HOURLY OPER. COST	RENT PER DAY	RENT PER WEEK	RENT PER MONTH	EQUIPMENT COST/DAY	
10	0010	**CONCRETE EQUIPMENT RENTAL** without operators R015433 -10							**10**
	0200	Bucket, concrete lightweight, 1/2 C.Y.	Ea.	.88	24.50	74	222	21.80	
	0300	1 C.Y.		.98	28.50	85.50	257	24.95	
	0400	1-1/2 C.Y.		1.24	38.50	115	345	32.90	
	0500	2 C.Y.		1.34	47	141	425	38.90	
	0580	8 C.Y.		6.49	265	795	2,375	210.90	
	0600	Cart, concrete, self-propelled, operator walking, 10 C.F.		2.86	58.50	176	530	58.10	
	0700	Operator riding, 18 C.F.		4.82	98.50	296	890	97.75	
	0800	Conveyer for concrete, portable, gas, 16" wide, 26' long		10.64	130	390	1,175	163.10	
	0900	46' long		11.02	155	465	1,400	181.15	
	1000	56' long		11.18	162	485	1,450	186.45	
	1100	Core drill, electric, 2-1/2 H.P., 1" to 8" bit diameter		1.57	56.50	170	510	46.55	
	1150	11 H.P., 8" to 18" cores		5.40	115	345	1,025	112.20	
	1200	Finisher, concrete floor, gas, riding trowel, 96" wide		9.66	148	445	1,325	166.30	
	1300	Gas, walk-behind, 3 blade, 36" trowel		2.04	24.50	73.50	221	31	
	1400	4 blade, 48" trowel		3.07	28	84.50	254	41.50	
	1500	Float, hand-operated (Bull float), 48" wide		.08	13.85	41.50	125	8.95	
	1570	Curb builder, 14 H.P., gas, single screw		14.04	285	855	2,575	283.30	
	1590	Double screw		15.04	340	1,025	3,075	325.35	
	1600	Floor grinder, concrete and terrazzo, electric, 22" path		3.04	190	570	1,700	138.30	
	1700	Edger, concrete, electric, 7" path		1.18	56.50	170	510	43.45	
	1750	Vacuum pick-up system for floor grinders, wet/dry		1.62	98.50	296	890	72.15	
	1800	Mixer, powered, mortar and concrete, gas, 6 C.F., 18 H.P.		7.42	127	380	1,150	135.35	
	1900	10 C.F., 25 H.P.		9.00	150	450	1,350	162	
	2000	16 C.F.		9.36	173	520	1,550	178.85	
	2100	Concrete, stationary, tilt drum, 2 C.Y.		7.23	243	730	2,200	203.85	
	2120	Pump, concrete, truck mounted, 4" line, 80' boom		29.88	1,000	3,025	9,075	844.05	
	2140	5" line, 110' boom		37.46	1,375	4,150	12,500	1,130	
	2160	Mud jack, 50 C.F. per hr.		6.45	123	370	1,100	125.60	
	2180	225 C.F. per hr.		8.55	145	435	1,300	155.40	
	2190	Shotcrete pump rig, 12 C.Y./hr.		13.97	218	655	1,975	242.75	
	2200	35 C.Y./hr.		15.80	242	725	2,175	271.40	
	2600	Saw, concrete, manual, gas, 18 H.P.		5.54	47.50	143	430	72.90	
	2650	Self-propelled, gas, 30 H.P.		7.90	78.50	235	705	110.15	
	2675	V-groove crack chaser, manual, gas, 6 H.P.		1.64	19	57	171	24.50	
	2700	Vibrators, concrete, electric, 60 cycle, 2 H.P.		.47	9.15	27.50	82.50	9.25	
	2800	3 H.P.		.56	11.65	35	105	11.50	
	2900	Gas engine, 5 H.P.		1.54	16.35	49	147	22.15	
	3000	8 H.P.		2.08	16.50	49.50	149	26.55	
	3050	Vibrating screed, gas engine, 8 H.P.		2.81	95	285	855	79.45	
	3120	Concrete transit mixer, 6 x 4, 250 H.P., 8 C.Y., rear discharge		50.72	615	1,850	5,550	775.80	
	3200	Front discharge		58.88	740	2,225	6,675	916.05	
	3300	6 x 6, 285 H.P., 12 C.Y., rear discharge		58.15	710	2,125	6,375	890.20	
	3400	Front discharge		60.59	750	2,250	6,750	934.70	
20	0010	**EARTHWORK EQUIPMENT RENTAL** without operators R015433 -10							**20**
	0040	Aggregate spreader, push type, 8' to 12' wide	Ea.	2.60	27.50	82.50	248	37.30	
	0045	Tailgate type, 8' wide		2.55	33	99	297	40.15	
	0055	Earth auger, truck mounted, for fence & sign posts, utility poles		13.85	460	1,375	4,125	385.85	
	0060	For borings and monitoring wells		42.65	690	2,075	6,225	756.20	
	0070	Portable, trailer mounted		2.30	33	99	297	38.20	
	0075	Truck mounted, for caissons, water wells		85.39	2,900	8,725	26,200	2,428	
	0080	Horizontal boring machine, 12" to 36" diameter, 45 H.P.		22.77	188	565	1,700	295.15	
	0090	12" to 48" diameter, 65 H.P.		31.25	330	985	2,950	447	
	0095	Auger, for fence posts, gas engine, hand held		.45	6.35	19	57	7.40	
	0100	Excavator, diesel hydraulic, crawler mounted, 1/2 C.Y. cap.		21.72	450	1,350	4,050	443.75	
	0120	5/8 C.Y. capacity		29.04	590	1,775	5,325	587.30	
	0140	3/4 C.Y. capacity		32.66	700	2,100	6,300	681.25	
	0150	1 C.Y. capacity		41.21	700	2,100	6,300	749.70	

For customer support on your Facilities Maintenance & Repair Costs with RSMeans data, call 800.448.8182.

01 54 33 | Equipment Rental

		UNIT	HOURLY OPER. COST	RENT PER DAY	RENT PER WEEK	RENT PER MONTH	EQUIPMENT COST/DAY	
20	0200	1-1/2 C.Y. capacity	Ea.	48.59	865	2,600	7,800	908.70	**20**
	0300	2 C.Y. capacity		56.58	1,050	3,125	9,375	1,078	
	0320	2-1/2 C.Y. capacity		82.64	1,300	3,900	11,700	1,441	
	0325	3-1/2 C.Y. capacity		120.12	2,150	6,475	19,400	2,256	
	0330	4-1/2 C.Y. capacity		151.63	2,750	8,275	24,800	2,868	
	0335	6 C.Y. capacity		192.39	3,400	10,200	30,600	3,579	
	0340	7 C.Y. capacity		175.20	3,225	9,650	29,000	3,332	
	0342	Excavator attachments, bucket thumbs		3.40	250	750	2,250	177.20	
	0345	Grapples		3.14	215	645	1,925	154.10	
	0346	Hydraulic hammer for boom mounting, 4,000 ft. lb.		13.48	375	1,125	3,375	332.85	
	0347	5,000 ft. lb.		15.95	465	1,400	4,200	407.60	
	0348	8,000 ft. lb.		23.54	675	2,025	6,075	593.35	
	0349	12,000 ft. lb.		25.72	800	2,400	7,200	685.75	
	0350	Gradall type, truck mounted, 3 ton @ 15' radius, 5/8 C.Y.		43.44	860	2,575	7,725	862.55	
	0370	1 C.Y. capacity		59.40	1,175	3,550	10,700	1,185	
	0400	Backhoe-loader, 40 to 45 H.P., 5/8 C.Y. capacity		11.90	207	620	1,850	219.20	
	0450	45 H.P. to 60 H.P., 3/4 C.Y. capacity		18.02	293	880	2,650	320.20	
	0460	80 H.P., 1-1/4 C.Y. capacity		20.36	375	1,125	3,375	387.85	
	0470	112 H.P., 1-1/2 C.Y. capacity		32.99	590	1,775	5,325	618.90	
	0482	Backhoe-loader attachment, compactor, 20,000 lb.		6.44	150	450	1,350	141.50	
	0485	Hydraulic hammer, 750 ft. lb.		3.68	103	310	930	91.45	
	0486	Hydraulic hammer, 1,200 ft. lb.		6.55	198	595	1,775	171.40	
	0500	Brush chipper, gas engine, 6" cutter head, 35 H.P.		9.17	110	330	990	139.35	
	0550	Diesel engine, 12" cutter head, 130 H.P.		23.67	335	1,000	3,000	389.40	
	0600	15" cutter head, 165 H.P.		26.59	400	1,200	3,600	452.70	
	0750	Bucket, clamshell, general purpose, 3/8 C.Y.		1.40	41	123	370	35.80	
	0800	1/2 C.Y.		1.52	50.50	152	455	42.55	
	0850	3/4 C.Y.		1.64	57.50	173	520	47.75	
	0900	1 C.Y.		1.70	62.50	188	565	51.20	
	0950	1-1/2 C.Y.		2.79	87	261	785	74.50	
	1000	2 C.Y.		2.92	95	285	855	80.40	
	1010	Bucket, dragline, medium duty, 1/2 C.Y.		.82	24.50	74	222	21.35	
	1020	3/4 C.Y.		.78	26	78	234	21.85	
	1030	1 C.Y.		.80	27	81.50	245	22.70	
	1040	1-1/2 C.Y.		1.26	41.50	125	375	35.10	
	1050	2 C.Y.		1.29	45	135	405	37.35	
	1070	3 C.Y.		2.08	66	198	595	56.25	
	1200	Compactor, manually guided 2-drum vibratory smooth roller, 7.5 H.P.		7.22	207	620	1,850	181.75	
	1250	Rammer/tamper, gas, 8"		2.21	46.50	140	420	45.65	
	1260	15"		2.63	53.50	160	480	53.05	
	1300	Vibratory plate, gas, 18" plate, 3,000 lb. blow		2.13	23.50	70.50	212	31.10	
	1350	21" plate, 5,000 lb. blow		2.62	32.50	98	294	40.55	
	1370	Curb builder/extruder, 14 H.P., gas, single screw		14.03	283	850	2,550	282.25	
	1390	Double screw		15.04	340	1,025	3,075	325.30	
	1500	Disc harrow attachment, for tractor		.47	79.50	239	715	51.60	
	1810	Feller buncher, shearing & accumulating trees, 100 H.P.		39.20	815	2,450	7,350	803.55	
	1860	Grader, self-propelled, 25,000 lb.		33.35	760	2,275	6,825	721.80	
	1910	30,000 lb.		32.85	660	1,975	5,925	657.85	
	1920	40,000 lb.		51.89	1,275	3,825	11,500	1,180	
	1930	55,000 lb.		66.93	1,650	4,925	14,800	1,520	
	1950	Hammer, pavement breaker, self-propelled, diesel, 1,000 to 1,250 lb.		28.40	475	1,425	4,275	512.20	
	2000	1,300 to 1,500 lb.		42.80	975	2,925	8,775	927.40	
	2050	Pile driving hammer, steam or air, 4,150 ft. lb. @ 225 bpm		12.15	565	1,700	5,100	437.20	
	2100	8,750 ft. lb. @ 145 bpm		14.35	810	2,425	7,275	599.75	
	2150	15,000 ft. lb. @ 60 bpm		14.68	840	2,525	7,575	622.40	
	2200	24,450 ft. lb. @ 111 bpm		15.69	935	2,800	8,400	685.50	
	2250	Leads, 60' high for pile driving hammers up to 20,000 ft. lb.		3.67	85.50	256	770	80.55	
	2300	90' high for hammers over 20,000 ft. lb.		5.45	152	455	1,375	134.60	

For customer support on your Facilities Maintenance & Repair Costs with RSMeans data, call 800.448.8182.

601

01 54 33	Equipment Rental	UNIT	HOURLY OPER. COST	RENT PER DAY	RENT PER WEEK	RENT PER MONTH	EQUIPMENT COST/DAY	
20	2350	Diesel type hammer, 22,400 ft. lb.	Ea.	17.82	475	1,425	4,275	427.55	20
	2400	41,300 ft. lb.		25.68	600	1,800	5,400	565.45	
	2450	141,000 ft. lb.		41.33	950	2,850	8,550	900.65	
	2500	Vib. elec. hammer/extractor, 200 kW diesel generator, 34 H.P.		41.37	690	2,075	6,225	745.95	
	2550	80 H.P.		73.03	1,000	3,000	9,000	1,184	
	2600	150 H.P.		135.14	1,925	5,775	17,300	2,236	
	2800	Log chipper, up to 22" diameter, 600 H.P.		46.26	665	2,000	6,000	770.10	
	2850	Logger, for skidding & stacking logs, 150 H.P.		43.53	835	2,500	7,500	848.25	
	2860	Mulcher, diesel powered, trailer mounted		18.04	220	660	1,975	276.35	
	2900	Rake, spring tooth, with tractor		14.72	360	1,075	3,225	332.75	
	3000	Roller, vibratory, tandem, smooth drum, 20 H.P.		7.81	150	450	1,350	152.45	
	3050	35 H.P.		10.13	252	755	2,275	232.05	
	3100	Towed type vibratory compactor, smooth drum, 50 H.P.		25.27	365	1,100	3,300	422.15	
	3150	Sheepsfoot, 50 H.P.		25.64	375	1,125	3,375	430.15	
	3170	Landfill compactor, 220 H.P.		70.01	1,575	4,750	14,300	1,510	
	3200	Pneumatic tire roller, 80 H.P.		12.92	390	1,175	3,525	338.35	
	3250	120 H.P.		19.39	640	1,925	5,775	540.10	
	3300	Sheepsfoot vibratory roller, 240 H.P.		62.21	1,375	4,125	12,400	1,323	
	3320	340 H.P.		83.82	2,075	6,250	18,800	1,921	
	3350	Smooth drum vibratory roller, 75 H.P.		23.34	635	1,900	5,700	566.70	
	3400	125 H.P.		27.61	715	2,150	6,450	650.90	
	3410	Rotary mower, brush, 60", with tractor		18.79	350	1,050	3,150	360.30	
	3420	Rototiller, walk-behind, gas, 5 H.P.		2.14	89	267	800	70.50	
	3422	8 H.P.		2.81	103	310	930	84.50	
	3440	Scrapers, towed type, 7 C.Y. capacity		6.44	123	370	1,100	125.50	
	3450	10 C.Y. capacity		7.20	165	495	1,475	156.65	
	3500	15 C.Y. capacity		7.40	190	570	1,700	173.20	
	3525	Self-propelled, single engine, 14 C.Y. capacity		133.29	2,575	7,725	23,200	2,611	
	3550	Dual engine, 21 C.Y. capacity		141.37	2,350	7,050	21,200	2,541	
	3600	31 C.Y. capacity		187.85	3,500	10,500	31,500	3,603	
	3640	44 C.Y. capacity		232.68	4,500	13,500	40,500	4,561	
	3650	Elevating type, single engine, 11 C.Y. capacity		61.87	1,125	3,350	10,100	1,165	
	3700	22 C.Y. capacity		114.62	2,325	6,950	20,900	2,307	
	3710	Screening plant, 110 H.P. w/5' x 10' screen		21.13	385	1,150	3,450	399.05	
	3720	5' x 16' screen		26.68	490	1,475	4,425	508.40	
	3850	Shovel, crawler-mounted, front-loading, 7 C.Y. capacity		218.65	3,800	11,400	34,200	4,029	
	3855	12 C.Y. capacity		336.90	5,275	15,800	47,400	5,855	
	3860	Shovel/backhoe bucket, 1/2 C.Y.		2.69	70.50	212	635	63.95	
	3870	3/4 C.Y.		2.67	79.50	238	715	68.90	
	3880	1 C.Y.		2.76	88	264	790	74.85	
	3890	1-1/2 C.Y.		2.95	103	310	930	85.65	
	3910	3 C.Y.		3.44	140	420	1,250	111.50	
	3950	Stump chipper, 18" deep, 30 H.P.		6.90	212	635	1,900	182.20	
	4110	Dozer, crawler, torque converter, diesel 80 H.P.		25.25	450	1,350	4,050	472	
	4150	105 H.P.		34.34	560	1,675	5,025	609.70	
	4200	140 H.P.		41.27	840	2,525	7,575	835.20	
	4260	200 H.P.		63.16	1,300	3,925	11,800	1,290	
	4310	300 H.P.		80.73	1,975	5,925	17,800	1,831	
	4360	410 H.P.		106.74	2,350	7,075	21,200	2,269	
	4370	500 H.P.		133.38	2,900	8,725	26,200	2,812	
	4380	700 H.P.		230.17	5,300	15,900	47,700	5,021	
	4400	Loader, crawler, torque conv., diesel, 1-1/2 C.Y., 80 H.P.		29.54	590	1,775	5,325	591.30	
	4450	1-1/2 to 1-3/4 C.Y., 95 H.P.		30.27	675	2,025	6,075	647.15	
	4510	1-3/4 to 2-1/4 C.Y., 130 H.P.		47.76	1,025	3,100	9,300	1,002	
	4530	2-1/2 to 3-1/4 C.Y., 190 H.P.		57.79	1,225	3,700	11,100	1,202	
	4560	3-1/2 to 5 C.Y., 275 H.P.		71.41	1,475	4,450	13,400	1,461	
	4610	Front end loader, 4WD, articulated frame, diesel, 1 to 1-1/4 C.Y., 70 H.P.		16.62	273	820	2,450	297	
	4620	1-1/2 to 1-3/4 C.Y., 95 H.P.		20.00	320	965	2,900	352.95	

For customer support on your Facilities Maintenance & Repair Costs with RSMeans data, call 800.448.8182.

01 54 33 | Equipment Rental

			UNIT	HOURLY OPER. COST	RENT PER DAY	RENT PER WEEK	RENT PER MONTH	EQUIPMENT COST/DAY	
20	4650	1-3/4 to 2 C.Y., 130 H.P.	Ea.	21.07	385	1,150	3,450	398.55	20
	4710	2-1/2 to 3-1/2 C.Y., 145 H.P.		29.53	490	1,475	4,425	531.20	
	4730	3 to 4-1/2 C.Y., 185 H.P.		32.09	515	1,550	4,650	566.70	
	4760	5-1/4 to 5-3/4 C.Y., 270 H.P.		53.19	915	2,750	8,250	975.55	
	4810	7 to 9 C.Y., 475 H.P.		91.17	1,750	5,275	15,800	1,784	
	4870	9 to 11 C.Y., 620 H.P.		131.92	2,625	7,850	23,600	2,625	
	4880	Skid-steer loader, wheeled, 10 C.F., 30 H.P. gas		9.57	163	490	1,475	174.55	
	4890	1 C.Y., 78 H.P., diesel		18.44	410	1,225	3,675	392.50	
	4892	Skid-steer attachment, auger		.75	137	410	1,225	87.95	
	4893	Backhoe		.74	118	355	1,075	76.90	
	4894	Broom		.71	123	370	1,100	79.65	
	4895	Forks		.15	26.50	80	240	17.25	
	4896	Grapple		.72	107	320	960	69.80	
	4897	Concrete hammer		1.05	177	530	1,600	114.40	
	4898	Tree spade		.60	100	300	900	64.80	
	4899	Trencher		.65	108	325	975	70.20	
	4900	Trencher, chain, boom type, gas, operator walking, 12 H.P.		4.18	50	150	450	63.40	
	4910	Operator riding, 40 H.P.		16.69	360	1,075	3,225	348.50	
	5000	Wheel type, diesel, 4' deep, 12" wide		68.71	935	2,800	8,400	1,110	
	5100	6' deep, 20" wide		87.58	2,150	6,475	19,400	1,996	
	5150	Chain type, diesel, 5' deep, 8" wide		16.30	350	1,050	3,150	340.40	
	5200	Diesel, 8' deep, 16" wide		89.66	1,875	5,600	16,800	1,837	
	5202	Rock trencher, wheel type, 6" wide x 18" deep		47.12	700	2,100	6,300	797	
	5206	Chain type, 18" wide x 7' deep		104.70	3,075	9,200	27,600	2,678	
	5210	Tree spade, self-propelled		13.69	390	1,175	3,525	344.55	
	5250	Truck, dump, 2-axle, 12 ton, 8 C.Y. payload, 220 H.P.		23.95	247	740	2,225	339.60	
	5300	Three axle dump, 16 ton, 12 C.Y. payload, 400 H.P.		44.63	350	1,050	3,150	567.05	
	5310	Four axle dump, 25 ton, 18 C.Y. payload, 450 H.P.		50.00	510	1,525	4,575	705	
	5350	Dump trailer only, rear dump, 16-1/2 C.Y.		5.74	147	440	1,325	133.95	
	5400	20 C.Y.		6.20	165	495	1,475	148.60	
	5450	Flatbed, single axle, 1-1/2 ton rating		19.05	71.50	214	640	195.20	
	5500	3 ton rating		23.12	102	305	915	245.95	
	5550	Off highway rear dump, 25 ton capacity		62.86	1,425	4,250	12,800	1,353	
	5600	35 ton capacity		67.10	1,550	4,675	14,000	1,472	
	5610	50 ton capacity		84.12	1,775	5,325	16,000	1,738	
	5620	65 ton capacity		89.83	1,925	5,800	17,400	1,879	
	5630	100 ton capacity		121.61	2,850	8,550	25,700	2,683	
	6000	Vibratory plow, 25 H.P., walking		6.79	60.50	181	545	90.55	
40	0010	**GENERAL EQUIPMENT RENTAL** without operators							40
	0020	Aerial lift, scissor type, to 20' high, 1200 lb. capacity, electric	Ea.	3.49	50.50	151	455	58.15	
	0030	To 30' high, 1,200 lb. capacity		3.78	67.50	202	605	70.70	
	0040	Over 30' high, 1,500 lb. capacity		5.15	122	365	1,100	114.20	
	0070	Articulating boom, to 45' high, 500 lb. capacity, diesel		9.95	275	825	2,475	244.60	
	0075	To 60' high, 500 lb. capacity		13.70	510	1,525	4,575	414.60	
	0080	To 80' high, 500 lb. capacity		16.10	560	1,675	5,025	463.80	
	0085	To 125' high, 500 lb. capacity		18.40	775	2,325	6,975	612.20	
	0100	Telescoping boom to 40' high, 500 lb. capacity, diesel		11.27	320	965	2,900	283.15	
	0105	To 45' high, 500 lb. capacity		12.54	315	945	2,825	289.35	
	0110	To 60' high, 500 lb. capacity		16.41	535	1,600	4,800	451.25	
	0115	To 80' high, 500 lb. capacity		21.33	835	2,500	7,500	670.65	
	0120	To 100' high, 500 lb. capacity		28.80	860	2,575	7,725	745.40	
	0125	To 120' high, 500 lb. capacity		29.25	840	2,525	7,575	739	
	0195	Air compressor, portable, 6.5 CFM, electric		.91	12.65	38	114	14.85	
	0196	Gasoline		.65	18	54	162	16.05	
	0200	Towed type, gas engine, 60 CFM		9.46	51.50	155	465	106.70	
	0300	160 CFM		10.51	53	159	475	115.85	
	0400	Diesel engine, rotary screw, 250 CFM		12.12	118	355	1,075	167.95	
	0500	365 CFM		16.05	142	425	1,275	213.35	

Note codes: 0020 — R015433-10; 0070 — R015433-15

01 54 33 | Equipment Rental

		UNIT	HOURLY OPER. COST	RENT PER DAY	RENT PER WEEK	RENT PER MONTH	EQUIPMENT COST/DAY
0550	450 CFM	Ea.	20.01	177	530	1,600	266.05
0600	600 CFM		34.20	240	720	2,150	417.60
0700	750 CFM		34.72	252	755	2,275	428.80
0930	Air tools, breaker, pavement, 60 lb.		.57	10.35	31	93	10.75
0940	80 lb.		.57	10.65	32	96	10.90
0950	Drills, hand (jackhammer), 65 lb.		.67	17.35	52	156	15.80
0960	Track or wagon, swing boom, 4" drifter		54.83	940	2,825	8,475	1,004
0970	5" drifter		63.49	1,100	3,325	9,975	1,173
0975	Track mounted quarry drill, 6" diameter drill		102.22	1,625	4,875	14,600	1,793
0980	Dust control per drill		1.04	25	75.50	227	23.45
0990	Hammer, chipping, 12 lb.		.60	27	81	243	21
1000	Hose, air with couplings, 50' long, 3/4" diameter		.07	10	30	90	6.55
1100	1" diameter		.08	13	39	117	8.40
1200	1-1/2" diameter		.22	35	105	315	22.75
1300	2" diameter		.24	40	120	360	25.90
1400	2-1/2" diameter		.36	56.50	170	510	36.85
1410	3" diameter		.42	37.50	112	335	25.75
1450	Drill, steel, 7/8" x 2'		.08	12.50	37.50	113	8.20
1460	7/8" x 6'		.12	19	57	171	12.30
1520	Moil points		.03	4.73	14.20	42.50	3.05
1525	Pneumatic nailer w/accessories		.48	31	92.50	278	22.35
1530	Sheeting driver for 60 lb. breaker		.04	7.50	22.50	67.50	4.85
1540	For 90 lb. breaker		.13	10.15	30.50	91.50	7.15
1550	Spade, 25 lb.		.50	7.15	21.50	64.50	8.30
1560	Tamper, single, 35 lb.		.59	39.50	119	355	28.55
1570	Triple, 140 lb.		.89	59.50	179	535	42.90
1580	Wrenches, impact, air powered, up to 3/4" bolt		.43	12.85	38.50	116	11.10
1590	Up to 1-1/4" bolt		.58	23	69.50	209	18.50
1600	Barricades, barrels, reflectorized, 1 to 99 barrels		.03	5.50	16.55	49.50	3.55
1610	100 to 200 barrels		.02	4.27	12.80	38.50	2.75
1620	Barrels with flashers, 1 to 99 barrels		.03	6.20	18.55	55.50	4
1630	100 to 200 barrels		.03	4.95	14.85	44.50	3.20
1640	Barrels with steady burn type C lights		.05	8.15	24.50	73.50	5.30
1650	Illuminated board, trailer mounted, with generator		3.29	135	405	1,225	107.30
1670	Portable barricade, stock, with flashers, 1 to 6 units		.03	6.15	18.50	55.50	4
1680	25 to 50 units		.03	5.75	17.25	52	3.70
1685	Butt fusion machine, wheeled, 1.5 HP electric, 2" - 8" diameter pipe		2.64	203	610	1,825	143.10
1690	Tracked, 20 HP diesel, 4" - 12" diameter pipe		11.27	525	1,575	4,725	405.15
1695	83 HP diesel, 8" - 24" diameter pipe		51.47	2,525	7,575	22,700	1,927
1700	Carts, brick, gas engine, 1,000 lb. capacity		2.95	69.50	208	625	65.20
1800	1,500 lb., 7-1/2' lift		2.92	65	195	585	62.40
1822	Dehumidifier, medium, 6 lb./hr., 150 CFM		1.19	74	222	665	53.95
1824	Large, 18 lb./hr., 600 CFM		2.20	140	420	1,250	101.60
1830	Distributor, asphalt, trailer mounted, 2,000 gal., 38 H.P. diesel		11.03	340	1,025	3,075	293.20
1840	3,000 gal., 38 H.P. diesel		12.90	365	1,100	3,300	323.25
1850	Drill, rotary hammer, electric		1.12	27	80.50	242	25.05
1860	Carbide bit, 1-1/2" diameter, add to electric rotary hammer		.03	5.10	15.30	46	3.30
1865	Rotary, crawler, 250 H.P.		136.18	2,225	6,650	20,000	2,419
1870	Emulsion sprayer, 65 gal., 5 H.P. gas engine		2.78	103	310	930	84.20
1880	200 gal., 5 H.P. engine		7.25	173	520	1,550	161.95
1900	Floor auto-scrubbing machine, walk-behind, 28" path		5.64	365	1,100	3,300	265.15
1930	Floodlight, mercury vapor, or quartz, on tripod, 1,000 watt		.46	22	66.50	200	16.95
1940	2,000 watt		.59	27.50	82	246	21.15
1950	Floodlights, trailer mounted with generator, one - 300 watt light		3.56	76	228	685	74.05
1960	two 1000 watt lights		4.50	84.50	254	760	86.80
2000	four 300 watt lights		4.26	96.50	290	870	92.05
2005	Foam spray rig, incl. box trailer, compressor, generator, proportioner		25.54	515	1,550	4,650	514.30
2015	Forklift, pneumatic tire, rough terr., straight mast, 5,000 lb., 12' lift, gas		18.65	212	635	1,900	276.20

40

For customer support on your Facilities Maintenance & Repair Costs with RSMeans data, call 800.448.8182.

01 54 33 | Equipment Rental

		UNIT	HOURLY OPER. COST	RENT PER DAY	RENT PER WEEK	RENT PER MONTH	EQUIPMENT COST/DAY
2025	8,000 lb, 12' lift	Ea.	22.75	350	1,050	3,150	392
2030	5,000 lb., 12' lift, diesel		15.45	237	710	2,125	265.60
2035	8,000 lb, 12' lift, diesel		16.75	268	805	2,425	295
2045	All terrain, telescoping boom, diesel, 5,000 lb., 10' reach, 19' lift		17.25	410	1,225	3,675	383
2055	6,600 lb., 29' reach, 42' lift		21.10	400	1,200	3,600	408.80
2065	10,000 lb., 31' reach, 45' lift		23.10	465	1,400	4,200	464.75
2070	Cushion tire, smooth floor, gas, 5,000 lb. capacity		8.25	76.50	230	690	112
2075	8,000 lb. capacity		11.37	89	267	800	144.30
2085	Diesel, 5,000 lb. capacity		7.75	83.50	250	750	112
2090	12,000 lb. capacity		12.05	130	390	1,175	174.40
2095	20,000 lb. capacity		17.25	150	450	1,350	228.05
2100	Generator, electric, gas engine, 1.5 kW to 3 kW		2.58	10.15	30.50	91.50	26.70
2200	5 kW		3.22	12.65	38	114	33.35
2300	10 kW		5.93	33.50	101	305	67.65
2400	25 kW		7.41	84.50	254	760	110.05
2500	Diesel engine, 20 kW		9.21	74	222	665	118.05
2600	50 kW		15.95	96.50	289	865	185.35
2700	100 kW		28.60	133	400	1,200	308.80
2800	250 kW		54.35	255	765	2,300	587.85
2850	Hammer, hydraulic, for mounting on boom, to 500 ft. lb.		2.90	90.50	271	815	77.40
2860	1,000 ft. lb.		4.61	135	405	1,225	117.85
2900	Heaters, space, oil or electric, 50 MBH		1.46	8	24	72	16.50
3000	100 MBH		2.72	11.85	35.50	107	28.85
3100	300 MBH		7.92	40	120	360	87.40
3150	500 MBH		13.16	45	135	405	132.25
3200	Hose, water, suction with coupling, 20' long, 2" diameter		.02	3.13	9.40	28	2.05
3210	3" diameter		.03	4.10	12.30	37	2.70
3220	4" diameter		.03	4.80	14.40	43	3.10
3230	6" diameter		.11	17	51	153	11.10
3240	8" diameter		.27	45	135	405	29.15
3250	Discharge hose with coupling, 50' long, 2" diameter		.01	1.33	4	12	.90
3260	3" diameter		.01	2.22	6.65	19.95	1.40
3270	4" diameter		.02	3.47	10.40	31	2.25
3280	6" diameter		.06	8.65	26	78	5.70
3290	8" diameter		.24	40	120	360	25.90
3295	Insulation blower		.83	6	18.05	54	10.25
3300	Ladders, extension type, 16' to 36' long		.18	31.50	95	285	20.45
3400	40' to 60' long		.64	112	335	1,000	72.10
3405	Lance for cutting concrete		2.21	64	192	575	56.05
3407	Lawn mower, rotary, 22", 5 H.P.		1.06	25	75	225	23.45
3408	48" self-propelled		2.90	127	380	1,150	99.20
3410	Level, electronic, automatic, with tripod and leveling rod		1.05	61.50	185	555	45.40
3430	Laser type, for pipe and sewer line and grade		2.18	153	460	1,375	109.40
3440	Rotating beam for interior control		.90	62	186	560	44.40
3460	Builder's optical transit, with tripod and rod		.10	17	51	153	11
3500	Light towers, towable, with diesel generator, 2,000 watt		4.27	98	294	880	92.95
3600	4,000 watt		4.51	102	305	915	97.10
3700	Mixer, powered, plaster and mortar, 6 C.F., 7 H.P.		2.06	21	62.50	188	28.95
3800	10 C.F., 9 H.P.		2.24	33.50	101	305	38.15
3850	Nailer, pneumatic		.48	32.50	97	291	23.25
3900	Paint sprayers complete, 8 CFM		.85	59.50	179	535	42.60
4000	17 CFM		1.60	107	320	960	76.80
4020	Pavers, bituminous, rubber tires, 8' wide, 50 H.P., diesel		32.02	550	1,650	4,950	586.20
4030	10' wide, 150 H.P.		95.91	1,875	5,650	17,000	1,897
4050	Crawler, 8' wide, 100 H.P., diesel		87.85	2,000	5,975	17,900	1,898
4060	10' wide, 150 H.P.		104.28	2,275	6,825	20,500	2,199
4070	Concrete paver, 12' to 24' wide, 250 H.P.		87.88	1,625	4,900	14,700	1,683
4080	Placer-spreader-trimmer, 24' wide, 300 H.P.		117.87	2,475	7,425	22,300	2,428

For customer support on your Facilities Maintenance & Repair Costs with RSMeans data, call 800.448.8182.

605

01 54 33 | Equipment Rental

		UNIT	HOURLY OPER. COST	RENT PER DAY	RENT PER WEEK	RENT PER MONTH	EQUIPMENT COST/DAY	
40	4100	Pump, centrifugal gas pump, 1-1/2" diam., 65 GPM	Ea.	3.93	52.50	157	470	62.90	40
	4200	2" diameter, 130 GPM		4.99	64.50	194	580	78.75	
	4300	3" diameter, 250 GPM		5.13	63.50	191	575	79.25	
	4400	6" diameter, 1,500 GPM		22.31	197	590	1,775	296.45	
	4500	Submersible electric pump, 1-1/4" diameter, 55 GPM		.40	17.65	53	159	13.80	
	4600	1-1/2" diameter, 83 GPM		.45	20.50	61	183	15.75	
	4700	2" diameter, 120 GPM		1.65	25	75.50	227	28.30	
	4800	3" diameter, 300 GPM		3.04	45	135	405	51.35	
	4900	4" diameter, 560 GPM		14.80	163	490	1,475	216.35	
	5000	6" diameter, 1,590 GPM		22.14	200	600	1,800	297.15	
	5100	Diaphragm pump, gas, single, 1-1/2" diameter		1.13	57.50	172	515	43.45	
	5200	2" diameter		3.99	70.50	212	635	74.30	
	5300	3" diameter		4.06	74.50	224	670	77.30	
	5400	Double, 4" diameter		6.05	155	465	1,400	141.40	
	5450	Pressure washer, 5 GPM, 3,000 psi		3.88	53.50	160	480	63.05	
	5460	7 GPM, 3,000 psi		4.95	63	189	565	77.40	
	5500	Trash pump, self-priming, gas, 2" diameter		3.83	23.50	71	213	44.80	
	5600	Diesel, 4" diameter		6.70	94.50	284	850	110.40	
	5650	Diesel, 6" diameter		16.90	165	495	1,475	234.20	
	5655	Grout pump		18.75	272	815	2,450	313	
	5700	Salamanders, L.P. gas fired, 100,000 BTU		2.89	14.15	42.50	128	31.65	
	5705	50,000 BTU		1.67	11.65	35	105	20.35	
	5720	Sandblaster, portable, open top, 3 C.F. capacity		.60	26.50	79.50	239	20.70	
	5730	6 C.F. capacity		1.01	39.50	118	355	31.65	
	5740	Accessories for above		.14	23	69	207	14.90	
	5750	Sander, floor		.77	19.35	58	174	17.75	
	5760	Edger		.52	15.85	47.50	143	13.65	
	5800	Saw, chain, gas engine, 18" long		1.76	22.50	67.50	203	27.55	
	5900	Hydraulic powered, 36" long		.78	67.50	202	605	46.65	
	5950	60" long		.78	69.50	208	625	47.85	
	6000	Masonry, table mounted, 14" diameter, 5 H.P.		1.32	81	243	730	59.20	
	6050	Portable cut-off, 8 H.P.		1.82	32.50	97.50	293	34	
	6100	Circular, hand held, electric, 7-1/4" diameter		.23	4.93	14.80	44.50	4.80	
	6200	12" diameter		.24	7.85	23.50	70.50	6.60	
	6250	Wall saw, w/hydraulic power, 10 H.P.		3.30	33.50	101	305	46.60	
	6275	Shot blaster, walk-behind, 20" wide		4.75	272	815	2,450	201	
	6280	Sidewalk broom, walk-behind		2.25	80	240	720	65.95	
	6300	Steam cleaner, 100 gallons per hour		3.35	80	240	720	74.80	
	6310	200 gallons per hour		4.35	96.50	290	870	92.75	
	6340	Tar kettle/pot, 400 gallons		16.53	76.50	230	690	178.20	
	6350	Torch, cutting, acetylene-oxygen, 150' hose, excludes gases		.45	14.85	44.50	134	12.50	
	6360	Hourly operating cost includes tips and gas		20.98				167.85	
	6410	Toilet, portable chemical		.13	22.50	67	201	14.45	
	6420	Recycle flush type		.16	27.50	83	249	17.90	
	6430	Toilet, fresh water flush, garden hose,		.19	33	99	297	21.35	
	6440	Hoisted, non-flush, for high rise		.16	27	81	243	17.45	
	6465	Tractor, farm with attachment		17.42	300	900	2,700	319.35	
	6480	Trailers, platform, flush deck, 2 axle, 3 ton capacity		1.69	21.50	65	195	26.55	
	6500	25 ton capacity		6.25	138	415	1,250	133	
	6600	40 ton capacity		8.06	197	590	1,775	182.50	
	6700	3 axle, 50 ton capacity		8.75	218	655	1,975	200.95	
	6800	75 ton capacity		11.11	290	870	2,600	262.90	
	6810	Trailer mounted cable reel for high voltage line work		5.90	282	845	2,525	216.20	
	6820	Trailer mounted cable tensioning rig		11.71	560	1,675	5,025	428.65	
	6830	Cable pulling rig		73.99	3,125	9,375	28,100	2,467	
	6850	Portable cable/wire puller, 8,000 lb. max. pulling capacity		3.71	205	615	1,850	152.70	
	6900	Water tank trailer, engine driven discharge, 5,000 gallons		7.18	153	460	1,375	149.45	
	6925	10,000 gallons		9.78	208	625	1,875	203.25	

For customer support on your Facilities Maintenance & Repair Costs with RSMeans data, call 800.448.8182.

01 54 33 | Equipment Rental

		UNIT	HOURLY OPER. COST	RENT PER DAY	RENT PER WEEK	RENT PER MONTH	EQUIPMENT COST/DAY	
40	6950	Water truck, off highway, 6,000 gallons	Ea.	71.97	810	2,425	7,275	1,061	**40**
	7010	Tram car for high voltage line work, powered, 2 conductor		6.90	152	455	1,375	146.20	
	7020	Transit (builder's level) with tripod		.10	17	51	153	11	
	7030	Trench box, 3,000 lb., 6' x 8'		.56	93.50	281	845	60.70	
	7040	7,200 lb., 6' x 20'		.72	120	360	1,075	77.75	
	7050	8,000 lb., 8' x 16'		1.08	180	540	1,625	116.65	
	7060	9,500 lb., 8' x 20'		1.21	225	675	2,025	144.65	
	7065	11,000 lb., 8' x 24'		1.27	212	635	1,900	137.15	
	7070	12,000 lb., 10' x 20'		1.49	255	765	2,300	164.95	
	7100	Truck, pickup, 3/4 ton, 2 wheel drive		9.26	59.50	179	535	109.90	
	7200	4 wheel drive		9.51	76	228	685	121.70	
	7250	Crew carrier, 9 passenger		12.70	88	264	790	154.35	
	7290	Flat bed truck, 20,000 lb. GVW		15.31	128	385	1,150	199.45	
	7300	Tractor, 4 x 2, 220 H.P.		22.32	208	625	1,875	303.55	
	7410	330 H.P.		32.43	285	855	2,575	430.40	
	7500	6 x 4, 380 H.P.		36.20	330	990	2,975	487.60	
	7600	450 H.P.		44.36	400	1,200	3,600	594.90	
	7610	Tractor, with A frame, boom and winch, 225 H.P.		24.81	283	850	2,550	368.50	
	7620	Vacuum truck, hazardous material, 2,500 gallons		12.83	300	900	2,700	282.65	
	7625	5,000 gallons		13.06	425	1,275	3,825	359.50	
	7650	Vacuum, HEPA, 16 gallon, wet/dry		.85	17.15	51.50	155	17.10	
	7655	55 gallon, wet/dry		.78	26.50	80	240	22.25	
	7660	Water tank, portable		.73	155	465	1,400	98.85	
	7690	Sewer/catch basin vacuum, 14 C.Y., 1,500 gallons		17.37	640	1,925	5,775	523.95	
	7700	Welder, electric, 200 amp		3.83	16	48	144	40.20	
	7800	300 amp		5.56	19.35	58	174	56.10	
	7900	Gas engine, 200 amp		8.97	23	69	207	85.60	
	8000	300 amp		10.16	24	72	216	95.65	
	8100	Wheelbarrow, any size		.06	10.35	31	93	6.70	
	8200	Wrecking ball, 4,000 lb.		2.51	72	216	650	63.25	
50	0010	**HIGHWAY EQUIPMENT RENTAL** without operators							**50**
	0050	Asphalt batch plant, portable drum mixer, 100 ton/hr.	Ea.	88.67	1,500	4,475	13,400	1,604	
	0060	200 ton/hr.		102.29	1,600	4,775	14,300	1,773	
	0070	300 ton/hr.		120.22	1,875	5,600	16,800	2,082	
	0100	Backhoe attachment, long stick, up to 185 H.P., 10-1/2' long		.37	24.50	74	222	17.75	
	0140	Up to 250 H.P., 12' long		.41	27.50	83	249	19.90	
	0180	Over 250 H.P., 15' long		.57	37.50	113	340	27.15	
	0200	Special dipper arm, up to 100 H.P., 32' long		1.16	77	231	695	55.50	
	0240	Over 100 H.P., 33' long		1.45	96.50	290	870	69.60	
	0280	Catch basin/sewer cleaning truck, 3 ton, 9 C.Y., 1,000 gal.		35.50	410	1,225	3,675	528.95	
	0300	Concrete batch plant, portable, electric, 200 C.Y./hr.		24.26	540	1,625	4,875	519.05	
	0520	Grader/dozer attachment, ripper/scarifier, rear mounted, up to 135 H.P.		3.16	61.50	184	550	62.10	
	0540	Up to 180 H.P.		4.15	92.50	278	835	88.80	
	0580	Up to 250 H.P.		5.87	148	445	1,325	135.95	
	0700	Pvmt. removal bucket, for hyd. excavator, up to 90 H.P.		2.17	56.50	169	505	51.10	
	0740	Up to 200 H.P.		2.31	72	216	650	61.70	
	0780	Over 200 H.P.		2.53	88.50	265	795	73.20	
	0900	Aggregate spreader, self-propelled, 187 H.P.		50.75	715	2,150	6,450	836	
	1000	Chemical spreader, 3 C.Y.		3.18	46.50	139	415	53.20	
	1900	Hammermill, traveling, 250 H.P.		67.55	2,250	6,750	20,300	1,890	
	2000	Horizontal borer, 3" diameter, 13 H.P. gas driven		5.43	57.50	172	515	77.85	
	2150	Horizontal directional drill, 20,000 lb. thrust, 78 H.P. diesel		27.66	685	2,050	6,150	631.30	
	2160	30,000 lb. thrust, 115 H.P.		34.00	1,050	3,125	9,375	897	
	2170	50,000 lb. thrust, 170 H.P.		48.74	1,325	3,975	11,900	1,185	
	2190	Mud trailer for HDD, 1,500 gallons, 175 H.P., gas		25.58	157	470	1,400	298.65	
	2200	Hydromulcher, diesel, 3,000 gallon, for truck mounting		17.48	255	765	2,300	292.90	
	2300	Gas, 600 gallon		7.52	107	320	960	124.15	
	2400	Joint & crack cleaner, walk behind, 25 H.P.		3.17	52.50	158	475	56.95	

R015433 -10

For customer support on your Facilities Maintenance & Repair Costs with RSMeans data, call 800.448.8182.

607

01 54 33 | Equipment Rental

		UNIT	HOURLY OPER. COST	RENT PER DAY	RENT PER WEEK	RENT PER MONTH	EQUIPMENT COST/DAY	
50	2500	Filler, trailer mounted, 400 gallons, 20 H.P.	Ea.	8.37	220	660	1,975	198.95	**50**
	3000	Paint striper, self-propelled, 40 gallon, 22 H.P.		6.78	163	490	1,475	152.25	
	3100	120 gallon, 120 H.P.		19.28	410	1,225	3,675	399.25	
	3200	Post drivers, 6" I-Beam frame, for truck mounting		12.45	390	1,175	3,525	334.60	
	3400	Road sweeper, self-propelled, 8' wide, 90 H.P.		36.01	690	2,075	6,225	703.10	
	3450	Road sweeper, vacuum assisted, 4 C.Y., 220 gallons		58.45	650	1,950	5,850	857.60	
	4000	Road mixer, self-propelled, 130 H.P.		46.37	800	2,400	7,200	851	
	4100	310 H.P.		75.23	2,100	6,275	18,800	1,657	
	4220	Cold mix paver, incl. pug mill and bitumen tank, 165 H.P.		95.25	2,250	6,725	20,200	2,107	
	4240	Pavement brush, towed		3.44	96.50	290	870	85.50	
	4250	Paver, asphalt, wheel or crawler, 130 H.P., diesel		94.52	2,200	6,600	19,800	2,076	
	4300	Paver, road widener, gas, 1' to 6', 67 H.P.		46.80	940	2,825	8,475	939.40	
	4400	Diesel, 2' to 14', 88 H.P.		56.55	1,125	3,350	10,100	1,122	
	4600	Slipform pavers, curb and gutter, 2 track, 75 H.P.		58.01	1,225	3,650	11,000	1,194	
	4700	4 track, 165 H.P.		35.79	815	2,450	7,350	776.35	
	4800	Median barrier, 215 H.P.		58.61	1,300	3,900	11,700	1,249	
	4901	Trailer, low bed, 75 ton capacity		10.74	273	820	2,450	249.90	
	5000	Road planer, walk behind, 10" cutting width, 10 H.P.		2.46	33.50	100	300	39.70	
	5100	Self-propelled, 12" cutting width, 64 H.P.		8.28	117	350	1,050	136.25	
	5120	Traffic line remover, metal ball blaster, truck mounted, 115 H.P.		46.70	785	2,350	7,050	843.60	
	5140	Grinder, truck mounted, 115 H.P.		51.04	810	2,425	7,275	893.35	
	5160	Walk-behind, 11 H.P.		3.57	54.50	164	490	61.35	
	5200	Pavement profiler, 4' to 6' wide, 450 H.P.		217.23	3,425	10,300	30,900	3,798	
	5300	8' to 10' wide, 750 H.P.		332.58	4,525	13,600	40,800	5,381	
	5400	Roadway plate, steel, 1" x 8' x 20'		.09	14.65	44	132	9.50	
	5600	Stabilizer, self-propelled, 150 H.P.		41.26	640	1,925	5,775	715.05	
	5700	310 H.P.		76.41	1,700	5,075	15,200	1,626	
	5800	Striper, truck mounted, 120 gallon paint, 460 H.P.		48.89	490	1,475	4,425	686.10	
	5900	Thermal paint heating kettle, 115 gallons		7.73	26.50	80	240	77.85	
	6000	Tar kettle, 330 gallon, trailer mounted		12.31	60	180	540	134.50	
	7000	Tunnel locomotive, diesel, 8 to 12 ton		29.85	600	1,800	5,400	598.75	
	7005	Electric, 10 ton		29.33	685	2,050	6,150	644.70	
	7010	Muck cars, 1/2 C.Y. capacity		2.30	26	77.50	233	33.95	
	7020	1 C.Y. capacity		2.52	33.50	101	305	40.35	
	7030	2 C.Y. capacity		2.66	37.50	113	340	43.90	
	7040	Side dump, 2 C.Y. capacity		2.88	46.50	140	420	51.05	
	7050	3 C.Y. capacity		3.86	51.50	154	460	61.70	
	7060	5 C.Y. capacity		5.64	66.50	199	595	84.90	
	7100	Ventilating blower for tunnel, 7-1/2 H.P.		2.14	51	153	460	47.75	
	7110	10 H.P.		2.43	53.50	160	480	51.40	
	7120	20 H.P.		3.55	69.50	208	625	70	
	7140	40 H.P.		6.16	91.50	275	825	104.25	
	7160	60 H.P.		8.71	98.50	295	885	128.70	
	7175	75 H.P.		10.40	153	460	1,375	175.20	
	7180	200 H.P.		20.85	300	905	2,725	347.75	
	7800	Windrow loader, elevating		54.10	1,350	4,025	12,100	1,238	
60	0010	**LIFTING AND HOISTING EQUIPMENT RENTAL** without operators							**60**
	0150	Crane, flatbed mounted, 3 ton capacity	Ea.	14.45	195	585	1,750	232.65	
	0200	Crane, climbing, 106' jib, 6,000 lb. capacity, 410 fpm		39.84	1,775	5,350	16,100	1,389	
	0300	101' jib, 10,250 lb. capacity, 270 fpm		46.57	2,275	6,800	20,400	1,733	
	0500	Tower, static, 130' high, 106' jib, 6,200 lb. capacity at 400 fpm		45.29	2,075	6,200	18,600	1,602	
	0520	Mini crawler spider crane, up to 24" wide, 1,990 lb. lifting capacity		12.54	535	1,600	4,800	420.30	
	0525	Up to 30" wide, 6,450 lb. lifting capacity		14.57	635	1,900	5,700	496.55	
	0530	Up to 52" wide, 6,680 lb. lifting capacity		23.17	775	2,325	6,975	650.40	
	0535	Up to 55" wide, 8,920 lb. lifting capacity		25.87	860	2,575	7,725	722	
	0540	Up to 66" wide, 13,350 lb. lifting capacity		35.03	1,325	4,000	12,000	1,080	
	0600	Crawler mounted, lattice boom, 1/2 C.Y., 15 tons at 12' radius		37.07	950	2,850	8,550	866.60	
	0700	3/4 C.Y., 20 tons at 12' radius		50.57	1,200	3,625	10,900	1,130	

Reference codes: R015433 -10 (for 0150 row); R312316 -45 (for 0300 row)

For customer support on your Facilities Maintenance & Repair Costs with RSMeans data, call 800.448.8182.

01 54 33 | Equipment Rental

		UNIT	HOURLY OPER. COST	RENT PER DAY	RENT PER WEEK	RENT PER MONTH	EQUIPMENT COST/DAY
60	0800	1 C.Y., 25 tons at 12' radius	Ea.	67.62	1,400	4,225	12,700	1,386
	0900	1-1/2 C.Y., 40 tons at 12' radius		66.51	1,425	4,300	12,900	1,392
	1000	2 C.Y., 50 tons at 12' radius		89.03	2,200	6,600	19,800	2,032
	1100	3 C.Y., 75 tons at 12' radius		75.49	1,875	5,650	17,000	1,734
	1200	100 ton capacity, 60' boom		86.17	1,975	5,950	17,900	1,879
	1300	165 ton capacity, 60' boom		106.42	2,275	6,850	20,600	2,221
	1400	200 ton capacity, 70' boom		138.63	3,175	9,550	28,700	3,019
	1500	350 ton capacity, 80' boom		182.75	4,075	12,200	36,600	3,902
	1600	Truck mounted, lattice boom, 6 x 4, 20 tons at 10' radius		39.88	915	2,750	8,250	869
	1700	25 tons at 10' radius		42.86	1,400	4,175	12,500	1,178
	1800	8 x 4, 30 tons at 10' radius		45.67	1,475	4,425	13,300	1,250
	1900	40 tons at 12' radius		48.70	1,550	4,625	13,900	1,315
	2000	60 tons at 15' radius		53.85	1,650	4,950	14,900	1,421
	2050	82 tons at 15' radius		59.60	1,775	5,350	16,100	1,547
	2100	90 tons at 15' radius		66.59	1,950	5,825	17,500	1,698
	2200	115 tons at 15' radius		75.12	2,175	6,525	19,600	1,906
	2300	150 tons at 18' radius		81.33	2,275	6,850	20,600	2,021
	2350	165 tons at 18' radius		87.30	2,425	7,275	21,800	2,153
	2400	Truck mounted, hydraulic, 12 ton capacity		29.59	390	1,175	3,525	471.75
	2500	25 ton capacity		36.46	485	1,450	4,350	581.70
	2550	33 ton capacity		50.82	900	2,700	8,100	946.60
	2560	40 ton capacity		49.62	900	2,700	8,100	937
	2600	55 ton capacity		53.94	915	2,750	8,250	981.50
	2700	80 ton capacity		75.94	1,475	4,400	13,200	1,487
	2720	100 ton capacity		75.18	1,550	4,675	14,000	1,536
	2740	120 ton capacity		103.12	1,825	5,500	16,500	1,925
	2760	150 ton capacity		110.25	2,050	6,125	18,400	2,107
	2800	Self-propelled, 4 x 4, with telescoping boom, 5 ton		15.18	230	690	2,075	259.45
	2900	12-1/2 ton capacity		21.48	335	1,000	3,000	371.85
	3000	15 ton capacity		34.52	535	1,600	4,800	596.20
	3050	20 ton capacity		24.09	660	1,975	5,925	587.75
	3100	25 ton capacity		36.80	615	1,850	5,550	664.40
	3150	40 ton capacity		45.03	650	1,950	5,850	750.25
	3200	Derricks, guy, 20 ton capacity, 60' boom, 75' mast		22.81	435	1,300	3,900	442.50
	3300	100' boom, 115' mast		36.15	750	2,250	6,750	739.20
	3400	Stiffleg, 20 ton capacity, 70' boom, 37' mast		25.48	565	1,700	5,100	543.85
	3500	100' boom, 47' mast		39.43	910	2,725	8,175	860.50
	3550	Helicopter, small, lift to 1,250 lb. maximum, w/pilot		99.44	3,525	10,600	31,800	2,916
	3600	Hoists, chain type, overhead, manual, 3/4 ton		.15	.33	1	3	1.35
	3900	10 ton		.79	6	18	54	9.90
	4000	Hoist and tower, 5,000 lb. cap., portable electric, 40' high		5.14	252	755	2,275	192.10
	4100	For each added 10' section, add		.12	19.65	59	177	12.75
	4200	Hoist and single tubular tower, 5,000 lb. electric, 100' high		6.98	350	1,050	3,150	265.80
	4300	For each added 6'-6" section, add		.21	34.50	103	310	22.25
	4400	Hoist and double tubular tower, 5,000 lb., 100' high		7.59	385	1,150	3,450	290.75
	4500	For each added 6'-6" section, add		.23	37.50	113	340	24.40
	4550	Hoist and tower, mast type, 6,000 lb., 100' high		8.26	400	1,200	3,600	306.10
	4570	For each added 10' section, add		.13	23	69	207	14.85
	4600	Hoist and tower, personnel, electric, 2,000 lb., 100' @ 125 fpm		17.55	1,075	3,200	9,600	780.40
	4700	3,000 lb., 100' @ 200 fpm		20.08	1,200	3,625	10,900	885.65
	4800	3,000 lb., 150' @ 300 fpm		22.28	1,350	4,075	12,200	993.30
	4900	4,000 lb., 100' @ 300 fpm		23.05	1,375	4,150	12,500	1,014
	5000	6,000 lb., 100' @ 275 fpm		24.77	1,450	4,350	13,100	1,068
	5100	For added heights up to 500', add	L.F.	.01	1.67	5	15	1.10
	5200	Jacks, hydraulic, 20 ton	Ea.	.05	2	6	18	1.60
	5500	100 ton		.40	12	36	108	10.40
	6100	Jacks, hydraulic, climbing w/50' jackrods, control console, 30 ton cap.		2.17	145	435	1,300	104.40
	6150	For each added 10' jackrod section, add		.05	3.33	10	30	2.40

01 54 33 | Equipment Rental

			UNIT	HOURLY OPER. COST	RENT PER DAY	RENT PER WEEK	RENT PER MONTH	EQUIPMENT COST/DAY	
60	6300	50 ton capacity	Ea.	3.49	232	695	2,075	166.95	60
	6350	For each added 10' jackrod section, add		.06	4	12	36	2.90	
	6500	125 ton capacity		9.13	610	1,825	5,475	438.05	
	6550	For each added 10' jackrod section, add		.62	41	123	370	29.55	
	6600	Cable jack, 10 ton capacity with 200' cable		1.82	122	365	1,100	87.60	
	6650	For each added 50' of cable, add		.22	14.65	44	132	10.55	
70	0010	**WELLPOINT EQUIPMENT RENTAL** without operators	R015433 -10						70
	0020	Based on 2 months rental							
	0100	Combination jetting & wellpoint pump, 60 H.P. diesel	Ea.	15.72	360	1,075	3,225	340.75	
	0200	High pressure gas jet pump, 200 H.P., 300 psi	"	33.93	305	920	2,750	455.45	
	0300	Discharge pipe, 8" diameter	L.F.	.01	.58	1.74	5.20	.40	
	0350	12" diameter		.01	.83	2.50	7.50	.60	
	0400	Header pipe, flows up to 150 GPM, 4" diameter		.01	.53	1.59	4.77	.40	
	0500	400 GPM, 6" diameter		.01	.62	1.86	5.60	.45	
	0600	800 GPM, 8" diameter		.01	.83	2.50	7.50	.60	
	0700	1,500 GPM, 10" diameter		.01	.88	2.64	7.90	.65	
	0800	2,500 GPM, 12" diameter		.03	1.70	5.10	15.30	1.20	
	0900	4,500 GPM, 16" diameter		.03	2.18	6.55	19.65	1.55	
	0950	For quick coupling aluminum and plastic pipe, add		.03	2.27	6.80	20.50	1.65	
	1100	Wellpoint, 25' long, with fittings & riser pipe, 1-1/2" or 2" diameter	Ea.	.07	4.52	13.55	40.50	3.25	
	1200	Wellpoint pump, diesel powered, 4" suction, 20 H.P.		7.02	207	620	1,850	180.20	
	1300	6" suction, 30 H.P.		9.42	257	770	2,300	229.35	
	1400	8" suction, 40 H.P.		12.76	350	1,050	3,150	312.10	
	1500	10" suction, 75 H.P.		18.83	410	1,225	3,675	395.65	
	1600	12" suction, 100 H.P.		27.32	660	1,975	5,925	613.55	
	1700	12" suction, 175 H.P.		39.09	725	2,175	6,525	747.75	
80	0010	**MARINE EQUIPMENT RENTAL** without operators	R015433 -10						80
	0200	Barge, 400 ton, 30' wide x 90' long	Ea.	17.69	1,150	3,475	10,400	836.50	
	0240	800 ton, 45' wide x 90' long		22.21	1,425	4,275	12,800	1,033	
	2000	Tugboat, diesel, 100 H.P.		29.66	230	690	2,075	375.25	
	2040	250 H.P.		57.58	415	1,250	3,750	710.70	
	2080	380 H.P.		125.36	1,250	3,750	11,300	1,753	
	3000	Small work boat, gas, 16-foot, 50 H.P.		11.38	46.50	139	415	118.85	
	4000	Large, diesel, 48-foot, 200 H.P.		74.90	1,325	3,975	11,900	1,394	

For customer support on your Facilities Maintenance & Repair Costs with RSMeans data, call 800.448.8182.

Crew No.	Bare Costs		In-House Costs		Incl. Subs O&P		Cost Per Labor-Hour		
Crew A-3H	Hr.	Daily	Hr.	Daily	Hr.	Daily	Bare Costs	In House	Incl. O&P
1 Equip. Oper. (crane)	$57.45	$ 459.60	$72.45	$ 579.60	$90.25	$ 722.00	$ 57.45	$ 72.45	$ 90.25
1 Hyd. Crane, 12 Ton (Daily)		708.90		708.90		779.79	88.61	88.61	97.47
8 L.H., Daily Totals		$1168.50		$1288.50		$1501.79	$146.06	$161.06	$187.72
Crew A-3I	Hr.	Daily	Hr.	Daily	Hr.	Daily	Bare Costs	In House	Incl. O&P
1 Equip. Oper. (crane)	$57.45	$ 459.60	$72.45	$ 579.60	$90.25	$ 722.00	$ 57.45	$ 72.45	$ 90.25
1 Hyd. Crane, 25 Ton (Daily)		785.85		785.85		864.43	98.23	98.23	108.05
8 L.H., Daily Totals		$1245.45		$1365.45		$1586.43	$155.68	$170.68	$198.30
Crew A-3J	Hr.	Daily	Hr.	Daily	Hr.	Daily	Bare Costs	In House	Incl. O&P
1 Equip. Oper. (crane)	$57.45	$ 459.60	$72.45	$ 579.60	$90.25	$ 722.00	$ 57.45	$ 72.45	$ 90.25
1 Hyd. Crane, 40 Ton (Daily)		1280.00		1280.00		1408.00	160.00	160.00	176.00
8 L.H., Daily Totals		$1739.60		$1859.60		$2130.00	$217.45	$232.45	$266.25
Crew A-3K	Hr.	Daily	Hr.	Daily	Hr.	Daily	Bare Costs	In House	Incl. O&P
1 Equip. Oper. (crane)	$57.45	$ 459.60	$72.45	$ 579.60	$90.25	$ 722.00	$ 53.33	$ 67.25	$ 83.78
1 Equip. Oper. (oiler)	49.20	393.60	62.05	496.40	77.30	618.40			
1 Hyd. Crane, 55 Ton (Daily)		1299.00		1299.00		1428.90			
1 P/U Truck, 3/4 Ton (Daily)		140.45		140.45		154.50	89.97	89.97	98.96
16 L.H., Daily Totals		$2292.65		$2515.45		$2923.80	$143.29	$157.22	$182.74
Crew A-3L	Hr.	Daily	Hr.	Daily	Hr.	Daily	Bare Costs	In House	Incl. O&P
1 Equip. Oper. (crane)	$57.45	$ 459.60	$72.45	$ 579.60	$90.25	$ 722.00	$ 53.33	$ 67.25	$ 83.78
1 Equip. Oper. (oiler)	49.20	393.60	62.05	496.40	77.30	618.40			
1 Hyd. Crane, 80 Ton (Daily)		2056.00		2056.00		2261.60			
1 P/U Truck, 3/4 Ton (Daily)		140.45		140.45		154.50	137.28	137.28	151.01
16 L.H., Daily Totals		$3049.65		$3272.45		$3756.49	$190.60	$204.53	$234.78
Crew A-3M	Hr.	Daily	Hr.	Daily	Hr.	Daily	Bare Costs	In House	Incl. O&P
1 Equip. Oper. (crane)	$57.45	$ 459.60	$72.45	$ 579.60	$90.25	$ 722.00	$ 53.33	$ 67.25	$ 83.78
1 Equip. Oper. (oiler)	49.20	393.60	62.05	496.40	77.30	618.40			
1 Hyd. Crane, 100 Ton (Daily)		2179.00		2179.00		2396.90			
1 P/U Truck, 3/4 Ton (Daily)		140.45		140.45		154.50	144.97	144.97	159.46
16 L.H., Daily Totals		$3172.65		$3395.45		$3891.80	$198.29	$212.22	$243.24
Crew A-16	Hr.	Daily	Hr.	Daily	Hr.	Daily	Bare Costs	In House	Incl. O&P
1 Maintenance Laborer	$30.80	$246.40	$40.10	$320.80	$49.65	$397.20	$30.80	$40.10	$49.65
1 Lawn Mower, Riding		99.20		99.20		109.12	12.40	12.40	13.64
8 L.H., Daily Totals		$345.60		$420.00		$506.32	$43.20	$52.50	$63.29
Crew A-17	Hr.	Daily	Hr.	Daily	Hr.	Daily	Bare Costs	In House	Incl. O&P
1 Maintenance Laborer	$30.80	$246.40	$40.10	$320.80	$49.65	$397.20	$30.80	$40.10	$49.65
1 Flatbed Truck, Gas, 1.5 Ton		195.20		195.20		214.72	24.40	24.40	26.84
8 L.H., Daily Totals		$441.60		$516.00		$611.92	$55.20	$64.50	$76.49
Crew A-18	Hr.	Daily	Hr.	Daily	Hr.	Daily	Bare Costs	In House	Incl. O&P
1 Maintenance Laborer	$30.80	$246.40	$40.10	$320.80	$49.65	$397.20	$30.80	$40.10	$49.65
1 Farm Tractor w/ Attachment		319.35		319.35		351.29	39.92	39.92	43.91
8 L.H., Daily Totals		$565.75		$640.15		$748.49	$70.72	$80.02	$93.56
Crew B-1	Hr.	Daily	Hr.	Daily	Hr.	Daily	Bare Costs	In House	Incl. O&P
1 Labor Foreman (outside)	$43.05	$ 344.40	$56.10	$ 448.80	$69.45	$ 555.60	$41.72	$54.37	$67.28
2 Laborers	41.05	656.80	53.50	856.00	66.20	1059.20			
24 L.H., Daily Totals		$1001.20		$1304.80		$1614.80	$41.72	$54.37	$67.28

Crew No.	Bare Costs		In-House Costs		Incl. Subs O&P		Cost Per Labor-Hour		

Crew B-17	Hr.	Daily	Hr.	Daily	Hr.	Daily	Bare Costs	In House	Incl. O&P
2 Laborers	$41.05	$ 656.80	$53.50	$ 856.00	$66.20	$1059.20	$45.19	$58.25	$72.25
1 Equip. Oper. (light)	51.65	413.20	65.15	521.20	81.15	649.20			
1 Truck Driver (heavy)	47.00	376.00	60.85	486.80	75.45	603.60			
1 Backhoe Loader, 48 H.P.		320.20		320.20		352.22			
1 Dump Truck, 8 C.Y., 220 H.P.		339.60		339.60		373.56	20.62	20.62	22.68
32 L.H., Daily Totals		$2105.80		$2523.80		$3037.78	$65.81	$78.87	$94.93

Crew B-18	Hr.	Daily	Hr.	Daily	Hr.	Daily	Bare Costs	In House	Incl. O&P
1 Labor Foreman (outside)	$43.05	$ 344.40	$56.10	$ 448.80	$69.45	$ 555.60	$41.72	$54.37	$67.28
2 Laborers	41.05	656.80	53.50	856.00	66.20	1059.20			
1 Vibrating Plate, Gas, 21"		40.55		40.55		44.60	1.69	1.69	1.86
24 L.H., Daily Totals		$1041.75		$1345.35		$1659.41	$43.41	$56.06	$69.14

Crew B-21	Hr.	Daily	Hr.	Daily	Hr.	Daily	Bare Costs	In House	Incl. O&P
1 Labor Foreman (outside)	$43.05	$ 344.40	$56.10	$ 448.80	$69.45	$ 555.60	$47.49	$61.36	$76.08
1 Skilled Worker	53.40	427.20	68.95	551.60	85.50	684.00			
1 Laborer	41.05	328.40	53.50	428.00	66.20	529.60			
.5 Equip. Oper. (crane)	57.45	229.80	72.45	289.80	90.25	361.00			
.5 S.P. Crane, 4x4, 5 Ton		129.72		129.72		142.70	4.63	4.63	5.10
28 L.H., Daily Totals		$1459.53		$1847.93		$2272.90	$52.13	$66.00	$81.17

Crew B-25B	Hr.	Daily	Hr.	Daily	Hr.	Daily	Bare Costs	In House	Incl. O&P
1 Labor Foreman (outside)	$43.05	$ 344.40	$56.10	$ 448.80	$69.45	$ 555.60	$45.90	$59.05	$ 73.25
7 Laborers	41.05	2298.80	53.50	2996.00	66.20	3707.20			
4 Equip. Oper. (medium)	55.10	1763.20	69.50	2224.00	86.55	2769.60			
1 Asphalt Paver, 130 H.P.		2076.00		2076.00		2283.60			
2 Tandem Rollers, 10 Ton		464.10		464.10		510.51			
1 Roller, Pneum. Whl., 12 Ton		338.35		338.35		372.19	29.98	29.98	32.98
96 L.H., Daily Totals		$7284.85		$8547.25		$10198.70	$75.88	$89.03	$106.24

Crew B-30	Hr.	Daily	Hr.	Daily	Hr.	Daily	Bare Costs	In House	Incl. O&P
1 Equip. Oper. (medium)	$55.10	$ 440.80	$69.50	$ 556.00	$86.55	$ 692.40	$ 49.70	$ 63.73	$ 79.15
2 Truck Drivers (heavy)	47.00	752.00	60.85	973.60	75.45	1207.20			
1 Hyd. Excavator, 1.5 C.Y.		908.70		908.70		999.57			
2 Dump Trucks, 12 C.Y., 400 H.P.		1134.10		1134.10		1247.51	85.12	85.12	93.63
24 L.H., Daily Totals		$3235.60		$3572.40		$4146.68	$134.82	$148.85	$172.78

Crew B-34P	Hr.	Daily	Hr.	Daily	Hr.	Daily	Bare Costs	In House	Incl. O&P
1 Pipe Fitter	$63.95	$ 511.60	$79.35	$ 634.80	$99.15	$ 793.20	$54.85	$69.27	$ 86.25
1 Truck Driver (light)	45.50	364.00	58.95	471.60	73.05	584.40			
1 Equip. Oper. (medium)	55.10	440.80	69.50	556.00	86.55	692.40			
1 Flatbed Truck, Gas, 3 Ton		245.95		245.95		270.55			
1 Backhoe Loader, 48 H.P.		320.20		320.20		352.22	23.59	23.59	25.95
24 L.H., Daily Totals		$1882.55		$2228.55		$2692.76	$78.44	$92.86	$112.20

Crew B-34Q	Hr.	Daily	Hr.	Daily	Hr.	Daily	Bare Costs	In House	Incl. O&P
1 Pipe Fitter	$63.95	$ 511.60	$79.35	$ 634.80	$99.15	$ 793.20	$55.63	$ 70.25	$ 87.48
1 Truck Driver (light)	45.50	364.00	58.95	471.60	73.05	584.40			
1 Equip. Oper. (crane)	57.45	459.60	72.45	579.60	90.25	722.00			
1 Flatbed Trailer, 25 Ton		133.00		133.00		146.30			
1 Dump Truck, 8 C.Y., 220 H.P.		339.60		339.60		373.56			
1 Hyd. Crane, 25 Ton		581.70		581.70		639.87	43.93	43.93	48.32
24 L.H., Daily Totals		$2389.50		$2740.30		$3259.33	$99.56	$114.18	$135.81

For customer support on your Facilities Maintenance & Repair Costs with RSMeans data, call 800.448.8182.

Crew No.	Bare Costs		In-House Costs		Incl. Subs O&P		Cost Per Labor-Hour		
Crew B-34R	Hr.	Daily	Hr.	Daily	Hr.	Daily	Bare Costs	In House	Incl. O&P
1 Pipe Fitter	$63.95	$ 511.60	$79.35	$ 634.80	$99.15	$ 793.20	$ 55.63	$ 70.25	$ 87.48
1 Truck Driver (light)	45.50	364.00	58.95	471.60	73.05	584.40			
1 Equip. Oper. (crane)	57.45	459.60	72.45	579.60	90.25	722.00			
1 Flatbed Trailer, 25 Ton		133.00		133.00		146.30			
1 Dump Truck, 8 C.Y., 220 H.P.		339.60		339.60		373.56			
1 Hyd. Crane, 25 Ton		581.70		581.70		639.87			
1 Hyd. Excavator, 1 C.Y.		749.70		749.70		824.67	75.17	75.17	82.68
24 L.H., Daily Totals		$3139.20		$3490.00		$4084.00	$130.80	$145.42	$170.17
Crew B-34S	Hr.	Daily	Hr.	Daily	Hr.	Daily	Bare Costs	In House	Incl. O&P
2 Pipe Fitters	$63.95	$1023.20	$79.35	$1269.60	$99.15	$1586.40	$ 58.09	$ 73.00	$ 91.00
1 Truck Driver (heavy)	47.00	376.00	60.85	486.80	75.45	603.60			
1 Equip. Oper. (crane)	57.45	459.60	72.45	579.60	90.25	722.00			
1 Flatbed Trailer, 40 Ton		182.50		182.50		200.75			
1 Truck Tractor, 6x4, 380 H.P.		487.60		487.60		536.36			
1 Hyd. Crane, 80 Ton		1487.00		1487.00		1635.70			
1 Hyd. Excavator, 2 C.Y.		1078.00		1078.00		1185.80	101.10	101.10	111.21
32 L.H., Daily Totals		$5093.90		$5571.10		$6470.61	$159.18	$174.10	$202.21
Crew B-34T	Hr.	Daily	Hr.	Daily	Hr.	Daily	Bare Costs	In House	Incl. O&P
2 Pipe Fitters	$63.95	$1023.20	$79.35	$1269.60	$99.15	$1586.40	$ 58.09	$ 73.00	$ 91.00
1 Truck Driver (heavy)	47.00	376.00	60.85	486.80	75.45	603.60			
1 Equip. Oper. (crane)	57.45	459.60	72.45	579.60	90.25	722.00			
1 Flatbed Trailer, 40 Ton		182.50		182.50		200.75			
1 Truck Tractor, 6x4, 380 H.P.		487.60		487.60		536.36			
1 Hyd. Crane, 80 Ton		1487.00		1487.00		1635.70	67.41	67.41	74.15
32 L.H., Daily Totals		$4015.90		$4493.10		$5284.81	$125.50	$140.41	$165.15
Crew B-35	Hr.	Daily	Hr.	Daily	Hr.	Daily	Bare Costs	In House	Incl. O&P
1 Labor Foreman (outside)	$43.05	$ 344.40	$56.10	$ 448.80	$69.45	$ 555.60	$51.22	$65.23	$81.11
1 Skilled Worker	53.40	427.20	68.95	551.60	85.50	684.00			
1 Welder (plumber)	63.15	505.20	78.35	626.80	97.95	783.60			
1 Laborer	41.05	328.40	53.50	428.00	66.20	529.60			
1 Equip. Oper. (crane)	57.45	459.60	72.45	579.60	90.25	722.00			
1 Equip. Oper. (oiler)	49.20	393.60	62.05	496.40	77.30	618.40			
1 Welder, Electric, 300 amp		56.10		56.10		61.71			
1 Hyd. Excavator, .75 C.Y.		681.25		681.25		749.38	15.36	15.36	16.90
48 L.H., Daily Totals		$3195.75		$3868.55		$4704.28	$66.58	$80.59	$98.01
Crew B-37	Hr.	Daily	Hr.	Daily	Hr.	Daily	Bare Costs	In House	Incl. O&P
1 Labor Foreman (outside)	$43.05	$ 344.40	$56.10	$ 448.80	$69.45	$ 555.60	$43.15	$55.88	$69.23
4 Laborers	41.05	1313.60	53.50	1712.00	66.20	2118.40			
1 Equip. Oper. (light)	51.65	413.20	65.15	521.20	81.15	649.20			
1 Tandem Roller, 5 Ton		152.45		152.45		167.69	3.18	3.18	3.49
48 L.H., Daily Totals		$2223.65		$2834.45		$3490.90	$46.33	$59.05	$72.73
Crew B-47H	Hr.	Daily	Hr.	Daily	Hr.	Daily	Bare Costs	In House	Incl. O&P
1 Skilled Worker Foreman (out)	$55.40	$ 443.20	$71.55	$ 572.40	$88.70	$ 709.60	$53.90	$69.60	$86.30
3 Skilled Workers	53.40	1281.60	68.95	1654.80	85.50	2052.00			
1 Flatbed Truck, Gas, 3 Ton		245.95		245.95		270.55	7.69	7.69	8.45
32 L.H., Daily Totals		$1970.75		$2473.15		$3032.15	$61.59	$77.29	$94.75
Crew B-55	Hr.	Daily	Hr.	Daily	Hr.	Daily	Bare Costs	In House	Incl. O&P
2 Laborers	$41.05	$ 656.80	$53.50	$ 856.00	$66.20	$1059.20	$42.53	$55.32	$ 68.48
1 Truck Driver (light)	45.50	364.00	58.95	471.60	73.05	584.40			
1 Truck-Mounted Earth Auger		756.20		756.20		831.82			
1 Flatbed Truck, Gas, 3 Ton		245.95		245.95		270.55	41.76	41.76	45.93
24 L.H., Daily Totals		$2022.95		$2329.75		$2745.97	$84.29	$97.07	$114.42

Crew No.	Bare Costs		In-House Costs		Incl. Subs O&P		Cost Per Labor-Hour		
Crew B-80	Hr.	Daily	Hr.	Daily	Hr.	Daily	Bare Costs	In House	Incl. O&P
1 Labor Foreman (outside)	$43.05	$ 344.40	$56.10	$ 448.80	$69.45	$ 555.60	$45.31	$58.42	$72.46
1 Laborer	41.05	328.40	53.50	428.00	66.20	529.60			
1 Truck Driver (light)	45.50	364.00	58.95	471.60	73.05	584.40			
1 Equip. Oper. (light)	51.65	413.20	65.15	521.20	81.15	649.20			
1 Flatbed Truck, Gas, 3 Ton		245.95		245.95		270.55			
1 Earth Auger, Truck-Mtd.		385.85		385.85		424.44	19.74	19.74	21.72
32 L.H., Daily Totals		$2081.80		$2501.40		$3013.78	$65.06	$78.17	$94.18
Crew B-80C	Hr.	Daily	Hr.	Daily	Hr.	Daily	Bare Costs	In House	Incl. O&P
2 Laborers	$41.05	$ 656.80	$53.50	$ 856.00	$66.20	$1059.20	$42.53	$55.32	$68.48
1 Truck Driver (light)	45.50	364.00	58.95	471.60	73.05	584.40			
1 Flatbed Truck, Gas, 1.5 Ton		195.20		195.20		214.72			
1 Manual Fence Post Auger, Gas		7.40		7.40		8.14	8.44	8.44	9.29
24 L.H., Daily Totals		$1223.40		$1530.20		$1866.46	$50.98	$63.76	$77.77
Crew D-1	Hr.	Daily	Hr.	Daily	Hr.	Daily	Bare Costs	In House	Incl. O&P
1 Bricklayer	$51.05	$408.40	$67.25	$538.00	$83.05	$ 664.40	$45.73	$60.23	$74.40
1 Bricklayer Helper	40.40	323.20	53.20	425.60	65.75	526.00			
16 L.H., Daily Totals		$731.60		$963.60		$1190.40	$45.73	$60.23	$74.40
Crew D-7	Hr.	Daily	Hr.	Daily	Hr.	Daily	Bare Costs	In House	Incl. O&P
1 Tile Layer	$47.85	$382.80	$60.50	$484.00	$75.30	$ 602.40	$42.80	$54.10	$67.35
1 Tile Layer Helper	37.75	302.00	47.70	381.60	59.40	475.20			
16 L.H., Daily Totals		$684.80		$865.60		$1077.60	$42.80	$54.10	$67.35
Crew D-8	Hr.	Daily	Hr.	Daily	Hr.	Daily	Bare Costs	In House	Incl. O&P
3 Bricklayers	$51.05	$1225.20	$67.25	$1614.00	$83.05	$1993.20	$46.79	$61.63	$76.13
2 Bricklayer Helpers	40.40	646.40	53.20	851.20	65.75	1052.00			
40 L.H., Daily Totals		$1871.60		$2465.20		$3045.20	$46.79	$61.63	$76.13
Crew E-11	Hr.	Daily	Hr.	Daily	Hr.	Daily	Bare Costs	In House	Incl. O&P
2 Painters, Struc. Steel	$44.65	$ 714.40	$63.65	$1018.40	$77.50	$1240.00	$45.50	$61.49	$75.59
1 Building Laborer	41.05	328.40	53.50	428.00	66.20	529.60			
1 Equip. Oper. (light)	51.65	413.20	65.15	521.20	81.15	649.20			
1 Air Compressor, 250 cfm		167.95		167.95		184.75			
1 Sandblaster, Portable, 3 C.F.		20.70		20.70		22.77			
1 Set Sand Blasting Accessories		14.90		14.90		16.39	6.36	6.36	7.00
32 L.H., Daily Totals		$1659.55		$2171.15		$2642.70	$51.86	$67.85	$82.58
Crew G-1	Hr.	Daily	Hr.	Daily	Hr.	Daily	Bare Costs	In House	Incl. O&P
1 Roofer Foreman (outside)	$47.05	$ 376.40	$68.50	$ 548.00	$83.10	$ 664.80	$42.14	$61.36	$74.44
4 Roofers Composition	45.05	1441.60	65.60	2099.20	79.60	2547.20			
2 Roofer Helpers	33.85	541.60	49.30	788.80	59.80	956.80			
1 Application Equipment		181.15		181.15		199.26			
1 Tar Kettle/Pot		178.20		178.20		196.02			
1 Crew Truck		154.35		154.35		169.79	9.17	9.17	10.09
56 L.H., Daily Totals		$2873.30		$3949.70		$4733.87	$51.31	$70.53	$84.53
Crew G-3	Hr.	Daily	Hr.	Daily	Hr.	Daily	Bare Costs	In House	Incl. O&P
2 Sheet Metal Workers	$60.95	$ 975.20	$76.70	$1227.20	$95.60	$1529.60	$51.00	$65.10	$80.90
2 Building Laborers	41.05	656.80	53.50	856.00	66.20	1059.20			
32 L.H., Daily Totals		$1632.00		$2083.20		$2588.80	$51.00	$65.10	$80.90
Crew G-5	Hr.	Daily	Hr.	Daily	Hr.	Daily	Bare Costs	In House	Incl. O&P
1 Roofer Foreman (outside)	$47.05	$ 376.40	$68.50	$ 548.00	$83.10	$ 664.80	$40.97	$59.66	$72.38
2 Roofers Composition	45.05	720.80	65.60	1049.60	79.60	1273.60			
2 Roofer Helpers	33.85	541.60	49.30	788.80	59.80	956.80			
1 Application Equipment		181.15		181.15		199.26	4.53	4.53	4.98
40 L.H., Daily Totals		$1819.95		$2567.55		$3094.47	$45.50	$64.19	$77.36

For customer support on your Facilities Maintenance & Repair Costs with RSMeans data, call 800.448.8182.

Crews - Maintenance

Crew No.	Bare Costs		In-House Costs		Incl. Subs O&P		Cost Per Labor-Hour		
	Hr.	Daily	Hr.	Daily	Hr.	Daily	Bare Costs	In House	Incl. O&P
Crew G-8									
1 Roofer Composition	$45.05	$ 360.40	$65.60	$ 524.80	$79.60	$ 636.80	$ 45.05	$ 65.60	$ 79.60
1 Telescoping Boom Lift, to 80'		670.65		670.65		737.72	83.83	83.83	92.21
8 L.H., Daily Totals		$1031.05		$1195.45		$1374.52	$128.88	$149.43	$171.81
Crew L-2	Hr.	Daily	Hr.	Daily	Hr.	Daily	Bare Costs	In House	Incl. O&P
1 Carpenter	$51.65	$413.20	$67.30	$538.40	$83.30	$ 666.40	$45.25	$59.40	$73.40
1 Carpenter Helper	38.85	310.80	51.50	412.00	63.50	508.00			
16 L.H., Daily Totals		$724.00		$950.40		$1174.40	$45.25	$59.40	$73.40
Crew L-4	Hr.	Daily	Hr.	Daily	Hr.	Daily	Bare Costs	In House	Incl. O&P
2 Skilled Workers	$53.40	$ 854.40	$68.95	$1103.20	$85.50	$1368.00	$48.55	$63.13	$78.17
1 Helper	38.85	310.80	51.50	412.00	63.50	508.00			
24 L.H., Daily Totals		$1165.20		$1515.20		$1876.00	$48.55	$63.13	$78.17
Crew Q-1	Hr.	Daily	Hr.	Daily	Hr.	Daily	Bare Costs	In House	Incl. O&P
1 Plumber	$63.15	$505.20	$78.35	$ 626.80	$97.95	$ 783.60	$56.83	$70.50	$88.13
1 Plumber Apprentice	50.50	404.00	62.65	501.20	78.30	626.40			
16 L.H., Daily Totals		$909.20		$1128.00		$1410.00	$56.83	$70.50	$88.13
Crew Q-2	Hr.	Daily	Hr.	Daily	Hr.	Daily	Bare Costs	In House	Incl. O&P
2 Plumbers	$63.15	$1010.40	$78.35	$1253.60	$97.95	$1567.20	$58.93	$73.12	$91.40
1 Plumber Apprentice	50.50	404.00	62.65	501.20	78.30	626.40	58.93	73.12	91.40
24 L.H., Daily Totals		$1414.40		$1754.80		$2193.60	$58.93	$73.12	$91.40
Crew Q-4	Hr.	Daily	Hr.	Daily	Hr.	Daily	Bare Costs	In House	Incl. O&P
1 Plumber Foreman (inside)	$63.65	$ 509.20	$78.95	$ 631.60	$98.70	$ 789.60	$60.11	$74.58	$93.22
1 Plumber	63.15	505.20	78.35	626.80	97.95	783.60			
1 Welder (plumber)	63.15	505.20	78.35	626.80	97.95	783.60			
1 Plumber Apprentice	50.50	404.00	62.65	501.20	78.30	626.40			
1 Welder, Electric, 300 amp		56.10		56.10		61.71	1.75	1.75	1.93
32 L.H., Daily Totals		$1979.70		$2442.50		$3044.91	$61.87	$76.33	$95.15
Crew Q-5	Hr.	Daily	Hr.	Daily	Hr.	Daily	Bare Costs	In House	Incl. O&P
1 Steamfitter	$63.95	$511.60	$79.35	$ 634.80	$99.15	$ 793.20	$57.55	$71.40	$89.22
1 Steamfitter Apprentice	51.15	409.20	63.45	507.60	79.30	634.40			
16 L.H., Daily Totals		$920.80		$1142.40		$1427.60	$57.55	$71.40	$89.22
Crew Q-6	Hr.	Daily	Hr.	Daily	Hr.	Daily	Bare Costs	In House	Incl. O&P
2 Steamfitters	$63.95	$1023.20	$79.35	$1269.60	$99.15	$1586.40	$59.68	$74.05	$92.53
1 Steamfitter Apprentice	51.15	409.20	63.45	507.60	79.30	634.40	59.68	74.05	92.53
24 L.H., Daily Totals		$1432.40		$1777.20		$2220.80	$59.68	$74.05	$92.53
Crew Q-7	Hr.	Daily	Hr.	Daily	Hr.	Daily	Bare Costs	In House	Incl. O&P
1 Steamfitter Foreman (inside)	$64.45	$ 515.60	$79.95	$ 639.60	$99.95	$ 799.60	$60.88	$75.53	$94.39
2 Steamfitters	63.95	1023.20	79.35	1269.60	99.15	1586.40			
1 Steamfitter Apprentice	51.15	409.20	63.45	507.60	79.30	634.40			
32 L.H., Daily Totals		$1948.00		$2416.80		$3020.40	$60.88	$75.53	$94.39
Crew Q-8	Hr.	Daily	Hr.	Daily	Hr.	Daily	Bare Costs	In House	Incl. O&P
1 Steamfitter Foreman (inside)	$64.45	$ 515.60	$79.95	$ 639.60	$99.95	$ 799.60	$60.88	$75.53	$94.39
1 Steamfitter	63.95	511.60	79.35	634.80	99.15	793.20			
1 Welder (steamfitter)	63.95	511.60	79.35	634.80	99.15	793.20			
1 Steamfitter Apprentice	51.15	409.20	63.45	507.60	79.30	634.40			
1 Welder, Electric, 300 amp		56.10		56.10		61.71	1.75	1.75	1.93
32 L.H., Daily Totals		$2004.10		$2472.90		$3082.11	$62.63	$77.28	$96.32

For customer support on your Facilities Maintenance & Repair Costs with RSMeans data, call 800.448.8182.

Crews - Maintenance

Crew No.	Bare Costs		In-House Costs		Incl. Subs O&P		Cost Per Labor-Hour		
Crew Q-9	Hr.	Daily	Hr.	Daily	Hr.	Daily	Bare Costs	In House	Incl. O&P
1 Sheet Metal Worker	$60.95	$487.60	$76.70	$613.60	$95.60	$764.80	$54.85	$69.03	$86.03
1 Sheet Metal Apprentice	48.75	390.00	61.35	490.80	76.45	611.60			
16 L.H., Daily Totals		$877.60		$1104.40		$1376.40	$54.85	$69.03	$86.03
Crew Q-10	Hr.	Daily	Hr.	Daily	Hr.	Daily	Bare Costs	In House	Incl. O&P
2 Sheet Metal Workers	$60.95	$975.20	$76.70	$1227.20	$95.60	$1529.60	$56.88	$71.58	$89.22
1 Sheet Metal Apprentice	48.75	390.00	61.35	490.80	76.45	611.60			
24 L.H., Daily Totals		$1365.20		$1718.00		$2141.20	$56.88	$71.58	$89.22
Crew Q-13	Hr.	Daily	Hr.	Daily	Hr.	Daily	Bare Costs	In House	Incl. O&P
1 Sprinkler Foreman (inside)	$62.00	$496.00	$77.15	$617.20	$96.35	$770.80	$58.55	$72.84	$91.00
2 Sprinkler Installers	61.50	984.00	76.50	1224.00	95.60	1529.60			
1 Sprinkler Apprentice	49.20	393.60	61.20	489.60	76.45	611.60			
32 L.H., Daily Totals		$1873.60		$2330.80		$2912.00	$58.55	$72.84	$91.00
Crew Q-14	Hr.	Daily	Hr.	Daily	Hr.	Daily	Bare Costs	In House	Incl. O&P
1 Asbestos Worker	$57.35	$458.80	$73.60	$588.80	$91.40	$731.20	$51.63	$66.25	$82.28
1 Asbestos Apprentice	45.90	367.20	58.90	471.20	73.15	585.20			
16 L.H., Daily Totals		$826.00		$1060.00		$1316.40	$51.63	$66.25	$82.28
Crew Q-15	Hr.	Daily	Hr.	Daily	Hr.	Daily	Bare Costs	In House	Incl. O&P
1 Plumber	$63.15	$505.20	$78.35	$626.80	$97.95	$783.60	$56.83	$70.50	$88.13
1 Plumber Apprentice	50.50	404.00	62.65	501.20	78.30	626.40			
1 Welder, Electric, 300 amp		56.10		56.10		61.71	3.51	3.51	3.86
16 L.H., Daily Totals		$965.30		$1184.10		$1471.71	$60.33	$74.01	$91.98
Crew Q-16	Hr.	Daily	Hr.	Daily	Hr.	Daily	Bare Costs	In House	Incl. O&P
2 Plumbers	$63.15	$1010.40	$78.35	$1253.60	$97.95	$1567.20	$58.93	$73.12	$91.40
1 Plumber Apprentice	50.50	404.00	62.65	501.20	78.30	626.40			
1 Welder, Electric, 300 amp		56.10		56.10		61.71	2.34	2.34	2.57
24 L.H., Daily Totals		$1470.50		$1810.90		$2255.31	$61.27	$75.45	$93.97
Crew Q-20	Hr.	Daily	Hr.	Daily	Hr.	Daily	Bare Costs	In House	Incl. O&P
1 Sheet Metal Worker	$60.95	$487.60	$76.70	$613.60	$95.60	$764.80	$55.89	$70.02	$87.34
1 Sheet Metal Apprentice	48.75	390.00	61.35	490.80	76.45	611.60			
.5 Electrician	60.05	240.20	74.00	296.00	92.60	370.40			
20 L.H., Daily Totals		$1117.80		$1400.40		$1746.80	$55.89	$70.02	$87.34
Crew R-3	Hr.	Daily	Hr.	Daily	Hr.	Daily	Bare Costs	In House	Incl. O&P
1 Electrician Foreman	$60.55	$484.40	$74.60	$596.80	$93.35	$746.80	$59.73	$73.93	$92.43
1 Electrician	60.05	480.40	74.00	592.00	92.60	740.80			
.5 Equip. Oper. (crane)	57.45	229.80	72.45	289.80	90.25	361.00			
.5 S.P. Crane, 4x4, 5 Ton		129.72		129.72		142.70	6.49	6.49	7.13
20 L.H., Daily Totals		$1324.33		$1608.33		$1991.30	$66.22	$80.42	$99.56
Crew R-26	Hr.	Daily	Hr.	Daily	Hr.	Daily	Bare Costs	In House	Incl. O&P
2 Electricians	$60.05	$960.80	$74.00	$1184.00	$92.60	$1481.60	$60.05	$74.00	$92.60
1 Articulating Boom Lift, to 45'		244.60		244.60		269.06	15.29	15.29	16.82
16 L.H., Daily Totals		$1205.40		$1428.60		$1750.66	$75.34	$89.29	$109.42

For customer support on your Facilities Maintenance & Repair Costs with RSMeans data, call 800.448.8182.

Travel Costs

The following table is used to estimate the cost of in-house staff or contractor's personnel to travel to and from the job site. The amount incurred must be added to the in-house total labor cost or contractor's billing application to calculate the total cost with travel.

In-House Staff–Travel Times versus Costs

Travel Time Versus Costs	Crew Abbr. Rate with Mark-ups	Skwk	Clab	Clam	Asbe	Bric	Carp	Elec	Pord	Plum	Rofc	Shee	Stpi
		$68.95	$53.50	$40.10	$73.60	$67.25	$67.30	$74.00	$55.70	$78.35	$65.00	$76.70	$79.35
Round Trip Travel Time (hours)	Crew Size												
0.50	1	34.48	26.75	20.05	36.80	33.63	33.65	37.00	27.85	39.18	32.50	38.35	39.68
	2	68.95	53.50	40.10	73.60	67.25	67.30	74.00	55.70	78.35	65.00	76.70	79.35
	3	103.43	80.25	60.15	110.40	100.88	100.95	111.00	83.55	117.53	97.50	115.05	119.03
0.75	1	51.71	40.13	30.08	55.20	50.44	50.48	55.50	41.78	58.76	48.75	57.53	59.51
	2	103.43	80.25	60.15	110.40	100.88	100.95	111.00	83.55	117.53	97.50	115.05	119.03
	3	155.14	120.38	90.23	165.60	151.31	151.43	166.50	125.33	176.29	146.25	172.58	178.54
1.00	1	68.95	53.50	40.10	73.60	67.25	67.30	74.00	55.70	78.35	65.00	76.70	79.35
	2	137.90	107.00	80.20	147.20	134.50	134.60	148.00	111.40	156.70	130.00	153.40	158.70
	3	206.85	160.50	120.30	220.80	201.75	201.90	222.00	167.10	235.05	195.00	230.10	238.05
1.50	1	103.43	80.25	60.15	110.40	100.88	100.95	111.00	83.55	117.53	97.50	115.05	119.03
	2	206.85	160.50	120.30	220.80	201.75	201.90	222.00	167.10	235.05	195.00	230.10	238.05
	3	310.28	240.75	180.45	331.20	302.63	302.85	333.00	250.65	352.58	292.50	345.15	357.08
2.00	1	137.90	107.00	80.20	147.20	134.50	134.60	148.00	111.40	156.70	130.00	153.40	158.70
	2	275.80	214.00	160.40	294.40	269.00	269.20	296.00	222.80	313.40	260.00	306.80	317.40
	3	413.70	321.00	240.60	441.60	403.50	403.80	444.00	334.20	470.10	390.00	460.20	476.10

Installing Contractors–Travel Times versus Costs

Travel Time Versus Costs	Crew Abbr. Rate with O & P	Skwk	Clab	Clam	Asbe	Bric	Carp	Elec	Pord	Plum	Rofc	Shee	Stpi
		$85.50	$66.20	$49.65	$91.40	$83.05	$83.30	$92.60	$69.10	$97.95	$79.60	$95.60	$99.15
Round Trip Travel Time (hours)	Crew Size												
0.50	1	42.75	33.10	24.83	45.70	41.53	41.65	46.30	34.55	48.98	39.80	47.80	49.58
	2	85.50	66.20	49.65	91.40	83.05	83.30	92.60	69.10	97.95	79.60	95.60	99.15
	3	128.25	99.30	74.48	137.10	124.58	124.95	138.90	103.65	146.93	119.40	143.40	148.73
0.75	1	64.13	49.65	37.24	68.55	62.29	62.48	69.45	51.83	73.46	59.70	71.70	74.36
	2	128.25	99.30	74.48	137.10	124.58	124.95	138.90	103.65	146.93	119.40	143.40	148.73
	3	192.38	148.95	111.71	205.65	186.86	187.43	208.35	155.48	220.39	179.10	215.10	223.09
1.00	1	85.50	66.20	49.65	91.40	83.05	83.30	92.60	69.10	97.95	79.60	95.60	99.15
	2	171.00	132.40	99.30	182.80	166.10	166.60	185.20	138.20	195.90	159.20	191.20	198.30
	3	256.50	198.60	148.95	274.20	249.15	249.90	277.80	207.30	293.85	238.80	286.80	297.45
1.50	1	128.25	99.30	74.48	137.10	124.58	124.95	138.90	103.65	146.93	119.40	143.40	148.73
	2	256.50	198.60	148.95	274.20	249.15	249.90	277.80	207.30	293.85	238.80	286.80	297.45
	3	384.75	297.90	223.43	411.30	373.73	374.85	416.70	310.95	440.78	358.20	430.20	446.18
2.00	1	171.00	132.40	99.30	182.80	166.10	166.60	185.20	138.20	195.90	159.20	191.20	198.30
	2	342.00	264.80	198.60	365.60	332.20	333.20	370.40	276.40	391.80	318.40	382.40	396.60
	3	513.00	397.20	297.90	548.40	498.30	499.80	555.60	414.60	587.70	477.60	573.60	594.90

617

Table H1010-101 General Contractor's Overhead

There are two distinct types of overhead on a construction project: project overhead and main office overhead. Project overhead includes those costs at a construction site not directly associated with the installation of construction materials. Examples of project overhead costs include the following:

1. Superintendent
2. Construction office and storage trailers
3. Temporary sanitary facilities
4. Temporary utilities
5. Security fencing
6. Photographs
7. Clean up
8. Performance and payment bonds

The above project overhead items are also referred to as General Requirements and therefore are estimated in Division 1. Division 1 is the first division listed in the CSI MasterFormat but it is usually the last division estimated. The sum of the costs in Divisions 1 through 49 is referred to as the sum of the direct costs.

All construction projects also include indirect costs. The primary components of indirect costs are the contractor's main office overhead and profit. The amount of the main office overhead expense varies depending on the following:

1. Owner's compensation
2. Project managers and estimator's wages
3. Clerical support wages
4. Office rent and utilities
5. Corporate legal and accounting costs
6. Advertising
7. Automobile expenses
8. Association dues
9. Travel and entertainment expenses

These costs are usually calculated as a percentage of annual sales volume. This percentage can range from 35% for a small contractor doing less than $500,000 to 5% for a large contractor with sales in excess of $100 million.

Table H1010-102 Main Office Expense

A general contractor's main office expense consists of many items not detailed in the front portion of the data set. The percentage of main office expense declines with increased annual volume of the contractor. Typical main office expense ranges from 2% to 20% with the median about 7.2% of total volume. This equals about 7.7% of direct costs. The following are approximate percentages of total overhead for different items usually included in a general contractor's main office overhead. With different accounting procedures, these percentages may vary.

Item	Typical Range			Average
Managers', clerical and estimators' salaries	40 %	to	55 %	48%
Profit sharing, pension and bonus plans	2	to	20	12
Insurance	5	to	8	6
Estimating and project management (not including salaries)	5	to	9	7
Legal, accounting and data processing	0.5	to	5	3
Automobile and light truck expense	2	to	8	5
Depreciation of overhead capital expenditures	2	to	6	4
Maintenance of office equipment	0.1	to	1.5	1
Office rental	3	to	5	4
Utilities including phone and light	1	to	3	2
Miscellaneous	5	to	15	8
Total				100%

For customer support on your Facilities Maintenance & Repair Costs with RSMeans data, call 800.448.8182.

Table H1010-201   Architectural Fees

Tabulated below are typical percentage fees by project size for good professional architectural service. Fees may vary from those listed depending upon degree of design difficulty and economic conditions in any particular area.

Rates can be interpolated horizontally and vertically. Various portions of the same project requiring different rates should be adjusted

proportionately. For alterations, add 50% to the fee for the first $500,000 of project cost and add 25% to the fee for project cost over $500,000.

Architectural fees tabulated below include Structural, Mechanical and Electrical Engineering Fees. They do not include the fees for special consultants such as kitchen planning, security, acoustical, interior design, etc.

Building Types	Total Project Size in Thousands of Dollars						
	100	250	500	1,000	5,000	10,000	50,000
Factories, garages, warehouses, repetitive housing	9.0%	8.0%	7.0%	6.2%	5.3%	4.9%	4.5%
Apartments, banks, schools, libraries, offices, municipal buildings	12.2	12.3	9.2	8.0	7.0	6.6	6.2
Churches, hospitals, homes, laboratories, museums, research	15.0	13.6	12.7	11.9	9.5	8.8	8.0
Memorials, monumental work, decorative furnishings	—	16.0	14.5	13.1	10.0	9.0	8.3

Table H1010-202   Engineering Fees

Typical **Structural Engineering Fees** based on type of construction and total project size. These fees are included in Architectural Fees.

Type of Construction	Total Project Size (in thousands of dollars)			
	$500	$500-$1,000	$1,000-$5,000	Over $5000
Industrial buildings, factories & warehouses	Technical payroll times 2.0 to 2.5	1.60%	1.25%	1.00%
Hotels, apartments, offices, dormitories, hospitals, public buildings, food stores		2.00%	1.70%	1.20%
Museums, banks, churches and cathedrals		2.00%	1.75%	1.25%
Thin shells, prestressed concrete, earthquake resistive		2.00%	1.75%	1.50%
Parking ramps, auditoriums, stadiums, convention halls, hangars & boiler houses		2.50%	2.00%	1.75%
Special buildings, major alterations, underpinning & future expansion		Add to above 0.5%	Add to above 0.5%	Add to above 0.5%

For complex reinforced concrete or unusually complicated structures, add 20% to 50%.

Table H1010-203   Mechanical and Electrical Fees

Typical **Mechanical and Electrical Engineering Fees** based on the size of the subcontract. The fee structure for both is shown below. These fees are included in Architectural Fees.

Type of Construction	Subcontract Size							
	$25,000	$50,000	$100,000	$225,000	$350,000	$500,000	$750,000	$1,000,000
Simple structures	6.4%	5.7%	4.8%	4.5%	4.4%	4.3%	4.2%	4.1%
Intermediate structures	8.0	7.3	6.5	5.6	5.1	5.0	4.9	4.8
Complex structures	10.1	9.0	9.0	8.0	7.5	7.5	7.0	7.0

For renovations, add 15% to 25% to applicable fee.

For customer support on your Facilities Maintenance & Repair Costs with RSMeans data, call 800.448.8182.

Table H1010-301 Builder's Risk Insurance

Builder's risk insurance is insurance on a building during construction. Premiums are paid by the owner or the contractor. Blasting, collapse and underground insurance would raise total insurance costs.

Table H1010-302 Performance Bond

This table shows the cost of a performance bond for a construction job scheduled to be completed in 12 months. Add 1% of the premium cost per month for jobs requiring more than 12 months to complete. The rates are "standard" rates offered to contractors that the bonding company considers financially sound and capable of doing the work. Preferred rates are offered by some bonding companies based upon financial strength of the contractor. Actual rates vary from contractor to contractor and from bonding company to bonding company. Contractors should prequalify through a bonding agency before submitting a bid on a contract that requires a bond.

Contract Amount	Building Construction Class B Projects			Highways & Bridges					
				Class A New Construction			Class A-1 Highway Resurfacing		
First $ 100,000 bid	$25.00 per M			$15.00 per M			$9.40 per M		
Next 400,000 bid	$ 2,500	plus	$15.00 per M	$ 1,500	plus	$10.00 per M	$ 940	plus	$7.20 per M
Next 2,000,000 bid	8,500	plus	10.00 per M	5,500	plus	7.00 per M	3,820	plus	5.00 per M
Next 2,500,000 bid	28,500	plus	7.50 per M	19,500	plus	5.50 per M	15,820	plus	4.50 per M
Next 2,500,000 bid	47,250	plus	7.00 per M	33,250	plus	5.00 per M	28,320	plus	4.50 per M
Over 7,500,000 bid	64,750	plus	6.00 per M	45,750	plus	4.50 per M	39,570	plus	4.00 per M

For customer support on your Facilities Maintenance & Repair Costs with RSMeans data, call 800.448.8182.

Table H1010-401 Workers' Compensation Insurance Rates by Trade

The table below tabulates the national averages for workers' compensation insurance rates by trade and type of building. The average "Insurance Rate" is multiplied by the "% of Building Cost" for each trade. This produces the "Workers' Compensation" cost by % of total labor cost, to be added for each trade by building type to determine the weighted average workers' compensation rate for the building types analyzed.

Trade	Insurance Rate (% Labor Cost) Range		Average	% of Building Cost Office Bldgs.	Schools & Apts.	Mfg.	Workers' Compensation Office Bldgs.	Schools & Apts.	Mfg.
Excavation, Grading, etc.	2.3 % to	19.2%	8.5%	4.8%	4.9%	4.5%	0.41%	0.42%	0.38%
Piles & Foundations	3.9 to	27.3	13.4	7.1	5.2	8.7	0.95	0.70	1.17
Concrete	3.5 to	25.6	11.8	5.0	14.8	3.7	0.59	1.75	0.44
Masonry	3.6 to	50.4	13.8	6.9	7.5	1.9	0.95	1.04	0.26
Structural Steel	4.9 to	43.8	21.2	10.7	3.9	17.6	2.27	0.83	3.73
Miscellaneous & Ornamental Metals	2.8 to	27.0	10.6	2.8	4.0	3.6	0.30	0.42	0.38
Carpentry & Millwork	3.7 to	31.9	13.0	3.7	4.0	0.5	0.48	0.52	0.07
Metal or Composition Siding	5.1 to	113.7	19.0	2.3	0.3	4.3	0.44	0.06	0.82
Roofing	5.7 to	103.7	29.0	2.3	2.6	3.1	0.67	0.75	0.90
Doors & Hardware	3.1 to	31.9	11.0	0.9	1.4	0.4	0.10	0.15	0.04
Sash & Glazing	3.9 to	24.3	12.1	3.5	4.0	1.0	0.42	0.48	0.12
Lath & Plaster	2.7 to	34.4	10.7	3.3	6.9	0.8	0.35	0.74	0.09
Tile, Marble & Floors	2.2 to	21.5	8.7	2.6	3.0	0.5	0.23	0.26	0.04
Acoustical Ceilings	1.9 to	29.7	8.5	2.4	0.2	0.3	0.20	0.02	0.03
Painting	3.7 to	36.7	11.2	1.5	1.6	1.6	0.17	0.18	0.18
Interior Partitions	3.7 to	31.9	13.0	3.9	4.3	4.4	0.51	0.56	0.57
Miscellaneous Items	2.1 to	97.7	11.2	5.2	3.7	9.7	0.58	0.42	1.09
Elevators	1.4 to	11.4	4.7	2.1	1.1	2.2	0.10	0.05	0.10
Sprinklers	2.0 to	16.4	6.7	0.5	—	2.0	0.03	—	0.13
Plumbing	1.5 to	14.6	6.3	4.9	7.2	5.2	0.31	0.45	0.33
Heat., Vent., Air Conditioning	3.0 to	17.0	8.3	13.5	11.0	12.9	1.12	0.91	1.07
Electrical	1.9 to	11.3	5.2	10.1	8.4	11.1	0.53	0.44	0.58
Total	1.4 % to	113.7%	—	100.0%	100.0%	100.0%	11.71%	11.15%	12.52%
		Overall Weighted Average	11.79%						

Table H1010-402 Workers' Compensation Insurance Rates by States

The table below lists the weighted average Workers' Compensation base rate for each state with a factor comparing this with the national average of 11.8%.

State	Weighted Average	Factor	State	Weighted Average	Factor	State	Weighted Average	Factor
Alabama	13.9%	129	Kentucky	11.3%	105	North Dakota	7.4%	69
Alaska	11.0	102	Louisiana	20.0	185	Ohio	6.2	57
Arizona	9.1	84	Maine	9.8	91	Oklahoma	8.7	81
Arkansas	5.7	53	Maryland	11.2	104	Oregon	9.0	83
California	23.6	219	Massachusetts	9.1	84	Pennsylvania	24.9	231
Colorado	8.1	75	Michigan	8.0	74	Rhode Island	10.6	98
Connecticut	17.0	157	Minnesota	16.2	150	South Carolina	20.3	188
Delaware	10.7	99	Mississippi	11.4	106	South Dakota	10.3	95
District of Columbia	9.1	84	Missouri	14.0	130	Tennessee	7.9	73
Florida	11.3	105	Montana	7.8	72	Texas	6.2	57
Georgia	32.9	305	Nebraska	13.4	124	Utah	7.4	69
Hawaii	9.2	85	Nevada	8.9	82	Vermont	10.8	100
Idaho	9.2	85	New Hampshire	11.3	105	Virginia	7.4	69
Illinois	20.1	186	New Jersey	15.4	143	Washington	8.9	82
Indiana	3.7	34	New Mexico	13.7	127	West Virginia	4.7	44
Iowa	12.1	112	New York	19.1	177	Wisconsin	13.3	123
Kansas	6.6	61	North Carolina	17.1	158	Wyoming	6.3	58
			Weighted Average for U.S. is	11.8% of payroll = 100%				

The weighted average skilled worker rate for 35 trades is 11.8%. For bidding purposes, apply the full value of Workers' Compensation directly to total labor costs, or if labor is 38%, materials 42% and overhead and profit 20% of total cost, carry 38/80 x 11.8% = 6.0% of cost (before overhead and profit)

into overhead. Rates vary not only from state to state but also with the experience rating of the contractor.

Rates are the most current available at the time of publication.

621

General Conditions — RH1020-300 Overhead & Miscellaneous Data

Unemployment Taxes and Social Security Taxes

State unemployment tax rates vary not only from state to state, but also with the experience rating of the contractor. The federal unemployment tax rate is 6.0% of the first $7,000 of wages. This is reduced by a credit of up to 5.4% for timely payment to the state. The minimum federal unemployment tax is 0.6% after all credits.

Social security (FICA) for 2019 is estimated at time of publication to be 7.65% of wages up to $128,400.

General Conditions — RH1020-400 Overtime

Overtime

One way to improve the completion date of a project or eliminate negative float from a schedule is to compress activity duration times. This can be achieved by increasing the crew size or working overtime with the proposed crew.

To determine the costs of working overtime to compress activity duration times, consider the following examples. Below is an overtime efficiency and cost chart based on a five, six, or seven day week with an eight through twelve hour day. Payroll percentage increases for time and one half and double times are shown for the various working days.

Days per Week	Hours per Day	Production Efficiency					Payroll Cost Factors	
		1st Week	2nd Week	3rd Week	4th Week	Average 4 Weeks	@ 1-1/2 Times	@ 2 Times
5	8	100%	100%	100%	100%	100%	1.000	1.000
	9	100	100	95	90	96	1.056	1.111
	10	100	95	90	85	93	1.100	1.200
	11	95	90	75	65	81	1.136	1.273
	12	90	85	70	60	76	1.167	1.333
6	8	100	100	95	90	96	1.083	1.167
	9	100	95	90	85	93	1.130	1.259
	10	95	90	85	80	88	1.167	1.333
	11	95	85	70	65	79	1.197	1.394
	12	90	80	65	60	74	1.222	1.444
7	8	100	95	85	75	89	1.143	1.286
	9	95	90	80	70	84	1.183	1.365
	10	90	85	75	65	79	1.214	1.429
	11	85	80	65	60	73	1.240	1.481
	12	85	75	60	55	69	1.262	1.524

General Conditions — RH1030-100 General

Sales Tax by State

State sales tax on materials is tabulated below (5 states have no sales tax). Many states allow local jurisdictions, such as a county or city, to levy additional sales tax.

Some projects may be sales tax exempt, particularly those constructed with public funds.

State	Tax (%)	State	Tax (%)	State	Tax (%)	State	Tax (%)
Alabama	4	Illinois	6.25	Montana	0	Rhode Island	7
Alaska	0	Indiana	7	Nebraska	5.5	South Carolina	6
Arizona	5.6	Iowa	6	Nevada	6.85	South Dakota	4.5
Arkansas	6.5	Kansas	6.5	New Hampshire	0	Tennessee	7
California	7.25	Kentucky	6	New Jersey	6.625	Texas	6.25
Colorado	2.9	Louisiana	4.45	New Mexico	5.125	Utah	5.95
Connecticut	6.35	Maine	5.5	New York	4	Vermont	6
Delaware	0	Maryland	6	North Carolina	4.75	Virginia	5.3
District of Columbia	5.75	Massachusetts	6.25	North Dakota	5	Washington	6.5
Florida	6	Michigan	6	Ohio	5.75	West Virginia	6
Georgia	4	Minnesota	6.875	Oklahoma	4.5	Wisconsin	5
Hawaii	4	Mississippi	7	Oregon	0	Wyoming	4
Idaho	6	Missouri	4.225	Pennsylvania	6	Average	5.10 %

For customer support on your Facilities Maintenance & Repair Costs with RSMeans data, call 800.448.8182.

Table H1040-101   Steel Tubular Scaffolding

On new construction, tubular scaffolding is efficient up to 60' high or five stories. Above this it is usually better to use hung scaffolding if construction permits. Swing scaffolding operations may interfere with tenants. In this case, the tubular is more practical at all heights.

In repairing or cleaning the front of an existing building, the cost of tubular scaffolding per S.F. of building front goes up as the height increases above the first tier. The first tier cost is relatively high due to leveling and alignment.

The minimum efficient crew for erection is three workers. For heights over 50', a crew of four is more efficient. Use two or more on top and two at the bottom for handing up or hoisting. Four workers can erect and dismantle about nine frames per hour up to five stories. From five to eight stories, they will average six frames per hour. With 7' horizontal spacing, this will run about 400 S.F. and 265 S.F. of wall surface, respectively. Time for

placing planks must be added to the above. On heights above 50', five planks can be placed per labor-hour.

The cost per 1,000 S.F. of building front in the table below was developed by pricing the materials required for a typical tubular scaffolding system eleven frames long and two frames high. Planks were figured five wide for standing plus two wide for materials.

Frames are 5' wide and usually spaced 7' O.C. horizontally. Sidewalk frames are 6' wide. Rental rates will be lower for jobs over three months' duration.

For jobs under twenty-five frames, add 50% to rental cost. These figures do not include accessories which are listed separately below. Large quantities for long periods can reduce rental rates by 20%.

Item	Unit	Monthly Rent	No. of Pieces	Rental per Month
			Per 1,000 S.F. of Building Front	
5' Wide Standard Frame, 6'-4" High	Ea.	$ 5.40	24	$129.60
Leveling Jack & Plate		1.78	24	42.72
Cross Brace		0.89	44	39.16
Side Arm Bracket, 21"		1.78	12	21.36
Guardrail Post		0.89	12	10.68
Guardrail, 7' section		0.89	22	19.58
Stairway Section		31.50	2	63.00
Walk-Through Frame Guardrail		2.22	2	4.44
			Total	$330.54
			Per C.S.F., 1 Use/Mo.	$ 33.05

Scaffolding is often used as falsework over 15' high during construction of cast-in-place concrete beams and slabs. Two-foot wide scaffolding is generally used for heavy beam construction. The span between frames depends upon the load to be carried, with a maximum span of 5'.

Heavy duty shoring frames with a capacity of 10,000#/leg can be spaced up to 10' O. C., depending upon form support design and loading.

Scaffolding used as horizontal shoring requires less than half the material required with conventional shoring.

On new construction, erection is done by carpenters.

Rolling towers supporting horizontal shores can reduce labor and speed the job. For maintenance work, catwalks with spans up to 70' can be supported by the rolling towers.

For customer support on your Facilities Maintenance & Repair Costs with RSMeans data, call 800.448.8182.

General: The following information on current city cost indexes is calculated for over 930 zip code locations in the United States and Canada. Index figures for both material and installation are based upon the 30 major city average of 100 and represent the cost relationship on July 1, 2018.

In addition to index adjustment, the user should consider:
1. productivity
2. management efficiency
3. competitive conditions
4. automation

5. restrictive union practices
6. unique local requirements
7. regional variations due to specific building codes

The weighted-average index is calculated from about 66 materials, 6 equipment types and 21 building trades. The component contribution of these in a model building is tabulated in Table J1010-011 below.

If the systems component distribution of a building is unknown, the weighted-average index can be used to adjust the cost for any city.

Table J1010-011 Labor, Material, and Equipment Cost Distribution for Weighted Average Listed by System Division

Division No.	Building System	Percentage	Division No.	Building System	Percentage
A	Substructure	6.1%	D10	Services: Conveying	4.0%
B10	Shell: Superstructure	19.2	D20-40	Mechanical	22.5
B20	Exterior Closure	12.2	D50	Electrical	11.9
B30	Roofing	2.5	E	Equipment & Furnishings	2.1
C	Interior Construction	15.8	G	Site Work	3.7
				Total weighted average (Div. A-G)	100.0%

How to Use the Component Indexes: Table J1010-012 below shows how the average costs obtained from this data set for each division should be adjusted for your particular building in your city. The example is a building adjusted for New York, NY. Indexes for other cities are tabulated in Division RJ1030. These indexes should also be used when compiling data on the form in Division K1010.

Table J1010-012 Adjustment of "Book" Costs to a Particular City

| Systems Division Number | System Description | New York, NY | | | | | | | |
		City Cost Index	Cost Factor	Book Cost	Adjusted Cost	City Cost Index	Cost Factor	Book Cost	Adjusted Cost
A	Substructure	141.9	1.419	$ 107,600	$ 152,700				
B10	Shell: Superstructure	132.7	1.327	297,300	394,500				
B20	Exterior Closure	144.7	1.447	201,500	291,600				
B30	Roofing	133.6	1.336	44,400	59,300				
C	Interior Construction	140.9	1.409	29,100	41,000				
D10	Services: Conveying	114.5	1.145	57,500	65,800				
D20-40	Mechanical	140.7	1.407	356,000	500,900				
D50	Electrical	150.7	1.507	199,200	300,200				
E	Equipment	109.2	1.092	30,100	32,900				
G	Site Work	112.5	1.125	84,700	95,300				
Total Cost, (Div. A-G)				$1,407,400	$1,934,200				

Alternate Method: Use the weighted-average index for your city rather than the component indexes. **For example:**

(Total weighted-average index A-G)/100 x (total cost) = Adjusted city cost
New York, NY 1.379 x $1,407,400 = $1,940,800

For customer support on your Facilities Maintenance & Repair Costs with RSMeans data, call 800.448.8182.

City to City: To convert known or estimated costs from one city to another, cost indexes can be used as follows:

$$\text{Unknown City Cost} = \text{Known Cost} \times \frac{\text{Unknown City Index}}{\text{Known City Index}}$$

For example: If the building cost in Boston, MA, is $2,000,000, how much would a duplicated building cost in Los Angeles, CA?

$$\text{L.A. Cost} = \$2,000,000 \times \frac{\text{(Los Angeles) } 112.3}{\text{(Boston) } 113.9} = \$1,971,905$$

The table below lists both the RSMeans City Cost Index based on January 1, 1993 = 100 as well as the computed value of an index based on January 1, 2019 costs. Since the January 1, 2019 figure is estimated, space is left to write in the actual index figures as they become available through the quarterly *RSMeans Construction Cost Indexes*. To compute the actual index based on January 1, 2019 = 100, divide the Quarterly City Cost Index for a particular year by the actual January 1, 2019 Quarterly City Cost Index. Space has been left to advance the index figures as the year progresses.

Table J1020-011 Historical Cost Indexes

Year	Historical Cost Index Jan. 1, 1993 = 100 Est.	Historical Cost Index Jan. 1, 1993 = 100 Actual	Current Index Based on Jan. 1, 2019 = 100 Est.	Current Index Based on Jan. 1, 2019 = 100 Actual	Year	Historical Cost Index Jan. 1, 1993 = 100 Actual	Current Index Based on Jan. 1, 2019 = 100 Est.	Current Index Based on Jan. 1, 2019 = 100 Actual	Year	Historical Cost Index Jan. 1, 1993 = 100 Actual	Current Index Based on Jan. 1, 2019 = 100 Est.	Current Index Based on Jan. 1, 2019 = 100 Actual
Oct 2019*					July 2004	143.7	63.2		July 1986	84.2	37.1	
July 2019*					2003	132	58.1		1985	82.6	36.3	
April 2019*					2002	128.7	56.6		1984	82.0	36.1	
Jan 2019*	227.3		100.0	100.0	2001	125.1	55.0		1983	80.2	35.3	
July 2018		222.9	98.1		2000	120.9	53.2		1982	76.1	33.5	
2017		213.6	94.0		1999	117.6	51.7		1981	70.0	30.8	
2016		207.3	91.2		1998	115.1	50.6		1980	62.9	27.7	
2015		206.2	90.7		1997	112.8	49.6		1979	57.8	25.4	
2014		204.9	90.1		1996	110.2	48.5		1978	53.5	23.5	
2013		201.2	88.5		1995	107.6	47.3		1977	49.5	21.8	
2012		194.6	85.6		1994	104.4	45.9		1976	46.9	20.6	
2011		191.2	84.1		1993	101.7	44.7		1975	44.8	19.7	
2010		183.5	80.7		1992	99.4	43.7		1974	41.4	18.2	
2009		180.1	79.2		1991	96.8	42.6		1973	37.7	16.6	
2008		180.4	79.4		1990	94.3	41.5		1972	34.8	15.3	
2007		169.4	74.5		1989	92.1	40.5		1971	32.1	14.1	
2006		162.0	71.3		1988	89.9	39.5		1970	28.7	12.6	
2005		151.6	66.7		1987	87.7	38.6		1969	26.9	11.8	

To find the **current cost** from a project built previously in either the same city or a different city, the following formula is used:

$$\text{Present Cost (City X)} = \frac{\text{Current HCI} \times \text{CCI (City X)}}{\text{Previous HCI} \times \text{CCI (City Y)}} \times \text{Former Cost (City Y)}$$

For example: Find the construction cost of a building to be built in San Francisco, CA, as of January 1, 2019 when the identical building cost $500,000 in Boston, MA, on July 1, 1968.

$$\text{Jan. 1, 2019 (San Francisco)} = \frac{\text{(San Francisco) } 128.5 \times 129.1}{\text{(Boston) } 24.9 \times 115.2} \times \$500,000 = \$2,891,500$$

Note: The City Cost Indexes for Canada can be used to convert U.S. national averages to local costs in Canadian dollars.

To Project Future Construction Costs: Using the results of the last five years average percentage increase as a basis, an average increase of 1.6% could be used.

The historical index figures above are compiled from the Means Construction Index Service.

Example:

To estimate and compare the cost of a building in Toronto, ON in 2019 with the known cost of $600,000 (US$) in New York, NY in 2019:

INDEX Toronto = 110.1

INDEX New York = 132.1

$$\frac{\text{INDEX Toronto}}{\text{INDEX New York}} \times \text{Cost New York} = \text{Cost Toronto}$$

$$\frac{110.1}{132.1} \times \$600,000 = .834 \times \$600,000 = \$500,076$$

The construction cost of the building in Toronto is $500,076 (CN$).

*Historical Cost Index updates and other resources are provided on the following website: **http://info.thegordiangroup.com/RSMeans.html**

How to Use the City Cost Indexes

What you should know before you begin

RSMeans City Cost Indexes (CCI) are an extremely useful tool for when you want to compare costs from city to city and region to region.

This publication contains average construction cost indexes for 317 U.S. and Canadian cities covering over 930 three-digit zip code locations, as listed directly under each city.

Keep in mind that a City Cost Index number is a percentage ratio of a specific city's cost to the national average cost of the same item at a stated time period.

In other words, these index figures represent relative construction factors (or, if you prefer, multipliers) for material and installation costs, as well as the weighted average for Total In Place costs for each Uniformat II Element. Installation costs include both labor and equipment rental costs.

The 30 City Average Index is the average of 30 major U.S. cities and serves as a national average.

Index figures for both material and installation are based on the 30 major city average of 100 and represent the cost relationship as of July 1, 2018. The index for each element is computed from representative material and labor quantities for that element. The weighted average for each city is a weighted total of the components listed above it. It does not include relative productivity between trades or cities.

As changes occur in local material prices, labor rates, and equipment rental rates (including fuel costs), the impact of these changes should be accurately measured by the change in the City Cost Index for each particular city (as compared to the 30 city average).

Therefore, if you know (or have estimated) building costs in one city today, you can easily convert those costs to expected building costs in another city.

In addition, by using the Historical Cost Index, you can easily convert national average building costs at a particular time to the approximate building costs for some other time. The City Cost Indexes can then be applied to calculate the costs for a particular city.

Quick calculations

Location Adjustment Using the City Cost Indexes:

$$\frac{\text{Index for City A}}{\text{Index for City B}} \times \text{Cost in City B} = \text{Cost in City A}$$

Time Adjustment for the National Average Using the Historical Cost Index:

$$\frac{\text{Index for Year A}}{\text{Index for Year B}} \times \text{Cost in Year B} = \text{Cost in Year A}$$

Adjustment from the National Average:

$$\frac{\text{Index for City A}}{100} \times \text{National Average Cost} = \text{Cost in City A}$$

Since each of the other RSMeans data sets contains many different items, any *one* item multiplied by the particular city index may give incorrect results. However, the larger the number of items compiled, the closer the results should be to actual costs for that particular city.

The City Cost Indexes for Canadian cities are calculated using Canadian material and equipment prices and labor rates in Canadian dollars. Therefore, indexes for Canadian cities can be used to convert U.S. national average prices to local costs in Canadian dollars.

How to use this section

1. Compare costs from city to city.

In using the RSMeans Indexes, remember that an index number is not a fixed number but a ratio: It's a percentage ratio of a building component's cost at any stated time to the national average cost of that same component at the same time period. Put in the form of an equation:

$$\frac{\text{Specific City Cost}}{\text{National Average Cost}} \times 100 = \text{City Index Number}$$

Therefore, when making cost comparisons between cities, do not subtract one city's index number from the index number of another city and read the result as a percentage difference. Instead, divide one city's index number by that of the other city. The resulting number may then be used as a multiplier to calculate cost differences from city to city.

The formula used to find cost differences between cities for the purpose of comparison is as follows:

$$\frac{\text{City A Index}}{\text{City B Index}} \times \text{City B Cost (Known)} = \text{City A Cost (Unknown)}$$

In addition, you can use RSMeans CCI to calculate and compare costs division by division between cities using the same basic formula. (Just be sure that you're comparing similar divisions.)

2. Compare a specific city's construction costs with the national average.

When you're studying construction location feasibility, it's advisable to compare a prospective project's cost index with an index of the national average cost.

For example, divide the weighted average index of construction costs of a specific city by that of the 30 City Average, which = 100.

$$\frac{\text{City Index}}{100} = \text{\% of National Average}$$

As a result, you get a ratio that indicates the relative cost of construction in that city in comparison with the national average.

3. Convert U.S. national average to actual costs in Canadian City.

$$\frac{\text{Index for Canadian City}}{100} \times \text{National Average Cost} = \text{Cost in Canadian City in \$ CAN}$$

4. Adjust construction cost data based on a national average.

When you use a source of construction cost data which is based on a national average (such as RSMeans cost data), it is necessary to adjust those costs to a specific location.

$$\frac{\text{City Index}}{100} \times \frac{\text{Cost Based on}}{\text{National Average Costs}} = \frac{\text{City Cost}}{\text{(Unknown)}}$$

5. When applying the City Cost Indexes to demolition projects, use the appropriate division installation index. For example, for removal of existing doors and windows, use the Division 8 (Openings) index.

What you might like to know about how we developed the Indexes

The information presented in the CCI is organized according to the Uniformat II classification system.

To create a reliable index, RSMeans researched the building type most often constructed in the United States and Canada. Because it was concluded that no one type of building completely represented the building construction industry, nine different types of buildings were combined to create a composite model.

The exact material, labor, and equipment quantities are based on detailed analyses of these nine building types, and then each quantity is weighted in proportion to expected usage. These various material items, labor hours, and equipment rental rates are thus combined to form a composite building representing as closely as possible the actual usage of materials, labor, and equipment in the North American building construction industry.

The following structures were chosen to make up that composite model:

1. Factory, 1 story
2. Office, 2–4 stories
3. Store, Retail
4. Town Hall, 2–3 stories
5. High School, 2–3 stories
6. Hospital, 4–8 stories
7. Garage, Parking
8. Apartment, 1–3 stories
9. Hotel/Motel, 2–3 stories

For the purposes of ensuring the timeliness of the data, the components of the index for the composite model have been streamlined. They currently consist of:

- specific quantities of 66 commonly used construction materials;
- specific labor-hours for 21 building construction trades; and
- specific days of equipment rental for 6 types of construction equipment (normally used to install the 66 material items by the 21 trades.) Fuel costs and routine maintenance costs are included in the equipment cost.

Material and equipment price quotations are gathered quarterly from cities in the United States and Canada. These prices and the latest negotiated labor wage rates for 21 different building trades are used to compile the quarterly update of the City Cost Index.

The 30 major U.S. cities used to calculate the national average are:

Atlanta, GA
Baltimore, MD
Boston, MA
Buffalo, NY
Chicago, IL
Cincinnati, OH
Cleveland, OH
Columbus, OH
Dallas, TX
Denver, CO
Detroit, MI
Houston, TX
Indianapolis, IN
Kansas City, MO
Los Angeles, CA

Memphis, TN
Milwaukee, WI
Minneapolis, MN
Nashville, TN
New Orleans, LA
New York, NY
Philadelphia, PA
Phoenix, AZ
Pittsburgh, PA
St. Louis, MO
San Antonio, TX
San Diego, CA
San Francisco, CA
Seattle, WA
Washington, DC

What the CCI does not indicate

The weighted average for each city is a total of the divisional components weighted to reflect typical usage. It does not include the productivity variations between trades or cities.

In addition, the CCI does not take into consideration factors such as the following:

- managerial efficiency
- competitive conditions
- automation
- restrictive union practices
- unique local requirements
- regional variations due to specific building codes

ALABAMA

DIV. NO.	BUILDING SYSTEMS	BIRMINGHAM			HUNTSVILLE			MOBILE			MONTGOMERY			TUSCALOOSA		
		MAT.	INST.	TOTAL	MAT.	INST.	TOTAL	MAT.	INST.	TOTAL	MAT.	INST.	TOTAL	MAT.	INST.	TOTAL
A	Substructure	99.2	74.0	84.5	94.4	73.2	82.1	90.9	72.7	80.3	90.4	72.6	80.0	97.1	73.5	83.4
B10	Shell: Superstructure	99.4	82.9	93.1	100.5	82.6	93.7	98.4	82.5	92.4	97.5	82.2	91.7	100.4	82.7	93.7
B20	Exterior Closure	84.5	69.6	77.8	85.6	67.8	77.6	85.5	62.2	75.0	83.6	62.5	74.1	85.2	68.5	77.7
B30	Roofing	96.0	67.1	84.4	95.1	66.2	83.5	93.9	66.2	82.8	92.4	66.4	82.0	95.2	66.8	83.8
C	Interior Construction	95.3	67.4	83.6	97.3	67.5	84.8	93.0	65.0	81.3	92.4	64.9	80.9	97.3	67.2	84.7
D10	Services: Conveying	100.0	85.7	95.7	100.0	84.8	95.5	100.0	85.9	95.8	100.0	85.1	95.6	100.0	85.1	95.6
D20 - 40	Mechanical	100.0	66.4	86.4	100.1	63.5	85.3	100.0	62.8	84.9	100.0	65.0	85.8	100.1	66.5	86.5
D50	Electrical	95.3	61.2	77.9	92.1	65.7	78.7	99.4	58.1	78.4	99.9	73.9	86.7	91.7	61.2	76.2
E	Equipment & Furnishings	100.0	65.6	98.1	100.0	65.1	98.0	100.0	66.8	98.1	100.0	66.1	98.1	100.0	65.5	98.1
G	Site Work	92.7	92.7	92.7	87.1	92.2	90.5	97.3	88.6	91.5	95.4	88.0	90.4	87.8	92.4	90.9
A-G	WEIGHTED AVERAGE	96.5	71.7	85.9	96.5	71.4	85.7	96.1	68.8	84.1	95.0	71.4	85.2	96.0	71.5	85.0

ALASKA / ARIZONA

DIV. NO.	BUILDING SYSTEMS	ANCHORAGE			FAIRBANKS			JUNEAU			FLAGSTAFF			MESA/TEMPE		
		MAT.	INST.	TOTAL	MAT.	INST.	TOTAL	MAT.	INST.	TOTAL	MAT.	INST.	TOTAL	MAT.	INST.	TOTAL
A	Substructure	110.8	120.5	116.5	115.7	120.7	118.6	122.4	120.5	121.3	93.8	76.3	83.6	91.0	76.0	82.2
B10	Shell: Superstructure	117.5	110.5	114.8	119.9	110.7	116.4	118.0	110.5	115.2	98.1	71.6	88.0	99.7	71.9	89.1
B20	Exterior Closure	144.3	118.7	132.8	133.3	118.9	126.8	132.9	118.7	126.5	113.4	61.3	90.0	101.0	61.9	83.4
B30	Roofing	169.7	118.4	149.1	181.1	119.7	156.4	185.9	118.4	158.8	98.8	71.3	87.8	98.6	69.5	86.9
C	Interior Construction	126.6	117.5	122.8	131.5	118.0	125.8	128.4	117.5	123.8	101.0	67.9	87.1	96.7	67.0	84.3
D10	Services: Conveying	100.0	108.4	102.5	100.0	108.5	102.5	100.0	108.4	102.5	100.0	84.7	95.5	100.0	85.1	95.6
D20 - 40	Mechanical	100.6	106.1	102.8	100.4	110.3	104.4	100.6	106.1	102.8	100.2	77.2	90.9	100.0	77.2	90.8
D50	Electrical	113.8	108.8	111.3	116.7	108.8	112.7	100.4	108.8	104.7	101.7	62.2	81.6	97.2	64.4	80.5
E	Equipment & Furnishings	100.0	115.8	100.9	100.0	115.8	100.9	100.0	115.8	100.9	100.0	70.2	98.3	100.0	68.6	98.2
G	Site Work	120.1	131.2	127.5	121.0	131.3	127.9	141.5	131.2	134.6	89.9	94.9	93.3	91.6	96.0	94.6
A-G	WEIGHTED AVERAGE	117.5	113.6	115.8	118.2	114.7	116.7	116.5	113.6	115.3	101.0	71.7	88.4	98.6	72.0	87.2

ARIZONA / ARKANSAS

DIV. NO.	BUILDING SYSTEMS	PHOENIX			PRESCOTT			TUCSON			FORT SMITH			JONESBORO		
		MAT.	INST.	TOTAL	MAT.	INST.	TOTAL	MAT.	INST.	TOTAL	MAT.	INST.	TOTAL	MAT.	INST.	TOTAL
A	Substructure	90.9	77.0	82.8	92.0	75.9	82.6	89.1	76.4	81.7	86.5	72.8	78.5	84.1	75.4	79.0
B10	Shell: Superstructure	100.9	73.8	90.6	97.9	71.3	87.8	100.0	72.1	89.4	98.2	71.3	88.0	92.2	77.3	86.5
B20	Exterior Closure	101.1	64.5	84.6	98.8	61.2	81.9	95.1	61.3	79.9	84.2	62.5	74.5	82.6	44.2	65.4
B30	Roofing	99.8	70.4	88.0	97.4	71.5	87.0	99.6	69.6	87.5	101.7	62.6	86.0	106.1	60.4	87.8
C	Interior Construction	99.8	69.3	87.1	99.8	67.0	86.1	93.7	68.0	82.9	93.4	63.0	80.7	92.9	54.5	76.9
D10	Services: Conveying	100.0	86.4	96.0	100.0	84.7	95.5	100.0	85.1	95.6	100.0	80.4	94.2	100.0	66.0	89.9
D20 - 40	Mechanical	99.9	78.6	91.2	100.2	77.0	90.8	100.0	77.1	90.7	100.1	49.1	79.5	100.4	52.2	80.9
D50	Electrical	103.0	62.2	82.2	101.4	64.4	82.6	99.5	59.5	79.2	93.3	58.4	75.6	99.4	60.9	79.9
E	Equipment & Furnishings	100.0	71.6	98.4	100.0	68.9	98.3	100.0	69.8	98.3	100.0	62.9	97.9	100.0	57.0	97.6
G	Site Work	92.1	94.7	93.8	78.3	94.9	89.4	87.5	96.0	93.2	83.3	89.7	87.6	101.8	107.2	105.4
A-G	WEIGHTED AVERAGE	100.0	73.0	88.4	98.7	71.7	87.1	97.6	71.4	86.4	95.1	63.6	81.6	94.7	62.7	81.0

ARKANSAS / CALIFORNIA

DIV. NO.	BUILDING SYSTEMS	LITTLE ROCK			PINE BLUFF			TEXARKANA			ANAHEIM			BAKERSFIELD		
		MAT.	INST.	TOTAL	MAT.	INST.	TOTAL	MAT.	INST.	TOTAL	MAT.	INST.	TOTAL	MAT.	INST.	TOTAL
A	Substructure	86.5	73.3	78.8	81.8	73.2	76.8	88.4	71.2	78.4	89.2	130.1	113.0	93.6	128.9	114.1
B10	Shell: Superstructure	98.1	72.0	88.2	99.0	71.9	88.7	94.0	68.8	84.4	102.6	122.6	110.2	97.5	121.8	106.8
B20	Exterior Closure	83.5	62.3	74.0	95.2	62.5	80.5	83.6	43.8	65.7	94.4	138.1	114.1	90.4	136.9	111.3
B30	Roofing	96.0	62.9	82.7	97.5	63.0	83.6	98.1	60.6	83.0	107.1	138.3	119.6	104.1	128.9	114.1
C	Interior Construction	91.9	64.5	80.5	95.2	63.0	81.7	98.7	54.2	80.1	101.7	134.2	115.3	94.6	132.5	110.4
D10	Services: Conveying	100.0	80.4	94.2	100.0	80.4	94.2	100.0	67.0	90.2	100.0	117.9	105.3	100.0	118.1	105.4
D20 - 40	Mechanical	99.8	52.6	80.7	100.0	52.6	80.8	100.0	55.6	82.1	100.0	131.8	112.8	100.1	129.0	111.8
D50	Electrical	100.4	66.8	83.3	94.0	58.9	76.2	95.7	58.9	77.0	90.2	111.0	100.8	107.1	110.7	109.0
E	Equipment & Furnishings	100.0	63.4	97.9	100.0	63.0	97.9	100.0	57.0	97.6	100.0	137.4	102.1	100.0	135.9	102.0
G	Site Work	90.2	90.0	90.1	88.8	89.7	89.4	98.4	91.8	94.0	101.2	110.8	107.6	97.2	108.6	104.9
A-G	WEIGHTED AVERAGE	95.4	65.9	82.8	96.7	64.6	82.9	95.6	60.4	80.5	98.9	126.9	110.9	98.0	125.3	109.7

CALIFORNIA

DIV. NO.	BUILDING SYSTEMS	FRESNO			LOS ANGELES			OAKLAND			OXNARD			REDDING		
		MAT.	INST.	TOTAL	MAT.	INST.	TOTAL	MAT.	INST.	TOTAL	MAT.	INST.	TOTAL	MAT.	INST.	TOTAL
A	Substructure	95.7	132.9	117.4	88.7	130.7	113.2	102.2	140.5	124.5	99.1	129.0	116.5	109.3	133.0	123.1
B10	Shell: Superstructure	98.4	125.8	108.8	95.8	124.3	106.7	101.8	131.4	113.1	95.5	122.3	105.7	111.2	126.1	116.8
B20	Exterior Closure	95.6	143.4	117.1	100.9	139.9	118.4	114.1	154.3	132.2	93.0	136.2	112.4	119.4	139.4	128.4
B30	Roofing	95.2	132.7	110.3	96.9	135.9	112.6	105.8	157.0	126.4	104.6	136.5	117.4	132.5	139.5	135.3
C	Interior Construction	94.5	145.6	115.9	102.1	134.5	115.7	98.9	162.9	125.7	92.1	133.8	109.6	113.2	149.0	128.2
D10	Services: Conveying	100.0	123.1	106.9	100.0	118.9	105.6	100.0	124.4	107.3	100.0	118.1	105.4	100.0	123.1	106.9
D20 - 40	Mechanical	100.2	130.3	112.3	99.9	131.9	112.8	100.2	168.0	127.6	100.1	131.8	112.9	100.4	130.3	112.5
D50	Electrical	96.8	108.0	102.5	98.6	130.5	114.8	101.5	161.0	131.8	101.0	117.5	109.4	103.9	122.4	113.3
E	Equipment & Furnishings	100.0	154.0	103.0	100.0	138.0	102.1	100.0	172.1	104.1	100.0	137.5	102.1	100.0	155.0	103.1
G	Site Work	101.4	108.0	105.9	96.0	112.7	107.1	119.8	110.7	113.8	101.7	106.7	105.0	128.4	107.5	114.4
A-G	WEIGHTED AVERAGE	97.7	129.3	111.2	98.7	130.4	112.3	102.7	151.2	123.5	97.2	127.1	110.0	109.1	131.5	118.7

CALIFORNIA

DIV. NO.	BUILDING SYSTEMS	RIVERSIDE			SACRAMENTO			SAN DIEGO			SAN FRANCISCO			SAN JOSE		
		MAT.	INST.	TOTAL	MAT.	INST.	TOTAL	MAT.	INST.	TOTAL	MAT.	INST.	TOTAL	MAT.	INST.	TOTAL
A	Substructure	92.9	129.7	114.3	88.5	136.1	116.2	94.7	122.2	110.7	113.3	141.7	129.8	107.8	138.4	125.6
B10	Shell: Superstructure	103.5	122.6	110.8	95.9	124.2	106.7	97.1	119.3	105.6	108.0	136.3	118.8	101.2	134.0	113.7
B20	Exterior Closure	91.0	137.8	112.1	104.4	144.2	122.3	99.9	130.6	113.7	123.1	152.7	136.4	112.3	149.8	129.1
B30	Roofing	107.3	138.3	119.7	115.2	140.2	125.2	101.2	119.9	108.7	111.4	155.9	129.3	106.2	156.7	126.5
C	Interior Construction	101.9	134.2	115.4	105.1	150.2	123.9	101.8	125.2	111.6	102.7	164.5	128.5	96.0	162.7	123.9
D10	Services: Conveying	100.0	117.9	105.3	100.0	123.7	107.0	100.0	116.0	104.8	100.0	124.4	107.3	100.0	123.8	107.1
D20 - 40	Mechanical	100.0	131.8	112.8	100.1	129.9	112.2	99.9	128.4	111.4	100.1	185.2	134.5	100.0	168.6	127.7
D50	Electrical	90.0	113.5	101.9	97.0	122.4	109.9	100.6	102.7	101.7	101.6	176.3	139.6	98.9	169.9	135.0
E	Equipment & Furnishings	100.0	137.5	102.1	100.0	157.5	103.2	100.0	124.5	101.4	100.0	172.8	104.1	100.0	172.5	104.1
G	Site Work	100.0	108.8	105.9	101.6	116.7	111.7	104.4	108.0	106.8	120.0	114.2	116.1	132.5	102.2	112.2
A-G	WEIGHTED AVERAGE	98.8	127.1	110.9	100.1	132.7	114.1	99.6	120.8	108.7	106.3	158.1	128.5	102.1	151.7	123.4

For customer support on your Facilities Maintenance & Repair Costs with RSMeans data, call 800.448.8182.

DIV. NO.	BUILDING SYSTEMS	CALIFORNIA SANTA BARBARA MAT.	INST.	TOTAL	STOCKTON MAT.	INST.	TOTAL	VALLEJO MAT.	INST.	TOTAL	COLORADO COLORADO SPRINGS MAT.	INST.	TOTAL	DENVER MAT.	INST.	TOTAL
A	Substructure	98.9	129.3	116.6	95.1	133.7	117.6	94.5	140.5	121.3	109.3	74.2	88.8	113.5	75.7	91.5
B10	Shell: Superstructure	95.8	122.2	105.9	100.8	126.6	110.6	99.5	128.3	110.5	96.8	73.8	88.0	99.9	73.7	89.9
B20	Exterior Closure	92.1	135.7	111.7	100.4	139.9	118.1	92.6	150.9	118.8	100.0	65.6	84.5	100.5	65.8	84.9
B30	Roofing	101.4	136.0	115.2	110.2	138.4	121.5	107.9	153.5	126.2	106.9	71.9	92.8	104.5	71.1	91.1
C	Interior Construction	92.8	133.8	110.0	102.5	150.3	122.5	105.7	163.2	129.7	99.3	65.5	85.1	103.4	65.3	87.4
D10	Services: Conveying	100.0	117.1	105.1	100.0	123.1	106.9	100.0	123.4	107.0	100.0	87.2	96.2	100.0	86.6	96.0
D20 - 40	Mechanical	100.1	131.8	112.9	100.0	130.3	112.2	100.2	147.9	119.5	100.2	73.8	89.5	100.0	73.6	89.4
D50	Electrical	94.2	114.2	104.3	96.0	114.3	105.3	94.5	127.9	111.4	101.4	74.9	87.9	103.2	80.2	91.5
E	Equipment & Furnishings	100.0	137.3	102.1	100.0	157.1	103.2	100.0	170.4	104.0	100.0	65.8	98.1	100.0	66.0	98.1
G	Site Work	101.6	108.7	106.3	102.5	107.5	105.9	103.9	116.5	112.3	101.4	91.2	94.6	107.0	102.4	103.9
A-G	WEIGHTED AVERAGE	96.5	126.7	109.4	100.3	130.8	113.4	99.5	141.5	117.5	100.0	73.0	88.4	101.8	74.4	90.0

DIV. NO.	BUILDING SYSTEMS	COLORADO FORT COLLINS MAT.	INST.	TOTAL	GRAND JUNCTION MAT.	INST.	TOTAL	GREELEY MAT.	INST.	TOTAL	PUEBLO MAT.	INST.	TOTAL	CONNECTICUT BRIDGEPORT MAT.	INST.	TOTAL
A	Substructure	118.1	75.8	93.4	111.6	78.1	92.1	103.0	77.6	88.2	102.7	73.7	85.8	106.7	121.2	115.2
B10	Shell: Superstructure	97.5	74.0	88.6	103.5	76.0	93.0	94.4	75.3	87.2	102.7	74.4	92.0	91.7	120.1	102.5
B20	Exterior Closure	107.4	63.4	87.6	116.6	66.1	93.9	98.4	63.6	82.8	95.7	63.9	81.4	101.5	128.9	113.8
B30	Roofing	106.5	71.1	92.3	106.8	72.3	92.9	105.7	71.8	92.1	106.2	70.1	91.7	97.8	124.3	108.5
C	Interior Construction	98.1	68.4	85.7	105.0	72.0	91.2	97.4	71.7	86.6	99.2	66.1	85.4	95.5	122.4	106.7
D10	Services: Conveying	100.0	86.5	96.0	100.0	87.8	96.4	100.0	86.5	96.0	100.0	88.0	96.4	100.0	112.9	103.8
D20 - 40	Mechanical	100.1	73.3	89.2	99.9	75.8	90.1	100.1	73.3	89.2	99.9	73.8	89.4	100.1	118.5	107.5
D50	Electrical	98.0	77.2	87.4	96.1	51.5	73.4	98.0	77.2	87.4	97.0	64.8	80.6	96.0	107.6	101.9
E	Equipment & Furnishings	100.0	71.0	98.4	100.0	74.0	98.5	100.0	75.9	98.6	100.0	65.1	98.0	100.0	116.9	101.0
G	Site Work	113.4	95.5	101.4	135.2	96.6	109.4	98.7	95.5	96.6	126.1	89.0	101.3	101.4	102.6	102.2
A-G	WEIGHTED AVERAGE	101.1	73.7	89.4	104.5	72.2	90.7	98.3	74.7	88.2	100.5	71.4	88.0	97.6	118.4	106.5

DIV. NO.	BUILDING SYSTEMS	CONNECTICUT BRISTOL MAT.	INST.	TOTAL	HARTFORD MAT.	INST.	TOTAL	NEW BRITAIN MAT.	INST.	TOTAL	NEW HAVEN MAT.	INST.	TOTAL	NORWALK MAT.	INST.	TOTAL
A	Substructure	102.4	121.1	113.3	103.7	121.1	113.9	103.5	121.1	113.8	104.5	121.3	114.3	105.6	121.6	114.9
B10	Shell: Superstructure	90.8	119.9	101.9	94.6	120.0	104.3	88.4	119.9	100.4	88.8	120.0	100.6	91.4	120.3	102.4
B20	Exterior Closure	97.8	128.9	111.8	96.3	128.9	110.9	98.7	128.9	112.3	112.2	128.9	119.7	97.7	130.0	112.2
B30	Roofing	98.0	119.7	106.7	102.7	119.7	109.5	98.0	120.1	106.8	98.1	120.5	107.1	98.0	124.6	108.7
C	Interior Construction	95.5	122.0	106.6	94.2	122.4	106.0	95.5	122.0	106.6	95.5	122.4	106.7	95.5	122.0	106.6
D10	Services: Conveying	100.0	112.9	103.8	100.0	112.9	103.8	100.0	112.9	103.8	100.0	112.9	103.8	100.0	112.9	103.8
D20 - 40	Mechanical	100.1	118.5	107.5	100.0	118.5	107.5	100.1	118.5	107.5	100.1	118.5	107.5	100.1	118.5	107.5
D50	Electrical	96.0	106.2	101.2	95.7	109.3	102.6	96.1	106.2	101.3	96.0	108.6	102.4	96.0	107.4	101.8
E	Equipment & Furnishings	100.0	116.8	100.9	100.0	117.0	101.0	100.0	116.8	100.9	100.0	116.9	101.0	100.0	116.9	100.9
G	Site Work	100.3	102.6	101.8	97.0	102.6	100.7	100.5	102.6	101.9	100.4	103.3	102.4	101.1	102.6	102.1
A-G	WEIGHTED AVERAGE	96.8	118.1	105.9	97.2	118.5	106.4	96.4	118.1	105.7	98.1	118.5	106.9	97.0	118.6	106.3

DIV. NO.	BUILDING SYSTEMS	CONNECTICUT STAMFORD MAT.	INST.	TOTAL	WATERBURY MAT.	INST.	TOTAL	D.C. WASHINGTON MAT.	INST.	TOTAL	DELAWARE WILMINGTON MAT.	INST.	TOTAL	FLORIDA DAYTONA BEACH MAT.	INST.	TOTAL
A	Substructure	106.8	121.7	115.5	106.7	121.2	115.1	108.4	82.9	93.5	102.8	108.3	106.0	93.2	69.9	79.6
B10	Shell: Superstructure	91.7	120.6	102.7	91.7	120.0	102.4	102.5	87.7	96.9	102.6	115.6	107.5	99.6	77.7	91.3
B20	Exterior Closure	98.0	130.0	112.4	98.0	128.9	111.9	101.1	81.6	92.3	95.3	104.3	99.4	84.0	61.8	74.0
B30	Roofing	98.0	124.6	108.7	98.0	120.5	107.0	102.8	84.6	95.5	103.7	114.9	108.2	100.9	67.0	87.3
C	Interior Construction	95.5	122.4	106.7	95.4	122.4	106.7	98.6	74.2	88.4	88.3	104.2	94.9	92.5	60.9	79.3
D10	Services: Conveying	100.0	112.9	103.8	100.0	112.9	103.8	100.0	99.2	99.8	100.0	106.0	101.8	100.0	85.4	95.7
D20 - 40	Mechanical	100.1	118.6	107.5	100.1	118.5	107.5	100.1	90.3	96.1	100.1	122.0	108.9	99.9	76.7	90.6
D50	Electrical	96.0	153.8	125.4	95.6	109.7	102.8	100.6	102.4	101.5	97.3	112.4	105.0	95.5	60.3	77.6
E	Equipment & Furnishings	100.0	119.1	101.1	100.0	117.0	101.0	100.0	73.8	98.5	100.0	100.7	100.0	100.0	61.7	97.8
G	Site Work	101.7	102.6	102.3	101.2	102.6	102.1	104.1	95.6	98.4	104.9	113.8	110.9	110.6	88.4	95.7
A-G	WEIGHTED AVERAGE	97.2	125.2	109.2	97.1	118.6	106.3	101.0	87.7	95.3	98.2	112.3	104.2	96.3	70.3	85.2

DIV. NO.	BUILDING SYSTEMS	FLORIDA FORT LAUDERDALE MAT.	INST.	TOTAL	JACKSONVILLE MAT.	INST.	TOTAL	MELBOURNE MAT.	INST.	TOTAL	MIAMI MAT.	INST.	TOTAL	ORLANDO MAT.	INST.	TOTAL
A	Substructure	91.8	66.2	76.9	93.8	68.2	78.9	105.6	70.6	85.2	91.6	66.6	77.0	103.7	70.4	84.3
B10	Shell: Superstructure	97.0	77.5	89.6	98.6	76.2	90.1	109.1	78.9	97.6	97.3	77.1	89.6	99.7	78.7	91.7
B20	Exterior Closure	85.4	64.4	76.0	83.9	59.3	72.8	85.1	62.0	74.7	84.5	58.0	72.6	83.8	61.9	74.0
B30	Roofing	101.3	61.0	85.1	101.2	62.1	85.5	101.4	65.4	87.0	101.2	61.9	85.4	105.9	67.0	90.3
C	Interior Construction	93.1	64.7	81.2	92.5	59.6	78.8	91.9	63.1	79.8	94.4	61.8	80.8	95.8	61.3	81.4
D10	Services: Conveying	100.0	85.0	95.5	100.0	80.5	94.2	100.0	85.5	95.7	100.0	85.4	95.7	100.0	85.4	95.7
D20 - 40	Mechanical	100.0	68.2	87.1	99.9	62.3	84.7	99.9	75.3	90.0	100.0	62.5	84.8	100.1	57.1	82.7
D50	Electrical	92.7	69.1	80.7	95.2	63.4	79.1	96.6	64.3	80.1	96.5	72.3	84.2	99.8	64.7	81.9
E	Equipment & Furnishings	100.0	65.0	98.0	100.0	60.3	97.8	100.0	62.0	97.9	100.0	65.2	98.0	100.0	62.0	97.9
G	Site Work	91.3	76.1	81.1	110.6	88.4	95.7	118.7	88.6	98.5	92.6	76.4	81.7	112.0	88.7	96.4
A-G	WEIGHTED AVERAGE	95.3	69.5	84.2	96.1	66.5	83.4	99.2	71.2	87.2	95.8	67.4	83.7	97.9	67.1	84.7

DIV. NO.	BUILDING SYSTEMS	FLORIDA PANAMA CITY MAT.	INST.	TOTAL	PENSACOLA MAT.	INST.	TOTAL	ST. PETERSBURG MAT.	INST.	TOTAL	TALLAHASSEE MAT.	INST.	TOTAL	TAMPA MAT.	INST.	TOTAL
A	Substructure	98.8	67.7	80.7	113.3	70.4	88.3	98.9	69.1	81.5	96.9	66.6	79.3	97.3	71.5	82.3
B10	Shell: Superstructure	100.3	78.2	91.9	104.1	79.1	94.6	100.7	79.9	92.8	98.5	75.1	89.6	99.7	81.3	92.7
B20	Exterior Closure	91.0	60.7	77.4	98.4	60.4	81.3	104.6	56.3	82.9	85.7	58.2	73.4	86.9	62.1	75.8
B30	Roofing	101.5	56.7	83.5	101.4	62.6	85.9	101.1	59.1	84.2	98.8	61.9	84.0	101.4	64.4	86.5
C	Interior Construction	92.1	63.6	80.2	91.9	62.4	79.6	93.1	59.8	79.2	95.8	56.6	79.4	94.3	61.4	80.5
D10	Services: Conveying	100.0	78.2	93.5	100.0	83.2	95.0	100.0	79.2	93.8	100.0	81.6	94.5	100.0	83.3	95.0
D20 - 40	Mechanical	99.9	51.5	80.4	99.9	63.2	85.1	100.0	55.7	82.1	100.0	66.6	86.5	100.0	60.0	83.8
D50	Electrical	94.3	57.2	75.4	97.9	51.9	74.5	93.0	62.1	77.3	99.9	57.2	78.2	92.7	64.7	78.5
E	Equipment & Furnishings	100.0	63.2	97.9	100.0	63.4	97.9	100.0	60.7	97.8	100.0	56.1	97.5	100.0	61.2	97.8
G	Site Work	123.2	87.6	99.3	124.1	88.1	100.0	105.2	87.9	93.6	104.7	88.1	93.6	105.2	88.5	94.0
A-G	WEIGHTED AVERAGE	97.6	64.2	83.3	100.3	66.4	85.8	99.0	65.2	84.5	97.3	65.7	83.7	96.8	68.1	84.5

629

For customer support on your Facilities Maintenance & Repair Costs with RSMeans data, call 800.448.8182.

GEORGIA

DIV. NO.	BUILDING SYSTEMS	ALBANY MAT.	INST.	TOTAL	ATLANTA MAT.	INST.	TOTAL	AUGUSTA MAT.	INST.	TOTAL	COLUMBUS MAT.	INST.	TOTAL	MACON MAT.	INST.	TOTAL
A	Substructure	88.7	71.5	78.7	108.3	77.3	90.2	101.5	76.4	86.9	88.4	71.6	78.6	88.3	75.4	80.8
B10	Shell: Superstructure	99.3	84.6	93.7	100.4	79.1	92.3	97.9	76.6	89.8	99.0	84.8	93.6	94.9	85.2	91.2
B20	Exterior Closure	84.1	70.3	77.9	93.9	71.1	83.7	88.2	70.5	80.2	84.1	70.3	77.9	88.0	70.4	80.1
B30	Roofing	98.7	67.9	86.0	101.6	72.0	89.7	99.9	70.6	88.2	98.6	68.7	86.6	97.1	71.1	86.6
C	Interior Construction	88.9	70.0	81.0	97.9	73.5	87.7	95.1	72.7	85.8	88.9	70.0	81.0	85.0	70.0	78.8
D10	Services: Conveying	100.0	85.2	95.6	100.0	88.1	96.5	100.0	85.1	95.6	100.0	85.2	95.6	100.0	85.3	95.6
D20 - 40	Mechanical	99.9	69.9	87.8	100.0	72.0	88.7	100.1	63.6	85.4	100.0	66.2	86.3	100.0	69.3	87.6
D50	Electrical	95.3	63.4	79.1	96.6	72.6	84.4	97.6	69.8	83.4	95.5	69.4	82.2	94.0	64.2	78.9
E	Equipment & Furnishings	100.0	67.7	98.2	100.0	72.7	98.5	100.0	74.7	98.6	100.0	68.0	98.2	100.0	68.0	98.2
G	Site Work	101.1	78.7	86.1	98.1	95.3	96.2	94.4	95.3	95.0	101.0	78.9	86.2	102.0	94.2	96.8
A-G	WEIGHTED AVERAGE	95.2	72.6	85.5	99.0	75.6	89.0	97.1	72.7	86.6	95.1	72.7	85.5	93.9	74.0	85.4

GEORGIA / HAWAII / IDAHO

DIV. NO.	BUILDING SYSTEMS	SAVANNAH MAT.	INST.	TOTAL	VALDOSTA MAT.	INST.	TOTAL	HONOLULU MAT.	INST.	TOTAL	STATES & POSS., GUAM MAT.	INST.	TOTAL	BOISE MAT.	INST.	TOTAL
A	Substructure	92.3	72.1	80.6	88.0	60.5	72.0	141.1	120.8	129.3	166.5	75.0	113.2	92.8	84.6	88.0
B10	Shell: Superstructure	96.8	83.8	91.9	97.8	74.4	88.9	135.3	114.7	127.4	155.5	71.0	123.4	102.3	81.4	94.3
B20	Exterior Closure	81.6	70.2	76.5	91.2	77.3	85.0	126.8	119.9	123.7	145.4	37.7	97.0	107.9	84.6	97.4
B30	Roofing	96.6	68.1	85.1	99.0	63.8	84.8	150.9	120.2	138.6	156.5	63.8	119.3	97.5	85.5	92.7
C	Interior Construction	93.4	71.3	84.1	86.2	51.5	71.7	128.1	127.6	127.9	152.4	49.1	109.2	94.9	75.8	86.9
D10	Services: Conveying	100.0	84.8	95.5	100.0	84.8	95.5	100.0	111.1	103.3	100.0	70.1	91.1	100.0	90.0	97.0
D20 - 40	Mechanical	100.1	67.1	86.7	100.0	69.3	87.6	100.4	112.0	105.1	103.1	34.3	75.3	100.0	74.8	89.8
D50	Electrical	99.1	70.2	84.4	93.8	52.9	73.0	112.8	124.3	118.7	162.9	36.8	98.8	97.3	69.8	83.3
E	Equipment & Furnishings	100.0	71.7	98.4	100.0	35.1	96.3	100.0	124.7	101.4	100.0	46.2	97.0	100.0	78.4	98.8
G	Site Work	100.9	80.3	87.1	109.7	78.7	89.0	158.6	107.6	124.5	187.1	102.2	130.3	88.8	96.9	94.2
A-G	WEIGHTED AVERAGE	95.6	73.1	86.0	95.3	66.2	82.8	120.9	118.3	119.8	138.8	52.6	101.9	99.7	79.4	91.0

IDAHO / ILLINOIS

DIV. NO.	BUILDING SYSTEMS	LEWISTON MAT.	INST.	TOTAL	POCATELLO MAT.	INST.	TOTAL	TWIN FALLS MAT.	INST.	TOTAL	CHICAGO MAT.	INST.	TOTAL	DECATUR MAT.	INST.	TOTAL
A	Substructure	102.6	87.7	93.9	95.1	84.4	88.8	97.7	83.9	89.6	117.9	145.7	134.1	94.5	110.3	103.7
B10	Shell: Superstructure	98.9	87.6	94.6	109.0	81.1	98.4	109.5	80.6	98.5	105.3	152.0	123.1	104.5	116.5	109.1
B20	Exterior Closure	115.1	85.6	101.9	104.8	83.0	95.0	113.0	79.5	98.0	99.2	165.3	128.9	79.8	119.5	97.6
B30	Roofing	153.1	85.5	125.9	97.3	74.8	88.3	98.2	82.2	91.7	93.6	150.6	116.5	100.7	112.7	105.5
C	Interior Construction	131.5	81.7	110.7	95.7	75.8	87.4	97.6	75.2	88.2	101.8	164.8	128.1	96.9	116.5	105.1
D10	Services: Conveying	100.0	96.2	98.9	100.0	90.0	97.0	100.0	89.0	96.7	100.0	123.8	107.1	100.0	103.6	101.1
D20 - 40	Mechanical	100.7	86.9	95.1	99.9	74.8	89.7	99.9	69.6	87.6	99.8	138.0	115.3	99.9	96.4	98.5
D50	Electrical	87.4	80.1	83.7	94.7	67.6	80.9	90.3	69.4	79.7	97.4	134.1	116.1	94.9	89.8	92.3
E	Equipment & Furnishings	100.0	82.0	99.0	100.0	78.2	98.8	100.0	78.2	98.8	100.0	161.1	103.4	100.0	114.3	100.8
G	Site Work	96.0	91.8	93.2	89.2	96.9	94.4	96.0	96.6	96.4	103.5	103.9	103.7	94.3	101.0	98.7
A-G	WEIGHTED AVERAGE	106.9	85.7	97.8	100.7	78.6	91.2	101.9	77.2	91.3	101.7	146.1	120.7	97.2	107.0	101.4

ILLINOIS

DIV. NO.	BUILDING SYSTEMS	EAST ST. LOUIS MAT.	INST.	TOTAL	JOLIET MAT.	INST.	TOTAL	PEORIA MAT.	INST.	TOTAL	ROCKFORD MAT.	INST.	TOTAL	SPRINGFIELD MAT.	INST.	TOTAL
A	Substructure	91.0	112.0	103.2	110.1	141.3	128.2	96.9	113.7	106.7	97.9	125.3	113.9	92.8	109.9	102.7
B10	Shell: Superstructure	101.1	124.5	110.0	102.0	144.0	118.0	97.4	121.5	106.5	97.6	137.7	112.8	101.8	116.3	107.3
B20	Exterior Closure	74.3	125.8	97.4	96.8	159.1	124.7	97.5	124.1	109.4	87.6	144.5	113.2	84.8	120.8	101.0
B30	Roofing	94.9	111.4	101.6	98.0	147.3	117.8	99.2	115.5	105.7	101.6	131.5	113.6	102.9	114.6	107.6
C	Interior Construction	90.6	114.3	100.5	98.1	162.8	125.1	95.2	122.0	106.4	95.2	133.6	111.2	98.9	116.7	106.4
D10	Services: Conveying	100.0	103.3	101.0	100.0	120.3	106.0	100.0	105.3	101.6	100.0	115.1	104.5	100.0	104.1	101.2
D20 - 40	Mechanical	99.9	99.3	99.7	99.9	133.0	113.3	99.9	102.4	100.9	100.0	118.7	107.6	99.9	102.4	101.0
D50	Electrical	92.1	102.1	97.2	96.5	136.5	116.9	96.1	93.7	94.9	96.4	128.7	112.8	97.4	84.2	90.7
E	Equipment & Furnishings	100.0	107.5	100.4	100.0	161.7	103.5	100.0	116.2	100.9	100.0	128.2	101.6	100.0	114.1	100.8
G	Site Work	101.2	101.5	101.4	98.8	102.0	100.9	99.8	100.5	100.3	99.5	101.6	100.9	98.0	101.3	100.2
A-G	WEIGHTED AVERAGE	94.4	111.3	101.6	99.7	142.2	117.9	97.8	111.5	103.7	96.8	128.7	110.5	97.9	107.7	102.1

INDIANA

DIV. NO.	BUILDING SYSTEMS	ANDERSON MAT.	INST.	TOTAL	BLOOMINGTON MAT.	INST.	TOTAL	EVANSVILLE MAT.	INST.	TOTAL	FORT WAYNE MAT.	INST.	TOTAL	GARY MAT.	INST.	TOTAL
A	Substructure	101.1	82.9	90.5	96.6	82.2	88.2	94.9	91.4	92.9	105.0	80.6	90.8	104.2	109.5	107.3
B10	Shell: Superstructure	99.4	84.9	93.9	99.4	76.6	90.7	93.1	83.4	89.4	100.3	82.4	93.5	100.0	108.5	103.2
B20	Exterior Closure	85.3	76.6	81.4	94.5	72.6	84.7	92.8	79.1	86.6	86.5	73.7	80.7	85.8	110.7	97.0
B30	Roofing	110.0	77.7	97.0	96.8	79.3	89.8	101.1	85.4	94.8	109.7	79.9	97.7	108.4	107.0	107.8
C	Interior Construction	92.5	78.0	86.5	94.7	80.4	88.7	91.0	79.6	86.3	92.4	74.9	85.1	92.0	112.8	100.7
D10	Services: Conveying	100.0	90.4	97.2	100.0	86.4	96.2	100.0	94.9	98.5	100.0	90.8	97.3	100.0	104.3	101.3
D20 - 40	Mechanical	99.9	78.7	91.4	99.8	78.9	91.3	100.0	80.4	92.0	99.9	73.8	89.4	99.9	105.1	102.0
D50	Electrical	86.5	85.2	85.8	99.2	86.6	92.8	95.2	83.4	89.2	87.2	76.7	81.8	98.1	110.7	104.5
E	Equipment & Furnishings	100.0	79.8	98.9	100.0	81.2	98.9	100.0	79.5	98.8	100.0	75.7	98.6	100.0	111.4	100.6
G	Site Work	100.0	93.6	95.7	87.8	92.3	90.8	93.2	121.5	112.2	101.8	93.5	96.2	100.7	97.3	98.4
A-G	WEIGHTED AVERAGE	95.9	81.8	89.8	97.7	80.3	90.3	95.4	84.8	90.9	96.4	78.2	88.6	97.2	103.8	102.0

INDIANA / IOWA

DIV. NO.	BUILDING SYSTEMS	INDIANAPOLIS MAT.	INST.	TOTAL	MUNCIE MAT.	INST.	TOTAL	SOUTH BEND MAT.	INST.	TOTAL	TERRE HAUTE MAT.	INST.	TOTAL	CEDAR RAPIDS MAT.	INST.	TOTAL
A	Substructure	99.5	86.4	91.9	100.5	82.3	89.9	101.4	83.6	91.0	93.0	89.8	91.1	105.0	87.9	95.0
B10	Shell: Superstructure	96.9	80.1	90.5	101.3	84.5	94.9	102.0	93.5	98.8	93.3	83.1	89.4	91.7	90.2	91.2
B20	Exterior Closure	93.6	78.8	87.0	90.1	76.7	84.1	86.1	77.9	82.4	100.4	75.6	89.3	91.9	82.0	87.5
B30	Roofing	99.7	82.1	92.7	99.3	79.2	91.2	102.6	81.4	94.1	101.1	82.8	93.8	105.2	83.1	96.3
C	Interior Construction	99.8	84.1	93.3	89.3	77.7	84.4	92.5	79.9	87.2	91.3	78.9	86.1	99.5	85.3	93.6
D10	Services: Conveying	100.0	91.2	97.4	100.0	89.2	96.8	100.0	91.9	97.6	100.0	92.1	97.6	100.0	95.1	98.5
D20 - 40	Mechanical	99.9	79.7	91.7	99.8	78.6	91.2	99.9	76.7	90.5	100.0	78.2	91.2	100.1	83.0	93.2
D50	Electrical	101.9	87.0	94.3	91.0	75.9	83.3	97.7	86.0	91.8	93.5	85.5	89.4	97.9	82.0	89.8
E	Equipment & Furnishings	100.0	86.0	99.2	100.0	79.6	98.9	100.0	77.5	98.7	100.0	78.6	98.8	100.0	85.6	99.2
G	Site Work	100.8	91.1	94.3	88.2	92.3	90.9	98.9	93.8	95.5	95.1	121.9	113.1	99.5	95.9	97.1
A-G	WEIGHTED AVERAGE	98.7	83.0	92.0	96.1	80.3	89.3	97.4	83.6	91.5	96.2	83.8	90.9	97.4	85.8	92.4

For customer support on your Facilities Maintenance & Repair Costs with RSMeans data, call 800.448.8182.

Cost Indexes

IOWA

DIV. NO.	BUILDING SYSTEMS	COUNCIL BLUFFS			DAVENPORT			DES MOINES			DUBUQUE			SIOUX CITY		
		MAT.	INST.	TOTAL	MAT.	INST.	TOTAL	MAT.	INST.	TOTAL	MAT.	INST.	TOTAL	MAT.	INST.	TOTAL
A	Substructure	107.5	82.0	92.7	102.1	99.0	100.3	97.5	90.5	93.4	101.3	86.3	92.5	105.7	79.6	90.5
B10	Shell: Superstructure	94.9	86.6	91.7	91.2	103.0	95.7	93.5	92.8	93.3	89.2	89.1	89.1	91.6	83.8	88.6
B20	Exterior Closure	93.4	78.8	86.8	90.6	95.1	92.6	82.6	86.8	84.5	91.3	73.1	83.1	89.3	70.2	80.8
B30	Roofing	104.5	76.9	93.4	104.6	95.1	100.8	96.7	86.3	92.5	104.8	81.1	95.3	104.6	73.8	92.2
C	Interior Construction	95.0	77.1	87.5	97.8	97.5	97.7	94.4	87.5	91.5	96.8	81.9	90.6	98.5	74.1	88.3
D10	Services: Conveying	100.0	91.3	97.4	100.0	97.6	99.3	100.0	95.6	98.7	100.0	94.0	98.2	100.0	93.3	98.0
D20 - 40	Mechanical	100.1	74.4	89.7	100.1	96.0	98.5	99.8	84.7	93.7	100.1	77.3	90.9	100.1	78.1	91.2
D50	Electrical	103.2	82.3	92.6	95.9	89.7	92.8	105.4	84.4	94.7	101.7	78.8	90.1	97.9	75.8	86.7
E	Equipment & Furnishings	100.0	74.5	98.6	100.0	97.4	99.9	100.0	85.0	99.2	100.0	83.8	99.1	100.0	73.9	98.5
G	Site Work	101.6	92.9	95.8	98.1	99.4	98.9	101.5	99.8	100.4	97.8	93.1	94.7	108.0	97.2	100.7
A-G	WEIGHTED AVERAGE	98.2	80.8	90.7	96.5	96.9	96.7	96.0	88.4	92.8	96.5	81.9	90.3	97.1	78.7	89.2

IOWA / KANSAS

DIV. NO.	BUILDING SYSTEMS	WATERLOO			DODGE CITY			KANSAS CITY			SALINA			TOPEKA		
		MAT.	INST.	TOTAL	MAT.	INST.	TOTAL	MAT.	INST.	TOTAL	MAT.	INST.	TOTAL	MAT.	INST.	TOTAL
A	Substructure	109.6	82.7	93.9	112.8	83.1	95.5	93.4	98.1	96.1	101.1	79.8	88.7	97.2	84.0	89.5
B10	Shell: Superstructure	93.5	86.0	90.6	95.0	88.7	92.6	95.7	102.3	98.2	92.4	86.8	90.3	93.7	90.6	92.5
B20	Exterior Closure	87.0	78.1	83.0	102.7	67.7	86.9	92.1	98.5	95.0	102.8	59.5	83.4	90.4	69.2	80.9
B30	Roofing	104.2	80.5	94.7	99.6	71.0	88.1	92.0	100.2	95.3	99.0	63.8	84.9	96.4	77.5	88.8
C	Interior Construction	92.8	72.7	84.4	92.9	63.1	80.4	89.4	97.3	92.7	92.2	57.9	77.9	98.2	68.4	85.7
D10	Services: Conveying	100.0	94.8	98.5	100.0	89.9	97.0	100.0	94.2	98.3	100.0	88.4	96.5	100.0	83.5	95.1
D20 - 40	Mechanical	100.0	81.9	92.7	100.1	72.1	88.8	99.8	98.8	99.4	100.1	70.9	88.3	100.0	74.7	89.7
D50	Electrical	95.7	63.1	79.1	93.8	69.5	81.5	101.4	97.9	99.6	93.6	71.9	82.5	100.1	73.6	86.6
E	Equipment & Furnishings	100.0	65.9	98.1	100.0	59.2	97.7	100.0	96.9	99.8	100.0	53.2	97.4	100.0	66.7	98.1
G	Site Work	108.3	97.3	100.9	110.0	96.0	100.6	95.7	94.3	94.8	100.7	96.6	98.0	97.0	93.7	94.8
A-G	WEIGHTED AVERAGE	96.2	79.3	89.0	98.3	75.4	88.5	96.0	98.6	97.1	96.9	72.8	86.6	97.0	77.7	88.7

KANSAS / KENTUCKY

DIV. NO.	BUILDING SYSTEMS	WICHITA			BOWLING GREEN			LEXINGTON			LOUISVILLE			OWENSBORO		
		MAT.	INST.	TOTAL	MAT.	INST.	TOTAL	MAT.	INST.	TOTAL	MAT.	INST.	TOTAL	MAT.	INST.	TOTAL
A	Substructure	96.9	76.2	84.9	86.4	79.8	82.5	95.7	84.7	89.3	91.5	79.6	84.6	89.3	90.5	90.0
B10	Shell: Superstructure	93.3	84.0	89.8	94.9	80.4	89.4	95.1	83.3	90.6	94.9	80.8	89.5	89.0	83.6	86.9
B20	Exterior Closure	91.8	54.7	75.1	93.4	73.1	84.3	79.1	73.8	76.7	78.2	69.1	74.2	99.5	80.7	91.1
B30	Roofing	97.9	60.8	83.0	88.1	81.5	85.4	104.1	77.1	93.2	100.0	73.9	89.5	101.1	78.7	92.1
C	Interior Construction	98.0	57.6	81.1	89.8	76.6	84.2	86.1	74.3	81.1	87.0	74.7	81.9	89.1	77.0	84.0
D10	Services: Conveying	100.0	87.6	96.3	100.0	86.6	96.0	100.0	92.5	97.8	100.0	88.7	96.6	100.0	97.2	99.2
D20 - 40	Mechanical	99.8	68.5	87.2	100.0	78.8	91.4	100.1	78.0	91.2	100.0	77.2	90.8	100.0	78.8	91.4
D50	Electrical	97.5	71.8	84.5	94.0	76.5	85.1	92.8	76.5	84.5	95.5	76.5	85.9	93.5	75.5	84.4
E	Equipment & Furnishings	100.0	53.6	97.4	100.0	78.1	98.8	100.0	73.2	98.5	100.0	78.0	98.8	100.0	77.3	98.7
G	Site Work	95.8	97.7	97.1	79.5	94.3	89.4	94.3	97.8	96.7	88.3	94.2	92.3	93.4	121.8	112.4
A-G	WEIGHTED AVERAGE	96.7	70.8	85.6	94.5	78.9	87.8	93.4	79.6	87.5	93.2	77.7	86.6	94.6	83.1	89.7

LOUISIANA

DIV. NO.	BUILDING SYSTEMS	ALEXANDRIA			BATON ROUGE			LAKE CHARLES			MONROE			NEW ORLEANS		
		MAT.	INST.	TOTAL	MAT.	INST.	TOTAL	MAT.	INST.	TOTAL	MAT.	INST.	TOTAL	MAT.	INST.	TOTAL
A	Substructure	91.8	69.6	78.9	96.4	75.7	84.3	96.7	72.2	82.5	91.6	68.7	78.3	94.7	73.7	82.5
B10	Shell: Superstructure	93.3	67.2	83.4	97.6	73.3	88.3	90.7	68.6	82.3	93.2	66.7	83.2	99.0	65.8	86.3
B20	Exterior Closure	94.0	62.6	79.9	87.2	65.5	77.4	91.2	67.7	80.6	91.9	61.0	78.0	99.5	65.4	84.2
B30	Roofing	98.6	67.3	86.1	97.0	70.1	86.2	96.5	69.4	85.7	98.6	66.4	85.7	95.1	69.7	84.9
C	Interior Construction	99.0	60.9	83.1	94.7	70.4	84.5	97.0	67.7	84.8	98.9	59.6	82.5	99.7	68.7	86.7
D10	Services: Conveying	100.0	83.3	95.0	100.0	86.9	96.1	100.0	86.9	96.1	100.0	83.2	95.0	100.0	87.5	96.3
D20 - 40	Mechanical	100.0	64.7	85.7	100.0	65.5	86.1	100.1	65.9	86.3	100.0	63.3	85.2	100.1	63.8	85.4
D50	Electrical	92.6	63.0	77.5	96.5	59.5	77.7	94.6	67.6	80.8	94.3	58.7	76.2	100.6	71.3	85.7
E	Equipment & Furnishings	100.0	60.6	97.8	100.0	74.2	98.5	100.0	69.9	98.3	100.0	60.3	97.8	100.0	72.7	98.5
G	Site Work	104.0	92.1	96.0	103.1	90.4	94.6	101.8	91.0	94.5	104.0	92.0	95.9	100.7	93.2	95.6
A-G	WEIGHTED AVERAGE	96.7	66.6	83.8	96.6	69.7	85.1	95.8	69.8	84.6	96.6	65.1	83.1	99.4	69.5	86.6

LOUISIANA / MAINE

DIV. NO.	BUILDING SYSTEMS	SHREVEPORT			AUGUSTA			BANGOR			LEWISTON			PORTLAND		
		MAT.	INST.	TOTAL	MAT.	INST.	TOTAL	MAT.	INST.	TOTAL	MAT.	INST.	TOTAL	MAT.	INST.	TOTAL
A	Substructure	95.7	69.0	80.1	93.3	92.4	92.8	80.0	92.7	87.4	85.2	92.7	89.6	93.5	92.7	93.1
B10	Shell: Superstructure	98.0	66.8	86.1	92.7	90.1	91.7	82.4	90.5	85.5	86.1	90.5	87.8	92.9	90.5	92.0
B20	Exterior Closure	88.3	58.5	74.9	96.0	89.4	93.1	105.0	92.0	99.1	94.9	92.0	93.6	95.4	92.0	93.9
B30	Roofing	96.8	66.5	84.6	106.3	72.6	92.8	103.0	80.0	93.7	102.8	80.0	93.6	106.1	80.0	95.6
C	Interior Construction	98.6	60.1	82.5	96.6	74.3	87.3	93.0	82.3	88.5	95.9	82.3	90.2	96.5	82.3	90.5
D10	Services: Conveying	100.0	86.9	96.1	100.0	102.2	100.6	100.0	105.3	101.6	100.0	105.3	101.6	100.0	105.3	101.6
D20 - 40	Mechanical	99.9	64.1	85.4	100.0	74.1	89.5	100.1	71.6	88.6	100.1	71.6	88.6	100.0	71.6	88.5
D50	Electrical	99.9	67.1	83.2	96.7	78.6	87.5	94.5	71.1	82.6	96.2	75.3	85.6	97.9	75.3	86.4
E	Equipment & Furnishings	100.0	61.4	97.8	100.0	76.7	98.7	100.0	74.3	98.6	100.0	74.5	98.6	100.0	74.5	98.6
G	Site Work	105.5	92.0	96.4	87.9	98.7	95.1	91.1	100.0	97.1	88.1	100.0	96.1	90.4	100.0	96.8
A-G	WEIGHTED AVERAGE	97.8	66.4	84.3	96.7	83.1	90.9	94.3	83.5	89.7	94.7	84.1	90.1	96.9	84.1	91.4

MARYLAND / MASSACHUSETTS

DIV. NO.	BUILDING SYSTEMS	BALTIMORE			HAGERSTOWN			BOSTON			BROCKTON			FALL RIVER		
		MAT.	INST.	TOTAL	MAT.	INST.	TOTAL	MAT.	INST.	TOTAL	MAT.	INST.	TOTAL	MAT.	INST.	TOTAL
A	Substructure	108.9	84.5	94.7	95.9	86.7	90.5	95.2	131.9	116.6	90.2	124.6	110.3	88.4	122.0	108.0
B10	Shell: Superstructure	103.2	90.2	98.2	97.8	95.1	96.8	94.6	136.2	110.4	89.8	128.0	104.3	89.4	122.1	101.8
B20	Exterior Closure	99.7	77.1	89.5	92.2	88.0	90.3	103.7	146.4	122.9	97.6	136.5	115.1	97.8	134.0	114.1
B30	Roofing	103.1	82.8	94.9	102.1	84.9	95.2	113.7	136.1	122.7	104.3	127.3	113.5	104.2	123.1	111.8
C	Interior Construction	100.2	78.8	91.2	96.3	82.7	90.6	99.5	145.0	118.5	94.9	134.2	111.3	94.8	133.3	110.9
D10	Services: Conveying	100.0	89.9	97.0	100.0	91.4	97.5	100.0	114.8	104.4	100.0	110.6	103.2	100.0	111.2	103.3
D20 - 40	Mechanical	100.0	83.8	93.5	99.9	84.9	93.8	100.1	127.5	111.2	100.3	106.1	102.6	100.3	105.8	102.5
D50	Electrical	100.4	89.0	94.6	97.7	81.0	89.2	100.8	131.1	116.2	98.0	100.1	99.1	97.9	100.1	99.0
E	Equipment & Furnishings	100.0	79.2	98.8	100.0	79.3	98.8	100.0	136.5	102.1	100.0	121.8	101.2	100.0	121.7	101.2
G	Site Work	101.4	97.2	98.6	92.6	93.9	93.5	94.6	101.3	99.1	91.2	102.9	99.0	90.1	103.0	98.8
A-G	WEIGHTED AVERAGE	101.2	85.0	94.2	97.5	87.0	93.0	99.4	133.3	113.9	96.1	119.2	106.0	95.9	117.4	105.1

631

MASSACHUSETTS

DIV. NO.	BUILDING SYSTEMS	HYANNIS MAT.	INST.	TOTAL	LAWRENCE MAT.	INST.	TOTAL	LOWELL MAT.	INST.	TOTAL	NEW BEDFORD MAT.	INST.	TOTAL	PITTSFIELD MAT.	INST.	TOTAL
A	Substructure	80.3	121.9	104.5	95.5	123.8	111.9	89.9	124.6	110.1	81.8	122.1	105.3	91.8	109.1	101.9
B10	Shell: Superstructure	84.8	121.8	98.9	91.5	125.9	104.6	90.2	124.4	103.2	88.0	122.1	101.0	90.5	109.5	97.7
B20	Exterior Closure	94.8	134.3	112.5	102.5	136.8	117.9	92.7	135.6	112.0	97.1	136.1	114.6	92.9	113.1	102.0
B30	Roofing	103.7	124.0	111.9	102.3	127.3	112.3	102.0	128.7	112.7	104.1	123.1	111.7	102.1	104.8	103.2
C	Interior Construction	91.2	133.1	108.7	92.4	133.8	109.7	97.2	134.4	112.8	94.7	133.3	110.8	97.2	111.6	103.2
D10	Services: Conveying	100.0	110.6	103.2	100.0	110.6	103.2	100.0	111.8	103.5	100.0	111.2	103.3	100.0	101.1	100.3
D20 - 40	Mechanical	100.3	106.5	102.8	100.1	121.9	108.9	100.1	123.4	109.5	100.3	105.7	102.5	100.1	96.3	98.5
D50	Electrical	95.7	100.1	97.9	96.9	125.9	111.6	97.3	122.2	110.0	98.8	100.1	99.4	97.3	98.5	97.9
E	Equipment & Furnishings	100.0	121.8	101.2	100.0	123.0	101.3	100.0	123.0	101.3	100.0	121.7	101.3	100.0	104.7	100.3
G	Site Work	88.9	103.1	98.4	92.8	102.9	99.6	91.2	103.1	99.2	88.6	103.0	98.3	92.5	101.5	98.6
A-G	WEIGHTED AVERAGE	93.4	117.5	103.7	96.7	125.8	109.2	95.8	125.4	108.5	95.3	117.6	104.9	96.0	105.1	99.9

MASSACHUSETTS / MICHIGAN

DIV. NO.	BUILDING SYSTEMS	SPRINGFIELD MAT.	INST.	TOTAL	WORCESTER MAT.	INST.	TOTAL	ANN ARBOR MAT.	INST.	TOTAL	DEARBORN MAT.	INST.	TOTAL	DETROIT MAT.	INST.	TOTAL
A	Substructure	92.2	116.3	106.2	91.9	123.4	110.3	85.6	101.8	95.0	84.4	102.2	94.8	91.2	104.5	98.9
B10	Shell: Superstructure	92.7	115.8	101.5	92.7	125.2	105.0	100.6	110.7	104.4	100.4	111.0	104.4	103.7	100.5	102.5
B20	Exterior Closure	92.8	120.2	105.1	92.6	133.4	110.9	92.3	99.5	95.5	92.2	101.2	96.2	100.0	101.1	100.5
B30	Roofing	102.0	109.8	105.1	102.0	119.8	109.2	110.0	100.9	106.4	108.2	103.8	106.4	106.3	107.1	106.6
C	Interior Construction	97.1	124.7	108.7	97.2	131.0	111.3	92.9	105.0	98.0	92.9	104.3	97.6	97.6	107.1	101.6
D10	Services: Conveying	100.0	102.9	100.9	100.0	104.7	101.4	100.0	93.9	98.2	100.0	94.4	98.3	100.0	102.1	100.5
D20 - 40	Mechanical	100.1	100.6	100.3	100.1	106.8	102.8	100.1	93.4	97.4	100.1	102.3	101.0	100.0	103.7	101.5
D50	Electrical	97.3	95.2	96.3	97.3	106.6	102.1	98.5	105.5	102.1	98.5	98.4	98.5	102.5	101.3	101.9
E	Equipment & Furnishings	100.0	121.2	101.2	100.0	119.1	101.1	100.0	105.3	100.3	100.0	105.0	100.3	100.0	109.0	100.5
G	Site Work	91.8	101.9	98.6	91.7	102.9	99.2	80.1	96.2	90.9	79.8	96.3	90.8	92.3	101.0	98.1
A-G	WEIGHTED AVERAGE	96.4	110.4	102.4	96.4	118.5	105.9	97.2	101.7	99.1	97.0	102.9	99.5	100.2	103.0	101.4

MICHIGAN

DIV. NO.	BUILDING SYSTEMS	FLINT MAT.	INST.	TOTAL	GRAND RAPIDS MAT.	INST.	TOTAL	KALAMAZOO MAT.	INST.	TOTAL	LANSING MAT.	INST.	TOTAL	MUSKEGON MAT.	INST.	TOTAL
A	Substructure	84.8	91.1	88.5	89.7	84.2	86.5	87.5	84.7	85.8	89.6	88.2	93.0	86.3	84.6	85.4
B10	Shell: Superstructure	100.9	103.3	101.9	99.5	83.4	93.4	101.4	83.0	94.4	101.6	100.9	101.3	99.3	83.7	93.3
B20	Exterior Closure	92.3	88.8	90.7	87.3	75.7	82.1	88.9	79.3	84.6	87.6	87.2	87.4	84.2	76.0	80.5
B30	Roofing	107.5	86.3	99.0	103.5	72.3	91.0	101.7	81.2	93.5	106.6	85.3	98.0	100.6	72.8	89.4
C	Interior Construction	92.7	84.6	89.3	94.9	77.3	87.5	86.0	77.1	82.3	94.6	77.4	87.4	83.8	77.3	81.1
D10	Services: Conveying	100.0	91.6	97.5	100.0	98.2	99.5	100.0	98.4	99.5	100.0	92.9	97.9	100.0	98.2	99.5
D20 - 40	Mechanical	100.1	85.1	94.1	100.1	81.4	92.5	100.1	80.0	92.0	100.0	85.5	94.1	99.9	81.4	92.4
D50	Electrical	98.5	91.3	94.8	100.3	85.0	92.5	94.7	75.9	85.2	99.1	89.8	94.4	95.2	74.7	84.8
E	Equipment & Furnishings	100.0	82.9	99.0	100.0	73.3	98.5	100.0	74.3	98.6	100.0	73.8	98.5	100.0	74.5	98.6
G	Site Work	71.2	95.8	87.7	92.3	86.1	88.2	93.8	86.1	88.7	91.5	96.0	94.5	91.2	86.1	87.8
A-G	WEIGHTED AVERAGE	96.9	90.8	94.3	97.1	81.6	90.5	95.6	80.7	89.2	97.9	88.7	93.9	94.1	80.3	88.2

MICHIGAN / MINNESOTA

DIV. NO.	BUILDING SYSTEMS	SAGINAW MAT.	INST.	TOTAL	DULUTH MAT.	INST.	TOTAL	MINNEAPOLIS MAT.	INST.	TOTAL	ROCHESTER MAT.	INST.	TOTAL	SAINT PAUL MAT.	INST.	TOTAL
A	Substructure	84.0	89.6	87.2	98.2	99.8	99.1	101.4	114.7	109.1	97.9	100.2	99.2	99.2	112.9	107.2
B10	Shell: Superstructure	100.6	100.2	101.1	93.1	109.6	99.4	93.7	121.8	104.4	92.9	111.7	100.1	93.2	120.1	103.4
B20	Exterior Closure	92.4	81.8	87.7	91.4	108.5	99.1	102.3	122.4	111.4	91.5	108.4	99.1	93.8	122.3	106.6
B30	Roofing	108.5	83.4	98.4	105.6	103.4	104.7	103.1	119.3	109.6	109.5	90.1	101.7	103.9	120.2	110.5
C	Interior Construction	91.7	81.9	87.6	99.0	102.4	100.4	101.1	119.3	108.8	95.9	98.3	96.9	95.5	119.9	105.7
D10	Services: Conveying	100.0	90.6	97.2	100.0	96.4	98.9	100.0	106.4	101.9	100.0	100.9	100.3	100.0	105.6	101.7
D20 - 40	Mechanical	100.1	80.0	92.0	99.8	96.5	98.4	99.9	113.0	105.2	99.9	93.5	97.3	99.9	112.7	105.1
D50	Electrical	96.6	86.6	91.5	97.4	101.8	99.6	105.4	113.0	109.3	101.3	92.4	96.7	102.1	113.0	107.6
E	Equipment & Furnishings	100.0	81.5	99.0	100.0	95.5	99.7	100.0	114.6	100.8	100.0	96.0	99.8	100.0	114.5	100.8
G	Site Work	73.7	95.7	88.4	99.2	102.8	101.6	94.9	108.4	103.9	96.7	102.3	100.4	92.8	103.7	100.1
A-G	WEIGHTED AVERAGE	96.5	87.3	92.6	97.1	100.3	99.5	99.7	116.5	106.9	97.1	100.3	98.4	97.2	115.8	105.2

MINNESOTA / MISSISSIPPI

DIV. NO.	BUILDING SYSTEMS	ST. CLOUD MAT.	INST.	TOTAL	BILOXI MAT.	INST.	TOTAL	GREENVILLE MAT.	INST.	TOTAL	JACKSON MAT.	INST.	TOTAL	MERIDIAN MAT.	INST.	TOTAL
A	Substructure	95.4	111.5	104.8	110.7	69.9	87.0	103.0	68.5	82.9	101.9	70.0	83.3	105.6	69.5	84.6
B10	Shell: Superstructure	91.1	117.8	101.3	95.5	75.2	87.8	95.0	78.0	88.5	98.7	75.8	90.0	92.5	75.3	85.9
B20	Exterior Closure	88.5	121.2	103.2	89.9	60.6	76.7	107.9	68.1	90.0	92.9	59.9	78.1	87.8	60.1	75.3
B30	Roofing	104.2	112.8	107.6	97.2	64.4	84.0	96.9	61.9	82.8	95.3	63.4	82.5	96.8	63.6	83.5
C	Interior Construction	86.3	116.3	98.8	93.1	62.1	80.2	93.8	60.0	79.6	93.3	63.1	80.7	90.6	61.0	78.2
D10	Services: Conveying	100.0	99.9	100.0	100.0	74.3	92.4	100.0	73.7	92.2	100.0	73.7	92.2	100.0	74.1	92.3
D20 - 40	Mechanical	99.5	110.9	104.1	100.0	56.5	82.4	100.0	58.2	83.1	100.0	59.0	83.4	100.0	58.7	83.3
D50	Electrical	101.7	113.0	107.4	99.7	55.0	77.0	95.7	56.1	75.6	101.3	56.1	78.3	98.7	57.0	77.5
E	Equipment & Furnishings	100.0	111.0	100.6	100.0	65.7	98.1	100.0	64.1	98.0	100.0	67.0	98.1	100.0	64.2	98.0
G	Site Work	93.8	105.8	101.8	108.6	89.4	95.7	105.6	89.4	94.7	102.8	89.4	93.8	104.7	89.5	94.5
A-G	WEIGHTED AVERAGE	94.4	114.0	102.8	97.3	64.6	83.3	98.6	66.0	84.6	98.0	65.4	84.0	95.6	65.1	82.5

MISSOURI

DIV. NO.	BUILDING SYSTEMS	CAPE GIRARDEAU MAT.	INST.	TOTAL	COLUMBIA MAT.	INST.	TOTAL	JOPLIN MAT.	INST.	TOTAL	KANSAS CITY MAT.	INST.	TOTAL	SPRINGFIELD MAT.	INST.	TOTAL
A	Substructure	90.8	85.7	87.9	88.3	87.9	88.1	104.6	83.1	92.1	101.2	103.1	102.3	94.2	83.7	88.1
B10	Shell: Superstructure	96.5	96.9	96.7	92.2	101.5	95.7	91.2	90.1	90.8	95.6	109.1	100.8	97.4	92.2	95.4
B20	Exterior Closure	100.0	80.7	91.3	104.2	87.8	96.8	89.9	82.5	86.6	95.4	104.3	99.4	93.3	84.4	89.3
B30	Roofing	99.2	86.6	94.2	96.6	85.8	92.2	92.6	83.3	88.9	92.6	105.1	97.6	101.1	76.5	91.3
C	Interior Construction	98.7	79.5	90.7	90.7	82.4	87.2	97.4	77.1	88.9	97.1	103.2	99.6	96.9	80.4	90.0
D10	Services: Conveying	100.0	93.4	98.0	100.0	96.8	99.1	100.0	82.0	94.7	100.0	98.6	99.6	100.0	94.4	98.3
D20 - 40	Mechanical	100.0	99.8	99.9	99.9	98.1	99.2	100.1	72.7	89.0	100.0	103.0	101.2	100.0	71.0	88.2
D50	Electrical	96.5	100.9	98.7	93.7	81.3	87.4	91.9	67.7	79.6	102.3	101.0	101.7	97.5	70.5	83.7
E	Equipment & Furnishings	100.0	80.0	98.9	100.0	77.7	98.7	100.0	74.9	98.6	100.0	101.6	100.1	100.0	76.5	98.7
G	Site Work	92.1	93.9	93.3	103.6	97.8	99.7	104.1	97.2	99.5	97.1	98.0	97.7	102.6	95.8	98.0
A-G	WEIGHTED AVERAGE	98.1	91.8	95.4	96.2	91.3	94.1	95.9	79.7	88.9	98.1	103.6	100.5	97.7	81.0	90.5

632

DIV. NO.	BUILDING SYSTEMS	MISSOURI						MONTANA								
		ST. JOSEPH			ST. LOUIS			BILLINGS			BUTTE			GREAT FALLS		
		MAT.	INST.	TOTAL	MAT.	INST.	TOTAL	MAT.	INST.	TOTAL	MAT.	INST.	TOTAL	MAT.	INST.	TOTAL
A	Substructure	97.6	96.1	96.7	98.6	103.0	101.2	108.4	76.2	89.7	117.1	76.1	93.2	122.3	75.9	95.3
B10	Shell: Superstructure	92.2	105.7	97.3	102.3	113.0	106.4	109.1	79.0	97.6	105.6	79.4	95.6	109.7	78.8	97.9
B20	Exterior Closure	92.3	93.1	92.6	95.7	110.6	102.4	94.4	73.5	85.0	90.3	80.6	85.9	94.5	73.4	85.0
B30	Roofing	93.0	90.4	91.9	97.6	105.5	100.7	105.5	69.2	90.9	105.2	71.2	91.6	105.9	70.7	91.8
C	Interior Construction	97.1	95.4	96.4	99.7	102.6	100.9	95.5	67.8	83.9	94.6	70.5	84.5	98.0	67.3	85.2
D10	Services: Conveying	100.0	96.8	99.0	100.0	102.4	100.7	100.0	93.8	98.2	100.0	93.8	98.2	100.0	93.8	98.2
D20 - 40	Mechanical	100.1	89.4	95.8	100.0	106.1	102.5	100.0	73.2	89.2	100.1	71.5	88.5	100.1	69.6	87.8
D50	Electrical	100.8	77.7	89.0	98.9	100.9	99.9	98.5	69.9	84.0	105.6	70.9	88.0	97.9	69.1	83.2
E	Equipment & Furnishings	100.0	90.1	99.4	100.0	101.2	100.1	100.0	63.7	98.0	100.0	63.6	97.9	100.0	62.8	97.9
G	Site Work	99.1	91.6	94.1	97.5	98.7	98.3	95.0	97.3	96.5	101.2	96.2	97.8	104.7	97.1	99.6
A-G	WEIGHTED AVERAGE	96.8	92.8	95.1	99.6	105.8	102.3	100.8	75.0	89.7	100.7	76.2	90.2	102.1	74.0	90.1

DIV. NO.	BUILDING SYSTEMS	MONTANA						NEBRASKA								
		HELENA			MISSOULA			GRAND ISLAND			LINCOLN			NORTH PLATTE		
		MAT.	INST.	TOTAL	MAT.	INST.	TOTAL	MAT.	INST.	TOTAL	MAT.	INST.	TOTAL	MAT.	INST.	TOTAL
A	Substructure	100.7	75.9	86.3	93.9	76.5	83.8	110.3	79.9	92.6	92.2	83.7	87.3	111.2	77.2	91.4
B10	Shell: Superstructure	104.3	79.1	94.7	98.3	80.5	91.5	95.3	85.2	91.5	93.1	87.9	91.1	96.6	84.2	91.9
B20	Exterior Closure	91.8	80.5	86.7	93.9	80.7	88.0	94.8	75.4	86.1	86.7	78.5	83.1	90.1	78.2	84.8
B30	Roofing	100.4	71.2	88.7	104.3	70.6	90.8	105.1	80.5	95.2	100.7	81.8	93.1	98.7	79.0	90.8
C	Interior Construction	98.3	70.3	86.6	94.3	69.4	83.9	89.4	71.2	81.8	98.9	79.5	90.8	91.3	77.2	85.4
D10	Services: Conveying	100.0	93.8	98.2	100.0	93.8	98.2	100.0	90.3	97.1	100.0	90.3	97.1	100.0	60.3	88.2
D20 - 40	Mechanical	100.1	71.5	88.5	100.1	69.7	87.8	100.0	81.3	92.5	99.9	81.3	92.4	99.9	75.3	90.0
D50	Electrical	104.7	70.9	87.5	103.4	69.7	86.3	94.2	67.3	80.5	106.4	67.3	86.5	92.4	67.3	79.6
E	Equipment & Furnishings	100.0	63.6	97.9	100.0	63.4	97.9	100.0	68.1	98.2	100.0	76.3	98.7	100.0	78.1	98.8
G	Site Work	94.2	96.2	95.5	84.0	95.9	92.0	112.1	94.9	100.6	99.0	94.9	96.2	106.3	94.3	98.3
A-G	WEIGHTED AVERAGE	100.1	76.1	89.8	97.9	75.7	88.4	97.0	78.5	89.1	97.1	81.1	90.2	96.6	77.3	88.3

DIV. NO.	BUILDING SYSTEMS	NEBRASKA			NEVADA									NEW HAMPSHIRE		
		OMAHA			CARSON CITY			LAS VEGAS			RENO			MANCHESTER		
		MAT.	INST.	TOTAL	MAT.	INST.	TOTAL	MAT.	INST.	TOTAL	MAT.	INST.	TOTAL	MAT.	INST.	TOTAL
A	Substructure	93.6	81.4	86.5	98.6	89.3	93.2	97.3	107.6	103.3	103.0	89.5	95.1	98.3	100.9	99.8
B10	Shell: Superstructure	93.5	81.2	88.9	103.7	91.0	98.9	112.6	104.7	109.6	108.6	91.4	102.0	93.6	94.8	94.0
B20	Exterior Closure	88.3	80.1	84.6	102.7	71.6	88.7	104.7	99.4	102.3	107.2	71.6	91.2	94.1	99.8	96.7
B30	Roofing	101.2	80.1	92.8	111.3	82.4	99.7	121.9	101.3	113.6	107.3	82.4	97.3	105.3	110.3	107.3
C	Interior Construction	99.1	77.3	90.0	96.7	77.5	88.7	96.2	106.4	100.4	95.6	77.6	88.1	95.8	96.9	96.3
D10	Services: Conveying	100.0	89.1	96.8	100.0	105.2	101.5	100.0	101.3	100.4	100.0	105.2	101.5	100.0	106.1	101.8
D20 - 40	Mechanical	99.9	76.8	90.6	100.1	78.7	91.4	100.1	101.3	100.6	100.0	78.7	91.4	100.0	86.3	94.4
D50	Electrical	102.7	83.4	92.9	98.9	90.4	94.6	100.6	110.4	105.6	96.8	90.4	93.6	91.3	77.1	84.1
E	Equipment & Furnishings	100.0	75.1	98.6	100.0	77.4	98.7	100.0	106.2	100.3	100.0	77.4	98.7	100.0	92.4	99.6
G	Site Work	90.3	94.3	93.0	88.4	97.6	94.6	82.3	100.1	94.2	78.9	97.6	91.4	88.5	100.8	96.7
A-G	WEIGHTED AVERAGE	96.9	80.8	90.0	100.4	84.2	93.5	102.7	104.1	103.3	101.5	84.2	94.1	96.2	93.0	94.8

DIV. NO.	BUILDING SYSTEMS	NEW HAMPSHIRE						NEW JERSEY								
		NASHUA			PORTSMOUTH			CAMDEN			ELIZABETH			JERSEY CITY		
		MAT.	INST.	TOTAL	MAT.	INST.	TOTAL	MAT.	INST.	TOTAL	MAT.	INST.	TOTAL	MAT.	INST.	TOTAL
A	Substructure	90.1	100.8	96.3	81.5	100.7	92.7	87.2	130.3	112.3	80.3	136.7	113.1	78.4	135.8	111.8
B10	Shell: Superstructure	90.5	94.7	92.1	85.5	95.1	89.1	100.1	123.5	109.0	93.6	133.7	108.8	96.1	131.7	109.6
B20	Exterior Closure	95.1	99.6	97.1	91.3	95.9	93.3	94.8	135.4	113.1	99.7	145.6	120.3	90.7	145.4	115.3
B30	Roofing	104.4	110.3	106.8	103.9	108.9	105.9	98.9	135.3	113.5	109.6	146.1	124.3	108.8	145.5	123.6
C	Interior Construction	97.8	96.1	97.1	95.3	95.9	95.6	94.5	145.2	115.7	96.5	156.1	121.4	93.8	156.4	120.0
D10	Services: Conveying	100.0	106.1	101.8	100.0	105.8	101.7	100.0	113.4	104.0	100.0	131.6	109.4	100.0	131.6	109.4
D20 - 40	Mechanical	100.1	86.3	94.5	100.1	85.7	94.3	99.9	129.1	111.7	100.1	137.9	115.4	100.1	138.3	115.5
D50	Electrical	93.1	77.1	85.0	91.6	77.1	84.2	99.1	135.3	117.5	96.2	141.9	119.4	100.2	142.5	121.7
E	Equipment & Furnishings	100.0	92.4	99.6	100.0	92.3	99.6	100.0	141.3	102.3	100.0	150.0	102.8	100.0	150.1	102.8
G	Site Work	90.2	100.7	97.2	85.0	101.2	95.9	87.9	104.4	98.9	108.9	106.2	107.1	94.9	106.4	102.6
A-G	WEIGHTED AVERAGE	95.9	92.8	94.6	93.3	92.2	92.8	97.5	130.7	111.8	97.3	139.7	115.4	96.3	139.4	114.8

DIV. NO.	BUILDING SYSTEMS	NEW JERSEY						NEW MEXICO								
		NEWARK			PATERSON			TRENTON			ALBUQUERQUE			FARMINGTON		
		MAT.	INST.	TOTAL	MAT.	INST.	TOTAL	MAT.	INST.	TOTAL	MAT.	INST.	TOTAL	MAT.	INST.	TOTAL
A	Substructure	96.4	136.7	119.9	90.7	136.7	117.5	99.9	129.5	117.1	94.9	76.4	84.1	97.2	76.4	85.1
B10	Shell: Superstructure	101.4	133.7	113.7	94.8	133.6	109.7	102.5	121.4	109.7	101.2	80.8	93.4	100.1	80.8	92.7
B20	Exterior Closure	93.3	145.6	116.8	92.3	145.6	116.2	95.0	135.5	113.2	92.8	60.8	78.4	98.5	60.8	81.6
B30	Roofing	111.0	146.1	125.1	109.5	137.2	120.6	99.8	136.4	114.5	102.0	71.5	89.7	102.3	71.5	89.9
C	Interior Construction	95.4	156.1	120.8	98.3	156.1	122.4	94.8	148.4	117.2	95.0	66.7	83.2	95.6	66.7	83.5
D10	Services: Conveying	100.0	131.6	109.4	100.0	131.6	109.4	100.0	113.5	104.0	100.0	87.2	96.2	100.0	87.2	96.2
D20 - 40	Mechanical	100.1	137.9	115.4	100.1	138.3	115.5	100.1	133.7	113.7	100.1	68.7	87.4	99.9	68.7	87.3
D50	Electrical	104.5	142.5	123.8	100.2	141.9	121.4	103.0	131.8	117.7	88.9	85.8	87.3	86.9	85.8	86.4
E	Equipment & Furnishings	100.0	150.0	102.8	100.0	150.0	102.8	100.0	142.8	102.4	100.0	69.1	98.3	100.0	69.1	98.3
G	Site Work	108.8	106.2	107.1	106.7	106.4	106.5	87.6	105.6	99.7	91.3	102.0	98.5	98.8	102.0	100.9
A-G	WEIGHTED AVERAGE	99.6	139.7	116.8	97.7	139.5	115.6	99.2	131.4	113.0	97.1	75.0	87.6	97.7	75.0	88.0

DIV. NO.	BUILDING SYSTEMS	NEW MEXICO									NEW YORK					
		LAS CRUCES			ROSWELL			SANTA FE			ALBANY			BINGHAMTON		
		MAT.	INST.	TOTAL	MAT.	INST.	TOTAL	MAT.	INST.	TOTAL	MAT.	INST.	TOTAL	MAT.	INST.	TOTAL
A	Substructure	93.8	69.0	79.4	98.0	76.4	85.4	97.7	76.4	85.3	86.3	110.7	100.5	102.8	100.7	101.6
B10	Shell: Superstructure	98.7	75.2	89.7	101.6	80.8	93.7	96.8	80.8	90.7	91.2	119.2	101.9	95.7	119.0	104.6
B20	Exterior Closure	83.3	60.0	72.8	109.3	60.8	87.5	86.4	60.8	74.9	88.1	115.8	100.5	91.5	107.0	98.5
B30	Roofing	88.6	66.0	79.5	101.8	71.5	89.6	104.6	71.5	91.3	107.2	111.1	108.8	109.4	94.4	103.4
C	Interior Construction	95.9	65.9	83.4	94.9	66.7	83.1	97.9	66.7	84.9	94.5	107.4	99.9	92.5	96.2	94.0
D10	Services: Conveying	100.0	84.4	95.4	100.0	87.2	96.2	100.0	87.2	96.2	100.0	102.2	100.7	100.0	98.9	99.7
D20 - 40	Mechanical	100.3	68.5	87.4	99.9	68.7	87.3	100.0	68.7	87.4	100.1	111.8	104.8	100.6	98.7	99.8
D50	Electrical	90.5	85.8	88.1	89.8	85.8	87.8	101.6	85.8	93.6	99.6	107.6	103.6	100.3	100.9	100.6
E	Equipment & Furnishings	100.0	68.4	98.2	100.0	69.1	98.3	100.0	69.1	98.3	100.0	105.1	100.3	100.0	91.2	99.5
G	Site Work	98.4	81.0	86.7	100.8	102.0	101.6	101.7	102.0	101.9	82.0	104.2	96.9	95.5	92.1	93.2
A-G	WEIGHTED AVERAGE	95.6	71.7	85.4	99.5	75.0	89.0	97.6	75.0	87.9	95.0	111.5	102.1	97.3	102.8	99.7

633

NEW YORK

DIV. NO.	BUILDING SYSTEMS	BUFFALO			HICKSVILLE			NEW YORK			RIVERHEAD			ROCHESTER		
		MAT.	INST.	TOTAL	MAT.	INST.	TOTAL	MAT.	INST.	TOTAL	MAT.	INST.	TOTAL	MAT.	INST.	TOTAL
A	Substructure	111.8	116.0	114.3	98.1	156.3	132.0	94.8	167.8	137.3	99.8	160.5	135.2	102.2	105.3	104.0
B10	Shell: Superstructure	102.9	112.8	106.7	101.4	169.7	127.4	95.9	178.8	127.4	102.4	165.2	126.3	103.8	113.6	107.5
B20	Exterior Closure	111.8	120.2	115.6	103.0	177.4	136.4	99.7	186.4	138.7	104.4	175.4	136.3	95.3	106.0	100.2
B30	Roofing	106.1	112.7	108.8	108.6	158.8	128.7	111.7	169.1	134.7	109.6	155.2	127.9	105.4	101.4	103.8
C	Interior Construction	99.1	119.3	107.5	92.3	168.7	124.3	99.7	193.3	138.8	92.6	164.8	122.8	98.8	103.2	100.7
D10	Services: Conveying	100.0	104.9	101.5	100.0	131.6	109.4	100.0	136.4	110.8	100.0	117.9	105.3	100.0	100.3	100.1
D20 - 40	Mechanical	100.0	99.7	99.9	99.8	163.2	125.4	100.1	175.3	130.5	99.9	157.9	123.4	100.0	90.1	96.0
D50	Electrical	103.7	102.8	103.3	99.6	144.4	122.4	97.7	186.7	142.9	101.2	138.2	120.0	102.0	92.1	97.0
E	Equipment & Furnishings	100.0	117.7	101.0	100.0	159.1	103.3	100.0	194.8	105.3	100.0	158.7	103.3	100.0	100.9	100.1
G	Site Work	105.7	101.8	103.1	114.5	125.0	121.6	105.5	111.7	109.6	116.1	124.5	121.7	93.0	108.5	103.4
A-G	WEIGHTED AVERAGE	103.0	110.0	106.0	99.8	160.5	125.8	99.0	176.2	132.1	100.5	156.8	124.6	100.3	101.3	100.8

NEW YORK

DIV. NO.	BUILDING SYSTEMS	SCHENECTADY			SYRACUSE			UTICA			WATERTOWN			WHITE PLAINS		
		MAT.	INST.	TOTAL	MAT.	INST.	TOTAL	MAT.	INST.	TOTAL	MAT.	INST.	TOTAL	MAT.	INST.	TOTAL
A	Substructure	94.1	110.6	103.7	97.9	100.5	99.4	91.0	98.2	95.2	100.1	102.7	101.6	83.2	146.2	119.9
B10	Shell: Superstructure	91.4	119.2	102.0	97.4	110.4	102.3	94.7	107.8	99.7	95.9	111.7	101.9	84.8	161.3	113.9
B20	Exterior Closure	90.8	115.8	102.0	96.6	102.9	99.4	94.0	101.2	97.3	101.9	106.8	104.1	89.0	159.3	120.6
B30	Roofing	101.8	111.1	105.6	103.8	95.5	100.5	92.3	95.7	93.6	92.6	99.0	95.1	109.5	149.5	125.6
C	Interior Construction	94.0	107.4	99.6	92.4	92.7	92.5	94.3	90.2	92.6	93.0	95.3	93.9	95.6	162.7	123.6
D10	Services: Conveying	100.0	102.2	100.7	100.0	98.0	99.4	100.0	92.3	97.7	100.0	99.1	99.7	100.0	126.7	107.9
D20 - 40	Mechanical	100.1	107.7	103.2	100.2	96.6	98.8	100.2	94.6	98.0	100.2	89.8	96.0	100.3	144.9	118.3
D50	Electrical	98.6	107.6	103.2	100.3	103.5	101.9	98.2	103.5	100.9	100.2	92.4	96.2	89.6	170.1	130.5
E	Equipment & Furnishings	100.0	105.1	100.3	100.0	89.3	99.4	100.0	86.0	99.2	100.0	92.6	99.6	100.0	152.5	103.0
G	Site Work	83.0	104.2	97.2	94.0	102.8	99.9	71.9	102.3	92.3	79.5	102.8	95.1	98.0	110.3	106.2
A-G	WEIGHTED AVERAGE	95.5	110.6	101.9	97.8	100.8	99.1	95.9	98.9	97.2	97.7	99.2	98.3	93.3	153.5	119.1

DIV. NO.	BUILDING SYSTEMS	NEW YORK YONKERS			NORTH CAROLINA ASHEVILLE			CHARLOTTE			DURHAM			FAYETTEVILLE		
		MAT.	INST.	TOTAL	MAT.	INST.	TOTAL	MAT.	INST.	TOTAL	MAT.	INST.	TOTAL	MAT.	INST.	TOTAL
A	Substructure	91.5	144.9	122.6	101.0	69.8	82.8	103.9	69.6	83.9	102.2	71.8	84.5	104.8	71.3	85.3
B10	Shell: Superstructure	92.4	160.3	118.2	99.0	78.9	91.3	100.5	79.0	92.3	111.9	79.1	99.4	116.0	78.9	101.9
B20	Exterior Closure	95.4	158.5	123.7	83.7	64.3	75.0	84.1	63.3	74.8	84.6	64.3	75.5	84.0	63.3	74.7
B30	Roofing	109.9	149.1	125.6	100.2	64.6	85.9	94.4	63.9	82.2	105.4	64.6	89.1	99.6	63.9	85.3
C	Interior Construction	99.3	160.4	124.8	89.4	61.3	77.6	92.9	61.0	79.6	92.0	61.3	79.2	89.9	61.0	77.8
D10	Services: Conveying	100.0	124.1	107.2	100.0	86.4	96.0	100.0	85.9	95.8	100.0	86.4	96.0	100.0	85.9	95.8
D20 - 40	Mechanical	100.3	144.8	118.3	100.5	61.5	84.7	99.9	61.9	84.6	100.6	61.5	84.8	100.3	60.8	84.3
D50	Electrical	95.8	159.8	128.4	101.0	59.3	79.8	100.1	61.8	80.6	95.2	58.8	76.7	100.6	58.8	79.3
E	Equipment & Furnishings	100.0	147.9	102.7	100.0	60.2	97.8	100.0	60.4	97.8	100.0	60.2	97.8	100.0	60.1	97.8
G	Site Work	102.7	110.2	107.8	94.9	80.3	85.1	96.8	80.6	85.9	99.9	89.5	92.9	94.5	89.4	91.1
A-G	WEIGHTED AVERAGE	97.3	151.2	120.5	96.3	67.0	83.8	97.0	67.3	84.3	99.3	67.7	85.7	100.0	67.3	86.0

DIV. NO.	BUILDING SYSTEMS	NORTH CAROLINA GREENSBORO			RALEIGH			WILMINGTON			WINSTON-SALEM			NORTH DAKOTA BISMARCK		
		MAT.	INST.	TOTAL	MAT.	INST.	TOTAL	MAT.	INST.	TOTAL	MAT.	INST.	TOTAL	MAT.	INST.	TOTAL
A	Substructure	101.5	71.8	84.2	104.4	71.4	85.2	100.9	69.5	82.6	103.4	71.8	85.0	102.9	84.7	92.3
B10	Shell: Superstructure	105.8	79.1	95.6	100.4	78.9	92.2	98.6	78.9	91.1	103.9	79.1	94.5	97.0	86.4	93.0
B20	Exterior Closure	83.5	64.3	74.9	79.6	63.3	72.2	79.6	63.3	72.3	83.6	64.3	75.0	87.5	84.0	85.9
B30	Roofing	105.2	64.6	88.9	99.5	63.9	85.2	100.2	63.9	85.6	105.2	64.6	88.9	105.2	83.9	96.6
C	Interior Construction	92.1	61.3	79.2	92.4	61.0	79.3	89.9	61.0	77.8	92.1	61.3	79.2	101.6	69.1	88.0
D10	Services: Conveying	100.0	83.3	95.0	100.0	85.9	95.8	100.0	85.9	95.8	100.0	86.4	96.0	100.0	91.6	97.5
D20 - 40	Mechanical	100.5	61.5	84.7	100.0	60.8	84.1	100.5	60.8	84.5	100.5	61.5	84.7	99.9	75.5	90.1
D50	Electrical	94.3	59.3	76.5	97.6	57.7	77.3	101.3	57.2	78.9	94.3	59.3	76.5	97.5	74.4	85.7
E	Equipment & Furnishings	100.0	60.2	97.8	100.0	60.1	97.8	100.0	60.1	97.8	100.0	60.2	97.8	100.0	69.9	98.3
G	Site Work	99.7	89.6	92.9	99.9	89.5	92.9	96.1	80.2	85.5	100.1	89.6	93.1	100.2	98.3	98.9
A-G	WEIGHTED AVERAGE	97.7	67.7	84.8	96.4	67.1	83.8	95.9	66.3	83.2	97.4	67.8	84.7	98.2	80.0	90.4

DIV. NO.	BUILDING SYSTEMS	NORTH DAKOTA FARGO			GRAND FORKS			MINOT			OHIO AKRON			CANTON		
		MAT.	INST.	TOTAL	MAT.	INST.	TOTAL	MAT.	INST.	TOTAL	MAT.	INST.	TOTAL	MAT.	INST.	TOTAL
A	Substructure	98.8	84.8	90.6	107.5	84.7	94.2	107.4	84.7	94.2	103.7	91.0	96.3	104.3	84.7	92.9
B10	Shell: Superstructure	95.0	86.7	91.9	94.6	86.3	91.5	94.7	86.4	91.5	96.0	84.2	91.5	95.2	76.8	88.8
B20	Exterior Closure	93.9	88.0	91.2	94.9	82.7	89.4	92.7	84.0	88.8	95.7	90.1	93.2	95.4	79.0	88.0
B30	Roofing	106.3	84.7	97.6	106.8	83.8	97.5	106.5	83.9	97.4	103.1	93.9	99.4	104.2	90.1	98.5
C	Interior Construction	97.9	69.9	86.2	96.8	69.7	85.5	96.4	69.0	84.9	105.6	85.9	97.3	102.0	74.2	90.4
D10	Services: Conveying	100.0	91.6	97.5	100.0	91.6	97.5	100.0	91.6	97.5	100.0	93.3	98.0	100.0	91.8	97.6
D20 - 40	Mechanical	99.9	75.6	90.1	100.2	73.8	89.5	100.2	73.6	89.4	100.0	89.8	95.9	100.0	79.6	91.8
D50	Electrical	102.1	71.1	86.3	98.9	69.9	84.2	101.9	74.4	87.9	96.5	83.9	90.1	95.8	87.1	91.4
E	Equipment & Furnishings	100.0	69.8	98.3	100.0	69.7	98.3	100.0	69.9	98.3	100.0	83.3	99.1	100.0	74.2	98.5
G	Site Work	97.3	98.3	98.0	108.9	98.3	101.8	106.6	98.3	101.1	100.4	99.4	99.8	100.6	99.1	99.6
A-G	WEIGHTED AVERAGE	98.2	80.3	90.5	98.4	79.0	90.1	98.3	79.6	90.3	99.5	88.3	94.7	98.9	81.4	91.4

OHIO

DIV. NO.	BUILDING SYSTEMS	CINCINNATI			CLEVELAND			COLUMBUS			DAYTON			LORAIN		
		MAT.	INST.	TOTAL	MAT.	INST.	TOTAL	MAT.	INST.	TOTAL	MAT.	INST.	TOTAL	MAT.	INST.	TOTAL
A	Substructure	96.2	84.3	89.3	101.5	93.9	97.1	100.2	82.9	90.1	89.4	85.0	86.8	100.2	88.6	93.5
B10	Shell: Superstructure	95.0	81.5	89.9	96.7	87.4	93.2	95.7	80.2	89.8	93.1	79.0	87.7	95.8	82.6	90.7
B20	Exterior Closure	91.1	80.5	86.3	99.1	95.8	97.6	91.6	85.0	88.6	86.0	77.6	82.2	93.8	93.1	93.4
B30	Roofing	98.8	82.4	92.2	100.2	97.8	99.2	93.6	83.9	89.7	105.1	82.4	96.0	104.1	92.5	99.4
C	Interior Construction	98.0	80.9	90.9	98.8	88.9	94.6	97.1	77.5	88.9	98.8	77.7	90.0	102.0	79.4	92.5
D10	Services: Conveying	100.0	90.8	97.3	100.0	99.4	99.8	100.0	91.1	97.3	100.0	87.8	96.4	100.0	95.5	98.6
D20 - 40	Mechanical	100.0	77.2	90.8	100.0	92.1	96.8	100.0	85.9	94.3	100.8	83.0	93.6	100.0	90.4	96.1
D50	Electrical	95.4	72.9	84.0	96.1	93.0	94.5	98.4	82.2	90.1	94.1	77.5	85.7	95.9	80.1	87.9
E	Equipment & Furnishings	100.0	81.9	99.0	100.0	86.9	99.3	100.0	77.1	98.7	100.0	79.3	98.8	100.0	72.5	98.5
G	Site Work	98.2	99.1	98.8	98.6	97.6	97.9	101.1	92.5	95.3	98.4	99.5	99.1	99.5	99.5	99.5
A-G	WEIGHTED AVERAGE	96.9	80.7	89.9	98.7	92.2	95.9	97.4	83.2	91.3	95.9	81.3	89.6	98.4	86.8	93.4

634

Cost Indexes

DIV. NO.	BUILDING SYSTEMS	OHIO SPRINGFIELD MAT.	INST.	TOTAL	TOLEDO MAT.	INST.	TOTAL	YOUNGSTOWN MAT.	INST.	TOTAL	OKLAHOMA ENID MAT.	INST.	TOTAL	LAWTON MAT.	INST.	TOTAL
A	Substructure	90.9	84.9	87.4	97.0	90.2	93.0	103.0	87.4	93.9	86.2	70.6	77.1	84.5	71.0	76.6
B10	Shell: Superstructure	93.4	79.0	87.9	94.9	87.1	91.9	95.9	80.6	90.1	92.4	62.0	80.8	94.9	62.0	82.4
B20	Exterior Closure	85.6	77.2	81.8	90.8	89.3	90.1	95.2	86.6	91.4	91.7	57.3	76.2	88.4	57.3	74.5
B30	Roofing	105.0	82.1	95.8	90.7	92.0	91.2	104.3	90.3	98.7	101.9	64.9	87.1	101.6	64.9	86.9
C	Interior Construction	98.0	77.6	89.5	96.2	87.3	92.4	102.0	80.5	93.0	94.6	57.8	79.2	96.2	57.8	80.2
D10	Services: Conveying	100.0	87.7	96.3	100.0	93.7	98.1	100.0	92.2	97.7	100.0	81.9	94.6	100.0	82.0	94.6
D20 - 40	Mechanical	100.8	82.8	93.5	100.1	95.1	98.0	100.0	85.8	94.3	100.2	67.8	87.1	100.2	67.8	87.1
D50	Electrical	94.1	82.2	88.1	98.6	106.0	102.3	95.9	76.2	85.9	94.4	70.9	82.4	96.0	70.9	83.3
E	Equipment & Furnishings	100.0	79.5	98.8	100.0	87.1	99.3	100.0	76.5	98.7	100.0	56.4	97.5	100.0	56.4	97.5
G	Site Work	98.8	99.5	99.2	98.6	95.9	96.8	100.2	99.2	99.6	103.1	90.7	97.7	98.8	96.6	97.3
A-G	WEIGHTED AVERAGE	95.9	81.8	89.8	96.7	92.8	95.0	98.7	84.0	92.4	95.5	66.5	83.1	95.9	66.6	83.3

DIV. NO.	BUILDING SYSTEMS	OKLAHOMA MUSKOGEE MAT.	INST.	TOTAL	OKLAHOMA CITY MAT.	INST.	TOTAL	TULSA MAT.	INST.	TOTAL	OREGON EUGENE MAT.	INST.	TOTAL	MEDFORD MAT.	INST.	TOTAL
A	Substructure	86.5	70.3	77.0	86.1	74.1	79.1	92.2	71.4	80.1	109.0	100.9	104.3	111.4	100.8	105.2
B10	Shell: Superstructure	93.6	69.1	84.3	90.2	64.2	80.3	97.0	71.0	87.1	103.2	97.0	100.8	102.9	96.8	100.6
B20	Exterior Closure	91.9	51.9	74.0	90.0	57.6	75.5	87.9	58.5	74.7	90.2	101.5	95.3	94.5	101.5	97.7
B30	Roofing	97.5	64.9	84.4	93.4	66.4	82.6	97.5	66.2	85.0	115.9	101.2	110.0	116.7	93.7	107.5
C	Interior Construction	92.7	51.5	75.5	94.9	63.3	81.7	93.4	57.6	78.5	97.2	95.4	96.5	99.6	95.0	97.7
D10	Services: Conveying	100.0	80.7	94.3	100.0	81.9	94.6	100.0	80.7	94.3	100.0	98.6	99.6	100.0	98.6	99.6
D20 - 40	Mechanical	100.2	62.7	85.1	100.1	67.8	87.0	100.2	65.1	86.0	100.0	100.8	100.3	100.0	107.5	103.0
D50	Electrical	93.6	68.8	81.0	101.5	70.9	86.0	95.5	68.7	81.9	98.7	93.7	96.2	102.2	80.5	91.2
E	Equipment & Furnishings	100.0	56.6	97.6	100.0	64.8	98.0	100.0	57.4	97.6	100.0	95.9	99.8	100.0	95.2	99.7
G	Site Work	90.3	92.2	91.6	96.7	97.1	97.0	96.1	91.4	93.0	103.9	103.3	103.5	111.5	103.3	106.0
A-G	WEIGHTED AVERAGE	95.1	64.4	81.9	95.2	68.2	83.6	96.0	67.1	83.6	99.8	98.5	99.3	101.3	97.8	99.8

DIV. NO.	BUILDING SYSTEMS	OREGON PORTLAND MAT.	INST.	TOTAL	SALEM MAT.	INST.	TOTAL	PENNSYLVANIA ALLENTOWN MAT.	INST.	TOTAL	ALTOONA MAT.	INST.	TOTAL	ERIE MAT.	INST.	TOTAL
A	Substructure	111.3	100.9	105.2	105.6	100.9	102.9	91.8	107.7	101.0	97.4	93.2	95.0	97.7	95.2	96.2
B10	Shell: Superstructure	104.7	97.0	101.7	108.4	97.0	104.1	96.3	116.2	103.9	91.9	103.9	96.5	92.9	105.4	97.7
B20	Exterior Closure	90.3	101.5	95.4	91.9	101.5	96.2	93.5	98.0	95.5	85.8	87.4	86.5	79.9	90.5	84.6
B30	Roofing	115.9	97.8	108.6	112.3	98.2	106.6	103.8	113.6	107.7	102.5	93.0	98.7	103.1	91.4	98.4
C	Interior Construction	96.1	95.8	96.0	98.1	95.7	97.1	91.6	109.7	99.1	86.6	89.8	87.9	88.3	91.0	89.4
D10	Services: Conveying	100.0	98.6	99.6	100.0	98.6	99.6	100.0	100.0	100.0	100.0	97.9	99.4	100.0	99.3	99.8
D20 - 40	Mechanical	100.0	104.7	101.9	100.0	107.5	103.1	100.2	117.1	107.1	99.8	85.5	94.0	99.8	95.7	98.1
D50	Electrical	98.9	101.5	100.2	106.6	93.7	100.0	99.6	98.1	98.8	90.0	110.3	100.3	91.7	95.4	93.6
E	Equipment & Furnishings	100.0	96.2	99.8	100.0	95.9	99.8	100.0	113.2	100.7	100.0	85.1	99.2	100.0	87.1	99.3
G	Site Work	107.2	103.3	104.6	98.7	103.3	101.8	92.4	101.4	98.4	94.2	101.2	98.8	91.2	101.5	98.1
A-G	WEIGHTED AVERAGE	100.2	100.4	100.3	101.7	99.9	100.9	96.7	108.4	101.7	93.2	95.1	94.0	93.1	96.2	94.5

DIV. NO.	BUILDING SYSTEMS	PENNSYLVANIA HARRISBURG MAT.	INST.	TOTAL	PHILADELPHIA MAT.	INST.	TOTAL	PITTSBURGH MAT.	INST.	TOTAL	READING MAT.	INST.	TOTAL	SCRANTON MAT.	INST.	TOTAL
A	Substructure	96.4	97.2	96.8	96.6	127.9	114.8	95.8	101.6	99.2	85.2	105.2	96.8	94.3	97.9	96.4
B10	Shell: Superstructure	101.9	108.3	104.4	100.2	124.0	109.3	100.7	104.6	102.2	96.6	118.7	105.0	98.4	109.4	102.6
B20	Exterior Closure	91.1	87.4	89.4	101.4	131.3	114.9	99.7	100.6	100.1	96.5	97.1	96.7	93.7	96.9	95.1
B30	Roofing	95.8	104.9	99.4	103.4	137.5	117.1	100.9	97.3	99.4	108.4	108.2	108.3	103.7	93.9	99.7
C	Interior Construction	94.6	89.0	92.3	99.6	141.5	117.1	99.1	100.5	99.7	88.0	92.7	90.0	92.4	92.0	92.2
D10	Services: Conveying	100.0	98.2	99.5	100.0	117.1	105.1	100.0	102.0	100.6	100.0	99.3	99.8	100.0	97.9	99.4
D20 - 40	Mechanical	100.1	93.7	97.5	100.2	140.0	116.3	99.9	98.3	99.2	100.2	109.2	103.8	100.2	98.2	99.4
D50	Electrical	98.2	87.9	93.0	100.4	160.5	131.0	100.8	110.8	105.9	99.5	93.6	96.5	99.6	97.6	98.6
E	Equipment & Furnishings	100.0	87.0	99.3	100.0	142.6	102.4	100.0	97.4	99.9	100.0	85.1	99.2	100.0	86.2	99.2
G	Site Work	88.9	99.5	96.0	96.6	100.0	98.9	100.4	100.0	100.1	98.2	112.4	107.7	93.1	101.4	98.6
A-G	WEIGHTED AVERAGE	97.8	94.9	96.5	100.1	135.3	115.2	99.9	102.2	100.9	96.5	104.0	99.7	97.4	98.9	98.1

DIV. NO.	BUILDING SYSTEMS	PENNSYLVANIA YORK MAT.	INST.	TOTAL	PUERTO RICO SAN JUAN MAT.	INST.	TOTAL	RHODE ISLAND PROVIDENCE MAT.	INST.	TOTAL	SOUTH CAROLINA CHARLESTON MAT.	INST.	TOTAL	COLUMBIA MAT.	INST.	TOTAL
A	Substructure	88.7	97.7	94.0	114.6	39.0	70.6	92.1	116.3	106.2	103.8	70.9	84.7	105.2	70.9	85.2
B10	Shell: Superstructure	95.9	108.6	100.8	109.0	32.1	79.7	92.4	116.0	101.4	101.9	79.6	93.4	100.2	79.6	92.4
B20	Exterior Closure	95.7	89.8	93.0	95.4	21.2	62.1	94.6	122.3	107.1	88.0	67.1	78.6	82.2	67.1	75.4
B30	Roofing	92.7	106.0	98.0	130.9	26.7	89.1	103.9	119.3	110.1	95.7	63.0	82.5	91.6	62.9	80.1
C	Interior Construction	87.7	90.2	88.8	171.4	21.0	108.5	95.3	122.9	106.8	94.2	66.8	82.8	92.5	66.6	81.7
D10	Services: Conveying	100.0	98.9	99.7	100.0	19.5	76.1	100.0	104.4	101.3	100.0	71.3	91.5	100.0	71.3	91.5
D20 - 40	Mechanical	100.1	94.8	98.0	103.5	17.9	68.9	100.1	113.2	105.4	100.5	58.3	83.4	100.0	57.7	82.9
D50	Electrical	92.3	84.9	88.5	120.6	16.9	67.9	97.4	97.9	97.6	95.9	58.3	76.8	96.2	61.1	78.4
E	Equipment & Furnishings	100.0	87.0	99.3	100.0	19.4	95.5	100.0	118.9	101.1	100.0	64.3	98.0	100.0	64.5	98.0
G	Site Work	84.5	99.7	94.6	129.3	89.7	102.8	86.5	102.4	97.2	106.4	88.1	94.1	106.4	88.1	94.1
A-G	WEIGHTED AVERAGE	94.9	95.3	95.1	117.8	27.2	79.0	96.2	113.7	103.7	98.0	67.6	84.9	96.5	67.9	84.2

DIV. NO.	BUILDING SYSTEMS	SOUTH CAROLINA FLORENCE MAT.	INST.	TOTAL	GREENVILLE MAT.	INST.	TOTAL	SPARTANBURG MAT.	INST.	TOTAL	SOUTH DAKOTA ABERDEEN MAT.	INST.	TOTAL	PIERRE MAT.	INST.	TOTAL
A	Substructure	93.4	70.8	80.3	93.8	70.7	80.4	94.0	71.0	80.6	109.0	82.7	93.7	108.1	73.6	88.0
B10	Shell: Superstructure	98.2	79.2	90.9	98.8	79.5	91.4	99.0	79.6	91.6	95.2	80.4	89.6	95.9	76.7	88.6
B20	Exterior Closure	88.0	67.1	78.6	85.0	67.1	77.0	85.9	67.1	77.5	96.6	72.1	85.6	93.4	75.1	85.2
B30	Roofing	96.0	63.0	82.8	95.9	63.0	82.7	96.0	63.0	82.7	103.1	82.3	94.8	104.8	75.5	93.0
C	Interior Construction	91.0	66.6	80.8	91.9	66.6	81.3	92.3	66.6	81.5	94.6	68.7	83.8	97.2	46.4	76.0
D10	Services: Conveying	100.0	71.3	91.5	100.0	71.3	91.5	100.0	71.3	91.5	100.0	87.8	96.4	100.0	89.8	97.0
D20 - 40	Mechanical	100.5	57.3	83.2	100.5	57.7	83.2	100.5	57.7	83.2	100.1	55.4	82.0	100.0	78.4	91.3
D50	Electrical	94.0	61.1	77.3	96.0	61.3	78.3	96.0	61.3	78.3	100.2	65.0	82.3	103.6	48.9	75.8
E	Equipment & Furnishings	100.0	64.5	98.0	100.0	64.5	98.0	100.0	64.5	98.0	100.0	76.0	98.7	100.0	36.4	96.4
G	Site Work	114.0	87.9	96.5	109.4	88.2	95.2	109.2	88.2	95.1	100.7	98.1	98.9	99.4	97.2	98.0
A-G	WEIGHTED AVERAGE	96.2	67.8	84.0	96.2	67.9	84.1	96.4	67.9	84.2	98.3	71.5	86.8	98.7	69.4	86.2

635

City Cost Indexes — RJ1030-010 Building Systems

SOUTH DAKOTA / TENNESSEE

DIV. NO.	BUILDING SYSTEMS	RAPID CITY MAT.	INST.	TOTAL	SIOUX FALLS MAT.	INST.	TOTAL	CHATTANOOGA MAT.	INST.	TOTAL	JACKSON MAT.	INST.	TOTAL	JOHNSON CITY MAT.	INST.	TOTAL
A	Substructure	105.9	77.9	89.6	91.9	86.1	88.6	98.5	70.4	82.1	97.9	65.5	79.0	85.2	65.8	73.9
B10	Shell: Superstructure	96.1	80.2	90.0	93.9	87.2	91.3	95.8	76.5	88.4	96.2	73.1	87.4	90.1	75.1	84.4
B20	Exterior Closure	95.1	76.7	86.9	85.6	78.9	82.6	96.8	60.0	80.3	97.4	45.8	74.2	122.6	48.9	89.5
B30	Roofing	103.6	82.0	94.9	101.7	86.4	94.9	95.7	63.0	82.6	96.8	56.3	80.5	91.8	55.0	77.0
C	Interior Construction	96.4	62.5	82.2	98.4	79.5	90.5	100.4	58.2	82.8	92.8	44.8	72.8	100.1	57.9	82.5
D10	Services: Conveying	100.0	89.8	97.0	100.0	89.8	97.0	100.0	70.5	91.2	100.0	65.3	89.7	100.0	79.2	93.8
D20-40	Mechanical	100.1	78.5	91.4	100.0	71.6	88.5	100.1	60.6	84.2	100.1	60.0	83.9	99.9	56.8	82.5
D50	Electrical	96.7	48.9	72.4	100.4	65.0	82.4	101.2	84.9	92.9	99.4	52.8	75.7	91.8	41.8	66.4
E	Equipment & Furnishings	100.0	49.9	97.2	100.0	75.9	98.6	100.0	59.2	97.7	100.0	41.0	96.7	100.0	62.2	97.9
G	Site Work	98.5	97.3	97.7	90.2	99.7	96.5	103.4	98.2	99.9	99.4	98.2	98.6	109.3	84.6	92.8
A-G	WEIGHTED AVERAGE	98.0	73.3	87.4	96.3	79.2	89.0	98.9	69.6	86.3	97.5	59.7	81.3	99.1	59.9	82.3

TENNESSEE / TEXAS

DIV. NO.	BUILDING SYSTEMS	KNOXVILLE MAT.	INST.	TOTAL	MEMPHIS MAT.	INST.	TOTAL	NASHVILLE MAT.	INST.	TOTAL	ABILENE MAT.	INST.	TOTAL	AMARILLO MAT.	INST.	TOTAL
A	Substructure	92.7	69.7	79.3	96.5	76.5	84.8	93.7	72.9	81.6	88.4	68.5	76.9	88.3	65.3	74.9
B10	Shell: Superstructure	95.2	76.8	88.2	91.3	77.9	86.2	101.3	76.7	92.0	99.7	66.7	87.1	93.1	64.0	82.0
B20	Exterior Closure	86.6	53.4	71.7	97.0	59.4	80.1	94.4	59.8	78.9	84.4	62.3	74.5	85.8	61.5	74.9
B30	Roofing	89.4	63.5	79.0	88.5	70.5	81.3	94.2	65.1	82.5	99.4	64.5	85.4	98.0	62.6	83.8
C	Interior Construction	95.4	61.0	81.0	99.3	63.9	84.5	98.3	63.5	83.8	93.5	61.2	80.0	97.3	54.2	79.3
D10	Services: Conveying	100.0	83.2	95.0	100.0	83.8	95.2	100.0	83.9	95.2	100.0	82.9	94.9	100.0	76.6	93.0
D20-40	Mechanical	99.9	61.8	84.5	100.0	71.9	88.6	100.0	75.5	90.1	100.3	53.3	81.3	100.0	52.2	80.7
D50	Electrical	97.4	56.3	76.5	103.2	65.1	83.9	94.1	62.8	78.2	96.1	55.3	75.4	99.8	60.3	79.7
E	Equipment & Furnishings	100.0	63.9	98.0	100.0	66.6	98.1	100.0	64.9	98.0	100.0	61.8	97.9	100.0	52.4	97.3
G	Site Work	91.4	87.5	88.8	86.9	94.6	92.1	102.6	97.2	99.0	93.7	91.7	92.4	92.6	91.0	91.6
A-G	WEIGHTED AVERAGE	95.6	65.2	82.6	97.3	71.1	86.1	98.4	71.1	86.7	96.0	62.8	81.8	95.7	61.1	80.9

TEXAS

DIV. NO.	BUILDING SYSTEMS	AUSTIN MAT.	INST.	TOTAL	BEAUMONT MAT.	INST.	TOTAL	CORPUS CHRISTI MAT.	INST.	TOTAL	DALLAS MAT.	INST.	TOTAL	EL PASO MAT.	INST.	TOTAL
A	Substructure	94.2	65.6	77.5	94.9	69.9	79.9	115.5	65.9	86.7	94.5	72.4	81.6	83.3	68.3	74.6
B10	Shell: Superstructure	102.8	63.7	87.9	102.4	69.3	89.8	103.5	73.2	92.0	101.1	75.5	91.3	97.5	66.2	85.6
B20	Exterior Closure	87.8	61.3	75.9	98.2	63.4	82.5	86.0	62.6	75.5	99.6	62.4	82.9	82.1	63.3	73.7
B30	Roofing	93.7	64.2	81.9	93.7	65.2	82.3	100.2	63.0	85.3	89.0	67.1	80.2	94.9	64.3	82.7
C	Interior Construction	96.5	54.4	78.9	94.9	58.9	79.8	100.6	54.7	81.4	96.7	62.9	82.6	92.8	58.9	78.6
D10	Services: Conveying	100.0	80.9	94.3	100.0	82.4	94.8	100.0	82.8	94.9	100.0	83.8	95.2	100.0	80.5	94.2
D20-40	Mechanical	100.0	60.0	83.8	100.1	64.4	85.7	100.1	49.8	79.8	100.0	60.7	84.1	99.9	65.9	86.2
D50	Electrical	93.7	58.7	75.9	98.9	64.9	81.6	90.5	65.0	77.5	98.5	61.3	79.6	94.8	51.7	72.9
E	Equipment & Furnishings	100.0	54.1	97.4	100.0	55.9	97.5	100.0	53.5	97.4	100.0	64.0	98.0	100.0	58.1	97.6
G	Site Work	96.5	90.7	92.6	90.0	96.0	94.0	137.1	86.5	103.2	106.4	96.5	99.8	91.5	92.4	92.1
A-G	WEIGHTED AVERAGE	97.4	62.7	82.5	98.8	67.1	85.2	99.7	63.0	84.0	99.1	67.7	85.6	94.6	64.7	81.8

TEXAS

DIV. NO.	BUILDING SYSTEMS	FORT WORTH MAT.	INST.	TOTAL	HOUSTON MAT.	INST.	TOTAL	LAREDO MAT.	INST.	TOTAL	LUBBOCK MAT.	INST.	TOTAL	ODESSA MAT.	INST.	TOTAL
A	Substructure	89.4	68.7	77.4	96.4	67.5	79.6	91.6	65.7	76.5	92.4	67.4	77.8	88.6	68.7	77.0
B10	Shell: Superstructure	101.9	66.7	88.5	105.6	69.6	91.9	101.3	64.4	87.3	102.7	74.1	91.8	99.2	66.7	86.8
B20	Exterior Closure	87.8	61.0	75.8	97.5	63.5	82.2	92.0	61.4	78.3	83.9	63.6	74.8	84.4	61.8	74.3
B30	Roofing	89.9	65.3	80.0	87.1	66.7	78.9	96.2	64.1	83.3	88.4	64.5	78.8	99.4	63.5	85.0
C	Interior Construction	96.9	60.7	81.8	101.5	60.0	81.8	94.0	53.9	77.2	98.7	56.2	80.9	93.5	60.6	79.7
D10	Services: Conveying	100.0	82.9	94.9	100.0	84.9	95.5	100.0	80.5	94.2	100.0	83.3	95.0	100.0	76.6	93.0
D20-40	Mechanical	99.9	56.6	82.4	100.1	65.5	86.1	100.1	60.2	84.0	99.7	54.3	81.3	100.3	53.7	81.4
D50	Electrical	97.0	60.0	78.2	100.7	68.1	84.1	92.3	58.5	75.1	94.9	61.6	77.9	96.2	60.3	78.0
E	Equipment & Furnishings	100.0	62.2	97.9	100.0	59.2	97.7	100.0	53.0	97.4	100.0	53.9	97.4	100.0	62.2	97.9
G	Site Work	97.9	92.4	94.2	102.0	94.3	96.8	99.9	90.7	93.7	114.6	90.5	98.4	93.7	92.4	92.9
A-G	WEIGHTED AVERAGE	97.4	64.1	83.1	100.7	67.9	86.6	97.1	62.7	82.4	97.5	64.4	83.3	96.0	63.3	82.0

TEXAS / UTAH

DIV. NO.	BUILDING SYSTEMS	SAN ANTONIO MAT.	INST.	TOTAL	WACO MAT.	INST.	TOTAL	WICHITA FALLS MAT.	INST.	TOTAL	LOGAN MAT.	INST.	TOTAL	OGDEN MAT.	INST.	TOTAL
A	Substructure	94.3	66.7	78.2	82.9	68.2	74.3	86.5	68.5	76.0	92.2	78.2	84.0	91.7	78.2	83.8
B10	Shell: Superstructure	103.9	62.8	88.2	101.3	65.5	87.7	102.0	66.7	88.6	101.6	78.6	92.8	102.1	78.6	93.2
B20	Exterior Closure	96.5	61.3	80.7	86.3	62.1	75.4	86.5	61.8	75.4	109.6	67.9	90.8	96.1	67.9	83.4
B30	Roofing	92.2	66.3	81.8	93.2	64.9	81.8	93.2	63.8	81.4	99.8	71.9	88.6	98.5	71.9	87.8
C	Interior Construction	100.2	54.6	81.1	79.5	60.4	71.5	79.7	62.0	72.3	92.4	65.6	81.2	91.5	65.6	80.7
D10	Services: Conveying	100.0	82.7	94.9	100.0	82.9	94.9	100.0	76.6	93.0	100.0	85.8	95.8	100.0	85.8	95.8
D20-40	Mechanical	100.1	61.1	84.3	100.1	60.1	83.9	100.1	53.3	81.2	99.9	69.9	87.8	99.9	69.9	87.8
D50	Electrical	93.7	61.2	77.2	99.8	55.4	77.2	101.5	55.3	78.0	95.7	70.5	82.9	96.0	70.5	83.0
E	Equipment & Furnishings	100.0	53.2	97.4	100.0	61.9	97.9	100.0	61.9	97.9	100.0	67.7	98.2	100.0	67.7	98.2
G	Site Work	98.4	93.5	95.2	95.6	92.3	93.4	96.5	92.4	93.8	99.3	93.9	95.7	88.0	93.9	92.0
A-G	WEIGHTED AVERAGE	99.3	63.5	83.9	94.3	63.9	81.3	94.9	62.7	81.1	99.4	73.1	88.1	97.5	73.1	87.0

UTAH / VERMONT / VIRGINIA

DIV. NO.	BUILDING SYSTEMS	PROVO MAT.	INST.	TOTAL	SALT LAKE CITY MAT.	INST.	TOTAL	BURLINGTON MAT.	INST.	TOTAL	RUTLAND MAT.	INST.	TOTAL	ALEXANDRIA MAT.	INST.	TOTAL
A	Substructure	93.3	77.8	84.3	97.5	78.1	86.2	106.3	95.0	99.7	93.5	94.9	94.3	103.7	79.4	89.5
B10	Shell: Superstructure	100.6	78.6	92.2	106.2	78.6	95.7	95.4	89.9	93.3	91.3	89.9	90.8	103.2	88.7	97.7
B20	Exterior Closure	111.1	67.9	91.7	117.5	67.9	95.2	100.0	87.2	94.3	91.8	87.2	89.8	92.7	74.6	84.6
B30	Roofing	101.8	71.9	89.8	106.3	71.9	92.5	106.1	86.6	98.3	99.1	86.6	94.1	102.3	80.7	93.6
C	Interior Construction	94.7	65.6	82.5	93.4	65.7	81.8	99.2	84.9	93.2	97.5	84.9	92.2	95.1	71.1	85.0
D10	Services: Conveying	100.0	85.8	95.8	100.0	85.8	95.8	100.0	91.9	97.6	100.0	91.9	97.6	100.0	89.7	96.9
D20-40	Mechanical	99.9	69.9	87.8	100.1	69.9	87.8	100.1	69.4	87.7	100.2	69.4	87.8	100.4	87.2	95.1
D50	Electrical	96.3	70.3	83.1	98.6	70.5	84.3	98.7	55.0	76.5	98.9	55.0	76.6	96.2	96.7	96.4
E	Equipment & Furnishings	100.0	67.7	98.2	100.0	67.7	98.2	100.0	78.8	98.8	100.0	78.8	98.8	100.0	69.2	98.3
G	Site Work	96.0	92.3	93.5	87.9	93.8	91.9	94.0	101.1	98.7	93.1	100.8	98.3	115.6	91.6	99.5
A-G	WEIGHTED AVERAGE	99.8	72.9	88.3	102.0	73.1	89.6	99.1	80.6	91.2	96.3	80.5	89.5	99.3	84.2	92.8

636

VIRGINIA

DIV. NO.	BUILDING SYSTEMS	ARLINGTON MAT.	INST.	TOTAL	NEWPORT NEWS MAT.	INST.	TOTAL	NORFOLK MAT.	INST.	TOTAL	PORTSMOUTH MAT.	INST.	TOTAL	RICHMOND MAT.	INST.	TOTAL
A	Substructure	104.6	78.9	89.7	101.2	73.5	85.0	106.1	74.0	87.4	99.5	64.1	78.9	97.9	81.4	88.2
B10	Shell: Superstructure	102.0	88.7	96.9	100.9	81.5	93.5	103.7	81.7	95.3	99.2	75.1	90.1	101.6	88.1	96.5
B20	Exterior Closure	102.6	74.6	90.0	94.0	64.3	80.7	93.4	64.7	80.5	96.4	50.8	75.9	90.9	66.9	80.1
B30	Roofing	104.4	80.7	94.9	103.0	71.8	90.5	100.6	72.0	89.1	103.0	62.5	86.7	100.6	76.4	90.9
C	Interior Construction	94.2	71.2	84.6	92.8	61.6	79.7	93.7	61.6	80.2	91.4	49.6	73.9	95.7	78.3	88.4
D10	Services: Conveying	100.0	86.9	96.1	100.0	82.2	94.7	100.0	82.2	94.7	100.0	67.7	90.4	100.0	85.0	95.5
D20 - 40	Mechanical	100.4	87.2	95.1	100.5	63.5	85.5	100.1	66.0	86.3	100.5	63.2	85.4	100.0	68.5	87.2
D50	Electrical	93.9	99.3	96.6	92.4	64.7	78.3	94.8	61.9	78.1	90.8	61.9	76.1	96.9	70.0	83.2
E	Equipment & Furnishings	100.0	69.3	98.3	100.0	59.2	97.7	100.0	59.1	97.7	100.0	45.5	96.9	100.0	83.2	99.1
G	Site Work	124.7	89.4	101.0	108.0	91.2	96.7	104.7	92.3	96.4	106.3	90.5	95.7	106.4	91.2	96.2
A-G	WEIGHTED AVERAGE	100.1	84.3	93.3	97.9	69.7	85.8	98.8	70.0	86.5	97.4	63.1	82.7	98.3	76.4	88.9

VIRGINIA / WASHINGTON

DIV. NO.	BUILDING SYSTEMS	ROANOKE MAT.	INST.	TOTAL	EVERETT MAT.	INST.	TOTAL	RICHLAND MAT.	INST.	TOTAL	SEATTLE MAT.	INST.	TOTAL	SPOKANE MAT.	INST.	TOTAL
A	Substructure	113.2	80.0	93.9	101.9	106.7	104.7	87.2	87.5	87.4	108.2	107.4	107.7	89.9	87.2	88.3
B10	Shell: Superstructure	105.1	86.2	97.9	108.7	99.9	105.3	90.4	86.8	89.0	110.9	102.2	107.6	93.0	86.3	90.5
B20	Exterior Closure	91.8	65.7	80.1	98.2	104.3	101.0	99.4	83.6	92.3	110.8	104.6	108.0	99.8	83.6	92.6
B30	Roofing	103.7	76.0	92.6	113.9	107.4	111.3	158.2	86.9	129.5	109.1	106.5	108.1	154.5	87.2	127.5
C	Interior Construction	94.5	69.6	84.1	106.1	99.5	103.3	113.8	80.5	99.9	109.0	101.4	105.8	113.0	80.7	99.5
D10	Services: Conveying	100.0	80.4	94.2	100.0	100.5	100.1	100.0	95.9	98.8	100.0	100.9	100.3	100.0	95.9	98.8
D20 - 40	Mechanical	100.4	65.2	86.2	100.1	102.8	101.2	100.6	110.9	104.8	100.1	112.5	105.1	100.5	85.5	94.4
D50	Electrical	95.8	56.1	75.6	104.5	101.5	102.9	82.6	95.1	89.0	103.7	115.4	109.6	81.2	78.9	80.0
E	Equipment & Furnishings	100.0	71.6	98.4	100.0	100.2	100.0	100.0	80.5	98.9	100.0	103.5	100.2	100.0	79.7	98.9
G	Site Work	106.9	89.4	95.2	93.2	111.1	105.2	105.6	90.4	95.4	99.9	110.2	106.8	105.0	90.4	95.2
A-G	WEIGHTED AVERAGE	99.7	71.6	87.7	103.4	102.6	103.1	99.6	92.2	96.4	106.0	107.4	106.6	99.9	84.5	93.3

WASHINGTON / WEST VIRGINIA

DIV. NO.	BUILDING SYSTEMS	TACOMA MAT.	INST.	TOTAL	VANCOUVER MAT.	INST.	TOTAL	YAKIMA MAT.	INST.	TOTAL	CHARLESTON MAT.	INST.	TOTAL	HUNTINGTON MAT.	INST.	TOTAL
A	Substructure	103.1	106.6	105.1	112.4	98.3	104.2	107.9	96.3	101.1	99.0	92.4	95.1	106.8	94.1	99.4
B10	Shell: Superstructure	109.9	99.6	106.0	109.6	96.5	104.6	109.4	90.7	102.3	95.8	98.1	96.7	100.1	100.5	100.2
B20	Exterior Closure	99.1	102.5	100.6	104.2	92.2	98.8	97.4	85.8	92.2	83.5	91.8	87.2	85.9	94.7	89.9
B30	Roofing	113.6	104.7	110.0	113.7	98.6	107.6	113.8	88.0	103.4	102.4	88.4	96.8	106.5	89.4	99.6
C	Interior Construction	105.9	99.5	103.2	101.5	89.3	96.4	105.2	92.2	99.8	94.5	91.3	93.1	92.5	93.8	93.1
D10	Services: Conveying	100.0	100.9	100.3	100.0	96.7	99.0	100.0	97.0	99.1	100.0	92.5	97.8	100.0	92.7	97.8
D20 - 40	Mechanical	100.1	102.8	101.2	100.2	100.0	100.1	100.1	109.8	104.0	100.1	94.3	97.8	100.7	92.6	97.4
D50	Electrical	104.3	100.1	102.2	109.5	98.5	103.9	107.4	95.1	101.2	98.1	86.2	92.0	96.8	91.8	94.3
E	Equipment & Furnishings	100.0	100.1	100.0	100.0	93.2	99.6	100.0	99.3	100.0	100.0	88.4	99.3	100.0	91.0	99.5
G	Site Work	96.1	111.0	106.1	105.6	97.8	100.4	98.9	109.5	106.0	97.7	91.8	93.7	102.9	91.8	95.5
A-G	WEIGHTED AVERAGE	103.8	102.1	103.1	104.8	96.2	101.1	104.0	96.6	100.8	96.1	92.5	94.6	97.5	94.3	96.1

WEST VIRGINIA / WISCONSIN

DIV. NO.	BUILDING SYSTEMS	PARKERSBURG MAT.	INST.	TOTAL	WHEELING MAT.	INST.	TOTAL	EAU CLAIRE MAT.	INST.	TOTAL	GREEN BAY MAT.	INST.	TOTAL	KENOSHA MAT.	INST.	TOTAL
A	Substructure	102.9	90.0	95.4	103.0	91.6	96.4	98.5	101.5	100.3	102.1	102.6	102.4	108.8	107.4	108.0
B10	Shell: Superstructure	101.5	95.8	99.3	101.6	99.1	100.6	92.9	103.8	97.1	96.2	104.0	99.1	104.2	106.2	105.0
B20	Exterior Closure	90.0	87.1	88.7	100.0	87.1	94.2	86.7	99.5	92.4	101.1	101.3	101.2	93.0	109.8	100.6
B30	Roofing	103.8	89.6	98.1	104.2	89.1	98.2	103.7	97.0	101.0	105.9	100.5	103.7	100.2	106.7	102.8
C	Interior Construction	93.7	88.5	91.5	94.2	87.8	91.5	97.2	98.1	97.6	95.6	101.9	98.2	92.9	111.7	100.8
D10	Services: Conveying	100.0	92.0	97.6	100.0	86.2	95.9	100.0	96.0	98.8	100.0	97.4	99.2	100.0	98.1	99.4
D20 - 40	Mechanical	100.4	90.3	96.3	100.4	91.8	96.9	100.1	90.2	96.1	100.3	85.3	94.2	100.3	98.1	99.4
D50	Electrical	96.3	90.4	93.3	93.6	91.5	92.5	103.6	86.5	94.9	98.0	83.2	90.5	100.1	99.5	99.8
E	Equipment & Furnishings	100.0	87.8	99.3	100.0	85.3	99.2	100.0	96.5	99.8	100.0	99.4	100.0	100.0	108.7	100.5
G	Site Work	109.2	91.7	97.5	109.9	91.5	97.6	96.4	104.0	101.5	99.0	100.1	99.7	101.7	103.7	103.0
A-G	WEIGHTED AVERAGE	98.2	90.7	95.0	99.3	91.5	96.0	96.5	96.7	96.7	98.7	95.8	97.5	99.4	106.4	101.6

WISCONSIN / WYOMING

DIV. NO.	BUILDING SYSTEMS	LA CROSSE MAT.	INST.	TOTAL	MADISON MAT.	INST.	TOTAL	MILWAUKEE MAT.	INST.	TOTAL	RACINE MAT.	INST.	TOTAL	CASPER MAT.	INST.	TOTAL
A	Substructure	90.0	100.9	96.4	100.4	102.8	101.8	93.4	107.3	101.5	100.0	108.2	104.8	104.1	73.4	86.2
B10	Shell: Superstructure	90.5	101.8	94.8	103.6	101.6	102.8	99.8	102.8	101.0	102.4	106.2	103.8	99.4	73.5	89.5
B20	Exterior Closure	83.3	97.4	89.6	92.1	102.1	96.6	100.1	111.7	105.3	93.9	109.8	101.1	94.9	66.0	81.9
B30	Roofing	103.0	95.8	100.1	96.8	103.0	99.3	102.2	108.9	104.9	99.8	105.2	102.0	103.5	66.7	88.7
C	Interior Construction	95.7	99.1	97.1	96.5	101.4	98.5	100.7	111.9	105.4	94.9	111.5	101.8	101.2	56.4	82.5
D10	Services: Conveying	100.0	94.4	98.3	100.0	98.7	99.6	100.0	101.9	100.6	100.0	98.1	99.4	100.0	99.4	99.8
D20 - 40	Mechanical	100.1	90.1	96.0	100.0	97.4	98.9	100.0	105.6	102.3	100.1	98.2	99.3	100.0	71.3	88.4
D50	Electrical	103.9	86.5	95.1	100.7	96.3	98.5	99.3	101.6	100.5	99.5	100.4	100.0	96.7	61.8	78.9
E	Equipment & Furnishings	100.0	96.5	99.8	100.0	97.1	99.8	100.0	109.1	100.5	100.0	108.7	100.5	100.0	49.7	97.2
G	Site Work	90.3	104.0	99.4	94.3	106.9	102.8	93.3	95.5	94.7	95.9	107.5	103.7	96.3	94.6	95.1
A-G	WEIGHTED AVERAGE	95.2	95.9	95.5	99.1	100.3	99.6	99.6	105.9	102.3	98.8	105.0	101.5	99.3	69.5	86.5

WYOMING / CANADA

DIV. NO.	BUILDING SYSTEMS	CHEYENNE MAT.	INST.	TOTAL	ROCK SPRINGS MAT.	INST.	TOTAL	CALGARY, ALBERTA MAT.	INST.	TOTAL	EDMONTON, ALBERTA MAT.	INST.	TOTAL	HALIFAX, NOVA SCOTIA MAT.	INST.	TOTAL
A	Substructure	99.0	78.1	86.9	99.6	75.9	85.8	136.1	105.0	118.0	131.9	104.9	116.2	112.9	88.4	98.6
B10	Shell: Superstructure	100.3	77.0	91.4	98.4	75.0	89.5	134.3	101.8	121.9	139.1	101.6	124.8	135.6	90.7	118.5
B20	Exterior Closure	97.6	65.7	83.3	120.8	59.2	93.1	139.3	87.9	116.2	137.7	87.8	115.4	130.4	84.2	109.6
B30	Roofing	98.1	68.9	86.4	99.9	71.0	88.3	128.8	99.2	116.9	140.7	99.2	124.1	134.3	85.4	114.7
C	Interior Construction	101.5	67.2	87.1	103.3	59.8	85.1	102.3	96.4	99.8	98.9	96.3	97.9	100.2	83.3	93.2
D10	Services: Conveying	100.0	91.9	97.6	100.0	87.2	96.2	131.1	93.7	120.0	131.1	93.6	120.0	131.1	65.5	111.6
D20 - 40	Mechanical	99.9	71.3	88.4	99.9	70.5	88.0	104.9	91.6	99.5	104.8	91.6	99.5	104.5	80.9	95.0
D50	Electrical	98.1	62.7	80.1	95.2	78.5	86.7	102.0	96.2	99.1	105.7	96.2	100.9	108.2	84.4	96.1
E	Equipment & Furnishings	100.0	66.1	98.1	100.0	62.2	97.9	131.1	96.7	129.2	131.1	96.7	129.2	131.1	82.4	128.4
G	Site Work	92.0	94.6	93.7	91.3	94.1	93.2	124.1	125.6	125.1	121.7	125.3	124.1	103.9	109.1	107.4
A-G	WEIGHTED AVERAGE	99.6	72.1	87.8	101.9	71.6	88.9	119.0	97.6	109.8	119.7	97.5	110.2	117.1	85.8	103.7

637

Cost Indexes

DIV. NO.	BUILDING SYSTEMS	CANADA														
		HAMILTON, ONTARIO			KITCHENER, ONTARIO			LAVAL, QUEBEC			LONDON, ONTARIO			MONTREAL, QUEBEC		
		MAT.	INST.	TOTAL	MAT.	INST.	TOTAL	MAT.	INST.	TOTAL	MAT.	INST.	TOTAL	MAT.	INST.	TOTAL
A	Substructure	120.5	101.7	109.6	109.3	95.3	101.1	114.7	87.1	98.6	124.5	98.9	109.6	125.0	97.4	108.9
B10	Shell: Superstructure	137.2	101.8	123.7	126.0	94.1	113.8	116.3	84.7	104.2	138.2	100.7	123.9	143.9	98.1	126.4
B20	Exterior Closure	120.8	99.4	111.1	102.6	95.7	99.5	120.8	75.9	100.6	128.2	97.5	114.4	122.2	86.3	106.1
B30	Roofing	124.8	99.8	114.8	115.3	96.3	107.7	114.6	88.3	104.1	124.5	97.2	113.5	118.7	97.7	110.3
C	Interior Construction	98.3	94.6	96.8	91.1	88.1	89.9	98.1	82.5	91.6	97.2	89.8	94.1	99.6	91.4	96.2
D10	Services: Conveying	131.1	91.6	119.4	131.1	89.4	118.7	131.1	76.8	115.0	131.1	91.3	119.3	131.1	80.9	116.2
D20 - 40	Mechanical	104.9	90.5	99.1	104.0	89.2	98.1	104.1	86.1	96.8	104.9	87.8	98.0	105.1	80.1	95.0
D50	Electrical	106.3	99.6	102.9	107.7	97.1	102.3	105.4	66.8	85.8	103.5	97.1	100.3	107.2	83.1	95.0
E	Equipment & Furnishings	131.1	91.8	128.6	131.1	85.5	128.6	131.1	80.5	128.3	131.1	86.5	128.6	131.1	89.2	128.8
G	Site Work	109.7	114.7	113.1	96.1	104.9	102.0	98.5	98.7	98.6	104.0	114.5	111.0	113.7	111.9	112.5
A-G	WEIGHTED AVERAGE	116.1	98.1	108.4	109.4	93.4	102.5	110.7	81.9	98.3	116.8	95.6	107.7	118.1	89.9	106.0

DIV. NO.	BUILDING SYSTEMS	CANADA														
		OSHAWA, ONTARIO			OTTAWA, ONTARIO			QUEBEC, QUEBEC			REGINA, SASKATCHEWAN			SASKATOON, SASKATCHEWAN		
		MAT.	INST.	TOTAL	MAT.	INST.	TOTAL	MAT.	INST.	TOTAL	MAT.	INST.	TOTAL	MAT.	INST.	TOTAL
A	Substructure	132.2	91.1	108.3	123.1	99.7	109.5	121.4	97.4	107.4	136.1	100.9	115.6	111.4	94.9	101.8
B10	Shell: Superstructure	122.8	91.5	110.9	138.3	101.8	124.4	137.9	99.0	123.1	145.6	99.6	128.1	110.2	91.5	103.1
B20	Exterior Closure	110.9	90.3	101.6	120.6	99.5	111.2	119.8	88.8	105.9	133.7	87.3	112.9	118.4	86.3	104.0
B30	Roofing	116.3	89.1	105.3	131.7	98.6	118.4	118.7	97.7	110.3	139.5	89.2	119.3	116.7	87.9	105.1
C	Interior Construction	94.1	90.1	92.4	101.8	88.5	96.2	101.0	91.3	97.0	105.9	94.2	101.0	98.1	93.8	96.3
D10	Services: Conveying	131.1	89.0	118.6	131.1	89.9	118.9	131.1	80.7	116.1	131.1	67.8	112.3	131.1	66.5	111.9
D20 - 40	Mechanical	104.0	100.0	102.4	104.9	89.5	98.7	104.8	80.1	94.8	104.4	88.8	98.1	104.3	88.7	98.0
D50	Electrical	108.8	87.6	98.0	103.4	97.9	100.6	104.5	83.1	93.6	107.6	93.7	100.6	107.4	93.7	100.4
E	Equipment & Furnishings	131.1	85.8	128.6	131.1	85.6	128.6	131.1	89.2	128.8	131.1	94.2	129.1	131.1	94.0	129.1
G	Site Work	107.4	103.8	105.0	108.3	114.4	112.4	116.1	111.8	113.2	126.8	119.0	121.6	111.8	97.7	102.4
A-G	WEIGHTED AVERAGE	111.6	92.9	103.6	116.9	96.4	108.1	116.4	90.3	105.2	122.1	94.2	110.1	109.6	90.8	101.5

DIV. NO.	BUILDING SYSTEMS	CANADA														
		ST CATHARINES, ONTARIO			ST JOHNS, NEWFOUNDLAND			THUNDER BAY, ONTARIO			TORONTO, ONTARIO			VANCOUVER, BRITISH COLUMBIA		
		MAT.	INST.	TOTAL	MAT.	INST.	TOTAL	MAT.	INST.	TOTAL	MAT.	INST.	TOTAL	MAT.	INST.	TOTAL
A	Substructure	105.6	97.3	100.8	135.3	95.9	112.3	111.8	96.4	102.8	121.0	106.5	112.6	114.8	101.4	107.0
B10	Shell: Superstructure	117.0	95.9	109.0	142.6	95.6	124.7	118.6	94.8	109.5	137.6	106.2	125.7	131.4	103.3	120.7
B20	Exterior Closure	102.3	97.9	100.4	135.1	89.1	114.4	105.6	97.8	102.1	120.7	105.6	113.9	121.9	88.2	106.8
B30	Roofing	115.3	100.8	109.5	141.5	98.3	124.2	115.5	98.0	108.5	132.5	106.8	122.2	138.0	89.6	118.6
C	Interior Construction	89.7	93.6	91.3	102.3	81.7	93.6	91.5	92.4	91.9	96.0	99.3	97.4	102.6	91.7	98.0
D10	Services: Conveying	131.1	66.2	111.8	131.1	66.7	112.0	131.1	66.7	112.0	131.1	93.6	120.0	131.1	91.8	119.4
D20 - 40	Mechanical	104.0	89.1	98.0	104.5	84.8	96.5	104.0	89.3	98.1	104.8	97.2	101.7	104.8	77.1	93.6
D50	Electrical	109.3	97.8	103.5	105.8	83.4	94.4	107.7	96.5	102.0	103.2	100.0	101.6	101.7	80.5	90.9
E	Equipment & Furnishings	131.1	92.8	129.0	131.1	84.0	128.5	131.1	90.4	128.8	131.1	96.9	129.2	131.1	91.4	128.9
G	Site Work	96.9	101.3	99.9	120.5	111.5	114.5	101.6	101.2	101.3	115.8	114.2	114.8	113.4	130.6	124.9
A-G	WEIGHTED AVERAGE	107.3	94.3	101.7	120.8	88.8	107.0	108.5	93.7	102.1	115.8	102.4	110.1	115.4	91.6	105.2

DIV. NO.	BUILDING SYSTEMS	CANADA														
		WINDSOR, ONTARIO			WINNIPEG, MANITOBA											
		MAT.	INST.	TOTAL	MAT.	INST.	TOTAL	MAT.	INST.	TOTAL	MAT.	INST.	TOTAL	MAT.	INST.	TOTAL
A	Substructure	107.6	96.1	100.9	136.2	78.4	102.5									
B10	Shell: Superstructure	117.8	94.8	109.0	145.8	79.5	120.5									
B20	Exterior Closure	102.2	97.1	99.9	131.9	65.4	102.0									
B30	Roofing	115.3	97.2	108.1	132.4	70.4	107.5									
C	Interior Construction	90.3	90.9	90.5	103.3	64.7	87.1									
D10	Services: Conveying	131.1	66.2	111.8	131.1	63.7	111.1									
D20 - 40	Mechanical	104.0	89.2	98.0	104.8	64.3	88.5									
D50	Electrical	112.5	97.9	105.1	107.0	63.2	84.7									
E	Equipment & Furnishings	131.1	88.8	128.8	131.1	64.7	127.4									
G	Site Work	92.7	101.2	98.4	115.4	115.0	115.2									
A-G	WEIGHTED AVERAGE	107.8	93.5	101.7	121.1	71.2	99.7									

For customer support on your Facilities Maintenance & Repair Costs with RSMeans data, call 800.448.8182.

Costs shown in RSMeans cost data publications are based on national averages for materials and installation. To adjust these costs to a specific location, simply multiply the base cost by the factor and divide by 100 for that city. The data is arranged alphabetically by state and postal zip code numbers. For a city not listed, use the factor for a nearby city with similar economic characteristics.

STATE/ZIP	CITY	MAT.	INST.	TOTAL
ALABAMA				
350-352	Birmingham	96.5	71.7	85.9
354	Tuscaloosa	96.6	71.5	85.8
355	Jasper	97.1	71.0	85.9
356	Decatur	96.6	70.1	85.2
357-358	Huntsville	96.5	71.4	85.7
359	Gadsden	96.6	70.6	85.5
360-361	Montgomery	95.6	71.4	85.2
362	Anniston	95.2	67.4	83.3
363	Dothan	95.6	72.9	85.9
364	Evergreen	95.2	71.4	85.0
365-366	Mobile	96.1	68.8	84.4
367	Selma	95.3	72.4	85.5
368	Phenix City	96.1	72.0	85.8
369	Butler	95.5	71.6	85.3
ALASKA				
995-996	Anchorage	117.5	113.6	115.8
997	Fairbanks	118.2	114.7	116.7
998	Juneau	116.5	113.6	115.3
999	Ketchikan	128.4	113.6	122.0
ARIZONA				
850,853	Phoenix	100.0	73.0	88.4
851,852	Mesa/Tempe	98.6	72.0	87.2
855	Globe	99.5	71.8	87.6
856-857	Tucson	97.6	71.4	86.4
859	Show Low	99.7	71.9	87.8
860	Flagstaff	101.0	71.7	88.4
863	Prescott	98.7	71.7	87.1
864	Kingman	97.2	71.8	86.3
865	Chambers	97.2	74.4	87.4
ARKANSAS				
716	Pine Bluff	96.7	64.6	82.9
717	Camden	94.8	59.6	79.7
718	Texarkana	95.6	60.4	80.5
719	Hot Springs	94.2	61.0	79.9
720-722	Little Rock	95.4	65.9	82.8
723	West Memphis	94.3	65.5	81.9
724	Jonesboro	94.7	62.7	81.0
725	Batesville	92.6	60.0	78.7
726	Harrison	94.0	59.1	79.0
727	Fayetteville	91.5	60.7	78.3
728	Russellville	92.7	59.6	78.5
729	Fort Smith	95.1	63.6	81.6
CALIFORNIA				
900-902	Los Angeles	98.7	130.4	112.3
903-905	Inglewood	94.2	129.6	109.4
906-908	Long Beach	95.9	129.6	110.3
910-912	Pasadena	94.8	129.4	109.7
913-916	Van Nuys	97.9	129.4	111.4
917-918	Alhambra	96.8	129.4	110.8
919-921	San Diego	99.6	120.8	108.7
922	Palm Springs	96.9	127.1	109.8
923-924	San Bernardino	94.5	126.8	108.4
925	Riverside	98.8	127.1	110.9
926-927	Santa Ana	96.5	127.0	109.6
928	Anaheim	98.9	126.9	110.9
930	Oxnard	97.2	127.1	110.0
931	Santa Barbara	96.5	126.7	109.4
932-933	Bakersfield	98.0	125.3	109.7
934	San Luis Obispo	97.5	126.6	110.0
935	Mojave	94.7	125.2	107.7
936-938	Fresno	97.7	129.3	111.2
939	Salinas	98.3	135.7	114.3
940-941	San Francisco	106.3	158.1	128.5
942,956-958	Sacramento	100.1	132.7	114.1
943	Palo Alto	98.5	151.4	121.2
944	San Mateo	100.9	150.2	122.1
945	Vallejo	99.5	141.5	117.5
946	Oakland	102.7	151.2	123.5
947	Berkeley	102.3	151.2	123.3
948	Richmond	101.8	145.2	120.4
949	San Rafael	103.9	149.3	123.4
950	Santa Cruz	104.0	136.0	117.7

STATE/ZIP	CITY	MAT.	INST.	TOTAL
CALIFORNIA (CONT'D)				
951	San Jose	102.1	151.7	123.4
952	Stockton	100.3	130.8	113.4
953	Modesto	100.2	129.4	112.7
954	Santa Rosa	100.8	148.5	121.2
955	Eureka	102.1	133.2	115.4
959	Marysville	101.2	131.3	114.1
960	Redding	109.1	131.5	118.7
961	Susanville	108.8	131.2	118.4
COLORADO				
800-802	Denver	101.8	74.4	90.0
803	Boulder	97.7	75.9	88.3
804	Golden	99.8	73.7	88.6
805	Fort Collins	101.1	73.7	89.4
806	Greeley	98.3	74.7	88.2
807	Fort Morgan	98.3	72.9	87.4
808-809	Colorado Springs	100.0	73.0	88.4
810	Pueblo	100.5	71.4	88.0
811	Alamosa	102.3	68.8	87.9
812	Salida	102.0	66.6	86.8
813	Durango	102.7	65.0	86.6
814	Montrose	101.4	69.3	87.6
815	Grand Junction	104.5	72.2	90.7
816	Glenwood Springs	102.5	65.4	86.6
CONNECTICUT				
060	New Britain	96.4	118.1	105.7
061	Hartford	97.2	118.5	106.4
062	Willimantic	97.0	118.2	106.1
063	New London	93.6	118.4	104.2
064	Meriden	95.5	118.4	105.4
065	New Haven	98.1	118.5	106.9
066	Bridgeport	97.6	118.4	106.5
067	Waterbury	97.1	118.6	106.3
068	Norwalk	97.0	118.6	106.3
069	Stamford	97.2	125.2	109.2
D.C.				
200-205	Washington	101.0	87.7	95.3
DELAWARE				
197	Newark	98.3	112.3	104.3
198	Wilmington	98.2	112.3	104.2
199	Dover	98.2	112.3	104.2
FLORIDA				
320,322	Jacksonville	96.1	66.5	83.4
321	Daytona Beach	96.3	70.3	85.2
323	Tallahassee	97.3	65.7	83.7
324	Panama City	97.6	64.2	83.3
325	Pensacola	100.3	66.4	85.8
326,344	Gainesville	97.8	64.9	83.7
327-328,347	Orlando	97.9	67.1	84.7
329	Melbourne	99.2	71.2	87.2
330-332,340	Miami	95.8	67.4	83.7
333	Fort Lauderdale	95.3	69.5	84.2
334,349	West Palm Beach	94.3	66.0	82.2
335-336,346	Tampa	96.8	68.1	84.5
337	St. Petersburg	99.0	65.2	84.5
338	Lakeland	96.2	67.1	83.8
339,341	Fort Myers	95.6	68.3	83.9
342	Sarasota	98.7	67.0	85.1
GEORGIA				
300-303,399	Atlanta	99.0	75.6	89.0
304	Statesboro	98.9	65.8	84.7
305	Gainesville	97.4	66.5	84.2
306	Athens	96.8	67.9	84.4
307	Dalton	98.7	71.0	86.8
308-309	Augusta	97.1	72.7	86.6
310-312	Macon	93.9	74.0	85.4
313-314	Savannah	95.6	73.1	86.0
315	Waycross	95.4	68.4	83.8
316	Valdosta	95.3	66.2	82.8
317,398	Albany	95.2	72.6	85.5
318-319	Columbus	95.1	72.7	85.5

639

STATE/ZIP	CITY	MAT.	INST.	TOTAL
HAWAII				
967	Hilo	116.3	118.3	117.1
968	Honolulu	120.9	118.3	119.8
STATES & POSS.				
969	Guam	138.8	52.6	101.9
IDAHO				
832	Pocatello	100.7	78.6	91.2
833	Twin Falls	101.9	77.2	91.3
834	Idaho Falls	99.2	78.2	90.2
835	Lewiston	106.9	85.7	97.8
836-837	Boise	99.7	79.4	91.0
838	Coeur d'Alene	106.8	84.4	97.2
ILLINOIS				
600-603	North Suburban	99.8	144.1	118.8
604	Joliet	99.7	142.2	117.9
605	South Suburban	99.8	144.0	118.7
606-608	Chicago	101.7	146.1	120.7
609	Kankakee	96.4	136.1	113.4
610-611	Rockford	96.8	128.7	110.5
612	Rock Island	94.9	102.1	98.0
613	La Salle	96.1	128.9	110.2
614	Galesburg	95.9	109.5	101.8
615-616	Peoria	97.8	111.5	103.7
617	Bloomington	95.2	112.2	102.5
618-619	Champaign	98.7	109.8	103.5
620-622	East St. Louis	94.4	111.3	101.6
623	Quincy	96.3	105.1	100.1
624	Effingham	95.7	109.2	101.5
625	Decatur	97.2	107.0	101.4
626-627	Springfield	97.9	107.7	102.1
628	Centralia	93.3	113.0	101.8
629	Carbondale	93.1	109.8	100.2
INDIANA				
460	Anderson	95.9	81.8	89.8
461-462	Indianapolis	98.7	83.0	92.0
463-464	Gary	97.2	108.3	102.0
465-466	South Bend	97.4	83.6	91.5
467-468	Fort Wayne	96.4	78.2	88.6
469	Kokomo	94.1	81.0	88.4
470	Lawrenceburg	92.6	78.5	86.5
471	New Albany	93.9	77.2	86.7
472	Columbus	96.1	80.4	89.3
473	Muncie	96.1	80.3	89.3
474	Bloomington	97.7	80.3	90.3
475	Washington	94.6	85.4	90.7
476-477	Evansville	95.4	84.8	90.9
478	Terre Haute	96.2	83.8	90.9
479	Lafayette	95.8	80.1	89.1
IOWA				
500-503,509	Des Moines	96.0	88.4	92.8
504	Mason City	94.7	76.2	86.8
505	Fort Dodge	94.9	74.1	86.0
506-507	Waterloo	96.2	79.3	89.0
508	Creston	95.2	83.1	90.0
510-511	Sioux City	97.1	78.7	89.2
512	Sibley	96.1	63.3	82.1
513	Spencer	97.7	63.6	83.1
514	Carroll	94.9	81.3	89.1
515	Council Bluffs	98.2	80.8	90.7
516	Shenandoah	95.4	83.8	90.4
520	Dubuque	96.5	81.9	90.3
521	Decorah	95.9	75.9	87.3
522-524	Cedar Rapids	97.4	85.8	92.4
525	Ottumwa	95.7	76.3	87.4
526	Burlington	95.1	86.3	91.3
527-528	Davenport	96.5	96.9	96.7
KANSAS				
660-662	Kansas City	96.0	98.6	97.1
664-666	Topeka	97.0	77.7	88.7
667	Fort Scott	94.8	78.2	87.7
668	Emporia	94.9	77.2	87.3
669	Belleville	96.6	72.1	86.1
670-672	Wichita	96.7	70.8	85.6
673	Independence	97.0	77.9	88.8
674	Salina	96.9	72.8	86.6
675	Hutchinson	92.4	73.1	84.1
676	Hays	96.3	73.5	86.5
677	Colby	97.1	75.7	87.9

STATE/ZIP	CITY	MAT.	INST.	TOTAL
KANSAS (CONT'D)				
678	Dodge City	98.3	75.4	88.5
679	Liberal	96.2	73.7	86.5
KENTUCKY				
400-402	Louisville	93.2	77.7	86.6
403-405	Lexington	93.4	79.6	87.5
406	Frankfort	95.5	78.7	88.3
407-409	Corbin	91.1	79.6	86.2
410	Covington	94.4	79.8	88.1
411-412	Ashland	93.2	92.0	92.7
413-414	Campton	94.4	79.9	88.2
415-416	Pikeville	95.8	87.6	92.3
417-418	Hazard	93.8	80.6	88.1
420	Paducah	92.5	83.8	88.8
421-422	Bowling Green	94.5	78.9	87.8
423	Owensboro	94.6	83.1	89.7
424	Henderson	92.2	82.4	88.0
425-426	Somerset	91.7	77.8	85.7
427	Elizabethtown	91.3	76.7	85.1
LOUISIANA				
700-701	New Orleans	99.4	69.5	86.6
703	Thibodaux	96.2	66.4	83.4
704	Hammond	93.8	64.5	81.2
705	Lafayette	95.6	68.9	84.1
706	Lake Charles	95.8	69.8	84.6
707-708	Baton Rouge	96.6	69.7	85.1
710-711	Shreveport	97.8	66.4	84.3
712	Monroe	96.6	65.1	83.1
713-714	Alexandria	96.7	66.6	83.8
MAINE				
039	Kittery	91.3	84.6	88.5
040-041	Portland	96.9	84.1	91.4
042	Lewiston	94.7	84.1	90.1
043	Augusta	96.7	83.1	90.9
044	Bangor	94.3	83.5	89.7
045	Bath	93.2	82.6	88.6
046	Machias	92.6	79.7	87.1
047	Houlton	92.8	79.7	87.2
048	Rockland	91.9	83.1	88.1
049	Waterville	93.1	83.1	88.8
MARYLAND				
206	Waldorf	97.7	87.4	93.3
207-208	College Park	97.7	88.6	93.8
209	Silver Spring	96.9	87.4	92.8
210-212	Baltimore	101.2	85.0	94.2
214	Annapolis	100.5	83.1	93.0
215	Cumberland	97.1	85.5	92.1
216	Easton	98.8	76.0	89.0
217	Hagerstown	97.5	87.0	93.0
218	Salisbury	99.3	68.5	86.1
219	Elkton	96.0	87.6	92.4
MASSACHUSETTS				
010-011	Springfield	96.4	110.4	102.4
012	Pittsfield	96.0	105.1	99.9
013	Greenfield	94.3	110.9	101.4
014	Fitchburg	93.0	118.7	104.0
015-016	Worcester	96.4	118.5	105.9
017	Framingham	92.4	127.0	107.2
018	Lowell	95.8	125.4	108.5
019	Lawrence	96.7	125.8	109.2
020-022, 024	Boston	99.4	133.3	113.9
023	Brockton	96.1	119.2	106.0
025	Buzzards Bay	91.1	117.7	102.5
026	Hyannis	93.4	117.5	103.7
027	New Bedford	95.3	117.6	104.9
MICHIGAN				
480,483	Royal Oak	95.2	100.6	97.5
481	Ann Arbor	97.2	101.7	99.1
482	Detroit	100.2	103.0	101.4
484-485	Flint	96.9	90.8	94.3
486	Saginaw	96.5	87.3	92.6
487	Bay City	96.6	87.5	92.7
488-489	Lansing	97.9	88.7	93.9
490	Battle Creek	95.3	82.5	89.8
491	Kalamazoo	95.6	80.7	89.2
492	Jackson	93.8	93.0	93.5
493,495	Grand Rapids	97.1	81.6	90.5
494	Muskegon	94.1	80.3	88.2

For customer support on your Facilities Maintenance & Repair Costs with RSMeans data, call 800.448.8182.

STATE/ZIP	CITY	MAT.	INST.	TOTAL
MICHIGAN (CONT'D)				
496	Traverse City	93.2	78.3	86.8
497	Gaylord	94.3	81.3	88.8
498-499	Iron Mountain	96.2	83.0	90.6
MINNESOTA				
550-551	Saint Paul	97.2	115.8	105.2
553-555	Minneapolis	99.7	116.5	106.9
556-558	Duluth	97.1	102.7	99.5
559	Rochester	97.1	100.3	98.4
560	Mankato	95.1	98.0	96.4
561	Windom	93.7	92.3	93.1
562	Willmar	93.3	99.9	96.1
563	St. Cloud	94.4	114.0	102.8
564	Brainerd	95.0	98.6	96.5
565	Detroit Lakes	96.7	92.2	94.8
566	Bemidji	96.0	95.4	95.8
567	Thief River Falls	95.6	89.4	92.9
MISSISSIPPI				
386	Clarksdale	95.2	54.4	77.7
387	Greenville	98.6	66.0	84.6
388	Tupelo	96.5	58.2	80.1
389	Greenwood	96.5	54.2	78.3
390-392	Jackson	98.0	65.4	84.0
393	Meridian	95.6	65.1	82.5
394	Laurel	97.2	56.9	79.9
395	Biloxi	97.3	64.6	83.3
396	McComb	95.4	54.5	77.9
397	Columbus	97.0	58.4	80.5
MISSOURI				
630-631	St. Louis	99.6	105.8	102.3
633	Bowling Green	97.5	96.0	96.9
634	Hannibal	96.5	94.0	95.4
635	Kirksville	100.2	88.8	95.3
636	Flat River	98.5	96.2	97.5
637	Cape Girardeau	98.1	91.8	95.4
638	Sikeston	96.9	89.4	93.7
639	Poplar Bluff	96.3	89.2	93.3
640-641	Kansas City	98.1	103.6	100.5
644-645	St. Joseph	96.8	92.8	95.1
646	Chillicothe	94.5	95.4	94.9
647	Harrisonville	94.0	102.6	97.7
648	Joplin	95.9	79.7	88.9
650-651	Jefferson City	96.3	90.8	94.0
652	Columbia	96.2	91.3	94.1
653	Sedalia	96.6	93.7	95.4
654-655	Rolla	94.3	96.6	95.3
656-658	Springfield	97.7	81.0	90.5
MONTANA				
590-591	Billings	100.8	75.0	89.7
592	Wolf Point	100.5	77.8	90.8
593	Miles City	98.4	77.9	89.6
594	Great Falls	102.1	74.0	90.1
595	Havre	99.5	75.7	89.3
596	Helena	100.1	76.1	89.8
597	Butte	100.7	76.2	90.2
598	Missoula	97.9	75.7	88.4
599	Kalispell	97.4	76.2	88.3
NEBRASKA				
680-681	Omaha	96.9	80.8	90.0
683-685	Lincoln	97.1	81.1	90.2
686	Columbus	95.6	82.7	90.1
687	Norfolk	96.9	79.2	89.3
688	Grand Island	97.0	78.5	89.1
689	Hastings	96.7	80.0	89.5
690	McCook	96.6	74.8	87.3
691	North Platte	96.6	77.3	88.3
692	Valentine	98.8	70.8	86.8
693	Alliance	98.9	74.7	88.5
NEVADA				
889-891	Las Vegas	102.7	104.1	103.3
893	Ely	101.6	95.6	99.0
894-895	Reno	101.5	84.2	94.1
897	Carson City	100.4	84.2	93.5
898	Elko	100.3	87.4	94.8
NEW HAMPSHIRE				
030	Nashua	95.9	92.8	94.6
031	Manchester	96.2	93.0	94.8

STATE/ZIP	CITY	MAT.	INST.	TOTAL
NEW HAMPSHIRE (CONT'D)				
032-033	Concord	96.1	92.4	94.5
034	Keene	93.0	89.2	91.3
035	Littleton	93.0	80.1	87.4
036	Charleston	92.5	88.6	90.8
037	Claremont	91.7	88.6	90.3
038	Portsmouth	93.3	92.2	92.8
NEW JERSEY				
070-071	Newark	99.6	139.7	116.8
072	Elizabeth	97.3	139.7	115.4
073	Jersey City	96.3	139.4	114.8
074-075	Paterson	97.7	139.5	115.6
076	Hackensack	95.9	139.6	114.6
077	Long Branch	95.6	133.6	111.9
078	Dover	96.1	139.6	114.7
079	Summit	96.2	139.7	114.8
080,083	Vineland	95.8	132.3	111.4
081	Camden	97.5	130.7	111.8
082,084	Atlantic City	96.4	132.6	111.9
085-086	Trenton	99.2	131.4	113.0
087	Point Pleasant	97.7	133.3	113.0
088-089	New Brunswick	98.3	137.7	115.2
NEW MEXICO				
870-872	Albuquerque	97.1	75.0	87.6
873	Gallup	97.3	75.0	87.7
874	Farmington	97.7	75.0	88.0
875	Santa Fe	97.6	75.0	87.9
877	Las Vegas	95.8	75.0	86.9
878	Socorro	95.5	75.0	86.7
879	Truth/Consequences	95.2	71.7	85.2
880	Las Cruces	95.6	71.7	85.4
881	Clovis	98.0	74.9	88.1
882	Roswell	99.5	75.0	89.0
883	Carrizozo	100.2	75.0	89.4
884	Tucumcari	98.7	74.9	88.5
NEW YORK				
100-102	New York	99.0	176.2	132.1
103	Staten Island	94.9	177.8	130.5
104	Bronx	93.3	176.8	129.1
105	Mount Vernon	93.5	151.3	118.3
106	White Plains	93.3	153.5	119.1
107	Yonkers	97.3	151.2	120.5
108	New Rochelle	93.9	146.0	116.2
109	Suffern	93.6	133.5	110.7
110	Queens	99.8	178.9	133.8
111	Long Island City	101.5	178.9	134.7
112	Brooklyn	101.8	178.9	134.9
113	Flushing	102.0	178.9	135.0
114	Jamaica	100.2	178.9	134.0
115,117,118	Hicksville	99.8	160.5	125.8
116	Far Rockaway	102.2	178.9	135.1
119	Riverhead	100.5	156.8	124.6
120-122	Albany	95.0	111.5	102.1
123	Schenectady	95.5	110.6	101.9
124	Kingston	98.9	136.0	114.8
125-126	Poughkeepsie	98.1	140.2	116.1
127	Monticello	97.4	136.9	114.3
128	Glens Falls	90.6	108.7	98.4
129	Plattsburgh	95.6	99.0	97.0
130-132	Syracuse	97.8	100.8	99.1
133-135	Utica	95.9	98.9	97.2
136	Watertown	97.7	99.2	98.3
137-139	Binghamton	97.3	102.8	99.7
140-142	Buffalo	103.0	110.0	106.0
143	Niagara Falls	98.8	110.2	103.7
144-146	Rochester	100.3	101.3	100.8
147	Jamestown	97.8	97.8	97.8
148-149	Elmira	97.6	103.9	100.3
NORTH CAROLINA				
270,272-274	Greensboro	97.7	67.7	84.8
271	Winston-Salem	97.4	67.8	84.7
275-276	Raleigh	96.4	67.1	83.8
277	Durham	99.3	67.7	85.7
278	Rocky Mount	95.3	67.5	83.4
279	Elizabeth City	96.1	69.2	84.6
280	Gastonia	97.0	67.7	84.4
281-282	Charlotte	97.0	67.3	84.3
283	Fayetteville	100.0	67.3	86.0
284	Wilmington	95.9	66.3	83.2
285	Kinston	94.3	67.1	82.7

For customer support on your Facilities Maintenance & Repair Costs with RSMeans data, call 800.448.8182.

STATE/ZIP	CITY	MAT.	INST.	TOTAL
NORTH CAROLINA (CONT'D)				
286	Hickory	94.7	68.3	83.4
287-288	Asheville	96.3	67.0	83.8
289	Murphy	95.5	66.2	82.9
NORTH DAKOTA				
580-581	Fargo	98.2	80.3	90.5
582	Grand Forks	98.4	79.0	90.1
583	Devils Lake	98.3	80.2	90.5
584	Jamestown	98.3	79.6	90.3
585	Bismarck	98.2	80.0	90.4
586	Dickinson	99.0	79.1	90.5
587	Minot	98.3	79.6	90.3
588	Williston	97.4	79.0	89.6
OHIO				
430-432	Columbus	97.4	83.2	91.3
433	Marion	93.4	88.5	91.3
434-436	Toledo	96.7	92.8	95.0
437-438	Zanesville	94.0	87.1	91.0
439	Steubenville	95.3	92.1	94.0
440	Lorain	98.4	86.8	93.4
441	Cleveland	98.7	92.2	95.9
442-443	Akron	99.5	88.3	94.7
444-445	Youngstown	98.7	84.0	92.4
446-447	Canton	98.9	81.4	91.4
448-449	Mansfield	96.5	86.4	92.1
450	Hamilton	95.9	81.5	89.7
451-452	Cincinnati	96.9	80.7	89.9
453-454	Dayton	95.9	81.3	89.6
455	Springfield	95.9	81.8	89.8
456	Chillicothe	95.4	90.8	93.4
457	Athens	98.2	86.4	93.2
458	Lima	98.3	83.8	92.1
OKLAHOMA				
730-731	Oklahoma City	95.2	68.2	83.6
734	Ardmore	93.8	66.2	82.0
735	Lawton	95.9	66.6	83.3
736	Clinton	95.0	66.1	82.6
737	Enid	95.5	66.5	83.1
738	Woodward	93.9	63.4	80.8
739	Guymon	94.9	64.8	82.0
740-741	Tulsa	96.0	67.1	83.6
743	Miami	92.8	66.5	81.5
744	Muskogee	95.1	64.4	81.9
745	McAlester	92.4	61.9	79.3
746	Ponca City	93.2	65.2	81.2
747	Durant	93.2	66.3	81.7
748	Shawnee	94.6	66.1	82.4
749	Poteau	92.4	66.1	81.1
OREGON				
970-972	Portland	100.2	100.4	100.3
973	Salem	101.7	99.9	100.9
974	Eugene	99.8	98.5	99.3
975	Medford	101.3	97.8	99.8
976	Klamath Falls	101.5	97.8	99.9
977	Bend	100.7	99.8	100.3
978	Pendleton	96.8	101.7	98.9
979	Vale	94.5	87.9	91.7
PENNSYLVANIA				
150-152	Pittsburgh	99.9	102.2	100.9
153	Washington	96.9	102.7	99.4
154	Uniontown	97.2	102.2	99.4
155	Bedford	98.2	94.3	96.5
156	Greensburg	98.2	98.7	98.4
157	Indiana	97.1	100.1	98.4
158	Dubois	98.7	98.3	98.5
159	Johnstown	98.2	94.2	96.5
160	Butler	91.3	103.1	96.4
161	New Castle	91.4	100.3	95.2
162	Kittanning	91.8	100.4	95.5
163	Oil City	91.3	100.3	95.1
164-165	Erie	93.1	96.2	94.5
166	Altoona	93.2	95.1	94.0
167	Bradford	94.9	99.5	96.8
168	State College	94.5	97.0	95.6
169	Wellsboro	95.5	95.2	95.4
170-171	Harrisburg	97.8	94.9	96.5
172	Chambersburg	94.9	91.6	93.4
173-174	York	94.9	95.3	95.1
175-176	Lancaster	93.5	96.8	94.9

STATE/ZIP	CITY	MAT.	INST.	TOTAL
PENNSYLVANIA (CONT'D)				
177	Williamsport	92.1	94.6	93.2
178	Sunbury	94.3	95.0	94.6
179	Pottsville	93.4	98.4	95.6
180	Lehigh Valley	95.0	113.6	103.0
181	Allentown	96.7	108.4	101.7
182	Hazleton	94.4	98.3	96.1
183	Stroudsburg	94.3	107.3	99.9
184-185	Scranton	97.4	98.9	98.1
186-187	Wilkes-Barre	94.1	97.9	95.7
188	Montrose	93.8	99.2	96.1
189	Doylestown	94.0	129.2	109.1
190-191	Philadelphia	100.1	136.3	115.2
193	Westchester	95.8	129.9	110.4
194	Norristown	94.8	130.0	109.9
195-196	Reading	96.5	104.0	99.7
PUERTO RICO				
009	San Juan	117.8	27.2	79.0
RHODE ISLAND				
028	Newport	94.6	113.7	102.8
029	Providence	96.2	113.7	103.7
SOUTH CAROLINA				
290-292	Columbia	96.5	67.9	84.2
293	Spartanburg	96.4	67.9	84.2
294	Charleston	98.0	67.6	84.9
295	Florence	96.2	67.8	84.0
296	Greenville	96.2	67.9	84.1
297	Rock Hill	96.0	67.3	83.7
298	Aiken	96.9	67.9	84.4
299	Beaufort	97.5	56.0	79.7
SOUTH DAKOTA				
570-571	Sioux Falls	96.3	79.2	89.0
572	Watertown	97.1	70.8	85.8
573	Mitchell	96.0	62.4	81.6
574	Aberdeen	98.3	71.5	86.8
575	Pierre	98.7	69.4	86.2
576	Mobridge	96.6	64.4	82.8
577	Rapid City	98.0	73.3	87.4
TENNESSEE				
370-372	Nashville	98.4	71.1	86.7
373-374	Chattanooga	98.9	69.6	86.3
375,380-381	Memphis	97.3	71.1	86.1
376	Johnson City	99.1	59.9	82.3
377-379	Knoxville	95.6	65.2	82.6
382	McKenzie	96.2	56.0	79.0
383	Jackson	97.5	59.7	81.3
384	Columbia	94.7	67.9	83.2
385	Cookeville	96.1	57.3	79.5
TEXAS				
750	McKinney	98.0	65.1	83.9
751	Waxahachie	98.1	65.1	83.9
752-753	Dallas	99.1	67.7	85.6
754	Greenville	98.2	64.5	83.8
755	Texarkana	97.5	63.6	83.0
756	Longview	98.4	62.6	83.0
757	Tyler	98.6	63.2	83.5
758	Palestine	95.1	62.0	80.9
759	Lufkin	95.5	64.2	82.1
760-761	Fort Worth	97.4	64.1	83.1
762	Denton	97.5	64.4	83.3
763	Wichita Falls	94.9	62.7	81.1
764	Eastland	94.0	61.8	80.2
765	Temple	92.5	60.0	78.5
766-767	Waco	94.3	63.9	81.3
768	Brownwood	97.4	60.2	81.4
769	San Angelo	97.1	60.8	81.6
770-772	Houston	100.7	67.9	86.6
773	Huntsville	99.4	64.3	84.4
774	Wharton	100.8	65.8	85.8
775	Galveston	98.5	68.1	85.5
776-777	Beaumont	98.8	67.1	85.2
778	Bryan	95.5	66.4	83.1
779	Victoria	100.5	64.4	85.0
780	Laredo	97.1	62.7	82.4
781-782	San Antonio	99.3	63.5	83.9
783-784	Corpus Christi	99.7	63.0	84.0
785	McAllen	100.6	57.9	82.3
786-787	Austin	97.4	62.7	82.5

For customer support on your Facilities Maintenance & Repair Costs with RSMeans data, call 800.448.8182.

STATE/ZIP	CITY	MAT.	INST.	TOTAL
TEXAS (CONT'D)				
788	Del Rio	100.6	62.0	84.0
789	Giddings	97.0	61.9	82.0
790-791	Amarillo	95.7	61.1	80.9
792	Childress	95.8	61.8	81.2
793-794	Lubbock	97.5	64.4	83.3
795-796	Abilene	96.0	62.8	81.8
797	Midland	98.1	63.7	83.3
798-799,885	El Paso	94.6	64.7	81.8
UTAH				
840-841	Salt Lake City	102.0	73.1	89.6
842,844	Ogden	97.5	73.1	87.0
843	Logan	99.4	73.1	88.1
845	Price	100.0	71.7	87.9
846-847	Provo	99.8	72.9	88.3
VERMONT				
050	White River Jct.	95.8	81.0	89.4
051	Bellows Falls	94.2	93.6	94.0
052	Bennington	94.6	90.0	92.6
053	Brattleboro	95.0	93.6	94.4
054	Burlington	99.1	80.6	91.2
056	Montpelier	97.7	85.3	92.4
057	Rutland	96.3	80.5	89.5
058	St. Johnsbury	95.8	80.5	89.3
059	Guildhall	94.5	80.2	88.3
VIRGINIA				
220-221	Fairfax	99.2	84.2	92.8
222	Arlington	100.1	84.3	93.3
223	Alexandria	99.3	84.2	92.8
224-225	Fredericksburg	97.9	80.6	90.5
226	Winchester	98.6	81.3	91.2
227	Culpeper	98.4	84.9	92.6
228	Harrisonburg	98.7	66.0	84.7
229	Charlottesville	99.1	68.0	85.8
230-232	Richmond	98.3	76.4	88.9
233-235	Norfolk	98.8	70.0	86.5
236	Newport News	97.9	69.7	85.8
237	Portsmouth	97.4	63.1	82.7
238	Petersburg	97.9	72.4	87.0
239	Farmville	97.1	62.8	82.4
240-241	Roanoke	99.7	71.6	87.7
242	Bristol	97.9	59.4	81.4
243	Pulaski	97.5	68.3	85.0
244	Staunton	98.3	65.6	84.3
245	Lynchburg	98.4	73.7	87.8
246	Grundy	97.8	59.2	81.2
WASHINGTON				
980-981,987	Seattle	106.0	107.4	106.6
982	Everett	103.4	102.6	103.1
983-984	Tacoma	103.8	102.1	103.1
985	Olympia	100.9	102.3	101.5
986	Vancouver	104.8	96.2	101.1
988	Wenatchee	104.1	91.0	98.5
989	Yakima	104.0	96.6	100.8
990-992	Spokane	99.9	84.5	93.3
993	Richland	99.6	92.2	96.4
994	Clarkston	98.9	83.9	92.5
WEST VIRGINIA				
247-248	Bluefield	96.7	90.2	93.9
249	Lewisburg	98.4	90.5	95.0
250-253	Charleston	96.1	92.5	94.6
254	Martinsburg	96.4	86.1	92.0
255-257	Huntington	97.5	94.3	96.1
258-259	Beckley	94.9	92.1	93.7
260	Wheeling	99.3	91.5	96.0
261	Parkersburg	98.2	90.7	95.0
262	Buckhannon	98.1	90.7	94.9
263-264	Clarksburg	98.7	91.4	95.5
265	Morgantown	98.5	91.5	95.5
266	Gassaway	98.0	90.5	94.8
267	Romney	97.9	89.2	94.2
268	Petersburg	97.7	87.7	93.4
WISCONSIN				
530,532	Milwaukee	99.6	105.9	102.3
531	Kenosha	99.4	104.6	101.6
534	Racine	98.8	105.0	101.5
535	Beloit	98.7	101.4	99.9
537	Madison	99.1	100.3	99.6

STATE/ZIP	CITY	MAT.	INST.	TOTAL
WISCONSIN (CONT'D)				
538	Lancaster	96.9	96.7	96.8
539	Portage	95.3	100.5	97.5
540	New Richmond	94.7	97.0	95.7
541-543	Green Bay	98.7	95.8	97.5
544	Wausau	94.1	95.8	94.8
545	Rhinelander	97.3	94.6	96.1
546	La Crosse	95.2	95.9	95.5
547	Eau Claire	96.8	96.5	96.7
548	Superior	94.5	98.7	96.3
549	Oshkosh	94.8	94.2	94.5
WYOMING				
820	Cheyenne	99.6	72.1	87.8
821	Yellowstone Nat'l Park	97.6	73.2	87.1
822	Wheatland	98.8	68.1	85.6
823	Rawlins	100.3	72.7	88.5
824	Worland	98.3	71.8	87.0
825	Riverton	99.4	69.7	86.6
826	Casper	99.3	69.5	86.5
827	Newcastle	98.1	72.3	87.1
828	Sheridan	100.9	70.6	87.9
829-831	Rock Springs	101.9	71.6	88.9
CANADIAN FACTORS (reflect Canadian currency)				
ALBERTA				
	Calgary	119.0	97.6	109.8
	Edmonton	119.7	97.5	110.2
	Fort McMurray	122.1	90.2	108.4
	Lethbridge	117.2	89.7	105.4
	Lloydminster	111.6	86.5	100.8
	Medicine Hat	111.7	85.8	100.6
	Red Deer	112.2	85.8	100.8
BRITISH COLUMBIA				
	Kamloops	113.0	85.7	101.3
	Prince George	113.9	85.0	101.5
	Vancouver	115.4	91.6	105.2
	Victoria	113.7	84.2	101.0
MANITOBA				
	Brandon	120.9	72.4	100.1
	Portage la Prairie	111.6	70.9	94.1
	Winnipeg	121.1	71.2	99.7
NEW BRUNSWICK				
	Bathurst	110.4	63.9	90.5
	Dalhousie	109.8	64.1	90.2
	Fredericton	117.9	71.2	97.9
	Moncton	110.6	68.1	92.4
	Newcastle	110.5	64.5	90.8
	Saint John	113.0	75.6	96.9
NEWFOUNDLAND				
	Corner Brook	125.0	70.2	101.5
	St. John's	120.8	88.8	107.0
NORTHWEST TERRITORIES				
	Yellowknife	130.4	85.0	110.9
NOVA SCOTIA				
	Bridgewater	111.2	72.2	94.5
	Dartmouth	121.8	72.1	100.5
	Halifax	117.1	85.8	103.7
	New Glasgow	119.9	72.1	99.4
	Sydney	118.1	72.1	98.4
	Truro	110.8	72.2	94.2
	Yarmouth	119.8	72.1	99.4
ONTARIO				
	Barrie	116.0	89.3	104.6
	Brantford	113.1	93.1	104.5
	Cornwall	113.1	89.6	103.0
	Hamilton	116.1	98.1	108.4
	Kingston	114.1	89.7	103.6
	Kitchener	109.4	93.4	102.5
	London	116.8	95.6	107.7
	North Bay	122.8	87.4	107.7
	Oshawa	111.6	92.9	103.6
	Ottawa	116.9	96.4	108.1
	Owen Sound	116.1	87.6	103.9
	Peterborough	113.1	89.4	102.9
	Sarnia	113.2	93.7	104.8

For customer support on your Facilities Maintenance & Repair Costs with RSMeans data, call 800.448.8182.

STATE/ZIP	CITY	MAT.	INST.	TOTAL
ONTARIO (CONT'D)				
	Sault Ste. Marie	108.6	90.0	100.6
	St. Catharines	107.3	94.3	101.7
	Sudbury	106.9	93.1	101.0
	Thunder Bay	108.5	93.7	102.1
	Timmins	113.2	87.5	102.2
	Toronto	115.8	102.4	110.1
	Windsor	107.8	93.5	101.7
PRINCE EDWARD ISLAND				
	Charlottetown	120.1	61.8	95.1
	Summerside	122.3	59.4	95.3
QUEBEC				
	Cap-de-la-Madeleine	110.7	82.1	98.4
	Charlesbourg	110.7	82.1	98.4
	Chicoutimi	110.3	87.3	100.4
	Gatineau	110.4	81.9	98.2
	Granby	110.6	81.8	98.3
	Hull	110.5	81.9	98.2
	Joliette	110.9	82.1	98.5
	Laval	110.7	81.9	98.3
	Montreal	118.1	89.9	106.0
	Quebec City	116.4	90.3	105.2
	Rimouski	110.3	87.3	100.4
	Rouyn-Noranda	110.4	81.9	98.1
	Saint-Hyacinthe	109.8	81.9	97.8
	Sherbrooke	110.6	81.9	98.3
	Sorel	110.9	82.1	98.5
	Saint-Jerome	110.4	81.9	98.2
	Trois-Rivieres	120.7	82.0	104.1
SASKATCHEWAN				
	Moose Jaw	109.0	65.6	90.4
	Prince Albert	108.2	63.9	89.2
	Regina	122.1	94.2	110.1
	Saskatoon	109.6	90.8	101.5
YUKON				
	Whitehorse	131.9	70.5	105.6

For customer support on your Facilities Maintenance & Repair Costs with RSMeans data, call 800.448.8182.

L1010-101 Minimum Design Live Loads in Pounds per S.F. for Various Building Codes

Occupancy or Use	Uniform (psf)	Concentrated (lbs)
1. Access floor systems		
Office use	50	2000
Computer use	100	2000
2. Armories and drill rooms	100	–
3. Assembly areas		
Fixed seats (fastened to floor)	60	
Follow spot, projections and control rooms	50	
Lobbies	100	
Movable seats	100	–
Stage floors	150	
Platforms (assembly)	100	
Other assembly areas	100	
4. Balconies and decks	Same as occupancy served	–
5. Catwalks	40	300
6. Comices	60	–
7. Corridors		
First floor	100	–
Other floors	Same as occupancy served except as indicated	
8. Dining rooms and restaurants	100	–
9. Elevator machine room grating (on an area of 2 in by 2 in)	–	300
10. Finish light floorplate construction (on area of 1 in by 1 in)	–	200
11. Fire escapes	100	
On single-family dwellings only		–
12. Garages (passenger vehicles only)	40	Note a
13. Hospitals		
Corridors above first floor	80	1000
Operating rooms, laboratories	60	1000
Patient rooms	40	1000
14. Libraries		
Corridors above first floor	80	1000
Reading rooms	60	1000
Stacks	150, Note b	1000
15. Manufacturing		
Heavy	250	3000
Light	125	2000
16. Marquees	75	–
17. Office buildings		
Corridors above first floor	80	2000
Lobbies and first-floor corridors	100	2000
Offices	50	2000
18. Penal institutions		
Cell blocks	40	
Corridors	100	
19. Recreational uses:		
Bowling alleys, poolrooms, and similar uses	75	
Dance halls and ballrooms	100	
Gymnasiums	100	
Reviewing stands, grandstands and bleachers	100	
Stadiums and arenas with fixed seats (fastened to floor)	60	
20. Residential		
One- and two- family dwellings		
Uninhabitable attics without storage, Note c	10	
Uninhabitable attics with storage, Note c, d, e	20	
Habitable attics and sleeping areas, Note e	30	
All other areas	0	
Hotels and multifamily dwellings		
Private rooms and corridors serving them	40	
Public rooms and corridors serving them	100	

645

L1010-101   Minimum Design Live Loads in Pounds per S.F. for Various Building Codes (cont.)

Occupancy or Use	Uniform (psf)	Concentrated (lbs)
21. Roofs		
All roof surfaces subject to maintenance workers		300
Awnings and canopies:		
Fabric construction supported by a skeleton structure	5	–
All other construction	20	
Ordinary flat, pitched and curved roofs (not occupiable)	20	
Where primary roof members are exposed to a work floor, at single panel point of lower cord of roof trusses or any point along primary structural members supporting roofs:		
Over manufacturing, storage warehouses, and repair garages		2000
All other primary roof members		300
Occupiable roofs:		
Roof gardens	100	
Assembly areas	100	
All other similar areas	Note f	Note f
22. Schools		
Classrooms	40	1000
Corridors	80	1000
First-floor corridors	100	1000
23. Scuttles, skylight ribs and accessible ceilings	–	200
24. Sidewalks, vehicular driveways and yards, subject to trucking	250	8000, Note g
25. Stairs and exits		
One- and two- family dwellings	40	300, Note h
All others	100	3000, Note h
26. Storage warehouses		
Heavy	250	–
Light	125	
27. Stores		
Retail		
First floor	100	1000
Upper floors	75	1000
Wholesale, all floors	125	1000
28. Walkways and elevated platforms (other than exitways)	60	–
29. Yards and terraces, pedestrians	100	–

Notes:

a. Floors in garages or portions of buildings used for the storage of motor vehicles shall be designed for the uniformly distributed live loads of Table 1607.1 or the following concentrated loads: (1) for garages restricted to passenger vehicles accommodating not more than nine passengers, 3,000 pounds acting on an area of 4.5 in. by 4.5 in.; (2) for mechanical parking structures without slab or deck that are used for storing passenger vehicles only, 2,250 pounds per wheel.

b. The loading applies to stack room floors that support nonmobile, double-faced library books stacks, subject to the following limitations:
  1. The nominal bookstack unit height shall not exceed 90 inches;
  2. The nominal shelf depth shall not exceed 12 inches for each face; and
  3. Parallel rows of double-faced book stacks shall be separated by aisles not less than 36 inches wide.

c. Uninhabitable attics without storage are those where the maximum clear height between the joists and rafters is less than 42 inches, or where there are not two or more adjacent trusses with web configurations capable of accommodating an assumed rectangle 42 inches in height by 24 inches in width, or greater, within the plane of the trusses. This live load need not be assumed to act concurrently with any other live load requirements.

d. Uninhabitable attics with storage are those where the maximum clear height between the joists and rafters is 42 inches or greater, or where there are two or more adjacent trusses with web configurations capable of accommodating an assumed rectangle 42 inches in height by 24 inches in width, or greater, within the plane of the trusses.

  The live load need only be applied to those portions of the joists or truss bottom chords where both of the following conditions are met:
    i. The attic area is accessible from an opening not less than 20 inches in width by 30 inches in length that is located where the clear height in the attic is a minimum of 30 inches; and
    ii. The slopes of the joists or truss bottom chords are no greater than two units vertical in 12 units horizontal.

  The remaining portions of the joists or truss bottom chords shall be designed for a uniformly distributed concurrent live load of not less than 10 lb/ft^2.

e. Attic spaces served by stairways other than the pull-down type shall be designed to support the minimum live load specified for habitable attics and sleeping rooms.

f. Areas of occupiable roofs, other than roof gardens and assembly areas, shall be designed for approporate loads as approved by the building official.

g. The concentrated wheel load shall be applied on an area of 4.5 inches by 4.5 inches.

h. The minimum concentrated load on stair treads shall be applied on an area of 2 in by 2 in. This load need not be assumed to act concurrently with the uniform load.

Excerpted from the 2012 *International Building Code*, Copyright 2011. Washington, D.C.: International Code Council. Reproduced with permission. All rights reserved. www.ICCSAFE.org

Table L1010-201   Design Weight Per S.F. for Walls and Partitions

Type	Wall Thickness	Description	Weight Per S.F.	Type	Wall Thickness	Description	Weight Per S.F.
Brick	4"	Clay brick, high absorption Clay brick, medium absorption	34 lb. 39	Clay tile	2"	Split terra cotta furring Non load bearing clay tile	10 lb. 11
		Clay brick, low absorption	46		3"	Split terra cotta furring	12
		Sand-lime brick	38			Non load bearing clay tile	18
		Concrete brick, heavy aggregate Concrete brick, light aggregate	46 33		4"	Non load bearing clay tile Load bearing clay tile	20 24
	8"	Clay brick, high absorption Clay brick, medium absorption	69 79		6"	Non load bearing clay tile Load bearing clay tile	30 36
		Clay brick, low absorption	89		8"	Non load bearing clay tile	36
		Sand-lime brick	74			Load bearing clay tile	42
		Concrete brick, heavy aggregate Concrete brick, light aggregate	89 68		12"	Non load bearing clay tile Load bearing clay tile	46 58
	12"	Common brick Pressed brick	120 130	Gypsum block	2"	Hollow gypsum block Solid gypsum block	9.5 12
		Sand-lime brick	105		3"	Hollow gypsum block	10
		Concrete brick, heavy aggregate	130			Solid gypsum block	18
		Concrete brick, light aggregate	98		4"	Hollow gypsum block	15
	16"	Clay brick, high absorption	134			Solid gypsum block	24
		Clay brick, medium absorption	155		5"	Hollow gypsum block	18
		Clay brick, low absorption	173		6"	Hollow gypsum block	24
		Sand-lime brick	138	Structural facing tile	2"	Facing tile	15
		Concrete brick, heavy aggregate	174		4"	Facing tile	25
		Concrete brick, light aggregate	130		6"	Facing tile	38
Concrete block	4"	Solid conc. block, stone aggregate Solid conc. block, lightweight	45 34	Glass	4"	Glass block	18
		Hollow conc. block, stone aggregate	30		1"	Structural glass	15
		Hollow conc. block, lightweight	20	Plaster	1"	Gypsum plaster (1 side)	5
	6"	Solid conc. block, stone aggregate Solid conc. block, lightweight	50 37			Cement plaster (1 side) Gypsum plaster on lath	10 8
		Hollow conc. block, stone aggregate	42			Cement plaster on lath	13
		Hollow conc. block, lightweight	30	Plaster partition (2 finished faces)	2"	Solid gypsum on metal lath	18
	8"	Solid conc. block, stone aggregate Solid conc. block, lightweight	67 48			Solid cement on metal lath	25
						Solid gypsum on gypsum lath	18
		Hollow conc. block, stone aggregate	55			Gypsum on lath & metal studs	18
		Hollow conc. block, lightweight	38		3"	Gypsum on lath & metal studs	19
	10"	Solid conc. block, stone aggregate	84		4"	Gypsum on lath & metal studs	20
		Solid conc. block, lightweight	62		6"	Gypsum on lath & wood studs	18
		Hollow conc. block, stone aggregate Hollow conc. block, lightweight	55 38	Concrete	6"	Reinf concrete, stone aggregate Reinf. concrete, lightweight	75 36-60
	12"	Solid conc. block, stone aggregate Solid conc. block, lightweight	108 72		8"	Reinf. concrete, stone aggregate Reinf. concrete, lightweight	100 48-80
		Hollow conc. block, stone aggregate Hollow conc. block, lightweight	85 55		10"	Reinf. concrete, stone aggregate Reinf. concrete, lightweight	125 60-100
Drywall	6"	Drywall on wood studs	10		12"	Reinf concrete stone aggregate Reinf concrete, lightweight	150 72-120

Table L1010-202   Design Weight per S.F. for Roof Coverings

Type		Description	Weight lb. Per S.F.	Type	Wall Thickness	Description	Weight lb. Per S.F.
Sheathing	Gypsum	1" thick	4	Metal	Aluminum	Corr. & ribbed, .024" to .040"	.4-.8
	Wood	¾" thick	3		Copper	or tin	1.5-2.5
Insulation	per 1"	Loose	.5		Steel	Corrugated, 29 ga. to 12 ga.	.6-5.0
		Poured in place	2	Shingles	Asphalt	Strip shingles	1.7-2.8
		Rigid	1.5		Clay	Tile	8-16
Built-up	Tar & gravel	3 ply felt	5.5		Slate	¼" thick	9.5
		5 ply felt	6.5		Wood		2-3

647

Table L1010-203 Design Weight per Square Foot for Floor Fills and Finishes

Type		Description	Weight lb. per S.F.	Type		Description	Weight lb. per S.F.
Floor fill	per 1"	Cinder fill	5	Wood	Single	On sleepers, light concrete fill	16
		Cinder concrete	9		7/8"	On sleepers, stone concrete fill	25
		Lightweight concrete	3-9		Double	On sleepers, light concrete fill	19
		Stone concrete	12		7/8"	On sleepers, stone concrete fill	28
		Sand	8		3"	Wood block on mastic, no fill	15
		Gypsum	6			Wood block on ½" mortar	16
Terrazzo	1"	Terrazzo, 2" stone concrete	25		per 1"	Hardwood flooring (25/32")	4
Marble	and mortar	on stone concrete fill	33			Underlayment (Plywood per 1")	3
Resilient	1/16"-1/4"	Linoleum, asphalt, vinyl tile	2	Asphalt	1-1/2"	Mastic flooring	18
Tile	3/4"	Ceramic or quarry	10		2"	Block on ½" mortar	30

Table L1010-204 Design Weight Per Cubic Foot for Miscellaneous Materials

Type		Description	Weight lb. per C.F.	Type		Description	Weight lb. per C.F.
Bituminous	Coal, piled	Anthracite	47-58	Masonry	Ashlar	Granite	165
		Bituminous	40-54			Limestone, crystalline	165
		Peat, turf, dry	47			Limestone, oolitic	135
		Coke	75			Marble	173
	Petroleum	Unrefined	54			Sandstone	144
		Refined	50		Rubble, in mortar	Granite	155
		Gasoline	42			Limestone, crystalline	147
	Pitch		69			Limestone, oolitic	138
	Tar	Bituminous	75			Marble	156
Concrete	Plain	Stone aggregate	144			Sandstone & Bluestone	130
		Slag aggregate	132		Brick	Pressed	140
		Expanded slag aggregate	100			Common	120
		Haydite (burned clay agg.)	90			Soft	100
		Vermiculite & perlite, load bearing	70-105		Cement	Portland, loose	90
		Vermiculite & perlite, non load bear	35-50			Portland set	183
	Reinforced	Stone aggregate	150		Lime	Gypsum, loose	53-64
		Slag aggregate	138		Mortar	Set	103
		Lightweight aggregates	30-120	Metals	Aluminum	Cast, hammered	165
Earth	Clay	Dry	63		Brass	Cast, rolled	534
		Damp, plastic	110		Bronze	7.9 to 14% Sn	509
		and gravel, dry	100		Copper	Cast, rolled	556
	Dry	Loose	76		Iron	Cast, pig	450
		Packed	95			Wrought	480
	Moist	Loose	78		Lead		710
		Packed	96		Monel		556
	Mud	Flowing	108		Steel	Rolled	490
		Packed	115		Tin	Cast, hammered	459
	Riprap	Limestone	80-85		Zinc	Cast rolled	440
		Sandstone	90	Timber	Cedar	White or red	24.2
		Shale	105		Fir	Douglas	23.7
	Sand & gravel	Dry, loose	90-105			Eastern	25
		Dry, packed	100-120		Maple	Hard	44.5
		Wet	118-120			White	33
Gases	Air	0C., 760 mm.	.0807		Oak	Red or Black	47.3
	Gas	Natural	.0385			White	47.3
Liquids	Alcohol	100%	49		Pine	White	26
	Water	4°C., maximum density	62.5			Yellow, long leaf	44
		Ice	56			Yellow, short leaf	38
		Snow, fresh fallen	8		Redwood	California	26
		Sea water	64		Spruce	White or black	27

For customer support on your Facilities Maintenance & Repair Costs with RSMeans data, call 800.448.8182.

Table L1010-225 Design Weight Per S.F. for Structural Floor and Roof Systems

Type Slab		Description	weight columns (Slab Depth in Inches)											
			1"	12.5	2"	25	3"	37.5	4"	50	5"	62.5	6"	75
Concrete Slab	Reinforced	Stone aggregate	1"	12.5	2"	25	3"	37.5	4"	50	5"	62.5	6"	75
		Lightweight sand aggregate		9.5		19		28.5		38		47.5		57
		All lightweight aggregate		9.0		18		27.0		36		45.0		54
	Plain, nonreinforced	Stone aggregate		12.0		24		36.0		48		60.0		72
		Lightweight sand aggregate		9.0		18		27.0		36		45.0		54
		All lightweight aggregate		8.5		17		25.5		34		42.5		51
Concrete Waffle	19" x 19"	5" wide ribs @ 24" O.C.	6+3	77	8+3	92	10+3	100	12+3	118				
			6+4½	96	8+4½	110	10+4½	119	12+4½	136				
	30" x 30"	6" wide ribs @ 36" O.C.	8+3	83	10+3	95	12+3	109	14+3	118	16+3	130	20+3	155
			8+4½	101	10+4½	113	12+4½	126	14+4½	137	16+4½	149	20+4½	173
Concrete Joist	20" wide form	5" wide rib	8+3	60	10+3	67	12+3	74	14+3	81				
		6" wide rib	↓	63	↓	70	↓	78	↓	86	16+3	94	20+3	111
		7" wide rib									↓	99	↓	118
		5" wide rib	8+4½	79	10+4½	85	12+4½	92	14+4½	99				
		6" wide rib	↓	82	↓	89	↓	97	↓	104	16+4½	113	20+4½	130
		7" wide rib									↓	118	↓	136
	30" wide form	5" wide rib	8+3	54	10+3	58	12+3	63	14+3	68				
		6" wide rib	↓	56	↓	61	↓	67	↓	72	16+3	78	20+3	91
		7" wide rib									↓	83	↓	96
		5" wide rib	8+4½	72	10+4½	77	12+4½	82	14+4½	87				
		6" wide rib	↓	75	↓	80	↓	85	↓	91	16+4½	97	20+4½	109
		7" wide rib									↓	101	↓	115
Wood Joists	Incl. Subfloor	12" O.C.	2x6	6	2x8	6	2x10	7	2x12	8	3x8	8	3x12	11
		16" O.C.	↓	5	↓	6	↓	6	↓	7	↓	7	↓	9

Table L1010-226 Superimposed Dead Load Ranges

Component	Load Range (PSF)
Ceiling	5-10
Partitions	20-30
Mechanical	4-8

For customer support on your Facilities Maintenance & Repair Costs with RSMeans data, call 800.448.8182.

Table L1010-301   Design Loads for Structures for Wind Load

Wind Loads: Structures are designed to resist the wind force from any direction. Usually 2/3 is assumed to act on the windward side, 1/3 on the leeward side.

For more than 1/3 openings, add 10 psf for internal wind pressure or 5 psf for suction, whichever is critical.

For buildings and structures, use psf values from Table L1010-301.

For glass over 4 S.F. use values in Table L1010-302 after determining 30′ wind velocity from Table L1010-303.

Type Structure	Height Above Grade	Horizontal Load in Lb. per S.F.
Buildings	Up to 50 ft.	15
	50 to 100 ft.	20
	Over 100 ft.	20 + .025 per ft.
Ground signs & towers	Up to 50 ft.	15
	Over 50 ft.	20
Roof structures		30
Glass	See Table below	

Table L1010-302   Design Wind Load in PSF for Glass at Various Elevations

Height From Grade	Velocity in Miles per Hour and Design Load in Pounds per S.F.																			
	Vel.	PSF	Vel.	PSF	Vel.	PSF	Vel.	PSF	Vel.	PSF	Vel.	PSF	Vel.	PSF	Vel.	PSF	Vel.	PSF	Vel.	PSF
To 10 ft.	42	6	46	7	49	8	52	9	55	10	59	11	62	12	66	14	69	15	76	19	83	22
10-20	52	9	58	11	61	11	65	14	70	16	74	18	79	20	83	22	87	24	96	30	105	35
20-30*	60	12	67	14	70	16	75	18	80	20	85	23	90	26	95	29	100	32	110	39	120	46
30-60	66	14	74	18	77	19	83	22	88	25	94	28	99	31	104	35	110	39	121	47	132	56
60-120	73	17	82	21	85	12	92	27	98	31	104	35	110	39	116	43	122	48	134	57	146	68
120-140	81	21	91	26	95	29	101	33	108	37	115	42	122	48	128	52	135	48	149	71	162	84
240-480	90	26	100	32	104	35	112	40	119	45	127	51	134	57	142	65	149	71	164	86	179	102
480-960	98	31	110	39	115	42	123	49	131	55	139	62	148	70	156	78	164	86	180	104	197	124
Over 960	98	31	110	39	115	42	123	49	131	55	139	62	148	70	156	78	164	86	180	104	197	124

*Determine appropriate wind at 30′ elevation Fig. L1010-303 below.

Table L1010-303   Design Wind Velocity at 30 Ft. Above Ground

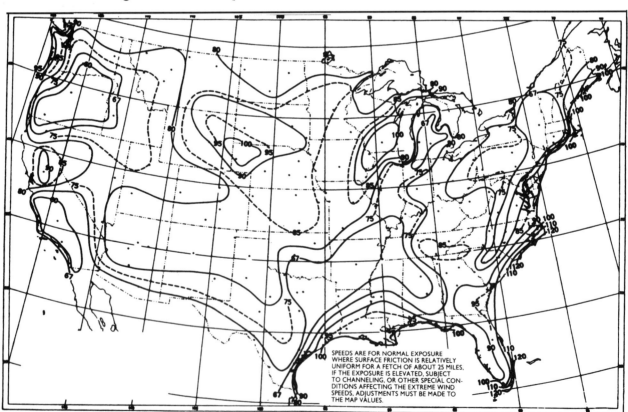

SPEEDS ARE FOR NORMAL EXPOSURE WHERE SURFACE FRICTION IS RELATIVELY UNIFORM FOR A FETCH OF ABOUT 25 MILES. IF THE EXPOSURE IS ELEVATED, SUBJECT TO CHANNELING, OR OTHER SPECIAL CONDITIONS AFFECTING THE EXTREME WIND SPEEDS, ADJUSTMENTS MUST BE MADE TO THE MAP VALUES.

For customer support on your Facilities Maintenance & Repair Costs with RSMeans data, call 800.448.8182.

Table L1010-401 Snow Load in Pounds Per Square Foot on the Ground

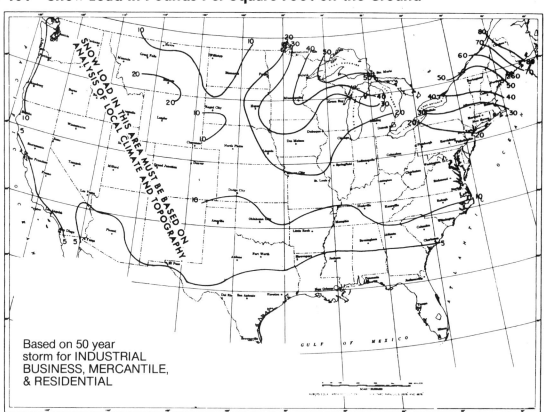

Based on 50 year
storm for INDUSTRIAL
BUSINESS, MERCANTILE,
& RESIDENTIAL

Table L1010-402 Snow Load in Pounds Per Square Foot on the Ground

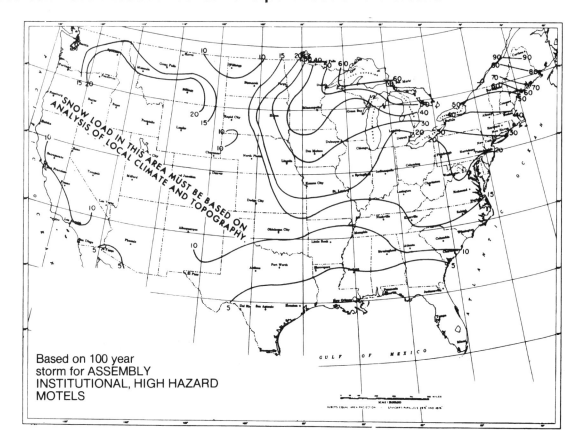

Based on 100 year
storm for ASSEMBLY
INSTITUTIONAL, HIGH HAZARD
MOTELS

651

Table L1010-403 Snow Loads

To convert the ground snow loads on the previous page to roof snow loads, the ground snow loads should be multiplied by the following factors depending upon the roof characteristics.

Note, in all cases $\frac{\alpha - 30}{50}$ is valid only for > 30 degrees.

Description	Sketch	Formula	Angle	Conversion Factor = C_f	
				Sheltered	Exposed
Simple flat and shed roofs	Load Diagram	For $\alpha > 30°$ $C_f = 0.8 - \frac{\alpha - 30}{50}$	0° to 30°	0.8	0.6
			40°	0.6	0.45
			50°	0.4	0.3
			60°	0.2	0.15
			70° to 90°	0	0
				Case I	Case II
Simple gable and hip roofs	Case I Load Diagram Case II Load Diagram	$C_f = 0.8 - \frac{\alpha - 30}{50}$ $C_f = 1.25\left(0.8 - \left(\frac{\alpha - 30}{50}\right)\right)$	10°	0.8	—
			20°	0.8	—
			30°	0.8	1.0
			40°	0.6	0.75
			50°	0.4	0.5
			60°	0.2	0.25
Valley areas of Two span roofs	Case I Case II	$\beta = \frac{\alpha_1 + \alpha_2}{2}$ $C_f = 0.8 - \frac{\alpha - 30}{50}$	$\beta \le 10°$ use Case I only $\beta > 10°$ $\beta < 20°$ use Case I & II $\beta \ge 20°$ use Case I, II & III		
Lower level of multi level roofs (or on an adjacent building not more than 15 ft. away)		$C_f = 15\frac{h}{g}$ h = difference in roof height in feet g = ground snow load in psf w = width of drift For h < 5, w = 10 h >15, w = 30	When $15\frac{h}{g} < .8$, use 0.8 When $15\frac{h}{g} > 3.0$, use 3.0		

Summary of above:

1. For flat roofs or roofs up to 30°, use 0.8 x ground snow load for the roof snow load.
2. For roof pitches in excess of 30°, conversion factor becomes lower than 0.8.
3. For exposed roofs there is a further 25% reduction of conversion factor.
4. For steep roofs a more highly loaded half span must be considered.
5. For shallow roof valleys conversion factor is 0.8.
6. For moderate roof valleys, conversion factor is 1.0 for half the span.
7. For steep roof valleys, conversion factor is 1.5 for one quarter of the span.
8. For roofs adjoining vertical surfaces the conversion factor is up to 3.0 for part of the span.
9. If snow load is less than 30 psf, use water load on roof for clogged drain condition.

Table L1020-101 Floor Area Ratios

The table below lists commonly used gross to net area and net to gross area ratios expressed in % for various building types.

Building Type	Gross to Net Ratio	Net to Gross Ratio	Building Type	Gross to Net Ratio	Net to Gross Ratio
Apartment	156	64	School Buildings (campus type)		
Bank	140	72	Administrative	150	67
Church	142	70	Auditorium	142	70
Courthouse	162	61	Biology	161	62
Department Store	123	81	Chemistry	170	59
Garage	118	85	Classroom	152	66
Hospital	183	55	Dining Hall	138	72
Hotel	158	63	Dormitory	154	65
Laboratory	171	58	Engineering	164	61
Library	132	76	Fraternity	160	63
Office	135	75	Gymnasium	142	70
Restaurant	141	70	Science	167	60
Warehouse	108	93	Service	120	83
			Student Union	172	59

The gross area of a building is the total floor area based on outside dimensions.

The net area of a building is the usable floor area for the function intended and excludes such items as stairways, corridors, and mechanical rooms.

In the case of a commercial building, it might be considered the "leasable area."

For customer support on your Facilities Maintenance & Repair Costs with RSMeans data, call 800.448.8182.

Table L1020-201 Partition/Door Density

Building Type		Stories	Partition/Density	Doors	Description of Partition
Apartments		1 story	9 SF/LF	90 SF/door	Plaster, wood doors & trim
		2 story	8 SF/LF	80 SF/door	Drywall, wood studs, wood doors & trim
		3 story	9 SF/LF	90 SF/door	Plaster, wood studs, wood doors & trim
		5 story	9 SF/LF	90 SF/door	Plaster, metal studs, wood doors & trim
		6-15 story	8 SF/LF	80 SF/door	Drywall, metal studs, wood doors & trim
Bakery		1 story	50 SF/LF	500 SF/door	Conc. block, paint, door & drywall, wood studs
		2 story	50 SF/LF	500 SF/door	Conc. block, paint, door & drywall, wood studs
Bank		1 story	20 SF/LF	200 SF/door	Plaster, wood studs, wood doors & trim
		2-4 story	15 SF/LF	150 SF/door	Plaster, metal studs, wood doors & trim
Bottling Plant		1 story	50 SF/LF	500 SF/door	Conc. block, drywall, metal studs, wood trim
Bowling Alley		1 story	50 SF/LF	500 SF/door	Conc. block, wood & metal doors, wood trim
Bus Terminal		1 story	15 SF/LF	150 SF/door	Conc. block, ceramic tile, wood trim
Cannery		1 story	100 SF/LF	1000 SF/door	Drywall on metal studs
Car Wash		1 story	18 SF/LF	180 SF/door	Concrete block, painted & hollow metal door
Dairy Plant		1 story	30 SF/LF	300 SF/door	Concrete block, glazed tile, insulated cooler doors
Department Store		1 story	60 SF/LF	600 SF/door	Drywall, metal studs, wood doors & trim
		2-5 story	60 SF/LF	600 SF/door	30% concrete block, 70% drywall, wood studs
Dormitory		2 story	9 SF/LF	90 SF/door	Plaster, concrete block, wood doors & trim
		3-5 story	9 SF/LF	90 SF/door	Plaster, concrete block, wood doors & trim
		6-15 story	9 SF/LF	90 SF/door	Plaster, concrete block, wood doors & trim
Funeral Home		1 story	15 SF/LF	150 SF/door	Plaster on concrete block & wood studs, paneling
		2 story	14 SF/LF	140 SF/door	Plaster, wood studs, paneling & wood doors
Garage Sales & Service		1 story	30 SF/LF	300 SF/door	50% conc. block, 50% drywall, wood studs
Hotel		3-8 story	9 SF/LF	90 SF/door	Plaster, conc. block, wood doors & trim
		9-15 story	9 SF/LF	90 SF/door	Plaster, conc. block, wood doors & trim
Laundromat		1 story	25 SF/LF	250 SF/door	Drywall, wood studs, wood doors & trim
Medical Clinic		1 story	6 SF/LF	60 SF/door	Drywall, wood studs, wood doors & trim
		2-4 story	6 SF/LF	60 SF/door	Drywall, metal studs, wood doors & trim
Motel		1 story	7 SF/LF	70 SF/door	Drywall, wood studs, wood doors & trim
		2-3 story	7 SF/LF	70 SF/door	Concrete block, drywall on metal studs, wood paneling
Movie Theater	200-600 seats	1 story	18 SF/LF	180 SF/door	Concrete block, wood, metal, vinyl trim
	601-1400 seats		20 SF/LF	200 SF/door	Concrete block, wood, metal, vinyl trim
	1401-22000 seats		25 SF/LF	250 SF/door	Concrete block, wood, metal, vinyl trim
Nursing Home		1 story	8 SF/LF	80 SF/door	Drywall, metal studs, wood doors & trim
		2-4 story	8 SF/LF	80 SF/door	Drywall, metal studs, wood doors & trim
Office		1 story	20 SF/LF	200-500 SF/door	30% concrete block, 70% drywall on wood studs
		2 story	20 SF/LF	200-500 SF/door	30% concrete block, 70% drywall on metal studs
		3-5 story	20 SF/LF	200-500 SF/door	30% concrete block, 70% movable partitions
		6-10 story	20 SF/LF	200-500 SF/door	30% concrete block, 70% movable partitions
		11-20 story	20 SF/LF	200-500 SF/door	30% concrete block, 70% movable partitions
Parking Ramp (Open)		2-8 story	60 SF/LF	600 SF/door	Stair and elevator enclosures only
Parking Garage		2-8 story	60 SF/LF	600 SF/door	Stair and elevator enclosures only
Pre-engineered	Steel	1 story	0		
	Store	1 story	60 SF/LF	600 SF/door	Drywall on metal studs, wood doors & trim
	Office	1 story	15 SF/LF	150 SF/door	Concrete block, movable wood partitions
	Shop	1 story	15 SF/LF	150 SF/door	Movable wood partitions
	Warehouse	1 story	0		
Radio & TV Broadcasting		1 story	25 SF/LF	250 SF/door	Concrete block, metal and wood doors
& TV Transmitter		1 story	40 SF/LF	400 SF/door	Concrete block, metal and wood doors
Self Service Restaurant		1 story	15 SF/LF	150 SF/door	Concrete block, wood and aluminum trim
Cafe & Drive-in Restaurant		1 story	18 SF/LF	180 SF/door	Drywall, metal studs, ceramic & plastic trim
Restaurant with seating		1 story	25 SF/LF	250 SF/door	Concrete block, paneling, wood studs & trim
Supper Club		1 story	25 SF/LF	250 SF/door	Concrete block, paneling, wood studs & trim
Bar or Lounge		1 story	24 SF/LF	240 SF/door	Plaster or gypsum lath, wooded studs
Retail Store or Shop		1 story	60 SF/LF	600 SF/door	Drywall metal studs, wood doors & trim
Service Station	Masonry	1 story	15 SF/LF	150 SF/door	Concrete block, paint, door & drywall, wood studs
	Metal panel	1 story	15 SF/LF	150 SF/door	Concrete block, paint, door & drywall, wood studs
	Frame	1 story	15 SF/LF	150 SF/door	Drywall, wood studs, wood doors & trim
Shopping Center	(strip)	1 story	30 SF/LF	300 SF/door	Drywall, metal studs, wood doors & trim
	(group)	1 story	40 SF/LF	400 SF/door	50% concrete block, 50% drywall, wood studs
		2 story	40 SF/LF	400 SF/door	50% concrete block, 50% drywall, wood studs
Small Food Store		1 story	30 SF/LF	300 SF/door	Concrete block drywall, wood studs, wood trim
Store/Apt. above	Masonry	2 story	10 SF/LF	100 SF/door	Plaster, metal studs, wood doors & trim
	Frame	2 story	10 SF/LF	100 SF/door	Plaster, metal studs, wood doors & trim
	Frame	3 story	10 SF/LF	100 SF/door	Plaster, metal studs, wood doors & trim
Supermarkets		1 story	40 SF/LF	400 SF/door	Concrete block, paint, drywall & porcelain panel
Truck Terminal		1 story	0		
Warehouse		1 story	0		

654

Table L1020-301 Occupancy Determinations

Function of Space	SF/Person Required
Accessory storage areas, mechanical equipment rooms	300
Agriculture Building	300
Aircraft Hangars	500
Airport Terminal	
Baggage claim	20
Baggage handling	300
Concourse	100
Waiting areas	15
Assembly	
Gaming floors (Keno, slots, etc.)	11
Exhibit gallery and museum	30
Assembly w/ fixed seats	load determined by seat number
Assembly w/o fixed seats	
Concentrated (chairs only-not fixed)	7
Standing space	5
Unconcentrated (tables and chairs)	15
Bowling centers, allow 5 persons for each lane including 15 feet of runway, and for additional areas	7
Business areas	100
Courtrooms-other than fixed seating areas	40
Day care	35
Dormitories	50
Educational	
Classroom areas	20
Shops and other vocational room areas	50
Exercise rooms	50
Fabrication and manufacturing areas where hazardous materials are used	200
Industrial areas	100
Institutional areas	
Inpatient treatment areas	240
Outpatient areas	100
Sleeping areas	120
Kitchens commercial	200
Library	
Reading rooms	50
Stack area	100
Mercantile	
Areas on other floors	60
Basement and grade floor areas	30
Storage, stock, shipping areas	300
Parking garages	200
Residential	200
Skating rinks, swimming pools	
Rink and pool	50
Decks	15
Stages and platforms	15
Warehouses	500

Excerpted from the 2012 *International Building Code*, Copyright 2011. Washington, D.C.: International Code Council. Reproduced with permission. All rights reserved. www.ICCSAFE.org

For customer support on your Facilities Maintenance & Repair Costs with RSMeans data, call 800.448.8182.

Table L1020-302   Length of Exitway Access Travel (ft.)

Occupancy Type	Without Sprinkler System (feet)	With Sprinkler System (feet)
A, E, F-1, M, R, S-1	200	250
I-1	Not Permitted	250
B	200	300
F-2, S-2, U	300	400
H-1	Not Permitted	75
H-2	Not Permitted	100
H-3	Not Permitted	150
H-4	Not Permitted	175
H-5	Not Permitted	200
I-2, I-3, I-4	Not Permitted	200

Note:

Refer to the 2012 *International Building Code* Section 1016 Exit Access Travel Distance for any exceptions or additions to the information above.

Excerpted from the 2012 *International Building Code*, Copyright 2011. Washington, D.C.: International Code Council. Reproduced with permission. All rights reserved. www.ICCSAFE.org

Table L1020-303 Capacity Per Unit Egress Width*

Use Group	Without Fire Suppression System (Inches per Person)*		With Fire Suppression System (Inches per Person)*	
	Stairways	Doors, Ramps and Corridors	Stairways	Doors, Ramps and Corridors
Assembly, Business, Educational, Factory Industrial, Mercantile, Residential, Storage	0.3	0.2	0.2	0.15
Institutional—1	0.3	0.2	0.2	0.15
Institutional—2	—	0.7	0.3	0.2
Institutional—3	0.3	0.2	0.2	0.15
High Hazard	0.7	0.4	0.3	0.2

* 1″ = 25.4 mm

For customer support on your Facilities Maintenance & Repair Costs with RSMeans data, call 800.448.8182.

Table L1030-401　Resistances ("R") of Building and Insulating Materials

Material	Wt./Lbs. per C.F.	R per Inch	R Listed Size	Material	Wt./Lbs. per C.F.	R per Inch	R Listed Size
Air Spaces and Surfaces				Insulation Loose Fill			
Enclosed non-reflective spaces, E=0.82				Cellulose	2.3	3.13	
50° F mean temp., 30°/10° F diff.					3.2	3.70	
.5"			.90/.91	Mineral fiber, 3.75" to 5" thick	2-5		11
.75"			.94/1.01	6.5" to 8.75" thick			19
1.50"			.90/1.02	7.5" to 10" thick			22
3.50"			.91/1.01	10.25" to 13.75" thick			30
Inside vert. surface (still air)			0.68	Perlite	5-8	2.70	
Outside vert. surface (15 mph wind)			0.17	Vermiculite	4-6	2.27	
				Wood fiber	2-3.5	3.33	
Building Boards				Masonry Brick, Common	120	0.20	
Asbestos cement, 0.25" thick	120		0.06	Face	130	0.11	
Gypsum or plaster, 0.5" thick	50		0.45	Cement mortar	116	0.20	
Hardboard regular	50	1.37		Clay tile, hollow			
Tempered	63	1.00		1 cell wide, 3" width			0.80
Laminated paper	30	2.00		4" width			1.11
Particle board	37	1.85		2 cells wide, 6" width			1.52
	50	1.06		8" width			1.85
	63	0.85		10" width			2.22
Plywood (Douglas Fir), 0.5" thick	34		0.62	3 cells wide, 12" width			2.50
Shingle backer, .375" thick	18		0.94	Concrete, gypsum fiber	51	0.60	
Sound deadening board, 0.5" thick	15		1.35	Lightweight	120	0.19	
Tile and lay-in panels, plain or					80	0.40	
acoustical, 0.5" thick	18		1.25		40	0.86	
Vegetable fiber, 0.5" thick	18		1.32	Perlite	40	1.08	
	25		1.14	Sand and gravel or stone	140	0.08	
Wood, hardwoods	48	0.91		Concrete block, lightweight			
Softwoods	32	1.25		3 cell units, 4"-15 lbs. ea.			1.68
Flooring Carpet with fibrous pad			2.08	6"-23 lbs. ea.			1.83
With rubber pad			1.23	8"-28 lbs. ea.			2.12
Cork tile, 1/8" thick			0.28	12"-40 lbs. ea.			2.62
Terrazzo			0.08	Sand and gravel aggregates,			
Tile, resilient			0.05	4"-20 lbs. ea.			1.17
Wood, hardwood, 0.75" thick			0.68	6"-33 lbs. ea.			1.29
Subfloor, 0.75" thick			0.94	8"-38 lbs. ea.			1.46
				12"-56 lbs. ea.			1.81
Glass				Plastering Cement Plaster,			
Insulation, 0.50" air space			2.04	Sand aggregate	116	0.20	
Single glass			0.91	Gypsum plaster, Perlite aggregate	45	0.67	
				Sand aggregate	105	0.18	
Insulation Blanket or Batt, mineral, glass				Vermiculite aggregate	45	0.59	
or rock fiber, approximate thickness				Roofing			
3.0" to 3.5" thick			11	Asphalt, felt, 15 lb.			0.06
3.5" to 4.0" thick			13	Rolled roofing	70		0.15
6.0" to 6.5" thick			19	Shingles	70		0.44
6.5" to 7.0" thick			22	Built-up roofing .375" thick	70		0.33
8.5" to 9.0" thick			30	Cement shingles	120		0.21
Boards				Vapor-permeable felt			0.06
Cellular glass	8.5	2.63		Vapor seal, 2 layers of			
Fiberboard, wet felted				mopped 15 lb. felt			0.12
Acoustical tile	21	2.70		Wood, shingles 16"-7.5" exposure			0.87
Roof insulation	17	2.94		Siding			
Fiberboard, wet molded				Aluminum or steel (hollow backed)			
Acoustical tile	23	2.38		oversheathing			0.61
Mineral fiber with resin binder	15	3.45		With .375" insulating backer board			1.82
Polystyrene, extruded,				Foil backed			2.96
cut cell surface	1.8	4.00		Wood siding, beveled, ½" x 8"			0.81
smooth skin surface	2.2	5.00					
	3.5	5.26					
Bead boards	1.0	3.57					
Polyurethane	1.5	6.25					
Wood or cane fiberboard, 0.5" thick			1.25				

For customer support on your Facilities Maintenance & Repair Costs with RSMeans data, call 800.448.8182.

Table L1030-501   Weather Data and Design Conditions

City	Latitude (1) °	Latitude (1) 1'	Winter Temperatures (1) Med. of Annual Extremes	Winter Temperatures (1) 99%	Winter Temperatures (1) 97½%	Winter Degree Days (2)	Summer (Design Dry Bulb) Temperatures and Relative Humidity 1%	Summer (Design Dry Bulb) Temperatures and Relative Humidity 2½%	Summer (Design Dry Bulb) Temperatures and Relative Humidity 5%
UNITED STATES									
Albuquerque, NM	35	0	5.1	12	16	4,400	96/61	94/61	92/61
Atlanta, GA	33	4	11.9	17	22	3,000	94/74	92/74	90/73
Baltimore, MD	39	2	7	14	17	4,600	94/75	91/75	89/74
Birmingham, AL	33	3	13	17	21	2,600	96/74	94/75	92/74
Bismarck, ND	46	5	-32	-23	-19	8,800	95/68	91/68	88/67
Boise, ID	43	3	1	3	10	5,800	96/65	94/64	91/64
Boston, MA	42	2	-1	6	9	5,600	91/73	88/71	85/70
Burlington, VT	44	3	-17	-12	-7	8,200	88/72	85/70	82/69
Charleston, WV	38	2	3	7	11	4,400	92/74	90/73	87/72
Charlotte, NC	35	1	13	18	22	3,200	95/74	93/74	91/74
Casper, WY	42	5	-21	-11	-5	7,400	92/58	90/57	87/57
Chicago, IL	41	5	-8	-3	2	6,600	94/75	91/74	88/73
Cincinnati, OH	39	1	0	1	6	4,400	92/73	90/72	88/72
Cleveland, OH	41	2	-3	1	5	6,400	91/73	88/72	86/71
Columbia, SC	34	0	16	20	24	2,400	97/76	95/75	93/75
Dallas, TX	32	5	14	18	22	2,400	102/75	100/75	97/75
Denver, CO	39	5	-10	-5	1	6,200	93/59	91/59	89/59
Des Moines, IA	41	3	-14	-10	-5	6,600	94/75	91/74	88/73
Detroit, MI	42	2	-3	3	6	6,200	91/73	88/72	86/71
Great Falls, MT	47	3	-25	-21	-15	7,800	91/60	88/60	85/59
Hartford, CT	41	5	-4	3	7	6,200	91/74	88/73	85/72
Houston, TX	29	5	24	28	33	1,400	97/77	95/77	93/77
Indianapolis, IN	39	4	-7	-2	2	5,600	92/74	90/74	87/73
Jackson, MS	32	2	16	21	25	2,200	97/76	95/76	93/76
Kansas City, MO	39	1	-4	2	6	4,800	99/75	96/74	93/74
Las Vegas, NV	36	1	18	25	28	2,800	108/66	106/65	104/65
Lexington, KY	38	0	-1	3	8	4,600	93/73	91/73	88/72
Little Rock, AR	34	4	11	15	20	3,200	99/76	96/77	94/77
Los Angeles, CA	34	0	36	41	43	2,000	93/70	89/70	86/69
Memphis, TN	35	0	10	13	18	3,200	98/77	95/76	93/76
Miami, FL	25	5	39	44	47	200	91/77	90/77	89/77
Milwaukee, WI	43	0	-11	-8	-4	7,600	90/74	87/73	84/71
Minneapolis, MN	44	5	-22	-16	-12	8,400	92/75	89/73	86/71
New Orleans, LA	30	0	28	29	33	1,400	93/78	92/77	90/77
New York, NY	40	5	6	11	15	5,000	92/74	89/73	87/72
Norfolk, VA	36	5	15	20	22	3,400	93/77	91/76	89/76
Oklahoma City, OK	35	2	4	9	13	3,200	100/74	97/74	95/73
Omaha, NE	41	2	-13	-8	-3	6,600	94/76	91/75	88/74
Philadelphia, PA	39	5	6	10	14	4,400	93/75	90/74	87/72
Phoenix, AZ	33	3	27	31	34	1,800	109/71	107/71	105/71
Pittsburgh, PA	40	3	-1	3	7	6,000	91/72	88/71	86/70
Portland, ME	43	4	-10	-6	-1	7,600	87/72	84/71	81/69
Portland, OR	45	4	18	17	23	4,600	89/68	85/67	81/65
Portsmouth, NH	43	1	-8	-2	2	7,200	89/73	85/71	83/70
Providence, RI	41	4	-1	5	9	6,000	89/73	86/72	83/70
Rochester, NY	43	1	-5	1	5	6,800	91/73	88/71	85/70
Salt Lake City, UT	40	5	0	3	8	6,000	97/62	95/62	92/61
San Francisco, CA	37	5	36	38	40	3,000	74/63	71/62	69/61
Seattle, WA	47	4	22	22	27	5,200	85/68	82/66	78/65
Sioux Falls, SD	43	4	-21	-15	-11	7,800	94/73	91/72	88/71
St. Louis, MO	38	4	-3	3	8	5,000	98/75	94/75	91/75
Tampa, FL	28	0	32	36	40	680	92/77	91/77	90/76
Trenton, NJ	40	1	4	11	14	5,000	91/75	88/74	85/73
Washington, DC	38	5	7	14	17	4,200	93/75	91/74	89/74
Wichita, KS	37	4	-3	3	7	4,600	101/72	98/73	96/73
Wilmington, DE	39	4	5	10	14	5,000	92/74	89/74	87/73
ALASKA									
Anchorage	61	1	-29	-23	-18	10,800	71/59	68/58	66/56
Fairbanks	64	5	-59	-51	-47	14,280	82/62	78/60	75/59
CANADA									
Edmonton, Alta.	53	3	-30	-29	-25	11,000	85/66	82/65	79/63
Halifax, N.S.	44	4	-4	1	5	8,000	79/66	76/65	74/64
Montreal, Que.	45	3	-20	-16	-10	9,000	88/73	85/72	83/71
Saskatoon, Sask.	52	1	-35	-35	-31	11,000	89/68	86/66	83/65
St. John, Nwf.	47	4	1	3	7	8,600	77/66	75/65	73/64
Saint John, N.B.	45	2	-15	-12	-8	8,200	80/67	77/65	75/64
Toronto, Ont.	43	4	-10	-5	-1	7,000	90/73	87/72	85/71
Vancouver, B.C.	49	1	13	15	19	6,000	79/67	77/66	74/65
Winnipeg, Man.	49	5	-31	-30	-27	10,800	89/73	86/71	84/70

(1) Handbook of Fundamentals, ASHRAE, Inc., NY 1989
(2) Local Climatological Annual Survey, USDC Env. Science Services Administration, Asheville, NC

For customer support on your Facilities Maintenance & Repair Costs with RSMeans data, call 800.448.8182.

Table L1030-502   Maximum Depth of Frost Penetration in Inches

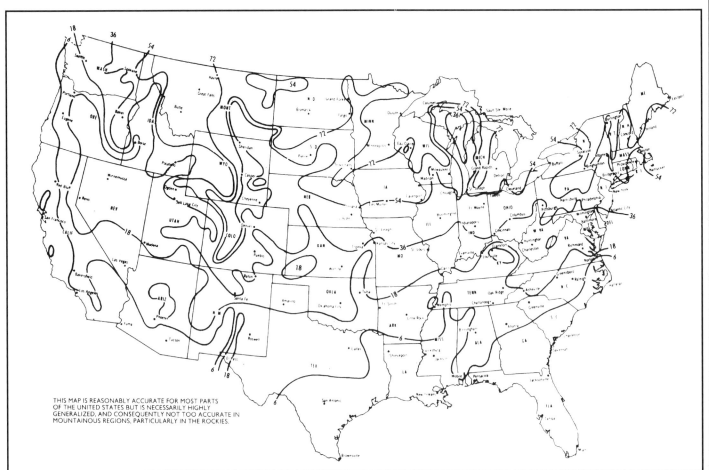

THIS MAP IS REASONABLY ACCURATE FOR MOST PARTS
OF THE UNITED STATES BUT IS NECESSARILY HIGHLY
GENERALIZED, AND CONSEQUENTLY NOT TOO ACCURATE IN
MOUNTAINOUS REGIONS, PARTICULARLY IN THE ROCKIES.

For customer support on your Facilities Maintenance & Repair Costs with RSMeans data, call 800.448.8182.

Table L1040-101   Fire-Resisting Ratings of Structural Elements (in hours)

Description of the Structural Element	No. 1 Fireproof		No. 2 Non Combustible			No. 3 Exterior Masonry Wall			No. 4 Frame	
			Protected		Unprotected	Heavy Timber	Ordinary		Protected	Unprotected
							Protected	Unprotected		
	1A	1B	2A	2B	2C	3A	3B	3C	4A	4B
Exterior, Bearing Walls	4	3	2	1½	1	2	2	2	1	1
Nonbearing Walls	2	2	1½	1	1	2	2	2	1	1
Interior Bearing Walls and Partitions	4	3	2	1	0	2	1	0	1	0
Fire Walls and Party Walls	4	3	2	2	2	2	2	2	2	2
Fire Enclosure of Exitways, Exit Hallways and Stairways	2	2	2	2	2	2	2	2	1	1
Shafts other than Exitways, Hallways and Stairways	2	2	2	2	2	2	2	2	1	1
Exitway access corridors and Vertical separation of tenant space	1	1	1	1	0	1	1	0	1	0
Columns, girders, trusses (other than roof trusses) and framing:										
Supporting more than one floor	4	3	2	1	0	—	1	0	1	0
Supporting one floor only	3	2	1½	1	0	—	1	0	1	0
Structural members supporting wall	3	2	1½	1	0	1	1	0	1	0
Floor construction including beams	3	2	1½	1	0	—	1	0	1	0
Roof construction including beams, trusses and framing arches and roof deck 15' or less in height to lowest member	2	1½	1	1	0	—	1	0	1	0

Note: a. Codes include special requirements and exceptions that are not included in the table above.
b. Each type of construction has been divided into sub-types which vary according to the degree of fire resistance required. Sub-types (A) requirements are more severe than those for sub-types (B).
c. Protected construction means all structural members are chemically treated, covered or protected so that the unit has the required fire resistance.

Type No. 1, Fireproof Construction — Buildings and structures of fireproof construction are those in which the walls, partitions, structural elements, floors, ceilings, roofs, and the exitways are protected with approved noncombustible materials to afford the fire-resistance rating specified in Table L1040-101; except as otherwise specifically regulated. Fire-resistant treated wood may be used as specified.

Type No. 2, Noncombustible Construction — Buildings and structures of noncombustible construction are those in which the walls, partitions, structural elements, floors, ceilings, roofs and the exitways are approved noncombustible materials meeting the fire-resistance rating requirements specified in Table L1040-101; except as modified by the fire limit restrictions. Fire-retardant treated wood may be used as specified.

Type No. 3, Exterior Masonry Wall Construction — Buildings and structures of exterior masonry wall construction are those in which the exterior, fire and party walls are masonry or other approved noncombustible materials of the required fire-resistance rating and structural properties. The floors, roofs, and interior framing are wholly or partly wood or metal or other approved construction. The fire and party walls are ground-supported; except that girders and their supports, carrying walls of masonry shall be protected to afford the same degree of fire-resistance rating of the supported walls. All structural elements have the required fire-resistance rating specified in Table L1040-101.

Type No. 4, Frame Construction — Buildings and structures of frame construction are those in which the exterior walls, bearing walls, partitions, floor and roof construction are wholly or partly of wood stud and joist assemblies with a minimum nominal dimension of two inches or of other approved combustible materials. Fire stops are required at all vertical and horizontal draft openings in which the structural elements have required fire-resistance ratings specified in Table L1040-101.

For customer support on your Facilities Maintenance & Repair Costs with RSMeans data, call 800.448.8182.

Table L1040-201 Fire Resistance Ratings

Fire Hazard for Fire Walls

The degree of fire hazard of buildings relating to their intended use is defined by "Fire Rating" the occupancy type. Such a rating system is listed in Table L1040-201 below. This type of rating determines the requirements for fire walls and fire separation walls (exterior fire exposure). For mixed use occupancy, use the higher Fire Rating requirement of the components.

Group	Fire-Resistance Rating (hours)
A, B, E, H-4[1], I, R-1, R-2, U	3
F-1[2], H-3[1], H-5[1], M, S-1	3
H-1[1], H-2[1]	4
F-2[3], S-2, R-3, R-4	2

Note: *The difference in "Fire Hazards" is determined by their occupancy and use.

1. High Hazard: Industrial and storage buildings in which the combustible contents might cause fires to be unusually intense or where explosives, combustible gases or flammable liquids are manufactured or stored.

2. Moderate Hazard: Mercantile buildings, industrial and storage buildings in which combustible contents might cause fires of moderate intensity.

3. Low Hazard: Business buildings that ordinarily do not burn rapidly.

Note:

In Type II or V construction, walls shall be permitted to have a 2-hour fire-resistance rating.

Table 706.4

Excerpted from the 2012 *International Building Code,* Copyright 2011. Washington, D.C.: International Code Council. Reproduced with permission. All rights reserved. www.ICCSAFE.org

Table L1040-202 Interior Finish Classification

Flame Spread for Interior Finishes

The flame spreadability of a material is the burning characteristic of the material relative to the fuel contributed by its combustion and the density of smoke developed. The flame spread classification of a material is based on a ten minute test on a scale of 0 to 100. Cement asbestos board is assigned a rating of 0 and select red oak flooring a rating of 100.

The three classes are listed in Table L1040-202.

The flame spread ratings for interior finish walls and ceilings shall not be greater than the Class listed in Table L1040-203.

Finish Class	Flame Spread Index	Smoke Developed Index
A	0-25	0-450
B	26-75	0-450
C	76-200	0-450

Section 803.1.1 Interior wall and ceiling finish materials

Excerpted from the 2012 *International Building Code,* Copyright 2011. Washington, D.C.: International Code Council. Reproduced with permission. All rights reserved. www.ICCSAFE.org

Table L1040-203 Interior Finish Requirements by Class

Group	Sprinklered			Nonsprinklered		
	Interior exit stairways, interior exit ramps and exit passageways Note a, b	Corridors and enclosure for exit access stairways and exit access ramps	Rooms and enclosed spaces Note c	Interior exit stairways, interior exit ramps and exit passageways Note a, b	Corridors and enclosure for exit access stairways and exit access ramps	Rooms and enclosed spaces Note c
A-1 & A-2	B	B	C	A	A (d)	B (e)
A-3 (f), A-4, A-5	B	B	C	A	A (d)	C
B, E, M, R-1	B	C	C	A	B	C
R-4	B	C	C	A	B	B
F	C	C	C	B	C	C
H	B	B	C	A	A	B
I-1	B	C	C	A	B	B
I-2	B	B	B (h, i)	A	A	B
I-3	A	A (j)	C	A	A	B
I-4	B	B	B (h, i)	A	A	B
R-2	C	C	C	B	B	C
R-3	C	C	C	C	C	C
S	C	C	C	B	B	C
U	No Restrictions			No Restrictions		

Notes:

a. Class C interior finish materials shall be permitted for wainscotting or paneling of not more than 1,000 square feet of applied surface area in the grade lobby where applied directly to a noncombustible base or over furring strips applied to a noncombustible base and fireblocked as required by IBC Section 803.11.1.

b. In other Group I-2 occupancies in buildings less than three stories above grade plane of other than Group I-3, Class B interior finish for nonsprinklered buildings and Class C interior finish for sprinklered buildings shall be permitted in interior exit stairways and ramps.

c. Requirements for rooms and enclosed spaces shall be based upon spaces enclosed by partitions. Where a fire-resistance rating is required for structural elements, the enclosing partitions shall extend from the floor to the ceiling. Partitions that do not comply with this shall be considered enclosed spaces and the rooms or spaces on both sides shall be considered one. In determining the applicable requirements for rooms and enclosed spaces, the specific occupancy thereof shall be the governing factor regardless of the group classification of the building or structure.

d. Lobby areas in Group A-1, A-2, and A-3 occupancies shall not be less than Class B materials.

e. Class C interior finish materials shall be permitted in places of assembly with an occupant load of 300 persons or less.

f. For places of religious worship, wood used for ornamental purposes, trusses, paneling or chancel furnishing shall be permitted.

g. Class B material is required where the building exceeds two stories.

h. Class C interior finish materials shall be permitted in administrative spaces.

i. Class C interior finish materials shall be permitted in rooms with a capacity of four persons or less.

j. Class B materials shall be permitted as wainscotting extending not more than 48 inches above the finished floor in corridors and exit access stairways and ramps.

Section 803.9

Excerpted from the 2012 *International Building Code*, Copyright 2011. Washington, D.C.: International Code Council. Reproduced with permission. All rights reserved. www.ICCSAFE.org

For customer support on your Facilities Maintenance & Repair Costs with RSMeans data, call 800.448.8182.

Description: This table is primarily for converting customary U.S. units in the left hand column to SI metric units in the right hand column.

In addition, conversion factors for some commonly encountered Canadian and non-SI metric units are included.

Table L1090-101 Metric Conversion Factors

If You Know		Multiply By		To Find
Length	Inches	x 25.4[a]	=	Millimeters
	Feet	x 0.3048[a]	=	Meters
	Yards	x 0.9144[a]	=	Meters
	Miles (statute)	x 1.609	=	Kilometers
Area	Square inches	x 645.2	=	Square millimeters
	Square feet	x 0.0929	=	Square meters
	Square yards	x 0.8361	=	Square meters
Volume (Capacity)	Cubic inches	x 16,387	=	Cubic millimeters
	Cubic feet	x 0.02832	=	Cubic meters
	Cubic yards	x 0.7646	=	Cubic meters
	Gallons (U.S. liquids)[b]	x 0.003785	=	Cubic meters[c]
	Gallons (Canadian liquid)[b]	x 0.004546	=	Cubic meters[c]
	Ounces (U.S. liquid)[b]	x 29.57	=	Milliliters[c, d]
	Quarts (U.S. liquid)[b]	x 0.9464	=	Liters[c, d]
	Gallons (U.S. liquid)[b]	x 3.785	=	Liters[c, d]
Force	Kilograms force[d]	x 9.807	=	Newtons
	Pounds force	x 4.448	=	Newtons
	Pounds force	x 0.4536	=	Kilograms force[d]
	Kips	x 4448	=	Newtons
	Kips	x 453.6	=	Kilograms force[d]
Pressure, Stress, Strength (Force per unit area)	Kilograms force per square centimeter[d]	x 0.09807	=	Megapascals
	Pounds force per square inch (psi)	x 0.006895	=	Megapascals
	Kips per square inch	x 6.895	=	Megapascals
	Pounds force per square inch (psi)	x 0.07031	=	Kilograms force per square centimeter[d]
	Pounds force per square foot	x 47.88	=	Pascals
	Pounds force per square foot	x 4.882	=	Kilograms force per square meter[d]
Flow	Cubic feet per minute	x 0.4719	=	Liters per second
	Gallons per minute	x 0.0631	=	Liters per second
	Gallons per hour	x 1.05	=	Milliliters per second
Bending Moment Or Torque	Inch-pounds force	x 0.01152	=	Meter-kilograms force[d]
	Inch-pounds force	x 0.1130	=	Newton-meters
	Foot-pounds force	x 0.1383	=	Meter-kilograms force[d]
	Foot-pounds force	x 1.356	=	Newton-meters
	Meter-kilograms force[d]	x 9.807	=	Newton-meters
Mass	Ounces (avoirdupois)	x 28.35	=	Grams
	Pounds (avoirdupois)	x 0.4536	=	Kilograms
	Tons (metric)	x 1000	=	Kilograms
	Tons, short (2000 pounds)	x 907.2	=	Kilograms
	Tons, short (2000 pounds)	x 0.9072	=	Megagrams[e]
Mass per Unit Volume	Pounds mass per cubic foot	x 16.02	=	Kilograms per cubic meter
	Pounds mass per cubic yard	x 0.5933	=	Kilograms per cubic meter
	Pounds mass per gallon (U.S. liquid)[b]	x 119.8	=	Kilograms per cubic meter
	Pounds mass per gallon (Canadian liquid)[b]	x 99.78	=	Kilograms per cubic meter
Temperature	Degrees Fahrenheit	(F-32)/1.8	=	Degrees Celsius
	Degrees Fahrenheit	(F+459.67)/1.8	=	Degrees Kelvin
	Degrees Celsius	C+273.15	=	Degrees Kelvin

[a]The factor given is exact
[b]One U.S. gallon = 0.8327 Canadian gallon
[c]1 liter = 1000 milliliters = 1000 cubic centimeters
  1 cubic decimeter = 0.001 cubic meter

[d]Metric but not SI unit
[e]Called "tonne" in England and
  "metric ton" in other metric countries

For customer support on your Facilities Maintenance & Repair Costs with RSMeans data, call 800.448.8182.

Life Cycle Costing (LCC) for Evaluating Economic Performance of Building Investments

Introduction

A complete analysis of maintenance costs requires consideration of life cycle costs, or LCC, as it is often called. Under current conditions, facilities managers have to wisely spend their limited operating and maintenance budgets. When deciding how to allocate limited budgets, maintenance professionals are constantly faced with the economic decision of when to stop maintaining and start replacing. Once the decision to replace has been made, LCC analysis can enhance the choice of alternatives as well.

The American Society of Testing and Materials (ASTM) Subcommittee E06.81 on Building Economics has developed a series of standards for evaluating the economic performance of building investments. These standards, which have been endorsed by the American Association of Cost Engineers (AACE), are intended to provide uniformity in both the terminology and calculation methods utilized for Life Cycle Costing. Without these standards as a basis for calculations, life cycle cost comparisons could be misleading or incomplete. This chapter is based on the current ASTM standards which are listed in the references at the end of this section.

Definition of Life Cycle Costing

(ASTM Standard Terminology of Building Economics E833-89)

A technique of economic evaluation that sums over a given study period the cost of initial investment less resale value, replacements, operations, energy use, and maintenance and repair of an investment decision. The costs are expressed in either lump sum present value terms or in equivalent uniform annual values.

This technique for life cycle costing is important to the facilities manager, as it is a valuable tool for evaluating investment alternatives and supports his or her decision-making.

To understand Life Cycle Costing, one should be familiar with the notions of compounding, discounting, present value, and equivalent uniform annual value.

Compounding

Definition of Compounding *(Webster's New Collegiate Dictionary)*

The process of computing the value of an original principal sum based on interest calculated on the sum of the original principal and accrued interest.

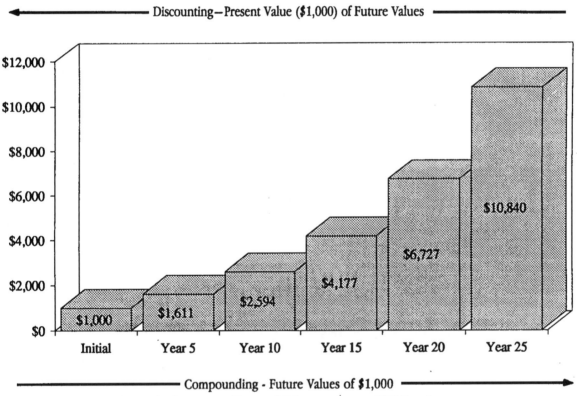

Discounting—Present Value ($1,000) of Future Values

$12,000

$10,000

$8,000

$6,000

$4,000

$2,000

$0

Initial $1,000 · Year 5 $1,611 · Year 10 $2,594 · Year 15 $4,177 · Year 20 $6,727 · Year 25 $10,840

Compounding - Future Values of $1,000

Figure 1: Compounding and Discounting at 10% Per Annum

For customer support on your Facilities Maintenance & Repair Costs with RSMeans data, call 800.448.8182.

Figure 1 illustrates compounding. As shown, a principal sum of $1,000 at a compound interest rate of 10% per annum increases to $1,611 in 5 years, $2,594 in 10 years, $4,177 in 15 years, $6,727 in 20 years, and $10,840 in 25 years; this process (compounding) is usually well understood in that it represents the amount accumulated in a bank account for a specific amount at a given interest rate and calculated on the sum of the original principal and accrued interest.

Discounting

ASTM Definition of Discounting

A technique for converting cash flows that occur over time to equivalent amounts at a common time (base time).

It is also evident from Figure 1 that if one had to pay out or meet an obligation of either $1,611 in 5 years, $2,594 in 10 years, $4,177 in 15 years, $6,727 in 20 years, or $10,840 in 25 years, the amount to be invested initially at 10% interest would be $1,000. This amount is referred to as the present value (PV) of a future amount and it is calculated by "discounting" a future value (F) in a specific year at a given rate of interest.

Present Value

ASTM Definition of Present Value

The value of a benefit or cost found by discounting future cash flows to the base time. (Syn. present worth)

The basic mathematics for calculating present values are relatively simple:

Given:

P = Present sum of money
F = Future sum of money
i = Interest or discount rate (expressed as a decimal and not a percentage)
n = Number of years

The future value (F) of a present sum of money (P) may be computed as shown in Table 1, a relatively simple process.

Year	Future Value (F)
1	$P(1+i)$
2	$P(1+i)(1+i) = P(1+i)^2$
3	$P(1+i)^2(1+i) = P(1+i)^3$
n	$P(1+i)^n$

Table 1: The Future Value of a Present Sum of Money

Since $F = P(1+i)^n$

Therefore $P = \dfrac{F}{(1+1)^n} = F \times \dfrac{1}{(1+i)^n}$

Example

What is the present value (P) of an anticipated maintenance expense of $300 in year 3 (F) if the interest rate is 10%?

$$P = F \times \frac{1}{(1+i)^n} = 300 \times \frac{1}{(1+0.10)^n} = 300 \times \frac{1}{1.33}$$

$$= 300 \times 0.75$$
$$= \$225$$

From an interest standpoint, this tells us that $225 is the amount that would have to be deposited today into an account paying 10% interest per annum in order to provide $300 at the end of year 3 to meet the anticipated expense. From the economist's point of view, discounting using the investor's minimum acceptable rate of return (10% in this case) makes the economic outlook the same, whether it is the future amount of $300 or a present value of $225.

Note that $(1+i)^n$ is commonly referred to as the *Single Compound Amount (SCA)* factor, and $1/(1+i)^n$ as the *Single Present Value Factor (SPV)*. Both factors may be readily found in standard discount factor tables for a wide range of interest rates and time periods, thus simplifying calculations. The use of these factors in calculating life cycle costs by the formula method is discussed later in this section.

A life cycle cost analysis can be performed in *constant dollars* (dollars of constant purchasing power, net of general inflation) or *current dollars* (actual prices inclusive of general inflation). If you use constant dollars, you must use a "real" discount rate (i.e., a rate which reflects the time value of money, net of general inflation) and differential escalation rates (also net of general inflation) for future cost items. Constant dollars are generally tied to a specific year. If you use current dollars, the discount rates and escalation rate must include general inflation (i.e., they must be "nominal" or "market" rates). In this chapter, all costs are expressed in current dollars and rates in nominal terms. When taxes are included in the analysis, the current dollar approach is generally preferred.

The Cash Flow Method for Calculating Life Cycle Costs

There are basically two approaches to calculating life cycle costs—the *cash flow* method and the *formula* method which, when applicable, is somewhat more simple.

To illustrate the cash flow method, a relatively simple example follows.

Example

A facility manager is considering the purchase of maintenance equipment required for a four-year period for which initial costs, energy costs, and maintenance costs of alternative proposals vary.

The financial criteria on which the economic evaluation will be based are the following:

Interest/Discount Rate	– 10%
Energy Escalation Rate	– 8% per annum
Labor and Material Escalation Rate	– 4% per annum
Period of Study	– 4 years

The following data is provided in one of the alternative proposals to be analyzed:

- Initial Capital Cost – $1,000
- Maintenance Costs – A fixed annual cost of $100 per year quoted by the supplier (no escalation to be considered)
- Annual Energy Costs – Initially $100 per year and subject to 8% annual escalation – i.e., $108 for year one.
- Salvage Value – At the end of the four-year period, the equipment has no further useful life and the supplier agrees to purchase it as scrap for $50.

What is the present value of this alternative proposal?

The steps to resolve this problem are the following:

Step 1 – Prepare a Cash Flow Diagram.

Step 2 – Establish a Time Schedule of Costs.

Step 3 – Calculate Annual Net Cash Flows.

Step 4 – Calculate Present Value Factors.

Step 5 – Calculate Present Values of Annual Cash Flows and Life Cycle Costs.

Step 1: Prepare a Cash Flow Diagram

To better understand the problem, cash flow diagrams are often prepared. Revenues are noted as vertical lines above a horizontal time axis, and disbursements are noted as vertical lines below it.

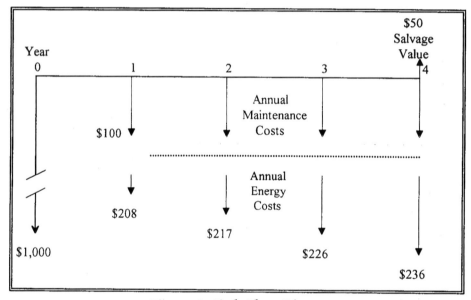

Figure 2: Cash Flow Diagram

For customer support on your Facilities Maintenance & Repair Costs with RSMeans data, call 800.448.8182.

Step 2: Establish a Time Schedule of Costs

Table 2 below presents revenues and disbursements for each year of the study:

A Year	B Capital Costs	C Maintenance Costs	D Energy Cost (8% escalation)
0	1,000	–	–
1	–	100	108
2	–	100	117
3	–	100	126
4	(50)	100	136

Table 2

Note that the disbursements in this example are considered positive, and revenues for the salvage value are negative.

Step 3: Calculate Annual Net Cash Flows

Table 3 below provides the net cash flow (NCF) for each year, which could be a disbursement or a revenue.

A Year	B Capital Costs	C Maintenance Costs	D Energy Cost (8% escalation)	E NCF (B + C + D)
0	1,000	–	–	1,000
1	–	100	108	208
2	–	100	117	217
3	–	100	126	226
4	(50)	100	136	186

Table 3

Step 4: Calculate Present Value Factors

As in a previous example, present value factors based on 10% interest will be calculated for each year to convert the annual net cash flows to present values.

Year n	PV Factor	$\dfrac{1}{(1 + i)^n}$
0		1
1		$\dfrac{1}{(1+ 0.1)^1}$ = 0.91
2		$\dfrac{1}{(1+ 0.1)^2}$ = 0.83
3		$\dfrac{1}{(1+ 0.1)^3}$ = 0.75
4		$\dfrac{1}{(1+ 0.1)^4}$ = 0.68

Table 4

Step 5: Calculate Present Values of Annual Cash Flows and Life Cycle Costs

Annual net cash flows are converted to present values of the base year. Their sum (Total Present Value) will represent the life cycle costs of the alternative.

Year n	NCF	PV Factor	PV $
0	1,000	1	1,000
1	208	0.91	189
2	217	0.83	180
3	226	0.75	170
4	186	0.68	126
PV Life Cycle Cost	–	–	1,665

Table 5

The total life cycle cost of this alternative in terms of today's dollar (if this year is the base year) is $1,665. This means that if a present sum of $1,665 were deposited today at an interest rate of 10%, all expenses could be paid over a four-year period, at which time the bank balance would be zero.

Once the costs of the other proposals have been calculated in present value dollars, the facilities manager will be in a position to quantify the differences in life cycle costs and decide which proposal represents the lowest cost of ownership in the long run.

The approach outlined is a practical one when considering the purchase of equipment such as elevators, controls, etc.; quotations should not only include capital costs, but also the costs of long-term maintenance contracts which could make a higher initial cost proposal lower over the life cycle of the equipment.

The Formula Method for Calculating Life Cycle Costs

It is also possible to solve the previous problem utilizing the formula method if the problem is adaptable to this technique. Business calculators or discount factor tables such as those published by ASTM can be employed to generate the present value factors, and results may be obtained more quickly.

In some cases, the complexity of the calculations (such as when taxes on net revenues are considered) will necessitate the "cash flow" method, with calculations done manually or using a spreadsheet program.

With the formula method approach, the calculations for the life cycle cost of the maintenance equipment alternative analyzed would be as follows:

1.	PV of Initial Investment	$1,000
2.	PV of Maintenance Costs	
	Annual Cost (A) x UPV factor	
	100 x 3.17	$ 317
3.	PV of Energy Costs	
	Base Year Cost (A) x UPV*	$ 382
	100 x 3.82	
4.	Salvage (Scrap) Value	
	Future Value (F) x SPV	
	(50) x 0.68	(34)
5.	Life Cycle Cost / Total Present Value	$1,665

Note: The above factors can be found in Discount Factor Tables.

UPV–Uniform Present Value
SPV–Single Present Value
UPV*–Uniform Present Value Modified (*)

The calculated life cycle cost of $1,665 is identical to that obtained with the cash flow method, but is arrived at with less effort.

Equivalent Uniform Annual Value

ASTM Definition of Equivalent Uniform Annual Value

A uniform annual amount equivalent to the project costs or benefits, taking into account the time value of money throughout the study period. (Syn. annual value)

As noted previously in the definition of Life Cycle Costs, these may be expressed as a lump sum present value with a base time reference, or in equivalent uniform annual values (EUAV) over the study period; decision-makers may prefer one or the other, or require both for the analysis of investment alternatives.

To express life cycle costs in annual values, it is necessary to first determine the lump sum present value of an alternative and multiply it by the Uniform Capital Recovery (UCR) factor given the interest rate and time period. This process is identical to calculating annual payments for a mortgage.

Example

The present value life cycle cost for one of the maintenance equipment alternatives analyzed amounted to $1,665; what is the equivalent uniform annual value "A"?

Given:	P	(a principal sum	
		or present value	
		life cycle cost)	= $1,665
	i		= 10%
	n		= 4 years
	UCR Factor		= 0.32 (from standard factor
			tables for i)
			= 10%, n = 4 years
	A		= equivalent uniform annual
			value – to be determined
Solution:	A		= P x UCR
			= $1,665 x 0.32
			= $530

The life cycle cost of the alternative may be expressed as a lump sum, $1,665, or an annual value of $530 based on a four-year time period.

Other Methods of Evaluating Economic Performance

While calculating life cycle costs is a helpful tool for the facilities manager evaluating alternatives, there are other approaches for evaluating economic performance of building investments. Among these other methods are *payback, net benefits,* and *internal rate of return;* these are described in detail in an ASTM compilation of *Building Economics Standards* which are endorsed by the American Association of Cost Engineers (AACE). These standards, as well as adjuncts available, such as discount factor tables and a life cycle cost analysis computer program, are included in the References at the end of this section.

ASTM has also developed a Standard Classification of Building Elements and Related Site Work entitled UNIFORMAT II (Designation E-1557-93). One of the many applications for this classification for facilities managers is its use to structure operating and maintenance cost data by building elements or systems, thus facilitating life cycle costing studies. This classification can also be the basis of life cycle cost plans for new facilities that should be developed during the design period with the assistance of the facility manager.

References (Annual Book of ASTM Standards Vol. 04.07)

ASTM Standards

E 833	Terminology of Building Economics
E 917	Practice for Measuring Life Cycle Costs of Buildings and Building Systems
E 964	Practice for Measuring Benefit-to-Cost and Savings-to-Investment Ratios for Buildings and Building Systems
E 1057	Practice for Measuring Internal Rates of Return and Adjusted Internal Rates of Return for Investments in Buildings and Building Systems
E 1074	Practice for Measuring Net Benefits for Investments in Buildings and Building Systems
E 1121	Practice for Measuring Payback for Investments in Buildings and Building Systems
E 1185	Guide for Selecting Economic Methods for Evaluating Investments in Buildings and Building Systems
E 1369	Guide for Selecting Techniques for Treating Uncertainty and Risk in the Economic Evaluation of Buildings and Building Systems
E 1557	Standard Classification for Building Elements and Related Site Work–UNIFORMAT II

ASTM Adjuncts

Discount Factor Tables, Adjunct to Practices E 917, E 964, E 1057, and E 1074.

Computer Program and User's Guide to Building Maintenance, Repair, and Replacement Database for Life Cycle Cost Analysis, Adjunct to Practices E 917, E 964, E 1057, and E 1121.

Note: The Standards noted above, other than E 1557, are available in one publication entitled *ASTM Standards on Building Economics, Second Edition.*

Abbreviations

A	Area Square Feet; Ampere	Brk., brk	Brick	Csc	Cosecant
AAFES	Army and Air Force Exchange Service	brkt	Bracket	C.S.F.	Hundred Square Feet
		Brs.	Brass	CSI	Construction Specifications Institute
ABS	Acrylonitrile Butadiene Stryrene; Asbestos Bonded Steel	Brz.	Bronze		
		Bsn.	Basin	CT	Current Transformer
A.C., AC	Alternating Current;	Btr.	Better	CTS	Copper Tube Size
	Air-Conditioning;	BTU	British Thermal Unit	Cu	Copper, Cubic
	Asbestos Cement;	BTUH	BTU per Hour	Cu. Ft.	Cubic Foot
	Plywood Grade A & C	Bu.	Bushels	cw	Continuous Wave
ACI	American Concrete Institute	BUR	Built-up Roofing	C.W.	Cool White; Cold Water
ACR	Air Conditioning Refrigeration	BX	Interlocked Armored Cable	Cwt.	100 Pounds
ADA	Americans with Disabilities Act	°C	Degree Centigrade	C.W.X.	Cool White Deluxe
AD	Plywood, Grade A & D	c	Conductivity, Copper Sweat	C.Y.	Cubic Yard (27 cubic feet)
Addit.	Additional	C	Hundred; Centigrade	C.Y./Hr.	Cubic Yard per Hour
Adh.	Adhesive	C/C	Center to Center, Cedar on Cedar	Cyl.	Cylinder
Adj.	Adjustable	C-C	Center to Center	d	Penny (nail size)
af	Audio-frequency	Cab	Cabinet	D	Deep; Depth; Discharge
AFFF	Aqueous Film Forming Foam	Cair.	Air Tool Laborer	Dis., Disch.	Discharge
AFUE	Annual Fuel Utilization Efficiency	Cal.	Caliper	Db	Decibel
AGA	American Gas Association	Calc	Calculated	Dbl.	Double
Agg.	Aggregate	Cap.	Capacity	DC	Direct Current
A.H., Ah	Ampere Hours	Carp.	Carpenter	DDC	Direct Digital Control
A hr.	Ampere-hour	C.B.	Circuit Breaker	Demob.	Demobilization
A.H.U., AHU	Air Handling Unit	C.C.A.	Chromate Copper Arsenate	d.f.t.	Dry Film Thickness
A.I.A.	American Institute of Architects	C.C.F.	Hundred Cubic Feet	d.f.u.	Drainage Fixture Units
AIC	Ampere Interrupting Capacity	cd	Candela	D.H.	Double Hung
Allow.	Allowance	cd/sf	Candela per Square Foot	DHW	Domestic Hot Water
alt., alt	Alternate	CD	Grade of Plywood Face & Back	DI	Ductile Iron
Alum.	Aluminum	CDX	Plywood, Grade C & D, exterior glue	Diag.	Diagonal
a.m.	Ante Meridiem			Diam., Dia	Diameter
Amp.	Ampere	Cefi.	Cement Finisher	Distrib.	Distribution
Anod.	Anodized	Cem.	Cement	Div.	Division
ANSI	American National Standards Institute	CF	Hundred Feet	Dk.	Deck
		C.F.	Cubic Feet	D.L.	Dead Load; Diesel
APA	American Plywood Association	CFM	Cubic Feet per Minute	DLH	Deep Long Span Bar Joist
Approx.	Approximate	CFRP	Carbon Fiber Reinforced Plastic	dlx	Deluxe
Apt.	Apartment	c.g.	Center of Gravity	Do.	Ditto
Asb.	Asbestos	CHW	Chilled Water; Commercial Hot Water	DOP	Dioctyl Phthalate Penetration Test (Air Filters)
A.S.B.C.	American Standard Building Code				
Asbe.	Asbestos Worker	C.I., CI	Cast Iron	Dp., dp	Depth
ASCE	American Society of Civil Engineers	C.I.P., CIP	Cast in Place	D.P.S.T.	Double Pole, Single Throw
A.S.H.R.A.E.	American Society of Heating, Refrig. & AC Engineers	Circ.	Circuit	Dr.	Drive
		C.L.	Carload Lot	DR	Dimension Ratio
ASME	American Society of Mechanical Engineers	CL	Chain Link	Drink.	Drinking
		Clab.	Common Laborer	D.S.	Double Strength
ASTM	American Society for Testing and Materials	Clam	Common Maintenance Laborer	D.S.A.	Double Strength A Grade
		C.L.F.	Hundred Linear Feet	D.S.B.	Double Strength B Grade
Attchmt.	Attachment	CLF	Current Limiting Fuse	Dty.	Duty
Avg., Ave.	Average	CLP	Cross Linked Polyethylene	DWV	Drain Waste Vent
AWG	American Wire Gauge	cm	Centimeter	DX	Deluxe White, Direct Expansion
AWWA	American Water Works Assoc.	CMP	Corr. Metal Pipe	dyn	Dyne
Bbl.	Barrel	CMU	Concrete Masonry Unit	e	Eccentricity
B&B, BB	Grade B and Better; Balled & Burlapped	CN	Change Notice	E	Equipment Only; East; Emissivity
		Col.	Column	Ea.	Each
B&S	Bell and Spigot	CO₂	Carbon Dioxide	EB	Encased Burial
B.&W.	Black and White	Comb.	Combination	Econ.	Economy
b.c.c.	Body-centered Cubic	comm.	Commercial, Communication	E.C.Y	Embankment Cubic Yards
B.C.Y.	Bank Cubic Yards	Compr.	Compressor	EDP	Electronic Data Processing
BE	Bevel End	Conc.	Concrete	EIFS	Exterior Insulation Finish System
B.F.	Board Feet	Cont., cont	Continuous; Continued, Container	E.D.R.	Equiv. Direct Radiation
Bg. cem.	Bag of Cement	Corkbd.	Cork Board	Eq.	Equation
BHP	Boiler Horsepower; Brake Horsepower	Corr.	Corrugated	EL	Elevation
		Cos	Cosine	Elec.	Electrician; Electrical
B.I.	Black Iron	Cot	Cotangent	Elev.	Elevator; Elevating
bidir.	bidirectional	Cov.	Cover	EMT	Electrical Metallic Conduit; Thin Wall Conduit
Bit., Bitum.	Bituminous	C/P	Cedar on Paneling		
Bit., Conc.	Bituminous Concrete	CPA	Control Point Adjustment	Eng.	Engine, Engineered
Bk.	Backed	Cplg.	Coupling	EPDM	Ethylene Propylene Diene Monomer
Bkrs.	Breakers	CPM	Critical Path Method		
Bldg., bldg	Building	CPVC	Chlorinated Polyvinyl Chloride	EPS	Expanded Polystyrene
Blk.	Block	C.Pr.	Hundred Pair	Eqhv.	Equip. Oper., Heavy
Bm.	Beam	CRC	Cold Rolled Channel	Eqlt.	Equip. Oper., Light
Boil.	Boilermaker	Creos.	Creosote	Eqmd.	Equip. Oper., Medium
bpm	Blows per Minute	Crpt.	Carpet & Linoleum Layer	Eqmm.	Equip. Oper., Master Mechanic
BR	Bedroom	CRT	Cathode-ray Tube	Eqol.	Equip. Oper., Oilers
Brg., brng.	Bearing	CS	Carbon Steel, Constant Shear Bar Joist	Equip.	Equipment
Brhe.	Bricklayer Helper			ERW	Electric Resistance Welded
Bric.	Bricklayer				

For customer support on your Facilities Maintenance & Repair Costs with RSMeans data, call 800.448.8182.

E.S.	Energy Saver	
Est.	Estimated	
esu	Electrostatic Units	
E.W.	Each Way	
EWT	Entering Water Temperature	
Excav.	Excavation	
excl	Excluding	
Exp., exp	Expansion, Exposure	
Ext., ext	Exterior; Extension	
Extru.	Extrusion	
f.	Fiber Stress	
F	Fahrenheit; Female; Fill	
Fab., fab	Fabricated; Fabric	
FBGS	Fiberglass	
F.C.	Footcandles	
f.c.c.	Face-centered Cubic	
f'c.	Compressive Stress in Concrete; Extreme Compressive Stress	
F.E.	Front End	
FEP	Fluorinated Ethylene Propylene (Teflon)	
F.G.	Flat Grain	
F.H.A.	Federal Housing Administration	
Fig.	Figure	
Fin.	Finished	
FIPS	Female Iron Pipe Size	
Fixt.	Fixture	
FJP	Finger jointed and primed	
Fl. Oz.	Fluid Ounces	
Flr.	Floor	
Flrs.	Floors	
FM	Frequency Modulation; Factory Mutual	
Fmg.	Framing	
FM/UL	Factory Mutual/Underwriters Labs	
Fdn.	Foundation	
FNPT	Female National Pipe Thread	
Fori.	Foreman, Inside	
Foro.	Foreman, Outside	
Fount.	Fountain	
fpm	Feet per Minute	
FPT	Female Pipe Thread	
Fr	Frame	
F.R.	Fire Rating	
FRK	Foil Reinforced Kraft	
FSK	Foil/Scrim/Kraft	
FRP	Fiberglass Reinforced Plastic	
FS	Forged Steel	
FSC	Cast Body; Cast Switch Box	
Ft., ft	Foot; Feet	
Ftng.	Fitting	
Ftg.	Footing	
Ft lb.	Foot Pound	
Furn.	Furniture	
FVNR	Full Voltage Non-Reversing	
FVR	Full Voltage Reversing	
FXM	Female by Male	
Fy.	Minimum Yield Stress of Steel	
g	Gram	
G	Gauss	
Ga.	Gauge	
Gal., gal.	Gallon	
Galv., galv	Galvanized	
GC/MS	Gas Chromatograph/Mass Spectrometer	
Gen.	General	
GFI	Ground Fault Interrupter	
GFRC	Glass Fiber Reinforced Concrete	
Glaz.	Glazier	
GPD	Gallons per Day	
gpf	Gallon per Flush	
GPH	Gallons per Hour	
gpm, GPM	Gallons per Minute	
GR	Grade	
Gran.	Granular	
Grnd.	Ground	
GVW	Gross Vehicle Weight	
GWB	Gypsum Wall Board	
H	High Henry	
HC	High Capacity	
H.D., HD	Heavy Duty; High Density	
H.D.O.	High Density Overlaid	
HDPE	High Density Polyethylene Plastic	
Hdr.	Header	
Hdwe.	Hardware	
H.I.D., HID	High Intensity Discharge	
Help.	Helper Average	
HEPA	High Efficiency Particulate Air Filter	
Hg	Mercury	
HIC	High Interrupting Capacity	
HM	Hollow Metal	
HMWPE	High Molecular Weight Polyethylene	
HO	High Output	
Horiz.	Horizontal	
H.P., HP	Horsepower; High Pressure	
H.P.F.	High Power Factor	
Hr.	Hour	
Hrs./Day	Hours per Day	
HSC	High Short Circuit	
Ht.	Height	
Htg.	Heating	
Htrs.	Heaters	
HVAC	Heating, Ventilation & Air-Conditioning	
Hvy.	Heavy	
HW	Hot Water	
Hyd.; Hydr.	Hydraulic	
Hz	Hertz (cycles)	
I.	Moment of Inertia	
IBC	International Building Code	
I.C.	Interrupting Capacity	
ID	Inside Diameter	
I.D.	Inside Dimension; Identification	
I.F.	Inside Frosted	
I.M.C.	Intermediate Metal Conduit	
In.	Inch	
Incan.	Incandescent	
Incl.	Included; Including	
Int.	Interior	
Inst.	Installation	
Insul., insul	Insulation/Insulated	
I.P.	Iron Pipe	
I.P.S., IPS	Iron Pipe Size	
IPT	Iron Pipe Threaded	
I.W.	Indirect Waste	
J	Joule	
J.I.C.	Joint Industrial Council	
K	Thousand; Thousand Pounds; Heavy Wall Copper Tubing, Kelvin	
K.A.H.	Thousand Amp. Hours	
kcmil	Thousand Circular Mils	
KD	Knock Down	
K.D.A.T.	Kiln Dried After Treatment	
kg	Kilogram	
kG	Kilogauss	
kgf	Kilogram Force	
kHz	Kilohertz	
Kip	1000 Pounds	
KJ	Kilojoule	
K.L.	Effective Length Factor	
K.L.F.	Kips per Linear Foot	
Km	Kilometer	
KO	Knock Out	
K.S.F.	Kips per Square Foot	
K.S.I.	Kips per Square Inch	
kV	Kilovolt	
kVA	Kilovolt Ampere	
kVAR	Kilovar (Reactance)	
KW	Kilowatt	
KWh	Kilowatt-hour	
L	Labor Only; Length; Long; Medium Wall Copper Tubing	
Lab.	Labor	
lat	Latitude	
Lath.	Lather	
Lav.	Lavatory	
lb.; #	Pound	
L.B., LB	Load Bearing; L Conduit Body	
L. & E.	Labor & Equipment	
lb./hr.	Pounds per Hour	
lb./L.F.	Pounds per Linear Foot	
lbf/sq.in.	Pound-force per Square Inch	
L.C.L.	Less than Carload Lot	
L.C.Y.	Loose Cubic Yard	
Ld.	Load	
LE	Lead Equivalent	
LED	Light Emitting Diode	
L.F.	Linear Foot	
L.F. Hdr.	Linear Feet of Header	
L.F. Nose	Linear Foot of Stair Nosing	
L.F. Rsr	Linear Foot of Stair Riser	
Lg.	Long; Length; Large	
L & H	Light and Heat	
LH	Long Span Bar Joist	
L.H.	Labor Hours	
L.L., LL	Live Load	
L.L.D.	Lamp Lumen Depreciation	
lm	Lumen	
lm/sf	Lumen per Square Foot	
lm/W	Lumen per Watt	
LOA	Length Over All	
log	Logarithm	
L-O-L	Lateralolet	
long.	Longitude	
L.P., LP	Liquefied Petroleum; Low Pressure	
L.P.F.	Low Power Factor	
LR	Long Radius	
L.S.	Lump Sum	
Lt.	Light	
Lt. Ga.	Light Gauge	
L.T.L.	Less than Truckload Lot	
Lt. Wt.	Lightweight	
L.V.	Low Voltage	
M	Thousand; Material; Male; Light Wall Copper Tubing	
M^2CA	Meters Squared Contact Area	
m/hr.; M.H.	Man-hour	
mA	Milliampere	
Mach.	Machine	
Mag. Str.	Magnetic Starter	
Maint.	Maintenance	
Marb.	Marble Setter	
Mat; Mat'l.	Material	
Max.	Maximum	
MBF	Thousand Board Feet	
MBH	Thousand BTU's per hr.	
MC	Metal Clad Cable	
MCC	Motor Control Center	
M.C.F.	Thousand Cubic Feet	
MCFM	Thousand Cubic Feet per Minute	
M.C.M.	Thousand Circular Mils	
MCP	Motor Circuit Protector	
MD	Medium Duty	
MDF	Medium-density fibreboard	
M.D.O.	Medium Density Overlaid	
Med.	Medium	
MF	Thousand Feet	
M.F.B.M.	Thousand Feet Board Measure	
Mfg.	Manufacturing	
Mfrs.	Manufacturers	
mg	Milligram	
MGD	Million Gallons per Day	
MGPH	Million Gallons per Hour	
MH, M.H.	Manhole; Metal Halide; Man-Hour	
MHz	Megahertz	
Mi.	Mile	
MI	Malleable Iron; Mineral Insulated	
MIPS	Male Iron Pipe Size	
mj	Mechanical Joint	
m	Meter	
mm	Millimeter	
Mill.	Millwright	
Min., min.	Minimum, Minute	

For customer support on your Facilities Maintenance & Repair Costs with RSMeans data, call 800.448.8182.

Misc.	Miscellaneous	PCM	Phase Contrast Microscopy	SBS	Styrene Butadiere Styrene
ml	Milliliter, Mainline	PDCA	Painting and Decorating	SC	Screw Cover
M.L.F.	Thousand Linear Feet		Contractors of America	SCFM	Standard Cubic Feet per Minute
Mo.	Month	P.E., PE	Professional Engineer;	Scaf.	Scaffold
Mobil.	Mobilization		Porcelain Enamel;	Sch., Sched.	Schedule
Mog.	Mogul Base		Polyethylene; Plain End	S.C.R.	Modular Brick
MPH	Miles per Hour	P.E.C.I.	Porcelain Enamel on Cast Iron	S.D.	Sound Deadening
MPT	Male Pipe Thread	Perf.	Perforated	SDR	Standard Dimension Ratio
MRGWB	Moisture Resistant Gypsum	PEX	Cross Linked Polyethylene	S.E.	Surfaced Edge
	Wallboard	Ph.	Phase	Sel.	Select
MRT	Mile Round Trip	P.I.	Pressure Injected	SER, SEU	Service Entrance Cable
ms	Millisecond	Pile.	Pile Driver	S.F.	Square Foot
M.S.F.	Thousand Square Feet	Pkg.	Package	S.F.C.A.	Square Foot Contact Area
Mstz.	Mosaic & Terrazzo Worker	Pl.	Plate	S.F. Flr.	Square Foot of Floor
M.S.Y.	Thousand Square Yards	Plah.	Plasterer Helper	S.F.G.	Square Foot of Ground
Mtd., mtd., mtd	Mounted	Plas.	Plasterer	S.F. Hor.	Square Foot Horizontal
Mthe.	Mosaic & Terrazzo Helper	plf	Pounds Per Linear Foot	SFR	Square Feet of Radiation
Mtng.	Mounting	Pluh.	Plumber Helper	S.F. Shlf.	Square Foot of Shelf
Mult.	Multi; Multiply	Plum.	Plumber	S4S	Surface 4 Sides
MUTCD	Manual on Uniform Traffic Control	Ply.	Plywood	Shee.	Sheet Metal Worker
	Devices	p.m.	Post Meridiem	Sin.	Sine
M.V.A.	Million Volt Amperes	Pntd.	Painted	Skwk.	Skilled Worker
M.V.A.R.	Million Volt Amperes Reactance	Pord.	Painter, Ordinary	SL	Saran Lined
MV	Megavolt	pp	Pages	S.L.	Slimline
MW	Megawatt	PP, PPL	Polypropylene	Sldr.	Solder
MXM	Male by Male	P.P.M.	Parts per Million	SLH	Super Long Span Bar Joist
MYD	Thousand Yards	Pr.	Pair	S.N.	Solid Neutral
N	Natural; North	P.E.S.B.	Pre-engineered Steel Building	SO	Stranded with oil resistant inside
nA	Nanoampere	Prefab.	Prefabricated		insulation
NA	Not Available; Not Applicable	Prefin.	Prefinished	S-O-L	Socketolet
N.B.C.	National Building Code	Prop.	Propelled	sp	Standpipe
NC	Normally Closed	PSF, psf	Pounds per Square Foot	S.P.	Static Pressure; Single Pole; Self-
NEMA	National Electrical Manufacturers	PSI, psi	Pounds per Square Inch		Propelled
	Assoc.	PSIG	Pounds per Square Inch Gauge	Spri.	Sprinkler Installer
NEHB	Bolted Circuit Breaker to 600V.	PSP	Plastic Sewer Pipe	spwg	Static Pressure Water Gauge
NFPA	National Fire Protection Association	Pspr.	Painter, Spray	S.P.D.T.	Single Pole, Double Throw
NLB	Non-Load-Bearing	Psst.	Painter, Structural Steel	SPF	Spruce Pine Fir; Sprayed
NM	Non-Metallic Cable	P.T.	Potential Transformer		Polyurethane Foam
nm	Nanometer	P. & T.	Pressure & Temperature	S.P.S.T.	Single Pole, Single Throw
No.	Number	Ptd.	Painted	SPT	Standard Pipe Thread
NO	Normally Open	Ptns.	Partitions	Sq.	Square; 100 Square Feet
N.O.C.	Not Otherwise Classified	Pu	Ultimate Load	Sq. Hd.	Square Head
Nose.	Nosing	PVC	Polyvinyl Chloride	Sq. In.	Square Inch
NPT	National Pipe Thread	Pvmt.	Pavement	S.S.	Single Strength; Stainless Steel
NQOD	Combination Plug-on/Bolt on	PRV	Pressure Relief Valve	S.S.B.	Single Strength B Grade
	Circuit Breaker to 240V.	Pwr.	Power	sst, ss	Stainless Steel
N.R.C., NRC	Noise Reduction Coefficient/	Q	Quantity Heat Flow	Sswk.	Structural Steel Worker
	Nuclear Regulator Commission	Qt.	Quart	Sswl.	Structural Steel Welder
N.R.S.	Non Rising Stem	Quan., Qty.	Quantity	St.; Stl.	Steel
ns	Nanosecond	Q.C.	Quick Coupling	STC	Sound Transmission Coefficient
NTP	Notice to Proceed	r	Radius of Gyration	Std.	Standard
nW	Nanowatt	R	Resistance	Stg.	Staging
OB	Opposing Blade	R.C.P.	Reinforced Concrete Pipe	STK	Select Tight Knot
OC	On Center	Rect.	Rectangle	STP	Standard Temperature & Pressure
OD	Outside Diameter	recpt.	Receptacle	Stpi.	Steamfitter, Pipefitter
O.D.	Outside Dimension	Reg.	Regular	Str.	Strength; Starter; Straight
ODS	Overhead Distribution System	Reinf.	Reinforced	Strd.	Stranded
O.G.	Ogee	Req'd.	Required	Struct.	Structural
O.H.	Overhead	Res.	Resistant	Sty.	Story
O&P	Overhead and Profit	Resi.	Residential	Subj.	Subject
Oper.	Operator	RF	Radio Frequency	Subs.	Subcontractors
Opng.	Opening	RFID	Radio-frequency Identification	Surf.	Surface
Orna.	Ornamental	Rgh.	Rough	Sw.	Switch
OSB	Oriented Strand Board	RGS	Rigid Galvanized Steel	Swbd.	Switchboard
OS&Y	Outside Screw and Yoke	RHW	Rubber, Heat & Water Resistant;	S.Y.	Square Yard
OSHA	Occupational Safety and Health		Residential Hot Water	Syn.	Synthetic
	Act	rms	Root Mean Square	S.Y.P.	Southern Yellow Pine
Ovhd.	Overhead	Rnd.	Round	Sys.	System
OWG	Oil, Water or Gas	Rodm.	Rodman	t.	Thickness
Oz.	Ounce	Rofc.	Roofer, Composition	T	Temperature; Ton
P.	Pole; Applied Load; Projection	Rofp.	Roofer, Precast	Tan	Tangent
p.	Page	Rohe.	Roofer Helpers (Composition)	T.C.	Terra Cotta
Pape.	Paperhanger	Rots.	Roofer, Tile & Slate	T & C	Threaded and Coupled
P.A.P.R.	Powered Air Purifying Respirator	R.O.W.	Right of Way	T.D.	Temperature Difference
PAR	Parabolic Reflector	RPM	Revolutions per Minute	TDD	Telecommunications Device for
P.B., PB	Push Button	R.S.	Rapid Start		the Deaf
Pc., Pcs.	Piece, Pieces	Rsr	Riser	T.E.M.	Transmission Electron Microscopy
P.C.	Portland Cement; Power Connector	RT	Round Trip	temp	Temperature, Tempered, Temporary
P.C.F.	Pounds per Cubic Foot	S.	Suction; Single Entrance; South	TFFN	Nylon Jacketed Wire

671

TFE	Tetrafluoroethylene (Teflon)	U.L., UL	Underwriters Laboratory	w/	With
T. & G.	Tongue & Groove;	Uld.	Unloading	W.C., WC	Water Column; Water Closet
	Tar & Gravel	Unfin.	Unfinished	W.F.	Wide Flange
Th., Thk.	Thick	UPS	Uninterruptible Power Supply	W.G.	Water Gauge
Thn.	Thin	URD	Underground Residential	Wldg.	Welding
Thrded	Threaded		Distribution	W. Mile	Wire Mile
Tilf.	Tile Layer, Floor	US	United States	W-O-L	Weldolet
Tilh.	Tile Layer, Helper	USGBC	U.S. Green Building Council	W.R.	Water Resistant
THHN	Nylon Jacketed Wire	USP	United States Primed	Wrck.	Wrecker
THW.	Insulated Strand Wire	UTMCD	Uniform Traffic Manual For Control	WSFU	Water Supply Fixture Unit
THWN	Nylon Jacketed Wire		Devices	W.S.P.	Water, Steam, Petroleum
T.L., TL	Truckload	UTP	Unshielded Twisted Pair	WT., Wt.	Weight
T.M.	Track Mounted	V	Volt	WWF	Welded Wire Fabric
Tot.	Total	VA	Volt Amperes	XFER	Transfer
T-O-L	Threadolet	VAT	Vinyl Asbestos Tile	XFMR	Transformer
tmpd	Tempered	V.C.T.	Vinyl Composition Tile	XHD	Extra Heavy Duty
TPO	Thermoplastic Polyolefin	VAV	Variable Air Volume	XHHW	Cross-Linked Polyethylene Wire
T.S.	Trigger Start	VC	Veneer Core	XLPE	Insulation
Tr.	Trade	VDC	Volts Direct Current	XLP	Cross-linked Polyethylene
Transf.	Transformer	Vent.	Ventilation	Xport	Transport
Trhv.	Truck Driver, Heavy	Vert.	Vertical	Y	Wye
Trlr	Trailer	V.F.	Vinyl Faced	yd	Yard
Trlt.	Truck Driver, Light	V.G.	Vertical Grain	yr	Year
TTY	Teletypewriter	VHF	Very High Frequency	Δ	Delta
TV	Television	VHO	Very High Output	%	Percent
T.W.	Thermoplastic Water Resistant	Vib.	Vibrating	~	Approximately
	Wire	VLF	Vertical Linear Foot	∅	Phase; diameter
UCI	Uniform Construction Index	VOC	Volatile Organic Compound	@	At
UF	Underground Feeder	Vol.	Volume	#	Pound; Number
UGND	Underground Feeder	VRP	Vinyl Reinforced Polyester	<	Less Than
UHF	Ultra High Frequency	W	Wire; Watt; Wide; West	>	Greater Than
U.I.	United Inch			Z	Zone

672

For customer support on your Facilities Maintenance & Repair Costs with RSMeans data, call 800.448.8182.

Other Data & Services

tradition of excellence in construction cost information
d services since 1942

Table of Contents
Cost Data Selection Guide
RSMeans Data Online
Training

more information visit our website at RSMeans.com

Unit prices according to the latest MasterFormat®

Cost Data Selection Guide

The following table provides definitive information on the content of each cost data publication. The number of lines of data provided in each unit price or assemblies division, as well as the number of crews, is listed for each data set. The presence of other elements such as reference tables, square foot models, equipment rental costs, historical cost indexes, and city cost indexes, is also indicated. You can use the table to help select the RSMeans data set that has the quantity and type of information you most need in your work.

Unit Cost Divisions	Building Construction	Mechanical	Electrical	Commercial Renovation	Square Foot	Site Work Landsc.	Green Building	Interior	Concrete Masonry	Open Shop	Heavy Construction	Light Commercial	Facilities Construction	Plumbing	Residential
1	584	406	427	531	0	516	200	326	467	583	521	273	1056	416	178
2	779	278	86	735	0	995	207	397	218	778	737	479	1222	285	274
3	1744	340	230	1265	0	1536	1041	354	2273	1744	1929	537	2027	316	444
4	961	21	0	921	0	726	180	615	1159	929	616	534	1176	0	448
5	1889	158	155	1093	0	852	1787	1106	729	1889	1025	979	1906	204	746
6	2453	18	18	2111	0	110	589	1528	281	2449	123	2141	2125	22	2661
7	1596	215	128	1634	0	580	763	532	523	1593	26	1329	1697	227	1049
8	2140	80	3	2733	0	255	1140	1813	105	2142	0	2328	2966	0	1552
9	2107	86	45	1931	0	309	455	2193	412	2048	15	1756	2356	54	1521
10	1089	17	10	685	0	232	32	899	136	1089	34	589	1180	237	224
11	1097	201	166	541	0	135	56	925	29	1064	0	231	1117	164	110
12	548	0	2	298	0	219	147	1551	14	515	0	273	1574	23	217
13	744	149	158	253	0	366	125	254	78	720	267	109	760	115	104
14	273	36	0	223	0	0	0	257	0	273	0	12	293	16	6
21	130	0	41	37	0	0	0	296	0	130	0	121	668	688	259
22	1165	7559	160	1226	0	1573	1063	849	20	1154	1682	875	7506	9416	719
23	1198	7001	581	940	0	157	901	789	38	1181	110	890	5240	1918	485
25	0	0	14	14	0	0	0	0	0	0	0	0	0	0	0
26	1512	491	10455	1293	0	811	644	1159	55	1438	600	1360	10236	399	636
27	94	0	447	101	0	0	0	71	0	94	39	67	388	0	56
28	143	79	223	124	0	0	28	97	0	127	0	70	209	57	41
31	1511	733	610	807	0	3266	289	7	1218	1456	3282	605	1570	660	614
32	838	49	8	905	0	4475	355	406	315	809	1891	440	1752	142	487
33	1248	1080	534	252	0	3040	38	0	239	525	3090	128	1707	2089	154
34	107	0	47	4	0	190	0	0	31	62	221	0	136	0	0
35	18	0	0	0	0	327	0	0	0	18	442	0	84	0	0
41	62	0	0	33	0	8	0	22	0	61	31	0	68	14	0
44	75	79	0	0	0	0	0	0	0	0	0	0	75	75	0
46	23	16	0	0	0	274	261	0	0	23	264	0	33	33	0
48	8	0	36	2	0	0	21	0	0	8	15	8	21	0	8
Totals	26136	19092	14584	20692	0	20952	10322	16446	8340	24902	16960	16134	51148	17570	12993

Assem Div	Building Construction	Mechanical	Electrical	Commercial Renovation	Square Foot	Site Work Landscape	Assemblies	Green Building	Interior	Concrete Masonry	Heavy Construction	Light Commercial	Facilities Construction	Plumbing	Asm Div	Residential
A		15	0	188	164	577	598	0	0	536	571	154	24	0	1	378
B		0	0	848	2554	0	5661	56	329	1976	368	2094	174	0	2	211
C		0	0	647	954	0	1334	0	1641	146	0	844	251	0	3	588
D		1057	941	712	1858	72	2538	330	824	0	0	1345	1104	1088	4	851
E		0	0	86	261	0	301	0	5	0	0	258	5	0	5	391
F		0	0	0	114	0	143	0	0	0	0	114	0	0	6	357
G		527	447	318	312	3378	792	0	0	535	1349	205	293	677	7	307
															8	760
															9	80
															10	0
															11	0
															12	0
Totals		1599	1388	2799	6217	4027	11367	386	2799	3193	2288	5014	1851	1765		3923

Reference Section	Building Construction Costs	Mechanical	Electrical	Commercial Renovation	Square Foot	Site Work Landscape	Assem.	Green Building	Interior	Concrete Masonry	Open Shop	Heavy Construction	Light Commercial	Facilities Construction	Plumbing	Resi.
Reference Tables	yes	yes	yes	yes	no	yes	yes	yes	yes	yes	yes	yes	yes	yes	yes	yes
Models					111			25					50			28
Crews	582	582	582	561		582		582	582	582	560	582	560	561	582	560
Equipment Rental Costs	yes	yes	yes	yes		yes		yes	yes	yes	yes	yes	yes	yes	yes	yes
Historical Cost Indexes	yes	yes	yes	yes	yes	yes	yes	yes	yes	yes	yes	yes	yes	yes	yes	no
City Cost Indexes	yes	yes	yes	yes	yes	yes	yes	yes	yes	yes	yes	yes	yes	yes	yes	yes

RSMeans data Online — Cloud-based access to North America's leading construction database

RSMeans Data Online

Cloud-based access to North America's leading construction cost database.

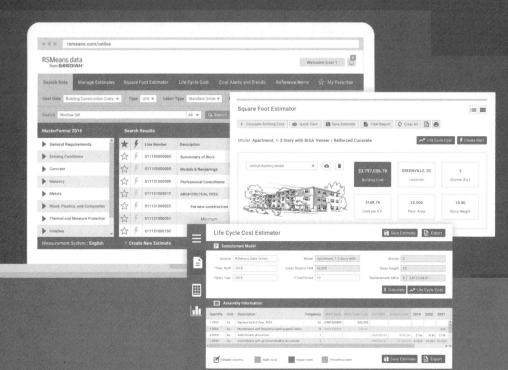

Plan your next budget with accurate prices and insight.

Estimate in great detail or at a conceptual level.

Validate unfamiliar costs or scopes of work.

Maintain existing buildings with schedules and costs.

RSMeans data
from G⦿RDIAN®

Learn more at rsmeans.com/online

rsmeans.com/core

RSMeans Data Online Core

The Core tier of RSMeans Data Online provides reliable construction cost data along with the tools necessary to quickly access costs at the material or task level. Users can create unit line estimates with RSMeans data from Gordian and share them with ease from the web-based application.

Key Features:

- Search
- Estimate
- Share

Data Available:

16 datasets and packages curated by project type

RSMeans Data Online Complete

The Complete tier of RSMeans Data Online is designed to provide the latest in construction costs with comprehensive tools for projects of varying scopes. Unit, assembly or square foot price data is available 24/7 from the web-based application. Harness the power of RSMeans data from Gordian with expanded features and powerful tools like the square foot model estimator and cost trends analysis.

rsmeans.com/complete

Key Features:

- Alerts
- Square Foot Estimator
- Trends

Data Available:

20 datasets and packages curated by project type

rsmeans.com/completeplus

RSMeans Data Online Complete Plus

The Complete Plus tier of RSMeans Data Online was developed to take project planning and estimating to the next level with predictive cost data. Users gain access to an exclusive set of tools and features and the most comprehensive database in the industry. Leverage everything RSMeans Data Online has to offer with the all-inclusive Complete Plus tier.

Key Features:

- Predictive Cost Data
- Life Cycle Costing
- Full History

Data Available:

Full library and renovation models

RSMeans data Online   Cloud-based access to North America's leading construction database

2019 Seminar Schedule
☎ 877-620-6245

Note: call for exact dates, locations, and details as some cities are subject to change.

Location	Dates	Location	Dates
Seattle, WA	January and August	San Francisco, CA	June
Dallas/Ft. Worth, TX	January	Bethesda, MD	June
Austin, TX	February	Dallas, TX	September
Jacksonville, FL	February	Raleigh, NC	October
Anchorage, AK	March and September	Baltimore, MD	November
Las Vegas, NV	March	Orlando, FL	November
Washington, DC	April and September	San Diego, CA	December
Charleston, SC	April	San Antonio, TX	December
Toronto	May		
Denver, CO	May		

Gordian also offers a suite of online RSMeans data self-paced offerings.
Check our website RSMeans.com/products/training.aspx for more information.

Facilities Construction Estimating

In this two-day course, professionals working in facilities management can get help with their daily challenges to establish budgets for all phases of a project.

Some of what you'll learn:
- Determining the full scope of a project
- Identifying the scope of risks and opportunities
- Creative solutions to estimating issues
- Organizing estimates for presentation and discussion
- Special techniques for repair/remodel and maintenance projects
- Negotiating project change orders

Who should attend: facility managers, engineers, contractors, facility tradespeople, planners, and project managers.

Mechanical & Electrical Estimating

This two-day course teaches attendees how to prepare more accurate and complete mechanical/electrical estimates, avoid the pitfalls of omission and double-counting, and understand the composition and rationale within the RSMeans mechanical/electrical database.

Some of what you'll learn:
- The unique way mechanical and electrical systems are interrelated
- M&E estimates—conceptual, planning, budgeting, and bidding stages
- Order of magnitude, square foot, assemblies, and unit price estimating
- Comparative cost analysis of equipment and design alternatives

Who should attend: architects, engineers, facilities managers, mechanical and electrical contractors, and others who need a highly reliable method for developing, understanding, and evaluating mechanical and electrical contracts.

Construction Cost Estimating: Concepts and Practice

This one or two day introductory course to improve estimating skills and effectiveness starts with the details of interpreting bid documents and ends with the summary of the estimate and bid submission.

Some of what you'll learn:
- Using the plans and specifications to create estimates
- The takeoff process—deriving all tasks with correct quantities
- Developing pricing using various sources; how subcontractor pricing fits in
- Summarizing the estimate to arrive at the final number
- Formulas for area and cubic measure, adding waste and adjusting productivity to specific projects
- Evaluating subcontractors' proposals and prices
- Adding insurance and bonds
- Understanding how labor costs are calculated
- Submitting bids and proposals

Who should attend: project managers, architects, engineers, owners' representatives, contractors, and anyone who's responsible for budgeting or estimating construction projects.

RSMeans data Training

ssessing Scope of Work for Facilities onstruction Estimating

his two-day practical training program addresses the vital portance of understanding the scope of projects in order produce accurate cost estimates for facility repair and modeling.

ome of what you'll learn:
Discussions of site visits, plans/specs, record drawings of facilities, and site-specific lists
Review of CSI divisions, including means, methods, materials, and the challenges of scoping each topic
Exercises in scope identification and scope writing for accurate estimating of projects
Hands-on exercises that require scope, take-off, and pricing

ho should attend: corporate and government estimators, anners, facility managers, and others who need to roduce accurate project estimates.

Maintenance & Repair Estimating for Facilities

This two-day course teaches attendees how to plan, budget, and estimate the cost of ongoing and preventive maintenance and repair for existing buildings and grounds.

Some of what you'll learn:
• The most financially favorable maintenance, repair, and replacement scheduling and estimating
• Auditing and value engineering facilities
• Preventive planning and facilities upgrading
• Determining both in-house and contract-out service costs
• Annual, asset-protecting M&R plan

Who should attend: facility managers, maintenance supervisors, buildings and grounds superintendents, plant managers, planners, estimators, and others involved in facilities planning and budgeting.

ractical Project Management for onstruction Professionals

n this two-day course, acquire the essential knowledge nd develop the skills to effectively and efficiently xecute the day-to-day responsibilities of the construction roject manager.

ome of what you'll learn:
General conditions of the construction contract
Contract modifications: change orders and construction hange directives
Negotiations with subcontractors and vendors
Effective writing: notification and communications
Dispute resolution: claims and liens

Who should attend: architects, engineers, owners' epresentatives, and project managers.

Life Cycle Cost Estimating for Facility Asset Managers

Life Cycle Cost Estimating will take the attendee through choosing the correct RSMeans database to use and then correctly applying RSMeans data to their specific life cycle application. Conceptual estimating through RSMeans new building models, conceptual estimating of major existing building projects through RSMeans renovation models, pricing specific renovation elements, estimating repair, replacement and preventive maintenance costs today and forward up to 30 years will be covered.

Some of what you'll learn:
• Cost implications of managing assets
• Planning projects and initial & life cycle costs
• How to use RSMeans data online

Who should attend: facilities owners and managers and anyone involved in the financial side of the decision making process in the planning, design, procurement, and operation of facility real assets.

Please bring a laptop with ability to access the internet.

Building Systems and the Construction Process

This one-day course was written to assist novices and those outside the ndustry in obtaining a solid understanding of the construction process - rom both a building systems and construction administration approach.

ome of what you'll learn:
Various systems used and how components come together to create a building
Start with foundation and end with the physical systems of the structure such as HVAC and Electrical
Focus on the process from start of design through project closeout

This training session requires you to bring a laptop computer to class.

Who should attend: building professionals or novices to help make the crossover to the construction industry; suited for anyone responsible for providing high level oversight on construction projects.

RSMeans data Training

Training for our Online Estimating Solution

Construction estimating is vital to the decision-making process at each state of every project. Our online solution works the way you do. It's systematic, flexible and intuitive. In this one-day class you will see how you can estimate any phase of any project faster and better.

Some of what you'll learn:
- Customizing our online estimating solution
- Making the most of RSMeans "Circle Reference" numbers
- How to integrate your cost data
- Generating reports, exporting estimates to MS Excel, sharing, collaborating and more

Also offered as a self-paced or on-site training program!

Training for our CD Estimating Solution

This one-day course helps users become more familiar with the functionality of the CD. Each menu, icon, screen, and function found in the program is explained in depth. Time is devoted to hands-on estimating exercises.

Some of what you'll learn:
- Searching the database using all navigation methods
- Exporting RSMeans data to your preferred spreadsheet format
- Viewing crews, assembly components, and much more
- Automatically regionalizing the database

This training session requires you to bring a laptop computer to class.

When you register for this course you will receive an outline for your laptop requirements.

Also offered as a self-paced or on-site training program!

Site Work Estimating with RSMeans data

This one-day program focuses directly on site work costs. Accurately scoping, quantifying, and pricing site preparation, underground utility work, and improvements to exterior site elements are often the most difficult estimating tasks on any project. Some of what you'll learn:
- Evaluation of site work and understanding site scope including: site clearing, grading, excavation, disposal and trucking of materials, backfill and compaction, underground utilities, paving, sidewalks, and seeding & planting.
- Unit price site work estimates—Correct use of RSMeans site work cost data to develop a cost estimate.
- Using and modifying assemblies—Save valuable time when estimating site work activities using custom assemblies.

Who should attend: Engineers, contractors, estimators, project managers, owner's representatives, and others who are concerned with the proper preparation and/or evaluation of site work estimates.

Please bring a laptop with ability to access the internet.

Facilities Estimating Using the CD

This two-day class combines hands-on skill-building with best estimating practices and real-life problems. You will learn key concepts, tips, pointers, and guidelines to save time and avoid cost oversights and errors.

Some of what you'll learn:
- Estimating process concepts
- Customizing and adapting RSMeans cost data
- Establishing scope of work to account for all known variables
- Budget estimating: when, why, and how
- Site visits: what to look for and what you can't afford to overlook
- How to estimate repair and remodeling variables

This training session requires you to bring a laptop computer to class.

Who should attend: facility managers, architects, engineers, contractors, facility tradespeople, planners, project managers, and anyone involved with JOC, SABRE or IDIQ.

Registration Information

Register early to save up to $100!!!

Register 45+ days before date of a class and save $50 off each class. This savings cannot be combined with any other promotional or discounting of the regular price of classes!

How to register

By Phone
Register by phone at 877-620-6245

Online
Register online at RSMeans.com/products/seminars.aspx

Note: Purchase Orders or Credits Cards are required to register.

Two-day seminar registration fee - $1,200.

One-Day Construction Cost Estimating or Building Systems and the Construction Process - $765.

Government pricing

All federal government employees save off the regular seminar price. Other promotional discounts cannot be combined with the government discount. Call 781-422-5115 for government pricing.

CANCELLATION POLICY:

If you are unable to attend a seminar, substitutions may be made at any time before the session starts by notifying the seminar registrar at 1-781-422-5115 or your sales representative.
If you cancel twenty-one (21) days or more prior to the seminar, there will be no penalty and your registration fees will be refunded. These cancellations must be received by the seminar registrar or your sales representative and will be confirmed to be eligible for cancellation.
If you cancel fewer than twenty-one (21) days prior to the seminar, you will forfeit the registration fee.
In the unfortunate event of an RSMeans cancellation, RSMeans will work with you to reschedule your attendance in the same seminar at a later date or will fully refund your registration fee. RSMeans cannot be responsible for any non-refundable travel expenses incurred by you or another as a result of your registration, attendance at, or cancellation of an RSMeans seminar.
Any on-demand training modules are not eligible for cancellation, substitution, transfer, return or refund.

AACE approved courses

Many seminars described and offered here have been approved for 14 hours (1.4 recertification credits) of credit by the AACE International Certification Board toward meeting the continuing education requirements for recertification as a Certified Cost Engineer/Certified Cost Consultant.

AIA Continuing Education

We are registered with the AIA Continuing Education System (AIA/CES) and are committed to developing quality learning activities in accordance with the CES criteria. Many seminars meet the AIA/CES criteria for Quality Level 2. AIA members may receive 14 learning units (LUs) for each two-day RSMeans course

Daily course schedule

The first day of each seminar session begins at 8:30 a.m. and ends at 4:30 p.m. The second day begins at 8:00 a.m. and ends at 4:00 p.m. Participants are urged to bring a hand-held calculator since many actual problems will be worked out in each session.

Continental breakfast

Your registration includes the cost of a continental breakfast and a morning and afternoon refreshment break. These informal segments allow you to discuss topics of mutual interest with other seminar attendees. (You are free to make your own lunch and dinner arrangements.)

Hotel/transportation arrangements

We arrange to hold a block of rooms at most host hotels. To take advantage of special group rates when making your reservation, be sure to mention that you are attending the RSMeans Institute data seminar. You are, of course, free to stay at the lodging place of your choice. (Hotel reservations and transportation arrangements should be made directly by seminar attendees.)

Important

Class sizes are limited, so please register as soon as possible.

Note: Pricing subject to change.